THERMAL PROCESSING AND QUALITY OF FOODS

Edited by

P. ZEUTHEN
Danish Meat Products Laboratory, Ministry of Agriculture, Copenhagen, Denmark

J. C. CHEFTEL
Université des Sciences et Techniques du Languedoc, Montpellier, France

C. ERIKSSON
SIK—The Swedish Food Institute, Gothenburg, Sweden

M. JUL
Danish Meat Products Laboratory, Ministry of Agriculture, Copenhagen, Denmark

H. LENIGER
Agricultural University Biotechnion, Wageningen, The Netherlands

P. LINKO
Helsinki University of Technology, Espoo, Finland

G. VARELA
Instituto de Nutrición, Ciudad Universitaria, Madrid, Spain

G. VOS
Commission of the European Communities, Brussels, Belgium

ELSEVIER APPLIED SCIENCE PUBLISHERS
LONDON and NEW YORK

ELSEVIER APPLIED SCIENCE PUBLISHERS LTD
Ripple Road, Barking, Essex, England

Sole Distributor in the USA and Canada
ELSEVIER SCIENCE PUBLISHING CO., INC.
52 Vanderbilt Avenue, New York, NY 10017, USA

British Library Cataloguing in Publication Data

Thermal processing and quality of foods.
1. Food—Preservation
2. Food—Thermal properties
I. Zeuthen, P.
664′.028 TP371.2

ISBN 0-85334-279-2

WITH 237 TABLES AND 322 ILLUSTRATIONS

Publication arrangements by: Commission of the European Communities, Directorate-General Information Market and Innovation, Luxembourg

EUR 9038

LEGAL NOTICE
Neither the Commission of the European Communities nor any person acting on behalf of the Commission is responsible for the use which might be made of the following information.

Printed in Great Britain by Galliard (Printers) Ltd, Great Yarmouth

THERMAL PROCESSING AND QUALITY OF FOODS

Proceedings of the concluding seminar held under the auspices of COST (European Cooperation in Scientific and Technical Research) on the thermal processing and quality of foods in Athens, 14–18 November 1983.

Editorial Note

The editing of these proceedings was delegated by the COST 91 COST–Community Coordinating Committee (CCCC) to the COST 91 'Executive Committee' which comprised:

P. Linko, *President, COST 91*
M. Jul, *Vice-President, COST 91, Chairman, Freezing and Thawing sub-group*
H. A. Leniger, *Vice-President, COST 91*
J. C. Cheftel, *Chairman, Extrusion Cooking sub-group*
C. E. Eriksson, *Chairman, Industrial Cooking sub-group*
V. Varela
G. Vos, *COST 91 Project Technical Secretary*
P. Zeuthen, *COST 91 Project Leader, with overall editorial responsibility*

Statements and opinions expressed in these Proceedings are those of the individual authors and not necessarily those of the editors, the publishers, the Commission or other participants in the COST 91 project.

Preface

Although food materials have been processed since ancient times, modern thermal processing and preservation techniques are of relatively recent origin. Today, the consumer spends more money on food than on any other item, and thermal processing has allowed him to be both geographically and climatically independent for his nutritional needs. Processing has made the year-round delivery of wholesome, nutritious and tasty foods possible. Nevertheless, some nutritional losses are unavoidable regardless whether the food is processed industrially or at home. On the other hand, thermal treatment allows the inactivation of many anti-nutritional factors and undesirable enzymes present in native food materials, resulting in an improved utilization of valuable nutritive components and in an improved shelf-life. Processing has also virtually eliminated food-borne diseases and poisonings, and has made possible the availability of a variety of foods with improved appearance, flavour and texture.

Much of this development has until recently been empirical, with the applications often preceding scientific knowledge. Novel technologies have been applied without sufficient information on the effects on quality and nutritional implications. Furthermore, during the last few years the availability and proper balance of minor nutrients, the effects of common salt in the diet, and the importance and availability of nutritive fibre have become a major concern. Yet little has been known of the effects of such factors during processing or of the influence of processing on such factors.

Realizing the importance of the improved availability of food processing-related data, Professor Erik von Sydow submitted in July 1974 a Swedish proposal to include food technology within the European Cooperation in Science and Technology programme COST. This was well received, and a formal proposal was presented in December 1974 to the Council of European Communities and in April 1975 COST Senior

Officials Committee established an *ad hoc* Working Party to draft a food technology research programme. This work led eventually to two separate projects, COST 90 on 'The Physical Properties of Foods', and COST 91 on 'The Effects of Thermal Processing on Quality and Nutritive Value of Food'. It was also agreed that the cooperation should make an important contribution to food and related industries, and to the consumer, and that it should benefit from international cooperation and should stimulate trade.

In November 1977 a historical seminar was held in Dublin as a brainstorming event to lay out the background for the project COST 91—historical in the sense that for the first time European experts representing different food disciplines discussed together research priorities. These discussions were later published as a book, *Food Quality and Nutrition*, edited by W. K. Downey (Applied Science Publishers, 1978). However, it took another two years of planning until the three-year COST 91 research programme was finally approved and the project was officially started. The project was divided into three sub-groups, extrusion cooking, industrial cooking and baking, and freezing and thawing. Realizing the already existing international commodity-oriented contacts and cooperation, the process-oriented interdisciplinary nature of the project was emphasized, and to my understanding this was the key to the success of the project. Furthermore, the importance of the systems approach, in which the total food chain from the raw material to the finished product delivered to the consumer, combining food processing with product development and distribution, was realized. After all, it is the optimization of the total food production system to meet consumer needs that is important. Throughout the duration of the project the principle of concerted action with research financed nationally by the participating countries has been followed, and the importance of the cooperation between universities, research institutes and the food industry stressed. The project created an atmosphere unique of its kind to allow scientists from many European countries to get together both at specialist and interdisciplinary seminars and meetings to exchange ideas and to discuss the latest research findings, and it brought together scientists and research institutes to solve important problems in cooperation.

The wish to hold an interdisciplinary concluding seminar had already been expressed in Dublin six years ago. During the week of 14–18 November 1983 European food experts got together again, this time in Athens, to report and to discuss the results of the project COST 91. Most of the papers presented at the seminar are included in this book, the impressive quantity of new valuable information speaking in itself of the success of the project. During the course of the project a number of other scientific and review papers have been published, some co-authored by scientists from different participating countries. These papers will be listed in the Final Report of the project. Still other papers resulting from the work carried out during the project are likely to be published later. But COST 91 has not only been a research effort. The cooperation has also made possible the preparation of a comprehensive terminology in several languages, recommendations related to vitamin analysis, guidelines for publications, etc. A new four-year project is currently under planning, expanding the European coordinated research effort in food technology to other high-temperature short-time processes, to chilling technology and to food biotechnology.

The importance of COST 91 as a European coordinated research effort in food technology was shown also by the fact that nearly everyone invited to contribute to the concluding seminar accepted. Undoubtedly many research institutes have learned to know each other better, and many new individual friendships have resulted. This success would not, however, have been possible without the continuing enthusiasm of the individual scientists and industrialists throughout the project, and not least of our expert project leader Peter Zeuthen and of the other friends and colleagues in the Executive Committee, to whom I am particularly indebted. Thanks are also due to all national organizations which have so generously supported the work reported in this book.

PEKKA LINKO
Chairman, COST 91 CCCC and Executive

Contents

SUB-GROUP 2: INDUSTRIAL COOKING

SUB-GROUP 4: NUTRITION

COST 91 - ORGANIZATION, GOALS AND PERFORMANCE

P. ZEUTHEN
Danish Meat Products Laboratory, Ministry of Agriculture,
and Food Technology Laboratory, Technical University of Denmark

1. INTRODUCTION

The work in this Programme, entitled "the effect of thermal (heat and cold) processing and distribution on the quality and nutritive value of foods" was initiated in 1980. The participating States were, besides the Community Member States, Finland, Spain, Sweden, Switzerland, and Yugoslavia. All research carried out under the Programme has been nationally funded. Only some travel and coordinating administration has been paid through common funding. Cooperation has taken place through a Project Leader and coordination of related projects in the participating countries, through workshops and coordinating meetings, a periodical newsletter distributed to the participants and extensive minutes of technical meetings, so that interim results could be reported quickly to institutes involved in individual project areas.

This book, the Proceedings of the Concluding Seminar of the Programme, contains most of the research accomplishments under the Programme. However, some reports of research under the Programme have already been reported, and others, not completed at the time of the Athens Seminar, i.e. November 1983, will be published by the executing institutes.

2. THE SUB-GROUPS

During the planning phase it was felt at an early stage that certain limitations would have to be made in order to make the Programme manageable. Further, that cooperation should be process-oriented rather than to be organized along commodity groups. It was also discussed that it could be relevant to include studies on the formation of toxic substances during certain processes. However, it was found more in accordance with the title of the project to concentrate on properties such as nutritive changes and losses of nutrients and quality brought about through processing, storage and distribution. Besides, other research projects within the European Community already deal with toxicological aspects.

The work was concentrated in the following areas and organized in four Sub-Groups: 1) Extrusion Cooking, 2) Industrial Cooking, incl. baking, 3) Freezing and Thawing, 4) Nutrition. This choice was made with that in view that the research areas selected should either constitute new processes or reflect areas of changing dietary habits.

The work in the Sub-Group on Extrusion Cooking has aimed at getting a better understanding of what happens in extrusion cooking when it operates under various conditions, e.g. biopolymers, so that extrusion cooking could be a process which was better understood, and not just a process mainly based on empiricism. Some projects have concentrated on optimization of design and operation during processing, whereas other projects tried new applications of extrusion cooking such as the use of the equipment for enzyme inactivation or for sterilization purposes.

The Sub-Group on Industrial Cooking has worked on boiling, frying and roasting processes, baking, and further treatment for catering. It has also worked out guidelines for quality assessment, descriptions of process modelling, and it has compiled a multi-lingual list of cooking terms. Details of these many subjects are reported in these proceedings; they all aim at getting a better understanding of heating and handling processes, especially in institutional food preparation. They are meant as a help for better understanding in further research in the areas in question, or take changing eating habits such as the increasing mass-feeding in consideration.

The Sub-Group on Freezing and Thawing of foods identified areas within these topics which have been inadequately studied. The work was divided in studies on the so-called TTT-PPP (time-temperature-tolerance and its dependence on product, process and packaging) concept of frozen foods, since only scant investigations have been made during the last two decennia. Besides, many deviations from the original theories in this area have been observed over the years. Also, new technologies in processing and new packaging materials as well as modified production methods of plant and animal raw materials have made many older data on keepability obsolete. The Sub-Group carried out large investigations on the conditions of frozen foods when they pass through the freezer chain today, i.e. the temperatures and residence times to which they are actually subjected during the various stages of storage, distribution, sale, and home use. Finally, research was carried out on industrial thawing, an area which hitherto has been fairly much neglected, and which is becoming increasingly important because of the growing use of frozen raw materials in industrial food processing.

The Sub-Group on Nutrition has had a dual task; it should partly advise on nutritional aspects of the investigations carried out in the other Sub-Groups, partly carry out its own research. This Sub-Group, therefore, worked out a "Nutritional Statement" for use in the other Sub-Groups. The Statement concludes that by and large the European consumer does not suffer from effects of inadequate nutrient intake but rather from wrong dietary compositions, such as too high intakes of lipids and too refined foods. However, it was also pointed out that certain groups such as the elderly stand at risk of inadequate intake of certain nutrients, if the diet is not carefully composed, a question which should be constantly monitored. Finally, the Sub-Group stressed the necessity of monitoring especially all new food processing methods and changing dietary habits in order to ensure that the intake of sufficient nutrients remains at a high level in the diet. The Sub-Group also took up specific subjects such as the influence of home cooking on the losses of nutrients, as well as arranging a large workshop on nutrient losses which may occur in various catering systems. Finally, a working party will be completing a proposal on recommended analytical methods for the determination of most vitamins in foods. This work will be published separately in 1984.

3. THE FOOD INDUSTRIES AND THE CONSUMER IN COST 91

During the course of this Programme, its relevance to industry and consumers has constantly been kept in mind. Industrial food technologists have participated in the work, not only in most of the workshops and the concluding seminar, also with concrete projects. This exchange of know-how and information has been most useful and has resulted in rapid transfer of the knowledge gained to practical food production. Similarly, consumer representatives have been invited to give their views on industrially processed foods on several occasions, including the concluding seminar. Basically, food research has one main objective, to optimize the quality of foods, i.e. consumers must be involved at the maximum extent possible.

4. BENEFIT TO RESEARCH OF COST 91

With many other international research activities in existence, it is appropriate to point out the particular advantages of this Programme. Throughout the work it has been the general feeling that national projects have profitted tremendously because they took place simultaneously, not only because they could supplement one another, but also because meetings and other means of exchange of ideas turned out to be very fruitful. Besides, it was seen several times that technologies and know-how in specific cases which were well-known and used in one region of Europe were less well understood or used in other regions; however, the knowledge could be transferred quickly and applied in these regions. Furthermore, many researchers in the Programme have expressed the usefulness of the impact of industry and the consumers in the Programme and have stressed that the mutual discussions have helped them focus their own projects better.

In conclusion it seems evident that the Programme has proved to be of great advantage, not only for the identification of gaps in knowledge on optimization of food processing methods with the purpose of improving the quality in all the meaning of this word, it has also furthered contacts among food researchers, industrialists, and the consumers, which, hopefully, will continue to exist long after the Programme has been terminated.

QUALITY OF PROCESSED FOODS

J. HAWTHORN
Professor of Food Science, The University of Strathclyde,
Glasgow, Scotland

Summary

'Quality' meaning 'degree of excellence' is a term of comparison. It therefore can only be defined and analysed in comparative terms. For those involved in food processing or in legislative processes aimed at controlling the actions of food processors, the term 'food quality' requires both definition and analysis. The paper considers those attributes which collectively comparise the concept, and relates them to the effects of processing. It suggests that, while the principal aim of the food processor is to sell his foodstuffs to the largest possible market, it is an error to suppose that the hedonic properties of foods (measured as colour, flavour and texture) are the sole or even the most important determinants of market size. Other quality attributes, even more difficult of measurement exert subtle influences on market acceptance, and may become of dominant significance in the formulation of food policies for large international markets such as the EEC.

1. INTRODUCTION

"There is a peculiar ceremony that takes place each day in all well-conducted British prisons. The prison governor, accompanied by his chief officer, passes through the precincts of the gaol and enters the kitchen where the day's meals are set out for his inspection. And the conscientious governor looks at the porridge, examines the cocoa, and the stew or the fish according to the day of the week, and the suet duff or the rice. He looks at the bread and perhaps pinches it. Then, unless something is remarkably amiss, he nods his approval and moves on."

"Now I would not for a moment belittle the judgement of a prison governor. In fact, almost twenty years ago, I had the duty of surveying the quality and the nutritional composition of the meals provided in long-term and short-term gaols, borstals, women's prisons - a spectrum of the penal institutions of the country, in fact - and, with some detailed proviso, confirmed the opinions of the respective heads of the establishments. Nevertheless, few normal individuals and almost no professional food technologists or directors of food firms would use the same criteria for judging the quality of food products intended for the open market. Yet the diet in Wormwood Scrubs or Wandsworth is wholesome and nourishing. That is to say, it is free from extraneous harmful materials and at the same time provides the appropriate nutrients in adequate amounts for health. There are, of course several reasons why these criteria alone are inadequate for normal civilised life" (1).

This quotation from a paper by Magnus Pyke (1959) sets my theme. Quality, as applied to foodstuffs means 'those attributes which render the food agreeable to the person who eats it' (2). Since tastes differ, the term "food quality" has no absolute meaning, and attempts to measure it in its totality cannot succeed. In the everyday world we use adjectives such as crispy, creamy, toothsome, tempting, tasty, horrid, revolting and the like to describe our reactions to a food or a food situation, and nouns such as goodness, freshness, purity, wholesomeness and richness, convey familiar day-to-day concepts which cannot be measured by physical or chemical parameters. They derive from the language of the senses and describe sensations which may be pleasant, unpleasant or neutral; but there are also overtones of our emotional response to these sensations in the context of use of such descriptors.

Furthermore, we place different values on different foods. By 'value' I mean the price the purchaser is prepared to pay. A litre of water in the middle of the Sahara desert is more valuable than a litre of water in, say, Paris. The difference between cost and price paid at the point of sale represents added value which the purchaser recognizes as one or another or a combination of characteristics such as appearance, flavour, purity, safety, nutritional value, convenience, price and (perhaps most important) previous experience of the product. On the integration of these depends the decision on whether or not to purchase.

Scientists are employed in the food industry to help the processor ensure that his product is right in all these respects at that critical instant in time when a purchase decision is made. Food scientists and technologists have traditionally assumed that the organoleptic properties of food, that is to say its colour, flavour and texture, are the dominant factors influencing that psychological mini-drama, the instant of decision to purchase. Much ink has been used to publish a host of techniques of measurement of organoleptic properties, and from the programme of the present meeting as seen in draft form, this is still the view of the majority. While not dissenting from this view, I wish to explore other more general aspects of food quality which in the long run may do more to influence long-term trends in food selection than the immediacy of hedonic choice.

2. ATTRIBUTES OF FOOD QUALITY

2.1 Safety and health

The desire for a long and healthy life is a universal human characteristic. Happily, there are widespread signs of an increase in life expectancy in industrialised societies. Thus in my own country of Scotland (whose health record so far as degenerative diseases are concerned is not good), the number of citizens in the over-85 age group was 19,800 in 1951 and 42,500 in 1978, an increase of around 115% against an increase in total population of about 2% (3). Indeed, in many countries in Europe the problem of caring for the elderly has become acute. General awareness of this phenomenon is associated with an equal awareness of the links between food and health, a situation encouraged by propaganda on healthier eating from government sources in many countries, coupled with media coverage of specific illnesses such as cancer and heart disease.

All this has produced what is no better than public obsessiveness about food safety. Anxieties are compounded by labelling regulations which require manufacturers to list additives either under EEC reference numbers which may suggest sinister secrecy to the purchaser, or as bad, under chemical names which conspire to confuse. Even positive factors such as declarations of nutrient content are presented (as required by law) in ways sometimes so tortuous as to be incomprehensible without the use of a calculator, even to me as a professional.

Thus suspicion is confounded. To Mrs. Everywoman looking at a label on a supermarket shelf, a declaration of "80 mg per 100 g of ascorbic acid" may suggest some nasty chemical, added by the manufacturer for nefarious purposes. It would illuminate, at the expense of a few extra words, to say "100 g of this product provides your recommended daily intake of vitamin C". The pedants and lawyers seek descriptions of scientific and legal precision for reasons they consider incontrovertible. But the user may be abused and confused by this very process of supposed user-protection. A sense of proportion seems essential, and we can easily lose this. The man in the street seems far more afraid of food additives than food-bourne infections. Yet proved illnesses from the use of legally permitted food additives are almost unknown, and where they have been tentatively identified they have been associated with rare and invidual idiosyncracies of allergies.

On the other hand, food-bourne infections are common, 10-12,000 cases being notified annually to the authorities in the UK and similar incidence rates are to be expected in other European countries. But there seems little general concern, and attempts at producing awareness and more care for food hygiene make too slow progress. These facts illustrate the lack of a due sense of proportion on the part of the public, whose desire for health and a long life makes them an easy prey to food fashions often born of magazine columnists and reared by commercial interests ready to exploit the public credulity. No food scientist worthy of the name would scorn some of the simple and wholesome foods offered in so-called health food shops although he might have reservations about their prices, sources and history. However, all food scientists must react at a philosophy which often (not always) infers magical properties to eccentric diets advocated by groups ill-qualified by training or experience to say anything about nutrition. The credulous, like the poor, will always be with us. To use science wisely to benefit our fellowcitizens is our objective. Perhaps an important part of our work is outside the laboratory and in education so that these selfsame fellowcitizens know the truth about the quality of their food supply. It seems clear to me now, and I venture to suggest that by the turn of the century, it will have become obvious to all, that food science must become a fundamental part of the school curriculum in all civilised societies. Only thus can our future citizens hope to optimise their diet for health and safety, the first and most important factors in any considerations of food quality.

2.2. Eating and well-being

I have argued elsewhere (4) that the sense of well-being we enjoy after a good meal is only partly due to the satisfaction of hunger. In part, I submit, it is due to the presence of traces of substances in many foods which have a pleasureable effect on our central nervous system. Such effects have been known since prehistoric times and have been exploited by all races of mankind. In dramatic form we call such substances drugs. It is seldom realised that a large proportion of the plants habitually used for human food contain drugs, but at such low concentrations that when consumed, we remain in control of our higher nervous centres. I describe such substances as the psychoactive constituents of food. Some we are all familiar with - the caffeine in coffee, the caffeine, theobromine and theophylline in tea, the ethanol found in food situations (as opposed to the ethanol deliberately consumed in beers, wines and spirits); all food scientists must be aware of the presence in substantial amounts of that powerful hallucinogen, myristicin, in nutmeg, but few seem to be aware that such psychoactive substances are to be found in many vegetables and fruits, in most root crops, in some species of fish and in cereals such as wheat and rye. Their identification in food is mostly a matter of straightforward chemistry (4) and their presence is beyond dispute. There is evidence (4) that some foods are deliberately sought for their stimulating properties. Coffee is an example familiar to us all, but outside of Japan, few know of the exotic sensations produced by eating fugu fish when skilfully prepared, or of the dangers of acute poisoning from eating the same fish prepared by a chef unaware of the extraordinary drug which it contains.

I have argued on the basis of much more extensive evidence than I can cite here that our enjoyment of food is only in part due to the satisfaction of hunger (4). In part it is due to the pleasureable sensations of these substances in our diet. If this thesis is correct, the presence of these substances which are relatively neglected by food researchers, are every bit as important and significant as the familiar attributes of colour, flavour and texture which continue (and rightly so) to be a major focus of attention.

An exception (which some will suggest throws doubt on this thesis) is mammalian meat, in which, to the best of my present knowledge, no psychoactive substances have been identified, yet which seems to play a peculiarly satisfying part in human food consumption. Furthermore, the primates as a group are essentially herbivorous, only man being a significant meat-eater. If "instinct" means "unlearned behaviour" there are evolutionary reasons for suggesting that meat-eating has played a highly significant role in the survival of the early human types from whom we are descended and also in the development of the modern human brain (5, 6). Some will argue that the evidence cited for this is suggestive but inconclusive. On the other hand, the evidence that humans can live long and healthy lives on a vegetarian diet is overwhelming. Equally, the evidence that meat consumption increases as increasing income provides greater freedom of choice is equally well established (7). It has been suggested that the drive to eat meat is a relic of human evolutionary development (5) and is 'instinctive' in the sense in which this word is used in this paper. Substantiation of this awaits further study. In the meantime, it is enough to note that for all ethnic groups so far studied, the inclusion of animal protein in reasonable amounts is regarded as an essential constituent in a diet of quality.

2.3 Psychological and emotional factors

That food plays an important role in the human emotions finds substantiation in everyday life. Great joys are marked by great feasts, and fasting has long been a traditional expression of grief. It is therefore not surprising that differing religious ceremonies have codified feasts and fasts within their separate disciplines in all parts of the globe and within almost all religious manifestations. Most religions have taboos or forbidden foods, some, like pork to the Jews, apparently having their origins in codes of hygiene, one of the earliest examples of these being the Jewish code described in the Book of Deuteronomy, chapter 14, verses 3-21.

This being so, it is not surprising that there is a sense in which food is sacred and to contaminate it is a form of desecration. We therefore react emotionally in the face of contamination which is physiologically harmless, such as a few rat hairs or insect fragments in a loaf of bread. Fortunately, so far as the general public is concerned, the rodent hairs and insect fragments are invisible without the use of special laboratory techniques to demonstrate their presence. Nevertheless 'pure' as a word applied to food has strong emotional overtones rooted in folk history and custom, and while to a chemist the only pure food is highly refined cane sugar, to a housewife and mother a major expression of love for her husband and family lies in cooking for them pure, wholesome food as she understands the terms.

Scientists are not exempt. Working as a young scientist in the food industry, I lived for a time in lodgings. In the particular city where I worked lodgings were difficult to find, yet I once left excellent rooms in an attractive district because my landlady, having put coal on her fire (it was long before central heating became common in Britain), served my meal leaving a thumbprint of coal-dust on the edge of the plate. I can still see the mark in my mind's eye and still remember my feeling of distaste. Yet I knew that even if I had swallowed a few milligrams of coal-dust it would have done me no harm. Nor was it because I had led an over-protected life, since I had often eaten roughly enough around a camp-fire. Thus we may be shocked by harmless contamination because it represents lack of respect for the special character given to food by human emotional attitudes, and also, perhaps, by the feeling that carelessness to harmless contamination also means indifference to the possibility of dangerous contamination. This concept of the special emotional context of eating can usefully be styled 'food aesthetics', and is an essential component in any consideration of food quality. Occasionally, certain forms of food processing offend aesthetic susceptibilities. The early history of milk pasteurisation affords interesting examples of this where its opponents objected on the grounds that the process destroyed the original "wholesomeness" and "goodness" of the milk as the cow gave it. All this despite the incontrovertible evidence that unpasteurised milk was causing 2000 deaths each year in the UK from tuberculosis (8).

Another psychological factor which influences ideas about food quality is age. This is not simply a matter of the diminishing sensitivity of flavour and odour receptors with advancing years, which certainly affects some but perhaps not all. There seems to be a generation gap where food is concerned. The modern European under about thirty years of age is not only developing a new lifestyle but new food habits to go with it. Furthermore, the old sex roles are changing. No longer is the housewife always in charge of the kitchen or in purchases in the supermarket. Despite the classical role of the mother as cook, lots of men enjoy cooking and some excel at it. Role reversals arising from the positions taken by the women's liberation movement have given men their chance and many have been quick to seize it. After all, all the great and famous chefs have been men. In a new sense, men are entering what has been a woman's world and are steadily exerting a new and lively influence upon criteria of food quality. I cannot offer a formal sociological study in support of this, but the phenomenum is obvious enough to those who read the cookery columns in newspapers and magazines or who watch trends in television programmes on this subject.

An additional and welcome factor which is accelerating these changes in taste is the influence, on the one hand, of ethnic minorities, and on the other, of foreign travel. The young intellectuals who are setting these trends, despite present problems of unemployment, are far more widely travelled by the age of twenty-five than were most of the previous generation by the age of forty. For those of a liberal or radical disposition, it has become fashionable to cultivate acquaintances and friendships across what were previously significant barriers of race and creed. All these factors and others which need not be detailed here are bringing about what can be described as a crossbreeding of cuisine - Indian, Chinese, Japanese, Malay, Indonesian - with European, all alike influencing ideas on what constitutes good food, or in other words, ideas on food quality.

2.4 Hedonic quality

The colour, flavour and texture of food, the familiar criteria of the tasting panel, reflect that mixture of sensations which begin when the smell of a dish first reaches us, when the sight of it pricks our appetite, when the taste of it reveals its texture and whether it stimulates pleasurable pain by its temperature or by the hot piquancy of peppers. The sensations do not end with tasting and chewing to release more flavour. Swallowing is also part of the process, because not only does the act bring into play further taste-buds located at the back of the tongue, but it also may send a little gust of aromatic air over the olfactory epithelium, thus intensifying the whole sensual experience.

Nor is the act of swallowing a neutral and unimportant one, the food being conveyed to the stomach by that wavelike movement of muscular contraction known as peristalsis. A man upside down and strung up by the heels can still eat and drink if offered food, if he should have a mind for it in this strange situation.

From its very nature then, a scientific taste panel is but the ghost of an eating situation. Rarely may the panel member even swallow the food for he would soon be sated. Never is the tasting done at a table, surrounded by friends, with the wine-bottle circulating and good conversation flowing as freely as the wine. Such circumstances make carefully subjective assessments impossible. But the feast with friends is real, the taste-panel situation an abstraction.

Yet so are most of our most useful measures. Five kilometres as measured on a map may mean a morning stroll on level paths well-suited to an elderly gentleman; in a rough mountain valley, five kilometres might tax the same gentleman to his limit. On a crevassed glacier five kilometres could be beyond the strength of a superbly fit young man of twenty-five. The measures are the same. The situations are different.

So with taste panels. They are maps of the gustatory world and as such, abstractions from the original. As is well known, taste panels can best be used to answer two kinds of questions. The first is the commonest 'Is sample A the same as or different from sample B? If different, describe the difference in precise terms.' The second 'Of samples A, B, C, etc., which do most people prefer?'. The first problem requires a relatively small number of specially trained tasters. The second, a large number of normal consumers. None of these tests measure our pleasure in eating although the data they generate may give us much information about preferences, about the hedonic consistency or variability of the output of a food factory, about variations in flavour response between the sexes, in health and disease, between sedentary and athletic occupations and between social groupings.

From their nature, organoleptic methods are crude indices when compared with the elegance and precision of physical and chemical measurements. Despite the use of statistical techniques to extract every possible shade of meaning from the voluminous data often generated by tasting panels, the results at best can but reflect the capricious nature of the operation of the human senses. How often over the past half-century must food technologists have dreamed of a quality meter which, by a magic pointer on a wondrous scale, would give an instant read-out of the flavour and texture properties of food products on the production line instead of having to stock-pile while slow, laborious and expensive organoleptic checks are made? Granted, some fairly rapid techniques have been developed for objective measurements of colour and texture but in the main, they are specific to a product or product group rather than general in application and are sometimes less precise than would be expected. Gas chromatography fingerprint techniques have been employed to follow flavour consistency in alcoholic drinks and other products where the pattern of peaks, rather than the identification and quantitation of individual components is the index against which product variation can be judged.

The safety of food, the sense of well-being or discomfort which it confers and its psychological and emotional connotations are factors in food selection which are every bit as important as its organoleptic properties. The control of the latter may be sufficient for practical purposes when established product types are being dealt with, but are insufficient when new products are under consideration.

3. ATTRIBUTES OF FOOD PROCESSING

Wheat and barley were first cultivated for human consumption about 10,000 years ago. The grinding of grain to flour is an extremely laborious process and the technology of mechanical engineering had its origins in attempts to use water and wind to ease the strain on human muscles. Similarly, chemical engineering began in Egypt with the development of products such as butter, cheese, wine and beers about 4,000 years ago (9). By Roman times gourmet cheeses were being sought after and smoked cheeses were especially favoured (10). The search for fine quality foods and wines is therefore not new and processing of a sort is at least as old as civilisation itself.

Modern food processing began as a 19th century response to the problems of providing food for the cities and towns whose populations had grown with uncontrolled rapidity to man the new factories of the industrial revolution. In the early days of the new food industries, unchecked by safety legislation or concepts of nationally-imposed food standards, quality in the overall sense of this collection of papers was not a major factor, but by the early decades of the present century two separate markets could be discerned. One was a mass market for products such as bread and preserved (mainly canned) fish, vegetables and fruits of which the then more exotic canned peaches, pears and pineapples predominated. This market was for comparatively low-priced foods produced with increasing degrees of mechanisation on a large scale, profitability depending on high turnover with low margins. The second was a gourmet market for which more exotic foods from all around the world were packed with care and with close attention to the kind of labelling which suggests sophistication and social superiority. These foods were often referred to as "quality" foods, thus both debasing the word and implying that other foods were inferior. This was unfortunate and for a time clouded early ideas of quality control and quality assurance, which, after all, are merely descriptions of the systematics of achieving as high a standard of quality as is compatible with the price at which a given food product is to be sold.

It would be expected that processing would generally lead to a loss of quality and this is, broadly speaking, the case. However, some processed flavours and textures came to be liked as something different and not necessarily inferior to the fresh raw material. For example, corned beef, canned salmon and some kinds of canned beans are quite unlike the original dishes they sought to simulate but they have come to be regarded as products in their own right and with their own collection of desirable properties. For most products, on the other hand, processing involves a loss of quality which must be kept to a minimum if the product is to be successful on a very competitive market.

During the past two decades, increasing disposable incomes in European countries have brought the two kinds of market closer together, many families shopping in both. This has had the further effect of making the mass market more discriminating and selective in its purchases and has put ever-increasing pressures on food technology to reduce quality losses on processing. An important result of this is the development of extended shelf-life products as opposed to foods preserved by conventional means for long-shelf life.

Of the three techniques available for this (chilled storage, low water activity and irradiation 'pasteurisation'), the former is at present the most widely used, because the technology is well-understood and most shops and supermarkets have existing means of exploiting it. Control of water activity has received much study during the past decade, and while the technique has so far found favour mainly in the production of petfoods, the concept has yet to make a major impact on the human food market. Irradiation at dose levels of up to ten kilograys is already approved for specific product types in a number of countries and the technique seems certain to come into widespread use within the next ten years. There is no reason, in principle, why combinations of these techniques should not be employed to extend the range of possibilities beyond the present practical limits, thus introducing a broader concept of high quality extended-life products of almost every conceivable type.

4. EDUCATION AND TRAINING

Most university courses in food science and technology pay a considerable amount of attention to food quality and food quality assurance. But classroom teaching is no substitute for practical experience. In the United Kingdom, the mid-seventies saw a wide-ranging debate on the need for a very high level qualification in this area. The Institute of Food Science and Technology of the UK set up a working group to study this question, and to recommend what action, if any, it considered to be appropriate. The committee proposed the establishment of a high level qualification to be known as the Mastership of Food Control whose holders should be entitled to designate themselves MFC. The Royal Institute of Chemistry and the Institute of Biology became engaged in the further discussion of these proposals and as a consequence, a Joint MFC Board was set up to implement them and to be responsible for their subsequent administration. Detailed regulations, syllabuses and study guides were prepared and a Board of Examiners (eight in all under the chairmanship of a chief examiner) was appointed. Details of the new programme was announced in a press release dated 1st December, 1977, the various arrangements becoming effective on 1st January, 1978.

The examination system falls into five parts, the first being a requirement of five years of appropriate industrial experience. The second is a formal written examination of four papers set on the published syllabus and requiring the candidates to show competence in the chemistry, biochemistry and properties of foods, in food microbiology, in food processing and in human nutrition. This is a formidable examination set at university honours degree standard, but holders of approved university degrees may be exempted. The third part is a further written examination, which may not be taken in the same year as the second and which also consists of four three-hour papers dealing with all aspects of food control and food legislation and enforcement. On completion of the first three parts, the candidate submits a dissertation on an approved aspect of food control and based on his or her own practical experience. Finally the candidate has to submit to an oral examination conducted by a panel drawn from the Board of Examiners (11).

With five sets of requirements to satisfy and with the high standards demanded by the regulations, an elite of food quality assurance specialists is being created at a level which the Joint Board believes to be the highest formal qualification of its kind in existence anywhere. As at 1st July 1983, forty-four candidates had satisfied the examiners, forty-one being resident in the United Kingdom, one in the Republic of Ireland, one in the United States and one in Canada (12). Since the qualification originated in the United Kingdom, it is not surprising that most of the holders are British. However, the examination system is open to citizens of any country whose competence in English is adequate for the purpose. There is nothing in principle to prevent the conduct of these examinations in other languages if the demand justifies it.

The control of food quality is of special significance in international trade. If present trends continue, the international use of agreed high-level qualifications of equivalent standard seems inevitable. The British initiative in this field is not intended to be exclusive, and if other countries wish to introduce a similar qualification for the benefit of their own food industries, we would wish to co-operate with them with the objective of facilitating international trade in foodstuffs. In the meantime the UK qualification is available to all competent to benefit from it.

REFERENCES

1. PYKE, M. (1959). Food quality - an attitude of mind. In 'Quality Control in the Food Industry' (J.M. Leitch, Ed.), Proc. Summer Course, Glasgow: University of Strathclyde.
2. HAWTHORN, J. (1967). The organisation of quality control. In 'Quality Control in the Food Industry' p.4 (S.M. Herschdoerfer, Ed.), London: Academic Press.
3. ANON. (1980). Annual Abstract of Statistics, p. 11-15, London: HMSO.
4. HAWTHORN, J. (1981). Pharmacologically active substances in food. In 'Criteria of Food Acceptance,' Zurich: Forster Verlag.
5. HAWTHORN, J. (1972). The carnivorous ape. SIK-Rapport Nr 302. Gothenburg: The Swedish Institute for Food Preservation Research.
6. HAWTHORN, J. (1976). A look forwards and backwards. SIK-Rapport Nr 399. Gothenburg: The Swedish Institute for Food Preservation Research.
7. ANON. (1964). Protein at the heart of the world food problem, p.24,25, Rome: FAO.
8. SMITH, J.A.B. (1978). A brief history of the Hannah Research Institute p.25, Ayr, Scotland: The Hannah Research Institute.
9. HAWTHORN, J. (1978). A history of milk in the food industry. Proc. Nutr. Soc., 37, 211-215.
10. BROTHWELL, D.R. and BROTHWELL, P. (1959). Food in Antiquity, p.52, London: Thames and Hudson.
11. ANON. (1983). Regulations, syllabuses and study guides for the Mastership in Food Control. (Obtainable from the Institute of Food Science and Technology, 20 Queensberry Place, London SW7 2DR).
12. ANON. (1983). Focus, 16, (3), 49. London: Institute of Food Science and Technology.

FOOD RESEARCH IN THE EEC REGIME

P. S. GRAY
Head of Division III/A/2
Directorate-General for internal market and industrial affairs
Commission of the European Communities

I am frequently asked why the EEC, which is an economic entity should concern itself with research and in particular with research in the food sector. Economics is basically concerned with allocation of resources and satisfying man's needs and aspirations and of the three basic needs food, clothing and shelter, food is the most fundamental, for without it our life expectancy can be measured in days and our other needs and aspirations are of little consequence. As food scientists you can congratulate yourselves in working in a product area which will never go out of fashion.

We are fortunate to be holding this symposium in Greece which was the cradle of western civilisation and two millenia later the onset of the industrial revolution brought with it a concentration of population in towns which demanded the creation of an industry for processing and a commerce for distributing food for our new urban society.

The process of industrial development which had been accelerated by two wars took a change of direction in the 1970s under the pressure of resource limitations, environmental impacts and the introduction of new technologies.

The population of the Community which had been rising is now virtually stable and we can expect the overall quantitative demand for food to stagnate and even slowly decline as the age structure of the population matures. One of the basic policies of the EEC Treaty the Common Agricultural Policy which had set out on a path of improving productivity and reducing deficits in many areas finding itself faced with growing surpluses. The food sector holds a key position in a rapidly changing society between the production regimes of the Common Agricultural Policy and the consumer. For the vast majority in western Europe the balance has swung from eating for survival towards eating for enjoyment or social reasons. The achievements of the food sector have created a paradoxical situation in which the very abundance and variety of food has led to a greater life expectancy where the under nutrition is having less of an impact on public health than overeating or unbalanced diet.

Food science has an important role to play in the feeding of an urban society, in the reconciliation between supply and demand and in the improvement of food quality and the nutritional requirements of the population as a whole.

It is not suprising that the Commission should regard research in the food field as of vital importance for the Community particularly at this stage of rapid mutation of the economy.

On 25 July 1983 the Council adopted a framework programme for a scientific and technical strategy for the period 1984-1987. This is a first attempt by the Community to devise a general research and development policy in which all the various subsequent individual actions can be set.

The programme identifies seven main objectives of policy:

- The promotion of agricultural competitivity
- The promotion of industrial competitivity
- Improvement of raw material resource management
- Improvement of energy resource management
- Reinforcing of development aid
- Improvement of living and working conditions
- Improvement of the application of the Scientific and Technical potential of the Community

Publicly financed research expenditure by the Member States amounted to some 26.5 thousand million ECU in 1982 compared with 38 thousand million in the USA and 7.5 thousand million in Japan.

The industrial expenditure in RD is about equal to that of the public sector in both USA and Europe whereas in Japan industrial financed research expenditure is more than double that in the public sector. It would be simplistic to say that the Community should emulate either one of our trading partners or competitors but these facts, should stimulate us to carefully examine the extent of public spending on research.

The economic and political situation in Europe is vastly different from that of either Japan or the USA. The diversity of the Community is a source of strength in terms of scientific inventiveness but a source of weakness if we do not realise the need to avoid the wastages of duplication and to exploit the advantages of cooperation.

It is clear that,at the sharp edge of industrial production, research along with manufacturing and marketing play a vital role in determining industrial competition at enterprise level and that in a mixed economy this aspect is best left to private business. The corollary is that publicly financed research cannot but benefit from cooperative effort and that much is to be gained by pooling resources in times of scarcity.

The COST (Cooperation in Science and Technology) framework gives the Community the possibility of enlarging this cooperation to other European OECD countries. The food sector where traditions and approaches vary so widely in Europe is an obvious candidate for this wider cooperation to which the success of COST 90 and 91 bears witness.

On the wider international scene the seven nations of the Versailles Summit, France, Canada, USA, Italy, Japan, the United Kingdom and Germany together with the EEC Heads of State identified food research as a vital sector in Technology Growth and Employment and set up an ad hoc Working Group with France and the United Kingdom as leaders.

The Working Group selected three themes for cooperation, food processing, the evaluation of additives and the problem of developing countries. For food processing they suggested that cooperation would be the most viable route. In the area of additives a coordination of fundamental research on the verification of test methods and a deeper understanding of the mechanisms of toxicity was proposed. For developing countries specific attention should be given to improving methods of production, a better use of raw materials, the development and application of storage of agricultural products and the reduction of food losses.

Within the EEC three administrative techniques of research cooperations are in use. Direct action involves projects carried out through the Communities' own joint research centre at Ispra which has extensive advanced facilities and staff, for example in domains such as analytical chemistry. Financing is on the basis of budgeting headings which may be specific but also may be for general technical support to the Commission's activities.

Indirect action research programmes involve a partial financing of research projects by the Community. On the basis of a carefully prepared proposal from the Commission the Council adopts a programme together with a budgetary committment spread over several years.

The programme, which has a designated general theme, such as raw materials, is divided into a number of main headings each of which are subdivided. Funds are allocated to each main heading and there is a possibility of transfer within certain circumscribed limits. Invitations to tender research projects from institutions or industry are published in the Official Journal and a consultative committee is set up to assist the Commission to choose the projects to be financed and to manage the programme. Funds are also set aside for the dissemination of information on the conclusion of the research programme.

Finally you will all be familiar with the method of concerted action used for COST 91 in which the activity of the Community is one of bringing together existing programmes in the Member States or other participating countries. The financing is directed towards administrative expenses of bringing participants together, organising meetings and the secretariat.

Concerted action requires the least funding at Community level and therefore is the approach to which it is easiest to obtain the agreement of the Member States. It is however limited to the coordination of programmes already in existence and if, as all too often happens, a deficiency is evident in the general pattern of the programme, it is not possible to remedy it. Indirect action has the advantage of presenting a coherent programme conceived at Community level but the consequent higher spending involves a transfer of research resources from national to Community level which can make agreement of the programme a more time consuming and difficult procedure. Finally direct action is best suited to those cases where expensive or complex facilities are needed which would possibly be under-utilised on a national basis.

In the food sector, so far, Community research has been entirely on the basis of concerted action. The COST 90 programme on the effect of physical properties of foodstuffs was adopted by the Council on 25 February 1978. It was aimed at three physical properties, Water Activity, Rheology and Thermal Properties. The programme was carried out by the EEC together with Finland, Sweden and Switzerland. The programme terminated originally in February 1981 but was extended to the end of November 1981 and a final seminar was held at Leuven in September to discuss the work of the project. Its primary objective was to acquire and disseminate more and better data on the physical properties of foodstuffs and further to try to develop a better understanding of the relationship between physical properties and other factors which affect their numerical value.

The achievements of the project went well beyond the original expectations showing in some cases high degrees of consistency in the measurement of physical properties and in others disturbing and unacceptable discrepancies between equally competent experimenters.

COST 90 bis was decided by the Council on the 22 November 1982, for a further programme of four years on 3 new physical properties in particular, mechanical properties, diffusional properties and electrical and optical properties. Two non Community states are participating Sweden and Switzerland.

Since Professor Linko has dealt with COST 91 in detail in his opening speech I will just recall its launching date on 22 October 1979.

In July this year the Commission proposed a further concerted action programme COST 91 bis to the Council on the effects of processing and distribution on the quality and nutritional value of food. This programme is proposed to run for a total of 780,000 ECUs from the Community budget and 15 million ECUs from national budgets.

There are three main themes

- various new heat treatments
- a study of quality and nutritional properties of food produced by biotechnology
- refrigeration and refrigerated storage

The programme was prepared using the usual consultative mechanisms available to the Commission starting with the Coordinating Committee of COST 91 and passing through the Committee for Scientific and Technical Research CREST.

This project can be seen as a transitional project between the previous COST projects and the larger research programme envisaged within the 1984-1987 framework for research strategy that I mentioned previously.

To put the framework in its perspective it is necessary to look back a little. In 1974 the total EC funds allocated to research were 70 million ECU. This has risen in 10 years to around 600 million ECU a year. The planned overall committment for the framework programme is 3750 million ECU over four years at constant 1982 values, representing an increase from present levels of 50% without allowing for inflation. In 1982 research received 2.6 percent of the EC budget and if the above target is met by 1987 it should be around 4 percent.

Such a projected increase may seem unrealistic at first sight especially when the Community is under severe budgetary pressure. However across the board cuts never make good policy. It is important to analyse the effectiveness of expenditure and to rationalise resource allocation. A very small percentage reduction in agricultural support expenditure which is currently some 70% of the Community budget would amply cover the projected increase in research appropriations.

Since the problem in the Community seems no longer how to increase production of basic foods but how to convert and use them this would be a logical policy objective.

We are at present in the early stages of developing the coordinated programme on food research and I can do no more than give you general indications. One important area would seem to be in the developing of more coordinated and better methods for safety evaluation. As Community law in the field of food is covering larger areas it would seem logical to parallel the harmonization of law with a coordination at research level. As a longer term objective development of a better understanding of toxicological mechanisms and in vitro test methods could fulfil a double aim of increasing consumer protection and reducing animal experiments.

The Commission has sent two Communications to the Council on biotechnology. Although biological techniques are among the earliest food processing techniques the impact of new technology will be changing many classical methods of production and introducing new areas in the food sector. Coordination of research in this field would seem to be a very important element of a food research programme.

Other processing technologies such as ultrafiltration and reverse osmosis through membranes are being increasingly used in the food sector and could be profitably coordinated at European level.

Meat processing and fish processing are also being looked at as areas of interest.

Finally the influence of oxidation on the quality, nutritive value and safety of food in particular the second or higher order interaction between the various components of food such as proteins, carbohydrates, lipids, minerals, vitamins and gases. A better knowledge of these interactions would explain a number of phenomena recurring in food, but this is such a vast field that a preparatory stage would be needed.

This list is by no means exhaustive but is long enough to highlight the need to carefully establish priorities. To do this we will need the comments of researchers, institutions, industry and governments and we are just starting this consultation process.

I started my talk with some comments about the economic background to the food industry. We could be facing even more drastic changes in the years to come if work patterns are changed by information technology and more leisure is created there will be further changes in the patterns of eating. The challenges will not be lacking, let us hope we can agree to meet them together.

DUBLIN 1977 - ATHENS 1983, WHAT WE HAVE ACCOMPLISHED

DR. S. NIELSEN
Programme Director
National Board for Science and Technology, Dublin

Summary

Nature's materials which supply man's food are complex, unstable and defy simple analysis. The whole range of scientific and technical tools is needed to reveal their complex structures and properties, to master their transformation and to understand their biological utilization. There are few industries where research is as multidisciplinary as in modern food research. Scientific and technical research is becoming increasingly important in the food industry, and calls for greater effort by all involved if solutions to the problems of world hunger are to be found. The challenge facing the food industry includes satisfying the needs of expanding populations, developing foods appropriate to the multinational requirements and resources of all consumer categories, preserving and transforming edible raw materials and establishing firms capable of the optimal production and distribution of the foods. The COST 91 concerted action programme has gone no small way towards addressing that challenge.

INTRODUCTION

In my paper here today, I start by recalling some background statistics, collected by the Commission in the late 1970's. These do not include the non-Community states, but are indicative of the general situation.

- o In many of our countries the food industry is the most important sector of manufacturing industry.
- o In the EEC the food industry had a turnover of 1500 billion ECU or 18% of all manufacturing industry, accounting for 10% of the value added, 9% of the employment or 2.2 million persons and involved some 12,900 firms.
- o It processed 2/3rds of the Community's agricultural output and provides the basis for numerous up and down stream industries.
- o The industry is served by over 400 laboratories and centres of which approximately 100 are industrial and the balance associated with universities and research centres.
- o There are some 1500 research workers in these laboratories spending about 250m ECU with industry contributing half of this.

In view of the foregoing statistics which indicate how important the food industry is in Europe it is perhaps surprising that it took so long to develop international collaboration. It is interesting to note that the seven themes developed for the COST framework for co-operation in science and technology included neither food nor agriculture. This accounts for the fact that agriculture eventually became the 80 series and food the 90 series in COST.

Major Achievement

As Chairman of COST during the 1975 - 1979 period I can vividly recall the frustrations experienced by scientists and technologists at the delays in getting the project going. Difficulties were encountered on reaching consensus on technical issues, there were intractable administrative problems to be resolved in addition to difficult legal and institutional issues concerning the relationship of the Community and non-Community States. The first major achievement was in fact getting COST 91 started, with the sub-group quickly establishing their programmes and priorities.

Sub-group I

Sub-group I held 5 major meetings, involving participation of 20 laboratories of which 9 were industrial and collaboration on some 51 projects. The excellent papers presented here at this seminar, is evidence of the progress which has been achieved in understanding and utilizing the technology of extrusion cooking. It is a powerful new technology in food processing and we have had interesting glimpses of new industries of the future associated with it.

Sub-group II

Sub-group II had 6 major meetings with many additional ones involving the particular process groups and the important working parties on Terminology; Quality Assessment and Process Modelling. But as we have seen here in Athens, this summation belies the scope of the activities or the enthusiasim shown or the progress achieved by the group.

Sub-group III

Sub-group III had 11 major meetings, with participation by 37 laboratories on 36 projects. Again, this reflects the extensive scope and wide interest in the area of freezing and thawing technology and the importance of an effective exchange of information.

Sub-group IV

Sub-group IV had perhaps the most difficult task of attempting to bring together the nutritional aspects of the work in other groups, as well as to contribute to the work of the other groups. It is clear that exchange of information has been excellent and that the development of the Food/Nutrition Data Banks, the Analytic Methodology manual and the publication of the Nutrition Statement have been major contributions. It seems though, that the scientific and technical aims are harder to define in this area and it is interesting to note the decision to integrate nutrition research rather than separate it, in the context of COST 91 bis.

Overall achievements

Because of the many achievements of the overall COST 91 programme and especially at the various sub-group levels, it is evident that COST 91 has been exceptionally strong in the areas of information exchange; data comparability; interdisciplinary co-operation; work exchange and joint projects. I have no doubt that other factors such as personnel mobility; shared facilities; industrial participation and synthesis validation will be evident in the proposed COST 91 follow-up, (COST 91 bis).

Conclusions

There is no doubt but that food science and technology has benefited from this programme. The prospects for industrial uptake are very encouraging from the large industrial representation at this seminar. The road from Dublin to Athens started in Sweden, through the foresightedness of Eric Von Sydow when he put proposals to COST in 1974. Through the work of Professor Linko and Professor Jul and their Nordic collaborators the main lines of COST 91 emerged. Finally, it took the successful seminar in Dublin in 1977, so well organised by Dr. Downey, to reach consensus on the main priorities for COST 91 and so signpost the route to this seminar here in Athens this week. To all these people and the many others associated with the execution of the programme a great debt of gratitude is due.

EXTRUSION COOKING

INTRODUCTION AND CONCLUSION

J C CHEFTEL
Universite des Sciences et Techniques du Languedoc, Montpellier, France

Continuous high temperature-short time processing of foods, combined with aseptic packaging, has been rated as the most significant innovation introduced in the food technology area within the last 20 years. HTST processes are applied mainly to liquid foods, such as milk. Other recent processes are being used or developed for solid foods or foods containing solid particles, e.g. sterilization or cooking in scraped-surface heat exchangers. Extrusion-cooking is an HTST process. Its main effects are cooking and texturization rather than preservation. Extrusion is usually performed on powder mixes of relatively low moisture contents and results in the formation of solid foods of porous structure and of distinct shape.

Although extrusion-cooking is carried out within one piece of equipment it cannot be considered as a unit operation, because it performs a large number of functions as the food mix or dough progresses along the screw(s) and barrel : transport, grinding, hydration, shearing, homogenization, mixing, dispersion of residence times, compression, degassing, thermal treatment (with partial melting and plastification of the mix, starch gelatinization, protein denaturation, destruction of microorganisms and various other chemical reactions), compaction, agglomeration, pumping, orientation of molecules or of aggregates, shaping, expansion, formation of porous and/or fibrous texture, partial drying. These functions are generally performed with small energy requirements and no waste or effluent production.

The extruder can be considered as a complex chemical reactor operating at high temperatures, with short residence times, under high pressures, high shear forces, on food mixes having a wide range of moisture contents and viscosities. Extruders can be used for producing textured food items but also to improve the functional properties of food ingredients (preparation of precooked flours or starches, reduction of the microbial charge of spices, etc.) or even to carry out chemical reactions (starch hydrolysis, transformation of casein into caseinate, etc.). Recent extruders with appropriate screw profiles develop specific temperature, pressure and shear profiles, and can be used to promote sequential reactions, especially if provision is made for the introduction of water, reagents, thermosensitive nutrients, etc., at various barrel locations, during the extrusion process.

Derived from the plastic extruders and from the screw press used for pasta production, the food extrusion cooker has swiftly moved ahead, from the production of pet foods and snacks to that of textured vegetable protein, flat bread, breakfast cereals, biscuits and a number of other products (see list of present and potential applications). Thus, in the 1980-1983 period, the COST 91 extrusion-cooking Subgroup has both witnessed,and contributed to impressive developments : new applications, new products, new machines, better quality control, better understanding of the basic scientific mechanisms underlying extrusion-cooking, either from a chemical engineering standpoint (measurement of process parameters, modeling) or with respect to the biochemical modifications of food constituents. This period of rapid growth and innovation, characteristic of a recently born technology, has been challenging and rewarding, and we feel that the 35 following scientific and technical papers give good support to the previous statements.

In addition, the activities of the COST 91 Subgroup have constituted an exercise in European and scientific collaboration. Open to all those involved in extrusion-cooking, the Subgroup has facilitated exchanges between specialists from National laboratories and from industrial firms, between engineers and nutritionists, between development experts and research scientists, between senior participants and Ph.D. students.

Through a relatively small number of meetings often held during symposia or professional courses, the Subgroup has enhanced personal acquaintance between most European investigators in this field. Together with regular information on current studies (through progress reports, discussions, preprints, news from industry, indications on new machines), this has promoted improved individual knowledge on extrusion-cooking, mutual visits of facilities, some exchange of scientists, better implementation and interpretation of experiments, the preparation of guidelines for publications, joint publications, including review papers. Better planning of future research has also probably been achieved thanks to new ideas, less duplication, stimulation through competition, comparison of data, validation of results, interdisciplinary discussion of problems. A number of collaborative research projects have been carried out, although often restricted to collaboration within one country or between geographically close regions.

It is also likely that the Subgroup activities have resulted in an increasing number of research contracts between public institutions and industrial firms and in extra financial support from National administrations.

Although COST programmes are intended as primers and not for self-perpetuation, we are convinced that more can be gained in terms of understanding, innovation and industry uptake from further promoting European collaboration in the well defined yet rapidly expanding field of extrusion-cooking. Comparison with other processing techniques, joint research on applications in developing countries, joint contracts, setting up of temporary common facilities, loan of equipment from manufacturers, specifications for building a low cost research extruder, and many other items of common interest can be proposed for future collaboration.

RECENT APPLICATIONS AND RESEARCH PERSPECTIVES IN THE FIELD OF EXTRUSION COOKING

J. M. HARPER
Vice President for Research
Colorado State University
Fort Collins, CO 80523, USA

Summary

The historical development of food extrusion is reviewed. Special emphasis is placed on recent applications including candy making, co-extruded products, extrusion efficiency, pretreatment of starch for enzyme hydrolysis, extrusion of oil seeds for solvent extraction and twin-screw extrusion applications. Suggested topics for further research include: (1) work on extrusion equipment to reduce wear and improve the measurement of process variables; (2) a better description of cooking reactions involving starch and protein; (3) understanding the nutritional and biological changes in foods caused by extrusion; and, (4) miscellaneous factors.

1. INTRODUCTION

Food extrusion has been practiced for nearly 50 years. It has developed from its initial roles of mixing and forming as applied to macaroni production to the point where the food extruder is now considered a high-temperature-short-time bioreactor which transforms a variety of raw ingredients into modified food ingredients and finished food products. The impetus for these developments has come from modern food processing's need for: (1) continuous processes having a high throughput; (2) energy efficiency which often involves processing relatively dry materials; (3) modified textural and flavor characteristics of foods; (4) control of the thermal effects on food constituents; and, (5) new and unique food products. Because of the versatility of extrusion, it has found ever increasing applications in a variety of food processes.

2. HISTORICAL APPLICATIONS

Initial applications of the food extruder involved its use to reform degermed cereal grits into macaroni, snacks and ready-to-eat cereals. The latter products required cooking, texturing and/or shaping, revealing the expanded capability of the extruder to perform a variety of valuable operations in a single processing step. During the 1940s, a number of single-screw expellers, which squeeze the oil from oil seeds, were developed and refined, replacing the use of much less efficient hydraulic presses previously employed for this purpose. The development of several new single-screw extruders expanded their application in the 1950s to commodity-type products such as dry pet foods, precooked cereal flours and heat treated cereals and oil seeds to enhance their value as animal feed constituents.

The decade of the 1960s saw a focused interest on the unique characteristics of extruded foods. A variety of extruded food products were developed having new properties and improved process economics. These included increasing numbers of RTE cereals and snacks, croutons for soups and salads, an expanding array of dry pet foods replacing canned or baked biscuits, the extrusion of defatted high protein flours to create textured meat-like food materials, and the processing of nutritious precooked food mixtures for infant feeding replacing the more expensive roller dried products. An expanding need for precooked cereals and starches required larger capacity machines where the economy of scale could be more easily realized. Extruders with a nominal capacity of five tons per hour were developed and placed into operation. Special purpose extrusion devices to debone meat scraps were also developed which employ a conical screw surrounded by a micro-perforated screen through which ground meat was forced leaving the coarser bone scraps behind. Continuous bread making processes required special purpose extruders to develop, proof and inject the bread dough into baking pans.

Further refinement of the food extrusion process and expansion of its applications occurred during the 1970s. Examples include the formulation of soft moist pet foods, with propylene glycol, sugars, acids, salts and alcohols to control the water activity in the final product, requiring an extruder as a mixer and pasteurizer. Co-extruded products containing more than one component such as egg rolls and ravioli were developed. Control of texture through the use of modified starches, emulsifiers, salts and pH adjustment was practiced. The use of two extruders in series, the first for cooking and the second for forming/steaming, resulted in products having unique textural, water hydration and density characteristics. Examples include the production of meat analogs from defatted soy protein which could be used to replace meat in a variety of prepared food products.

The use of twin-screw extruders for food processing was of interest in the 1970s with an expanding number of applications in the 1980s. Although twin-screw extruders can come in a variety of designs, the co-rotating intermeshing screw type has found the widest acceptance. These machines have improved conveying and mixing capabilities using interchangeable screw profiles created by slipping various screw sections onto parallel shafts, making them very flexible and amenable to a variety of process applications.

3. RECENT APPLICATIONS

The variety of food products being produced on food extruders continues to expand. Recent applications have gone far beyond the forming, cooking, expansion and texturizing of cereal and vegetable protein ingredients. The newer applications often utilize the extruder's ability to convey and heat viscous materials. An example would be the continuous production of candies either through the anhydrous melting of sucrose or the continuous cooking of sugar solutions. In many cases, twin-screw extruders are used for these applications because the point of energy addition and pressure profile to prevent boiling within the machine can be closely controlled by the configuration of the screw.

To increase product variety, the co-extrusion of food products has been an area of increased interest. Filled snacks which contain an oil emulsion filling, RTE cereals with a fruit gel or fruit flavored component, and pet foods containing multicolored textured pieces to simulate bones or other meat products are examples. Such processes require the synchronization of two extruders producing the contrasting parts of the product which are combined in a special die. When one component is more liquid, a variable speed positive displacement pump can be used to inject that component into or around the extruded piece.

To increase the efficiency of extrusion systems, greater attention is being placed on operating them at maximum capacity while utilizing a minimum amount of energy for the entire process. The optimal set of operating conditions are site specific due to differences in the cost of various forms of energy and money. Holay and Harper (8) have described the basic procedures to minimize energy costs. One cost extreme occurs with wetter extrusion, where steam additions to the feed mixture in a preconditioner or through the extruder barrel wall add most of the energy in the form of latent heat, thereby reducing the electrical energy required to turn the screw but increasing needs for drying capacity and energy. The other is a drier extrusion with increased electrical energy requirements to turn the screw because the dough is quite viscous, making it the major source of heat to the system. Under dry extrusion conditions, system drying requirements and related energy are reduced. Optimal conditions fall somewhere between these two extremes.

The cooking of starchy materials to increase their susceptibility to enzymatic hydrolysis has become an important preprocessing step for fermentation processes. The extruder has shown itself effective as a technique to alter starch under relatively dry conditions, thereby reducing the energy necessary using the normal high moisture process since the dry mass of the material being extruded is substantially less (11). One further advantage of the extrusion cooked material is a lower initial viscosity of the mash being hydrolyzed making it possible to avoid the expense of adding α-amylase to reduce the viscosity necessary to increase processing efficiency.

Extrusion of oil seeds, such as soy, has been shown to improve the solvent extraction process. During extrusion the cells are ruptured, exposing the oil while making the resulting meal more dense. The latter allows greater quantities of meal to be loaded into the extractor increasing its throughput. Extraction also is enhanced because of an increased solvent percolation rate through the bed even with concentrated miscella. With improved drainage rates and overall extractions, reductions in desolventizing costs and increases in oil yields result.

Another recent example of the application of extruders to the vegetable oil industry is their use to stabilize rice bran. Rice bran exists as a very light powdery material resulting from the polishing of rice. It contains approximately 20% oil but has a very strong natural lipase which hydrolyzes it to free fatty acids in a matter of hours once the bran is removed. Extrusion at 140°C inactivates the lipase (16) and densifies the product making solvent extraction feasible and creating a new vegetable oil source for Southeast Asia which generally has an edible oil deficit.

The baking lines required for the production of flat breads and crackers can now be replaced by the food extruder. The extrusion process is continuous, high capacity, sanitary, energy efficient and capital cost saving. The twin-screw extruder has been the most successful type of equipment for these operations. The processing of dry ingredients greatly reduces the drying/baking requirements following extrusion with only a short baking step necessary to increase flavor and reduce moisture. To provide pieces with precise shapes, the ribbons of product are run through gaging rolls to control thickness and width.

4. RESEARCH PERSPECTIVES

Research in food extrusion has been primarily confined to projects searching out new applications and defining the basic concepts of the process during the formative years of the field. Relatively little basic work in the area was undertaken until the 1970s. Academic and research institutes have only recently become interested in food extrusion and equipped their laboratories with the necessary equipment. In contrast, thermoplastics extrusion used to form and process many polymeric materials from the petrochemical industry had been receiving extensive theoretical treatment which led to process and product improvements. Even today, food extrusion has not developed into the science which characterizes much of the plastics extrusion field. Instead it remains largely empirical although increased understanding and analytical description of the food extrusion process is developing.

This section of the paper will summarize some research challenges which exist in the food extrusion field. In all respects, food extrusion is more complicated than plastics extrusion because of the variability of food materials, the use of multicomponent systems where the ingredients interact and change during processing, and the wide variety of raw materials which are extruded. To overcome these complexities, academic and research institutes should concentrate on basic extrusion work of general applicability and leave specific product development work to industry.

4.1 Extrusion Equipment

One of the continuing problems plaguing the users of extrusion equipment has been the wear of screws and barrel pieces over time. Worn components reduce the capacity of extruders, alter the properties of the products produced and increase the probability of surging of the extrusion equipment. Wear or erosion of metal parts is increased by the use of acids or salts as part of the raw ingredient mixture. Research, in collaboration with metallurgists, needs to test new hardenable corrosion resistant materials and develop improved and less expensive methods for rebuilding worn extrusion parts.

The theoretical analysis of single-screw extrusion theory shows that extruder throughput would be increased if slip could be induced at the root of the screw. Teflon coatings on screws demonstrate the phenomenon but do not have sufficient wear resistance to generate any commercial interest. New metals which contain embedded polymers appear promising as aids to increase slip at the root of the screw and require further evaluation.

Measurement of pertinent extrusion parameters continues to be a limiting and expensive proposition for many extrusion operations. Thermocouples can measure dough and barrel temperatures but the accuracy of these measurements are often questionable because of poor placement of the thermocouples, the influence of surrounding temperatures such as heated jackets and the mechanical fragility of their tip. Pressure measurement behind the die, essential for on-line interpretation of dough rheology, requires expensive devices which are easily damaged by exceeding pressure limitations, require frequent calibration and are often temperature sensitive. Although motor power can be measured with a recording watt meter, variations in efficiency of drives and bearings between the motor and screw make it difficult to determine the actual mechanical energy being added to the system. The list could be expanded to include improved measurement of water and steam rates. Further research is needed to give manufacturers and users improved recommendations on measurement of key extrusion variables.

Many extrusion processes are operated with minimal control and rely extensively on manual operator adjustments to maintain uniform processing conditions. The use of computers to monitor and adjust extrusion conditions based upon multivariable control algorithms is only in its infancy (14). Basic research to define effective control schemes needs to be undertaken to meet the requirements of improved extrusion operations and consistent product uniformity.

Several extruders allow an infinite variety of screw designs, varying screw lengths and barrel configurations. Empirical rules exist which allow product developers to make selections from these vast arrays of possibilities. Further research is required for users to knowledgeably select these mechanical features based on product characteristics such as particle density, size, friction factors, cooking requirements, changes required during processing, energy input, mixing, extrusion rate, etc.

Dies are currently designed empirically or through a series of trial and error steps. Finite element computational methods exist which should allow more precise die design and optimization if an adequate rheological model including elastic effects is known as a function of shear rate and temperature.

4.2 Cooking Reactions

The thermal processing of the raw food ingredients in an extruder occurs under high temperature and limited moisture conditions. The use of the differential scanning calorimeter (17) and X-ray diffraction (12) techniques are providing new answers to the changes occurring at conditions relevant to extrusion processes. To extend these studies over the entire range of conditions found in the extruder, good experimental procedures are needed which allow the independent control of time, temperature and shear environment used to produce experimental samples. Once these effects are understood and translated into mathematical representations, their use to optimally design, operate and control extrusion processes will follow.

Considerable work has been expended to study the changes in starch after it passes through extrusion processes (2, 6, 13). These data suggest that the loss of crystallinity in the starch granule occurs by a different mechanism than it does during classical high moisture gelatinization where the granule swells in excess water allowing the amylose extrudate to migrate into the water to form a gel. Mechanical disruption of the molecule in the intense shear fields within the extruder screw and die appears to be a more likely mechanism with only limited water available. Because of the molecular size and conformation, amylopectin appears to be affected to a greater degree than does the more linear amylose. Monoglycerides can interact with the amylose in extrudates to form complexes exhibiting unique properties (12). Finally, increasing shear rates through dies can change the texture of expanded cereal grits from an open cellular flinty structure to a soft, highly water absorbent product. These products have different mouth feel, water absorption and enzyme susceptibility characteristics. Further research on the changes occurring in starch in a low moisture extrudate as a function of time, temperature, shear or added ingredients will be important to the design of improved extruders and development and control of extruded products.

The restructuring of protein molecules has been pictured as an initial denaturation where the native protein unfolds allowing it to align in the streamlines of flow within the extruder screw channel (1). Crosslinking between adjacent molecules occurs in the form of both ionic (4) and covalent (15) bonds between appropriate reactive sites. Holay and Harper (9) have shown high shear rates through dies can disrupt these crosslinks after they are formed to significantly alter the structure of the product. These observations point to the need for research to describe the restructuring of protein in shear fields.

A fruitful approach to future research work would be the description of an extruder as a continuously stirred chemical reactor. Such a model coupled with appropriate reaction rate kinetics as influenced by temperature, shear, water activity and composition could serve as a powerful design tool for extruders. The environment existing within the extruder including temperature profiles, residence time distributions, and shear environments could be modeled using some of the techniques developed for thermoplastic extrusion coupled with the chemical reactor concept.

Observations from existing extrusion data would indicate that cooking reactions are greatly influenced by the composition of the feed materials. Often minor ingredients such as electrolytes, acids/bases, fat, surface active agents, amino acids, and reducing sugars can cause pronounced effects on the extruded product. Examples include the use of small amounts of sodium bicarbonate to increase flavor development and expansion of cereal based products, adding monoglycerides to control puffing of snack items, and altering water content to control product texture and flavor. Research to quantitatively elucidate the effects these minor ingredients have on the extrudate would be of general interest and benefit to the food industry.

4.3 Nutritional and Biological Factors

Extrusion has been used to beneficially alter the nutritional value of extruded foods and has been reviewed by Harper (7). Despite a considerable body of literature on the impact of extrusion on the general nutritional value of foods, the specific details are usually unknown. For example, loss of available lysine has been studied by Thompson et al. (18) but kinetic models for lysine loss have not been applied to actual extrusion data with any degree of accuracy. A research topic for the future would be the kinetic study of various enzymes' inactivation, antinutritional factor destruction such as trypsin inhibitors in soybeans and gossypol in cottonseed, aflotoxin inactivation, the formation of phytic acid complexes, microorganism destruction, vitamin loss, mineral binding, etc.

Extruded foods are known to have increased digestibility and make ideal foods for young children or others at nutritional risk (10). The optimal extrusion conditions to achieve maximum digestibility and nitrogen retention for protein-rich extruded foods needs to be studied and understood.

Many cereals represent good sources of fiber in the diet. With increased interest in high fiber diets, various sources of dietary and non-dietary fiber are being added to extruded foods (3). Little is known about the effect of extrusion on these fiber fractions and how their biological activity may be altered.

4.4 Miscellaneous Items

In an effort to change and control the physical structure of extruded products, new techniques which can be applied to the extruder need to be developed. For example, is it possible to employ blowing agents, materials which will release gas upon exiting from the die, to increase and control expansion? Could injection molding techniques be applied to the manufacture of food items to improve shape control and create new and unique forms of products? Are there techniques that would coat products upon leaving the die to avoid hard to control and expensive enrobing operations? What additives could be combined with food ingredients to alter texture or structure after extrusion? These and other potential adjuncts to the extrusion process can lead to a variety of new and novel food products.

4.5 Reporting Experimental Conditions

No specific recommendations have been developed concerning the data which should be reported to properly describe different types of extrusion devices used to perform experimental studies. Merely reporting the model of the extruder is an insufficient description of the physical structure of the machine. The inability to compare experimental data from a number of researchers is partially related to the lack of necessary data about the physical configuration of the extruder and detail about how the processing variables were measured. Table I gives a list of items which need to be specified in any research report. I'm encouraging this and other groups to review them in hopes of gaining a consensus on a reporting standard.

Table I: Reporting of Extrusion Processing Conditions

Extruder Barrel

- Make and model number
- Single
  - Dimensional drawing showing length, diameter and grooves

  Multiple
  - Identification numbers
  - Dimensional drawing showing individual barrel sections, length, diameter and grooves
- Clearance between screw and barrel

Extruder Screw

- Single
  - Identification number
  - Drawing showing pertinent dimensions for length, diameter, flight height(s), pitch, flight dimensions, etc.
- Multiple
  - Identification number
  - Drawing showing individual screw sections, dimensions of length, diameter, flight height(s), pitch, flight dimensions, etc.
  - Steam locks - dimensions
- Die Plate - crossectional drawing showing all pertinent dimensions and construction so void volume is completely described
- Die Inserts - crossectional drawing showing diameter, length, transitional area, number of inserts

Instrumentation
- Temperature - type, location
- Pressure - type, location
- Energy input - type and location

Feed Materials
- Complete formulation
- Analysis of any ingredients which are of variable or unpublished composition

REFERENCES

1. AGUILERA, J. M., KOSIKOWSKI, F. V. and HOOD, L. F. (1976). Ultrastructural changes occurring during thermoplastic extrusion of soybean grits. J. Food Sci. 41:1209.
2. ANDERSON, R. A., CONWAY, H. F. and PEPLINSKI, A. K. (1970). Gelatinization of corn grits by roll cooking, extrusion cooking, and steaming. Staerke 22(4)130.
3. ANDERSSON, Y., HEDLUND, B., JONSSON, L. and SVENSON, S. (1981). Extrusion cooking of a high-fiber cereal product with crispbread character. Cereal Chem. 58:370.
4. BURGESS, L. D. and STANLEY, D. W. (1976). A possible mechanism for thermal texturization of soybean protein. Can. Inst. Food Sci. Technol. J. 43:775.
5. CHEFTEL, J. C., LI-SUI-FONG, J. C., MOSSO, K. and ARNAULD, J. (1981). Maillard reactions during extrusion-cooking of protein-enriched biscuits. Prog. Food Nutr. 5:487.
6. CHIANG, B. Y. and JOHNSON, J. A. (1977). Gelatinization of starch in extruded products. Cereal Chem. 54(3):436.
7. HARPER, J. M. (1983). Effect of commercial processing and storage on nutrients. 6. Heat processing. Part 3. Extrusion processing. In: Nutritional Evaluation of Food Processing, 3rd Ed. E. Karmas and R. S. Harris, Eds. AVI Publishing Co., Inc., Westport, CT, USA.
8. HOLAY, S. H. and HARPER, J. M. (1979). Optimal energy usage in food extrusion. Paper No. 79-6508. Presented at Winter Meeting of ASAE, New Orleans, LA, USA. Dec. 11-14.
9. HOLAY, S. H. and HARPER, J. M. (1982). Influences of the extrusion shear environment on plant protein texturization. J. Food Sci. 47(6):1869.
10. JANSEN, G. R. and HARPER, J. M. (1980). Application of low-cost extrusion cooking to weaning foods in feeding programmes. Part 2. FAO Food Nutr. Quart. 6(2):15.
11. KORN, S. R. and HARPER, J. M. (1982). Extrusion of corn for ethanol fermentation. Biotechnol. Letters 47(7):417.
12. MERCIER, C. (1980). Structure and digestibility alterations of cereal starches by twin-screw extrusion-cooking. In: Food Process Engineering. Vol. I. Food Processing Systems. P. Linko, Y. Malkki, J. Olkku and J. Larinkari, Eds. Applied Science Publishers, London, UK. p. 795.
13. MERCIER, C. and FEILLET, P. (1975). Modification of carbohydrate components by extrusion-cooking of cereal products. Cereal Chem. 52(3):283.
14. OLKKU, J., HASSINEN, H., ANTILA, J., POHJANPOLO, H. and LINKO, P. (1980). Automation of HTST-extrusion cooker. In: Food Process Engineering. Vol. I. Food Processing Systems. P. Linko, Y. Malkki, J. Olkku and J. Larinkari, Eds. Applied Science Publishers, London, UK. p. 777.
15. RHEE, K. C., KUO, C. K. and LUSAS, E. W. (1981). Texturization. In: Protein Functionality in Foods. J. P. Cherry, Ed. ACS Symp. Ser. 147:51.
16. SAYRE, R. N., SAUNDERS, R. M., ENOCHIAN, R. V., SCHULTZ, W. G. and BEOGLE, E. C. (1982). Review of rice bran stabilization systems with emphasis on extrusion cooking. Cereal Foods World 27(7):318.
17. STEVENS, D. J. and ELTON, G. A. H. (1971). Thermal properties of the starch/water system. Part 1. Measurement of heat of gelatinization by differential scanning calorimetry. Staerke 23:8.
18. THOMPSON, D. R., WOLF, J. C. and REINECCIUS, G. A. (1976). Lysine retention in food during extrusion-like processes. Trans. ASAE 19(5):989.

A comparative study of the operational characteristics of single and twin screw extruders

D.J. van Zuilichem, B. Alblas, P.M. Reinders, W. Stolp.

Wageningen Agricultural University, Department of Food Science,
Food Process and Bio Engineering Group,
De Dreijen 12, 6703 BC Wageningen, The Netherlands.

ABSTRACT.

The main differences in operation of twin (t.s.e.) and single screw (s.s.e.) extruders are explained. For the s.s.e. designs the behaviour is illustrated considering the relation between theoretical throughput, throttle-ratio and the derived efficiency and optimum working points.

The practical situation for s.s.e.'s and t.s.e.'s is explained considering slip-phenomena and temperature and heating effects. The constraints and possibilities are given ; for the t.s.e. a case is worked out for a Cincinnati-design, processing a confectionary recipe. An elegant calculation scheme leads to a fair and accurate relation between the product temperature, the throughput, the wall temperature and the r.p.m. number, thus demonstrating the usefulness of a process engineering approach.

1. Introduction.

A single screw extruder is like a friction pump. When used for 'liquids' a more correct name is 'viscosity' pump. The plasticized food product wets the wall of the pump inner-cylinder and due to the mostly applied barrel - grooving the no slip condition applies. The moving surface drags the fluid. The mechanical efficiency is low, since a large part of the power supplied by the shaft is being dissipated as heat. Single screw extruders are thereby eminently suitable for processes in which the medium being transported must be heated.

Single screw extruders (s.s.e.) can generate high pressures, depending on the length L of the screw, the channel depth H, the pitch and the apparent viscosity η_{app} of the extruded product.

The mentioned drag flow effect can be realized in single and twin screw extruders (t.s.e.). The main difference between twin and single screw extruders is in the mechanism of conveying. As stated above, in a single screw extruder, the transport results primarily from the differences in the frictional and viscous forces at the contact locations screw/product and barrel/product.

In a t.s.e. with intermeshing screws the product is constrained and physically prevented from rotating with the screw. Now the friction is less important than in the s.s.e., although screw geometry can have some influence.

2. Extruder throughput.(Theoretical).

The throughput of a s.s.e. can be easily calculated. The momentum equation says :

$$\frac{d\,\tau_{yx}}{d\,y} = \frac{d\,P}{d\,x} \quad , \text{ with } \tau_{yx} = \mu \frac{d\,u}{d\,y} \quad , \text{(see Fig. 1).}$$

The velocity field is obtained by integration between the boundary conditions u = 0 for y = 0 and u = U for y = H. Now follows (see Figure 2) :

$$u = U\frac{y}{H} + \frac{H^2}{2\mu} \cdot \frac{d\,P}{d\,x}\left\{\left(\frac{y}{H}\right)^2 - \frac{y}{H}\right\} \tag{1}$$

The throughput is obtained by integration of the velocity u over the channel depth H and with w as :

$$Q_v = Q_{drag} + Q_{pressure} = \int_0^w \int_0^H u\,dy = \left(\tfrac{1}{2}\,UH - \frac{H^3}{12\mu} \cdot \frac{d\,P}{d\,x}\right) \cdot W \tag{2}$$

In this expression for the output Q_v we recognize the most important share of the drag flow $\frac{1}{2}$ UH which is often completed by the introduction of the correction factor F_{drag} and is written as

$\frac{UH + F_d}{2}$. For the remaining pressure flow component the factor F_p is introduced and follows the expression $\frac{H^3}{12\mu} \cdot \frac{d\,P}{d\,x} \cdot F_p$. Both factors F_d and F_p are functions of the channel dimensions H and w (width) as is demonstrated by Bernhardt in Figure 3. Introducing the throttle ratio a, which indicates to what extent the drag flow is reduced by the backflow (= pressure flow) :

$$a = \frac{H^2\,\Delta P}{6\mu U L} \tag{3}$$

then this ratio between pressure flow and drag flow influences strongly the flow profile , (see Figure 4). Now an efficiency η may be expressed in terms of the throttle ratio a (from equation 3) :

$$\eta = 3\,\frac{a\,(1-a)}{1+3a} \tag{4}$$

The pressure at which this efficiency reaches its maximum value for a given screw design and screw velocity U, is determined by solving $\frac{d\eta}{d\Delta P} = 0$, from which follows that :

$$\Delta P = \frac{2 \mu U L}{H^2} \qquad (5)$$

At this value for ΔP the throughput is

$$Q_v = \tfrac{1}{2} U H - \frac{H^3}{12\mu} \cdot \frac{2 \mu U L}{L H^2} = \frac{1}{2} U H - \frac{1}{6} U H = \frac{1}{3} U H$$

Inserting those values for ΔP in Q_v the maximum efficiency is $\eta_{max} = \frac{1}{3}$.

From equation 4 the value of η is found to be at $a = \frac{1}{3}$, which corresponds to the solution $\Delta P = \frac{2\mu UL}{H}$ from equation 5.

The useful pump work equals $Q_v \cdot \Delta P$. Apparently $2 Q_v \cdot \Delta P$ has been dissipated and this amount of energy must therefore reappear as sensible heat in the medium. For a temperature increase ΔT of the product in the die head of the extruder the amount of energy $Q_v . \rho c_p . \Delta T$ is needed. It is therefore that $Q_v . \rho c_p . \Delta T = 2 Q_v . \Delta P$, from which we see that

$$\Delta T = \frac{2 \Delta P}{\rho . c_p} \qquad (6)$$

(assuming no exchange of heat with the environment).

This temperature increase is a minimum, which occurs at maximum efficiency. For lower efficiencies the temperature increase ΔT will be larger.

3. Working point.

The pressure generated by the s.s.e. depends mainly on the screw compression ratio and the 'exit' conditions. Now the question arises, how much pressure is needed to force the product volume through the die. For the die the relationship between volume flow and pressure is written as

$$Q_v = \frac{k}{\mu} \Delta P \qquad (7)$$

In which k is a characteristic value for a die, reciprocal to a flow resistance. Now the die characteristic can be put in a diagram, together with the extruder characteristic out of equation 2, giving at the intermeshing points the so called 'working point' of the extruder, see Figure 5.

For a better understanding different working points of the extruder are given in this diagram of throughput versus pressure in Figure 6. In this figure 6 line 1 shows a fast decrease of the throughput at increasing pressure when extrusion is carried out at low pressure e.g. without die-head. Line 2 is plotted for the same channel dimensions but for a larger pumping zone. This means an unchanged drag flow, but a smaller effect of the pressure flow caused by the longer pumping zone. Line nr. 3 represents a shallower metering zone, causing a smaller throughput without die. At the same time a throughput is less sensitive to the pressure built up by the shallower channel.

Lines nr. 4 and 5 represent die characteristics. It is logic that a bigger die orifice gives a faster increase of the throughput at increasing pressure in comparison with a small die orifice.

4. Practical throughput/Slip consideration.

In practice there is a considerable difference between theoretical and experimental throughput, which is demonstrated in Figure 7 for maize grits in a single screw extruder at different compression ratio's e.g. working point

This is caused mainly by slip phenomena's. For food products, which always contain a certain amount of water, there always exists the possibility that a film of water is built up upon the barrel inside or screw outside upon which the bulk of the product will slide e.g. will slip.

It is as well possible that the slip layer is formed by oil or other unmixed phases.

The throttle ratio a from equation 3 can be expressed in the pitch angle θ from Fig. 1 by $a = \dfrac{H^2 Q_v \operatorname{tg} \theta}{6 k U L}$. From this expression the optimum pitch angle can be calculated via $dQ/d\theta = 0$ as $\sin \theta = \frac{1}{2}$, which means that the optimal pitch angle for a s.s.e therefore is $\theta = 30°$.

In the case of wall slip one may try to prevent the slip by grooving the barrel wall. Then the situation exists as given in Figure 8 where longitudinal barrel-grooving is showed in combination with an unrolled screw-pattern. Analysis of the slip phenomenon in this case and optimization of the pitch angle gives now $\theta_{opt} \cong 35°$. This means a considerable increase of the pitch in practical circumstances for most food materials in single screw extruders in order to overcome the negative effect of the slip.

This situation for t.s.e.'s is completely different, caused by the effect of intermeshing of screws in the various t.s.e.-designs. Intermeshing screws imply that more or less cloes C-shaped chambers, (see Figure 9) to a certain extend filled-up with product, are conveyed positively from feed zone to the die-head. This means that in principle slip at the barrel wall becomes irrelevant. The practical output of a t.s.e. can be calculated as the number of C-shaped chambers that are delivered at the die, thereupon multiplied with the chamber volume and is expressed by

$$Q_{t.s.e.} = 2 N m V \qquad (8)$$

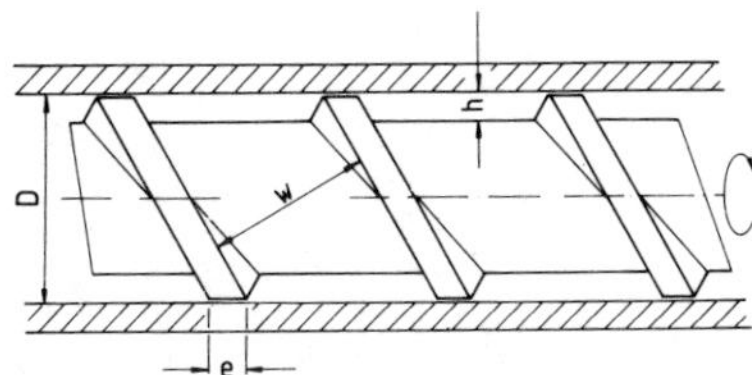

schematic diagram of a single screw extruder

fig. 1

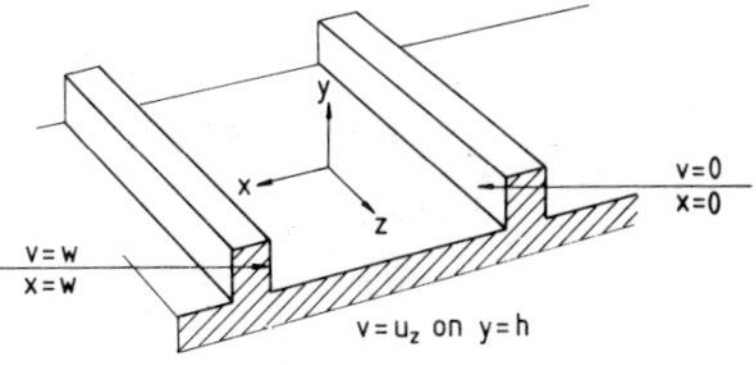

flattened out extruder channel to show coordinates and boundary conditions

fig. 2

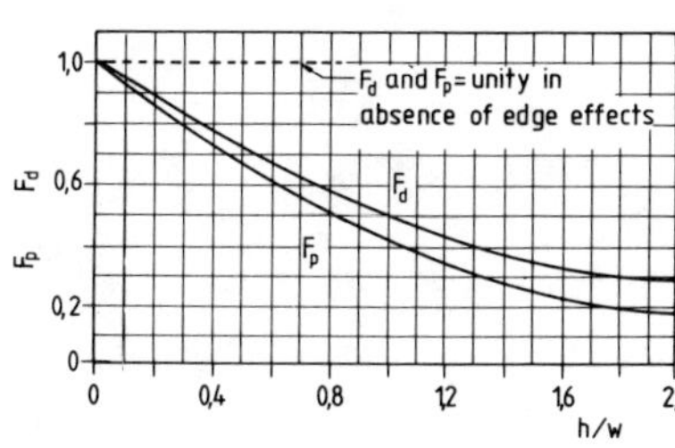

correctionfactors F_d and F_p as a function of height-width ratio h/w

fig. 3

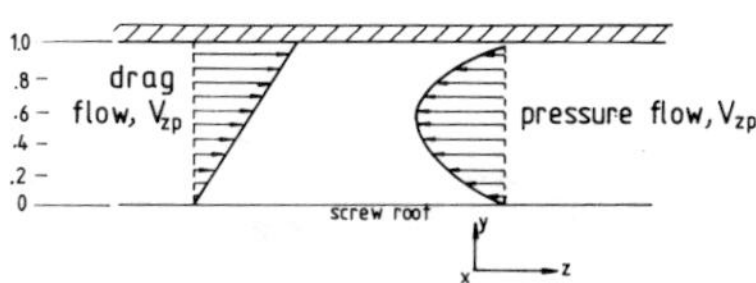

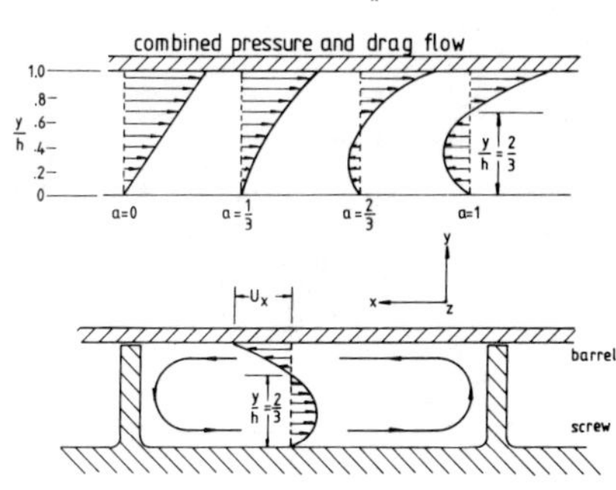

transversal velocity profile

fig. 4

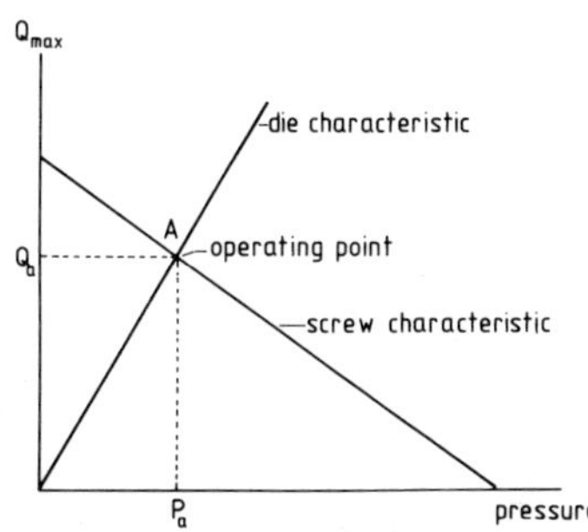

definition of operating point

fig. 5

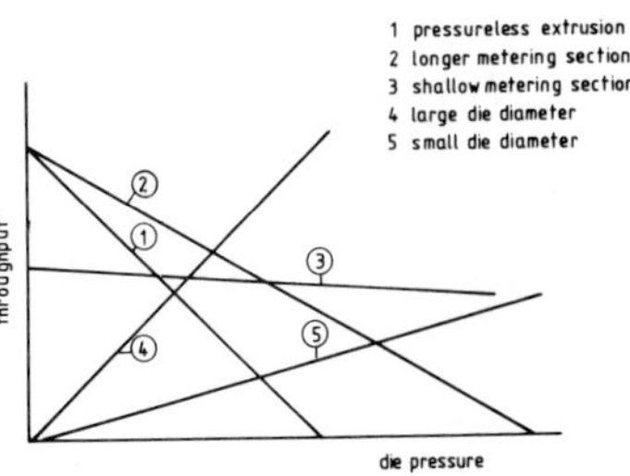

throughput related to die pressure

fig. 6

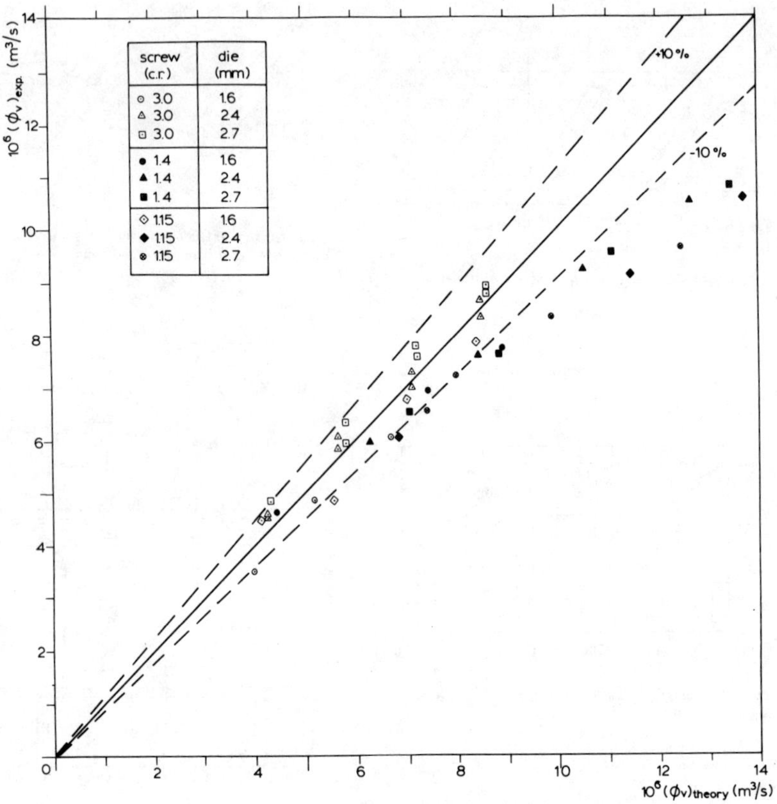

screw (c.r.)	die (mm)
⊙ 3.0	1.6
△ 3.0	2.4
⊡ 3.0	2.7
● 1.4	1.6
▲ 1.4	2.4
■ 1.4	2.7
◇ 1.15	1.6
◆ 1.15	2.4
⊗ 1.15	2.7

Fig.7 . Comparison of calculated and experimental flowrates for corn grits.

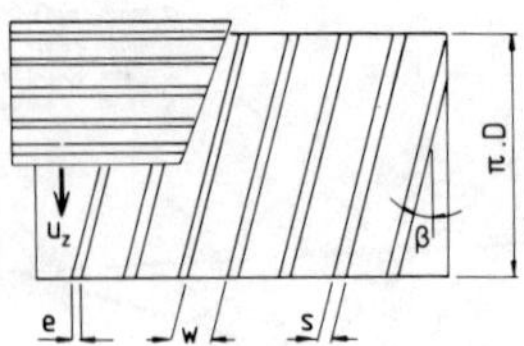

profiled barrel (axial) and screw
fig. 8

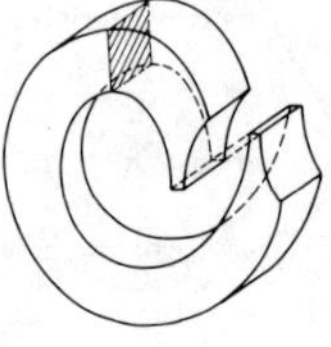

C-shaped chamber of a twin screw extruder
fig. 9

in which N represents the r.p.m. number, m the amount of screw starts and V stands for the volume of the C-shaped chambers. Of course the extent to which a designer has succeerded to determine and to control the various leakage gaps is responsible for the actual throughput of a t.s.e.

5. Temperature- and heating effects.

For a s.s.e. the temperature rise can be calculated via equation 6, taking into account all the restrictions made for this deviation. The overall impression that a s.s.e. causes the temperature rise by dissipation of motor power holds for a large group of materials, with respect to the coefficients of internal friction (power) or to the developed viscosity η_{app} (fluid paste). It is possible to make a good use of 'combined' heating, which means the use of barrel heaters extra upon the heat generated by the dissipation of mechanical power. Now we have to consider two dimensionless groups, in order to distinguish between the created heating effects. For this subject the Graetz-number and the Brinkman-number are used. The Graetz-number

$$Gz = \frac{U\ H^2}{a'\ L} \tag{9}$$

$(a' = \lambda/\rho\ c)$, gives the ratio of heat transport by convection to that by conduction via the barrel heating. The Brinkman-number gives the ration of heat generated in the flow dissipation of mechanical power to the amount of heat supplied by conduction with

$$Br = \frac{\mu\ U^2}{\lambda\ \Delta T} \tag{10}$$

For a t.s.e. most of the heat is taken up by the product by means of external (barrel) heating. This means that the Br-number should be below ~ 0.1. Taking the extrusion of sucrose-starch syrups as an example the Br-number is calculated to be below 10^{-3}. In a t.s.e. the heat flux $\emptyset_w$ through a boundary layer is described by

$$\emptyset_w = \alpha\ (T_{wall} - T_{mass})_z \tag{11}$$

at location z. The flux per unit length at diameter $D = 2\ R_w$ follows as

$$\emptyset_w = 2\ \Pi\ R_w\ \alpha_z\ (T_w - T_m)_z \tag{12}$$

The heat penetration takes place in a thin layer at the wall. This layer is scaped off by the screwflight and is mixed up in the bulk, etc. With the theory of Levêcque α can be calculated :

$$\alpha = 2\ \lambda_m\ \left(\frac{36\ a\ z}{\dot{\gamma}} \right)^{-1/3} \tag{13}$$

in which $\dot{\gamma}$ equals the velocity gradient at the wall. Now the Graetz-number must be greater than 50. Kwant defines the Gz-number as

$$Gz = \frac{\langle v_z \rangle D^2}{a.z} .$$

In the sucrose case Gz > 10^5, which means that the expression for α in equation 13 can be used. The velocity gradient

$$\dot{\gamma} = \left.\frac{d\,v}{d\,r}\right|_{r=R} \quad \text{is calculated using} \quad \left.\frac{d\,v}{d\,r}\right|_{r=R} = \frac{v_z}{\delta_o}$$

For the hydrodynamic boundary layer-thickness δ_o can be used the expression

$$\delta_o = 4{,}6 \sqrt{\frac{\eta\, z}{\rho\, v_z}} \text{ , which gives}$$

$$\dot{\gamma} = \frac{(\Pi\, D_o\, N)^{3/2} \,.\, z^{-1/2}}{4{,}6 \sqrt{\eta/\rho}} \qquad (14)$$

Now follows

$$\emptyset_w = 5{,}71 \left(\frac{\rho^3\, \lambda_m^4\, c_p^2}{\eta}\right)^{1/6} .N^{1/2} .\, R_w \,.\, R_s^{1/2} .\, (T_w - T_m)_z \,.\, z^{-1/2} \qquad (15)$$

In the calculation is further introduced the logarithmic averaged temperature difference $(T_w - T_m)_{ln}$. Taking into account that in the sucrose case a Cincinnati-t.s.e.- CM 45 (see Figure 10) was used with tempered screws, now can be derived a comparable heat flux expression for the screws and thus for the total heat input in the product $\emptyset_{w,tot}$:

$$\emptyset_{w,tot} = \emptyset_{w,wall} + 2\, \emptyset_{w,screw} .$$

This calculation can be performed and a correction for the Chamber filling degree (U) is introduced by

$U = \frac{\emptyset_m}{2\ m\ N\ v_m\ \rho}$. Multiplication with $U^{2/3}$ gives the final expression for $\emptyset_{w,total}$.

On the other hand the product needs to take up heat flux $\emptyset_w^*$ in order to bring the glucose syrup and the sucrose to die temperature and to melt the sucrose crystals.

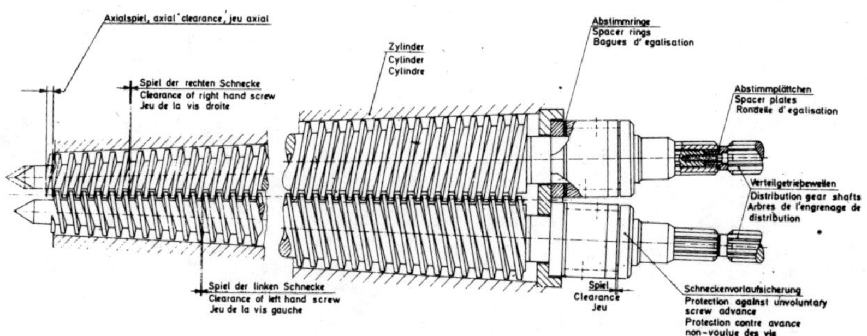

Fig.10. Example of a conical twin-screw extruder (Cincinnati)

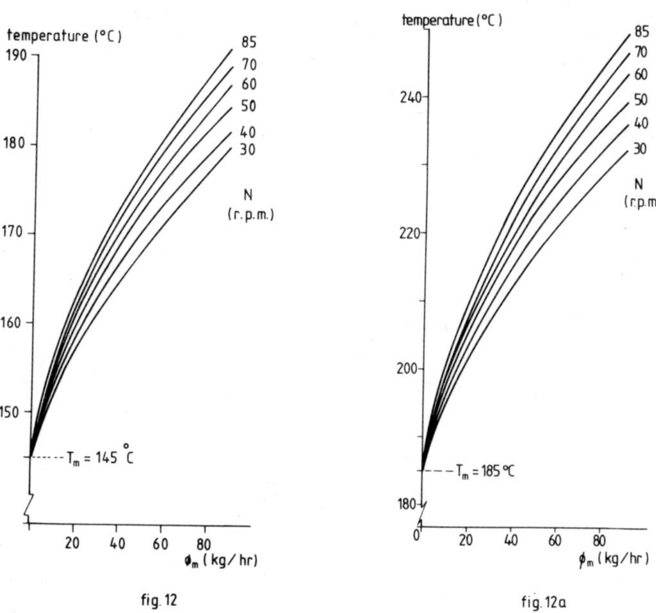

fig. 12

fig. 12a

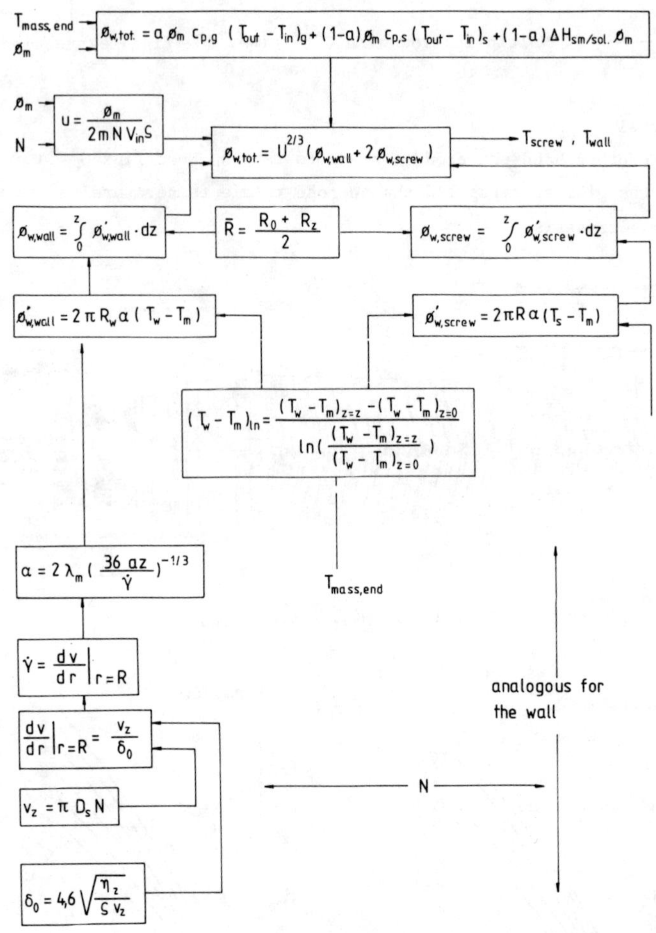

calculating scheme
fig.11

Thus follows :

$$\emptyset^*_{w,tot} = \{a\ \emptyset_m\ c_p\ (\Delta T)\}_g + \{1 - a)\ \emptyset_m\ c_p\ (\Delta T)\}_s +$$

$$+ \{(1-a)\ \Delta\ H_m \cdot \emptyset_m\}_s \qquad (16)$$

It is evident that equation 15 equals equation 16 ; giving after simple calculations the wanted relation between N, $\emptyset_m$, T_{wall} and T_{out}. The calculation scheme is given in Figure 11 as an overview. With this method follows that $T_{wall} \cong N^{1/6} . Q_m^{1/3} . T_{out}$, which can be recalculated in graphs, demonstrated in Figure 2, showing those relations. In practice the experiments prove these approach, as the maximum temperature differences in the die area are not greater than 10 °C.

6. Conclusion.

It can be concluded that the different operational characteristics of single and twin screw extruders should lead to tailor made calculation methods for the calculation of total shear, total heat input and total throughput. In this paper a survey of knowledge in this field is collected and where some doubts exist about the usefulness of the physical-technological apporach, some calculations are made, e.g. the heat householding of a counter rotating t.s.e.- CM 45 producing a well known confectionary product. There is still little knowledge in this field but our Process Engineering Group will continue to work out more of these cases.

REFERENCES.

1. Bernhardt, Ernest C., Processing of thermoplastic materials, Van Nostrand Reinhold Cy, 1959.
2. Schenkel, Gerhard., Kunstoff extrudertechnik. Carl Hanser Verlag, München, 1963.
3. Janssen, L.P.B.M., Van Zuilichem, D.J., Ontwikkelingen in extrusie. Nederlandse Vakgroep voor Extrusie, Delft, 1979.
4. Kroesser, F.W., Middleman, S., The calculation of screw characteristics.
5. Bruin, S., Van Zuilichem, D.J., Stolp, W. A review of fundamental and engineering aspects of extrusion of biopolymers in single screw extruders, Food and Nutrition Press, Vol. 1, January 1978.
6. Van Zuilichem, D.J. , Stolp, W. New twin screw extruder equipment for food extrusion. Proceedings Seminar Koch u. Extrudertechnik, ZDS Solingen-Gräfrath, BRD, 1982.
7. Van Zuilichem, D.J., Stolp, W. Extrusion of sucrose crystals. Proceedings Seminar Koch u. Extrudertechnik, ZDS Solingen-Gräfrath, BRD, 1982.
8. Janssen, L.P.B.M., Twin screw extrusion, Amsterdam (Elsevier), 1978.

PROCESS CONTROL AND AUTOMATION IN EXTRUSION COOKING

ANTILA, J.*), PIPATTI, R.*) and LINKO, P.**)

*) Technical Research Centre of Finland, Food Research Laboratory
SF-02150 Espoo 15

**) Helsinki University of Technology, Department of Chemistry,
Laboratory of Biochemistry and Food Technology, SF-02150 Espoo 15

Summary

Control, automation and the measurement of different extrusion cooking process parameters have been discussed, based on experimental work done with a Creusot-Loire BC-45 twin-screw extruder. Process automation applications and the advantages gained with computer process control have also been discussed. Applications are general and can be utilized as well in single screw as in double screw extrusion.

1. INTRODUCTION

Cooking extrusion has intensively been studied in general and also from the mathematical point of view (1...5). Computer analysis and manipulation has become a part of up-to-date analytical work. Results can be transformed in the form where they can be used, not only in product development work but also in process control and automation research. Adequate instrumentation and measurement of processing parameters are the basis of fruitful extrusion applications. This paper deals with problems and solutions involved in instrumentation and control. Also some examples of our recent automation research have been discussed. A more detailed report will be published elsewhere.

2. PROCESS EQUIPMENT

All experimental work was done with a French Creusot-Loire BC-45 twin-screw extruder. Temperature was monitored with 11 isolated measurement amplifiers. Thermocouples and PT-100 resistance sensors, thermally isolated ones for mass temperature, were used as temperature probes. Pressure measurements were made with a Dynisco PT-411 pressure transducer. Power and energy consumption at different stages of the process were recorded with kWh-meters equipped with pulse output of up to 9000 impulses/kWh, allowing measurements with high accuracy. Part of the information collection was done with a 12 channel measurement module, µMac-4000 (Analog Devices, USA) allowing for example transportation of data obtained via acustical modems and telephone lines. The control of process variables was first realized with a RCA-1802 µ-processor based microcomputer (4) and later, since 1981, with a commercially available fast ABC-800 microcomputer-based control system consisting of 64+128kb RAM-memory, a floppy disk memory of 2•160kb 8bit data, 4 D/A-converter output channels, 32 A/D-converter input channels and 16 pulse counter channels. Moreover there are 32 transistor input/output ports and 8 relay output ports, 2 RS-232-C ports and an IEEE-488 port for additional data transportation. High degree calculations and data manipulation could be carried out with a Cyber-170 central computer of the Research Centre via a RS-232-C connection.

3. PROBLEMS AND SOLUTIONS IN ON-LINE MEASUREMENTS

3.1 Temperature and pressure measurements

The most important temperature in extrusion process evaluations is the mass temperature, under which the raw material is actually processed. However, accurate measurements can be difficult, and usually a compromise between mass and barrel temperature measurement is accepted and the temperature measured 1...3 mm from the surface with an insulated temperature sensor. Even if we should succeed in mass temperature measurement, we are usually measuring only in one or in a few places along the reactor. When for example counter-screw elements are used, the actual reactor zone could be located far from the temperature probe and the barrel temperature may in some cases be found to characterize well the processing itself. Furthermore, barrel temperature is measured from a thermally stable large heat capacity. Thermocouples locating along the extruder screws give a good description of temperature distribution. A disadvantage can be that measurement arrangements possibly lead to complicated signal transportation through rotating slide-contacts which work, for example with metallic mercury. This kind of arrangement can be used only in certain extruder types in which the signal transportation can easily be arranged.

Because of the measurement difficulties, also pressure has to be often measured somewhere else than in the actual reactor zone of the extruder. Thus it has turned out that in some applications the importance both of pressure and mass-temperature information from the process can easily be overestimated. Pressure for example is not only influenced by processing variables but also by mass flow fluctuation, fouling of the dies, accidental changes in temperature of the processed mass, etc. However, although reliable and reproducible absolute pressure values are difficult to obtain, pressure readings describe the stability of the process reasonably well.

3.2 Power and energy consumption

The loading of extruder main motor, which in certain cases has a good correlation with pressure readings, reflects often "the total extrusion pressure" or the severity of material handling better than pressure alone. This electrical loading can be measured with power transformers, or by integrating the energy consumption with kWh-meters which, if they are equipped with pulse output, can easily be used also in power measurements by measuring the pulse frequency. Another way is to calibrate and calculate the electrical motor power consumption as a function of screw speed and electrical motor current. The electrical current again can be measured with instruments which often belong to the standard measuring capabilities of extruders. This function, which is specific for each extruder, was in our case:

$$\text{Power} = 1.324 + 2\cdot10^{-3}\cdot X\cdot Y - 3\cdot10^{-3}\cdot X \qquad (1)$$

where X = screw speed, 1/min
Y = electrical current of extruder main motor, A
R = 0.997 (correlation coefficient between measured and calculated values)

3.3 Other on-line measurements

Screw speed can be accurately measured with tachometers, which give an output pulse frequency or a voltage linearly proportional to the rotational speed. There are also many other possibilities for similar measurements with optical and other sensors.

Feed rate is usually the most critical extrusion process variable. It must be even, and all possible should be done to avoid any disturbances. In most extruder configurations, feed rate is not measured at all, but only the rotation or transportation speed of the feeding instrument is controlled with the assumption that there are not too much disturbance in the material flow. The most sophisticated measurement system is the derivative weighing in which a change of feed hopper weight is kept constant. Another suitable method is belt conveyer weighing.

The rate of the outcoming product describes well processing stability and is a good indicator of die-blocking. This can be measured with tachometer type instruments, which are driven by the out-flowing material. This signal can be used in controlling the process stages following extrusion.

4. OFF-LINE ANALYTICS

Product quality characteristics, including sensory evaluation are examples of off-line analytics, which can be applied in process optimization and automation. This type of feedback information comes to the production stage far too late to be used in on-line process control. When this type of off-line information can be expressed in the form of a mathematical model it can most effectively be utilized. Consequently, great emphasis in our extrusion research has been on the transformation of information into the mathematical form where product properties and production conditions have been related. In this way an abundant data basis has been collected to be used in process automation and product development work.

5. PROCESS MODELLING

In addition to the standard methods for studying the dynamic behaviour of extrusion processes we have also used radioactive tracers in order to investigate residence time distribution and the influence of process variables on material transportation (5). Results from tracer experiments have also been used in modelling of material flow inside the reactor. Plug flow, (PF) and continuous stirred tank reactors (CSTR) were differently combined in order to find the best process model. Experimental work was done with a computer program based on impulse response simulation. According to our simulation results, a model combining a plug flow and two CSTR units in a series described the extruder behaviour best. Equations (2) to (4) describe the unit reactors of the extruder model.

PF: $C_o(t) = C_i\cdot(t-\tau_{PF})$ (2)

CSTR1: $C_o(t) = C_i(t) - \tau_{CSTR1}\cdot dC_o(t)/dt$ (3)

CSTR2: $C_o(t) = C_i(t) - \tau_{CSTR2}\cdot dC_o(t)/dt$ (4)

where C_i = input concentration into reactor; C_o = output concentration from reactor; t = time; τ_n = time constant (the mean residence time in a reactor, index expresses the reactor type); d /dt = time derivative.

Regression analysis was used in estimating time constants (τ_n) as a function of processing conditions. Some dependencies of reactor τ_ns´ with multiple correlation coefficienfs varying from 0.95 to 0.997 are expressed in Fig.1 and an example of simulation in Fig. 2.

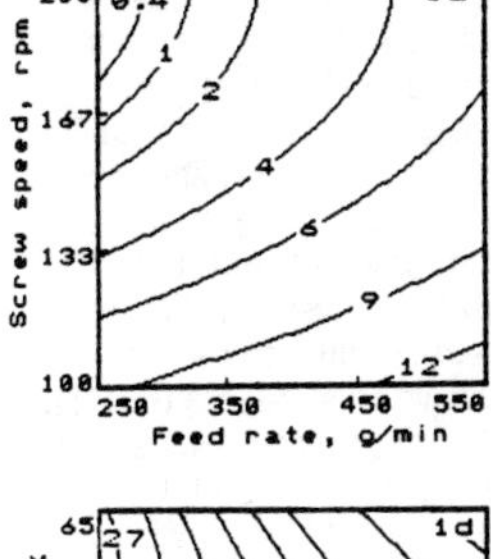

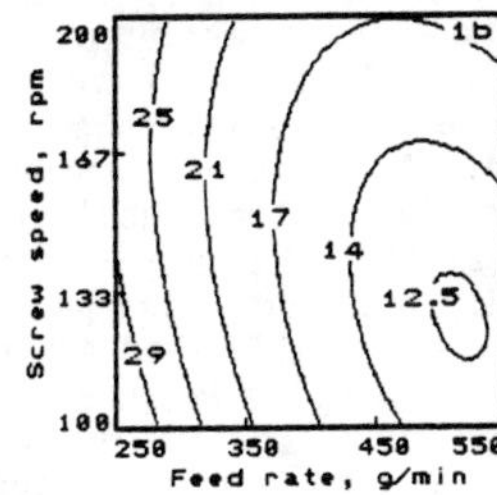

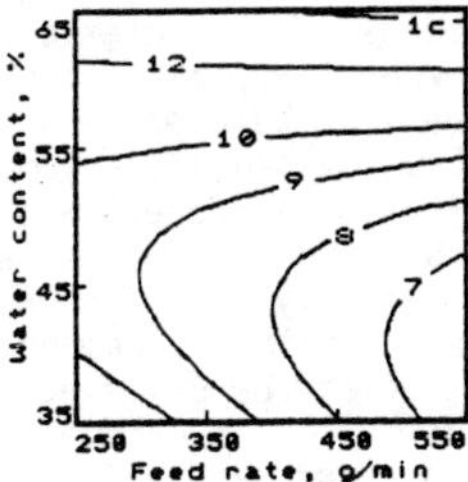

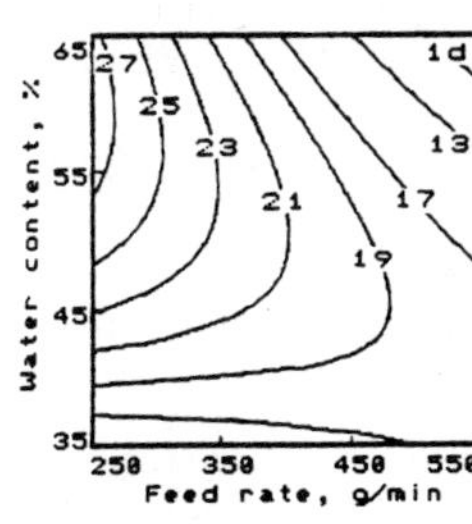

Fig. 1. Dependencies of τ_{PF} (a and c) and τ_{CSTR} (b and d) on feed rate and screw speed (a and b) or feed moisture (c and d).

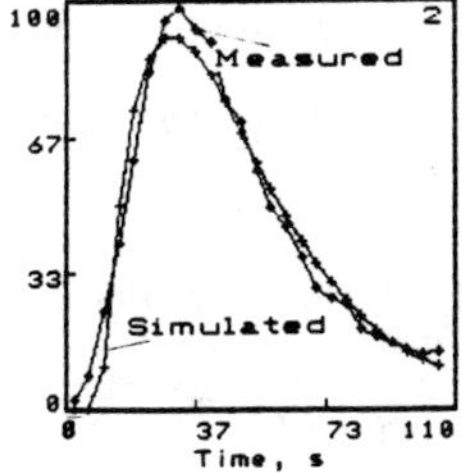

Fig. 2. Simulated and measured extruder output impulse responses.

6. PROCESS CONTROL AND AUTOMATION

Cooking extrusion is a process which practically always needs some degree of automation. Processing variables can be continuously remotely controlled, for example by a computer. Feed rate, screw speed and temperature profile of the reactor can be used as independent variables. If control is realized with a computer, also water content of feed can be controlled when feed moisture content differs from the raw material moisture. Also other process parameters can be measured to be used under on-line feedback control. We have shown that pressure measured in front of the dies can be stabilized to desired levels and held constant with a mean accuracy of a few bars. By employing both adaptive and standard PI, PID or 3-P control algorithms for pressure control we have obtained significant increase in process stabilization when pressure control variables, e.g. feed rate or screw speed have not been the critical processing variables.

In addition to the earlier mentioned controls also some product properties can be controlled through on-line measurements. In some cases measurements can be substituted by empirical mathematical functions relating the product properties to processing conditions. Reliability and flexibility will be highly increased when computer automation of processing parameter controls will be carried out parallelly with the conventional controllers. This allows manual control when the situation cannot be handled by a computer. Many advantages can be gained by extrusion process automation. As it is well known that extrusion is both an unstable process and sensitive to the stability of process parameters such as feed rate, it is clear that the optimal process handling can be realized by computer control. Accidental changes in process conditions can be optimally compensated, especially when empirical process information can be utilized. Recipe handling is an excellent other type of application of extrusion computer automation which can minimize errors in dosage and material handling. A computer can be taught to start and stop the processing, and thus also automatically change the product and the same computer can be used to control downstream processing like drying, packaging etc.

7. REFERENCES

1. LINKO, P., COLONNA, P. and MERCIER, C. (1982). High-temperature, short-time extrusion cooking. In "Advances in Cereal Science and Technology, Vol. 4" (Y. Pomeranz, ed.), A.A.C.C., St Paul, USA, pp. 145-235.
2. LINKO, P. (1983). Recent progress in the art of extrusion cooking. In "Progress in Food Engineering" (C. Cantarelli and C. Peri, eds.), Forster Verlag AG, Küsnacht, pp. 593 - 609.
3. OLKKU, J., HASSINEN, H., ANTILA, J. and POHJANPALO, H. (1980). Prozessdynamik und Automation der Hochtemperatur-Kurzzeit-Extrusion. Getreide, Mehl und Brot 34, 46 - 52.
4. OLKKU, J., HASSINEN, H., ANTILA, J. and LINKO, P. (1980). Automation of HTST-extrusion cooker. In "Food Process Engineering, Vol. 1" (Linko, P., Mälkki, Y., Olkku, J. and Larinkari, J. eds.) Applied Science Publishers Ltd., London, pp. 777 - 790.
5. OLKKU, J., ANTILA, J., HEIKKINEN, J. and LINKO, P. (1980). Residence time distribution in a twin-screw extruder. In "Food Process Engineering, Vol 1." (Linko, P., Mälkki, Y., Olkku, J. and Larinkari, J. eds.) Applied Science Publishers Ltd., London, pp. 791 - 794.

THE INTERACTION OF DESIGN AND OPERATING CONDITIONS IN EXTRUSION COOKING

L.P.B.M. Janssen
Laboratory for Physical Technology
Delft University of Technology
Delft, the Netherlands

Summary

Many papers on the performance of the pump zone of extruders have been published. Especially in food extrusion where grooved barrels are common the feed zone plays an important role. For a good understanding of the process the interactions between feed zone and pumping zone are vital. If these interactions are not taken into account an erroneous design may be the result and an instable operation of the extruder may occur.

1. INTRODUCTION

An extruder is a pump. In normal operation it is fed with solid material, a plastification or melding operation is normally achieved within a few diameters and in the last part of the extruder pressure is build up and homogenisation of the material occurs to achieve the operating goal: the production of a homogeneous stream of molten or plastizised material with desired physical properties.

It is striking to see that in extrusion research during the past decades a strong emphasis has been laid on the working of the last zone of the extruder: the pump zone. Various models have been published for pressure build up, throughput and flow profiles in this zone, both for single screw extruders as for various twin screw extruders. A further distinction can be made into models for different material rheology and models with different assumptions concerning operating conditions. Models for single screw extruders exist for isothermal operation, adiabatic operation, operation with imposed heat transfer and temperature dependent viscosity, all as well for newtonian as for power law liquids (a good overview of these models is given by Middleman (1)). However all these models concern melt extrusion, assuming that the pumpzone is an isolated unit, fed with a liquid at an unrestricted rate. Hardly any attempts have been made to understand the solids transport in the feed zone and to match the operating characteristics of these two zones.

In food extrusion where the friction coefficient of the solid material may change considerably from batch to batch the interaction of feed zone and pump zone is important and, in fact, determines the whole process. Many extruders are starved fed, a predetermined flow of material is metered into the feed port that, because of continuity, determines completely the throughput, independent of back pressure. In grooved feed zones, within normal operating ranges, the transport is hardly dependent on back pressure similar to for instance closely intermeshing twin screw extruders. These considerations change the concept for the description and understanding of the entire process as it occurs in an extruder.

2. The feed zone

In classical theories on the feed zone the walls are smooth (2). For calculations of the throughput of the feed zone a plug flow is generally assumed. A force balance over the plug shows that, apart from the forces of the pushing flight and any back pressure which may be present, the frictional forces developed between the plug and the several surfaces play an important role. For a stationary plug velocity the sum of the forces must be zero, from which the transport angle of the plug, relative to the barrel, can be calculated. From this force balance it is clear that the throughput of the solids conveying zone is severely dependent on the friction between the solid plug and the barrel and screw surfaces. In case of greater friction at the screw surfaces than at the barrel surface the output of the solids conveying zone stops completely since the material will then rotate with the screw without being pushed forward.

From the desire to enhance the output by increasing the tangential friction at the barrel surface the concept of axial grooves emerged in the literature in the late fifties (3). This has a very limited application in the processing of synthetic polymers but soon became wide spread in food extrusion. A good theory for food processing allowing the calculation of output and pressure is not yet available. However it is recognized that the operating characteristics of such feed zones allow the construction of feed controlled machines in which the output is to a large extent independent of pressure or temperature (4).

The pressure which can be attained in grooved feed zones is limited by thermal properties. The friction against the material in the grooves will generate heat causing a softening of the material in the grooves. This will reduce the tangential friction and the advantages of the grooves will be lost. From these arguments Ingen-Housz has constructed operating characteristics for grooved feed zones with experimental verifications (4) (see fig. 1).

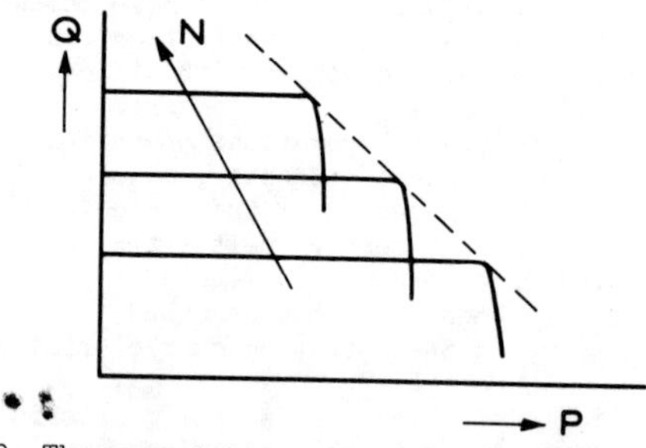

Fig. 1.
Operating characteristics of a grooved feed zone.

3. The pump zone

As stated before there exists an abundance of models for the pump zone of an extruder. However all these models are based on a judicious combination of drag flow and pressure flow. In its simplest form these models provide the well known equation

$$Q = AN + \frac{B}{\eta}\frac{dP}{dz} \qquad (1)$$

where A and B are constants, determined only by the screw geometry.

The characteristics of the feed zone play an important role on the working of the pump zone, since within normal operating conditions the feed zone determines the throughput. With a given throughput and a constant rotational rate the pressure term must be constant and independent on die pressure. For an isoviscous situation this even means that the pressure gradient is predetermined by the feed process, screw geometry, viscosity and rotational rate.

4. A combined model

In order to combine the interactions of the different zones of the extruder a parallel model is considered. The extruder is divided into three distinct zones each with their own characteristics (fig. 2).

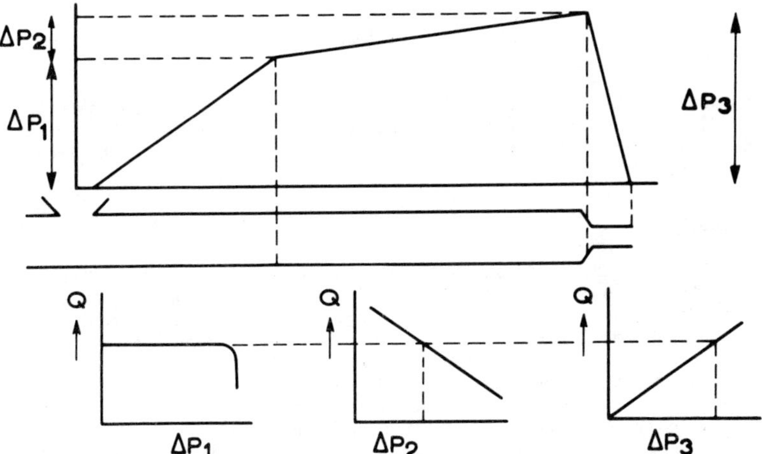

Fig. 2. Pressure built-up in an extruder and operating characteristics of the various zones.

Material enters the extruder in the feed zone where - within limits - the throughput is insensitive to the back pressure, then it passes the pump zone with characteristics according to equation 1 and finally it passes the die. A justification for a distinct division between feed zone and pump zone can be found in the fact that the melting zone in food extrusion is generally very short and its place in the extruder is found to be reasonable insensitive to operating conditions (5). Once melting occurs the typical characteristics of the grooved feed zone are lost.

The three zones must - because of continuity - have the same throughput. This implies at once that at a given viscosity the pressure drop in the die is not affected by the pump zone but completely determined by the feed zone. Therefore the die pressure is fixed at a given rotational speed. Under similar circumstances also the pressure gradient in the pump zone is fixed and therefore the pressure at the end of the feed zone is determined.

A graphical construction of the operating diagram is shown in fig. 3. At the left side of the diagram the characteristics of the feed zone are shown, the right hand side gives the characteristics of the pump zone. From

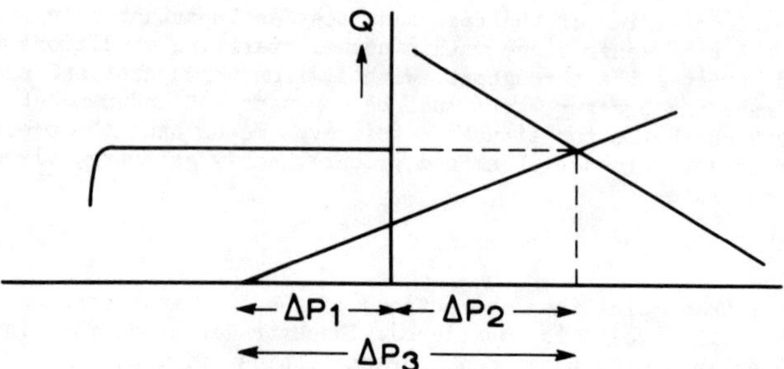

Figure 3. Combination of the three operating diagrams in the extruder
- Δp_1 pressure built-up in the feed zone
- Δp_2 pressure built-up in the pump zone
- Δp_3 pressure drop in the die

these two lines the pressure built-up in the pump zone can be constructed (Δp_2). Since the slope of the die characteristic is a known parameter we now are able to construct the die pressure (Δp_3) and from this the pressure built-up in the feed zone can be found (Δp_1).

5. Instabilities

It can be concluded from the foregoing that, although the throughput of the extruder is determined by the feed zone, the pressure built-up in the same feed zone actually depends strongly on die design and pump zone performance. Under normal conditions pressure fluctuations that occur at the die are damped out in the feed zone and do not have any influence on the throughput of the extruder. Only if the pressure built-up in the feed zone reaches extreme limits severe instabilities can occur.

It is easy to visualize that if the pressure built-up in the feed zone is too large its performance drops and no stable output at a desired rate can be obtained (fig. 4).

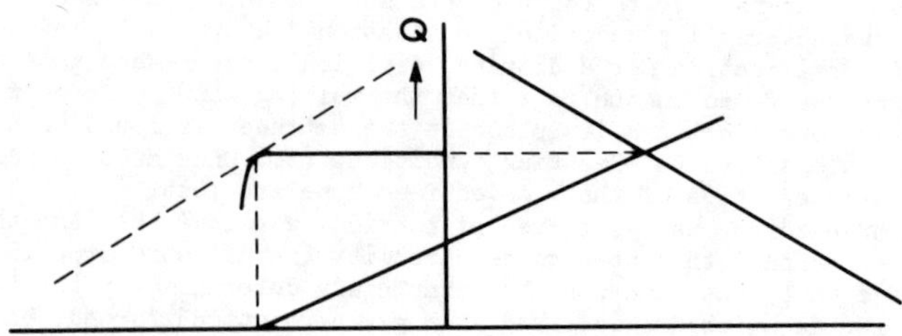

Figure 4. Instable operation, pressure built-up in the feed zone too large resulting in groove inefficiency.

On the other hand if the pressure distribution in the extruder is such that no pressure built-up in the feed zone occurs the solid bed may break. This solid bed break-up introduces instabilities in the throughput and melting process, that are a well known origin for surging (fig. 5).

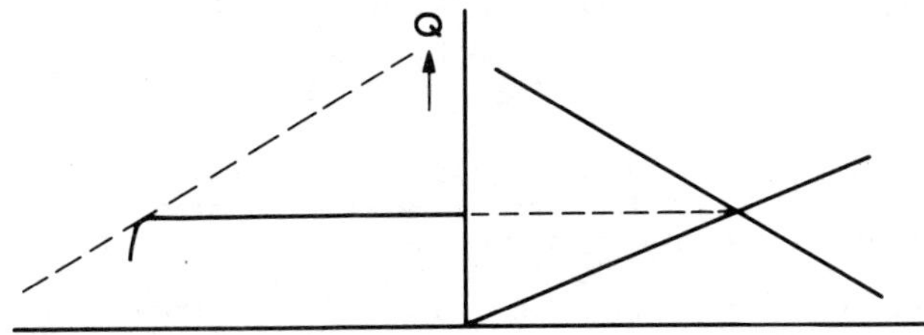

Figure 5. Instable operation, no pressure built-up in the feed zone resulting in solid bed break-up.

Conclusive remarks

The design of extruders has traditionally focussed on the pump zone. In this context it generally has been assumed that the main objective of the feed zone was to provide enough material to feed the pump zone. In contrary to most plastics extrusion processes where this argument has some validity, in food extrusion processes, because of the grooved barrel surface, no justification for this approach is present.

In cooker extruders the grooved feed zone determines the throughput of the machine. This zone acts as a stiff solids pump. The actual pressure built-up in this zone results from a balance between pressure drop in the die and the pressure built-up in the pump zone and may vary between limitations. Fluctuations in the pressure profile in the extruder are damped out in this feed zone. However if the pressure built-up in the feed zone exceeds a maximum value or drops to zero feed zone instabilities occur and no stable operation of the machine can be achieved.

Unfortunately the theory on grooved feed zones is hardly developed, especially for biological materials. Further research on this subject based on particle mechanics is needed in order to develop theories that provide a sound basis for the design of food extruders.

REFERENCES

1. MIDDLEMAN, S. (1977). Fundamentals of Polymer Processing. McGraw-Hill, New York.
2. DARNELL, W.H. and MOLL, E.A. SPE Journal 12 (1965) 20.
3. JACKSON, M.L., LAVACOT, F.J. and RICHARDS, H.R. Ind. Eng. Chem. 50 (1958) 1569.
4. INGEN-HOUSZ, J.F. Proc. Conf. Polymer Extrusion II, p.2.1, Plastics and Rubber Inst. London (1982).
5. BARTELS, P., M.Sc. thesis Wageningen (1981).

TWIN-SCREW EXTRUSION-COOKING OF CORN STARCH : FLOW PROPERTIES OF STARCH PASTES

B. LAUNAY and T. KONÉ
Département Science de l'Aliment, ENSIA, 91305 - MASSY (F).

Summary

Corn starch was processed with a twin screw extruder at various water contents, barrel temperatures and screw speeds. Starch macromolecular degradation was evaluated by intrinsic viscosity measurements in KOH solutions : it increases with screw speed and barrel temperature and it decreases with feed water content. The flow properties of pastes made with native or extruded starch have been studied in a large shear rate range. At low - and high - shear rates we have observed limiting Newtonian behaviours, separated by an intermediate plastic-like zone. The Cross equation has been found to fit very well the flow curves with an exponent close to 1 in every case. When the extrusion parameters vary, the low-shear rate Newtonian viscosities and the relaxation times of the pastes are related to the corresponding intrinsic viscosities through power laws. It is concluded that pastes flow properties are closely connected with the level of starch macromolecular degradation brought about by the extrusion-cooking process.

1. INTRODUCTION

Extrusion-cooking is a very efficient process for pregelatinizing starches. Several working parameters can be modified to change the end-products characteristics : screw design and composition, water content, barrel temperature and screw speed, among others. Therefore, extrusion-cooking is potentially more versatile than roll-cooking. Pregelatinized starches are used for their thickening and stabilizing properties, most often evaluated in terms of hot and cold paste consistencies. However, in previous works, we have shown that flow curves are better suited than empirical consistencies for this purpose (1, 2). Generally, extrusion-cooking of starches does not yield low molecular weight carbohydrates (3). However, we have assumed from rheological evidences that it does not merely destroy starch granular structure and that, in addition, some depolymerization should occur (1, 2). This assumption has been recently confirmed, with gel permeation chromatography and intrinsic viscosity measurements, for extruded manioc starch (4) and corn starch (5).

The aim of this study is to give a full account of the flow properties of starch pastes and to relate them to three important process parameters, viz. feed water content, barrel temperature and screw speed whose efficiency on starch depolymerization will be evaluated.

2. MATERIALS AND METHODS

Commercial corn starch (Rofec) has been purchased to Roquette (Lestrem, France) and extruded with a twin-screw pilot-plant apparatus (Creusot-Loire BC 45) : screw profile is given in figure 1.

SCREW COMPOSITION							
	FEED THROAT ↓						2DIES Ø
Length mm	200	100	50	100	50	100	4mm
Pitch mm	50	35	25	15	-15	15	

Figure 1 - The minus sign corresponds to the reverse pitch section.

Water content before extrusion varied from 16 % (moist basis) for starch without added water to 20, 23.5 or 26.5 % by adding continuously water with a pump. Barrel temperature in the last zone (T_3) was set between 148 and 191°C and screw speed was 150, 175 or 204 rpm. Feeding rate was 25 kg/h. After reaching steady state conditions a 10 - 15 min. running time was allowed before collecting samples. They were grinded and sieved (0.15 mm aperture) once, and occasionally twice, for separating "white spots" which did not gelatinize at 95°C; generally, about 10 % of the grinded extrudate was eliminated by sieving. The resulting sample was dispersed in cold water (total weight 450 g), placed in the Brabender Viscoamylograph bowl, heated to 95°C, maintained 10 min. at this temperature, and cooled to 65°C (heating and cooling rates 1.5°C/min.). Flow properties at 60°C were obtained without delay using a Rheometrics Fluids Rheometer fitted with coaxial cylinders ($r_1/r_2 = 0.92$) for low and intermediate shear rates and a Contraves Rheomat 30, fitted with MSA ($r_1/r_2 = 0.95$) or double shear MSO cylinders ($r_1/r_2 = r_3/r_4 = 0.98$), for high shear rates.

Starch depolymerization was estimated by intrinsic viscosity measurements, according to Colonna and Mercier (4). After dissolving the sample in 1 N KOH, efflux times were measured in 0.2 N KOH at 25°C with an Ubbelhode capillary viscometer associated with an automatic diluting unit (Fica). The Huggins equation was used to calculate $[\eta]_o$.

3. RESULTS

As shown in figure 2 the flow curves, when drawn as viscosity versus shear stress, display a quasi-vertical part strongly indicative of a yield stress σ_s.

Evans and Haisman (6), using a pasting method rather similar to ours, have also observed yield stresses with various starches, in particular $\sigma_s = 0.75$ Pa for a corn starch paste at 42 g/l and 60°C. At very low shear rates they used the Herschel-Bulkley equation (power law + yield stress) and they also suggested that the Cross equation (7) :

$$\eta = \eta_\infty + (\eta_o - \eta_\infty)/(1 + (\tau\dot{\gamma})^m) \qquad \text{(eq. 1)}$$

could be applied. However, they were unable to extrapolate the low-and high-shear rate Newtonian viscosities η_o and η_∞, respectively. On the contrary, it can be seen in figure 2 that η_o and η_∞ are closely approached and figure 3 indicates that equation 1 fits very well experimental data.

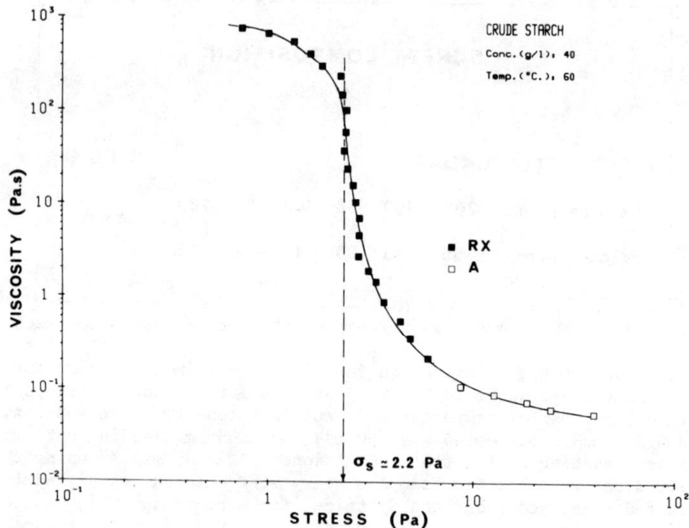

Figure 2 - Flow curve represented as viscosity versus shear stress and corresponding to native starch sample. RX : Rheometrics Fluids Rheometer, A : Rheomat 30 and MSA cylinders, σ_s : apparent yield stress.

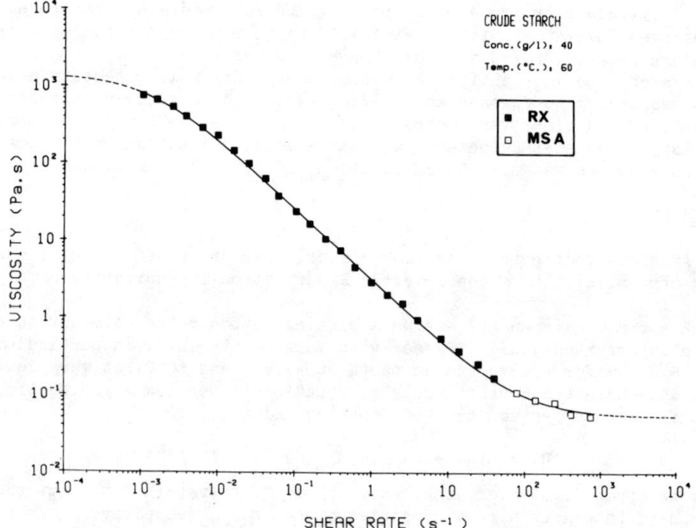

Figure 3 - Same results as in figure 2, with shear rate as variable. The curve has been calculated using equation 1.

Extruded starches give similar results with smaller - and generally not so well defined - σ_s and lower viscosities. As shown in figure 4, η_∞ is experimentally attained and the low shear rate η-values approximate to η_o. In every case, the Cross equation has provided very good results and Table I gives the values of the four parameters (η_o, η_∞ , relaxation time τ, exponent m), along with intrinsic viscosities $[\eta]_o$, for various extrusion conditions.

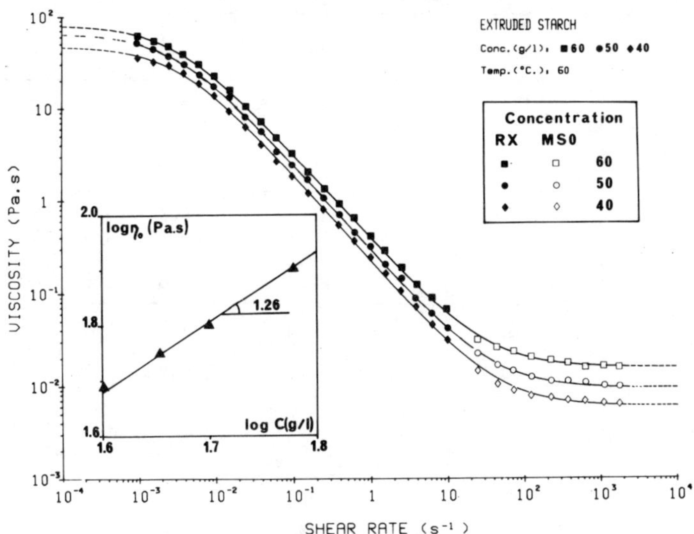

Figure 4 - Flow curves, corresponding to an extruded starch sample, fitted with equation 1. MSO : Rheomat 30 with MSO cylinders. For clarity, the curve C = 45 g/l has been omitted. Inset : power-law relationship between η_o and C.

It is seen in Table I that m is always close to 1, and that η_∞ has a small and almost constant value; $[\eta]_o$, as well as η_o and τ, decreases with increasing severity of treatment, that is to say when barrel temperature or screw speed increase or when starch water content decreases. Similar variations of $[\eta]_o$ with screw speed and barrel temperature were observed by Colonna and Mercier (4), but $[\eta]_o$ reached a limit for T_3 = 150°C. Starch depolymerization seems less pronounced that what has been obtained by Colonna and coworkers for manioc starch extruded with the same equipment at smaller screw speeds (4) and for corn starch processed with a BC 72 twin-screw extruder (Creusot-Loire) at 145 rpm (5) : 33 % reduction, at most, in $[\eta]_o$ instead of about 60 %. The difference may be attributed to the various screw configurations, especially in the reverse flight zone, used in these three studies.

Figure 5 shows that there is a close relationship between η_o and $[\eta]_o$ for extruded starches. There is also an analogous but looser relationship between τ and $[\eta]_o$ (fig. 6), but it has to be noticed that the values of τ are much less reproducible than the values of η_o. For the native starch sample, these results are not suitable to η_o (calculated value 139 Pas instead of 1411 Pas) but possibly to τ (790 s instead of 749 s).

TABLE I : intrinsic viscosities of native and extruded starches, flow parameters for starch pastes (40 g/l, 60°C) and mean relative deviations (MRD) between experimental and calculated flow curves (eq. 1). The central extrusion conditions were : 20 % water, setting temperature T_3 = 169°C, screw speed 150 rpm.

Sample		T_3(°C)	$[\eta]_o$(ml/g)	η_o(Pas)	η_∞(10^{-3}Pas)	τ(s)	m	MRD(%)
Native starch		/	204	1411	53	749	0.92	7.5
Water content (%)	16	175	134	22.4	5.6	157	1.00	6.9
	20	164	153	48.2	5.9	290	0.95	5.2
	23.5	173	149	38.1	4.7	168	1.00	6.3
	26.5	171	165	55.7	6.3	356	0.98	5.2
Barrel temp. T_3 (°C)	148		164	53.1	5.5	412	0.94	4.4
	169		152	43.6	5.1	435	0.99	2.3
	180		141	32.8	5.1	201	1.00	4.8
	191		137	24.3	4.6	218	1.00	3.6
Screw speed (rpm)	150	172	188	93.3	6.2	542	0.98	3.8
	175	171	158	41.8	4.1	258	0.98	2.5
	204	180	141	25.0	6.0	180	0.96	4.8

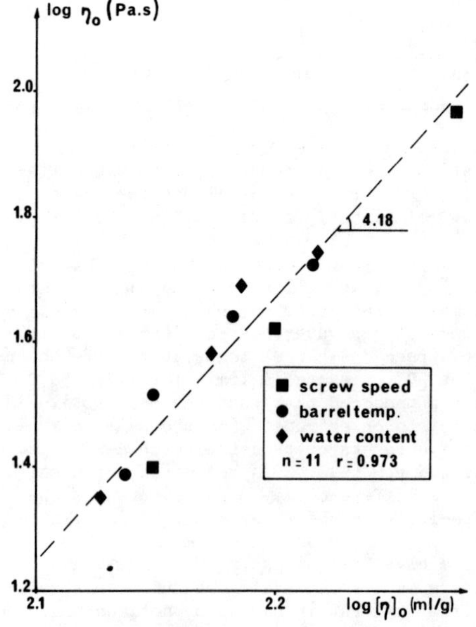

Figure 5 - Relationschip between η_o and $[\eta]_o$ for starches extruded in various conditions (c = 40 g/l, 60°C).

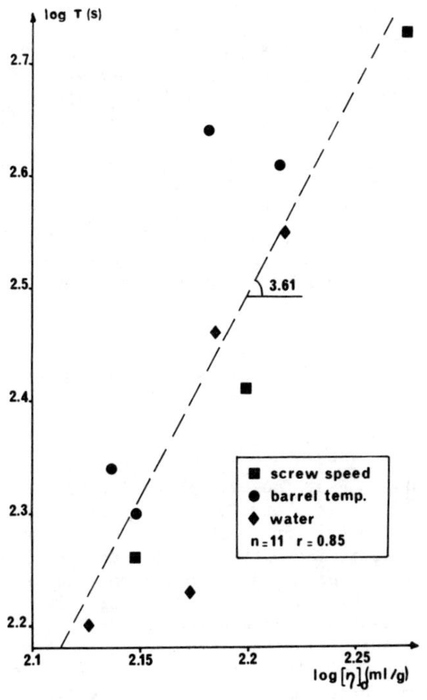

Figure 6 - Relationship between τ and $[\eta]_o$ (see fig. 5).

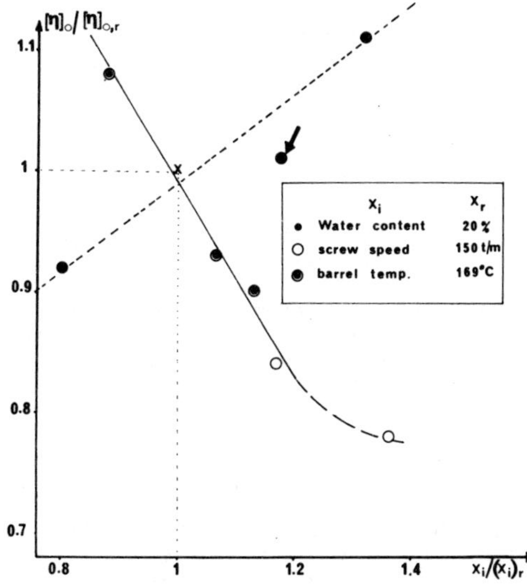

Figure 7 - $[\eta]_o$ as a function of extrusion parameters : both coordinates are referred to their values in central operating conditions (X). When T_3 was not a variable, a correction to T_3 = 169°C was done, using the results at various barrel temperatures : $[\eta]_o = [\eta]_{o,exp.} + 0.65\ (T_3 - 169)$. The arrowed point has not been taken into account (interrupted line).

From Table I and figures 5 and 6, it can be inferred that $[\eta]_o$ is a good explanatory parameter for paste flow properties of extruded starches and, therefore, $[\eta]_o$ can be used as an index of process efficiency. Figure 7 shows that, in a small variation range, $[\eta]_o$ depends approximately linearly on the operating variables. Screw speed and barrel temperature have similar effects on $[\eta]_o$, and a change in feed water content seems to be somewhat less effective.

4. CONCLUSION

Compared to our previous results (1, 2), we have been able to obtain the flow properties of starch pastes in a much larger shear rate range (more than 6 decades). A low - and a high - shear rate Newtonian zone have been found out, separated by a plastic - like domain; Cross equation fits very well the whole curve. The process parameters are closely related to the intrinsic viscosity of extruded starch which, in turn, determines paste flow properties. From figure 5 and from the inset in figure 4 the following equation can be calculated :

$$\eta_o = 2.961\ 10^{-10}\ C^{1.26}\ [\eta]_o^{\ 4.18} \qquad \text{(eq. 2)}$$

with C in g/l, $[\eta]_o$ in ml/g and η_o in Pas. The experimental concentration range is narrow but the value of the exponent (1.26) is practically identical to the one obtained by Doublier (9) for wheat starch pastes at 25°C and $c \leqslant 5$ g/l. This result tends to indicate that soluble macromolecules which have exuded from extruded starch granules did not form an entangled system, at least until c = 60 g/l; therefore, these macromolecules should be highly depolymerized compared with those exuding from native starch granules. The very large low-shear rate viscosities could be attributed to interactions in a deformable and highly expanded dispersed phase, as observed among others by Bagley and Christianson (10) for cooked wheat starch.

Acknowledgements : this work has been supported by the French Ministry of Industry and Research (effet des traitements thermiques sur les aliments). Mr P. Cantoni and Mrs M. Angé have contributed for technical assistance and typing, respectively.

REFERENCES

1. LAUNAY, B. and LISCH, J.M. (1975). In : Nouvelles voies d'utilisation et de valorisation de diverses céréales, légumineuses et produits amylacés par cuisson-extrusion (eds J.F. de la Guérivière and J. Grèbaut), Action concertée DGRST 71-7-2828, APRIA, Paris.
2. LAUNAY, B. and LISCH, J.M. (1983). Twin-screw extrusion-cooking of starches : flow behaviour of starch pastes, expansion and mechanical properties of extrudates. J. Fd Engineering, 2, 259-280.
3. MERCIER, C. and FEILLET, P. (1975). Modification of carbohydrate components by extrusion cooking of cereal products. Cereal Chem., 52, 283-297.
4. COLONNA, P. and MERCIER, C. (1983). Macromolecular modifications of manioc starch components by extrusion-cooking with and without lipids. Carbohyd. Polymers, 3, 87-108.
5. COLONNA, P., MELCION, J.P., VERGNES, B. and MERCIER, C. (1983). Flow, mixing and residence time distribution of maize starch within a twin-screw extruder with a longitudinally-split barrel. J. Cereal Sci., 1,

115-125.
6. EVANS, I.D. and HAISMAN, D.R. (1979). Rheology of gelatinized starch suspension. J. Text. Stud., 10, 347-370.
7. CROSS, M.M. (1965). Rheology of non-Newtonian fluids : a new flow equation for pseudoplastic systems. J. Colloid Sci., 20, 417-437.
8. DOUBLIER, J.L. and LAUNAY, B. (1977). Propriétés rhéologiques des solutions aqueuses de galactomannanes : rôle de la masse moléculaire moyenne et de la concentration. Cahiers du Groupe Français de Rhéologie, 4, 191-198.
9. DOUBLIER, J.L. (1981). Rheological studies on starch - Flow behaviour of wheat starch pastes. Stärke, 33, 415-420.
10. BAGLEY, E.R. and CHRISTIANSON, D.D. (1982). Swelling capacity of starch and its relationship to suspension viscosity. Effect of cooking time, temperature and concentration. J. Text. Stud., 13, 115-126.

MODELLING A TWIN SCREW CO-ROTATING EXTRUDER

W. A. Yacu
RHM Research Limited, Lincoln Road,
High Wycombe, Bucks. HP12 3QR., England.

ABSTRACT

This work describes the simulation of a twin screw co-rotating extruder. Based on a uni-directional analysis, the temperature and pressure profiles within the extruder are predicted from knowledge of feed and extruder characteristics. A non-Newtonian, non-isothermal viscosity model is used to describe the materials rheology. The analysis covers an evaluation of conditions within the solid conveying zone, the melt pumping zone and a melt shearing zone achieved by reverse screw elements. The total energy input (both mechanical and through the barrel) is also evaluated. The simulation work has proved satisfactory in the design and optimization of extruder operation.

1. INTRODUCTION

Very little work has been published about the mechanisms of material flow and energy transfer in intermeshing co-rotating extruders. Most of this work (3,7,19) treated the melt flow in a similar way to that of a single screw extruder as a combination of drag and pressure flows. Booy (2) added a third flow element due to the pushing action of the second screw in the intermeshing region. All these assumed a Newtonian and isothermal fluid. Another approach (14) has been the analysis of process response variables by multiple regression methods. While this approach achieves information about the specific process and product properties, it does not provide engineering understanding of the interactions between the extruder design, the material's characteristics and the operating conditions of the extruder. Greater understanding of the flow mechanism seems to exist in intermeshing counter-rotating extruders. This has been based on the axial movement of material in closed C-shaped chambers (3,9).

The present work is primarily concerned with the evaluation of the temperature and pressure profiles in the axial direction (parallel to the screw shaft). This then serves as the basic information required for the evaluation of the extruder performance.

A schematic diagram of the extruder is presented in Figure 1. The extruder analysis covers the solid conveying zone, the melt pumping zone and a melt shearing zone caused by reverse screw elements fitted at the end of the shaft before the die plate. The food melt rheology is described by a non-Newtonian , non-isothermal viscosity model taking into consideration the effect of moisture and fat content. Steady state behaviour and uniform conditions in the direction normal to the screw shaft are assumed. The melt flow is highly viscous and in the laminar flow regime. Gravity effects are negligable and are ignored. Heat losses to the screw shafts are neglected. Only the net flow in the axial direction is considered.

The material's temperature increases rapidly along the melt section causing a change in its viscosity. This prevents the straight forward calculation of the required pressure and hence the length of the melt pumping section. A trial and error approach is therefore adopted. This is based on starting with the best estimate of the melt pumping section length, evaluating the temperature profile first and then the pressure profile.

The pressure drop across the plate hole is evaluated independently using the product temperature achieved at that stage and the specific die characteristics. If this die pressure is higher or lower than that evaluated earlier at the die plate, the pumping section length is increased or decreased respectively and the evaluation procedure started again. This calculation continues until the acceptable accuracy tolerance is obtained.

2. THEORETICAL TREATMENT

2.1. Temperature Profile

2.1.1. The Solid Conveying Section

In this section the screws are partially empty (9), and no pressure is developed. Therefore, negligible viscous heat dissipation takes place. If re-quired heat can be transferred to the powder through the barrel; this is expected to take place by conduction. To simplify the analysis, convection heat transfer is assumed. A pseudo heat transfer coefficient U_s is used to replace the characteristic conductivity property (k/ℓ). This approach is realistic owing to the fact that mixing within the channel is reasonably good in co-rotating extruders. Constructing a heat balance across an element normal to the axial direction and solving for T with the boundary condition at $\chi = 0$ $T = T^{\circ}_f$ results

$$T = T_{bi} - (T_{bi} - T^{o}_{f})\, e^{-\frac{\nu U_s A}{Q C_{vs}}\cdot \chi} \qquad \ldots\ldots (1)$$

ν describes the degree of screw filling which increases with decreasing the screw pitch (9).

2.1.2. The Melt Pumping Section

The change of material from solid powder to fluid melt is assumed to take place in a step change. The screws are completely full in this section. In order to construct and evaluate a heat balance across an element in this zone, it is necessary first to describe the way the mechanical energy is consumed. Part of this energy is converted to thermal heat (including melting and heat of reaction, if any) and potential energy within the product. The kinetic energy is relatively small and can be ignored. Martelli (11) estimated the energy converted to thermal heat by viscous dissipation per screw channel in one screw turn. He assumed that heat was dissipated within the channel and due to leakage flow in the various gaps. These were (i) between the flight tip of the screw and the inside surface of the barrel, (ii) between the flight tip of one screw and the bottom of the channel of the other and (iii) between the flights of opposite screws parallel to each other.

The heat dissipation within the channel was estimated by:

$$Z_1 = \mu \gamma^2 V_p \qquad \ldots. (3)$$

where V_p is the channel volume and was estimated as

$$V_p = \frac{\pi^2}{2} D.D_e \, h \tan\theta_p \qquad .... (4)$$

D_e is the equivalent twin screw diameter estimated as

$$D_e = \frac{2}{\pi} (\pi D - \sqrt{2Dh}) \qquad .... (5)$$

$$\gamma = \frac{\pi D_e N}{h} \qquad .... (6)$$

therefore Z_1 =

$$Z_1 = \frac{\pi^4 D_e^3 D \tan\theta_p}{2h} \mu N^2 \qquad .... (7)$$

Based on a force balance, the power dissipation within a clearance space was derived (12,17), as

$$dZ = \frac{\mu u^2}{c} dA \qquad .... (8)$$

Martelli defined the relative product velocity (u) within the various clearances and calculated their resultant energy dissipation per one channel in one screw turn. For a screw with an m number of flight starts these can be described as follows:

(i) between the flight tip of the screw and the inside surface of the barrel

$$Z_2 = \frac{\pi^2 D^2 e C_e}{\alpha} m\mu N^2 \qquad .... (9)$$

C_e is the equivalent twin screw circumference and was estimated by

$$C_e = 2 (\pi D - \sqrt{2Dh})$$

(ii) between the flight tip of one screw and the bottom of the channel of the other

$$Z_3 = \frac{8\pi^2 I^3 e}{\varepsilon} m\mu N^2 \qquad .... (10)$$

(iii) between the flights of opposite screws parallel to each other

$$Z_4 = \frac{\pi^2 I^2 h \sqrt{D^2 - I^2}}{2\sigma} m\mu N^2 \qquad .... (11)$$

The total energy per channel per one screw turn was therefore expressed as:

$$Z_p = Z_1 + Z_2 + Z_3 + Z_4 \qquad .... (12)$$

$$= C_{1p} \mu_p N^2 \qquad .... (13)$$

where C_{1p} can be termed as the pumping section screw geometry factor and is described as:

$$C_{1p} = \frac{\pi^4 D_e^3 D \tan\theta_p}{2h} + m\left(\frac{\pi^2 D^2 e C_e}{\alpha} + \frac{8\pi^2 I^3 e}{\varepsilon} + \frac{\pi^2 I^2 h \sqrt{D^2 - I^2}}{2\sigma}\right) \qquad .. (14)$$

The amount of heat generated within an element of $\delta\chi$ thickness can be therefore evaluated as:

$$dZ_p = C_{1p}\mu_p N^2 \frac{\delta\chi}{\pi D \tan\theta_p} \quad \text{.... (15)}$$

The overall shear rate the product is exposed to while passing through the pumping zone (including that due to leakage flows) can be estimated by assimilating Equation 13 with Equation 3.

$$\mu_p \gamma_p{}^2 V_p = C_{1p}\mu_p N^2$$

$$\therefore \gamma_p = N\sqrt{\frac{C_{1p}}{V_p}} \quad \text{.... (16)}$$

The product apparent viscosity μ_p is described by power law and a dependence on temperature, moisture content and fat content as follows:

$$\mu_p = \mu_o\, e^{-a_1(MC-MC^o)}\, e^{-a_2 F}\, \gamma_p{}^{-n_1}\, e^{-bT}$$

$$= \mu_1\, \gamma_p{}^{-n_1}\, e^{-bT} \quad \text{.... (17)}$$

Further discussion of the viscosity and shear rate definition is given in Section 2.3. μ_1 is the initial viscosity after accounting for moisture and fat content. At a fixed screw speed, this can be further reduced to:

$$\mu_p = \mu_{2p} e^{-bT} \quad \text{.... (18)}$$

where $\mu_{2p} = \mu_1 \gamma_p^{-n_1}$

Constructing a heat balance across an element in the melt pumping zone

$$QC_{vm}T + U_m A(T_{bi} - T)d\chi + \frac{C_{1p}\mu_{2p}N^2 e^{-bT}}{\pi D \tan\theta_p} d\chi = QC_{vm}(T + dT)$$

or $$\frac{dT}{d\chi} = C_{2p}e^{-bT} + C_{3p}(T_{bi} - T) \quad \text{.... (19)}$$

where $$C_{2p} = \frac{C_{1p}\mu_{2p}N^2}{\pi D \tan\theta_p QC_{vm}} \quad \text{.... (20)}$$

and $$C_{3p} = \frac{U_m A}{QC_{vm}} \quad \text{.... (21)}$$

Equation 19 is a first order, non-linear differential equation. Coupled with the boundary condition: $T = T^\circ_m$ at $\chi = \chi^\circ_m$, it can be integrated numerically by the Simpson 3/8 rule (10).

2.1.3. The Reverse Element Section

The mechanism of fluid flow and heat generation within this section are extremely complicated and no relevant work could be cited. The existance of cross channels normal to the screw channel (in some cases) adds further to this complication. These cross channels are introduced in order to improve mixing and reduce the total shear stress. The net flow in this section is necessarily equal to that of the melt pumping zone. As a reasonable approximation, the heat generation by viscous dissipation is assumed to take place much in the same way as in the melt pumping section. The presence of cross channels is accounted for by modifying the screw geometry factor C_{1r} (equivalent to C_{1p} defined by Equation 14) in the following way:

$$C'_{1r} = C_{1r}\left(1 - \frac{m_1 BG}{\pi Dh}\right) \quad .... (22)$$

The screws remain full in this section. A heat balance across this section can therefore be described as follows:

$$QC_{vm}T + U_m A(T_{bi} - T)dX + \frac{C'_{1r}\mu_r N^2}{\pi D \tan\theta_r}dX - QdP = QC_{vm}(T + dT) \quad .... (23)$$

$-QdP$ is the energy dissipated due to pressure drop. The pressure gradient in this section is estimated from a formula having a similar structure to that of the fluid pressure gradient in the annulus between two concentric pipes (1):

$$\frac{dP}{dX} = -\frac{Q\mu_r}{K_1} \quad .... (24)$$

where K_1 is the screw conductance and is approximated by

$$K_1 = \frac{h^5 \tan\theta_r}{D(1 - m_1 BG/\pi Dh)} \quad .... (25)$$

The apparent viscosity in this section (μ_r) is described in a similar way to that of the melt pumping section:

$$\mu_r = \mu_1 \gamma_r^{-n} e^{-bT} \quad .... (26)$$

Substituting Equations 24 and 26 in Equation 23 and rearranging,

$$\frac{dT}{dX} = (C_{2r} + C_{4r})e^{-bT} + C_{3r}(T_{bi} - T) \quad .... (27)$$

where C_{2r} and C_{3r} are similar to C_{2p} and C_{3p} described by Equation 20 and 21 for the melt pumping section, and

$$C_{4r} = \frac{QD\mu_1\gamma_r^{-n}(1 - m_1 BG/\pi Dh)}{C_{vm} h^5 \tan\theta_r}$$

Coupled with the boundary condition: $T = T_p°$ at $X = X°_p$, Equation 27 is also solved numerically by the Simpson 3/8 rule intregration method. Experimentally, the mechanical energy input was found to be underestimated by Equations 19 and 27. Better predictions were obtained when the geometry factors C_{1p} and C'_{1r} were increased by a factor of 1.3.

2.2. Pressure Profile

The screws are partially empty in the solid conveying zone. Therefore no pressure is developed at this stage.

2.2.1. The Melt Pumping Section

Considerable disagreement exists in the published literature over the definition of flow mechanism and the boundary conditions (2,3,7,19). Burkhardt et al. (3) Denso and Hwang (7) and Wyman (19) calculated the volumetric flow rate by integrating the velocity vector v_z (in the channel direction) with respect to channel width and height in a similar way to that of a single screw extruder. Wyman (19) however concluded that the flow due to pressure gradient had the same direction as that of the drag flow. In other words the pressure gradient is negative in the z direction. This concept is hard to accept owing to the open nature of co-rotating screws and the existance of a continuous channel. Booy (2) added a positive displacement flow element in the axial direction to account for the pushing action of the second screw in the intermeshing region. Generally it can be stated that the theory of fluid flow and the development of the pressure profile in co-rotating extruders is still in its early development stages. This is more so when accounting for specific extruder geometry and the fact that extruded materials are generally non-Newtonian undergoing considerable temperature changes.

A subjective and quantitative analysis of the pressure profile is followed in this work. This is based on theoretical input verified by experimental results. The length of the pumping section and the pressure generated behind the die plate were measured experimentally. The formula adopted and tested is that of the maximum pressure gradient (when the discharge end is closed) derived by Wyman (19).

$$\left(\frac{dP}{dz}\right)_{max} = \frac{6\pi DN\mu_p \cos\theta_p}{h^2} \qquad \text{.... (28)}$$

Assuming that a continuous channel exists, the pressure profile in the axial direction $\frac{dP}{dX}$ can be calculated as:

$$\frac{dP}{dX} = \frac{dP}{dz} \cdot \frac{1}{\sin\theta_p} \cdot \frac{D_e}{D} \qquad \text{.... (29)}$$

for a small $\frac{h}{D}$ ratio Equation 28, therefore reduces to

$$\left(\frac{dP}{dX}\right)_{max} = \frac{12\pi DN\mu_p}{h^2 \tan\theta_p} \qquad \text{.... (30)}$$

Experimentallly Equation 30 was found to provide good prediction of the length and final pressure in the melt pumping zone when the constant value was replaced by 4. This is then solved for a non-Newtonian non-isothermal fluid whose apparent viscosity is defined by Equation 18. Substituting and rearranging:

$$\frac{dP}{dX} = K_2 e^{-bT} \qquad \text{.... (31)}$$

where $K_2 = \dfrac{4\pi DN\mu_{2p}}{h^2 \tan\theta_p}$

To simplify the solution of Equation 31, T is defined as a function of χ

$$T = T^\circ_m + g_p (\chi - \chi_m) \qquad \text{.... (32)}$$

Substituting Equations 32 in Equation 31 and solving for P with the boundary condition: P = 0 at $\chi = \chi^\circ_m$, the final pressure at $\chi = \chi^\circ_p$ is then calculated as:

$$P_p = \frac{K_2 e^{-bT^\circ_m}}{bg_p} (1 - e^{-bg_p (\chi^o{}_p - \chi^o{}_m)}) \qquad \text{.... (33)}$$

2.2.2. <u>The Reverse Element Section</u>

The pressure gradient in this section was defined by Equation 24 as:

$$\frac{dP}{dX} = \frac{-Q\mu_r}{K_1}$$

substituting for μ_r by Equation 26 and re-writing:

$$\frac{dP}{dX} = \frac{-Q\mu_1 \gamma_r^{-n} e^{-bT}}{K_1} \qquad \text{.... (34)}$$

Similarly, defining T as a function of χ

$$T = T^\circ_p + g_r (\chi - \chi^o{}_p) \qquad \text{.... (35)}$$

Substituting Equation 35 in Equation 34 and solving for P with the boundary condition: $P = P^\circ_p$ at $\chi = \chi^\circ_p$,

$$P = P^\circ_p - \frac{K_3}{bg_r} (1 - e^{-bg_r (\chi - \chi^o{}_p)}) \qquad \text{.... (36)}$$

where $K_3 = \dfrac{Q\mu_1 \gamma_r^{-n} e^{-bT^\circ_p}}{K_1}$

Substituting L for χ in Equation 36 the melt pressure at the end of this section can be evaluated.

2.2.3. <u>Pressure Drop Across the Die Hole</u>

This is calculated by:

$$\Delta P = \frac{-Q\mu_d}{K_d} \qquad \text{.... (37)}$$

From a forces balance across the die plate, K_d for a circular hole can be defined as:

$$K_d = \frac{\pi d^4}{132\ell} \quad \text{.... (38)}$$

Pressure drop due to entrance (end) effect is accounted for from experimental evaluation of the materials rheology (8).

2.3. Rheology Characteristics

2.3.1. Rheology Model

The rheology model for a wheat A-starch was determined using the twin screw extruder itself as a capillary viscometer. Pressure drop measurements were carried out across three capillaries of L/D ratio: 3, 4 and 6. The calculation procedure for capillary shear rate, end effect, the effect of moisture and fat is given in references (4,5,8, 15). The value of the various indices was determined by Multiple Regression Analysis to fit the following rheology equation. The shear rate element (γ^{-n}) definition is dealt with in the next section

$$\mu = \mu_o e^{-a_1(MC - MC^o)} e^{-a_2 F} \gamma^{-n} e^{-bT} \quad \text{.... (39)}$$

2.3.2. Definition of Shear Rate

The food material faces different kinds of shear regimes while passing through the melting zones of the extruder. It also undergoes considerable physical and chemical reactions which change its rheology irreversibly. Until more information is available about these reactions, emperical approximation seems the most practical way of describing the food rheology. From our work on the capillary rheology measurements using the extruder as a viscometer, it was found necessary to take into consideration the shear characteristics of the screws as well as those of the capillary. The rate of the physical/chemical reactions appeared to be related to the shear rate within the extruder. The definition of the shear rate in the various melt sections of the extruder is therefore important as it could indirectly explain some of the material phase changes.

2.3.2.1. The Melt Pumping Section

The overall shear rate in this section γ_p was defined by Equation 16. This was based on the summation of absolute values of the average shear rate within the channel and the various gaps. The effect of shear rate only on the apparent viscosity μ_p is predicted using the Ostwalde power law model

$$\mu_p = \mu^o \gamma^{-n_1}$$

or

$$\mu_p = \mu^o \left(N \sqrt{\frac{C_{1p}}{V_p}}\right)^{-n_1} \quad \text{.... (40)}$$

2.3.2.2. The Reverse Element Section

In this section the average shear rate was first described in a similar way to that of fluid flow in the annulus between two concentric pipes. However, due to the non-Newtonian nature of food melts, it was decided to account also for the average shear rate caused by the screw rotation, i.e.

$$\mu_r = \mu^o \gamma_p^{-n_2} \gamma_r^{-n_3} \qquad \text{.... (41)}$$

where $\gamma_r = \dfrac{QD(1 - m_1 BG/\pi Dh)}{h^4 \tan\theta_r}$

2.3.2.3. Within The Die Hole

As mentioned earlier, it was found necessary to account for the average shear rate caused by the screw rotation as well as the axial average shear rate caused by the fluid flow within the capillary.

$$\mu_d = \mu^o \gamma_p^{-n_2} \gamma_d^{-n_3} \qquad \text{.... (42)}$$

where

$$\gamma_d = \frac{32Q}{\pi d^3}$$

The values of n_1, n_2, n_3 were estimated from a series of experiments evaluating the material's rheology. Different screw geometry and configuration (with and without the reverse element) were used. The rheology results were correlated using multiple regression analysis.

2.4. Energy Requirement

2.4.1. Mechanical Energy Supplied by the Motor

This is calculated from knowledge of total viscous heat dissipation and potential energy developed within the extruder.

$$\int_{X_m^o}^{L} dE = \int_{X_m^o}^{X_p^o} \frac{C_{1p} N^2 \mu_p dX}{\pi D \tan\theta_p} + \int_{X_p^o}^{L} \frac{C'_{1r} N^2 \mu_r dX}{\pi D \tan\theta_r} + \int_{P_a}^{P_d} QdP \qquad \text{.... (43)}$$

μ_p and μ_r were defined by Equations 16 and 26 respectively. The temperature is then defined as a function of X in a similar way to that carried out for evaluating the pressure profile in Section 2.2. Solving Equation 43 for E with the boundary conditions: $T = T_m^o$ at $X = X_m^o$ and $T = T_p^o$ at $X = X_p^o$ resulted in

$$E = K_4 \left(1 - e^{-bg_p(X_p^o - X_m^o)}\right) + K_5 \left(1 - e^{-bg_r(L - X_p^o)}\right) + Q.P_d \qquad \text{.... (44)}$$

where $K_4 = \dfrac{C_{1p} N^2 \mu_{2p}\, e^{-bT^\circ_m}}{bg_p\ \pi D \tan\theta_p}$

and $K_5 = \dfrac{C'_{1r} N^2 \mu_{2r}\, e^{-bT^\circ_p}}{bg_r\ \pi D \tan\theta_r}$

Accounting for a motor efficiency factor ε, the total motor energy consumption E_T is:

$$E_T = \frac{E}{\varepsilon}$$

2.4.2. Energy Supplied or Taken by the Barrel

In most cases heat is supplied or taken by the barrel according to a controlled temperature set point. In some cases, multiple zone temperature control is provided. The calculation will be confined here, as an example to the melt pumping zone. The total heat flow through the barrel is the summation of heat flow across all the individual heating/cooling zones.

Assuming that the controlled temperature is the inside barrel surface temperature T_{bi} and is constant along the melt pumping zone. The logarithmic mean temperature difference is

$$\Delta T_{\ell m} = \frac{T^\circ_p - T^\circ_m}{\ell n\left(\dfrac{T^\circ_{bi} - T^\circ_m}{T_{bi} - T^\circ_p}\right)}$$

Heat transfer to the material H_{1p}

$$H_{1p} = \pi D_e L_p U_m \Delta T_{\ell m}$$

This heat is equal to that transferred across the thickness of the barrel (assuming that both the heating and/or the cooling source are placed at the outside barrel surface). The outside barrel temperature (T_{bo}) is calculated assuming a symmetrical cylinder barrel of an inside diameter D_e and outside diameter D_o as:

$$T_{bo} = T_{bi} + \frac{H_{1p}\ell n(D_o/D_e)}{2\pi k L_p}$$

Heat losses to the air, in the case of no insulation and no resistance to heat transfer between the heating bands and the barrel surface is estimated as:

$$H_{2p} = \pi D_o L_p U_a (T_{bo} - T_a)$$

Total heat supplied through this section of the barrel H_{tp} is then:

$$H_{tp} = H_{1p} + H_{2p}$$

Total energy supply per unit throughput (S) $= \dfrac{E_T + H_T}{M}$

3. RESULTS AND DISCUSSION

A computer programme in the BASIC language was written to carry out the trial and error calculation and predict the extruder performance. The results were compared with those experimentally measured at the same operating conditions. The measured variables (at steady state) included product pressure and temperature behind the die plate, barrel temperatures, screw speed and armature current of the DC motor. Owing to instrument measurement limitations, the material temperature and pressure could not be measured along the barrel. In some selected experiments, the powder feeder, the water supply and the extruder were stopped simultaneously, the extruder cooled and the screw shafts pulled out for inspection. The extruder geometry details are given in Table I.

Using the physical properties and the rheology model data listed in Table II, the predicted results were generally in good agreement with those measured. A sample of the predicted results is given in Table III and the corresponding temperature and pressure profiles in Figure 2. The ultimate test of such modelling exercise is its capability of predicting successfully the results of other workers. Such test was however not possible due to lack of published comprehensive data but hopefully be arranged in the future. From the obtained results the following points can be made:

(i) Uni-directional analysis of co-rotating twin screw extruders is realistic since good cross channel mixing is achieved by the relatively high screw speed, thus reducing product conductivity limitations.

(ii) Food material (in this case, wheat A-starch) under extrusion conditions appeared to behave as a non-Newtonian fluid and its viscosity could be described reasonably well by the power law. The definition of shear rate was very important as it affected the physical and chemical changes in the product.

(iii) The length of the melt pumping section was relatively short. In most cases of low moisture/fat concentration, it was only about one-two pitch length. This was in agreement with previous work carried out on a Creusot-Loire BC-72 extruder (6). The pressure gradient in the pumping section was much higher than that achieved by single screw extruders.

(iv) Under conditions of high viscosity and high operating speeds, the contribution of barrel heating was relatively small. This becomes smaller in commercial extruders since the throughput is roughly proportional to D^3 while the heat transfer area is proportional to D only.

(v) The presence of a reverse element at the end of the screw shaft was essential to achieve relatively high product temperatures at moderate screw speeds. It also had significant effects on the product textural properties.

(vi) In order to account for all the heat supplied to the product, a product specific heat value of 2.7 kj/kg°C was assumed. This is relatively high for low moisture cereal products (13) which suggests that some heat was consumed as heat of melting/reaction. Similar findings were reported (16) using a single screw extruder.

4. CONCLUSIONS

Co-rotating twin screw extrusion of food material was satisfactorily simulated. This was based on theoretical understanding and intuitive assumptions supported by experimental verification. Although food extrusion is a very complicated system, a uni-directional analysis proved very useful in the design and optimization of extruder operations. Considerable basic research work needs to be done to identify in detail: (i) The mechanism of fluid flow and energy transfer, (ii) The phase changes the food material is undergoing and their effect on its physical and chemical properties.

NOTATION

A	Surface area, m^2
a_1	Moisture coefficient of viscosity
a_2	Fat coefficient of viscosity
B	Width of the drilled cross channel in the reverse screw element, m
b	Temperature coefficient of viscosity 1/°C
C_1	Geometry factory, m^3
C_2	Viscous heat dissipation factor, °C/m
C_3	Barrel/melt convection heat transfer factor, °C/m
C_4	Potential energy factor, °C/m
C_v	Specific heat kj/m^3.°C
c	General clearance width, m
D	Screw diameter, m
d	Die hole diameter, m
E	Motor energy, W
e	Screw flight tip width, m
F	Fat concentration, % of total weight
G	Depth of the drilled cross channel in the reverse screw element, m
g	Temperature profile slope, °C/m
H	Barrel heat supply or loss, W
h	Screw channel depth, m
I	Centre to centre distance between the two screw shafts, m
K_1	The reverse screw element conductance factor, m^4
K_2,K_3	Pressure profile factor for the pumping section and the reverse screw section respectively, N/m^3
K_4,K_5	Viscous heat dissipation factor for the pumping section and the reverse screw section respectively, N.m/s
K_d	Die hole conductance, m^3
k	Thermal conductivity, W/m.°C
L	Barrel length, m
	Die plate hole length, m
M	Mass flow rate, kg/s
MC	Moisture concentration, % of total weight
m	No. of screw flight starts
m_1	No. of cross channels drilled in the reverse screw element
N	Screw speed, 1/s
n,n_1, n_2,n_3	Power low index
P	Pressure, N/m^2
Q	Volumetric flowrate m^3/s
S	Specific energy consumption, W.s/kg
T	Temperature, °C
U	Convection heat transfer coefficient W/m^2.°C
u	Relative material velocity in the clearance, m/s
V	Channel volume, m^3
v	Material velocity within the channel
x	Axial distance or coordinate, m
Z	Energy due to viscous heat dissipation, W
z	Coordinate parallel to the screw channel

α	Distance between the screw flight tip and the inside surface of the barrel, m
ε	Distance between the flight tip of one screw and the bottom channel of the other
σ	Distance between flights of opposite screws parallel to each other, m
ν	Screw filling ratio in the solid conveying section
μ	Viscosity, $N.s/m^2$
γ	Shear rate, I/s
ε	Extruder efficiency factor
θ	Screw helix angle, radians

Subscripts

a	Air
b	Barrel
d	Die hole
e	Equivalent
f	Feed material
i	Inside barrel surface
m	Melt
o	Outside barrel surface
°	Initial or reference condition
p	Pumping section
r	Reverse screw section
s	Solid (powder)
t,T	Total

ACKNOWLEDGEMENTS

The author wishes to thank: the Director of RHM Research Centre, Dr. J. Edelman, for allowing the publication of this manuscript; the Programme Manager Dr. G.L. Solomons; the Head of Process Development Dept., Mr. E.C. Pape and Dr. M.C. Jones of Aston University in Birmingham for encouragement and valuable discussions; Mr. J. Gourlay and Miss A. Horn for help with the experimental work; and Mrs. J. Wharton for her patience in typing the manuscript.

REFERENCES

1. Bird, R.B., Stewart, W.E. and Lightfoot, E.N., Tansport Phenomena, John Wiley & Son Inc., 1960
2. Booy, M.L., Polym. Eng. & Sci. 20, (18), 1980, 1220
3. Burkhardt, K., Hermann, H. and Jakopin, S., SPE ANTEC Technical Papers, Volume XXIV, Washington DC, 1978, 498
4. Cervone, N.W. & Harper, J.M., J. Food Process Eng. 2, 1978, 83
5. Chen. Y.C.J., Lewandowski, D. & Irwin, W.E., Ibid, 2, 1978, 97
6. Colona, P., Melcion, J.-P., Vergnes, B. & Mercier, C., J. of Cereal Science 1, 1983, 115
7. Denson, C.D. & Hwang, J.R., Polym. Eng. & Sci. 20 (14), 1980, 965
8. Harper, J.M., Extrusion of Food. Vol.1, CRC Press Inc., 1981
9. Janssen, L.P.B., Twin Screw Extrusion, Elsevier Sci. Pub. Co. 1978
10. Jenson, V.G. & Jeffreys, G.V., Mathematical Methods in Chemical Engineering, Academic Press 1977

11. Martelli, F.G., Twin Screw Extruders: A basic understanding. Van Nostrand, Reinhold, 1982
12. McKelevey, J.M., Polymer Processing. John Wiley & Sons Inc., 1962
13. Mohsenin, N.N., Thermal Processing of Food and Agricultural Materials, Gordon and Breach Science Publishers 1980
14. Olkko, J., Hassinen, H., Antilla, J., Pohjanpalo, H. and Linko, P., Food Process Engineering. Vol.1. Edited by Linko, P., Applied Sci. Publishers Ltd.
15. Remsen, C.H. and Clark, J.P., J. Food Process Eng. 2, 1978, 39
16. Sahagun, J.F. and Harper, J.M., J. Food Process Eng. 3, (4), 1980,199
17. Schenkel, G., Plastic Extrusion, Technology and Theory, London Illiffe Books Ltd. 1966
18. Tadmor, Z. and Klein, I., Engineering Principles of Plasticating Extruders, Van Nostrand, Reinhold 1970
19. Wyman, C.E., Polym. Eng. & Sci. 15 (8), 1975, 606

FIGURE 1 - SCHEMATIC DIAGRAMME OF THE EXTRUDER

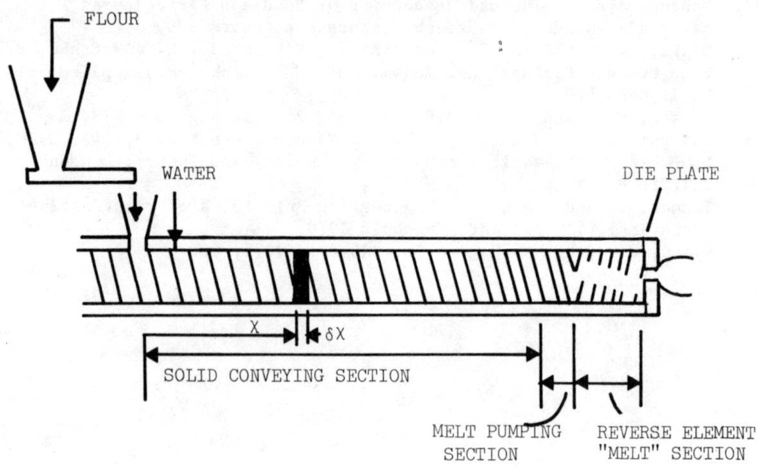

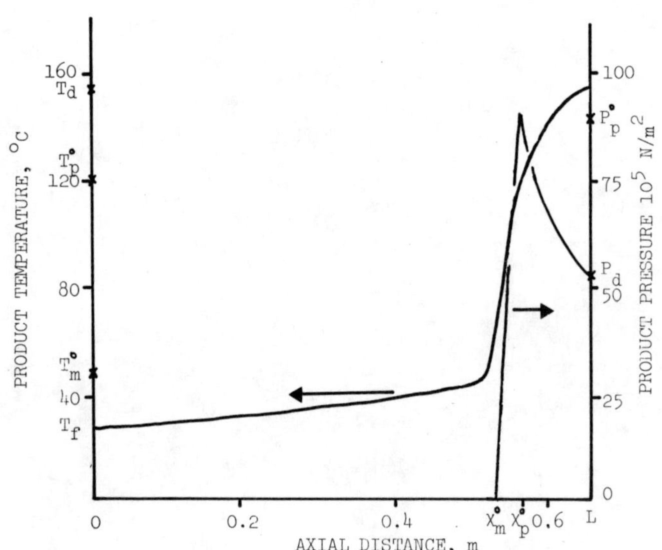

FIGURE 2 - PREDICTED TEMPERATURE AND PRESSURE PROFILES WITHIN THE EXTRUDER

Table I

Extruder Geometry Specifications

Co-rotating twin screw extruder

Barrel: Length = 0.66m; Inside Diam. = 55mm; Outside Diam. = 130mm
Screw Channel Depth = 7mm
Screw Shafts Centre-Centre Distance = 48mm

Screw Configuration

	Screw Pitch, mm	No. of Starts	Length, mm
Feed Section 1	55	2	16.5
Feed Section 2	40	2	33.0
Pump Section	30	2	8.5
Reverse Section	30	2	8.0

Screw Details in the Pumping and Reverse Sections

e	1.8
α	0.7
ε	0.7
σ	3.0

Reverse screw cross channel:	
Width (B)	10
Depth (G)	7
No. of Channels (m_1)	3

Die Hole: Circular of 4mm diameter and 12mm long

Table II

Physical Properties, Operating Conditions and Rheology Data Used in the Modelling Example

A. Physical Properties and Operating Conditions

Material: Wheat A-Starch, supplied by
Tenstar Aquitaine, Bordeaux, France

	Solid Conveying Section	Melt Sections
Specific heat, kj/kg°C	1.7	2.7
Density, kg/m^3	700	1,400
Heat transfer coefficient, W/m^2°C	30	350
Feed temperature, °C	25	
Air temperature, °C	30	30
Natural convection heat Transfer coefficient, W/m^2°C	10	20
Barrel inside surface temp., °C	100	160
Screw speed, rps	2.8	
Feed rate, kg/hr	36.0	
Moisture content, %	19.5	
Fat content (ground nut oil), %	0	

Table II cont..

B. Rheology Model Data

Initial viscosity μ_o, N.S./m^2	120,000
Reference moisture content MC°,%	6
Moisture coefficient a_1	0.07
Fat coefficient a_2	0.1
Temperature coefficient b	0.005
Shear rate (power law) index	
n_1	0.7
n_2	0.4
n_3	0.3

Table III

Predicted Extruder Performance Results

	Solid Conveying Section	Melt Pumping Section	Melt Reverse Section
Length, cm	53	5	8
Material final:			
Temp., °C	44	120	153
Pressure, 10^5 N/m^2	0	87	53
Viscosity, N.s/m^2		620	340 (314 within the die hole)
Barrel outside temp., °C	101	172	172
Energy supplied to the material, W-hr/kg			
i - By the motor (net)			
Thermal		45	19
Potential		1.6	-0.6
Motor efficiency at 65%			

Total (gross) energy supplied by the motor = 100 W-hr/kg

	Solid Conveying Section	Melt Pumping Section	Melt Reverse Section
ii - By the heating bands, W-hr/kg			
to the material	8.3		14.8
lost to the air	4.2		1.7
	12.5		16.5

Total heater band energy supply = 29 W-hr/kg

Total energy = 129 W-hr/kg

QUALITY ASPECTS IN EXTRUSION COOKING

Yngve Andersson
SIK-The Swedish Food Institute
Box 5401

S-402 29 Göteborg, Sweden

SUMMARY

This paper deals with aspects on sensory properties of extruded products. The importance of a proper selection of raw materials is discussed. Examples are given regarding e.g. the influence of fibre and fat contents of the raw materials on the quality of the products. It is also shown how different process parameters as e.g. extrusion temperature, screw speed, and water addition can change the product characteristics. Also, post-extrusion treatments of extruded materials are discussed.

INTRODUCTION

When food materials are processed in a cooker-extruder, they are subjected to high temperatures and pressures in combination with action of shearing forces. As a result of this treatment the material undergoes chemical and physical modifications. For instance, the chemical composition can be changed by "gelatinization" and breakdown of starch, denaturation of proteins, and complex formation between carbohydrates and proteins. Besides, a structure formation is achieved. Of course, such changes also will influence the appearance, the taste, and the texture of the extruded products. Figure 1 illustrates how such products can be divided into two groups, depending on whether the main purpose is to achieve a structure or to chemically modify the raw materials.

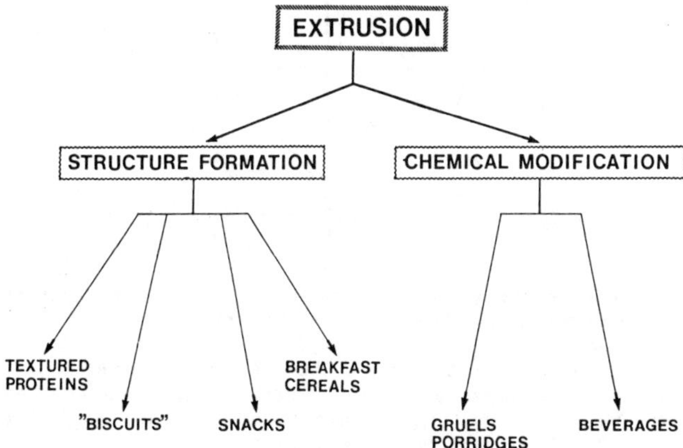

Figure 1 Main groups of products manufactured through extrusion cooking.

The desired properties of extruded foods are related to how they are intended to be used. In snacks and "biscuits", where the extruded material constitutes a complete food item, the sensory properties are extremely important for the consumer acceptance. In fact, in the advertisements the manufacturers in many cases stress that the consumers, when buying their products, get a specific taste, texture, or "sensation".

When extruded materials are mixed with other food components (e.g. milk powder, sugar, or minced meat), the sensory characteristics of the complete food are important. Examples are extruded carbohydrate-rich semi-manufactured products for gruels, porridges, and beverages and textured vegetable proteins for meat patties. Figure 2 shows how incorporation of textured soy proteins into meat patties influence two sensory properties, biting resistance and "beany" taste. From Figure 2 it can be seen that the biting resistance can be increased by addition of up to 8% textured soy protein. This may in some cases be considered as positive. However, when the amount of soy protein increases in a recipy, the typically "beany" soy flavour will become more intense, which is negative. This means that the incorporation level of the extruded components must be optimized with consideration to the sensory properties.

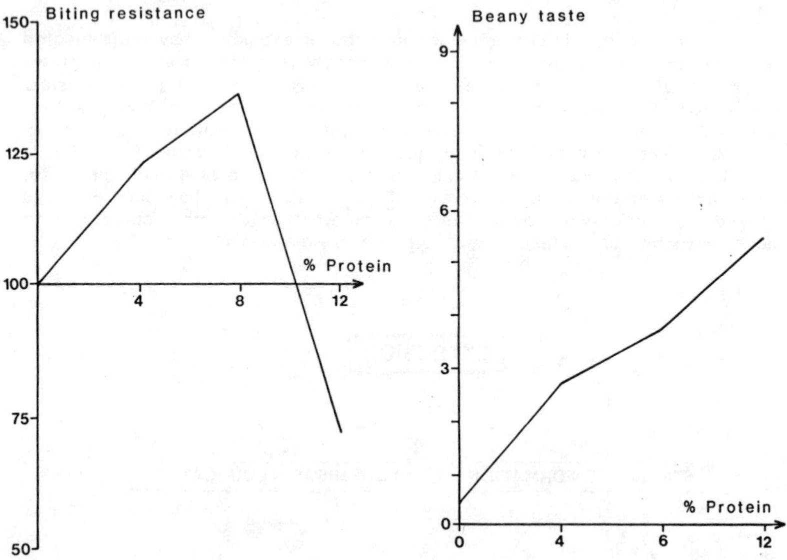

Figure 2 Biting resistance (a) and beany taste (b) in meat patties as function of the incorporation level of textured soy protein.

The sensory properties of extruded products can be influenced in various ways. Here, it will shortly be discussed how the sensory quality can be changed by selection of raw materials, changing of processing conditions, and/or post-treatments of the extruded items by drying, frying etc.

RAW MATERIAL

It it obvious that the selection of raw materials is very important for the properties of the extruded product. Appearance, taste, and texture will be influenced, if for instance a starchy raw material is mixed with protein or fibre. As an example I will relate to an investigation performed at SIK -The Swedish Food Institute (Andersson et al., 1981). We studied how the product properties were influenced by changing the proportions between the amounts of wheat starch and wheat bran used as raw materials for production of a "bread"- or "biscuit"-resembling, fibre-rich product. When the amount of wheat starch was increased and, consequently, the amount of wheat bran was decreased, the following was noticed:

A more open structure with larger "cells" was achieved; the colour became brighter, the hardness of the product decreased, the products became more gluey, the bitter taste increased, and the cereal taste decreased.

Also, different properties noticed in different varieties of the same raw material influence the product. The different behaviours during extrusion of corn with a high content of amylopectin and a low content of amylos ("waxy" corn) and corn with a low content of amylopectin and a high content of amylose are well-known. The waxy corn promotes puffing and give an extremely light, fragile product. Therefore, the relation between amylose and amylopectin can be used to control the texture of the extruded product. (Feldberg, 1969)

Another example of the possibility of regulating the texture is incorporation of gluten. Inclusion of gluten into soy proteins can yield tougher, chewier products if the gluten is prepared from hard wheat, while the use of gluten from soft wheat yields more tender, friable products. (Smith, 1975).

Also the fat content of the raw materials is important for the product quality. In order to investigate the influence of fat, two studies were performed at SIK-The Swedish Food Institute, which may be shortly reported. One study was carried out with oat flour. As opposed to corn and wheat flour, which are very easy to extrude and have a fat content of about 1.5 %, oat flour, which has a fat content of about 6%, is extremely difficult to extrude. It fails to expand in the extruder, coming out of the die as hard, compact pieces that disintegrate. There is also an uncontrollable increase in temperature during the extrusion process.

Suspecting the higher fat content in the oat flour to be the source of the problem, we decreased the fat content in the oat flour to about 1% by extraction with hexane. Then, when the fat-extracted material was extruded, the problems had vanished. The material flowed out of the extruder in strips, which had expanded some 250%, and the uncontrollable temperature increase had also disappeared.

In the case of oat flour the higher fat content was apparently a disadvantage. The reverse could also be true - better performance during the extrusion with a higher fat content. When up to 10 % corn oil was added to wheat starch, a slight increase in the degree of expansion was observed (from about 400% for pure wheat starch to about 440% for wheat starch with oil added). The oil content did not seem to influence the degree of expansion: 1.5 - 10 % corn oil resulted in about the same expansion. However, when 10 % corn oil was added, a part of the fat separated from the product and flowed out freely from the die.

PROCESS PARAMETER

Important process parameters for the product properties are i.e. water addition, retention time (which can be influenced by feed speed, screw speed, screw configuration, die size etc.), extrusion temperature, and pressure.

Generally, by changing the processing conditions the product properties can be regulated within very large ranges. A wheat flour processed at 105°C during a short time (retention-time 10-15 secs) has a typical, very mild cereal taste. In a microscope picture one can see that many starch granules still are unswollen. By the other hand, if the same wheat flour is processed at 200°C during a somewhat longer time (retention time 20-25 secs), the product will have a darker appearance and taste burned. The microscope picture shows that no unswollen starch granules exist any longer (Figure 3).

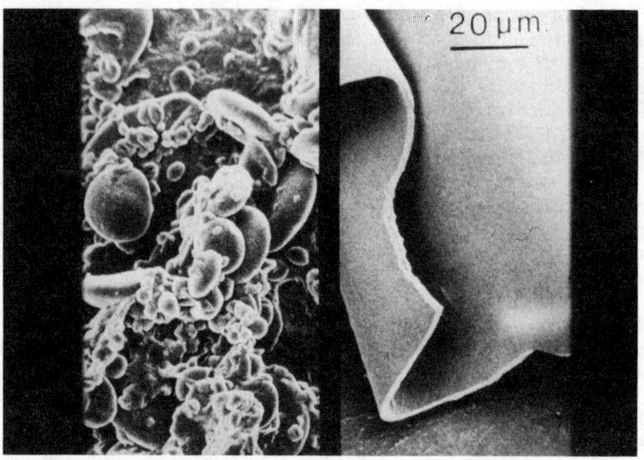

Figure 3 Scanning electron microscope picture of extruded wheat flour.
Left: Material treated at 105°C with 20% water added
Right: Material treated at 200°C with 0% water added
(Copywright: A-M Hermansson and M Langton, SIK-The Swedish Food Institute)

In the SIK investigation of fibre rich products mentioned earlier, we also studied the influence of screw speed and water addition on the product quality. When increasing the screw speed, the burnt taste increased, the cereal taste decreased, while no influence on appearance and texture was noticed. The increase in the burnt taste can possibly be attributed to the higher temperatures, which ought to occur locally in the screw channels, as a result of the higher shearing forces obtained when increasing the screw speed. When increasing the water addition during extrusion, the sensory properties of the products were influenced in the following ways: The structure became more open with larger "cells", the colour became brighter, the hardness decreased, and the stickiness increased, while no taste changes were observed.

POST-EXTRUSION TREATMENT

An after-treatment of the extruded material can also influence the properties. For instance, a too high addition level of water may give products that are tough and sticky. A drying step immediately after extrusion can change the texture so it becomes more crispy and friable. Often flavours (e.g. onion, garlic, barbecue, bacon) are added after extrusion, for instance by spraying a flavour solution on the surface. This procedure is often used for snacks. Another way to add flavours is to mix temperature stable flavours into the raw material mix.

For snacks several processes have been developed where a starchy material with about 30 % water is extruded as flat strips or compact pellets. The extruded half-products are dried under such conditions as to give the central parts of the products about 2.5 times as much moisture as the hard outer layers. Then, when frying in hot fat (180-210°C), crispy snacks are formed as a result of rapid vapourization of the moisture. (Harper, 1981)

CONCLUSIONS

The properties of extruded products are very important for the consumer acceptance. In order to be able to predict and control these properties it is essential that much work is devoted to explaining the extrusion process in detail. If we perform such investigations, and if we are able to find correlations between raw material properties, processing conditions, and product properties, we will also be able to optimize the process in regard to consumer acceptance (Figure 4).

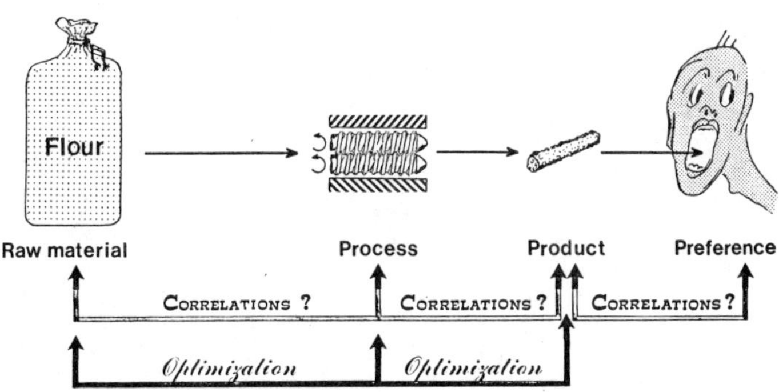

Figure 4. Schematic illustration of the extrusion process

REFERENCES

Andersson, Y. et al., 1981 — Extrusion cooking of a high-fiber cereal product with crispbread character. Cereal Chemistry 58(1981):5, p.370

Feldberg, C., 1969 — Extruded starch-based snacks. Cereal Sci. Today 14(1969), p.211

Harper, J.M., 1981 — Extrusion of foods. Volume II Boca Raton, Florida (CRC Press, Inc.) 1981, p.74

Smith, O.B., 1975 — Extrusion and forming: Creating new foods Fd.Eng. 47(1975):7 p.48

DECRISTALLISATION DU SACCHAROSE PAR CUISSON-EXTRUSION: APPLICATION EN CONFISERIE

M. BILLON
Ingénieur R & D
Département du matériel pour l'industrie alimentaire
Creusot-Loire 42700 Firminy - France

Résumé

Les produits de confiserie sont normalement fabriqués à partir de saccharose et de glucose, ces sucres se présentant à l'état amorphe. Pour décristalliser le saccharose, on commence généralement par le dissoudre dans de l'eau en présence de glucose. Il faut ensuite porter ce sirop à ébullition ménagée pour éliminer l'excès d'eau en évitant la reformation de fins cristaux.

Cette communication présente un procédé de décristallisation de sucre au moyen d'un cuiseur-extrudeur en présence de glucose avec une faible quantité d'eau dont l'excédent s'évapore à la sortie de la machine pour donner un produit directement utilisable en confiserie.

Abstract

Confectionery products are normally manufactured from sucrose and glucose, in the amorphous state. In order to ensure the decrystallization of sugar, sucrose is first dissolved in water in presence of glucose. Then, this syrup is boiled with care in order to evaporate the excess water, avoiding another formation of fine crystals.

This paper presents a process of sugar decrystallization by means of a cooker-extruder in presence of glucose with a small amount of water, of which the excess is evaporated at the outlet of the machine, giving a directly convenient product for confectionery.

1. INTRODUCTION

Les produits de confiserie sont normalement fabriqués à partir de saccharose, de glucose et d'eau. L'eau sert uniquement à dissoudre le saccharosequi est obtenu industriellement à l'état cristallin, afin d'assurer sa décristallisation. Généralement, cependant, on confectionne d'abord un sirop de glucose dans des récipients chauffés afin de faciliter la dissolution du saccharose qu'on y ajoute, puis on porte ce sirop à ébullition ménagée afin d'évaporer l'excès d'eau en évitant la reformation de fins cristaux. Cette méthode conventionnelle est donc pratiquée en discontinu et sous vide. On connaît des méthodes continues dans lesquelles l'ébullition s'effectue en couches fines de sorte que la température d'ébullition puisse être obtenue très rapidement et même à un degré plus élevé qu'en méthode conventionnelle. Par exemple, on utilise une vis qui tourne à l'intérieur d'un cylindre en laissant un espace entre le diamètre extérieur de la vis et le diamètre intérieur du cylindre pour permettre la formation d'une couche mince.

Dans toutes le méthodes connues, il est obligatoire de dissoudre auparavant le sucre dans de l'eau, puis d'évaporer celle-ci. On peut penser qu'un tel procédé n'est pas rationnel mais il est jusqu'ici indispensable puisqu'on ne peut decristalliser du saccharose par d'autres moyens.

L'emploi du cuiseur-extrudeur permet de décristalliser du sucre sans mise en solution préalable dans une importante quantité d'eau qui devra être évaporée ensuite.

2. DESCRIPTION DU PROCEDE.

Le procédé consiste à introduire un mélange de sucre cristallisé et de glucose dans un cuiseur-extrudeur à deux vis, muni d'une filière de sortie et d'un moyen de chauffage externe. Détaillons la description du cuiseur-extrudeur:

Chaque vis est composée d'un certain nombre d'éléments joints bord à bord et verrouillés sur un arbre cannelé. L'espace entre les vis est très faible (quelques 1/10 à 1 mm) et le profil conçu pour que les vis se nettoient réciproquement en tournant. Prenons l'exemple de la composition suivante pour des vis de 44,2 mm de diamètre. Nous avons d'amont en aval:

1/ Une section de longueur .... 200 mm
avec un pas de .... 50 mm
2/ Une section de longueur .... 100 mm
avec un pas de .... 35 mm
3/ Une section inverse de .... 50 mm
avec un pas de .... 15 mm
4/ Une série de disques
malaxeurs de longueur .... 50 mm
5/ Une section de longueur .... 200 mm
avec un pas de .... 15 mm

L'enveloppe, appelée "fourreau", qui est le cylindre entourant les vis, est munie à l'extérieur sur sa moitié en aval de deux induthermes ou bobines de chauffage par induction d'une puissance installée respective de 7 Kw et 3 Kw et se termine par deux filières de Ø = 2 mm.

L'alimentation du glucose dans le cuiseur-extrudeur peut se faire de deux manières: sous forme de sirop ou sous forme sèche, pulvérulente.

Dans le premier cas, le sirop est un produit commercial dont les caractéristiques sont les suivantes: D.E. 36 à 39 Matière sèche: 80%. Ce sirop est introduit par l'intermédiaire d'une pompe doseuse directement par une tubulure à l'entrée du fourreau.

Dans le second cas, le glucose sec est introduit par un doseur à poudre, séparé ou en mélange avec le saccharose, par l'ouverture d'entrée du fourreau, au moyen d'une trémie. Une certaine quantité d'eau est alors ajoutée au moyen de la pompe doseuse.

Dans les deux cas, le saccharose est introduit sous forme cristallisée en proportion égale avec le glucose.

Dans l'essai que nous décrivons, nous ajoutons 15 Kg/h de saccharose pour 15 Kg/h de glucose sec. La quantité d'eau ajoutée est de 2,4 litres/heure. La vitesse des vis est réglée à 150 tours/mn.

Si nous suivons la montée des températures, telle qu'elle est réglée et relevée par les sondes placées dans le métal du fourreau, nous obtenons les résultats suivants:

1ère moitié du fourreau: 90°c
1er indutherme : 180°c
2d indutherme : 180°c

Le produit fini sort à environ 130°c de la machine, se refroidit rapidement à 100°c par suite de l'évaporation puis progressivement jusqu'à la température ambiante.

Dans la première partie du fourreau, on est souvent obligé de maintenir la température au dessous de 100°c par une gaine de refroidissement, pour éviter une évaporation prématurée de l'eau. Cela est particulièrement sensible dans le cas d'alimentation du glucose sous forme pulvérulente, à cause de l'élévation mécanique intense de la température dans l'élément à pas inverse, dit "contre-filet". Une fois franchi ce contre-filet qui forme un bouchon, les constituants fondus dans une masse visqueuse ne peuvent plus revenir en arrière et suivent vers l'aval l'effet de pompage engendré par les deux vis co-pénétrantes.

3. RESULTATS ET COMMENTAIRES.

Après refroidissement, le produit est vitreux, limpide, incolore. Cet aspect est confirmé par examen au microscope binoculaite et au microscope polarisant. Le spectre de diffraction aux rayons X ne comporte rigoureusement aucune raie. Le halo obtenu témoigne d'un état parfaitement amorphe. L'humidité résiduelle de ces essais se situe entre 3 et 3,2 %.

Il nous faut maintenant suggérer une explication des phénomènes qui se déroulent dans la machine. Nous pouvons les décrire sans de trop grands risques d'erreur, compte-tenu des con naissances que nous avons acquises sur l'action des divers éléments de vis, notamment grâce à l'emploi de radio-éléments.

Si nous suivons la composition des vis telle que décrite plus haut, nous voyons que dans les 200 premiers mm les matières sont simplement transportées en mélange proportionnel sans transformation physique. Chaque sucre reste dans l'état où il a été introduit.

Dans la section suivante,le mélange subit une élévation de pression progressive qui culmine à la rencontre du contre-filet. C'est en effet dans ce dernier élément que la transformation a lieu. Il se produit à ce niveau quatre opérations:

- un mélange intense
- une rapide élévation de température d'au moins 40°c.
- une dispersion des temps de séjour faisant passer la moyenne à plus de 1 mn.
- une élévation de pression dûe à la fois à l'action mécanique et à la tension de vapeur d'eau.

Il s'ensuit donc: un broyage important des cristaux sous haute température (environ 150°c) et haute pression (environ 200 bars)pendant un temps suffisant pour produire à la fois une fusion et une dissolution dans l'eau présente dont l'excédent s'évapore en sortie de machine en revenant à la pression atmosphérique. Notre hypothèse actuelle est que le mécanisme de base est bien la dissolution du saccharose mais que l'action est accélérée par le mélange intime du saccharose et du glucose réalisé par le cuiseur-extrudeur, ainsi que par l'effet de broyage des vis.

En ajoutant aux matières premières les colorants et les arômes souhaités ou en modifiant les températures pour avoir un effet de caramélisation, on peut obtenir directement un produit apte à être moulé.

Quant à l'avantage économique de ce procédé, tout dépend de la différence des coûts entre procédé traditionnel et cuisson-extrusion. Dans nos essais, sur machine pilote dont on peut améliorer le rendement à l'échelle industrielle, nous avons consommé, tout compris, 0,2 Kw/h par Kg de produit obtenu. Nous pensons que la baisse relative des coûts de l'électricité nucléaire par rapport à ceux de l'énergie fossile jouera en faveur de la cuisson-extrusion.

RELATIONSHIP BETWEEN DIE-VISCOSITY, ULTRASTRUCTURE AND TEXTURE OF EXTRUDED SOYA PROTEINS

P. J. FRAZIER and A. CRAWSHAW
Dalgety Spillers Limited, Group Research Laboratory,
Research & Technology Centre, Cambridge, U.K.

Summary

Textured soya protein (TSP) was produced under standard conditions on a Brabender laboratory extruder from two different samples of soya grits of similar specification. Instron assessment of texture after retorting indicated marked differences in quality and the poor-texturing grits were shown to have a considerably higher apparent viscosity in the extruder die for the same conditions of temperature, moisture and shear rate. Examination of the TSP over a wide range of magnification by light, scanning electron and transmission electron microscopy showed that well textured material always had a greater homogeneity of structure, consistent with adequate plastication or 'melting' of the protein and laminar flow in the die, and supporting the existence of an upper viscosity limit for good texture.

1. INTRODUCTION

Extrusion cooking of defatted soya grits provides an economical method of converting their globular protein content into an insoluble network structure with a fibrous texture simulating meat. Such analogues have found wide application as meat extenders for human food and in canned petfoods where retort sterilisation poses a severe test of texture quality. Although many production problems have now been resolved, fundamental understanding of the mechanism of extrusion texturing is still relatively poor, as evidenced for example, by the great variability sometimes found in textured soya protein from feedstocks of apparently similar specification.

Our earlier studies of this problem (1) concluded that plastication of the protein was an essential stage in texturing and that achievement of a critical range of apparent viscosity in the extruder die was necessary for good TSP. Subsequently, we examined (2) seven extrusion process variables for their effect on TSP quality, using response surface techniques to define interactions and indicate regions of optimum processing conditions. This present paper returns to the problem of feedstock variability, examining the microstructure of good and poor TSP and relating it to viscosity and texture quality.

Microscopical techniques have been used extensively to examine the textural properties of food materials (3) but there are few reports on extruded soya. Cumming et al. (4, 5) used scanning electron microscopy (SEM) to observe structural changes between soya beans, defatted meal and TSP, and Aguilera et al. (6) followed changes in soya flour along the screw of a Wenger X5 extruder. Mixing prevailed until the last head, where most of the cooking occurred, strand formations were visible and the material was thought to be behaving like a 'melt'. Our observations on a Brabender extruder (7) showed compression and melting to occur more gradually along the barrel. A dense plastic mass formed in the die head with a pronounced laminar structure but fibres were only seen after the material expanded on exiting from the die.

Generally SEM has proved most useful in documenting gross morphological differences (8,9). However, light microscopy (LM) (10) and transmission electron microscopy (TEM) (9) have been found more valuable in relating protein and insoluble carbohydrate distribution to texture integrity.

2. EXPERIMENTAL

2.1 Materials

Two samples of defatted soya grits were used ('A' and 'B') from the same batches described previously (1) having respectively : protein, 52.3% and 51.9%; moisture, 9.4% and 9.0%; fat (by mixed solvent), 2.5% and 2.0%; ash, 5.9% and 6.7% and nitrogen solubility index (NSI), 43 and 32.

2.2 Extrusion Texturing

A Brabender 20DN ¾ inch single screw laboratory extruder was used (1,2) under the following process conditions to produce samples for microscopy : moisture 40% dsb; screw compression ratio 4:1; screw speed 250 rev/min; barrel temperature, feed end 125°C, die end 150°C; die temperature 150°C; die exit diameter 6mm. The resulting extrudate was dried for 2 hours at 70°C to a stable moisture content.

2.3 Viscosity Determination and Texture Assessment

Flow curves were constructed as before (1) from pressure drop measurements at 150°C for different screw speeds using a 10:1 length : diameter die head of 8mm internal diameter and making approximations for the pressure drop in the conical entrance region. The system was calibrated at 25°C using a silicone oil of known viscosity.

Texture assessment was carried out on samples of retorted TSP using an Ottawa Texture Measuring Cell (OTMS) ($50cm^2$ base, 8 wire grid) on an Instron as described previously (1,2).

2.4 Light Microscopy

Small pieces of dry-stabilised TSP were fixed overnight in a mixture of 3% glutaraldehyde, 1.5% formaldehyde and 1.5% acrolein buffered with 0.05M phosphate to pH 7.0. After washing in buffer, the specimens were dehydrated in a graded acetone series, embedded in epoxy resin and stained with periodic acid, Schiff/amido black (PAS/AB) for carbohydrate and protein.

2.5 Scanning Electron Microscopy

Samples of TSP were examined in three conditions: dry-stabilised; after rehydration in excess distilled water (7:1) overnight; and after canning and retorting as for texture assessment. All samples were frozen in liquid nitrogen and freeze dried. Dried specimens were fractured and small pieces mounted, fracture face up, on a specimen stub, sputter coated with 20nm gold-palladium and examined in a Cambridge Instruments S4 scanning microscope.

2.6 Transmission Electron Microscopy

Small pieces of dry-stabilised TSP (approx. $2mm^3$) were fixed and washed as above (2.4) and post-fixed in 1% OsO_4 for 2h. The specimens were dehydrated in a graded acetone series, placed in 70% methanol containing 2% saturated aqueous uranyl acetate for 16h at 4°C, passed through three changes of anhydrous acetone and embedded in epoxy resin. Thin sections were cut on an LKB ultratome, stained with lead citrate and examined in an AEI EM6B transmission electron microscope.

3. RESULTS AND DISCUSSION

3.1 Viscosity and Texture Measurement

As reported previously (1) soya 'A' textured well, producing an expanded, laminar structure which survived retorting. The Instron texture measurement (OTMS peak force) was between 220 and 200N for extrusion moisture contents from 36 to 42% dsb. Soya 'B' however, formed a very dense extrudate with little evidence of laminar structure and, on retorting, the product disintegrated completely, giving OTMS values below 100N. The flow curves in Figure 1 show clearly that, across the ranges of moisture from 36 to 44% dsb and shear rates from 20 to 50 s^{-1}, the poor-texturing soya protein 'B' always had a viscosity in the die at least 1000 Pa s higher than the good material 'A', regardless of shear thinning effects.

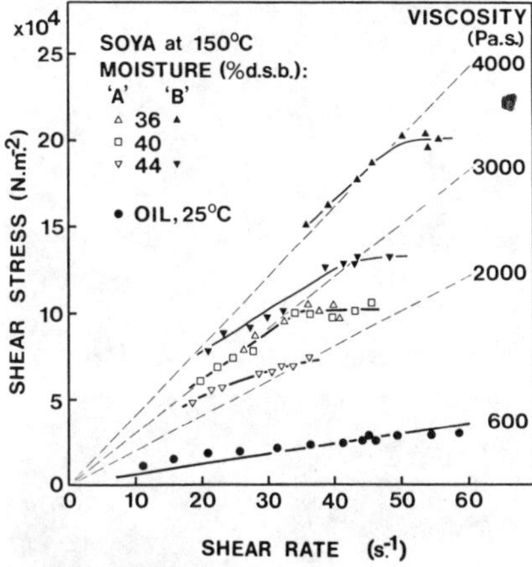

Figure 1. (left) Flow curves for good quality 'A' and poor quality 'B' plasticised soya protein in the extruder die at 150°C over a range of moisture levels; and calibration check using silicone oil of 600 Pa s viscosity at 25°C.

Figure 2. (below) Light micrographs (x 200) of sections of dry-stabilised TSP 'A' and 'B' stained to show carbohydrate (c) and protein (p).

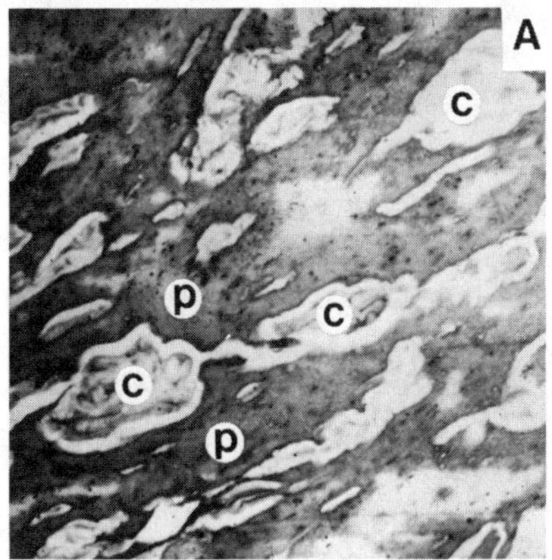

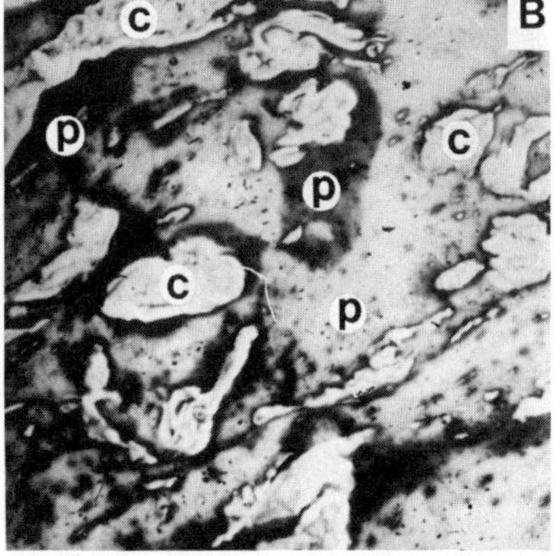

3.2 Light Microscopy

Typical sections of dry-stabilised TSP from soya grits 'A' and 'B' are shown in Figure 2. The PAS/AB stain resulted in a clear differentiation between the two major components of TSP, insoluble carbohydrate material showing as magenta (PAS +ve) against a blue-stained (AB) protein background.

In both TSP samples there was a sharp separation of the two components, the PAS +ve material forming lacunae of varying sizes within the protein-rich matrix. However, not all these cavities were carbohydrate bodies; many were empty and corresponded in size and distribution to the numerous air spaces which were present in both specimens. On a gross scale TSP 'A' showed generally a regular and more uniform protein background than TSP 'B', possibly indicating a more homogeneous protein 'melt' during extrusion of the better-textured material.

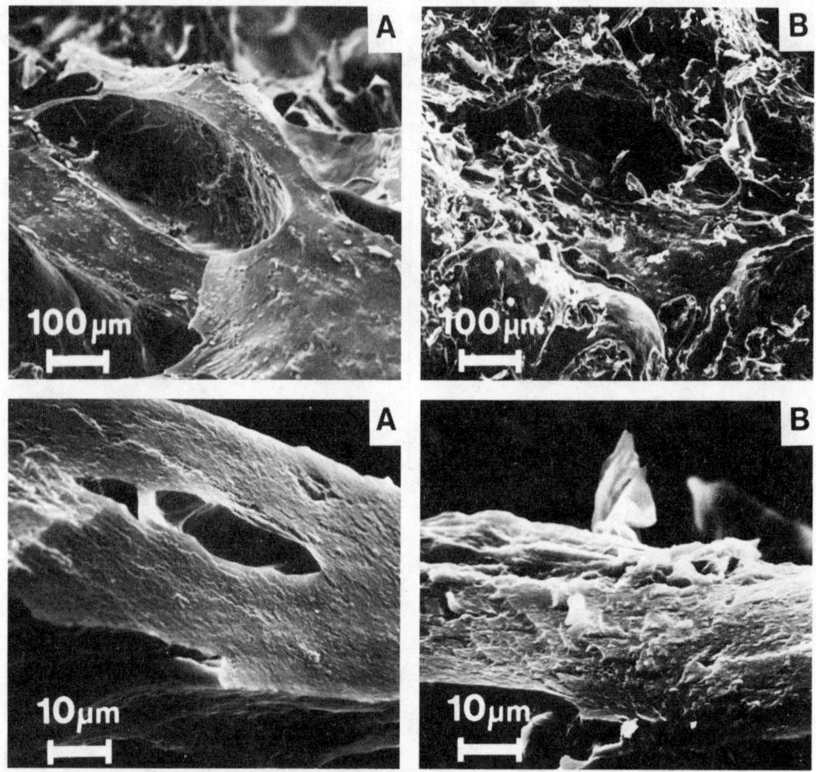

Figure 3. Scanning electron micrographs of dry-stabilised TSP 'A' and 'B' (Upper: x 60, lower: x 600)

3.3 Scanning Electron Microscopy

Examination of TSP in its dry-stabilised state from grits 'A' and 'B' produced scanning micrographs of which typical examples are shown in Figure 3. Both TSP samples were made up of a large, open network with walls of varying thickness. Holes were up to 1-1.5mm in diameter and walls were between 50 and 150µm thick. The fractured surfaces of the walls of the well-textured grits were relatively smooth, while their sides had a coarse fibrous appearance at low magnification. High magnification did not resolve any fine fibrillar structure however, rather the wall material appeared dense and compact but tending to split in places as though laminated. The poorly-textured grits while showing many areas of broadly similar structure to the good TSP, tended to have a more 'bitty' appearance with small flakes of material seen adhering to, or erupting from, the surface both at low and high magnification.

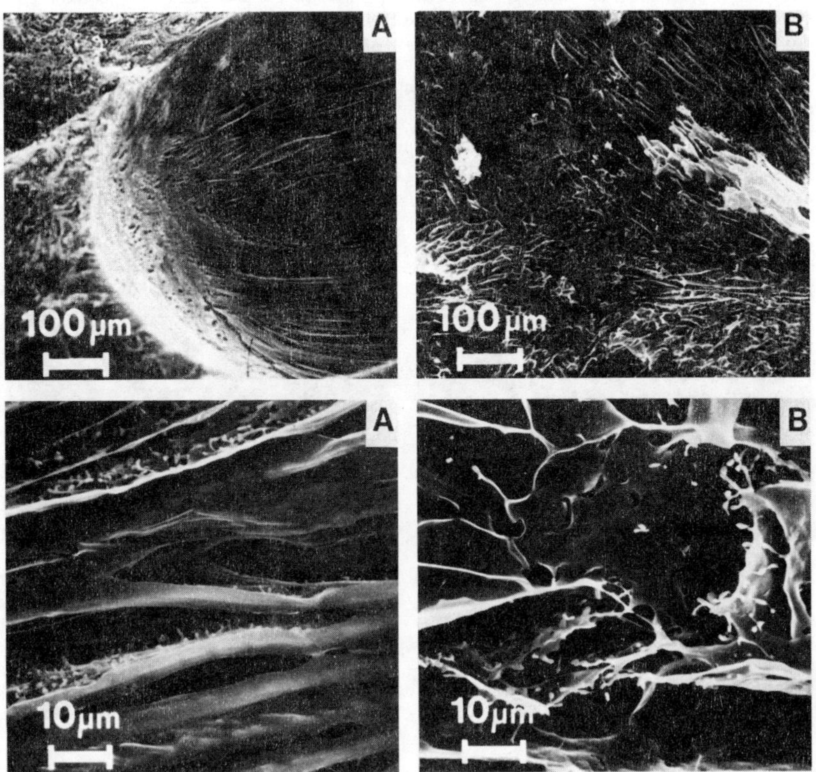

Figure 4. Scanning electron micrographs of rehydrated and retorted TSP 'A' and 'B'. (Upper: x 60, lower: x 600)

Rehydration of the TSP had the effect of opening up numerous splits in the thick walls around the larger holes but again no fine fibre structure was seen. Both TSP specimens appeared to consist of interconnected laminated sheets, having a more open structure than the dry-stabilised material. As before, the poorer textured sample was characterised by the presence of many fine flecks of material, widely distributed over the surface.

Figure 4 shows the effect of rehydration and retorting on the structure of good 'A' and poor 'B' TSP. The large holes seen clearly in the dry-stabilised, well textured TSP (Figure 3) were still clearly visible after retorting (Figure 4) but the walls were now seen to be made up of a large number of small interlocking fibres and sheets about 0.5 µm across and having, in places, a fine beaded sub-structure. Thus the extensive hydration associated with retorting had opened out the laminar structure, providing many cavities for holding free water, and giving a finely fenestrate appearance to the fractured surfaces of the main walls. (Because of the large amount of water present, and subsequently frozen, in the sample the structure would have been sensitive to ice crystal damage and this appearance may be partly a freezing artifact). The poorly textured material had an overall appearance of fine fenestrations similar to the good TSP except that there was very little evidence of interlocking fibres or sheets. Instead the material resembled a sponge-like, cellular mass with a sub-structure of many small beads visible at the edges of the cells.

In summary, the scanning electron microscopic appearance of TSP was consistent with the properties of a material formed as a viscous plastic melt. Thus the good TSP had a relatively smooth, homogeneous appearance indicating an adequate 'melt', while the poor TSP had evidently not formed a sufficiently homogeneous plastic mass, resulting in flecks of material readily sloughing off its surface instead of contributing to the laminar struture.

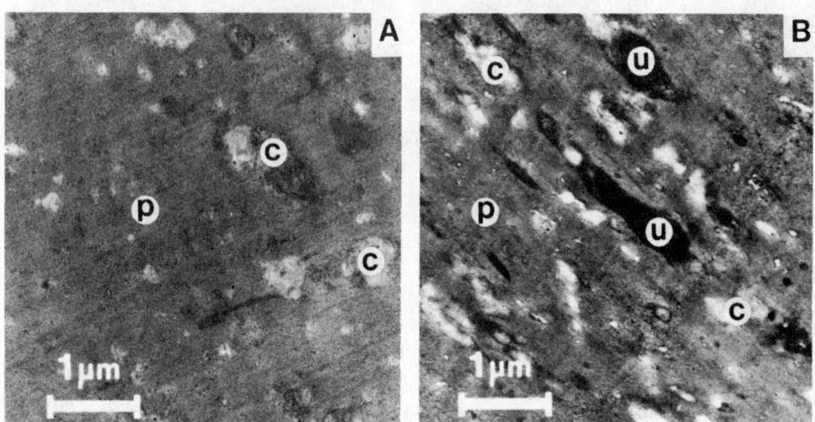

Figure 5. Transmission electron micrographs (x 10,000) of thin sections of dry-stabilised TSP 'A' and 'B' (c : carbohydrate, p : protein, u : unplasticised protein)

3.4 Transmission Electron Microscopy

Typical examples of the ultrastructural appearance of dry-stabilised TSP at high magnification are shown in Figure 5. There was little definitive or readily recognisable structure in either sample but there were regions of differing electron density which provided a means of identifying components. The bulk of the material was made up of a medium electron-dense component which formed the plasticised, continuous matrix of insoluble protein. Within this material there were electron dense bodies of varying size, probably representing the protein component in a separate (unplasticised) state. These bodies were both larger and very much more prevalent in the poorly textured sample than in the well textured material. Also contained within the continuous matrix were electron-transparent regions. Some of these may have been micro-air cavities, although it should be noted that all TEM samples were taken from air-cell wall material and contained few air spaces. Most electron transparent bodies therefore probably represented carbohydrate material. This was evenly distributed throughout both samples of TSP but the average particle size was smaller in the well-textured material. A few membranous fragments were also seen, associated with the carbohydrate material, particularly in the well textured sample.

Overall, these high magnification micrographs reinforce the conclusions from LM and SEM studies, that the well textured material shows greater homogeneity of structure consistent with the extrudate having been plasticised or 'melted' to a greater extent than the poorly textured material under the same extrusion conditions of temperature, moisture content, pressure and shear field.

4. CONCLUSIONS

Well-textured, retort-stable TSP with an OTMS retorted texture value above 200N, can be produced on a laboratory scale, single screw extruder when the plasticised protein flowing in the die registers an apparent viscosity below 3000 Pa s. Under these conditions TSP microstructure at LM, SEM and TEM levels shows a relatively smooth and homogeneous protein matrix, having the appearance of a 'melt', containing small, evenly distributed carbohydrate bodies. When apparent viscosities exceed this value under the same process conditions, as a result of differences in the feedstock, texture quality is markedly reduced and the microstructure is indicative of non-uniform or incomplete plastication. Further work is required to establish the biochemical reasons for differences in soya protein viscosity under identical plasticating conditions.

ACKNOWLEDGEMENTS

Our thanks are due to the U.K. Ministry of Agriculture, Fisheries and Food for financial support, to Mrs. L. E. Gardiner for assistance in operating the extruder and carrying out texture tests and to Dr. P. Echlin, University of Cambridge, for taking the light and electron micrographs.

REFERENCES

1. FRAZIER, P.J., CRAWSHAW, A., STIRRUP, J.E., DANIELS, N.W.R. and RUSSELL EGGITT, P.W. (1980). In : Food Process Engineering, Vol. I. (Linko, P., Malkki, Y., Olkku, J. and Larinkari, J., eds) Applied Science Publishers Ltd., 768-776.
2. FRAZIER, P.J., CRAWSHAW, A., DANIELS, N.W.R., and RUSSELL EGGITT, P.W. (1983). J. Food Engineering, 2, 79-103.
3. STANLEY, D.W. and TUNG, M.A. (1976). In : Rheology and Texture in Food Quality (de Man, J.M., Voisey, P.W., Rasper, V.F. and Stanley, D.W., eds) AVI Publishing Company, 48-78.
4. CUMMING, D.B., STANLEY, D.W. and de MAN, J.M., (1972). Can. Inst. Food Sci. Technol. J., 5, 124-128.
5. MAURICE, T.J. and STANLEY, D.W., (1978). Can. Inst. Food Sci. Technol. J., 11, 1-6.
6. AGUILERA, J.M., KOSIKOWSKI, F.V. and HOOD, L.F. (1976) J. Food Sci. 41, 1209-1213.
7. DANIELS, N.W.R. (1981). In : Utilization of Protein Resources (Stanley, D.W., Murray, E.D. and Lees, D.H., eds). Food and Nutrition Press Inc., 180-207.
8. TARANTO, M.V., CEGLA, G.F., BELL, K.R. and RHEE, K.C. (1978). J. Food Sci. 43, 767-771.
9. TARANTO, M.V., CEGLA, G.F. and RHEE, K.C. (1978). J. Food Sci. 43, 973-979.
10. CEGLA, G.F., TARANTO, M.V., BELL, K.R. and RHEE, K.C. (1978). J. Food Sci. 43, 775-779.

PHYSICAL AND FUNCTIONAL PROPERTIES OF WHEAT STARCH AFTER EXTRUSION-COOKING AND DRUM-DRYING

P. COLONNA, J.L. DOUBLIER, J.P. MELCION, F. de MONREDON and C. MERCIER
Institut National de la Recherche Agronomique
Centre de Recherches Agro-Alimentaires
NANTES - France

Summary

Native wheat starch was modified by twin screw extrusion-cooking under five different conditions of moisture and temperature and by drum-drying under two various conditions. The comparison of the transformation intensity was followed on the properties of the end-products. For both technology, the end-products were transformed into a continuous solid phase of melted starch, including variable amounts of air bubbles. In comparison with native starch no modification of the specific gravity was detectable, whereas the surface area of the products was more reduced by extrusion than by drum drying. Extrusion-cooking led to a macromolecular degradation of both amylose and amylopectin by random chain splitting, observed by intrinsic viscosity measurement, Sepharose CL-2B gel permeation chromatography and average molecular weight determination. In contrast, drum-drying as well as the extrusion-cooking under the so called "pasta" conditions, degraded only slightly starch components. The rheological behaviour of modified starches, characterized by Brabender viscoamylograms and flow curves, can be partly explained by their respective macromolecular degradation.

1. INTRODUCTION

Pregelatinized starches refer to cooked starches, when prepared by complete gelatinization and drying of the products (1) and where the destruction of the granular structure is the major physical event. Their main properties are water absorption and water solubility as soon as they are mixed with water,leading to instant starch slurries prior heating.Previously pregelatinized starch was prepared by drum-drying, which is recognized difficult to handle.Recently, extrusion-cooking, becoming popular, is more and more used for achieving such starch gelatinization which would be therefore a challenging technique. In contrast to drum-drying, this latter has been extensively studied (2, 3), mainly for the functional properties of the extruded products (4,5,6). Researches have been devoted to the physico chemical changes occuring during the treatment (6-10), whereas the comparison of the effects between extrusion-cooking, roll- and drum-drying have been rarely studied and only on cereals (11,12,13) and corn- soy- whey-mixtures (14). Although various authors (15,16) have suspected a molecular degradation during these two technologies, amylose and amylopectin depolymerizations had been shown to occur only up to now during extrusion of tuber starches (4,8).

The aim of this study is to compare morphological changes and chemical tranformations of wheat starch, when modified by extrusion-cooking and drum drying. Their characteristics will be used to understand their paste behaviour .

2. EXPERIMENTAL

Prime wheat starch was extruded on a semi-industrial twin-screw extruder (Creusot Loire BC45) in five different conditions of moisture and temperature, in order to obtain a large range of processing intensity. Feed rate (45 Kg h^{-1} on dry basis) and screw speed (∿270rpm) were constant. Drum-drying was carried out on monorolls at a speed of 10rpm, with a pressure of 10 bars, on starch suspension containing 450g per liter ; the 2 assays were different by a pre-cooking on one sample.

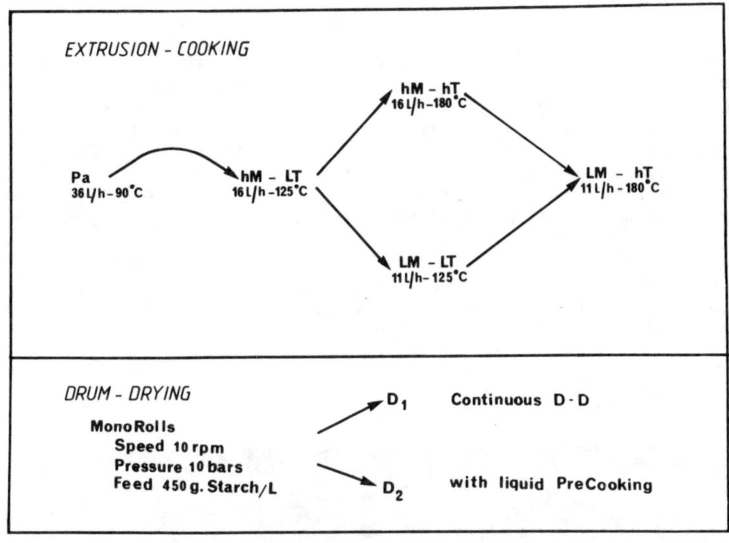

Fig. 1. Schematic diagram of operating conditions used for extrusion-cooking and drum-drying.

Every modified starch was ground and sieved to the 100-125μ fraction. Surface area, S ($cm^2.g^{-1}$), was calculated from argon and sorption measurements using a Quantasorb equipment by the dynamic method of Nelson and Eggersten (17), from the BET laws. Specific gravity, ρ ($g.cm^{-3}$) was determined by the pycnometer method (volume ∿ 25ml) using toluene at 25°C on 2.5g sample, previously ground to less than 50μ, in order to remove air bubbles.

Scanning electron microscopy was carried out, as described by Colonna et al. (18). On every treated sample starch, primary chemical structure was characterized by β-amylolysis, isoamylase debranching (8) and iodine binding capacity (19). Macromolecular degradation was followed by measurement of the intrinsic viscosity in 0.2N KOH and by gel permeation chromatography on Sepharose CL-2B eluted with 0.1N KOH (8).

Some starches were fractionated into amylose and amylopectin using thymol or butanol complex procedure (20). Average molecular weights were measured by the end-group determinations ($\overline{M}n$), the viscosimetry ($\overline{M}v$) and the light scattering ($\overline{M}w$) methodology (21).

Viscograms of 8.5% dispersions were performed with a Brabender viscograph type E on the 250cmg sensitivity. Rheological measurements on these dispersions were also carried out after cooking at 30°C with a co-axial cylinders viscometer (Rheomat 30) and the flow curves were plotted as described by Doublier (22). Before the viscosity measurements, dispersions were filtrated on sintered glass (#0) in order to avoid the presence of some clumps. It has been verified that this filtration was not changing significantly the final concentrations.

3. RESULTS AND DISCUSSION

For both processes, native birefringent starch granules (16 ± 1.4μ) are transformed into a continuous solid phase in which more or less air bubbles are entrapped. This character is not useful in all thickeners uses, since pregelatinized starches are always ground before utilization. For the sample extruded under our so called "pasta" conditions (Pa) (Fig. 2A) 20-40% of the extrudate volume is occupied by air bubbles (80-150μ diameter). Samples obtained by drum-drying(D1 and D2)(Fig.2B)look like flat and thin(∿150μ) flakes, with 40-60% of entrapped bubbles. Two faces must be distinguished : a bright one, in contact with the roll surface and a dull one. For the extruded samples (hM-LT, hM-hT, LM-LT and LM-hT) (Fig. 2C) a lattice of melted starch (5-10μ thin) is formed, with large air cells, whose diameter is within the range 0.5-5mm. This last characteristic is the main morphological difference between the four extruded samples.

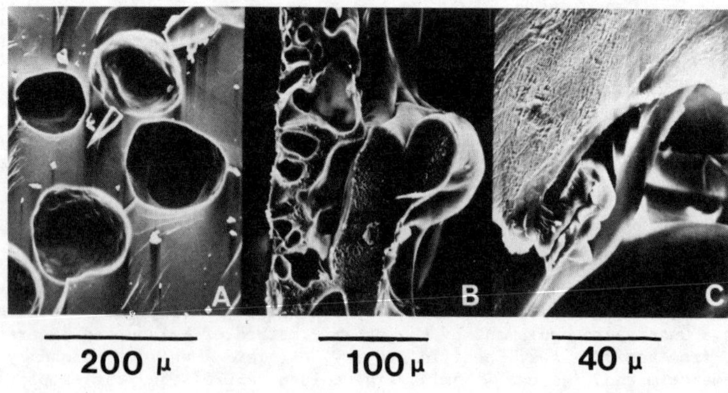

Fig. 2. Scanning electron micrographs of wheat starch samples. A : extruded "Pa" ; B : drum-dried D2 ; C : extruded LM-hT.

No variation in specific gravity (∿1.47-1.48g.cm^{-3}) is detectable (Table I). The specific surface for native starch is of 1220cm^2.g^{-1}, corresponding to a theoretical value (1320cm^2.g^{-1}) deduced from the granule size distribution and specific gravity. Surface area decreases greatly to 620-650cm^2.g^{-1} for drum-dried samples and to 270-480cm^2.g^{-1} for extruded ones. Since a solid flake of dimensions (100x100x10μ) should have a specific surface of 1600cm^2.g^{-1}, these results demonstrate that sieved ground starch are large particules (∿125μ), roughly cubic and impervious to argon. At the present time, there is a complete failure of methodology for describing such porous and solid structure in terms of porosity (accessibility to water) and hydrogen bonds strengths between macromolecules.

Table I. Solid phase properties of native and modified wheat starches

Starch samples	Specific gravity (g.cm^{-3})	Surface area (cm^2.g^{-1})
Native	1.48	1220
Extruded		
Pa	1.48	230
hM-LT	1.47	500
hM-hT	1.47	390
LM-LT	1.48	270
LM-hT	1.48	480
Drum-dried		
D1	1.47	650
D2	1.47	620

After thermal modification, no difference can be detected between the β-amylolysis limits (58.4-60.8%) and the α-1,6 linkages contents (3.6-4.0%) of all samples (Table II). The iodine binding capacities (IBC) are still within the range 5.2-6.1mgI$_2$ per 100mg of dry starch. As pure amylose has an IBC of 20.7, the apparent amylose contents are 25.1-29.5%. In contrast, from native starch (210ml.g^{-1})(Table II), the intrinsic viscosity of the treated starches decreases slightly (D1 : 188ml.g^{-1}) or strongly (LM-hT : 70ml.g^{-1}). Since the primary chemical structure is not modified by the 2 processes, the lowering of the intrinsic viscosity demonstrates that the macromolecular structure of wheat starch must be degraded. The behaviour of modified starches, compared to native starch, on gel permeation chromatography, carried out on Sepharose CL-2B, confirms this assumption (Fig. 3).

The relative amount of excluded material, corresponding to the amylopectin fraction with a λmax at 550nm, decreases from 82.1% for native starch to 77.9% for D1 and up to 8.5% for the extruded starches (Table II). Simultaneously to the decrease of the intrinsic viscosity of extruded samples, an amylopectin-like material (λmax 550nm) appears located at Kav between 0.05-0.30 and the amylose fraction, absorbing at λmax ∿ 620nm, appears later on the gel, up to Kav 0.60, indicating a lower molecular weight. Samples, treated by drum-drying, do not show any amylopectin-like material and the amylose is only slightly degraded . When extrusion-cooking is carried out under the conditions of "pasta" transformation (Pa sample), the process modifies slightly starch molecules, similarly to drum-drying.

Since the precited methods are unable to quantify any molecular weight decrease, starch fractionation, followed by direct average molecular weight determinations have been carried out on the native and three modified starches(LM-LT, LM-hT, and D2). $\overline{Mn}$ and $\overline{Mv}$ of amylose (Table III) decrease for every sample. The polydispersity ratio $\overline{Mv}/\overline{Mn}$ does not change when degradation increases from native amylose to samples D2, LM-LT and LM-hT Amylopec-

Table II. Molecular characteristics of native and modified wheat starches

Starch samples	Iodine binding capacity (mgI_2/100mg) (dry starch)	β-amylolysis limit (%)	α-1,6 linkages (%)	Intrinsic viscosity $[\eta]$ ($ml.g^{-1}$)	Elution profile on Sepharose CL-2B: Material with Kav<0.1 (%)	Kav of last fraction with λ-max ∿540-550nm	Kav of first fraction with λmax ∿ 610nm
Native	5.9	59.6	3.9	210	82.1	0.04	0.21
Extruded							
Pa	5.3	59.7	3.8	180	59.5	0.05	0.35
hM-LT	6.0	59.8	3.6	119	38.5	0.07	0.50
hM-hT	5.8	58.4	3.7	105	33.6	0.12	0.54
LM-LT	6.1	60.8	3.9	90	15.2	0.25	0.58
LM-hT	5.4	60.8	4.0	70	8.5	0.30	0.60
Drum-dried							
D1	5.2	60.3	3.7	188	77.9	0.05	0.23
D2	5.5	58.4	3.6	182	64.7	0.06	0.39

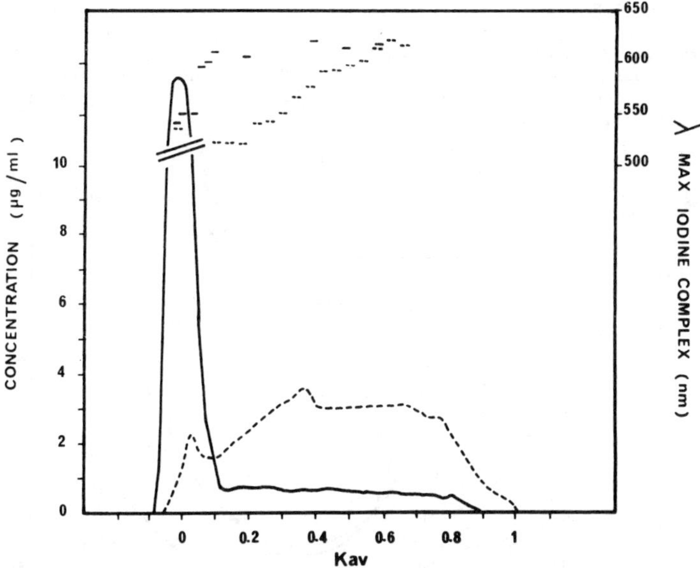

Fig. 3. Elution profiles from Sepharose CL-2B of native (——) and extruded LM-hT (---) wheat starches.

tin $\overline{Mw}$ decreases also, but with a higher intensity than amylose $\overline{Mn}$ and $\overline{Mv}$.

Table III. Average molecular weights ($g.mole^{-1}$) of native and modified wheat starch components

Starch samples	Amylose			Amylopectin	
	$\overline{Mn}\times10^{-3}$ (1)	$[\eta]$ (2)	$\overline{Mv}\times10^{-3}$ (3)	$[\eta]$ (2)	$\overline{Mw}\times10^{-3}$ (4)
Native	158	165	409	97	58,300
Extruded					
LM-LT	99	120	272	46	3,800
LM-hT	93	98	210	36	2,700
Drum-dried					
D2	153	152	368	75	22,700

(1) number average molecular weight ; (2) intrinsic viscosity ($ml.g^{-1}$) ; (3) viscosimetric average molecular weight ; (4) weight average molecular weight.

Although amylose and amylopectin are more or less degraded by both thermal processes, the macromolecular degradation is much more drastic with extrusion and particularly under the severe conditions of low moisture (161/h) and high temperature (180°C). The absence of oligosaccharides and the appearance of amylose polydispersity support arguments for a pure random chain scission (23,24). The mechanical shear degradation by extrusion-cooking imposes severe limitations on uses in case of effectiveness is determined by high molecular weight.

Viscograms of the extruded samples (hM-LT, hM-hT, LM-LT and LM-hT) display a low consistency in comparison to the drum-dried and the pasta-like one (Fig. 4).

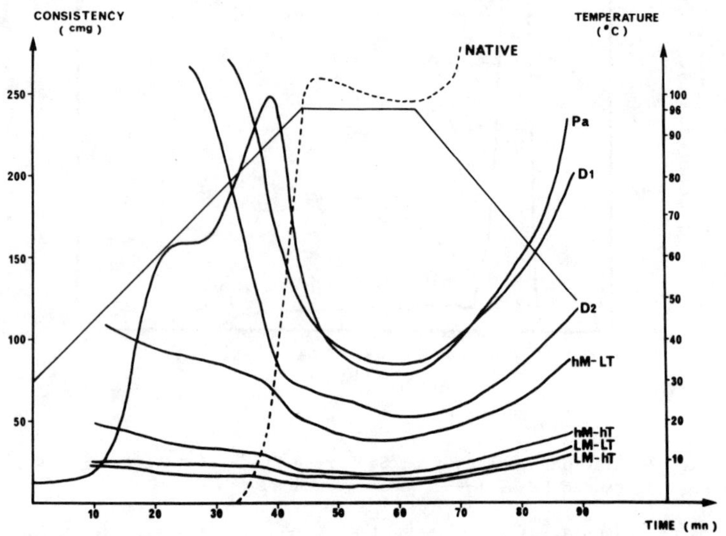

Fig. 4. Brabender consistency curves of native and modified wheat starch.

A large range of hot-consistencies (at 96°C) is observed between these four extruded starches : 10.1, 13.3, 17.9 and 37.4cmg for LM-hT, LM-LT, hM-hT and LM-LT respectively. As expected, drum-drying leads to profiles with a very high consistency after cold dispersion which decreases sharply on cooking. In the meantime, the porridge-like texture of cold dispersions is transformed into a smooth texture, ressembling to usual starch pastes. On another hand, pasta-like sample (Pa) is easily dispersed in cold water with a very low consistency and a highly viscous dispersion is obtained after heating : hot consistency is very close to this of D1 and cooling leads to a slight higher viscosity than for D1. Consistency profile is very peculiar with a maximum for about 85°C and a sharp decrease between 92°C and 96°C.

Flow curves of cold dispersion (Fig. 5), plotted in logarithmic coordinates, display a slope of about 0.6-0.7, indicating a highly shear-thinning behaviour. The viscosities are in the same order as on the viscograms with however a better accuracy. Apparent viscosities of LM-LT and LM-hT are very low : 0.21 and 0.36 Pascals x S, respectively at $57s^{-1}$.On the contrary hM-LT displays a cold flow behaviour close to this of one drum-dried sample (D2), 0.84 against 1.02 Pascals x S at this same shear rate, whereas

Pa is largely more viscous than D1 : 3.7 against 1.7 Pascals x S.

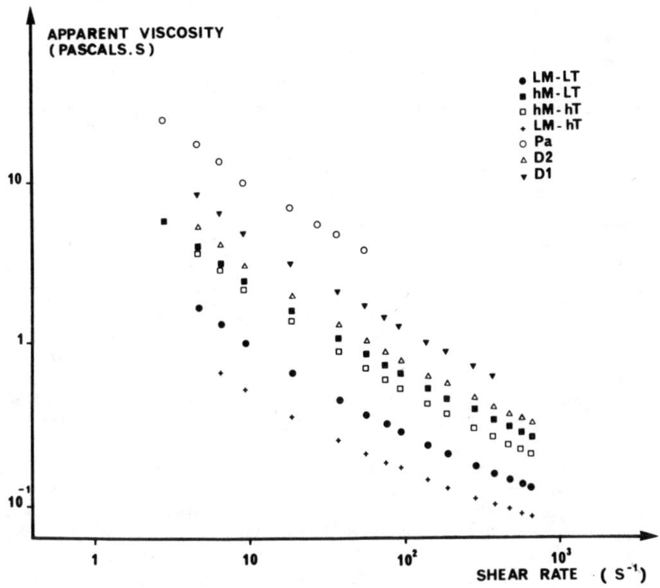

Fig. 5. Flow curves in logarithmic coordinates of modified starches (temperature : 30°C)

4. CONCLUSION

Extrusion-cooking depolymerizes wheat starch macromolecules with a variable intensity depending upon the process severity, the more drastic conditions being low moisture and high temperature. In contrast, drum-drying degrades only slightly starch components. Morphological modifications concern essentially the expansion in case of extrusion, but since air bubbles are not communicating, there is mainly a decrease of "accessible" surface area. At the present state, porosity measurements by gazeous adsorption are unable to describe functional properties of treated starches. The rheological behaviour of extruded and drum-dried starches are following the macromolecular degradation, observed by gel permeation and average molecular weight determination. By acting on the ratio moisture/temperature, it is possible to obtain by extrusion-cooking, various textures, expanded or not, with typical functional properties which, under peculiar conditions, can be similar to the effect of drum-drying, known as a less-versatile technology.

5. REFERENCES

1. POWELL, E.L. (1967). Production and use of pregelatinized starches. In Starch : Chemistry and Technology, vol. II. Edited by R.L. WHISTLER and E.F. PASCHALL, published by Academic Press.
2. HARPER, J.M. (1981). Extrusion of foods, vol. 1 and 2. Published by CRC Press, Boca Raton.
3. LINKO, P., COLONNA, P. and MERCIER C. (1981). "HTST-Extrusion cooking". In Advances in Cereal Science and Technology, vol. 4, 145-235, published by A.A.C.C.
4. MERCIER, C. (1977). Effect of extrusion-cooking on potato starch using a twin-screw french extruder. Stärke, 29, 48-52.
5. MERCIER, C. and FEILLET, P. (1975). Modification of Carbohydrate Components by Extrusion-Cooking of Cereal Products. Cereal Chem., 52, 283-297.
6. MERCIER, C., CHARBONNIERE, R., GALLANT, D. and GUILBOT, A. (1979). Structurl modification of various starches by extrusion-cooking with a twin-screw french extruder. In "Polysaccharides in Foods", 153-170. Edited by J.M.V. BLANSHARD and J. MITCHELL, published by Butterworths.
7. MERCIER, C., CHARBONNIERE, R., GREBAUT, J. and de la GUERIVIERE, J.F. (1980). Formation of amylose-lipid complexes by twin-screw extrusion-cooking of manioc starch. Cereal Chem., 57, 4-9.
8. COLONNA, P. and MERCIER, C. (1983). Macromolecular modifications of manioc starch components by extrusion-cooking with and without lipids. Carbohydrate Polymers, 3, 87-108.
9. GOMEZ, M.H. and AGUILERA, J.M. (1983). Changes in the starch fraction during extrusion-cooking of corn. J. Food Science, 48, 378-381.
10. OWUSU, J., VAN de VOORT, F.R. and STANLEY, D.W. (1983). Physicochemical changes in corn starch as a function of extrusion variables. Cereal Chem., 60, 319-324.
11. ANDERSON, R.A., CONWAY, H.F., PFEIFER,W.F. and GRIFFIN,E.L.(1969a). Cereal Sci. Today, 14, 4-12.
12. ANDERSON, R.A., CONWAY H.F., PFEIFER, W.F. and GRIFFIN, E.L. (1969b). Roll and extrusion-cooking of grain sorghum grits. Cereal Sci. Today, 14, 372-375 + 381.
13. ANDERSON, R.A., CONWAY, H.F. and PEPLINSKI, A.J. (1970). Gelatinization of corn grits by roll-cooking, extrusion-cooking and steaming. Stärke, 22, 130-135.
14. AGUILERA, J.M. and KOSIKOWSKI, F.V. (1978). Extrusion and roll-cooking of corn-soy-whey mixtures. J. Food Sci., 43, 225-227 + 230.
15. LORENZ, K. and JOHNSON, J.A. (1972). Starch hydrolysis under high temperatures and pressures. Cereal Chem., 49, 616-628.
16. CHIANG, C.Y. and JOHNSON, J.A. (1977). Gelatinization of starch in extruded products. Cereal Chem., 54, 436-443.
17. NELSON, F.M. and EGGERSTEN, F.R. (1958). Quoted in LOWELL, S. Introduction to powder surface area (1979) pp. 189. Edited by Wily , New-York.
18. COLONNA, P., GALLANT, D. and MERCIER, C. (1980). "Pisum sativum" and "Vicia faba" carbohydrates : studies of fractions obtained after dry and wet protein extraction processes. J. Food Sci., 45, 1629-1636.
19. LARSON, B.L., GILLES, K.A. and JENNESS, R. (1953). Amperometric method for determining the sorption of iodine by starch. Anal Chem.,25,802-804
20. BANKS, W. and GREENWOOD, C.T. (1967). The fractionation of laboratory isolated cereal starches using dimethyl sulfoxyde. Stärke, 19, 394-398.
21. COLONNA, P. and MERCIER, C. (1984). Macromolecular structure of wrinkled and smooth pea starch components. Carbohydr. Res.(in press).
22. DOUBLIER, J.L. (1981). Rheological studies on starch ; flow behaviour or wheat starch pastes. Stärke, 33, 415-420.

23. BANKS, W. and GREENWOOD, C.T. (1975). Starch and its components. Published by Edinburgh University Press.
24. JELLINEK, H.H.G. (1978). In Aspects of degradation and stabilization of polymers. Edited by H.H.H. JELLINEK, published by Elsevier Scientific Publishing Company.

MODIFICATION OF WHEAT FLOUR FOR THE PREPARATION OF FLOUR-CONFECTIONERY GOODS

K. SEILER

Federal Research Centre of Grain and Potato Processing
Institute of Baking Technology
D-4930 Detmold, Federal Republic of Germany

Summary

Since the Montzheimer report in 1931 it became commonly known that cake-making properties of wheat flours could be improved with chlorine treatment. Using chlorinated flour, it was possible to greatly increase the sugar and water content of a recipe to yield a high volume cake of fine grain, and light and tender eating properties. Today, both in the UK and USA, this type of wheat flour is employed for manufacturing special flour-confectionery products. Flour treatment by chlorine gas and its application for bakery purposes is forbidden in the Federal Republic of Germany. In order to be able to produce similar types of special cakes, German wheat flours were treated by physical methods such as kneading, shear and heat treatment to obtain flour with properties similar to the chlorinated UK or USA flours. Protein quality and starch gelatinization behaviour of two chosen German flours were modified by extrusion cooking with a Werner & Pfleiderer twin screw extruder. The resulting extruded flours, obtained under varying extrusion conditions, were dried, milled, and tested to their viscosity behaviour, water absorption capacity and baking properties in the manufacture of flour-confectionery goods. The results are presented in this paper.

1. INTRODUCTION

Chlorination of wheat flour is commonly used in the UK, USA, and in some other countries, to improve the cake-making properties (1, 6), chlorination is not permitted in most of the EEC countries. Cakes produced with flours which have been optimally treated with chlorine are higher in volume, have superior grain and more tender texture than those prepared from untreated flours (5). This treatment also increases the sugar and water carrying capacity of flours, permitting the use of "high-ratio" recipes (4), in which sugar content exceeds that of flour (11). Flour confectionery goods such as swiss-rolls, slab cakes and high-ratio cakes are made by this type of recipe (12).

Employing of ordinary flour in cake making in place of chlorinated flour allows a maximum level of sugar (and water) in a normal recipe of about 100 - 110 per cent of flour weight.

To avoid such difficulties, different attempts have been made in the past to find alternatives to chlorine mostly based on physical treatment of wheat flour. Little is known about these experiments, except an UK patent of 1968 and a recent paper published by KIM and ROTTIER in 1979 (3).

In the Federal Research Centre of Grain and Potato Processing in Detmold possibilities of physical flour treatment by means of kneading, shear and heating to get properties of German extruded wheat flours similar to chlorinated flours in England and USA were proved (Fig. 1).

2. RAW MATERIALS

The tests were carried out with three different flours. For comparison a UK chlorinated flour (No. 1) was used in a normal high-ratio cake recipe published by HODGE (2). The other two flours (No. 2 and 3) were typical German wheat flours, one of high and one of low protein content.

Table I : Wheat flours for manufacturing of flour-confectionery goods

flour characteristics		chlorinated flour	German flour	German flour
no.		1	2	3
type		-	550	550
protein content	[%]	9.6	12.1	10.5
sedimentation value	-	18	30	25
falling number	[s]	236	264	328
water absorption (500 Farin.Units)	[ml/100 g]	54.3	55.5	52.9

3. MACHINERY

The heat treatment, shear and kneading to affect wheat protein quality and to slightly gelatinized starch was carried out by a Werner & Pfleiderer twin screw extruder (8). The diameter D of the screw was 58 mm, the length of the five-segment-cylinder was 20 D, the agitator motor was of 20 kW, and the corresponding two-circuit oil heating system had a 2 x 12 kW heat capacity. The maximum throughput of the W&P-extruder was 100 kg/h. The combination-possibilities of the various screw elements permitted wide variation in process conditions (Fig. 2).

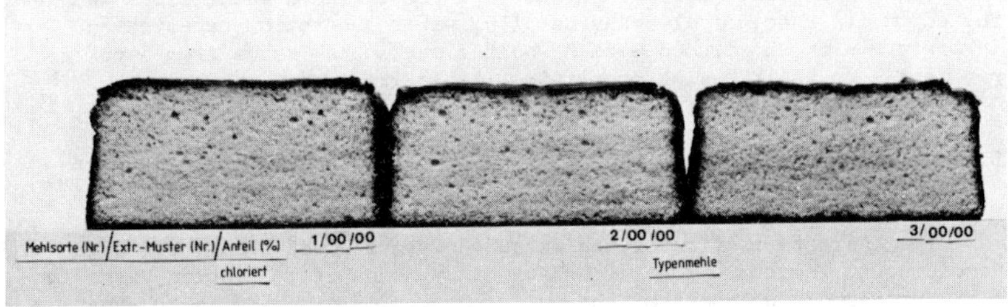

Fig. 1 : High-ratio cakes, produced from chlorinated flour (No.1) and German flours (No. 2) without additives of extruded flour

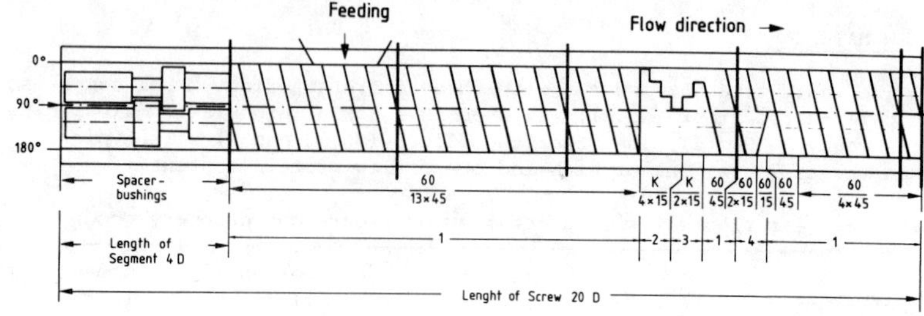

Fig. 2 : Five segment extruder barrel of the Werner & Pfleiderer extruder, showing the processing zone in the fourth segment

4. TECHNOLOGY

To weaken gluten, and to enable gradual gelatinization of starch of the German wheat flour, the combination possibilities of the segments of the two screws is of major importance (7). For our experiments we used an extruder-cylinder of a length of 16 D and arranged the screw segments onto the screw-root in such a way that within the processing zone the wheat-flour treatment was relatively mild. To receive best baking results with the extruded flour, the processing section of the mass was disposed into the fourth cylinder segment and combined mild kneading conditions with a short distance of kneading in opposite mass flow direction. The extrudate coming out of two 8 mm diameter dies was dried and milled with a Kamas mill to a maximum particle size of 400 µm (9). The extruded wheat flour was tested on its rheological behaviour (19, water absorption capacity and - in mixture with unextruded German wheat flour in the range from 5 to 20 per cent - on their baking properties (Table II).

5. INVESTIGATION AND EVALUATION OF EXTRUDED WHEAT FLOURS AND HIGH-RATIO CAKES

The characterisation of the extruded German wheat flour was done according to their rheological behaviour by means of a Rheotron-Viscosimeter. Furthermore we compared the water absorption capacity and the baking properties of different mixtures of extruded flours with the unextruded German wheat flours no. 2 and 3 (Table I). Typical rheological curves are shown in the Figures 2 and 4 (4). As special the specific baking volume of the high-ratio cake in ml/g was used. Table III summarizes the results and shows the comparison of the baked high-ratio cakes,manufactured by unextruded and extruded wheat flours.

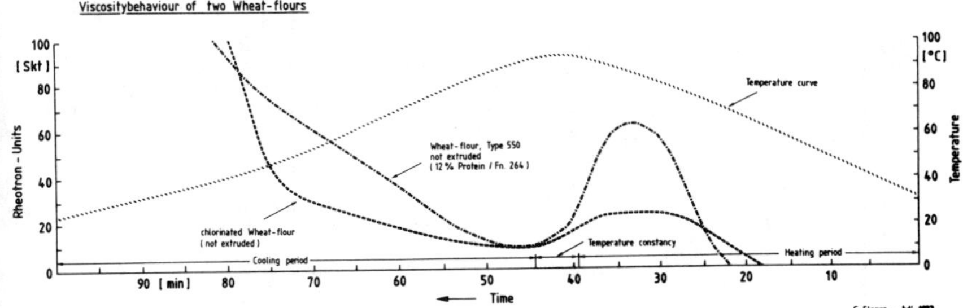

Fig. 3 : Viscosity diagram of chlorinated wheat flour and German high protein wheat flour type 550

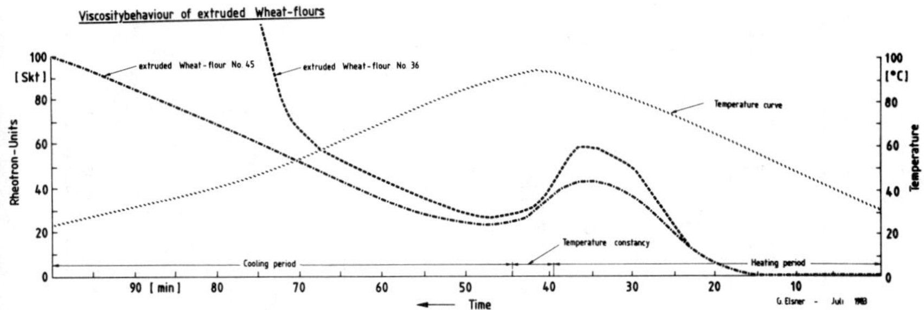

Fig. 4 : Viscosity diagram of extruded German high protein wheat flour, processed under mild extrusion conditions

Obviously the improved cake-making property of the chlorinated flour is shown in comparison to the German flours no.2 and 3. High-ratio cakes produced with unextruded flour no.2, supplemented with 5 to 10 per cent of extruded flour processed a very good specific volume and relatively good sensoric quality. The extremely mild treatment of the German flour of high protein content was important. Extruded flours, processed under severe kneading forces, high shear and high temperature resulted in unsuitable batters and a distinct decrease both in specific volume and organoleptic quality (Table III).

Wheat starch addition (30 %) with and without extruded flours to the high-ratio recipe gave no significant improvement in cake making properties. An improvement in organoleptic quality resulted when German unextruded and extruded flour siftings (150 µm) were used for high-ratio cake making, whereas the specific volume of the high-ratio cake decreased (Fig.5).

The use of semol instead of wheat flour as raw material for extrusion cooking resulted in an extruded product with very good baking properties regarding high-ratio-cakes. Semol extruded under mild conditions,

Table II : Rheological behaviour of the unextruded flours (Table I) in comparison to the extruded flours

sample no.	feed rate [g/min]	moisture content [%]	rpm [min^{-1}]	die diameter [mm]	temperature mass [°C]	temperature heating [°C]	pressure mass [bar]	viscosity start [Rh-U]	viscosity maximum [Rh-U]	viscosity end [Rh-U]
1								0	23	147
2								0	67	141
3								0	74	157
45*	555	35	100	8	45	12	28	0	44	98
36*	1000	35	180	8	60	12	28	0	57	291
26*	771	35	250	8	125	135	17	320	252	204

*) Raw material for this samples : German flour no. 2 (Table I)

Table III : Comparison of specific volume and sensoric evaluation of some flour-confectionery goods prepared without and with extruded flour

sample	spec.vol. [ml/g]	porosity		crumb structure	mouthfeeling
chlor. flour	2.64	rather uniform	fine/open	soft/moist	soft/short
high protein flour	2.53	rather uniform	middle/still open	still soft/moist	viscous/aggl.
low protein flour	2.48	rather uniform	swh.unregl./middle	rather firm/moist	viscous/aggl.
wheat flour plus 10 % extruded flour additive					
extruded flour -45-	2.57	rather uniform	middle/open	soft/moist	swh.viscous
extruded flour -36-	2.63	rather uniform	middle/open	soft/moist	swh.aggl.
extruded flour -26-	2.41	swh. uneven	rough/middle	rather firm/moist	agglomerated

gives an extruded 'flour' with a low degree of starch gelatinization and a not entirely denaturated structure of the wheat gluten. This extruded semola (extruded 'flour') - used in No. 36 - had a good cake baking strength and gave acceptable high-ratio-cakes without any addition of unextruded flour components.

Fig. 5 : High-ratio cake, manufactured by a mixture of German flours no. 2 and 3 with extruded flour (test no. 45)

REFERENCES

1. GOUGH, B.M., WHITEHOUSE, M.E. and GREENWOOD, C.T. (1978). The role and function of chlorine in the preparation of high-ratio cake flour. Critical reviews in food science and nutrition 9, 91-113
2. HODGE, D.G. (1975). Alternatives to chlorination for high-ratio cake flours. Baking Ind. J. 7, 12-19
3. KIM, J.C. and ROTTIER, W. (1979). Modifizierung von Aestivumweizen-Grieß durch Extrusion. Getreide Mehl und Brot 7, 188-190
4. KULP, K. (1972). Some effects of chlorine treatment of soft wheat flour. The Baker's Dig. 6, 26-30
5. KULP, K., TSEN, C.C. and DALY, D.J. (1972). Effects of chlorine on the starch component of soft wheat flour. Cereal Chem. 2, 194-200
6. LUDEWIG, H.-G. (1983). Mehlqualitäten für die Herstellung saftiger Kuchen. Brot u. Backwaren 7/8, 12-19
7. MERCIER, C. and FEILLET, P. (1975). Modification of carbohydrate component by extrusion cooking of cereal products. Cereal Chem.3,283-297
8. MILLAUER, C., WIEDMANN, W.M. and STROBEL, E. (1981). Verfahrenstechnik und Anwendung des selbstreinigenden, zweiwelligen Misch- und Kochextruders. Seine Wirkungsweise in Abhängigkeit verschiedener Prozeßparameter. Private Mitteilung (Hausdruck Fa. Werner&Pfleiderer, Stuttgart)
9. SEILER, K. (1980). Rohstoffe und Extrusion. Verhalten einiger Rohstoffe auf Getreidebasis während des Extrusionsprozesses. Gordian 9, 210-212; Gordian 10, 235 - 242
10. SEILER, K., WEIPERT, D. and SEIBEL, W. (1980). Viskositätsverhalten vermahlener Extrudate in Abhängigkeit von verschiedenen Parametern. Z. Lebensm. Technol. und Verfahrenstechn. 2, 37-42
11. TSEN, C.C., KULP, K. and DALY, D.J. (1971). Effects of chlorine on flour proteins, dough properties and cake quality. Cereal Chem. 3, 247 - 255

12. WEILAND, P.J. (1976). Weizenmehl für Feine Backwaren. Getreide Mehl und Brot 3, 82

ACKNOWLEDGEMENT

The author is indebted to the Forschungskreis der Ernährungsindustrie e.V., Hannover, for financial support of this work.

EXTRUDED PROTEIN-RICH ANIMAL BY-PRODUCTS WITH IMPROVED TEXTURE

K.H. Kristensen, P. Gry and F. Holm
Jutland Technological Institute, Aarhus, Denmark

Summary

In the meat and fish industry tremendous amounts of underutilized edible by-products are available. One of the processing techniques which can be utilized for up-grading the animal by-products is HTST extrusion cooking. By this technique it is possible to produce products with attractive textures, e.g. crispy chips, pasta products and textured products with meat structure.

1. INTRODUCTION

1.1 The Potential of Animal By-Products

The food industry is characterized by high raw material prices and a very low value increase from raw material to products. This fact is especially true for slaughterhouses, meat-processing industries and the fish industry. E.g. in Denmark the value increase for slaughterhouses is as low as 19% in comparison to around 48% for the rest of the industry.

For this reason the food industry has to decrease waste and up-grade by-products in order to be competitive. This is true in particular for the meat and fish industry where the amount of low value by-products is substantial.

According to Lawrie (6) the main underutilized tissues from Angus Carcase, e.g. with a living weight of 250 kg, are blood, stomachs and lungs (table I).

Table I. Yield of currently underutilized tissues from a typical Angus Carcase of 250 kg weight (6)

Tissue		Wt. kg	Protein %	Potential protein yield, kg
Blood		18.0	16	2.90
Stomachs	(Rumen/Reticulum)	7.0	10-13	0.90
	(Abomasum)	1.8	7-9	0.15
	(Omasum)	2.7	8.5-9.5	0.25
Lungs		3.2	16-17	0.50
Total		32.7		4.70

These underutilized edible tissues amount to around 100,000 tons in Great Britain so even when subtracting the weight of the white plasma fraction of the blood, which today is a valuable product for e.g. sausages, the amount of edible underutilized tissues is substantial.

For the Danish pig production the comparable figure for edible underutilized tissues is around 60,000 tons. If other by-products are included, e.g. deboned meat, skin and other organs, the figure increases significantly to close to 90,000 tons edible underutilized pig tissue.

In the fish industry waste also occurs. From the Danish landing of approximately 2 million tons of fish, by-products are available such as roe, liver and other parts, but mainly as fish-mince, produced in the fillet industry using bone separators. For example filleting of codfish (table II) results in only 36% product and probably at least 20% edible waste and underutilized tissues, including the fish-mince.

Table II. Component weight of codfish (7)

Component	Percentage whole weight
Head	21
Guts	7
Liver	5
Roe	4
Backbone, etc.	14
Fins and lungs	10
Skin	3
Fillet skinned	36

In general, between 1/5 and 1/3 of the fish caught for processing would be edible waste.

Globally, the production of meat is 120 million tons and 50 million tons of fish for human consumption (7). Using typical figures for waste and underutilized tissues, the total amount of these products is estimated at several million tons annually.

Although only part of the production will pass the industry and give rise to a well-defined waste figure, and although the habits throughout the world will vary according to the concept of waste, it will be clear to everybody that an up-grading of the animal by-products is a matter of great perspectives.

1.2 Utilization of Animal By-Products

Today the animal by-products are used mainly as animal feed and petfoods, but a smaller amount is used for human consumption. Because every country has its own traditions, products which are waste in one country are highly estimated speciality food products in other countries.

But generally speaking, most of the by-products from the meat and fish processing industries, e.g. stomachs, lungs, other organs, roe and greaves, have a low as is acceptability because of lack of functional properties, lack of tenderness, unacceptable taste, aroma and appearance and non-convenient processing requirements, but most of all because of lack of tradition.

On the other hand, most of the by-products have a satisfactory nutritional value ranging in protein between 12 and 23% and between 3 and 28% fat.

From an engineering point of view an up-grading of the by-products could involve a degradation of the meat followed by a building up of new fibres, structures or shapes to give texture and functionality to the products. Some products also have to be dehydrated mechanically or thermally and other products have to be defatted.

The most well-known techniques for up-grading of animal waste is production of protein concentrates and isolates using alkaline extraction followed by precipitation, membrane concentration and separation, hydrolysis or ion exchange. Several of these techniques have already reached production.

The results of these processes are protein powders without texture or any resemblance to meat although it is valuable products and fillers giving functionality to foods.

For some by-products the meat structure can be achieved by size reduction and gluing together the meat fibres, or by spinning of new fibres from a solution.

Special techniques are also known where the materials are changed into new structures which can involve meat texture or meat gels. These techniques are especially known from the Japanese fish industry where the products Surimi, Kamaboko and Marine beef are of special interest. In the Marine beef e.g. the texture of fine disintegrated fish tissue is performed by kneading the mince with salt under strict control of pH and temperature, resulting in gluing the muscle fibres together with solubilized actomyosine and coagulating the network and forming it by means of an extruder (10).

Utilization of HTST extrusion cooking of animal tissue either alone or in combination with starch, plant proteins and other ingredients is also known from the literature, in particular the patent literature (1, 2, 3, 9, 8). Both dry and semi-moist products can be produced.

In the extruder the possibility exists of producing both expanded products and meat-textured products. In the following I would like to give a few examples of the possibilities of HTST extrusion cooking of animal by-products. This work is done at the Jutland Technological Institute, Denmark.

2. EXTRUSION COOKING OF ANIMAL BY-PRODUCTS

2.1 Extrusion Cooking of Fish-Based Products

The aim of this work has been to utilize fish by-products for production of 1) an expanded fish snack containing 35-45% fish meat, 2) a dry fish soup and 3) pasta products containing fish. It was further the aim to demonstrate that the extrusion cooker can be utilized to produce nutritious fish products with good shelf lives. This could be of special interest to less developed countries where the infrastructures do not allow distribution of fresh fish.

In the following the raw materials used were minced meat from codfish and herring, but mackerel and sprat have been tested, too. The typical composition of the fish is shown in table III.

Table III. Chemical composition of fish

Fish	Water content %	Protein %	Fat %	Calories per 100 g
Cod	80.0	16-18	0-1	70-90
Herring	60-70	17-19	5-20	120-260
Mackerel	70.8	16-19	5-25	120-290
Sprat	60-70	17-19	5-18	120-220

To absorb water from the mince and to assist the expansion and ensure an adequate texture, the following cereals and potato products were used: Potato starch, potato flour, mashed potatoes, maize starch, wheat starch, and wheat flour. Further, crossbonded waxy maize starch, xantangum, carob seed flour and salt were used in combination with monoglycerides, calcium or sodium stearoyl-2-lactylate or citric acid esters of monoglycerides as well as antioxydants. The extrusion cookers used were a twin-screw Werner & Pfleiderer, Continua 37/5 and a single-screw Brabender, 20 DN.

The quality of the products was mainly investigated in a taste panel using a 0-10 scale giving 0 for completely unacceptable products and 10 for perfect products. The acceptability limit was 5 and the quality factors were texture and taste. The chemical analyses, e.g. Free Fatty Acids (FFA), peroxide number, anisidine number and TVN figures, could not be correlated to the organoleptic tests and for this reason omitted in this text. For all experiments the FFA decreased and the TVN figures increased for the products during storage at room temperature in the dark, packed in low oxygene barrier film (polyethylene).

Tables IV and V show some typical examples of experimental conditions for production of pellets intended for nuddles or expanded snacks. For snack use the pellets were expanded by deep fat frying or by heating in a microwave oven.

Table IV. Typical experimental conditions for production of fish pellets by extrusion cooking (twin-screw)

Extruder type: Werner & Pfleiderer, Continua 37/5, twin-screw
Number of sections: 5
Screw configuration: 6 mm D, 30 mm S 40, 90 mm S 60, 40 mm Mix, 670 mm S 40
Die: 0.7 mm x 20 mm

Test No.	Feed comp.	Section, °C 1	2	3	4	5	Die	RPM	Water feed l/h	Torque %	Pressure atm	Prod. temp. °C	Capacity kg/h
6	A	Cooling	50	-	81	84	76	105	7	99	138	99	10.4
8	B	Cooling	35	-	79	80	75	205	0	68	135	104	12.5

Composition A: 30% codfish, 26% potato starch, 26% wheat starch, 15% wheat flour, 3% salt

Composition B: 41% codfish, 22% potato starch, 22% wheat starch, 12% wheat flour, 3% salt

Added 25 ppm BHA and 25 ppm octylgallate

A and B are recipes for snacks

Table V. Typical experimental conditions for production of fish pellets by a single-screw extrusion cooker

Extruder type: Brabender 20 DN
Screw: Compression 1:2
Die: 0.7 mm x 20 mm

Test No.	Feed comp.	Zone, temperature °C 1	2	head	RPM	Torque mp	Capacity g/min.
25	C	70	120	94	75	2500	45
43	D	70	120	94	80	1450	35
41	E	70	120	94	40	2000	28

Feed composition C: 35% codfish, 31% potato starch, 24% potato flour, 6% crossbonded waxy maize starch, 3.5% salt, 1% calcium-stearoyl-2-lactylate

" " D: 30% codfish, 26% potato starch, 26% potato flour, 15% wheat flour, 3% salt, 40 ppm dodecylgallate

" " E: 30% codfish, 67% wheat flour, 3% salt

Feed C and D are for snack food, feed E for nuddles.

Using fish with a high fat content, e.g. herring, the fish content could not be higher than 20% in order to ensure a good quality. Using fish with less fat, it is possible to produce expanded snacks, soups and pasta with a content of up to near 50% fish. Figures 1 and 2 show the pellets and the expanded snacks.

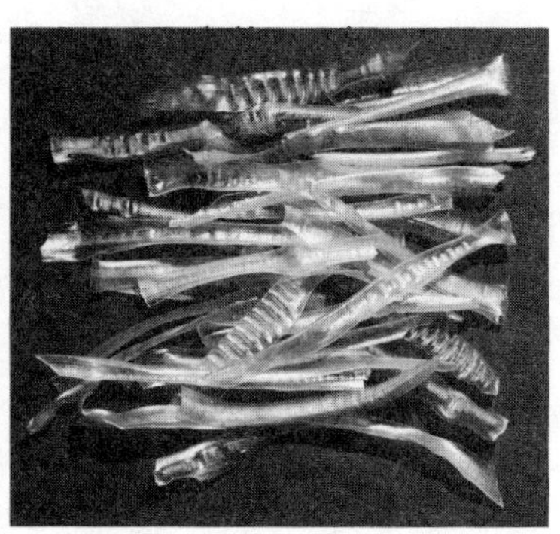

Figures 1 and 2
Pellets produced for snacks (left) and deep fat fried snacks (right)

The conclusions from the organoleptic tests are that a good quality fish mince can be used for the HTST extrusion processes. Even without anti-oxydants the pellets do not develop an objective rancid taste within one year if low fat fish is used, but the fish taste seems to intensify during storage.

For the recipes in general was found that addition of crossbonded waxy maize starch stabilized the flavour of expanded products. This effect was demonstrated on an expanded model snack by following the peroxide number. The relation is shown in figure 3.

The protective effect of crossbonded waxy maize starch can be utilized in extrusion of fat fish in particular, but the branched starch also seems to have a positive effect on homogeneity of fat and pressure build-up in the extruder (4).

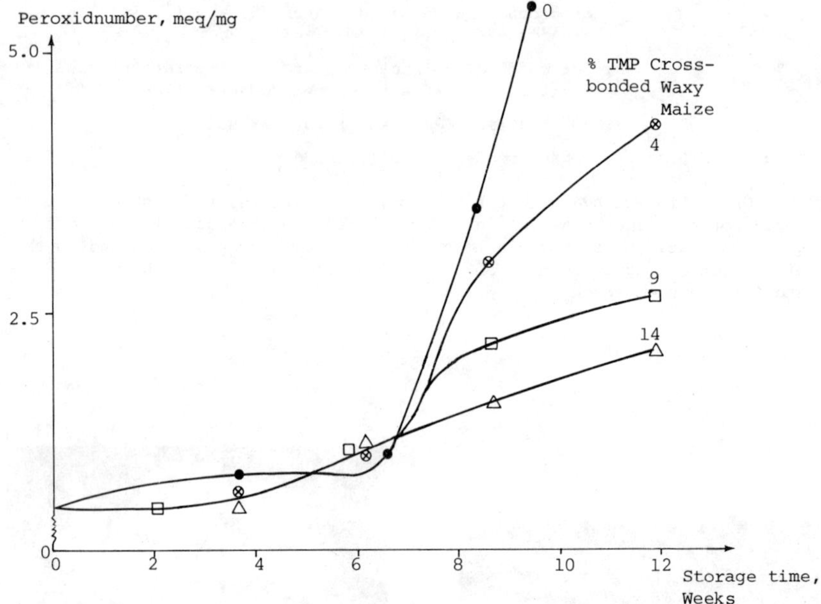

Figure 3. Relation between storage time of expanded product, peroxide number and per cent crossbonded waxy maize starch added (5).

2.2 Extrusion Cooking of Slaughterhouse By-Products

The aim of this work has been to utilize by-products, e.g. meat-mince from bone separators, lungs and stomachs from pigs and cattle, to produce protein chunks with meat texture which can be used in meat emulsions and other minced-meat products.

The extruder used was a twin-screw, co-rotating Werner & Pfleiderer C 37 pilot machine with up to eight heating and cooling sections. A typical process diagram is shown in figure 4, but the process chosen will depend on the raw material such as pressability and meat structure, fat content and water content.

In processing the by-products it is highly needed to control the degree of cutting, the pH and the temperature in order to optimize the dewatering and defatting in the pressing process and to ensure the building up of a new meat texture in the extruder.

Dependent on the raw material and the pressing operation, different ingredients can be added to control water content and pressure build-up in the extrusion cooker. Both different starches and proteins have been used. The textured product is finally dried to 14% water in a belt dryer.

In table VI the typical extrusion parameters used can be seen.

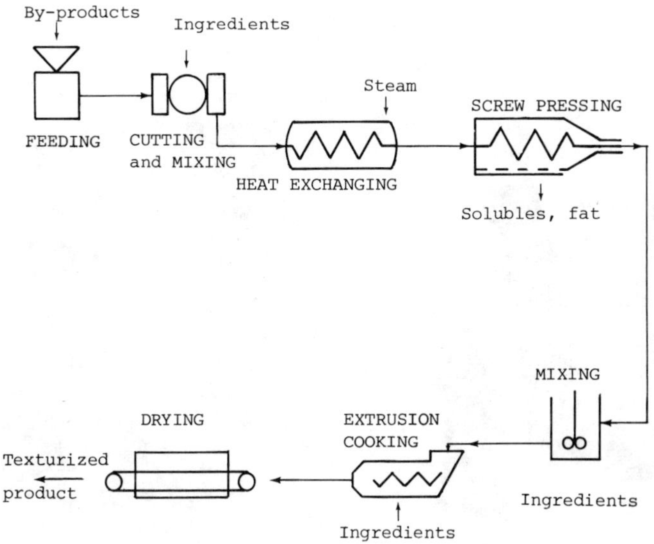

Figure 4. Process diagram for production of meat chunks from slaughterhouse by-products.

Table VI. Typical parameters for extrusion cooking of slaughterhouse by-products

Extruder type: Werner & Pfleiderer, Continua 37/5, twin-screw

Number of sections: 8

Die: 5 mm ø, 62 mm

Test No.	Section, °C 1 2 3 4 5 6 7 8	RPM	Torque %	Pressure atm	Temperature °C	Capacity kg/h
23	← cooling → 157 154	100	69	26	126	26.1
27	←cool.→ 150 151 - - -	180	46	8	130	16.2
28	←cool.→ 151 147 - - -	180	38	0	119	27.0

Composition: Presscake and potato starch, 2:1.

The resulting meat chunks can absorb water up to 200% of its own weight and after rehydration the chunks have a chemical composition very much like that of meat, e.g. 60% water, 12% protein, 1% fat and up to a total of 100% carbohydrates.

A product example is shown in figure 5.

Figure 5. Meat chunks from lungs, produced by extrusion cooking. Hydrated product right.

From these examples it appears that HTST extrusion processes can be used with great success for up-grading underutilized animal by-products.

Very often we are involved in product development based on this technique, e.g. up-grading of broken shrimps, chicken skin or extrusion cooking of whey proteins and krill.

The requirements for the extrusion process are in general a water content of the feed of 40-50%, less than 8% fat, a material temperature of 150°C and a low material pressure.

REFERENCES

(1) Buckley, K. et al. (1979). Gelatinized animal food product. U.S. patent 4, 143, 171.

(2) Ernst, T.J. (1977). Process for the production of a formed high-moisture pet food product. U.S. patent 4, 011, 346.

(3) Feldbrugge, A. et al. (1975). Process for preparing meatlike fibers. U.S. patent 3, 886, 299.

(4) Gry, P. and Kristensen, K.H. (1982). Extrusion-cooking of protein-enriched cereal mixtures for high-protein foods. 7th World Cereal and Bread Congress, ICC Symposium on HTST-Extrusion-Cooking, Prague.

(5) Kristensen, K.H. (1982). The influence of starches and starch-derivates on quality of extruded products. Symposium "Cooking and Extruding Techniques", Solingen.

References (continued)

(6) Lawrie, R.A. (1981). Recovery and utilization of Abattoire waste. IFST Annual Symposium - Food Technology in Europe. July 1981, 118-130.

(7) Mackie, I.M. (1983). New approaches in the use of fish proteins. Developments in food proteins - 2. Editor B.J.F. Hudson, Applied Science Publishers, 215-262.

(8) Meyer, R.H. et al. (1981). Composition containing animal parts for production of a fried snack food and method for production thereof. U.S. patent 4, 262, 028.

(9) Stupec, G.B. (1978). Reconstituted fried puffed pork skins. U.S. patent 4, 119, 742.

(10) Suzuki, T. (1981). Fish and krill protein: processing technology. Applied Science Publishers Ltd., London, chapter 4.

CONTINUOUS EXTRUSION PROCESSING OF STARCHY MATERIALS FOR THE PRODUCTION OF SYRUPS AND ETHANOL

P. LINKO, Y.-Y. LINKO and S. HAKULIN
Helsinki University of Technology, Department of Chemistry, Laboratory of Biochemistry and Food Technology, SF-02150 Espoo 15, Finland

Summary

Barley starch liquefied in a Creusot-Loire BC 45 twin-screw HTST-extrusion cooker was saccharified either to glucose syrup with *Aspergillus niger* amyloglucosidase, or to maltose syrup with barley-β-amylase and pullulanase.

The highest DE-values of ~ 96 (for 30% d.m.) and ~ 98 (for 10% d.m.) in glucose syrup after 24 hours saccharification were obtained after extrusion-liquefaction at 135°C, 60% water content with 1.5% (w/w d.m.) *Bacillus licheniformis* Termamyl α-amylase. The highest maltose content of 87.5% after 25 hours saccharification, was obtained after extrusion-cooking at 125°C, 19% water content without α-amylase to a DE ~ 1.3.

Extrusion cooked barley starch or crushed whole barley was saccharified with glucoamylase for 1-5 hours before the temperature was lowered to ~ 30°C, and *Saccharomyces cerevisiae* or *Zymomonas mobilis* centrifuged cell culture was added to initiate the ethanol production simultaneously with the completition of the saccharification. The maximum ethanol production of 109 g/l in 44 hours from barley starch syrup (22.5% d.m.) was obtained at low initial DE 58. For extruded crushed whole barley (23.5% d.m.), the best result, 90 g/l in 48 hours, was obtained when the barley was extruded at 135°C, 50% water content with 0.12% (w/w d.m.) Termamyl.

1. INTRODUCTION

Continuous extrusion cooking has been applied by the food and animal feed industries for a long time (1). Only recently, however, the possibility to use the HTST-extruder as a continuous reactor for thermomechanical pretreatment and enzymatic modification of starches has been investigated. We have preveously applied HTST-extrusion cooking in glucose syrup production (2,3) and observed that this technique may also be applied to obtain low DE thinned starches (4,5). The possibilities of using a twin-screw HTST-extrusion cooker for gelatinization and enzymatic conversion of starchy materials have been further investigated.

2. EXPERIMENTAL

A Creusot-Loire BC 45 twin-screw extruder was used for thinning of commercial barley starch or crushed whole barley at various moisture levels and temperatures mostly with addition of thermostable α-amylase Termamyl 60L or 120L. Glucoamylase was then used for the saccharification to glucose syrup, on barley-β-amylase either alone or with pullulanase for the saccharification to maltose syrup. For ethanol production saccharification with glucoamylase was started 1 hour before *S. cerevisiae* or *Z. mobilis* centrifuged cell culture was added.

Saccharification with glucoamylase was carried out at 60°C, pH 4.5, and with β-amylase and pullulanase at 55°C, pH 6.0. Ethanol production took place at 28°-30°C, pH 4.5.

Dry matter was determinated by heating samples mixed with sand at 105°C over-night. DE-values were determinated as described by Miles Laboratories, Inc. (6). Glucose was determinated by emploing the hexokinase/glucose-6-phosphate dehydrogenase system as described by Boeringer Mannheim (7), and maltose was determinated by high pressure liquid chromatography (Bio-Rad HPX-42 column). Ethanol was determinated by gas chromatography (1/4" 2m glass column with Chromosorb 101 80/100).

3. RESULTS AND DISCUSSION

3.1. Glucose syrup

Barley starch was liquefied in the extruder at various water contents (40-60%) and temperatures (105°-160°C) with the addition of 0-3% (w/w d.m.) Termamyl 120L α-amylase. Saccharification was carried out at various dry matter (10-45%) and glucoamylase levels (0.36-1.2% w/w d.m.).

DE-values after extrusion but before saccharification increased with increasing water content during extrusion-cooking. This is in good agreement with earlier results (3). The increase in water content during extrusion-cooking, with best results obtained at 60% moisture increased also final DE-value after subsequent saccharification. The DE-value thus obtained in 24 hours varied from about 93 to 96. The mass temperature during exrusion-cooking did not affect subsequent saccharification very much. Only the extrudate obtained at the relatively high 160°C mass temperature differed significantly from the others. The DE-value reached after 10 hours saccharification was only 76 for the sample extruded at 160°C, compared with DE 90 to 95 for all the other samples.

Extrusion-cooking of starch without or with small amounts of Termamyl (<0.36% w/w d.m.) resulted in a solid extrudate, difficult to mix homogeneously with water to a 30% d.m. slurry for saccharification. After 24 hours saccharification DE 87 was reached for starch extrusion-cooked without Termamyl, and DE 94 when the starch was extrusion-cooked with 1.5% (w/w d.m.) Termamyl, but immediately mixed with hot acetate buffer (pH 3.5) for α-amylase inactivation. DE-values before saccharification were 3.5, 4.7 and 26, respectively, and DE 98 was reached in 24 hours if the starch was extrusion-cooked with 1.5% (w/w d.m.) Termamyl without inactivating the α-amylase after extrusion (Figure 1.).

With 10-15% d.m. solutions the DE-values were over 95 as early as after 10 hours saccharification, decreasing thereafter with increasing dry matter content. Thus only a DE 92-93 was obtained at or above 30% d.m. substrate level. The saccharification time was reduced by increasing the

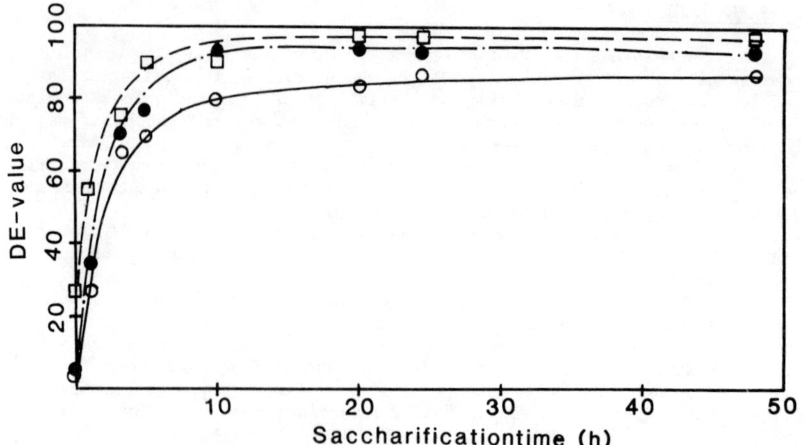

Figure 1. The effect of Termamyl during the extrusion on the subsequent saccharification. o——o extrusion without Termamyl, ●-·-● extrusion with Termamyl inactivated immediately after extrusion, □- - □ extrusion with Termamyl without inactivation.

glucoamylase amount. In 3 hours DE 72 was reached with 0.36% (w/w d.m.), DE 78 with 0.6% (w/w d.m.) and DE 89 with 1.2% (w/w d.m.) glucoamylase, but after 24 hours the differences were already insignificants, with all of the DE-values of order of 91-93. We have also been able to show with a Werner & Phleiderer continua 58 twin-screw extruder with a longer barrell that excellent results can be obtained if starch gelatinization, enzymatic liquefaction and saccharification all are sequentially initiated during extrusion cooking (8). A DE-value of 94 was obtained in a total processing time of 5 hours, and DE 97 in ~ 21 hours.

3.2. Maltose syrup

Barley starch was liquefied in extruder at various water contents (19-65%) and temperatures (105°-160°C) either with 3% (w/w d.m.) or without Termamyl 60L α-amylase. Saccharification was carried out with β-amylase either alone or with pullulanase.

The highest maltose content was obtained when the starch was extrusion cooked without α-amylase at 19% water content, and 125°C mass temperature to about DE 1.3. Saccharification was carried out with both β-amylase and pullulanase. The syrup contained 87.5% maltose, 8.4% maltotriose and 4.0% higher oligomers.

Table I shows the composition, as determinated by HPLC-techniques, of high maltose syrups produced from variously extrusion-cooked barley starch substrate (10% d.m.) saccharified with barley-β-amylase (ABM 1500 L) and pullulanase (ABM pulluzyme S 2000). One can see that the highest maltose content is obtained when the starch has been extrusion-cooked without α-amylase to a low DE-value. The effect of water content and mass temperature during extrusion-cooking are both rather insignificant.

Table I. Composition of various maltose syrups obtained

Extrusion	DE	Sacch.time (h)	DE	glucose (%)	maltose (%)	maltotriose (%)	higher oligomers
40% H_2O, 135°C without Termamyl	3.2	26	46.2	-	86.1	7.0	6.9
40% H_2O, 135°C with Termamyl	10.1	26	50.6	0.7	71.2	22.5	5.7
55% H_2O, 135°C with Termamyl	12.0	23	57.5	0.8	72.3	21.9	5.0
19% H_2O, 125°C without Termamyl	1.3	25	45.5	-	87.5	8.4	4.0

3.3. Ethanol

Extrusion-cooked barley starch or crushed whole barley was saccharified with glucoamylase for 1-5 hours before the temperature was lowered to 30°C and *S. cerevisiae* was added to start the ethanol production, allowing saccharification to continue simultaneously. Best results with barley starch were obtained when ethanol fermentation was started at the relatively low DE ~ 58 after 1 hour saccharification (9). The short saccharification time of 1 hour was used in all cases for extruded crushed whole barley.

At an initial dry matter content of 22.5% for the barley starch slurry an ethanol concentration of 109 g/l was reached in 44 hours. For extruded crushed barley the best results were obtained when the barley was extruded at 50% water content and 135°C mass temperature with the relatively low 0.12% (w/w d.m.) Termamyl 120L amount. The extrusion-cooked barley (23.5% d.m.) was saccharified for 1 hour with 0.36% (w/w d.m.) glucoamylase before 0.4% (w/w d.m.) *S. cerevisiae* was added. The ethanol concentration reached in 48 hours was 90 g/l, that is 3.2 kg barley was needed to get 1 kg 100% ethanol. Figure 2. shows that by increasing the *S. cerevisiae* amount the batch fermentation rate is also increasing, but the maximum ethanol amount was still reached in about 45 hours for all cases.

With *Z. mobilis* as much as 50 g ethanol/l was obtained in 7 hours compared with 30 g/l for *S. cerevisiae*, but the batch fermentation slowed down rapidly with further accumulation of ethanol.

4. REFERENCES

1. LINKO, P., COLONNA, P. and MERCIER, C., High-temperature, short-time extrusion cooking, Advances in Cereal Science and Technology 4 (1981) 145-235.
2. LINKO, Y.-Y., LINDROOS, A. and LINKO, P., Soluble and immobilized enzyme technology in bioconversion of barley starch, Enzyme Microb. Technol. 1 (1979) 273-278.

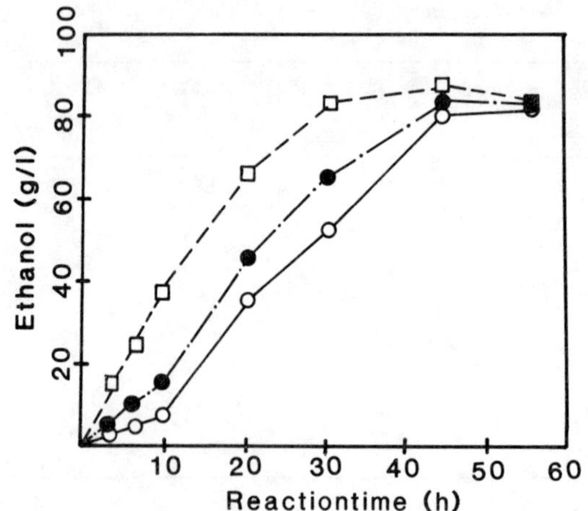

Figure 2. The effect of *S. cerevisiae* amount on ethanol production
o——o 4 g/kg d.m., ●-·-● 20 g/kg d.m., □--□ 72 g/kg d.m.

3. LINKO, Y.-Y., VUORINEN, H., OLKKU, J. and LINKO, P., The effect of HTST-extrusion on retention of cereal α-amylase activity and on enzymatic hydrolysis of barley starch, in 'Food Process Engineering' Vol. 2. 'Enzyme Engineering in Food Processing' (P. Linko and J. Larinkari, eds) Applied Science Publishers, London 1980, 210-223.
4. LINKO, P., HTST (High Temperature-Short Time)-Extruder als biochemischer Reaktor. Getreide Mehl u. Brot 36 (1982) 326-332.
5. LINKO, P., LINKO, Y.-Y. and OLKKU, J., Extrusion cooking and bioconversions, J. Food. Eng. 2 (1983) (in press).
6. MILES LABORATORIES INC., Technical Bulletin No 1-174 (1963) 20.
7. BOERINGER MANNHEIM, Test Combination Cat. No 139106.
8. HAKULIN, S., LINKO, Y.-Y., LINKO, P., SEILER, K. and SEIBEL, W., Enzymatic Conversion of Starch in Twin-Screw HTST-Extruder, Starch/Stärke 35 (1983) (in press).
9. LINKO, P., HAKULIN, S. and LINKO, Y.-Y., Extrusion cooking of starch for glucose syrup and ethanol, J. Cereal Sci. 1 (1983) (in press).

THE USE OF HIGH TEMPERATURE SHORT TIME EXTRUSION COOKING OF MALT IN BEER PRODUCTION

F. MEUSER, E. KRÜGER, B. VAN LENGERICH and E. GRONEICK
Technische Universität Berlin
Institut für Lebensmitteltechnologie - Getreidetechnologie -
Institut für Fermentation und Brauwesen

Summary

Since the processing techniques for the production of beer are limited by which raw materials may be used, an attempt was made to integrate HTST extrusion cooking into the malting process. Air-dried extruded malt had the typical taste and colour components usually found in kilned malt. Taste components in air-dried extruded malt were much more intensive than normal. This material was mashed in various ratios with air-dried malt and processed to beer. The worts and beers from these raw material mixtures were then compared with those produced conventionally from kilned malt. The results showed that it is possible to produce comparable products with acceptable extract yields from the raw materials. However, the colour of the worts from the various MIXes was darker than those from normal malt.

1. INTRODUCTION

Beer is conventionally brewed from malt. The malt provides the enzymes necessary for the breakdown of carbohydrates to fermentable sugars as well as contributing to the typical taste of the beer. Each step in the preparation of malt and beer has, therefore, a specific influence on the beer quality with the result that the production technique can hardly be changed. However, certain differences in the use of equipment and raw materials are not unusual.

In the Federal Republic of Germany, beer can only be brewed with water, malt, hops and yeast. It is not allowed to use other sources of carbohydrates or enzymes. This restriction in the use of raw materials imposes a limitation on other brewing possibilities and thereby makes the formation and development of certain characteristics of beer difficult. Moreover, the conventional methods of malting and brewing require a great deal of energy.

For example, the drying and kilning of the green malt (GM) necessary for the formation of the taste in beer makes up about 30% of the total cost of producing the malt (1). To date, these high costs of energy could only be reduced through the introduction of certain techniques to recover the energy (2). These techniques are so far advanced that any further savings on energy can only be made possible by modifications to the kilning technique. Through kilning, the moisture content is reduced to about 4% with a simultaneous formation of colour and taste components. This low moisture content is not necessary, by itself, for the storability of the product. Air-dried malt (AM) with a moisture content of 14% or below could be conveniently stored and milled. However, in such cases, the colour and taste components are absent. The fact that these components could be formed either wholly or in part by high temperature short time (HTST) extrusion cooking of the AM opens up prospects of another process technique which is worthy of investigation. This technique could possibly lead to energy savings and could be used to influence certain changes during kilning.

The aim of our investigations, therefore, was to integrate HTST extrusion cooking into the malting process. In addition to this, specific qualities of the air-dried extruded malt (AEM) were characterised and compared with conventionally cured malt (CM) and AM.

2. MATERIALS AND METHODS

2.1 Extrusion cooking

GM germinated for seven days was dried with warm air to a moisture content of 15%. The AM was then separated from the rootlets and then ground (particle size 1000 µm). Part of the AM was extruded (Extruder: Werner & Pfleiderer, Continua 37). The extrusion conditions were laid down by means of a fractional factorial experimental design (4). The AEM were studied in respect of specific product criteria (extract yield, viscosity and colour of the wort). The data were then evaluated by means of polynomial regression equations.

2.2 Mashing and brewing

Mashes, worts and beers were produced both on a laboratory and a pilot scale in accordance with industrial mashing norms. The AM and CM used in these experiments came from the same sample. - The development of the malt viscosity was studied through the simulation of the mashing process in the measuring cell of a viscosimeter. - The formation of maltose during mashing was enzymatically studied (5). The samples that were taken during the process were boiled to inactivate the enzymes. - The quality characteristics of the worts and beers were determined using standard methods (6). - The taste components were quantitatively determined and identified by means of a liquid-liquid extraction followed by gaschromatography and mass spectrometry (7).

3. RESULTS AND DISCUSSION

3.1 Design of a new brewing procedure

A comparison of the malts used showed that the enzymes formed in the GM were completely inactivated during extrusion (Fig. 1). The moisture content of the AEM was so low that it, like the CM and the AM, could be stored. Odour, taste and colour of the product were firmly established

Criterion / Product	Enzyme Activity	Moisture Content	Taste/Flavour Intensity	Colour (Malt)	Viscosity (Mash)
Green Malt	high	high	low	low	low
Air-dried Malt	high	medium	low	low	low
Malt	high	low	medium	medium	low
Air-dried Extruded Malt	none	low	high	high	high

Fig. 1. Distinctive Characteristics of Various Malts

through extrusion. They were much more pronounced in comparison with the CM. The viscosity of the mash from the AEM was very high, due to gelatinization of the starch during extrusion and the absence of amylolytic enzymes. From the resulting malts, it became clear that the AEM alone could not be used to produce worts. This made it necessary to mix the AM and the AEM in order to obtain a fermentable wort. The criteria for determining the ratio of both fractions in the mixture were worked out with the aid of the scheme in Figure 2. The utilization of HTST extrusion cooking could thus lead to various processing techniques in brewing. This is presented below (Fig. 3).

Criterion / Product	Enzyme Activity	Moisture Content	Taste/Flavour Intensity	Colour (Wort)
Air-dried Malt				
Mixture	medium	low	medium	medium
Air-dried Extruded Malt				

Fig. 2. Required Performance of the Mixture

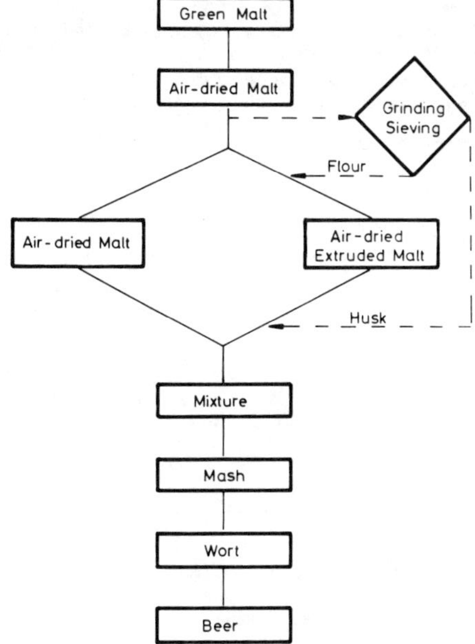

Fig. 3. The Use of HTST Extrusion Cooking in the Malting and Brewing Process

GM is dried to AM which can be stored, ground and extruded. Part of the AM is extruded to AEM which can then be mixed with ground AM. The mixture can be mashed and filtered and the resulting wort then fermented. A possible processing alternative lies in the separation of the husks from the AM to be extruded. This step can, among others, help to improve the performance of the extruder or impart special qualities to the AEM. It is necessary to mention at this point that a whole series of other processing methods are possible. However, these are not covered in the results presented below.

3.2 Extrusion cooking and mashing

As a first step, the possibility of processing a mixture (MIX) of AM and AEM to a fermentable wort was studied. The mash viscosity and the formation of maltose during mashing were measured. It was to be expected, on the one hand, that the mash viscosity of the MIX from AM and AEM, by virtue of the gelatinization of the starch resulting from the HTST extrusion cooking, would be appreciably higher than that of the mash conventionally prepared from CM. On the other hand, it was to be expected that the formation of maltose from the gelatinized starch would take place much faster. These two aspects are important in the wort processing technique with particular reference to mash concentration and duration of mashing.

Under the usual mash concentration of 25 kg solids/100 l water, the AEM, alone, formed a solid gel with a high viscosity. Exchanging 25% AEM for AM completely prevented gel formation. In the mashing phase at about 40°C the MIX was already liquefied through the influence of α-amylase to a suspension with a lower viscosity (Fig. 4). In the course of mashing, the viscosity of the mash from the MIX dropped drastically to that of the mash from the CM. The increase in viscosity due to the gelatinization of the starch from the AEM fraction was not so pronounced. The extent and duration of this rise in viscosity depended on the ratio of the MIX. Nevertheless, it was lower in all cases than that of the mash from the CM.

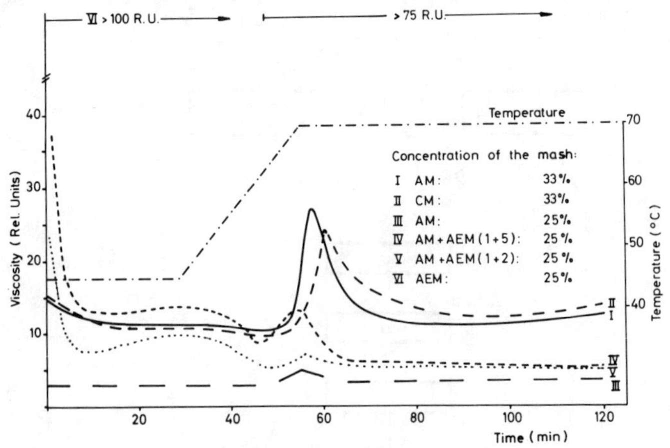

Fig.4. The Development of the Viscosity during Mashing

These differences in the gelatinization of the starch had an effect on the amylolytic degradation (Fig. 5). The higher the AEM fraction in the MIX, the more maltose was formed even at a lower temperature.

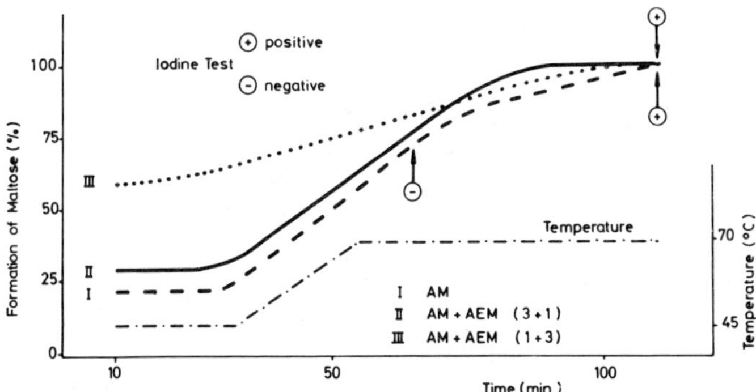

Fig.5. Enzymatic Hydrolysis of Starch during Mashing

Under the chosen experimental mashing ratio, the enzyme potential had no observable effect on the rate and extent of maltose formation. However, this clearly depended on the rate and degree of gelatinization of the starch in the mash. Although the maltose formation in the CM- mash took longer to start and occured at a slower rate than in the MIX- mash, the iodine test was already negative 10 min after reaching the chosen maximum mashing temperature. This showed that the amylose was quickly broken down to smaller polymers of α-glucans which had no effect on the iodine blue colouration. On the other hand, the iodine test for all the MIX-mashes even after the relatively long mashing time of 110 min was still positive. Of particular interest was the fact that the iodine blue colouration was no longer observable in the filtered worts. This implies that part of the starch from the AEM was not accessible for enzymatic degradation. Since the starch was, on the one hand, completely dissolved through mashing and, on the other hand, a higher filterable polymer was present, it was assumed that this phenomenon was due to the formation of amylose-lipid enclosure compounds. Such compounds are known to be formed between lipids and starches (8, 9). The presence of amylose-lipid complexes in the AEM had been established by us. The characterization of such enclosure compounds is an area for further research.

The incomplete degradation of starch necessitated the study of the effect of extrusion on the processability of the mash and the composition of the wort. Furthermore, the effect of extrusion on the extract yield as well as the viscosity and colour of the wort were determined. In order to differentiate the effect of extrusion conditions on these criteria, rather extreme ratios of AEM and AM (8 + 1) were chosen for the preparation of the mash.

The presentation of the relationship between the thermal and mechanical energy inputs and the extract yield showed that with increasing thermal energy input, presented here as a rise in the product temperature, the extract yield also increased (Fig. 6). A rise in the specific mechanical energy (SME), on the other hand, had the opposite effect. However, the highest extract yield could only be achieved when a relatively low SME was combined with a high thermal energy input.

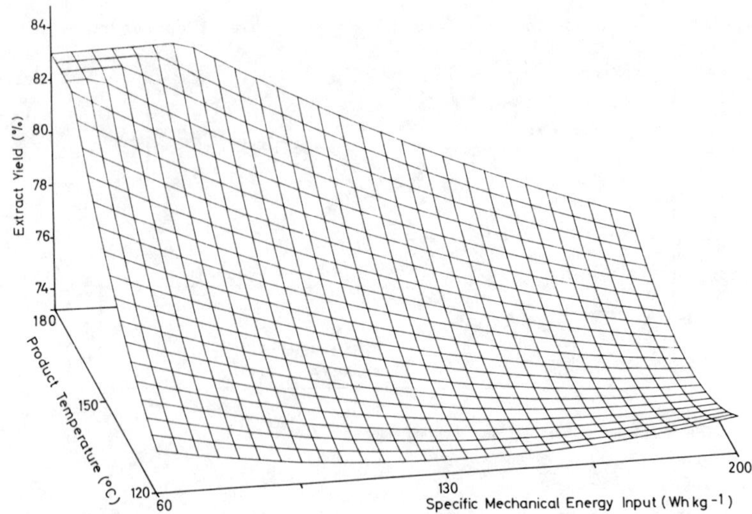

Fig.6. The Effect of the Energy Input on the Extract Yield of the Mixture

This type of energy input led to an increase in the viscosity of the wort, which in turn had a negative effect on the flux rate during filtration. The viscosity of the MIX-worts were, in all cases, higher than those of the AM or CM-worts. The viscosity of the MIX- worts must have been caused by macromolecular substances which were either only not partly, or not at all, degraded by the enzymes present in the malt. It was concluded from this that extrusion of the AM led to qualitatively and/or quantitatively different reaction compounds in comparison with those during kilning. The formation of colouring components points to this fact. The colour of the worts from all the MIX-mashes was darker than those from CM-mashes.

It became clear under the extrusion conditions that, in spite of the much shorter reaction time required and the higher moisture content of the mass, more colour and taste components were formed. The colour of the MIX-wort was deeper, the higher the product temperature (Fig. 7). The rise in colour intensity increased proportionately with the SME. An increase in the product temperature and the SME led to dark-coloured and bitter-tasting AEM.

A combination of the results with the aid of contour plots showed that, under the extrusion and mashing conditions, the MIX-worts could not match the conditions laid down for the CM-worts (Fig. 8). The colour and viscosity of the worts were too high. These wort charateristics produced from a MIX with more AEM could be improved upon by reducing the proportion of AEM in the MIX. The extract yield remains unchanged.

It should therefore be mentioned that this procedure could lead to a further saving of energy in the malting. The curves in Figure 9 show that a drop in the fraction of extruded material would lead to a reduction in energy costs for the manufacture of the starch-containing component. A ratio of 60 parts AM to 40 parts AEM would reduce energy costs below those when malting with energy recycling. This saving could be further improved by separating the husks before extrusion.

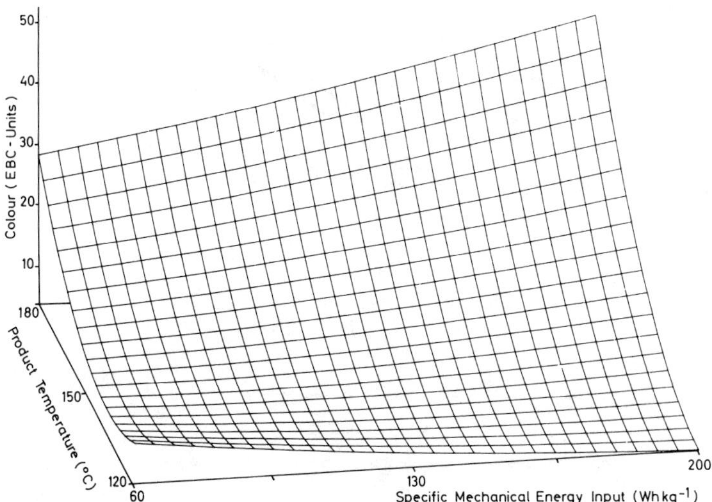

Fig. 7. The Effect of the Energy Input on the Colour of the Wort from the Mixture

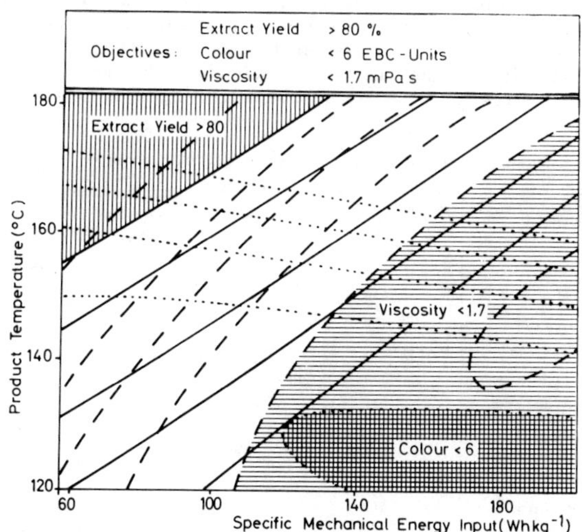

Fig. 8. Optimization Diagram for the Extrusion of Air-dried Malt

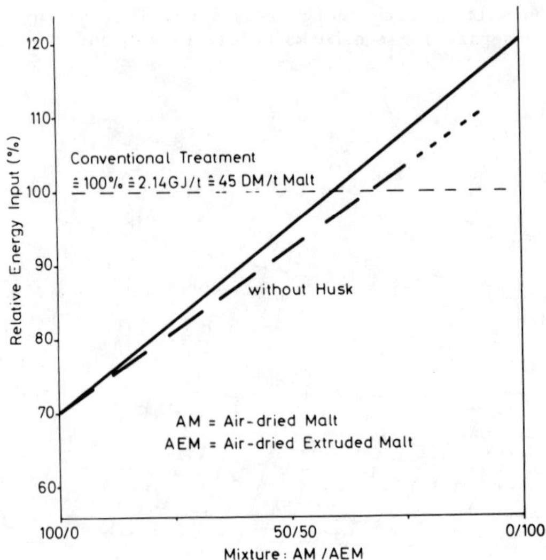

Fig.9. Comparison of Energy Input for Various Mixtures

3.3 Charateristics of wort and beer

The use of a greater fraction of AM in the MIX is possible because the presence of typical taste components in AEM could supplement their absence in the AM. For example, 4-vinylguaiacol, 2-acetylpyridin and 2-phenylethanol were increased through thermal treatment (Table 1). These substances also often occur in malts which are kilned at higher temperatures. This results in dark malt from which beer with a good aroma can be brewed.

Table I. Taste and Flavour Components of Various Raw Materials

Component (ppb)	Air-dried Malt	Malt	Air-dried Extruded Malt A*	Air-dried Extruded Malt B**
Hexanal	610	730	905	972
tr-2-Nonenal	100	170	30	50
3-Methional	15	15	165	215
2-Phenylethanol	490	690	240	275
4-Vinylguaiacol	75	90	750	960
2-Acetylpyridin	-	5	30	60
2-Pentylfuran	100	405	500	970

Energy Input by HTST Extrusion Cooking:

* SME = 55 Wh / kg ; Product Temperature = 180°C

** SME = 122 Wh / kg ; Product Temperature = 190°C

Nonetheless, substances which could be responsible for off-flavours were present, to some extent, in the AEM. These included such substances as hexanal, trans-2-nonenal and 3-methional. Their formation was, however, not uniform with the result that their effect could not be evaluated. In any case, it has been established beyond reasonable doubt that, the taste of beer can be altered over a wide range through the use of AEM. The ratio of AEM to AM would therefore play an important role in influencing the quality and character of the beer.

It is possible to produce beer from a MIX having the desired enzyme potential whilst restricting the taste and colour components to the desired limits.

For example, it was possible to produce a wort comparable to that from a CM malt with a 33% AEM in the MIX (Table 2). The extract yield was about the same. The viscosity of the MIX-worts was only a little higher, but still within the normal limits. All the malts gave clear worts. The macromolecular fraction of the nitrogen compounds in the MIX-worts was somewhat smaller than that in the AM and CM-worts. This difference was caused by the denaturation of the proteins during extrusion. The cooking colour of the MIX-worts was darker than that of the CM-wort. Because the colour could be affected by extrusion conditions and the MIX-ratio, the conditions necessary for production could be achieved by adjusting these parameters.

Table II. Distinctive Characteristics of the Raw Materials used for Brewing

Criterion	Raw Material			
	Air-dried Malt	Mixture I*	Mixture II**	Malt
Yield (% d.s.)	83.6	81.8	81.8	81.9
Haze (EBC)	1.8	<1.0	<1.0	2.0
Viscosity (cp)	1.513	1.530	1.527	1.513
N-Compounds (mg/100 g)				
Soluble	765	712	686	765
High-molecular	195	162	160	197
Boiled Wort Colour (EBC)	7.4	9.3	12.9	6.8
App. Fin. Attenuation (%)	80.1	77.6	79.2	79.9

Ratio of the Mixtures: 33% Air-dried Extruded Malt / 67% Air-dried Malt

Energy Input by HTST Extrusion Cooking:

* SME = 73 Wh / kg ; Product Temperature = 173°C

** SME = 55 Wh / kg ; Product Temperature = 179°C

For example the fermentability of a MIX-wort with a high AEM fraction, of the order of 75%, is presented in Table 3. In spite of the rather low product temperatures chosen for the extrusion - a step which reduced the formation of taste and colour components in the AEM - the beers showed no difference in fermenting behaviour.

Table III. Distinctive Characteristics of the Beer from Various Raw Materials

Criterion	Raw Material			
	Malt	Mixture I*	Mixture II**	Mixture III***
Original Gravity (%)	11.70	12.39	12.16	11.68
Ethanol (%)	3.89	4.26	3.93	3.82
App. Fin. Attenuation (%)	78.5	82.1	77.2	78.5
Final Attenuation (%)	64.5	66.5	62.6	63.5

Ratio of the Mixtures:

75% Air-dried Extruded Malt / 25% Air-dried Malt

Energy Input by HTST Extrusion Cooking:

* SME = 88 Wh / kg ; Product Temperature = 140°C

** SME = 84 Wh / kg ; Product Temperature = 139°C

*** SME = 107 Wh / kg ; Product Temperature = 135°C

This leads to the conclusion that HTST extrusion cooking could be integrated into the malting process. This represents a new processing technique in the production of beer. Of particular interest in this connection is the fact that a wide range of possibilities exists to adjust the taste qualities of beers.

REFERENCES

1. ECKART, P. Energieverbrauch bei der Malzherstellung. Der Doemensianer 23, Nr. 1 (1983), 11.
2. diagramm 77, Firmenschrift Bühler-Miag, Braunschweig, 1983.
3. Behandeln von angekeimten Getreidekörnern für die Brotherstellung. DP 2851053, Ausgabetag 25.9.1980.
4. MEUSER, F., VAN LENGERICH, B. and KÖHLER, F. Einfluß der Extrusionsparameter auf funktionelle Eigenschaften von Weizenstärke. starch/stärke 34 (1982), 366.
5. Methoden der enzymatischen Lebensmittelanalytik. Boehringer GmbH, Mannheim, 1980.
6. KRÜGER, E. and BIELIG, H.J. Betriebs- und Qualitätskontrolle in Brauerei und alkoholfreier Getränkeindustrie. Verlag Paul Parey, Berlin und Hamburg, 1976.
7. KOSSA, T., BAHRI, D. and TRESSL, R. Aromastoffe des Malzes und deren Beitrag zum Bieraroma. Monatsschrift für Brauerei 32 (1979), 249.
8. MERCIER, C., CHARBONNIERE, R., GREBAUT, J. and DE LA GUERIVIERE, J.F. Formation of amylose-lipid-complexes by twin-screw extrusion cooking of manioc starch. Cereal Chem. 57 (1980), 4.
9. MEUSER, F., VAN LENGERICH, B. and KÖHLER, F. Technologische und ernährungsphysiologische Aspekte zur Herstellung eines mit Speisetrebern angereicherten extrudierten Flachbrots. Lebensmittel-Technologie 16 (1983), 13.

EXTRUSION COOKING OF DAIRY ENRICHED PRODUCTS AND MODIFICATION OF DAIRY PROTEINS

C. MILLAUER
W. M. WIEDMANN
E. STROBEL
Werner & Pfleiderer, Stuttgart

Summary:

With milk ingredients, which are added to a cereal flour, the properties of extrusion-cooked products can be altered to a large extent. Apart from being an additive, dairy-protein is also suitable in pure form for the mechanical modification and also for the chemical reaction with lye in the extruder in order to produce entirely soluble caseinates.

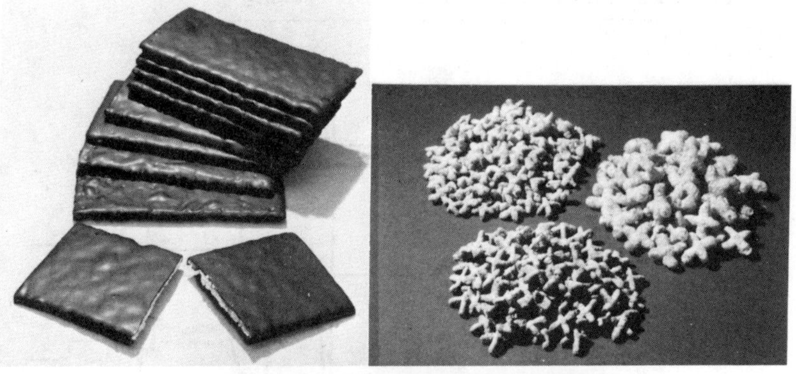

Fig. 1: Examples for extruded dairy enriched products

1. INTRODUCTION

Food extrusion was started about 50 years ago with single-screw extruders for extruding noodle doughs without gelatinization of the grain. Not until a long time after this was it possible to gelatinize the starch in cereals and to extrude snacks and other expanded products. Only recently has it been possible, particularly with twin-screw extruders, to start from flours and mixtures of flours and make semi-finished and finished products which used to be made on other plants with a considerably higher water content and therefore more uneconomically. For extruders, the addition of milk powder is very important as this improves the flavour protein content, texture and browning, although the process conditions in the extruder are permanently altered. As an example, see the products shown in Fig. 1, which were made on extruders and which contain milk components such as 3-4% whole milk powder for crisp bread, coated with chocolate with milk powder again, or dairy enriched breakfast cereals with casein, milk powder or milk sugar, and particularly products filled during extrusion which also contain a lot of milk powder in the filling compound.

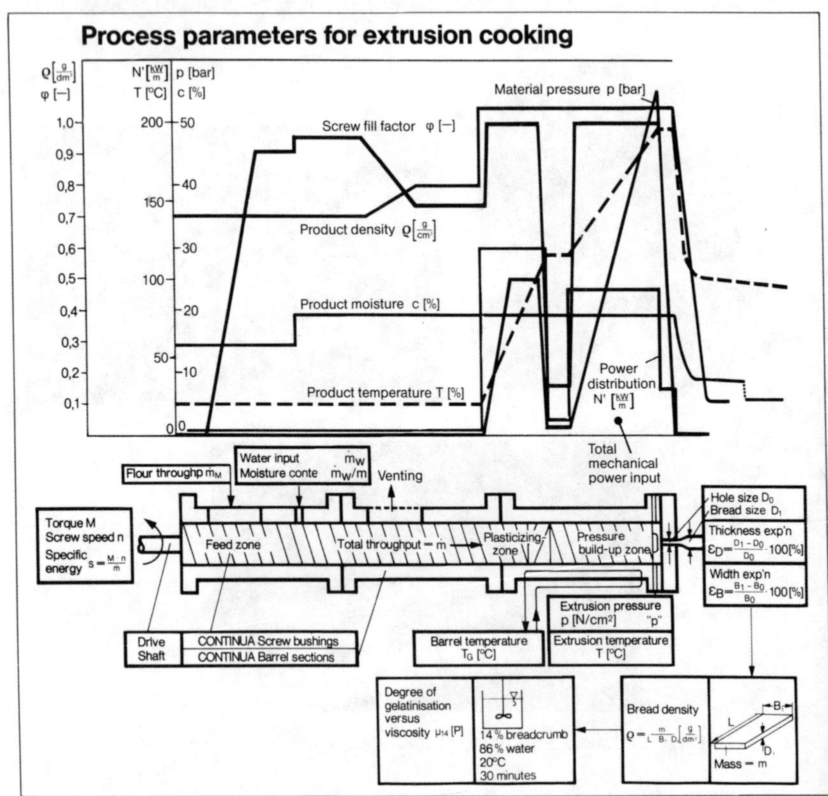

Fig. 2: Change of the process parameters in an extrusion cooker

2. PROCESSES

The centre-piece of such a process is of course the extruder, which must homogenize, gelatinize, disperse and carry out the reaction, structuring and shaping of the products. What happens now in such a machine? (Fig. 2).

The flour is fed to the extruder with additives like ballast, fats, spices, salt, sugars and some milk components. Then 0-15% water is added, so that the total water content is 14-30%, and is pre-mixed in the screws. This mix is then devolatilized at very low bulk weights to ex- pell the entrapped air so that it does not obstruct the feed area, which might reduce throughput. The screw fill factor thus drops and no back- pressure can occur. The pre-mixed material is conveyed into the plasticizing zones, which produce increased stock pressure in the barrel as a result of an increase in the fill factor, which again leads to an increase in shear and thus energy dissipation into the product with simultaneous temperature increase. The starch grain structure is thus destroyed and the water contributes to gelatinization.

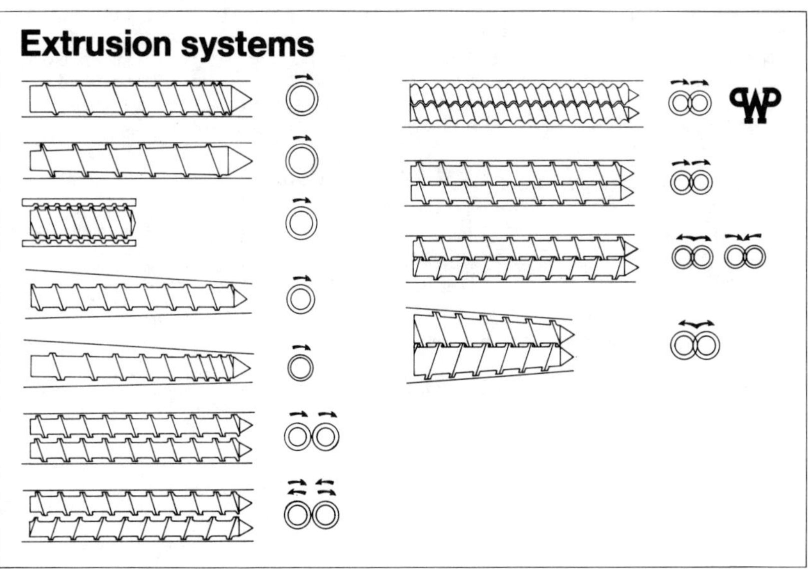

Fig. 3: Different single- and twin-screw extrusion systems

The plasticizing zone and pressure build-up zone are separated so that they can be controlled independently. The fill factor and the pressure thus fall abruptly between the two zones. The plasticized and gelatinized product is then compressed again in the pressure build-up zone, and thus the fill factor, the pressure and the temperature increase again until the material is extruded or formed into a strip through the die, which causes the pressure to fall to atmospheric pressure.

When the product is discharged into the open air, it relaxes, thus causing expansion and evaporation of about 10% water. The flexible material, which is under pressure in the extruder, foams when it relaxes and this foamy structure remains when the water and heat are removed. It does not matter whether the product is pressed through a flat die to make flat bread or other dies, for example to make pellets of various shapes by cutting the strands at the die-plate. Different extrusion systems (Fig. 3) are in use to fulfil different requirements.

3. FLAT BREAD EXTRUSION

Fig. 4 shows a flat bread plant based on a Continua twin-screw self-cleaning co-rotating mixing and cooking extruder for a throughput of about 500 kg/h roasted metering units. The Continua 120 has a 4-hole die and a combined take-off and cross-cutting unit. The four flat bread strands are guided by plastic lips as they leave the die at a speed of approx. 40 metres/min. and are conveyed to the cutting unit. After this, the product is dried from approx. 9% to 4% water content and roasted. The flat bread, which is at a temperature of approx. 120°C, is then cooled down to 70°C, stacked and conveyed to the packing unit.

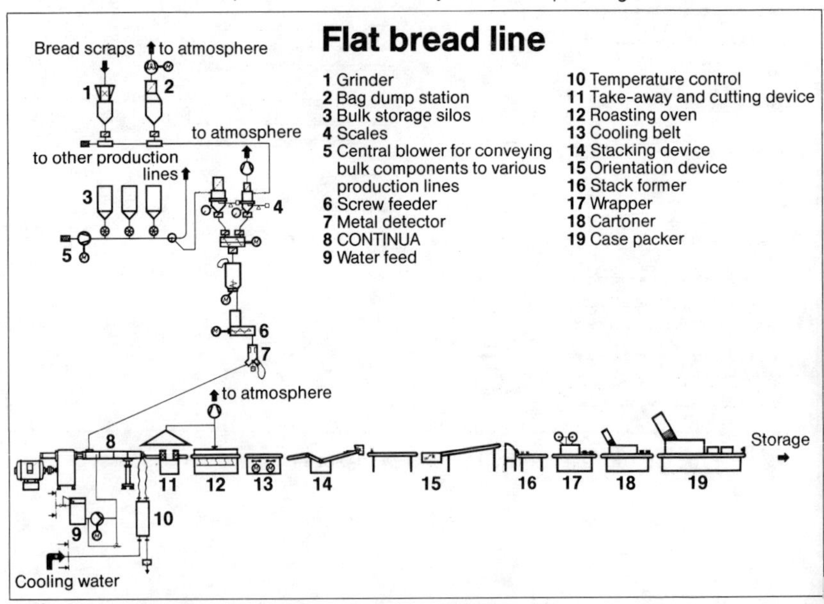

Fig. 4: Automatic production line for 1700000 slices of crisp bread per da

4. INFLUENCE OF WHOLE MILK POWDER

The use of whole milk powder (Fig. 5) in flat bread or other extruded products is extremely beneficial for the products themselves since, as the content of whole milk powder increases, the product becomes browner, tastier, more brittle, more biscuit-like and takes impressions more easily. On the other hand the influence on the process parameters must be considered. As the influence of the fat predominates, the energy and extrusion pressure drop slightly and the degree of expansion is about halved when the content of whole milk powder is 20%. With a lot of plasticizing zones, even low concentrations of fat create pulsating extrusion conditions. Normally, therefore, fat contents of 1% cannot be exceeded. Split plasticizing zones with stabilized intermediate conveying stages make it possible to add up to 20% whole milk powder, which equals 5% total fat content and is ample for most applications.

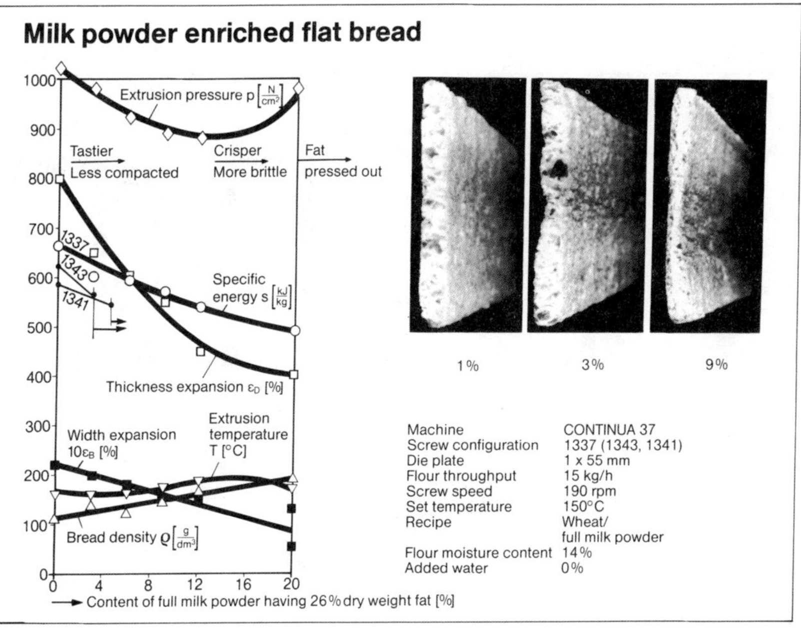

Fig. 5: Influence of milk powder on extrusion parameters and product quality

5. INFLUENCE OF CASEIN

In Fig. 6, two variants are shown for casein enrichment. In the left-hand graph, the added water content of 14% and total water content of 24%, which is necessary to texturize pure casein has been maintained, although for wheat flour the energy and pressure are then too low, the density too high and the product too hard. It is therefore necessary to adjust the added water-content between 0 and 14% when changing from pure wheat flour to pure casein powder. This gives the curves in the right-hand graph with higher energy and higher pressure for wheat extrusion than for casein extrusion. As the casein content increase, the density decreases as pure casein has a particularly strong and foamy expansion pattern. Immediately after the die (while hot), the density of the casein is only about half the value shown, as contraction occurs on cooling. Improvements in texture without demaging the flavour are only possible with low casein contents. The best grinding conditions for purely extruded casein are given by direct drying of the extrudate to preserve the foamy structure.

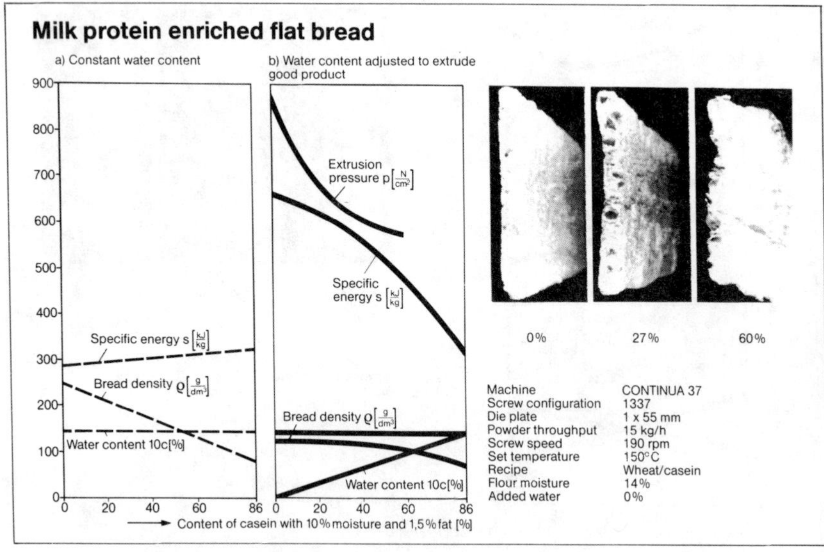

Fig. 6: Influence of Casein powder with constant or adjusted water content on extrusion parameters and product quality

6. REACTIONS OF CASEIN TO FORM CASEINATE

The casein coming from the decanter (Fig. 7) with approx. 50% dry compound and a pH value of 4.6 is volumetrically fed to the Continua by a twin-screw feed unit. In the machine, a 25% soda lye is also added by a metering piston pump and homogeneously mixed with the casein, which reacts to from caseinate. This takes place at throughputs of 1400-2000 kg/h casein, the caseinate leaving the extruder at about 80-90°C with a viscous consistency and being conveyed to a drum dryer for further drying.

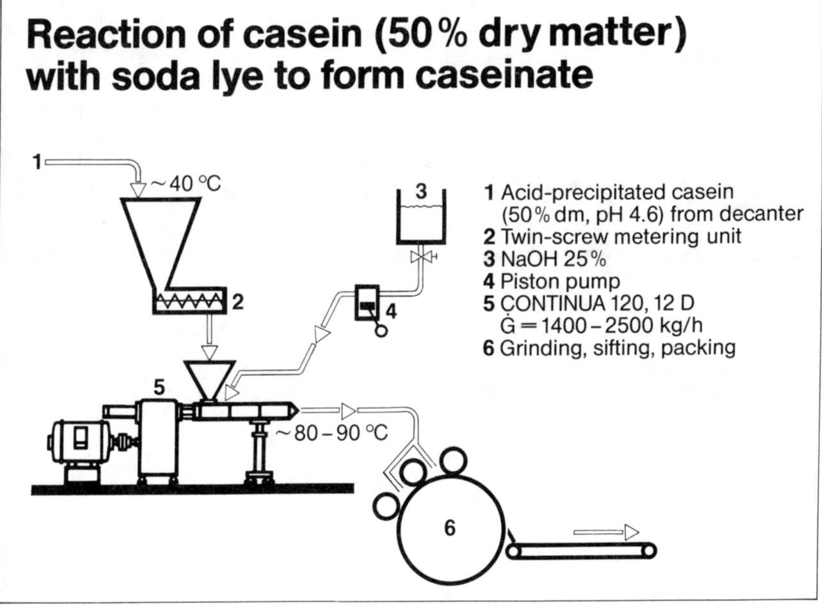

Fig. 7: Flow sheet of a production line for the chemical reaction of wet casein to sodium caseinate

It is however also possible to start with pre-dried casein with 6-20% moisture content (Fig. 8). Here too, the casein is fed into the extruder and then the soda lye (20-40%) added. The energy dissipation by shear increases the product temperature to about 120°C at the die which induces the reaction to form expanded and structured strands of caseinate which are perhaps post-dried and ground afterwards.

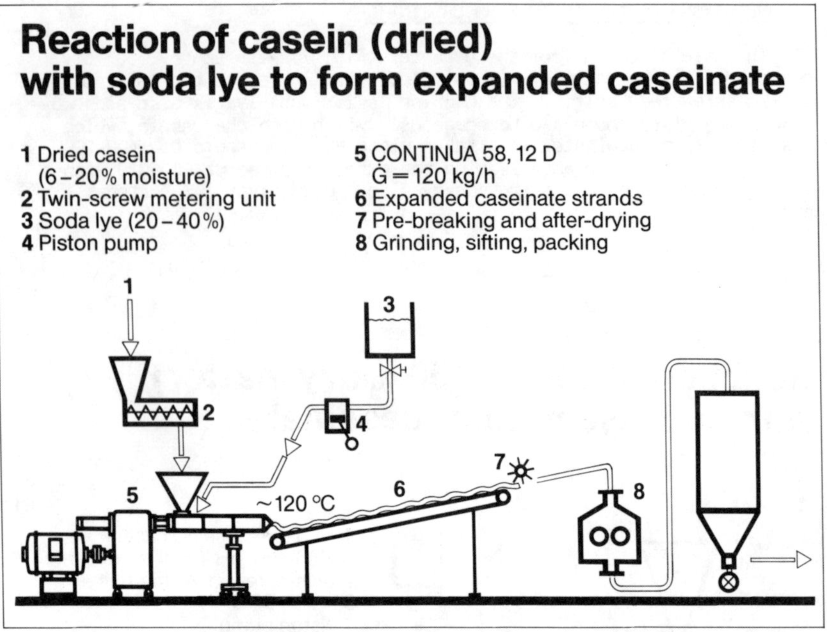

Reaction of casein (dried) with soda lye to form expanded caseinate

1 Dried casein (6 – 20 % moisture)
2 Twin-screw metering unit
3 Soda lye (20 – 40 %)
4 Piston pump
5 CONTINUA 58, 12 D $\dot{G}$ = 120 kg/h
6 Expanded caseinate strands
7 Pre-breaking and after-drying
8 Grinding, sifting, packing

~120 °C

Fig. 8: Flow sheet of a pilot plant for the chemical reaction of casein powder to sodium or calcium caseinate

Potassium- and sodium-caseinates are excellent water soluble and have a high capability of water absorption at low temperatures. They are stabilizing oil in water-emulsions and aqueous foams and can be applied in meat, toppings, dressings, sweets and dairy products.

Calcium caseinates are only colloidal soluble and are applied for example in doughs, baker's wares and as a brightener.

REFERENCES

1. WIEDMANN, W. M., STROBEL, E. and MILLAUER, C. (1981). The influence of recipe variations on process parameters and product quality taking flat bread as an example. WP-presentation "Food extrusion".
2. MILLAUER, C. (1983). Kochextrusion von Mehlen unter besonderer Berück- sichtigung des Einsatzes von Milchbestandteilen. Swiss Food, 1983.
3. SCHAER, H. U. (1983). Funktionelle Eigenschaften der Milchproteine und -proteinfraktionen. Swiss food Nr. 6 (1983).

PREPARATION D'ALIMENTS POUR ANIMAUX MARINS PAR CUISSON-EXTRUSION

(PREPARATION BY EXTRUSION-COOKING OF IMPROVED FEEDS FOR MARINE ANIMALS)

J.P. MELCION*, J. GUILLAUME**, J. MEHU*, R. METAILLER**, G. CUZON***

* I.N.R.A. rue de la Géraudière - 44072 NANTES CEDEX - FRANCE

** C.N.E.X.O., Centre Océanologique de Bretagne
B.P. 337 - 29273 BREST CEDEX - FRANCE

*** C.N.E.X.O., Centre Océanologique du Pacifique
B.P. 7004 - TARAVAO - TAHITI - POLYNESIE FRANCAISE

RESUME

Les animaux marins (poissons et crustacés) possèdent des comportements alimentaires variés et souvent mal connus. La cuisson-extrusion apparaît comme un moyen de conférer aux aliments qui leur sont destinés des caractéristiques particulières mesurées par l'aptitude à la réhydratation, la stabilité à l'eau, la consistance, l'aptitude ou non à flotter.

Dans nos conditions expérimentales, une stabilité à l'eau satisfaisante (76-89 p. 100) des aliments pour crevettes Pénéïdes est obtenue par extrusion à température (80-90°C) et hydratation (22 p. 100) modérées d'un mélange contenant du gluten de blé. A l'opposé, une expansion des aliments pour saumon et turbot sera recherchée en vue de favoriser l'absorption d'eau ou d'huiles fluides. L'expansion est obtenue par extrusion à température relativement élevée (180°C) d'une farine contenant des amidons prégélatinisés . La capacité de réhydratation est alors de 250 à 330 p. 100. Si l'aptitude à flotter des extrudats ainsi obtenus est connue et bien adaptée au comportement du saumon, le problème de la fabrication d'aliments destinés au turbot, à la fois réhydratables et coulants au fond de l'eau, est plus difficile à résoudre.

Comparée avec d'autres techniques, la cuisson-extrusion est coûteuse et risque de détruire certaines vitamines de l'aliment, mais elle semble, à l'heure actuelle, bien adaptée à la production d'aliments pour les salmonidés et les poissons plats.

SUMMARY

The feeding behaviours of marine animals are quite different and often bad known. Extrusion-cooking is a way to obtain unusual physical caracteristics of the feeds measured by the rehydration, the water stability, the consistency and the floating or sinking ability.

Extrusion-cooking at moderate temperatures (80-90°C) and water additions (22 p. 100) of a gluten containing meal gives favourable water stability (76-89 p. 100) to shrimp feeds.

At the opposite, salmon and turbot feeds need expansion to increase water and/or oil absorption.

Expansion would be given by extrusion at fairly high temperature (180°C) of a meal added with pregelatinized starches. Rehydration capacity of the product is between 250-330 p. 100.

A good flottability is usual with such a feed, well adapted to salmon and trout behaviour, but the problem of the sinking ability of a turbot feed is difficult to solve.

Compared with other techniques, extrusion-cooking is expensive and has a tremendous effect on the vitamins of the feed, but it seems to be actually well adapted to produce feeds for salmonids and flat fishes.

1. CARACTERISTIQUES DES ALIMENTS POUR ANIMAUX AQUATIQUES

Les aliments destinés aux animaux aquatiques sont soumis à des contraintes liées au milieu liquide rendant impossible l'emploi de solutions, suspensions ou mélanges très solubles.

Par ailleurs, pour les animaux domestiques les plus courants, les problèmes d'appétence sont rarement ardus, mais, pour certaines espèces aquatiques, l'acceptabilité peut revêtir une toute autre importance.

1.1. Caractéristiques physiques des aliments : rôle de la stabilité, de la flottabilité, de la taille et de la dureté des particules (tableau I)

1.1.1. En fonction du comportement alimentaire

Tous les aliments mis en forme sont sensés conserver la forme et la taille qu'on leur a données. Le délitement crée toujours des difficultés de manutention et de distribution et constitue une cause de pertes. Toutefois, ces inconvénients sont spécialement graves chez les animaux aquatiques où un agglomérat qui se désagrège dans l'eau peut très bien ne plus être consommable du tout.

Tableau I

CARACTERISTIQUES PHYSIQUES RECHERCHEES DES ALIMENTS POUR ANIMAUX AQUATIQUES

(Requested physical properties of feeds for aquatic species)

	"TRUITE DE MER" ET SAUMON ("sea-trout"and salmon)(a)	TURBOT (Turbot) (b)	BAR ET DAURADES (Sea bass and sea breams) (c)	SOLE (Sole) (d)	CREVETTE JAPONAISE (Japanese shrimp (e)
REHYDRATATION (Rehydration)	++	+		(+)	
STABILITE A L'EAU (Water stability)	+	++	+	+++	+++
APTITUDE A FLOTTER (Floating ability)	++	--	+	--	--
CONSISTANCE (Consistency)	-	-		-	

(a) Salmo Gairdneri, Onchorynchus Kisutch, Salmo salar, (b) Scophtalmus maximus, (c) Dicentrarchus labrax, Sparus sp., Chrysophrys sp., (d) Solea solea, (e) Penaeus Japonicus.

Le problème est cependant différent selon le comportement alimentaire de l'animal :

- Le prédateur actif, tel que la truite, le saumon ou le bar, se jette sur le granulé dès qu'il tombe dans l'eau, ou le saisit alors qu'il est en train de couler. Un prédateur de ce type apprend plus difficilement à récupérer une particule reposant déjà sur le fond, plus difficilement encore des débris résultant du délitement des granulés. Dès lors, le granulé doit rester entier quand il est plongé dans l'eau, mais cette résistance peut n'être que de courte durée.

- Le prédateur à l'affût, tel que le turbot, passe une partie de son temps inactif (reposant sur le fond par exemple) et se déplace moins facilement ; il attend que la particule soit à sa portée avant de la happer. Un nouvel impératif apparaît donc : les aliments doivent couler sinon ils ne sont pratiquement pas consommés.

- Certains animaux au comportement brouteur ou détritivore (sole, crevette par exemple) restent immobiles pendant une grande partie de la journée et ne recherchent vraiment leur nourriture que pendant leur phase active qui est généralement nocturne. Ces animaux dédaignent une particule qui tombe ; ils ne se déplaceront vers elle que plus tard, quand elle les attirera par le biais de stimulus olfactifs. Les aliments restent donc dans l'eau pendant une période qui peut atteindre plusieurs heures et la stabilité des agglomérats devient absolument nécessaire ; sans cela, une partie des nutriments est entraînée par lessivage et l'aliment devient déséquilibré, carencé ou inappétent.

1.1.2. <u>En fonction de l'âge des animaux</u>

La vie des animaux aquatiques (à l'exception des reptiles et mammifères marins) débute toujours par un stade larvaire et la taille des larves peut être extrêmement réduite : au stade actuel d'évolution des techniques, on est pratiquement incapable de nourrir avec des particules artificielles les jeunes larves qui chassent activement des proies mobiles de quelques centaines, voire quelques dizaines de micromètres.

Toutefois, dès que les animaux atteignent un stade plus avancé, situé généralement aux alentours de la métamorphose, le recours à des aliments inertes devient possible, pourvu que l'on respecte deux conditions :

- densité adéquate permettant, soit une certaine "flottaison" entre deux eaux (cas du turbot), soit une chute rapide vers le fond (cas de la sole et de la crevette) ;
- taille de la particule adaptée à celle de la bouche de la "post-larve".

Ces conditions impliquent le recours à des techniques sûres permettant d'obtenir des particules de très petite taille, homogènes, calibrées, de densité choisie, gardant leur cohésion pendant une durée pouvant atteindre quelques heures.

Par la suite, la taille des particules reste primordiale, sans créer de problème particulier, mais la flottabilité reste préoccupante : elle peut être recherchée (cas des salmonidés d'eau douce répugnant à prendre des granulés sur certains fonds) ou évitée (cas des turbots de grande taille qui attendent que l'aliment coule).

La dureté de la particule s'est avérée être une composante importante de l'appétence chez les post-larves et chez les adultes de certaines espèces comme le turbot qui refuse les miettes ou les granulés de consistance dure.

1.2. <u>Qualités organoleptiques - Rôle des attractants</u>

Les poissons autres que les prédateurs actifs sont généralement attirés vers leurs aliments par des substances chimiques douées de propriétés olfactives. Le rôle des attractants est également fondamental chez les jeunes poissons récemment métamorphosés. De ce fait, les granulés destinés à reposer un certain temps sur le fond jusqu'à ce que le poisson les trouve, de même que les miettes calibrées utilisées pour le "sevrage" des larves de poisson, doivent laisser diffuser des substances chimiques tout en gardant leur cohésion globale.

2. CARACTERISATION DES PROPRIETES PHYSIQUES DES ALIMENTS DES ANIMAUX MARINS

Les aliments des animaux terrestres sont classiquement caractérisés par leur forme, leurs dimensions adaptées à la taille de l'animal et leur texture : résistance aux chocs et à l'abrasion ou friabilité (1), résistance à l'écrasement ou dureté (2), densité en place.

Les aliments destinés aux animaux aquatiques seront l'objet de déterminations complémentaires simples: capacité de réhydratation, stabilité à l'eau, consistance après imbibition, degré de flottabilité, adaptées au comportement alimentaire des espèces considérées. Dans nos essais, nous avons surtout utilisé les deux premièrs critères.

2.1. Aptitude à la réhydratation

Les produits mis en forme sont plongés dans un excès d'eau douce à 20°C et la masse d'eau absorbée est mesurée au cours du temps. Un excès d'eau permet une absorption homogène par toute la surface de l'aliment.

L'aliment est disposé alors en mono couche dans des paniers grillagés (section 70 x 70 mm, maille de 2 mm) de manière à ce que les agglomérés ne soient pas en contact, avant d'être plongé dans l'eau. Après un temps t, les paniers sont égouttés sur papier filtre et pesés (3) ; le taux de réhydratation est exprimé par le rapport :

$$Rh = (\text{produit réhydraté}/\text{produit sec initial}) \times 100$$

Exprimés en coordonnées inverses, les courbes Rh = f(t) sont des droites de régression d'équation :

$$1/Rh = A + B/t$$

L'ordonnée à l'origine permet de connaître la capacité maximale théorique de réhydratation pour un temps infini $Rh\ \infty$, l'inverse de la pente est la vitesse de réhydratation.

Une simplification de la méthode est de mesurer la quantité d'eau absorbée au bout d'un temps fixé conventionnellement, 30 minutes par exemple, notée Rh 30 (tableau II).

Tableau II

EXEMPLE DE CARACTERISATION D'ALIMENTS EXTRUDES REHYDRATABLES (SAUMON)

(Caracterisation of rehydratable extruded feeds for salmon)

ALIMENTS COMMERCIAUX (commercial feeds)	REHYDRATATION (rehydration)			STABILITE A L'EAU (water stability)
	Rh 30 (p. 100)	Rh ∞ (p. 100)	Vitesse (speed)	St 60 (p. 100)
A	180	259	0,12	82,0
S	150	239	0,02	97,6

Limites de la méthode :

Il s'agit d'une mesure par défaut en raison du délitement et de la solubilisation de matière sèche dans le milieu lorsque le temps de séjour est trop long. Elle doit être complétée par une mesure de la stabilité de l'aliment dans l'eau.

2.2. Stabilité à l'eau

La méthode utilisée est directement dérivée de celle préconisée aux Etats-Unis (4), puis modifiée en Israël (5) avec des aliments pour poissons d'eau douce (poisson-chat, carpe).

Les aliments sont placés dans des paniers grillagés de 70 x 70 mm à maille de 2 mm de côté, puis plongés en eau douce ou en eau de mer à 20°C et ramenés à l'air à la cadence de 10 mouvements par minute. Le résidu est ensuite séché à l'étuve, jusqu'à poids constant, pesé et les résultats exprimés en p. 100 de la matière sèche initale. La mesure est effectuée sur 5 g d'aliment (6).

La perte en matière sèche augmente d'abord brutalement en fonction du temps, puis la pente de la courbe diminue au bout de 1 à 3 heures : conventionnellement, la perte en matière sèche au bout de 1 heure de séjour dans l'eau sera retenue pour le calcul de la stabilité.

Cas particulier des aliments réhydratables

En raison de l'existence d'une liaison inverse entre aptitude à la réhydratation et stabilité à l'eau, il peut être utile d'adapter la durée de la mesure de stabilité à celle nécessaire pour atteindre un taux de réhydratation donné. Les comparaisons entre aliments s'effectuent alors à réhydratation constante, mais à des temps variables.

Une autre solution est d'accélérer artificiellement la réhydratation des aliments jusqu'à ce que ceux-ci ne flottent plus avant d'effectuer la mesure de stabilité au moyen du vide. La réhydratation est plus élevée que celle spontanée en ½ heure , mais l'agitation créée par la mise sous vide, même pendant un temps court (1 à 2 mn) favorise une légère dissolution de certains composants de l'aliment (tableau III).

Tableau III

STABILITE A L'EAU D'UN ALIMENT EXTRUDE REHYDRATABLE (TURBOT) APRES REHYDRATATION SOUS VIDE (34 mm Hg)

(Water stability of a rehydratable extruded feed for turbot after rehydration under vacuum 34 mm Hg)

ALIMENTS n° (feeds n°)	TAUX EXPANSION (expansion ratio)	REHYDRATATION (p. 100) (rehydration)		STABILITE A L'EAU (p. 100) (water stability)	
		Rh 30	sous vide (under vacuum) 3 mn	en l'état (as such)	Après réhydratation sous vide (after rehydration under vacuum)
1	1,9	330	391	66,0	60,0
2	1,8	280	325	69,0	66,0
3	1,4	196	213	73,3	73,1

3. INTERET DE LA CUISSON-EXTRUSION

Dans ce contexte, plusieurs techniques de mise en forme ont été employées : agglomération classique en alimentation animale à l'aide de presses à filières tournantes (7), granulation par voie humide et séchage ultérieur (8), cuisson-extrusion. Sans permettre de résoudre tous les problèmes, cette dernière technique s'est révélée offrir un grand nombre de possibilités (9).

3.1. Aliments non réhydratables (cas des aliments pour crevette et sole)

L'extrudeur est utilisé comme appareil de mise en forme simple, similaire à celle des pâtes alimentaires.

Dans nos conditions expérimentales (extrudeur Creusot-Loire BC45 bi-vis co-rotatives,muni de filières de 2 mm) la farine contenant une notable proportion d'eau ajoutée au préalable (18-22 p. 100) est extrudée à température modérée (76-90°C). Le contact avec l'eau permet de développer un réseau protéique à partir du gluten de blé contenu dans l'aliment. Ce réseau est ensuite fixé par un séchage à l'air chaud (80°C) aussi limité que possible, qui élimine l'humidité excédentaire. D'autres protéines aisément coagulables peuvent être employées : farine de sang, poudre d'oeuf. Les colloïdes de nature polysaccharidiques sont généralement peu utiles : les gels qui auraient pu se former sont détruits par le cisaillement dans les conditions d'hydratation et de température dans lesquelles nous avons opéré.

Plusieurs points sont à considérer :

- la finesse de broyage : phénomène bien connu, une plus grande finesse accroît la stabilité à l'eau (10), d'autant plus que les filières d'extrusion sont de petit diamètre. Nous avons pu vérifier expérimentalement que l'optimum semble se situer entre 0,100 et 0,150 mm de diamètre moyen des particules en travaillant avec une filière de diamètre 2 mm.
- la séquence d'incorporation de l'eau : on cherche à préserver l'action du gluten en incorporant tout d'abord les matières grasses insaturées, préférentiellement adsorbées sur la farine, puis le gluten et enfin l'eau.
- la teneur en gluten et la quantité d'eau ajoutée : une teneur en gluten de 12 à 15 p. 100 apparaît nécessaire. A ce niveau, le rapport eau ajoutée/gluten présent est de 1,6 à 1,7.
- le séchage : la température et la durée d'un séchage en lit fluidisé ne semblent pas jouer de rôle important sur la stabilité à l'eau des aliments pour crevettes japonaises (tableau IV). On note également un effet légèrement favorable d'un temps de rétraction de l'extrudé à la sortie de l'appareil.

Au-delà de 22-25 p. 100 d'eau ajoutée et jusqu'à 40-43 p. 100, les mécanismes de liaison des particules alimentaires sont différents : il est nécessaire de faire alors appel à des agents de texture : gélifiants (alginates), propylène- glycol, etc..., en utilisant d'autres techniques. La cuisson-extrusion peut toutefois se concevoir pour cuire (et éventuellement stériliser) des déchets de viande ou de poisson frais dans un aliment semi-humide, dans la mesure où sont résolus les problèmes liés au conditionnement et à la distribution. Appliqué depuis longtemps à des espèces d'eau douce (poisson-chat) (11), ce schéma serait applicable aux aliments pour turbots où l'on recherche des produits plus denses que l'eau et de consistance proche de celle d'une proie vivante.

Tableau IV

INTENSITE DU SECHAGE ET STABILITE A L'EAU D'ALIMENTS POUR CREVETTES JAPONAISES

(Drying intensity and water stability of shrimp feeds)

	TEMPERATURES DE SECHAGE (°C) x TEMPS (mn) (drying temperatures (°C) x time (mn))				
	90 x 10 60 x 10	90 x 10	60 x 20	60 x 10	Air ambiant 20 x 24 h (room temperature)
Stabilité (p 100) (water stability)	86,6 ± 3,1	84,9 ± 2,5	86,1 ± 3,2	86,2 ± 3,4	88,9 ± 2,6

Conditions d'extrusion (extrusion conditions) :

Température fourreau (barrel temperature) ... : 70-75°C
Vitesse vis (screws rotation) .... : 270 rpm
Eau ajoutée (added water) .... : 21 p. 100
Débit alimentation (feeding input) .... : 95 kg.h-1

3.2. Aliments réhydratables (cas des aliments pour truites de mer, saumon, turbot)

La cuisson-extrusion est alors nécessaire à la fois pour améliorer la digestibilité de l'amidon présent dans l'aliment et pour expanser le produit qui devient poreux et absorbe l'eau plus rapidement.

Le taux et la fragilité des matières grasses présentes dans l'aliment n'autorisent pas des températures d'extrusion très élevées : il sera alors nécessaire d'utiliser des amidons soit déjà transformés, soit facilement attaquables.

3.2.1. Aliments réhydratables flottants (cas des aliments pour truites de mer et saumon)

La cuisson-extrusion autorise la production d'aliments expansés flottants de manière relativement aisée avec 10 à 15 p. 100 d'amidon transformé (amidons gélatinisés industriels, déchets de biscuiterie, etc...) dans la formule. Ce taux est compatible avec les besoins nutritionnels des salmonidés.

Dans nos conditions expérimentales, la capacité de réhydratation, liée au taux d'expansion et à la densité des produits extrudés a pu être contrôlée en agissant sur :

- le couple humidité-température : pour une température de fourreau de 185°C, et une farine à 25 p. 100 d'humidité, l'addition de 7 p. 100 d'eau se traduit par un accroissement de 15 à 25 p. 100 de la capacité de réhydratation sans toutefois modifier la stabilité à l'eau ;
- le débit d'alimentation : à humidité et température constante, un débit élevé accroît l'aptitude à la réhydratation ;
- la forme de la vis : une vis relativement longue (60 cm au lieu de 50), munie de disques malaxeurs en position intermédiaire permet d'obtenir des produits de caractéristiques favorables.

3.2.2. Aliments réhydratables non flottants (cas des aliments pour turbots)

Dans ce type d'aliment, l'expansion doit être limitée de manière à ce que le produit extrudé s'imbibe d'eau, puis coule le plus rapidement possible, tout en conservant sa stabilité.

Le problème n'est pas à l'heure actuelle résolu de manière satisfaisante : l'eau d'imbibition ne se substitue que partiellement à l'air occlus, sauf pour les aliments concassés se présentant sous forme de miettes où la surface de contact avec le liquide est plus importante.

Plusieurs voies d'approche sont possibles :

- L'air occlus dans les extrudats peut être éliminé mécaniquement, soit en fragilisant sa structure (fabrication de miettes destinées aux jeunes turbots), soit en les soumettant à l'action d'un vide modéré (34 mm Hg : tableau V) durant 1 à 3 minutes.

Tableau V

DUREE D'APPLICATION DU VIDE ET APTITUDE A FLOTTER D'ALIMENTS POUR TURBOTS

(Vacuum time and sinking ability of turbot feeds)

ALIMENT (feed)	TAUX EXPANSION DIAMETRALE (diameter expansion ratio)	EXTRUDES FLOTTANTS (p. 100) (floating extruded products at a given time)			
		A pression atmosphérique (at atmospheric pressure)	Sous vide partiel (under vacuum)		
		30 mn	1 mn	2 mn	3 mn
1	1,42	100	0	-	-
2	1,82	100	20	5	0
3	1,92	100	80	10	0

- Les propriétés des amidons incorporés à l'aliment peuvent aussi être mises à profit : dans nos conditions expérimentales, l'incorporation d'amidon d'amylo-maïs, riche en amylose hydrophobe et facilement complexable avec les acides gras présents, limite l'expansion et la réhydratation, mais semble favoriser la stabilité à l'eau des extrudats et leur aptitude à couler (tableau VI). Une solubilité limitée à froid de l'amidon employé pourrait être un facteur favorable.

D'autres auteurs (12) préconisent une cuisson du produit en tête d'extrudeur pour refroidir ensuite avant la filière. La densité du produit serait alors de 430 à 560 kg.m-3 au lieu de 250-400 kg.m-3 pour des aliments flottants obtenus en chauffant le produit dans la partie terminale du fourreau.

Tableau VI

PRESENCE D'AMYLOSE ET CARACTERISTIQUES D'ALIMENTS EXTRUDES POUR TURBOTS

(Amylose level and physical caracteristics of extruded turbot feeds)

TYPES D'AMIDONS (Starches of)	TAUX D'EXPANSION (expansion ratio)	ALIMENT FLOTTANT (p. 100) (floating feed) à/at 30mn	REHYDRATATION (p. 100) (rehydration) à/at 30 mn	STABILITE A L'EAU (p. 100) (water stability)
Amylo-maïs (amylo-maïze)	1,57	75	190	75,8
Maïs normal (normal maïze)	1,8	100	242	72,3
Maïs cireux (waxy maïze)	1,7	100	280	57,7

4. LIMITE DE LA CUISSON-EXTRUSION POUR LA FABRICATION DES ALIMENTS POUR ANIMAUX MARINS

4.1. Conséquences sur la composition de l'aliment

Un aliment composé complet doit apporter à l'animal tout ce qui est nécessaire à ses besoins : or, il est bien connu que les conditions de température et de cisaillement régnant dans un extrudeur sont susceptibles d'altérer la composition vitaminique de l'aliment.

Il est fait état de pertes importantes en vitamine C (32 p. 100) et thiamine (13 p. 100) au cours de l'extrusion à 135-155°C d'un aliment pour salmonidés (13). La pyridoxine, les acides folique et pantothénique paraissent plus stables.

Les composants majeurs de l'aliment semblent moins altérés ; nos essais ont montré que la disponibilité de la lysine d'un aliment pour saumon extrudé à 185°C n'était pas significativement réduite lorsque l'hydratation était supérieure à 25 p. 100. Les matières grasses, huiles fluides contenant des acides gras à longue chaîne et hautement insaturés sont plus sensibles, mais peuvent être ajoutées à la sortie de l'extrudeur ou avant l'utilisation.

4.2. Coût économique

La cuisson-extrusion est une technique onéreuse, non seulement par son coût direct (investissement, entretien, fluides), mais aussi par son coût annexe car elle est intégrée dans une chaîne de fabrication : broyage poussé, mélange, incorporation des liquides en amont, séchage et/ou enrobage en aval. Elle ne peut donc s'adresser qu'à des productions animales autorisant des marges élevées.

D'autre part, les aliments pour animaux marins s'adressent à une très large gamme d'espèces, dont les comportements alimentaires sont très dissemblables et de plus mal connus. Si un coût aberrant peut être admis parce que l'acceptation par l'animal est meilleure, le risque est élevé de voir se développer d'autres techniques de fabrication, moins coûteuses, au fur et à mesure de l'évolution des connaissances et des méthodes d'élevage.

5. CONCLUSION

Nos essais, bien que de portée limitée, confirment qu'il est possible d'obtenir des aliments dont les caractéristiques physiques sont en rapport avec les habitudes alimentaires des poissons et crustacés d'élevage. En particulier, la cuisson-extrusion paraît bien adaptée aux aliments pour les salmonidés et les poissons plats où la stabilité à l'eau et la capacité d'absorption d'eau douce, d'huiles et de substances attractantes sont absolument nécessaires en l'état actuel de nos connaissances.

BIBLIOGRAPHIE

(1) PFOST H.B., ALLEN R.N., 1962. A standard method of measuring pellet durability. Proc. Feed Prod. School, Kansas City, 12-14 Nov., 25-29.

(2) DELORT-LAVAL J., DREVET S., 1970. Méthode d'appréciation de la dureté des aliments agglomérés. Ind. Alim. Anim., 213, 43-54.

(3) MEHU J., 1980. Réhydratation et stabilité à l'eau des aliments pour salmonidés : mise au point méthodologique, application à quelques aliments obtenus par cuisson-extrusion. Mémoire ENITIAA, Nantes, sept., pp. 73.

(4) HASTINGS W.H., 1964. Fish feed processing research. Feedstuffs, 36 (21), 13.

(5) HEPHER B., 1968. A modification of Hastings'method for the determination of water stability of fish feed pellets. Recent developments in fish food technology. Working-party FAO-EIFAC/CECPI, Roma (Italy), 20-24 may, appendix 4, p. 48-54 (EIFAC T9).

(6) MELCION J.P., 1969. Carp feed processing. Working-party FAO-EIFAC/CECPI, Alvkarleö (Sweden), 24-28 nov., pp. 11.

(7) GOUBY F., 1982. Elaboration d'un aliment pour crevettes par agglomération. Mémoire ENITIAA, Nantes, sept., pp. 39.

(8) AQUACOP, 1978. Equipement pour fabriquer des granulés par voie humide destinés aux animaux marins. Symposium FAO, Hambourg (FRG), 20-23 Juin, p. 9 (EIFAC E 28).

(9) MEYERS S.P., ZEIN-ELDIN Z.P., 1972. Binders and pellet stability developpement of crustacean diets. Proc. 3rd World Mariculture Soc. Workshop, 351-364.

(10) LECENKO V., VALACKIJ O., 1972. L'influence du degré de finesse des aliments sur la qualité des agglomérés pour poissons (en russe). Mukomol'no Elevatornaja Prom, 1, 20-21.

(11) HUBLOU W.F., 1963. Oregon pellets. Prog. Fish Culturist, 25, 175-180.

(12) WOOD J., 1982. The recipe for success scaling the problem to make feeds for fish. Milling, 165 (2), 32-34.

(13) SLINGER S.J., RAZZAQUE A., CHO C.Y., 1978. Effect of feed processing and leaching on the losses of certain vitamins in fish diets. Symposium FAO, Hamburg (FRG), p. 18 (EIFAC E 70).

DESTRUCTION OF MICROORGANISMS AND TOXINS BY EXTRUSION-COOKING

C. VAN DE VELDE, D. BOUNIE, J.L. CUQ and J.C. CHEFTEL
Laboratoire de Biochimie et Technologie Alimentaires,
Université des Sciences et Techniques,
34060 Montpellier, France

Summary

1. A Liquid or freeze-dried inoculum of Bacillus stearothermophilus spores was introduced into a CL-BC 45 extruder operating at a true end temperature of 150-180°C with a biscuit formulation (14 % H_2O) and a residence time distribution (RTD) of 30-200 seconds. Spore reduction ratios varied from 3.10^{-5} to 7.10^{-8}, corresponding to commercial sterilization.
2. A peanut meal containing 250 µg aflatoxin B_1 per kg was extruded at 175-185°C in the presence of 20 % H_2O and 2-2.5 % NH_3, with an RTD of 100-140 seconds. TLC determination indicated a 85 % "destruction" of aflatoxin. The lysine content of the meal was but slightly reduced ($<$ 10 % loss).
3. RTD in the extruder were determined using pulse injections of erythrosin or of ^{51}Chromate. "Effective" minimum and mean residence times (RT) in the high temperature-high shear zone of the extruder were calculated. Effective mean RT were more affected by feed rate than by screw rotation, while the reverse was observed for effective minimum RT.

The extrusion-cooking of full-fat or defatted soybean has long been used to inactivate deterioration enzymes such as lipoxydase, and toxic proteins such as trypsin inhibitors and hemagglutinins (1,2). While it is well known that extruded foods possess low microbial loads, the precise sterilizing effects of extrusion-cooking have been little investigated.

1. DESTRUCTION OF HEAT-RESISTANT SPORES

In this laboratory, spores of Bacillus stearothermophilus FS 1518, which are among the most heat-resistant spores encountered in foods, are used as sterilization indicators. As previously published (3), a 3 ml liquid inoculum containing 10^{10} or 10^{11} spores is introduced as a pulse, within ca 1 s, into the inlet of an operating BC 45 Creusot-Loire extruder (screw profile FFFFRF). The food mix corresponds to a biscuit formulation containing 52 % starch, 18 % protein and 20 % sucrose. Its water content is equilibrated at 14 % w/w (a_w 0.75 at 20°C) prior to extrusion. Its pH is close to 6.5. Extrusion conditions are the following : feed rate of moist mix : 40 $kg.h^{-1}$; screw rotation : 80 rpm ; temperature of the food mix just before the die : 165 or 182°C. The extruded product is collected 5 seconds per 5 seconds as it comes out of the die, and the number of surviving spores is determined in each 5 s sample. The results are the average of 2 independent experiments.

The relationship between surviving spores in the extruded biscuit and time after spore introduction is shown in figures 1A and 1B. Also shown on these figures is the residence time distribution (RTD) of the food mix in the extruder, as determined by introducing 1 g of erythrosin as a pulse in the extruder inlet, and measuring the absorbance of the extruded samples.

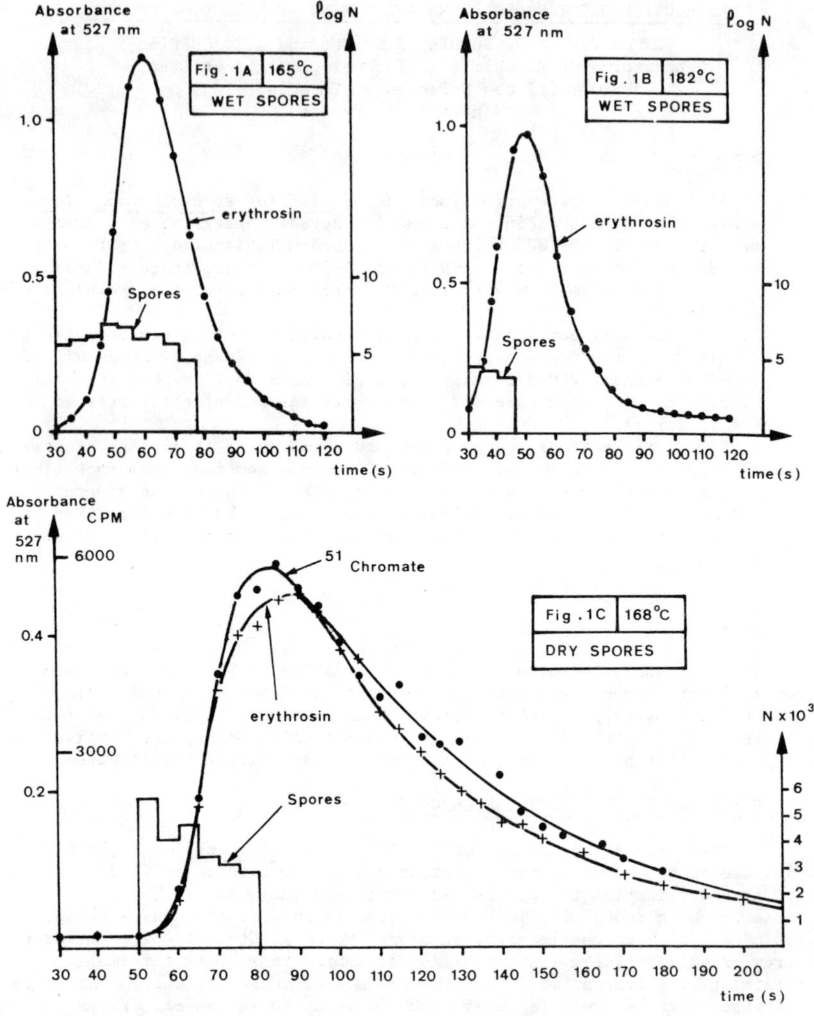

Figure 1. Residence time distribution of erythrosin, ^{51}Chromate or surviving spores during extrusion-cooking. N = number of surviving spores. Time is counted after introduction of erythrosin, chromate or spores in the extruder inlet.

Under the present conditions of extrusion, minimum and maximum residence times (RT) are found to be equal to 35 ± 5 s and to approximately 120 s, respectively. When extrusion is carried out at 165°C, surviving spores are found exclusively in the samples collected within 30 and 75 seconds after introduction of the spores. Samples collected before 30 s, i.e. before the minimum RT, are not yet contaminated. Samples collected between 75 s and 120 s (maximum RT) still contain some of the initially contaminated food particles, but these particles have remained long enough in the extruder for spore sterilization. When extrusion is carried out at 182°C, surviving spores are found only in samples collected between 30 and 45 s (figure 1 B). This means that a residence time of 45 s is then sufficient for spore sterilization. It can be seen that the "concentration" of surviving spores decreases in samples collected from 30 to 45 s, while the RTD indicates that the "concentration" of dead plus surviving spores increases during that period. This emphasizes the importance of residence time (and temperature) in determining sterilizing effects.

It is possible to calculate the total number of surviving spores in cumulated samples and to divide it by the number of spores in the inoculum. Overall reduction ratios (N/No) are equal to 2.9×10^{-5} and 7.4×10^{-8} at 165 and 185°C, respectively. These values correspond to commercial sterilization, as can be briefly discussed now. The decimal reduction time D of these spores of B. stearothermophilus, as determined in test tubes in a liquid medium at 121°C, was found to vary from 1.7 to 3.9 min depending on the spore samples. A D value of 2 min is assumed for calculation purposes. The overall reduction ratios obtained during extrusion, 4.54xD (= log (2.9×10^{-5}) x-1) (at 165°C) and 7.13xD (at 185°C) can thus be considered to correspond to treatments of $\simeq$ 9.1 and $\simeq$ 14.3 min at 121°C in a liquid medium, respectively. It is known that 4 min in such conditions reduce the number of Clostridium botulinum spores to $1/10^{12}$ and bring commercial sterilization.

It should be pointed out however that the introduction of a liquid suspension of spores may have led to an over evaluation of the sterilizing efficiency for two reasons : 1) high water content and a_w of the spores, at least initially, and therefore possibly lower heat resistance ; 2) increase in the water content of the food mix above 14 % around the spores when the liquid inoculum (containing 2.5 ml H_2O) is introduced in the extruder. Other extrusion tests were therefore carried out using either the liquid inoculum or a dry inoculum of 200 mg of freeze-dried spores (ca 10^{11} living spores). The dry spores were mixed with an equal weight of food mix for easier handling. The water content of the food mix was kept at 14 %. Extrusion conditions were modified as follows : the feed rate was decreased from 40 to 30 $kg.h^{-1}$, giving a minimum RT of 45 ± 5 s at 80 rpm, and the temperature of the food mix just before the die was set at 150 or 165-168°C.

The overall reduction ratios obtained in the four extrusion tests are indicated in table I.

Table I. Sterilizing effect of extrusion-cooking on wet and dry spores of B. stearothermophilus

Extrusion temperature	Overall spore reduction ratios (N/No)	
	Liquid inoculum (wet spores)	Powder inoculum (freeze-dried spores)
150°C	2.5×10^{-7}	3.6×10^{-7}
165 - 168°C	1.3×10^{-7}	3.6×10^{-7}

These results indicate 1) that overall spore reduction at 165-168°C appears to be more marked than previously observed at the same temperature. This can perhaps be explained by the increase in RT at the lower feed rate ; 2) that similar spore reductions are now observed at 150 and 165-168°C. This is difficult to explain but it should be recalled that the temperature profiles of the food mix along the screws are not known ; 3) that sterilizing effects are not significantly different with freeze-dried spores and with wet spores. This behavior may perhaps be attributed to a rapid equilibration of the water content of the spores at these high extrusion temperatures (with a_w values probably close to 1), or to the known fact that the thermal resistance of very heat-resistant spores does not significantly increase at low water contents.

It is of interest to examine figure 1C which illustrates extrusion cooking at 168°C with freeze-dried spores, and where the RTD has been determined with erythrosin and, independently, with a pulse introduction of 50 μl of sodium ^{51}Cr chromate followed by radioactivity measurements in the samples collected 5 s per 5 s. Erythrosin and radioactivity measurements indicate the same minimum RT of 55 $\pm$ 5 s. The first surviving spores appear in the sample collected from 50 to 55 s after the inoculation. It could be argued that the instantaneous sterilizing ratio of "surviving/surviving plus dead" spores in this 50 to 55 s sample is much less favorable than the overall reduction ratio of cumulated samples containing surviving spores. This would be even more marked with lower minimum RT. However it should also be considered that in order to be able to measure surviving spores in extruded samples (10 to 4000 spores per 100 mg extruded sample), it was necessary to use an inoculum of 10^{10} or 10^{11} spores over approximately 10 g of food mix. The usual microbial load of heat-resistant spores in food ingredients is close to 5 spores/g, i.e. at least 10^8 times smaller. It can therefore be safely assumed that most foods extruded at temperatures above 150°C are sterile immediately as they come out of the extruder provided that the minimum RT in the extruder is $\gg$ 40 s.

Such sterilizing effects are of interest in the manufacture of some rehydratable animal feeds or foods (weaning foods, instant drinks). Extrusion could also be used to reduce the microbial load of contaminated meals, flours, starches, sugars, spices, etc. Further experiments are needed to study the efficiency of an extruder as a scraped surface heat exchanger for the continuous processing and sterilization of high moisture foods such as vegetable or fruit purées, processed cheese, etc.

2. DESTRUCTION OF AFLATOXIN B1

The extruder can also be used as a continuous chemical/heat reactor for the detoxification of such food ingredients as peanut meal or corn flour contaminated by aflatoxin.

The extrusion experiments are carried out with peanut meal containing ca 250 μg aflatoxin B1/kg d.w. Prior grinding and homogenization of the meal were necessary in order to obtain a coefficient of variation of the aflatoxin content below 12 %. Results from the first series of experiments (4) indicate that when peanut meal is extruded at feed rates of 20-30 kg. h^{-1}, screw rotation of 30 rpm, water content during extrusion of 20 %, and meal temperatures (before the die) from 140 to 185°C, 23 to 66 %, respectively, of the initial aflatoxin are destroyed or, more precisely, escape extraction by chloroform-water and/or determination by T.L.C. The water content during extrusion, from 11 to 38 % by weight, was found not to influence the extent of aflatoxin destruction, as long as the temperature of the metal barrel was adjusted to maintain a constant temperature of the meal before the die of 170-175°C. It is well known that aflatoxin B1 is relatively heat-resistant but that destruction is facilitated by various alkaline or oxidizing agents. The combined effects of heat and ammonium hydroxide during extrusion-cooking at 175 or 185°C are shown in figure 2.

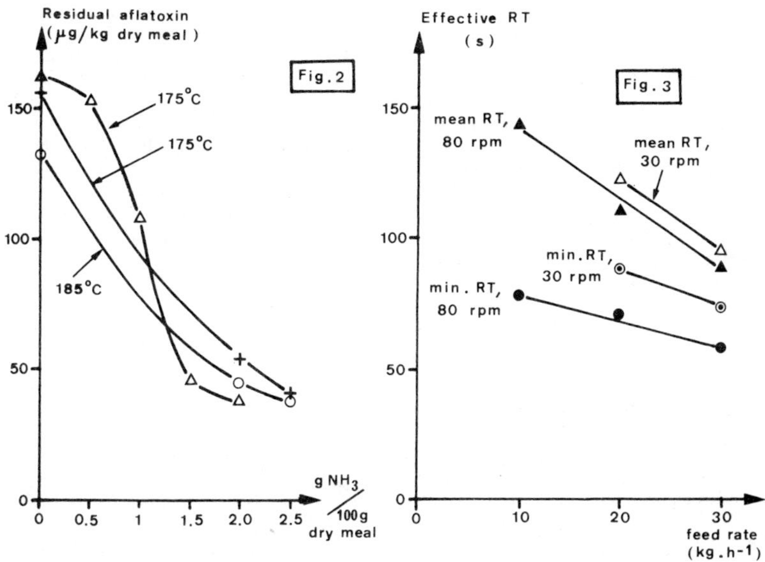

Figure 2. Destruction of aflatoxin B1 during extrusion of peanut meal with various NH_3 levels.
Figure 3. Effective mean and minimum residence times of peanut meal in the extruder as a function of feed rate and screw rotation.

The water content during extrusion was 20 %, the feed rate 30 $kg.h^{-1}$, the screw rotation 30 rpm. Under these conditions the minimum and maximum RT are respectively 100 and 240 s. The residual aflatoxin after extrusion in the absence of NH_4OH corresponds to a destruction of 36 to 54 % of the initial aflatoxin. About 85 % destruction were obtained when extrusion was carried out in the presence of 2-2.5 % NH_4OH (expressed as g NH_3 per 100 g dry meal). Both water and ammonia are partly removed from the textured ribbon of peanut meal as it comes out of the die since the moisture content decreases to 11 %, and the pH (measured after rehydration) to 8.4 (from an initial 9.6).

Preliminary experiments indicate that the protein nutritional value is not severely affected. The lysinoalanine content was determined in all the extruded samples of figure 2 and remained below 0.15 mmoles/100 g d.w., a negligible value corresponding to 1.4 % of the lysine content on a molar basis. The lysine content itself was little affected since a maximum loss of 13 % was observed, in the absence of ammonium hydroxide. Some desulfuration of cysteine into dehydroalanine may have occurred followed by further reaction with ammonia to form β-amino alanine. Ammonia may also protect against lysine loss and cross-linking formation by way of its reaction with reducing carbohydrates.

As illustrated by figure 2, a maximum extent of aflatoxin destruction of 87 % was obtained at 185°C and 2.5 % NH_4OH, under extrusion conditions where the minimum and maximum RTs were 100 and 240 s, respectively. When the mean RT was increased by about 25 % by reducing the feed rate from 30 to 20 $kg.h^{-1}$, a non significant increase (2 %) in the extent of aflatoxin destruction was observed. A non significant decrease (2 %) in aflatoxin destruction was noted also when the screw rotation was changed from 30 to 80 rpm and the RT therefore decreased by about 10 %. Since high capacity operation of a double screw extruder requires high feed rates and therefore high screw rotations, it would probably be of interest to use long barrel extruders with numerous reverse screw elements in order to increase RT.

Extrusion of contaminated meals with ammonium hydroxide is of practical interest even if the extent of aflatoxin destruction remains close to 85 %. Maximum permitted levels of aflatoxin B1 in animal feeds vary from 20 to 50 μg/kg. Chemical treatment of contaminated meals or flours is allowed only when they initially contain less than 500 μg aflatoxin B1/kg. The economics of extrusion treatments remain to be assessed, together with the long term effects of $NH_4\,OH$ on the extruder. It also remains to be determined whether the aflatoxin molecules are irreversibly broken down or just bound to meal constituents. Preliminary studies, including in vitro proteolysis and mutagenic activity of extruded peanut meal, have not yet given conclusive evidence.

3. RESIDENCE TIME DETERMINATIONS

Whether extrusion-cooking is used to destroy microorganisms or toxins, the extruder always acts as a continuous chemical reactor,and the residence time within the heat and mix zone becomes an important parameter. RTD curves obtained with erythrosin or ^{51}Cr (see figure 1C) can be processed to yield useful information. As a complement to minimum and maximum RT, median RT indicates the time when half the erythrosin or radioactivity has come out of the extruder. Mean RT can be computed from the RTD curve according to the following equation :

$$\bar{t} = \frac{\int_0^\infty t.\,c(t).dt}{\int_0^\infty c(t).dt}$$

, where t = exit age (or RT) of an unit particle,

and c(t) = concentration distribution of erythrosin or ^{51}Cr at the extruder outlet as a function of time following introduction of a pulse tracer at the extruder inlet.

The mean RT is also equal to the following ratio :

$$\frac{\text{cumulated residence times of all unit particles containing erythrosin or } ^{51}\text{Cr}}{\text{total number of unit particles}}$$

In addition, it is possible to calculate "effective" RTs by subtracting from the previously defined RTs the RT in the transport zone of the extruder (from the inlet to the reverse screw element). RT in the transport zone is calculated assuming plug flow transport (i.e. without dispersion) at the speed of the screw. Thus the "effective" mean RT represents the mean time spent in the heat and mix zone of the extruder.

RTD curves of erythrosin and ^{51}Cr have been determined with peanut meal (20 % moisture, 2 % NH_3) at different feed rates and screw rotations. The barrel temperatures were kept at 100, 162 and 182°C at position θ_1, θ_2 and θ_3, respectively. The various RTs are given in table II. Fair agreement (within ± 5 s) is observed between erythrosin and ^{51}Cr measurements. When effective minimum RT and effective mean RT are plotted against feed rates for 2 values of screw rotation, it is apparent (figure 3) that effective mean RT is more affected by feed rate than by screw rotation, while the reverse is true for effective minimum RT. Further experiments are in progress to determine if the extent of a zero order reaction taking place in the extruder is linearly proportional to the effective mean residence time.

Table II. Influence of feed rate and screw rotation on the residence times of peanut meal (with 2 % NH_3) in a Creusot-Loire BC 45 extruder

feed rate ($kg.h^{-1}$)	screw rotation (rpm)	mode of determination	min. RT(s)	median RT (s)	max. RT (s)	mean RT (s)	effective mean RT(s)
10	80	Erythrosin	90	145	280	152	142
10	80	^{51}Cr	85	147	310	155	145
20	30	Er	115	144	250	147	120
20	30	^{51}Cr	115	145	>250	151	124
20	80	Er	80	121	230	124	114
20	80	^{51}Cr	80	110	210	116	106
30	30	Er	100	122	>240	126	99
30	30	^{51}Cr	100	125	183	118	91
30	80	Er	70	90	165	102	92
30	80	^{51}Cr	65	89	155	94	84

REFERENCES

1. MUSTAKAS,G.C., GRIFFIN,E.L., ALLEN,L.E., and SMITH,O.B. (1964). J. Am. Oil Chem. Soc. 41, 607-614.
2. MUSTAKAS,G.C., ALBRECHT,W.J., BOOKWALTER,G.N., Mc GHEE,J.E., KWOLEK, W.F., and GRIFFIN,E.L. (1970). Food Technol. 24, 1290-1296.
3. BOUVERESSE,J.A., CERF,O., GUILBERT,S. and CHEFTEL,J.C. (1982). Lebensm. Wiss. u. Technol., 15, 135-138.
4. GREHAIGNE,B., CHOUVEL,H., PIÑA,M., GRAILLE,J., and CHEFTEL,J.C. (1983). Lebensm. Wiss. u. Technol. In Press.

THE EFFECT OF EXTRUSION-COOKING ON NUTRITIONAL VALUE

N.-G. ASP and I. BJOERCK
Department of Food Chemistry, Chemical Center, University of Lund
P.O.Box 740, S-220 07 LUND, Sweden

Summary

This paper gives a brief overview of effects on various nutrients during extrusion-cooking. Some recent results on starch availability to amylase *in-vitro*, rate of absorption *in-vivo*, and availability to fermentation by dental plaque bacteria are also presented, as well as studies on dietary fibre and fat content in extruded wheat products.

1. INTRODUCTION

The unique features of extrusion-cooking - high shear at elevated pressure and temperature - makes a thorough documentation of the effects on nutritional value essential. The process is very versatile and new applications are made currently for nutritious products such as weaning foods, breakfast cereals, bread substitutes and meat extenders. A more extensive review of the present knowledge on nutrient retention during extrusion was published recently (1).

2.1 STARCH

Slowly absorbed carbohydrates are considered beneficial in relation to diabetes and obesity due to a lower insulin need and a longer duration of satiety. Cooking and gelatinization of starch increases the availability to amylase. In addition, disruption of the organized structure appears to influence the availability of carbohydrates in a food item (2). In spite of the low water content during extrusion, complete gelatinization is usually obtained even at fairly moderate temperature (3), and the starch in extruded materials is generally highly available to enzymatic digestion.

In a study by Bjoerck et al. (4) extrusion-cooking of wheat products rendered the starch more available to salivary amylase *in-vitro* than boiling for 20 min. The plasma glucose respons in young rats after a gastric load containing an equivalent amount of starch, did not differ between boiling and extrusion cooking at mild conditions (Fig.1) However, with materials processed at severe conditions, the early plasma glucose respons was significantly higher compared with the boiled control. Thus, extrusion conditions can influence the glycemic respons to starch. A wheat flour product of crisp-bread character extruded at mild conditions gave a plasma glucose respons in rats very similar to that obtained with a soft bread baked from the same flour. Thus, at mild conditions, extrusion-cooking appears to influence the rate of *in-vivo* starch absorption similarly to baking and boiling.

In Fig.2 corresponding *in-vivo* data for drum-dried wheat flour are presented (4). Wheat flour drum-dried under realistic conditions produced a significantly lower plasma glucose respons than boiling. At more severe conditions, the difference was less pronounced. Boiling of drum-dried wheat flour increased the *in-vivo* availability above the level seen after boiling alone.

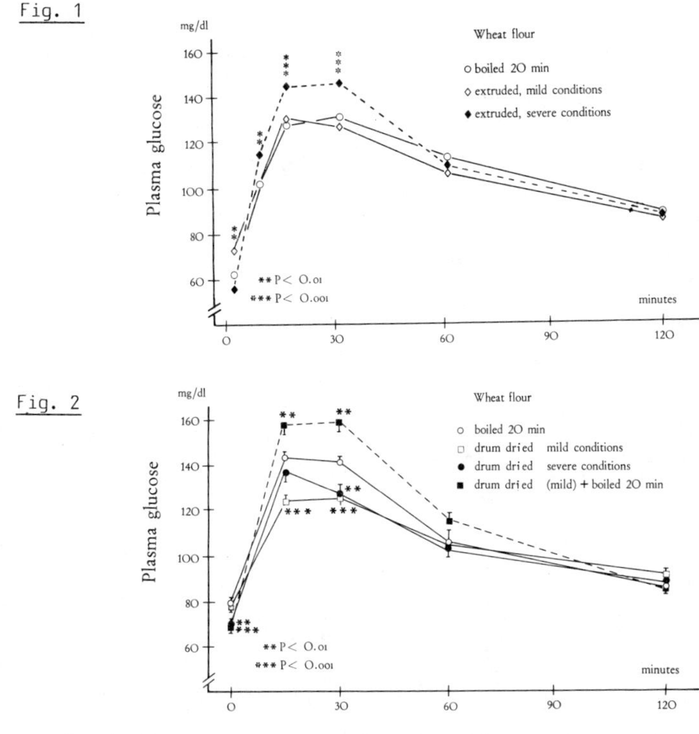

Due to the presence of α-amylase in saliva, an increased availability of starch might favour development of caries. Microorganisms in the dental plaque ferment dextrins and maltose produced in the oral cavity to organic acids, with a resulting pH drop at the tooth surface.

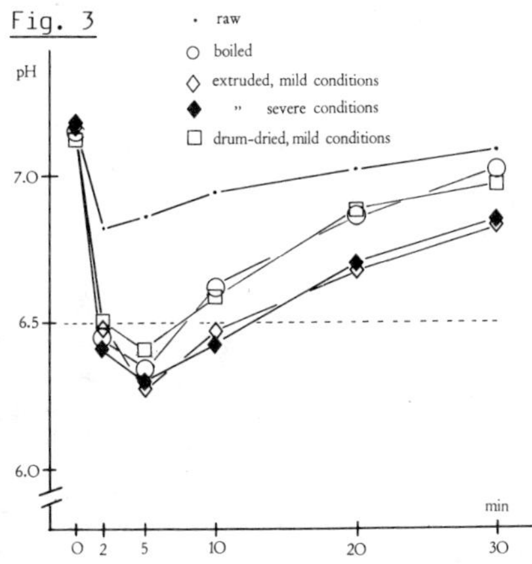

The effect on pH in human dental plaque after a 30s mouth rinse with various wheat flour suspensions are shown in Fig.3. With the method used for collection of plaque, a pH of 6.5 has been accepted as critical for development of caries. Thus, all the processed products must be considered more or less cariogenic. In contrast, raw wheat flour produced only a slight drop in pH. The pH drop after rinsing with extruded products was significantly more pronounced than with the boiled control, whereas drum-dried and boiled products gave a similar drop in plaque pH (4).

2.2 DIETARY FIBRE

The effect of extrusion-cooking on dietary fibre content was studied by Bjoerck et al. (5). Dietary fibre was analysed with an enzymatic gravimetric method recently published (6). After preincubation with a thermostable α-amylase (Termamyl), remaining protein and starch is solubilized with pepsin and pancreatin. The values obtained are corrected for remaining protein and ash. In wheat flour there was an increase in fibre content from 4 to 5% at severe extrusion conditions. However, if corrections were made also for remaining starch available to extensive amyloglucosidase digestion, the differences in total fibre before and after extrusion were no longer significant. Results obtained in raw and extruded whole grain wheat flour, using different enzyme systems are shown in Table I. With physiological enzymes only, the increase in fibre content due to processing was 6%. After correction for starch available to amyloglucosidase the difference was reduced to 2.6%. If Termamyl was included in the dietary fibre assay as described by Asp et al. (6), the difference in fibre content was only about 1%. Thus, although extrusion cooking makes the starch highly available to amylase, a small fraction is rendered more or less resistant to enzymatic degradation. The difference in fibre values with and without Termamyl is probably due to amylose-lipid complexes. This complexed starch is completely solubilized by Termamyl at high temperature (7).

Very few data exist on the impact of processing on the physiological properties of dietary fibre. Extrusion cooking of wheat flour solubilized the dietary fibre and increased the availability to fermentation in the rat colon (5). This could be expected to influence the fecal bulking capacity of the fibre. No such effect was observed with extruded whole grain wheat flour.

Table I. Enzymatic gravimetric assay of dietary fibre

Whole grain wheat flour		Dietary fibre (% dry basis) raw	extruded	
Physiological enzymes only		14.5	20.5	($P<0.001$)
	a	12.5	15.1	($P<0.01$)
Preincubation with Termamyl (6)		13.1	14.2	($P<0.05$)
	a	13.0	13.9	($P<0.05$)

a corrected for remaining starch (amyloglucosidase)
Creusot-Loire BC 45 twin screw extruder (166°C, 200 rpm, 20% H2O)
Feed rate 200 g/min; Die geometry - width 20 mm, height 2 mm, length 27 mm; screw configuration Feed-CCCR-Die.

2.3 PROTEIN

Available reports on the protein nutritional value after processing in single screw extruders indicate little change (8). Less biological data are available on the effects of twin screw extruders, which are able to operate over a wider range of moisture contents. Some raw vegetable proteins, e.g. soy, contain protease inhibitors that must be sufficiently inactivated to avoid blocking of proteolytic enzymes in the gut. Several authors have described the use of cooking extruders in processing full-fat soy flour with a high protein nutritional value (9).

Extrusion-cooking takes place at comparatively low water content which is known to favour the Maillard reaction. Wet heat treatment, such as boiling or autoclaving, is considered less detrimental. Materials containing low levels of reducing sugars are less prone to deterioration during heat treatment. However, during extrusion-cooking formation of reducing carbohydrates may occur through hydrolysis of sucrose (10) or starch (11), thus increasing reactivity. Lysine is by far the most reactive protein bound amino acid and also limiting in cereals. Thus, the lysine retention in extruded products is especially important. At severe conditions the availability of other amino acids such as cystine, arginine, histidine, aspartic acid and serine, may also be reduced (12). Present data suggest that a reduced residence time, a higher feed rate, a lower process temperature and an increased moisture content increases the protein nutritional value of the product.

Depending on the Maillard compounds formed, different chemical methods for determination of available lysine correlates more or less well with biological availability. At mild extrusion conditions in a Creusot-Loire twin screw extruder analysis of FDNB-lysine and even total lysine predicted the loss in biologically available lysine (13). However, at more severe conditions, the chemical methods underestimated lysine loss. This was mainly due to a lowered protein digestibility. In another study (Bjoerck, I., Nair, B. and Matoba, T., unpublished results) the reliability of an enzymatic method (14) for determination of available amino acids in extruded wheat products was investigated. This enzymatic method predicted the loss in biologically available lysine very well also in products with lowered digestibility. In addition to losses in lysine there were also losses in the availability of several other amino acids at severe conditions. It could also be demonstrated that an increase in moisture content from 13 to 18% at 210°C specifically improved the availability of lysine.

Another interesting aspect is the texturization of vegetable proteins for use as meat extenders. The mechanism behind texturization is not fully known, although hydrogen and disulphid bonds are believed to play a major role (15). These bonds should not reduce the availability to proteolytic enzymes. However, very few biological evaluations exist on the effects on protein nutritional value.

2.4 LIPIDS

The nutritional value of lipids may be affected by processing through different mechanisms such as oxidation, isomerization or hydrogenation. According to Maga (16), the ratio of unsaturated/saturated fatty acids was only slightly affected during extrusion and formation of trans fatty acids was very small. However, there are reports of a reduced extractability of lipids with organic solvents (17,18). Compared with other low moisture heat treatments (pelleting, popping, flaking) the reduction in extractable lipids was most pronounced with extrusion cooking. A decrease in extractable lipids has also been reported during processes such as dough mixing and baking (19).

Table II shows the fat recovery obtained with two different methods in raw and extruded wheat products (Bjoerck, I. and Siljestroem, unpublished results). The decrease in lipid content in extruded materials was considerable when chloroform/methanol extraction was used (20). In contrast, with acid hydrolysis prior to extraction with ether/petrolether (21), the fat content was very similar before and after extrusion.

Table II. Recovery of fat in extruded whole grain wheat flour products

	chloroform/ methanol (20)	acid hydrolysis (21)
Raw material	100	100
Extruded products		
(164 °C, 100 rpm, 20% H20)	37%	105%
(166 °C, 150 rpm, 20% H20)	35%	105%
(166 °C, 200 rpm, 20% H20)	43%	100%

Creusot-Loire BC45 twin screw extruder. Feed rate 200 g/min; Die geometry-width 20 mm, height 2 mm, length 27 mm. Screw config.Feed-CCR-Die.

2.5 MINERALS

The availability of trace elements e.g. Zn is lower in vegetable foods than in foods of animal origin. With an increasing use of cooker extruders for production of meat extenders from vegetable sources, it is essential to evaluate the bioavailability of trace elements in such products.

Until recently, minerals have been looked upon as nutritionally stable during processing. However, during e.g. baking, iron may become less soluble and thus less available for absorption (22). Toasting and boiling affect the ability of fibre components to bind minerals in-vitro (23). It was suggested that the increase in binding due to toasting of wheat bran is caused by formation of "lignin-like" substances during this process. Formation of Maillard products recovered in the lignin fraction has been reported during extrusion cooking (24).

When comparing mineral absorption from browned corn flakes and unbrowned corn grits in humans, no differences were found (25). However, Zn retention was reduced when feeding corn flakes. Further, the apparent availability of Zn, Cu and Mg in patients with ileostomy was lower with an extruded bread product than with the corresponding raw material (26). Thus, the effects of processing on mineral metabolism is an interesting field for further research.

Another aspect is "mineral fortification" due to screw wear. According to Maga and Sizer (27), the iron content in the product increased with increasing temperature. It was suggested that this iron addition might catalyze and aggravate vitamin C destruction during extrusion.

2.6 VITAMINS

Cereals constitute an important source of B-vitamins in the human diet which makes the fate of these vitamins during extrusion-cooking particularly important. However, in some products, e.g. weaning food, the retention of all vitamins must be considered. Within the B-complex, thiamin seems to be most sensitive, whereas the other vitamins are comparatively stable.

Extrusion conditions that improve retention of one vitamin may increase destruction of another one. For instance, an increase in extrusion temperature leads to an increased destruction of thiamin (28) and vitamin C (27), but may even improve the retention of riboflavin (28,29). Further, an increase in moisture content improves thiamin stability (29) but may aggravate riboflavin (30) and vitamin C loss (29). Increased screw speed increased the loss of thiamin, riboflavin (30) and vitamin C (31), whereas the opposite was observed for vitamin A (32). An increase in feed rate generally improves vitamin retention.

High sensory acceptability does not guarantee good vitamin retention. Under extrusion conditions suitable for production of breakfast cereals, the loss in thiamin and riboflavin was substantial, 90% and 50% respectively (28).

REFERENCES

1. BJOERCK, I. and ASP, N.-G. (1983). J Food Eng (in press)
2. HEATON, K.W. (1980). Medical aspects of dietary fibre - topics in gastroenterology (eds. Spiller, G.A. and Mc Pherson Kay, R.). Plenum Medical Book Comp. New York and London, pp 223-38.
3. RABE, E. et al. (1983). Zucker Süsswaren Wirtsch. 33, 158-62
4. BJOERCK, I. et al. (1983). J. Cereal Sci. (submitted)
5. BJOERCK, I. et al. (1983). J. Cereal Chem. (submitted)
6. ASP, N.-G. et al. (1982). J. Agric. Food Chem. 31, 476-482-
7. HOLM, J. et al. (1983). Die Stärke (in press).
8. HARPER, J. and JANSEN, G.R. (1981). Nutritious foods produced by low-cost technology LEC Report 10, Colorado State Univ., Fort Collins, Co.
9. MUSTAKAS, G.C. et al. (1964). J. Am. Oil Chem. Soc. 41, 607-14.
10. NOGUCHI, A. et al. (1982). Lebensm. Wiss. Techn. 15, 105-10
11. SAHAGUN, J.F. and HARPER, J.M. (1980). J. Food Proc. Eng. 3, 199-216.
12. KÖHLER, F. (1981). Veränderungen der ernährungsphysiologischen und physikalischen Eigenschaften von von Getreidemahlerzeugnissen durch Extrusion unter besonderer Berücksichtigung proteinangereicherter Produkte. Dissertation, Institut für Lebensmitteltechnologie, Berlin
13. BJOERCK, I. et al. (1983). J. Agric. Food Chem. 31, 488-492
14. MATOBA, T. et al. (1982). Agric. Biol. Chem. 46, 465-472.
15. JEUNINK, J. and CHEFTEL, J.-C. (1979). J. Food Sci. 44, 1322-5, 1328
16. MAGA, J.A. (1978). Lebensm. Wiss. Techn. 11, 183-4.
17. DELORT-LAVAL, J. and MERCIER, C. (1976). Ann. Zootech. 25, 3-12
18. NIERLE, W. et al. (1980). Getreide Mehl Brot 34, 73-8
19. CHUNG, O.K. et al. (1981). Cereal Chem. 58, 220-6.
20. Handbuch der Lebensmittelchemie, Vol IV. Fette und Lipoide (Lipids) Springer Verlag 1969, pp 419-20
21. AOAC 14.019 13th ed. Washington 1980, p 213
22. LEE, K. and CLYDESDALE, F.M. (1980). J. Food Sci. 45, 1500
23. CAMIRE, A.L. and CLYDESDALE, F.M. (1981). J. Food Sci. 46, 548-51
24. THEANDER, O. (1983). Advances in the chemical characterization and analytical determination of dietary fibre components. In: Dietary Fibre (eds. Birch, G.G. and Parker, K.J.) Applied Science Publishers, Ripple Rd, Barking, Essex, England, pp 77-93. England
25. JOHNSON, P.E. et al. (1983). The effect of browned and unbrowned corn products on absorption of zinc, iron and copper in humans. In: Acs Symp. Ser. 213, Washington D.C., pp 349-360
26. NÄVERT, B. et al. (1983). Betydelsen av extrudering för tillgänglighet av mineraler. Paper presented at Nordiskt Symposium om Kostfiber, Umeå 1-3 June, Sweden
27. MAGA, J.A. and SIZER, C.E. (1978). Lebensm.Wiss.Techn. 11, 192-4
28. BEETNET, G. et al. (1976). J. Milk Food Techn. 39, 244-5
29. HARPER, J.M. et al. (1977). Evaluation of low-cost extrusion cookers for use in LDC´s. LEC Report 2. Colorado State Univ.,Fort Collins, Co.
30. BEETNER, G. et al. (1974). J. Food Sci. 39, 207-8
31. MAGA, J.A. and COHEN, M.R. (1978). Lebensm. Wiss. Techn. 11, 195-7.
32. LEE, T-C. et al. (1978). AICHE Symp. Ser. 74(172), 192-5.

STABILITE DE QUELQUES PIGMENTS CAROTENOIDES EN CUISSON-EXTRUSION

C.BERSET - J.DEBONTRIDDER - C.MARTY
Laboratoire de Biochimie Industrielle Alimentaire - ENSIA-MASSY , FRANCE

Résumé

L'évaluation de la couleur des extrudats d'amidon de maïs, colorés par des pigments caroténoïdes, a été mesurée par colorimétrie tristimulaire au cours de la conservation des échantillons à température ambiante et à l'obscurité ou à la lumière. Quatre préparations commerciales ont été testées. Le β-apo-8'-caroténal et le β-carotène type II montrent tous deux une bonne résistance à la cuisson-extrusion, les pertes n'excédant pas respectivement 16 et 17% après 7 semaines de stockage à l'abri de la lumière. Le traitement thermique de la cuisson-extrusion induit l'isomérisation du β-carotène *all-trans* en deux composés présumés être le 15-*cis* et le 9-*cis*-β-carotène, ainsi que l'apparition de composés d'oxydation comme le 5,6-époxyde-β-carotène. La cinétique de dégradation des 3 isomères a été suivie dans les extrudats par chromatographie sur colonne et dosage spectrophotométrique.

Summary

The discoloration of extruded corn starch samples, colored by commercial powders of carotenoid pigments was followed by means of tristimulus colorimetry during storage at ambient temperature in darkness or under light. β-apo-8'-carotenal and β-carotene II showed good stability after extrusion-cooking, pigments losses being respectively 16% and 17% after 7 weeks of storage in darkness. Thermal treatment of extrusion-cooking induces the isomerisation of *all-trans* β-carotene probably into 15-*cis* and 9-*cis*-β-carotene and the appearance of oxydized compounds as β-carotene-5,6-epoxyde. Kinetics of degradation were established for the three isomers isolated from the snacks and purified by liquid chromatography.

1 . INTRODUCTION

Les études consacrées au comportement des colorants en cuisson-extrusion sont peu nombreuses et concernent essentiellement les colorants artificiels, lesquels offrent par rapport aux pigments naturels de meilleures garanties de stabilité. Or le retour à une législation plus stricte en matière de colorants artificiels a conduit les industriels de l'alimentation et les chercheurs à s'interesser de nouveau aux pigments naturels. Parmi ceux-ci les caroténoïdes offrent le double avantage d'être relativement résistants au chauffage et compétitifs sur le plan financier. En outre, certains d'entre eux sont des précurseurs de la vitamine A.

KONE et BERSET (3) ont étudié la cinétique de dégradation d'une préparation commerciale de β-carotène de synthèse, incorporée à de l'amidon de maïs et extrudée sous différentes conditions. Ils montrent qu'une température de cuisson-extrusion modérée (170°C) et un stockage des extrudats à l'abri de la lumière limitent la vitesse de décoloration ultérieure des produits. Néanmoins, la préparation commerciale utilisée dans cette étude ne semble pas offrir une résistance suffisante en cuisson-extrusion pour une utilisation industrielle.

Poursuivant ce travail, nous avons comparé la stabilité en cuisson-extrusion de divers types de préparations colorantes commerciales et nous avons suivi plus particulièrement l'évolution, dans les extrudats, de 3 isomères du β-carotène.

2. MATERIEL ET METHODES

Le cuiseur-extrudeur utilisé dans nos essais est un modèle bivis BC 45 de la société CREUSOT-LOIRE, équipé d'une filière de 4mm et d'un couteau rotatif. Les essais ont été réalisés à des températures comprises entre 170 et 185°C., mesurées prés de la filière.

Les extrudats obtenus à partir d'amidon de maïs mélangé à une préparation commerciale d'un pigment caroténoïde avaient la forme de billes d'environ 1cm de diamètre. Ils étaient conservés par fractions de 20g en boites métalliques à l'abri de la lumière et à la température du laboratoire, ou disposés en une monocouche et exposés en continu aux radiations d'un tube fluor "lumière du jour" de 30 watts. Ils étaient, dans ce second cas, retournés tous les jours.

La décoloration des extrudats était suivie par colorimétrie tristimulaire, sur un colorimètre DU COLOR NEOTEC équipé d'une tête à éclairement circonférentiel et les lectures étaient faites dans le système CIELAB (1) sur des poudres obtenues par broyage des extrudats dans un broyeur à billes et tamisées entre 200 et 400 µm. Nous avons choisi d'exprimer la décoloration des extrudats au cours du temps en terme de différence de couleur totale des échantillons entre le jour 0 de l'extrusion et un jour i:

$$\Delta E = (\Delta L^2 + \Delta a^2 + \Delta b^2)^{1/2}$$

avec: $\Delta L = L_0 - L_i$
$\Delta a = a_0 - a_i$
$\Delta b = b_0 - b_i$

La quantité de pigments résiduels dans les extrudats était déterminée par dosage spectrophotométrique aprés digestion enzymatique de l'amidon et extraction des pigments, selon la méthode de KONE et BERSET (3).

3. RESULTATS ET DISCUSSION

3.1. Dégradation de quelques pigments caroténoïdes au cours du stockage des extrudats.

3.1.1. Conservation à l'obscurité.

Le β-carotène, le β-apo-8'-caroténal et la canthaxanthine fournis par la société HOFFMANN-LAROCHE se présentaient sous forme de poudres hydrodispersables titrant 1% en pigments. Une modification récemment introduite dans la formulation du β-carotène CWS 1% nous a conduits à distinguer un β-carotène type I (ancienne formule) et un β-carotène type II (nouvelle formule).

Le tableau I donne le pourcentage de pigments détruits aprés cuisson-extrusion et la figure 1 présente l'évolution de la couleur des extrudats en fonction du temps de stockage, pour les quatre préparations commerciales

Tableau I : Pourcentage de pigments détruits immédiatement après cuisson-extrusion et après plusieurs semaines de conservation.

Pigment / Temps de stockage	β-carotène type I	β-carotène type II	β-apo-8'-caroténal	canthaxanthine
0	13 %	6 %	12 %	18 %
3 semaines	88 %	10 %	15 %	30 %
7 semaines	96 %	17 %	16 %	50 %

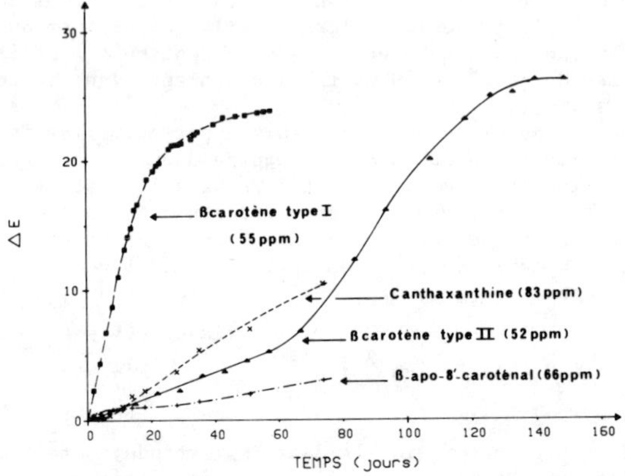

Figure 1 : Evolution de la couleur des extrudats d'amidon colorés par divers pigments caroténoïdes.

On constate que les pertes en pigments engendrées directement par la cuisson-extrusion sont faibles (inférieures à 20%) mais que l'évolution ultérieure de la couleur des extrudats dépend étroitement de la nature du colorant utilisé. Le β-apo-8'-caroténal apparait ainsi comme le pigment le plus stable, avec la vitesse de dégradation la plus faible. Ce résultat est intéressant, s'agissant d'un précurseur de vitamine A.

Le β-carotène type II, également stable pendant les deux premiers mois de stockage, donne, pour une même concentration, des snacks dont la teinte jaune est beaucoup plus prononcée que celle des extrudats colorés par le β-carotène type I. Cette différence de couleur (et peut être de résistance au traitement thermique) s'explique par une différence de composition des deux préparations comme le montre l'analyse des pigments des poudres I et II réalisée selon le mode opératoire suivant :

- Isolement d'une fraction A sur colonne d'alumine II-III éluée par un mélange hexane-éther éthylique 90:10 v/v.

- chromatographie de la fraction A sur alumine I et séparation de 3 isomères:
  - . isomère 1 : élué par le mélange hexane-éther ethylique 90:10 v/v. Maxima d'absorption dans l'hexane à 469 - 441 nm et pic " *cis*" à 344 nm.Ces caractéristiques spectrales sont celles du 15-*cis* - β-carotène (6,8).
  - . β -carotène *all-trans*: élué par le mélange hexane-éther éthylique 70:30 v/v. Maxima d'absorption dans l'hexane à 477 - 450 nm . Spectres IR et RMN (9).
  - . isomère 3 : élué par le mélange hexane-éther éthylique 60:40 v/v. Maxima d'absorption dans l'hexane à 471 - 443 nm. Pic "*cis*" à 337 nm. Ces caractéristiques spectrales sont celles du 9- *cis*- β -carotène (2).

Pour les dosages, nous avons utilisé les coefficients d'extinction donnés par GOODWIN (2) pour le β-carotène *all* - *trans* et les isomères 15-*cis* et 9 -*cis*.

Tableau II : Composition de la fraction A du β-carotène type I et type II

	Fraction A % pigments totaux	Composition centésimale de la fraction A β-carotène *all-trans* %	isomère 1 %	isomère 3 %
β-carotène type I	92	96	3	1
β-carotène type II	51	38	27	35

Dans la nouvelle formulation, la fraction A ne représente donc que la moitié des pigments totaux (exprimés ici en équivalents β-carotène = 2798) avec un composé *all-trans* déjà fortement isomérisé. Les autres pigments constituant la matière colorante du β-carotène type II n'ont pas été identifiés.

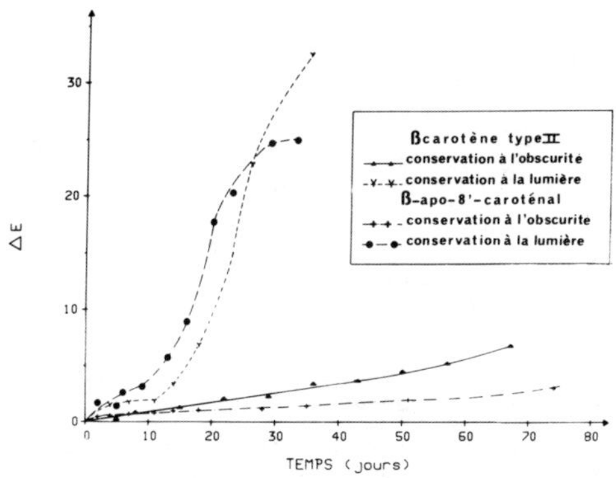

Figure 2 : Influence de la lumière sur la couleur des extrudats.

3.1.2. Influence de la lumière sur la stabilité de la couleur des extrudats

Cet essai a été conduit sur les deux préparations les plus stables, β-carotène type II et β-apo-8'-caroténal. La figure 2 montre l'accélération du processus de décoloration des extrudats sous l'action de la lumière, l'exposition aux radiations lumineuses réduisant de 4 à 5 fois la durée de vie du colorant.

Les courbes de photodestruction présentent un caractère de sigmoïdie marquée, caractéristique des réactions d'oxydation en chaînes qui se produisent sous l'effet des photons.

3.2. Etude des produits de dégradation du β-carotène apparus après cuisson-extrusion.

Les pigments extraits à partir des extrudats colorés par du β-carotène type I ont été chromatographiés sur colonne d'alumine II-III. Deux fractions sont en cours d'étude.

- Fraction A : éluée par hexane-éther éthylique 90:10 v/v. On note dans cette fraction une modification des proportions relatives des 3 isomères présents initialement dans la préparation commerciale. La teneur en β-carotène *all-trans* passe de 96 à 74% au profit des isomères 1 et 3 qui augmentent respectivement de 3 à 19% et de 1 à 7%. Il s'agit là d'un phénomène bien connu, l'isomérisation des caroténoïdes *all-trans* en composés *cis* se produisant fréquemment sous l'action du chauffage ou de la lumière(4,5,7).

L'évolution de ces trois isomères a été suivie pendant 30 jours après cuisson-extrusion. La figure 3 montre le profil des courbes de dégradation.

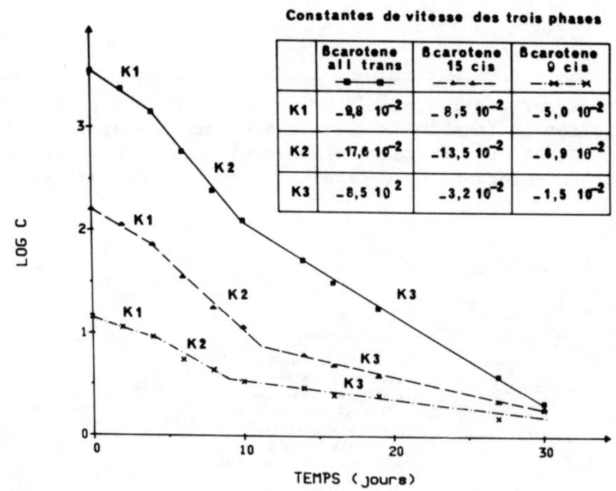

Constantes de vitesse des trois phases

	Bcarotene all trans	Bcarotene 15 cis	Bcarotene 9 cis
K1	-9,8 10-2	- 8,5 10-2	- 5,0 10-2
K2	-17,6 10-2	-13,5 10-2	- 6,9 10-2
K3	-8,5 10 2	-3,2 10-2	-1,5 10-2

Figure 3 : Cinétiques de dégradation de trois isomères du β-carotène après cuisson-extrusion.

La dégradation des 3 isomères s'opère simultanément et présente trois phases distinctes. Elle n'est pas totale après 30 jours. Les constantes de vitesse k données pour les isomères 1 et 3 sont des constantes apparentes puisque la dégradation observée est la résultante de deux processus

. accumulation par isomérisation du composé *all-trans*
. transformation en des formes plus dégradées.

La présence, dans les trois cas, d'une phase d'induction de quatre jours suivie d'une accélération de la vitesse de dégradation pourrait traduire l'existence de réactions d'autooxydation eventuellement induites par la formation de radicaux libres dans le support du colorant ou l'amidon de maïs. Toutefois, nos observations et l'allure générale du phénomène laissent supposer un mécanisme plus complexe dans lequel la structure des extrudats pourrait jouer un rôle. En effet cette structure n'est pas homogène, les extrudats présentant de place en place de petits noyaux vitreux, sièges d'une coloration plus durable. On aurait ainsi des régions amorphes, poreuses et perméables aux échanges gazeux, favorisant la destruction des pigments et des régions moins accessibles à l'oxygène, dans lesquelles la dégradation serait plus lente.

- Fraction B : éluée par hexane-éther éthylique 80:20 v/v. On y dénombre plusieurs bandes parmi lesquelles un composé a été purifié et étudié. Ses maxima dans l'alcool (474-446 nm) et dans l'hexane (473-445 nm) sont très proches de ceux rapportés par ZINSOU pour le 5,6-époxyde-β-carotène (9). En présence de HCl 0,1 M, ce composé en solution éthanolique présente un déplacement hypsochromique de 18 nm caractéristique selon GOODWIN (2) de l'isomérisation d'une fonction 5,6-époxyde en 5,8-époxyde. Ce composé n'avait pas été détecté dans la préparation avant cuisson-extrusion. L'absence de données sur son coefficient d'extinction n'a pas permis de le doser.

CONCLUSION

Des quatre préparations colorantes étudiées, nous retiendrons plus particulièrement le β-apo-8'-caroténal pour la coloration *in situ* des snacks expansés. Le β-carotène appelé dans cette étude type II est également très résistant dans les conditions de nos essais de cuisson mais il donne une couleur moins attrayante et sa composition reste encore mal définie. La canthaxanthine permet d'obtenir des extrudats de couleur agréable mais sa stabilité est plus faible. L'addition d'antioxydants pourrait peut-être permettre de l'améliorer. Dans tous les cas, les produits obtenus doivent être conservés à l'abri de la lumière.

Cette étude se poursuit actuellement avec des poudres susceptibles de colorer les snacks tout en les aromatisant. Des essais sont en cours avec le paprika, la tomate et l'annatto.

REFERENCES BIBLIOGRAPHIQUES

1. C.I.E. (1976). Colorimétrie. Recommandations officielles de la Commission Internationale de l'Eclairage. Supplément n° 2 à la publication CIE n° 15 (E.1.3.1.) 1971 / (TC.1.3.)
2. GOODWIN, T.W. (1980). The Biochemistry of the Carotenoids. Vol. 1 Plants. Chapman and Hall. London and New-york.
3. KONE,T. BERSET,C. (1982). Stabilité du β-carotène après cuisson-extrusion dans un produit modèle à base d'amidon de maïs. Sci. Alim., 2, 465-481.
4. LAND, P.G. (1962). Stability of plant pigments. Rec. Adv. Food Sci., 2, 50-56.
5. OGUNLESI, A.T., LEE, C.Y. (1979). Effects of thermal processing on the stereoisomerisation of major carotenoids and vitamin A value of carrots. Food Chem., 4, n° 4, 311-318.

6. WOJTKOWIAK, B., CHABANEL, M. (1977). Spectrochimie moléculaire. Lavoisier. Paris.
7. ZECHMEISTER, L. (1962). *Cis-trans* isomeric carotenoids, vitamin A and arylpolyenes. Academic Press. New-York.
8. ZINSOU, C. (1971). Degradation enzymatique et photooxydative du β -carotène. Phys. Vég. , 9, 149-167.
9. ZINSOU, C. (1973). Oxydation enzymatique et photochimique des caroténoïdes. Leur rôle dans la protection des chlorophylles. Thèse série A, n° 1177, Université Paris-Sud.

SYSTEM ANALYTICAL MODEL FOR THE EXTRUSION OF STARCHES

F. MEUSER and B. VAN LENGERICH
Technische Universität Berlin
Institut für Lebensmitteltechnologie
- Getreidetechnologie -

Summary

The functional characteristics of native starches can be changed under conditions of extrusion cooking, depending upon the operational variables used. The influence of these variables on the starch characteristics can be differentiated with the aid of a system analysis. A model is proposed which separates extrusion into its influencing and influenced parameters and allocates these according to their functional relationships to each other. The model distinguishes between influencing process and system parameters as well as influenced structural and target parameters. The process parameters correspond to the process variables which control the specific energy. The latter is characterised by the system parameters which for their part change the starch structure and therefore the structural parameters. Structural changes to the starch influence the characteristics of the starch itself, which are to be seen as target parameters. The operational variables are responsible for the relative magnitude of structural breakdown of the starch, upon which, amongst other things, its solubility and viscosity behaviour depends. The system analysis relationships were represented quantitatively using the results of an experimental test programme by means of regression equations. This makes it possible either to estimate the resulting starch characteristics given the operational variables and the system parameters, or to determine the required extrusion conditions given required characteristics of the starch.

1. INTRODUCTION

Starch shows a specific reaction behaviour under conditions of high temperature short time (HTST) extrusion cooking when energy is introduced into native starch granules. Various changes in the characteristics of the substance result from this, depending upon the type and magnitude of the energy transmitted. The processes in the extruder depend on a large number of variable machine and raw material related control parameters. These determine reactions which are so complex that differentiating between the influence of the individual variables on the changes in characteristics is only possible to a limited extent. Thus the use of extrusion technology for modifying the characteristics of starch is based predominantly upon empirically acquired knowledge. Scientific literature on extrusion has hitherto discribed principally the physical and chemical properties of the extruded products, which result from the use of specific extrusion conditions in varying machines. This approach permits consideration of the reaction behaviour of starch, under the influence of specific energy, as related to the operational variables. And yet it is precisely this which is of decisive significance in the optimization of extrusion technology as well as the modification of starch or starch containing raw materials. What is more, the result of the empirical application of the technology is bound to the design features of the extruder. Therefore, the results gained with one extruder cannot be transferred to another at will.

Our investigation stemmed from the outline given: it comprised the differentiation and experimental examination, by system analysis, of the influence of the operational variables in a twin-screw extruder on the extrusion process and its effect on starch.

2. MATERIAL AND METHODS

First a system analysis model is proposed which separates extrusion into its influencing and influenced parameters, and allocates them according to their functional relationships with each other. The model (Fig. 1) distinguishes between influencing process and system parameters as well as influenced structural and target parameters. The process parameters correspond to the operational variables, on which the introduction of specific energy depends. The latter is characterised by the system parameters which for their part change the structure of starch and thereby the structural parameters. The effects on the characteristics of the starch, which are to be seen as the target parameters, are due to changes in its structure (1).

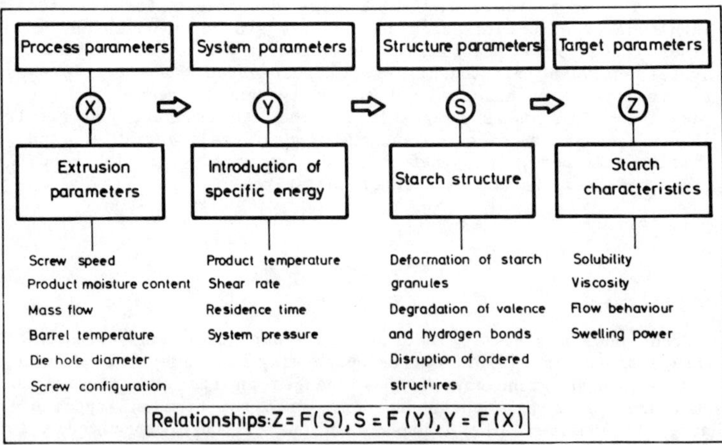

Fig. 1. System Analytical Model Used for the Extrusion of Starches

Experimental examination of the model was carried out by processing wheat starch (commercially available grade 1 starch) in a twin-screw extruder from Werner & Pfleiderer, type CONTINUA 37. The process length for all tests was 12 D with a screw diameter, D, of 37 mm. Operational variables were screw speed, mass flow, barrel temperature, die hole diameter, and product moisture content. The extrusion experiments were carried out by using a fractional factorial design in accordance with statistical experimental guidelines. This procedure had the advantage of shortening the scope of experiments by a third without putting any noticeable limitation on the results (2).

Polynomial regression calculation by means of least squares was used to determine the system analytical relationships between the process and system variables. The regression equations so obtained permitted the calculation of the relative magnitude of specific mechanical energy input (SME) and the product temperature for any combinations of extrusion variables tested in the process spectrum. These are graphically presented by means of a coordinate system for the limiting variables of die hole diameter, product moisture content and barrel temperature. The relationships between the forms of energy input and some of the structural parameters of the starch material could not be determined quantitatively because of the great amount of analytical work involved. This explains the qualitative characterization of selected extruded starches after processing at different levels of energy input. Indication of changes in the starch structure were determined by means of scanning electron microscopy. The degree of dissolution of the main valency bonds of the starch molecules was determined with the aid of HPLC. Changes in crystallinity and the formation of amylose-lipid complexes with lipids present in the starch were determined by x-ray diffraction spectrometry and differential scanning calorimetry. The last method was also used to determine the degree of gelatinization of the extruded starch. The thermal and mechanical energy input resulted in the fact that hydrogen bonds between the starch molecules and main valency bonds in the starch molecules were broken. Both resulted in increased solubility and reduced viscosity. As more of the main valency bonds were broken down, the reaction products became more easily soluble and less viscous. Consequently this rather easy means of determining the quality of the products was used to show changes in the starch structure. This fact was used to quantitatively establish the effect of the system variables on the structural variables.

The functional relationship between the thermal and mechanical energy input and the solubility as well as the cold paste viscosity of the ground extruded starch was established by means of a further polynomial regression analysis. The results were then presented as a functional equation in a three-dimensional diagram.

3. RESULTS

Figure 2 shows the effect of die hole diameter, product moisture content and the barrel temperature on the SME. The scope of the experiment was limited through the block in the diagramm. Each combination of extrusion parameters results in a definite SME. It could be seen that a lowering of the product moisture content was related to a higher SME. A lowering of the barrel temperature also led to a higher SME, as was the case when the die hole diameter was reduced. The highest SME values resulted when the extrusion took place simultaneously at low die hole diameter, low product moisture content and low barrel temperature. It could be concluded that, on the one hand, an infinite number of combinations of adjustments were possible in practice in order to obtain a definite SME. On the other hand, each extrusion parameter could have a limiting effect if a definite minimal energy input were necessary. For example, it was impossible to exceed an energy input of the order of 200 Wh/kg if the die hole diameter was greater than 2,7 mm. In this case, a variation in the other parameters, such as increasing the screw speed or changing the geometry of the screw by building in a backward movement element could be necessary in order to increase the SME.

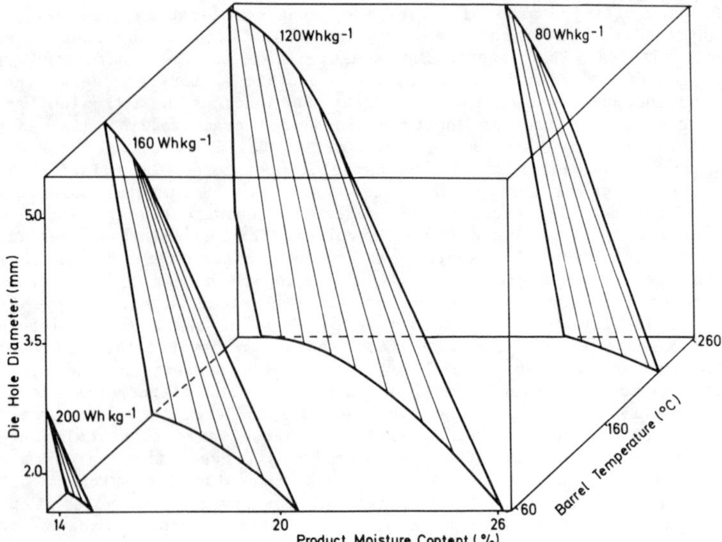

Fig. 2. The Effect of the Extrusion Parameters on the Specific Mechanical Energy Input

The calculated results on the effect of the extrusion parameters on the product temperature show that high product temperatures could be reached through low moisture content, small die hole diameter and high barrel temperature. Particularly remarkable was the effect of the moisture content on the product temperature due to increased frictional heat resulting from the increased energy input.

The specific thermal and mechanical energy inputs caused wide-ranging changes to the structural variables under the extrusion conditions. It showed that a partial structural disturbance of the starch granule only took place in case of less energy input. On the other hand, higher energy inputs resulted in a complete structural disturbance of the granules. A determination of the molecular weight distribution, by means of high pressure gel permeation chromatography of the different extruded starches showed that the main valency bonds of the starch molecules were broken down under the extrusion conditions. The most affected were the amylopectin molecules. With increasing mechanical energy input the molecules of the amylopectin fraction were increasingly broken down to smaller units which permeated with the amylose molecules under the conditions of the HPLC.

Furthermore, the ability of the products to diffract x-ray radiations was lost with increasing energy input. However, the x-ray diffraction diagrams suggest that the extrusion of wheat starch resulted in the formation of new crystalline components which in turn point to the formation of amylose-lipid complexes. Extruded potato starch did not show a similar peak.

The calorimetric studies showed that, depending on the magnitude of energy input, partial to complete gelatinization of the starch took place. Moreover, the DSC measurements confirmed the formation of amylose-lipid inclusion compounds. These were only dissociated when the temperature exceeded 110°C, and they reformed on cooling.

Changes in the quantifiable qualities of the extruded starch such as solubility and viscosity were indirectly used to show both the structural changes that took place and the relationship between the target parameters and the forms of energy input. It could be seen from the threedimensional graphical presentation of the functional equations (Fig. 3) that the solubility increased when either the SME or the temperature of the product or both variables simultaneously increased.

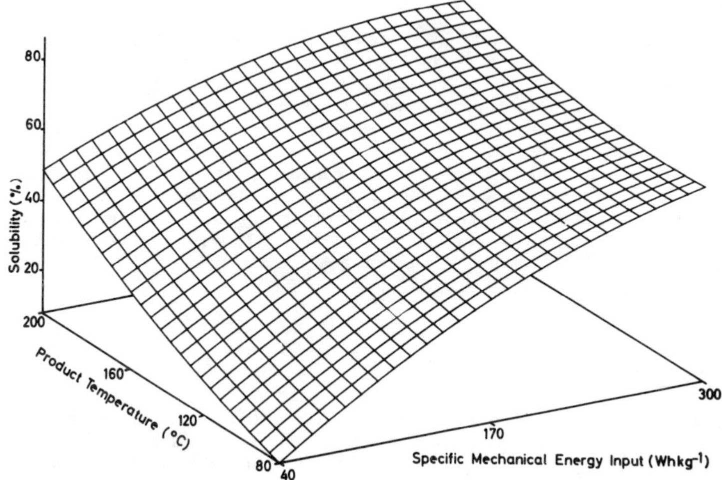

Fig. 3. The Effect of the Energy Input on the Solubility of the Extruded Starch

The energy input also had a clear effect on the viscosity of the pastes from the extruded starch. With increasing mechanical energy input at low product temperatures the viscosity decreased dramatically at first and then later less so. The temperature of the mass also had a remarkable influence on the viscosity. At higher mechanical energy inputs, the viscosity rose with increasing product temperature. This could be explained by a lowering of the structural viscosity of the plastified mass in the screw chamber due to increased temperature.

REFERENCES

1. MEUSER, F., VAN LENGERICH, B. and KÖHLER, F. Einfluß der Extrusionsparameter auf funktionelle Eigenschaften von Weizenstärke. starch/stärke 34 (1982), 366.
2. DAVIES, O. L. The Design and Analysis of Industrial Experiments. Oliver and Boyd, London, 1954.

POSSIBILITIES OF QUALITY OPTIMIZATION OF INDUSTRIALLY EXTRUDED FLAT BREADS

F. MEUSER and B. VAN LENGERICH
Technische Universität Berlin
Institut für Lebensmitteltechnologie
- Getreidetechnologie -

Summary

The quality of extruded flat bread was optimized on a technical scale extruder using a recently developed system analysis model. The described relations were calculated by means of polynomial regression analysis. Moreover, the results obtained were presented either in the form of equations or as contour plots. The effect of the individual extrusion parameters on the process or product criteria could then be read off easily. The contour plots permit a prediction of the required extrusion conditions to obtain specific product qualities. Consequently, it is possible in practice to influence extrusion conditions with the aim of either achieving definite product qualities or improving the product quality.

1. INTRODUCTION

The product qualities of extruded materials, produced on an industrial scale, are mostly empirically optimized. On account of the diverse variable parameters of the extruder, this procedure involves a great deal of material and personnel. Therefore we have considered these aspects as a priority for our reseach. We have solved the essential aspects of this problem with the aid of a system analysis model with particular reference to the extrusion of starch. This model is applicable in the solution of problems pertaining to HTST extrusion cooking, an example of which is the extrusion of flat bread. The model divides the extrusion process into the influencing and influenced variables which are allocated according to their functional relationships with one another and which would be described by means of polynomial regression equations (1). The evaluation of the calculated results made it possible to establish the effect of mechanical and thermal energy inputs on the product qualities. It was thus possible, through a combination of the calculated results for different known data, to optimize either a product or a process.

2. QUALITY IMPROVEMENT OF FLAT BREAD

For our optimization programme, a strategy was first developed (Fig. 1). Objectives which described the specific qualities of the desired products were then formulated. These were then assigned a definite number. Such variables as the expansion index, the specific weight and the width, all of which represent important packaging data, were used as end variables.

The following example describes the optimization of the end product qualities of a flat bread manufactured on a technical scale with an extruder, type Continua 120 (Werner & Pfleiderer). The limiting factor for the experiments was the quantity of raw material available. This amounted to 3 tons of wheat flour which made up the main component of the recipe. Because of this limitation, the maximum duration of the experimental work was limited to 12 h at a mass flow of 250 kg/h. Since each experiment had to last for at least 15 min in order to achieve a constant run of the extruder after changing certain process parameters, a maximum of 48 experiments could be carried out. The effect of the extrusion parameters, barrel temperature, screw speed, product moisture content and the die hole diameter on the end-product quality were studied. The composition of the raw material was kept constant.

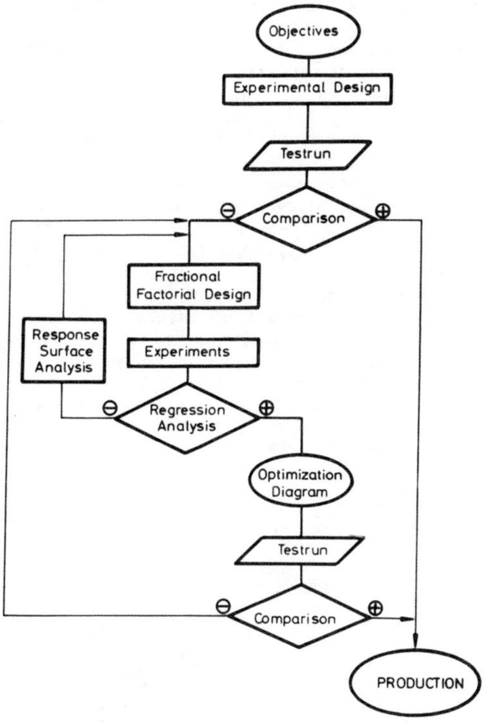

Fig.1. Strategy of the Optimization of Extruded Flat Bread

In addition to the above, a complete programme of a factorial design was developed. The four parameters, to be studied in three stages, were so chosen that the upper and lower values of each were the same distance from the middle in order to convert the values by means of a transformation calculation with the numbers -1.0; 0.0 and +1.0. The scope of the programme was such that 81 experiments should have been carried out but were reduced to a maximum of 48 with a view to limiting the number whilst at the same time maintaining the same degree of dependability. The programme was further reduced through a selective fractionating to the minimum of 37 experiments.

The next step that was followed was in accordance with the programme strategy. This was achieved with the aid of a trial experiment in which the extrusion parameters were kept at a middle level. The results so obtained then served as a basis for a detailed study of the working parameters of the extruder. Furthermore, a definite experimental sequence was laid down because this was of paramount importance in determining the duration of the individual experiments.

3. RESULTS

After carrying out the experiments in the main programme, a relationship was first established by means of polynomial regression calculation between the work parameters of the extruder and the relative magnitute of the energy inputs. With the aid of a further regression calculation, a functional relationship between the energy input and the resultant product qualities then was established. The equations of the latter regression calculations were each separately presented in a contour plot. The results showed that it was not possible, within the limits of the programme, to manufacture the desired product. The contour plots for the relationship between the expansion index and the forms of energy input showed that the expansion necessary for the desired product quality was excessively large. Thus the expansion index increased with the specific mechanical energy input (SME). However, it lessened with increasing product temperature. This meant that, to a chieve a desired product quality with an expansion index of 3.35 to 3.55 at a moderate SME, one had to reckon with a high product temperature.

Furthermore, the specific weight of between 102 to 106 mg/cm^3, prescribed for this section, could not be achieved. However, the contour plot showed that the specific weight could be increased by reducing the SME. On the other hand, the contour plot showed that the desired width from 64-66 mm could be achieved for the section under consideration.

It could be seen from the results that neither the expansion nor the specific weight could be achieved with the chosen die hole diameter.

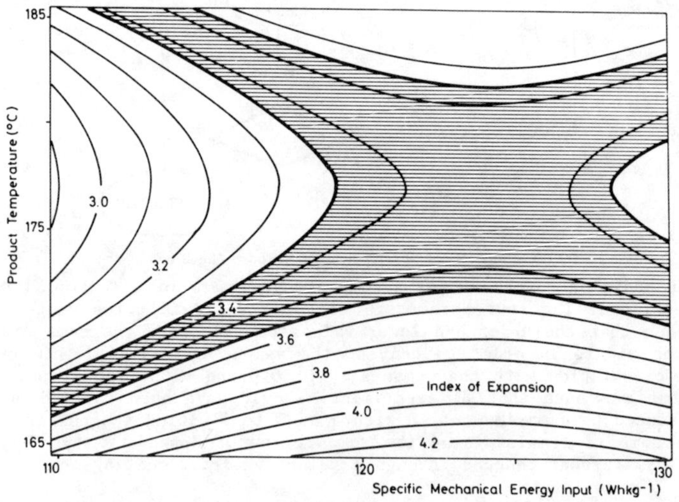

Fig. 2. The Effect of the Energy Input on the Expansion of the Extruded Material

Furthermore, it was established that the screw speed as well as the product moisture content of the sample had a much greater effect than the barrel temperature on the energy input. In the light of these facts, a supplementary experimental design was drawn up in which five experiments were to be carried out. Using an increased die hole diameter the parameters under consideration were the screw speed and the product moisture content of the sample, both of which were set at various levels.

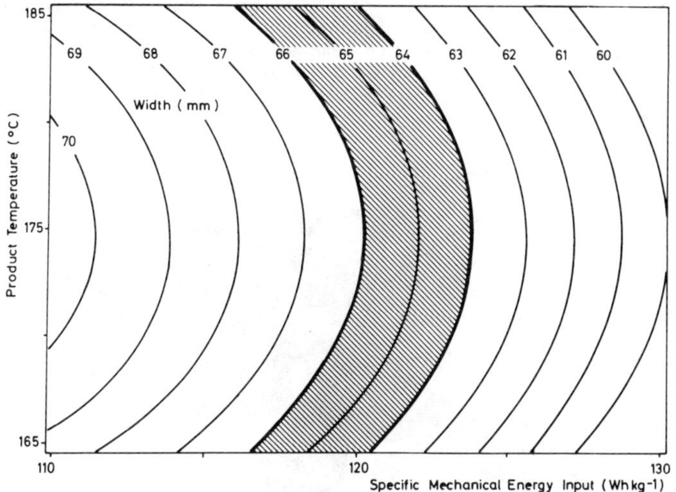

Fig. 3. The Effect of the Energy Input on the Width of the Flat Bread

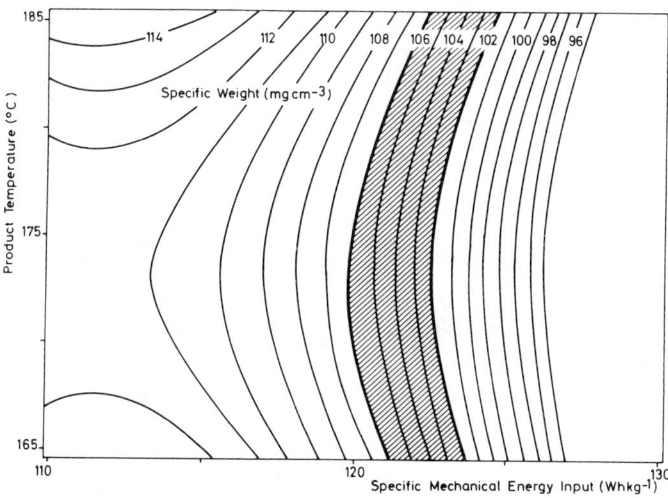

Fig. 4. The Effect of the Energy Input on the Specific Weight of the Flat Bread

The results showed that the product temperature increased with higher moisture contents when the screw velocity increased while the SME could be kept constant. The higher product temperature resulted in a more pronounced expansion of the extruded material, so that the desired expansion could be achieved (Fig. 2). Similar results were obtained for the width of the extruded material and its specific weight respectively (Fig. 3 and 4).

After obtaining these results, the contour plots for each parameter (Fig. 2-4) were combined into a single plot (Fig. 5). From such an optimization diagram, it was possible to read off the parameters necessary to produce the flat bread with the desired qualities using the flour and the extruder available. These were experimentally verified using parameters chosen from the optimization diagram.

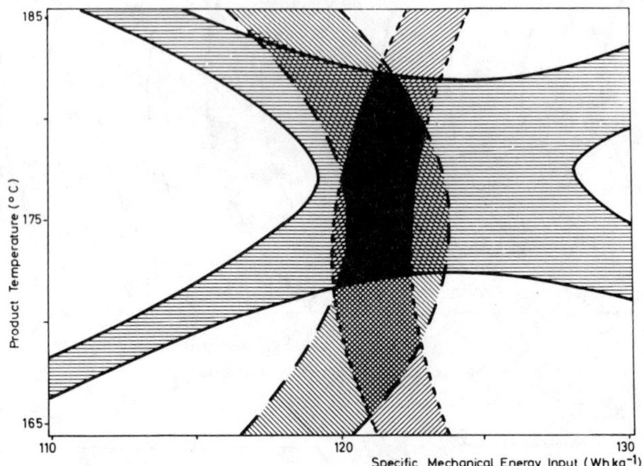

Fig.5. Optimization Diagram for the Flat Bread Production

REFERENCES

1. MEUSER, F., VAN LENGERICH, B. and KÖHLER, F. Einfluß der Extrusionsparameter auf funktionelle Eigenschaften von Weizenstärke. starch/stärke 34 (1982), 366.
2. MEUSER, F., VAN LENGERICH, B. and KÖHLER, F. Technologische und ernährungsphysiologische Aspekte zur Herstellung eines mit Speisetrebern angereicherten extrudierten Flachbrots. Lebensmittel-Technologie 16 (1983), 13.

DEGERMINATION OF SPICES IN AN EXTRUDER

P. Gry, F. Holm and K.H. Kristensen
Jutland Technological Institute, Aarhus, Denmark

Summary

By extrusion of pulverized natural spices in a Werner & Pfleiderer Continua 37 extruder we have succeeded in making products whose content of viable micro-organisms, inclusive of bacterial spores, is less than 1000 per gram. The spices used are black pepper and paprika. By comparative taste evaluations it has not been possible to find any significant difference in taste between unprocessed and extruded samples.

1. BACKGROUND

Natural spices contain large quantities of micro-organisms and spores, frequently in the order of several million per gram. When these spices are used in foods which are micro-biologically sensitive, e.g. meat-products, it is necessary to degerminate the spices to obtain sufficient quality and shelf life.

Most microbiological requirements for degerminated spices state that the total content of germs must be less than 10,000 per gram, and that the content of bacterial spores must be less than 1,000 per gram. To obtain these low figures, the spices are today treated with ethylene-oxide. This treatment is no longer regarded as being safe from a nutritional point of view and will be permitted only till acceptable alternatives are found.

A possible alternative is treatment with ionizing radiation which encounters emotional resistance from the consumers. Another possible alternative is to use extracts of spices (oleoresins). The extraction process is expensive, and the taste of the extracts differs considerably from that of the natural spices.

In the patent literature methods for degermination of spices through extrusion are described (1 and 2).

The purpose of this work is to develop extrusion processes which not only degerminate spices, but also treat spices so gently that aroma, taste and colour are not changed in an unacceptable way, and which will thus be an acceptable alternative to the well-known methods.

2. MATERIALS AND METHODS

For the experiments we have used pulverized black pepper from Sarawak and pulverized Hungarian Edelsüss paprika. Both spices have been delivered from Slagteriernes Fællesindkøbsforening A.m.b.A. We have decided to work with these spices because they are the most frequently used qualities of the spice groups pepper and paprika, and because pepper and paprika together constitute the greater part of the total spice consumption in Denmark.

For the extrusion is used a Werner & Pfleiderer Continua 37 pilot extruder which is a twin-screw extruder with self-cleaning screws. The extruder barrel has been built up of up to 8 sections. The temperature of each section can be adjusted by circulation of a hot or cold liquid. In our experiments we have used cooling of all sections. The screws are built up of screw elements which are placed on screw shafts. The screw elements are double-flights with a large or small pitch, right- or left-turned. Moreover, there are kneading elements of various length.

The grinding of the extruded spices has been carried out in a Bizerba coffee mill - for pepper with the setting 1 and for paprika with the setting 2 because these settings gave the particle size distributions most resembling those of the basic materials.

During the preliminary experiments we measured the total number of germs, fungi, bacteria and bacterial spores. It appeared that there were no fungi after the extrusion, and that the total number of germs, bacteria and bacterial spores was the same in the individual samples. After that we measured only the number of bacterial spores. We applied the method: ICC Draft Method for Aerobic Bacterial Spores.

The samples which gave bacterial spore counts lower than 1000 per gram have been evaluated sensorically. From these samples the best were selected for evaluation by trained taste panels. Pepper has been tasted in the concentration 0.4% in luncheon meat and 0.1% in neutral sauce.

Paprika has been tested in the concentration 0.2% in neutral sauce. The taste evaluations were made as triangle tests. The tastings took place between 1 and 8 weeks after the extrusion.

Moreover, storage experiments are being carried out. During these experiments both the unprocessed and the extruded spices are being kept in a gas and aroma-tight packing in the dark at $8^{o}C$ for six months.

The colour measurements have been made on a Hunterlab D25-9 according to the Lab-method.

3. RESULTS

In the extrusion experiments we used L/D ratios from 14 to 36, R.P.M. varied from 80 to 450, max. temperature varied from $50^{o}C$ to $180^{o}C$, max. pressures varied from 25 to 120 bars, torque varied from 30 to 80 Nm and capacity varied from 2 to 25 kg/hour.

The process conditions for the best pepper sample appear from table I. The process conditions for the best paprika sample appear from table II.

It was necessary to protect paprika with N_2-gas during the extrusion to avoid unacceptable colour changes. Some results of the colour measurements are shown in table III. Table III shows that N_2-covering gives a smaller decrease in the a value, and thereby in the red colour of the extruded sample.

The screws were built up in such a way that the greatest pressure, the greatest shear stresses and the highest temperature were obtained in the middle of the extruder. The later extruder sections were thus used for cooling to reduce the evaporation of volatile aroma components.

The shape of the extruded spices varies with die-design and granulation. In most cases, it will be necessary to grind the extruded spices. The ground extruded spices have a more uniform appearance than unprocessed spices.

As to black pepper, the taste panels could not establish any significant difference in taste (not even at the 90% level) between unprocessed and the best samples of extruder-degerminated black pepper neither in the freshly processed samples nor in the samples which had been stored for six months.

As to paprika, the taste panels could not find any significant difference in taste (90% level) between unprocessed and the best sample of extruder-processed paprika. The storage experiments have not yet been finished.

Optimum process conditions for black pepper		
Screw length:		836 mm
Length - diameter ratio:		22.6
Pitch of screw elements:		40 mm
except	90 mm pitch 60 in the feed zone 40 mm mixing elements just after feed zone 40 mm mixing elements 446 mm from the feed end	
Screw velocity:		400 R.P.M.
Dye-plate:		2 x 2.5 mm ø
Barrel temperature:		12-15°C
Max. product temperature:	abt.	150°C
Max. pressure:	abt.	80 bars
Max. torque:	abt.	40 Nm
Capacity:	abt.	20 kg/hour

Table I. Process variables for the best sample of extruder-degerminated black pepper.

Process variables for "Edelsüss" paprika		
Screw length:		1317 mm
Length - diameter ratio:		35.6
Pitch of screw elements:		40 mm
except	180 mm pitch 60 in the feed zone 40 mm mixing elements just after feed zone 30 mm counter-screwed element 413 mm from the feed end	
Screw velocity:		400 R.P.M.
Dye-plate:		2 x 3.5 mm ø
Barrel temperature:		16-23°C
Max. product temperature:	abt.	90°C
Max. pressure:	abt.	30 bars
Max. torque:	abt.	35 Nm
Capacity:	abt.	10 kg/hour
The paprika was processed under N_2-cover.		

Table II. Process variables for the best sample of extruder-degerminated paprika.

Colour measurements on paprika samples			
Sample	Unprocessed	Extruded	Extruded and N_2-covered
L	32.05	31.19	32.59
a	28.04	21.04	25.16
b	18.40	17.85	18.96

Table III.

4. CONCLUSION

The results have, so far, been very positive, and this is promising for the further work on degermination of nutmeg and possibly other spices. The method is not applicable to spices that must be used un-ground.

REFERENCES

(1) M.J. Bayusik and P.H. Chen (1978)
Method of sterilizing spices. U.S. patent 4, 210, 678

(2) H. Gremli und J.-R. Mor (1979)
Verfahren zur Sterilization von Gewürzen. Europäisches Patent 0-012-813-B1

APPLICATION OF EXTRUSION-COOKING TO CONVENIENCE AND DIETARY FOODS

G. PAPOTTO
Mapimpianti S.p.A. - Galliera Veneta (Padua) - ITALY

SUMMARY

This paper describes the properties of some foods obtained by different extrusion-cooking techniques. The involved technologies refer to two Mapimpianti extruder-cookers, one low shear (G), the other high shear (MPE) performing.
The author illustrates different convenience foods, whose nutritional properties are examined; products range from precooked flours to cereal flakes, from snack foods, protein- or fiber-enriched, to unconventional and precooked pasta products and to corn chips.

1. INTRODUCTION

In the recent years a growing attention of consumers towards convenience and dietary food has been observed. This attitude can be found both in industrialized and developing countries even though its premises are different. In fact, while in the developed countries the problems to be faced are a hasty life's style and excessive food intakes in the developing countries they are the acceptance of certain kinds of foods and nutritional deficiencies. Extrusion-cooking, one of the most versatile modern technologies, can meet these demands. In order to launch this new developed technology on th the market, the plant manufactures have underlined the multi-purpose utilization of extruder-cookers.
In our opinion, however, it cannot be disregarded that generally the multi-purpose machines don't allow the best results in each single production process. According to us, each extruder-cooker, different in type (single or twin screw) or in screw geometry or rotating rate, corresponds to a specific kind of chemical-physical transformation of food biopolymers and therefore it performs better in a particular kind of processing than in others.

2. EXPERIMENTAL

Mapimpianti, whose activity is principally addressed to the development of technologies for the production of convenience and dietary foods, has built two models of extruder-cookers that allow to prepare food products with different organoleptic and nutritional features.

The model G is a low shear single screw extruder-cooker, which rotates at low speed, with extrusion barrel equipped with a series of independently thermoregulated zones. The capacity of the G model ranges from 20 to 1000 kg/hr with screw rotation varying from 35 to 75 rpm. The cooking diagram can be preset directly on the control board and be automatically controlled. The extruder G can process any kind of food powder with moisture ranging between 25 and 45 % or more, depending on the nature of ingredients.

Basically the machine is designed to utilize virtually any form of meal or flour, either single or in mixture, which may be combined with non starch based ingredients such as sugar, proteinaceous additives, colourings, flavours etc.
The dry ingredients metered into the extruder-cooker from the feeding hopper are hydrated by micrometric tuning. The machine is equipped with a system designed to spray the water and floury ingredients enabling a very even moisture imbibition, which will come to completion during the holding time in the mixer, before feeding the extruder-cooker. The geometry of the screw provides for a gentle cooking/mixing of the material, passing through different temperature controlled zones.
A very stringent control of the temperature, preset on the control board, is realized by means of thermocouples, which feel the product passing through each cooking zone. In this way the operator may achieve the required cooking level of the product very easily.
The model MPE is a high shear single screw extruder-cooker, which rotates at high speed, (variable from 200 to 300 rpm), with variable configuration and length of screw and barrel, obtained by properly assembling and combining independent elements. The capacity ranges from 100 to 300 kg/hr. Further on the MPE extruder has the same sophisticated thermoregulation system as the G model. This unit can process a very extensive range of raw materials, preferably ranging between 10 and 35 % moisture content.
In this case the mixture is dosed directly into the cooking barrel, where it is properly hydrated by a water injection system micrometrically controlled.
When advancing inside the suitably configurated screw/barrel system, the material is mixed, melted and, at the outlet, a proper die provides the required shape to puffing product. Also in the field of direct expanded snack foods, we have exploited the great potentiality of the machine to develop snack products belonging to different classes, according to the type of raw materials entering the mixtures, the sweet or savoury coating, nutritional value etc.

3. RESULTS AND DISCUSSION

A) PRECOOKED FLOURS OF CEREAL AND LEGUMES

The legumes supplement the nutritional contents of cereals with significant amount of high-value protein, iron, thiamin, riboflavin and trace minerals.
Under this section we listed:

1.1 Pea-based instant soup
1.2 Rice and soya flours
1.3 Rice-based baby food

They're a mixture of peas, rice, and wheat flours with some flavours, processed in the G extruder-cooker, dried and ground to obtain the precooked products. Their properties are illustrated in a more detailed form in table 1.

Table 1.

sample	calories (100 g)	protein (%)	fat (%)	dispersibility powder:water
1.1	350	23.8	1.6	1 : 8
1.2	390	21.5	1.7	1 : 4.5
1.2	385	13.5	4.4	1 : 5.6

All the three samples show a good absorption rate both in cold or warm water, with just a slight tendency to syneresis for the sample <u>1.2.</u> The sample <u>1.3</u> didn't present any tendency to syneresis, also after a freezing cycle.
The values of dispersibility are to be intended as ratio powder/water giving an optimum of consistency.
For the sample <u>1.2</u> and <u>1.3</u> we report also the percentage of aminoacids calculated on the basis of total aminoacid content.

sample	aminoacids (% on total aminoacids content)													
	ASP	THR	SER	GLU	PRO	GLY	ALA	VAL	MET	LEU	TYR	PHE	LYS	ARG
<u>1.2</u>	27.3	2.9	6.9	17.5	--	14.8	7.41	3.06	0.39	6.12	1.06	3.17	4.71	4.51
<u>1.3</u>	24.5	3.16	6.33	18.8	--	16.4	11.3	5.08	--	9.15	2.26	1.81	3.39	--

B) CEREAL FLAKES READY-TO-EAT (RTE)

Ready-to-eat cereals are convenience foods which can be given a specific dietary characteristic by exploiting the versatility of the extruder-cooker G. The mixture, pre-cooked in the extruder-cooker G, is then formed into pellets in a former-extruder, flaked in a flaking mill and toasted for a few seconds, at high temperature, in a toasting oven. When utilizing different cereal flours, supplemented by soya protein, gluten, etc, flakes (20 % protein) can be produced. Bran-additioned mixtures result in dietary fiber-enriched flakes. The role of the dietary fiber is well-known as a factor in etiology of colon constipation, diverticulosis, colon cancer, hyperlipoproteinemia and diabetes, by effecting stool bulk and transit time, by lowering colon pressure, by reducing the exposure of the intestinal mucosa to possible carcinogenic substances in the feces, by increasing the excretion of cholesterol bound to bile acids and so on.
We report on this section:

<u>2.1</u> Bran flakes
<u>2.2</u> Protein- enriched flakes (soya and wheat protein based)
<u>2.3</u> Multi-grain flakes

As before table 2 gives their properties.

Table 2

sample	calories (per 100g)	protein(%)	fat (%)
<u>2.1</u>	300	13.9	2.8
<u>2.2</u>	330	20.2	1.8
<u>2.3</u>	330	7.1	1.5

Crude fiber of sample <u>2.1</u> is 6.1 %, almost all coming from wheat bran. The contents of amino-acids for samples <u>2.2</u> and <u>2.3</u> is as follows:

sample	aminoacidis (% on total aminoacid content)													
	ASP	THR	SER	GLU	PRO	GLY	ALA	VAL	MET	LEU	THR	PHE	LYS	ARG
<u>2.2</u>	23.2	2.49	8.70	16.6	--	17.0	11.5	3.31	--	9.12	1.10	2.21	1.52	3.31
<u>2.3</u>	24.1	1.80	7.88	18.2	--	19.6	11.5	3.38	1.13	8.78	1.35	1.13	1.13	--

C) PROTEIN- OR FIBER-ENRICHED SNACK FOODS

Mixture with a high protein or fiber content can be cooked in the extruder-cooker G and then formed into pellets and dried. The pellets are puffed by frying of baking, becoming light and crispy. Pellets with 40 % bran contents have been produced in Mapimpianti's Research Center. Mixture with high protein or fiber contents, passed through the extruder-cooker MPE, can be directly puffed at the die outlet. These products after a short drying and flavouring step are ready to be eaten.

Among the large amount of snacks obtained in our Research Center we've chosen as representative:

3.1 High fiber fish-shaped snacks

They are produced by the extruder-cooker G, processing a selected blend of wheat and rice flours, corn starch, wheat bran, sugar, salt and a little percentage of sodium bicarbonate. The pellets obtained are then fried and flavoured to give the finished product.

3.2 High fiber rings.

3.3 Sugar-coated protein-enriched rings

THese last two samples have been obtained on the extruder-cooker MPE; thus they are products directly puffed at the die outlet. They've been purposely developed to obtain a sweet appetizing and nutritional snack. The table shows some physico-chemical properties of these snacks.

Table 3.

sample	calories (per 100g)	protein(%)	fat (%)	bulk density(g/l)
3.1	470	7.2	25	---
3.2	350	12.8	3.1	95
3.3	400	13	1	135

The crude fiber for sample 3.1 and 3.2 was 3.9 % and 6.5 % respectively. The amino-acids contents for sample 3.3 were (% on total a. a. content): ASP 31.9, THR 2.67, SER 6.55, GLU 17.2, PRO --, GLY 16.2, ALA 7.76, VAL 2.86, MET 0.51, LEU 6.30, TYR 1.52, PHE 2.54, LYS 4.01, ARG --.

THe acceptability of protein recipes from people of underdeveloped countries is one of the difficulties met, trying to face the problem of the hungry in the world. We believe that protein enriched sweet snack foods can contribute to solve this problemn offering something of pleasant, especially for children.

D) CORN CHIPS

The attention towards the fat intakes has dramatically increased in the last years, especially in relation with their effects on the cardiovascular diseases. Moreover, the oil, stressed by frying, is suspected to contain carcinogenic substances. Due to this fact, there is a tendency to reduce fat content in snack foods, as we have done with these products, directly obtained from the extruder-cooker G.

Here we compare the results from "normal" corn chip with the ones obtained from the "low fat" corn chips baked in an oven, after extrusion-cooking and then flavoured in a tumbler.

Here are the results:

Table 4.

sample	calories (per 100g)	protein (%)	fat (%)
normal	509	5	29
low fat	420	6	12

E) UNCONVENTIONAL CEREAL AND LEGUME PASTA PRODUCTS

The extruder-cooker G enables the conglomeration and texturization of gluten-free flours to achieve cereal and legume pasta products which, when eaten with proper sauces, give a complete and balanced meal.With the same procedure, protein-free (particularly suitable in some renal disease therapy) or fiber-enriched pasta products can be made with acceptable fineness and stickness.

We listed:

5.1 Wheat and bean pasta
5.2 Lupine pasta
5.3 Potato pasta
5.4 Protein free pasta

Their properties are shown in table 5.

Table 5.

sample	calories (per 100g)	protein (%)	fat (%)
5.1	270	16.6	1.8
5.2	315	15.3	2.1
5.3	355	4.8	trace
5.4	370	0.5	0.3

For samples 5.1 and 5.2 we give also the contents in amino-acids:

sample	aminoacids (% on total aminoacid content)													
	ASP	THR	SER	GLU	PRO	GLY	ALA	VAL	MET	LEU	THR	PHE	LYS	ARG
5.1	31.1	3.11	6.07	14.6	0.83	11.9	6.80	5.14	0.31	5.45	1.24	2.70	5.45	4.98
5.2	28.6	2.67	6.55	17.2	--	16.2	7.76	2.86	0.51	6.30	1.52	2.54	4.01	--

4. CONCLUSION

Without claiming to have exhausted the subject, we have tried to illustrate the potentiality of extrusion-cooking as a technology for the preparation of foods with a substantial nutritional value. Particularly we have stressed the fact that it is possible to integrate the food nutritional characteristics in a most satisfactory way, by properly playing upon the ingredients and the production technologies.

We propose to investigate the biological value of this food, by means of biochemical and biological analyses.

EVALUATION OF THE BEHAVIOUR OF DIFFERENT TYPES OF STARCHES AND FLOURS AFTER PRECOOKING IN A HIGH SHEAR OR IN A LOW SHEAR EXTRUDER-COOKER

E. GUIDOLIN - V. SANGIOVANNI - L. VIRTUCIO - G. PAPOTTO
Mapimpianti S.p.A. - Galliera Veneta (Padua) - ITALY

1. INTRODUCTION

Although several experimental studies have been published in the recent years, little is still known about the behaviour, transformation and interaction of food components (particularly starch and protein) during the above mentioned process. In fact, on the one hand foods composition (especially in the case of mixtures) is very complex, on the other hand a great number of variables is involved in the process. With this work we want to demonstrate that the shear level reached in two single screw extruders having different geometries is fundamental in determining the kind of transformation induced on the processed material and the chemical-physical characteristics of the finished product.

2. MATERIALS AND METHODS

Raw commercial corn starch was purchased from SPAD (Cassano-Spinola, AL) and flour rice from INVERNIZZI (Novara). Laboratory analyses indicated that corn starch contained 0.31 % protein (N x 5.7), 0.5 % fat and 12.3 % moisture.
Rice flour contained 6.7 % protein, 1.1 % fat and 13.4 % moisture.
To process corn starch and rice flour two Mapimpianti extruder-cookers, namely MPE 100 and G 20, were used. The mod. G is a low shear single-screw extruder-cooker, which rotates at slow speed, with extrusion barrel equipped with a series of independent thermoregulated zones. The cooking diagram along the different barrel stages can be preset directly on the control board and be automatically controlled. The extruder G can process any kind of food powder with moisture ranging between 25 and 45 %. The model MPE is a high RPM single-screw extruder-cooker equipped with the same thermoregulating system as in the extruder G. It can process any kind of food powder with moisture ranging between 10 and 35 %.

Each sample was taken after 10 minutes running at stable conditions in each trial. The products were immediately dried, milled and packed in plastic bags to be analysed later. The water absorption index (WAI) was determined by the AACC method and the extent of gelatinization by the o-toluidine method as reported by Chiang and Johnson [Cereal Chemistry, 54 (3), 429 (1977)]
The amylograms were obtained on a Brabender amylograph, based on 45 g of substance diluted to 500 ml with a 350 cmg cartridge.
In table 1 we listed the process conditions:

Table 1 Samples of rice flour

Samples	G1	G2	M3	M4	M5	M6	M7	M8[a]
Temperature°C	140	140	90	100	110	120	130	100
Moisture %	35	25	35	35	35	35	35	35

Table 1/contd.

Samples of corn starch

Samples	G9	M10	M11	M12	M13	M14	M15[a]	M16[a]
Temperature °C	100	70	80	90	100	105	100	100
Moisture %	35	35	35	35	35	35	35	25

a. extruded at high pressure

3. RESULTS AND DISCUSSION

The gelatinization values of the various samples after their processing are shown in table 2:

Table 2

GELATINIZATION VALUES

RICE FLOUR EXTRUDATE		CORN STARCH EXTRUDATE	
Sample	Gelatinization %	Sample	Gelatinization %
G1	98.9	G9	99.4
G2	98.3	M10	97.6
M3	99.0	M11	98.3
M4	98.3	M12	98.9
M5	97.5	M13	99.1
M6	97.7	M14	98.6
M7	99.1	M15	98.2
M8	97.8	M16	97.9

All the extrudates appear totally gelatinized independently from the type of extruder and from the processing conditions used. Diagram 1 shows the initial part of the amylograph curve for corn starch, while diagram 2 illustrates the curve with rice flour.

In both cases the cold viscosity of the samples processed on the extruder-cooker G is markedly superior to that of the samples processed on the MPE extruder.

The difference is more considerable with rice flour, but also substantial with corn starch. By comparing samples M 15 and M 16, and samples G1 and G2, the effects of the moisture content of the mixture on viscosity can be observed.

As foreseen, a major viscosity can be noted in those products processed with higher moisture. Table 3 shows the WAI results.

Also in this case, the higher values are noted in those samples processed on the extruder-cooker G.

As for syneresis, the only two dispersions which did not show separation after 12 h of holding time at 20°C were the ones obtained using G9 and G1 samples.

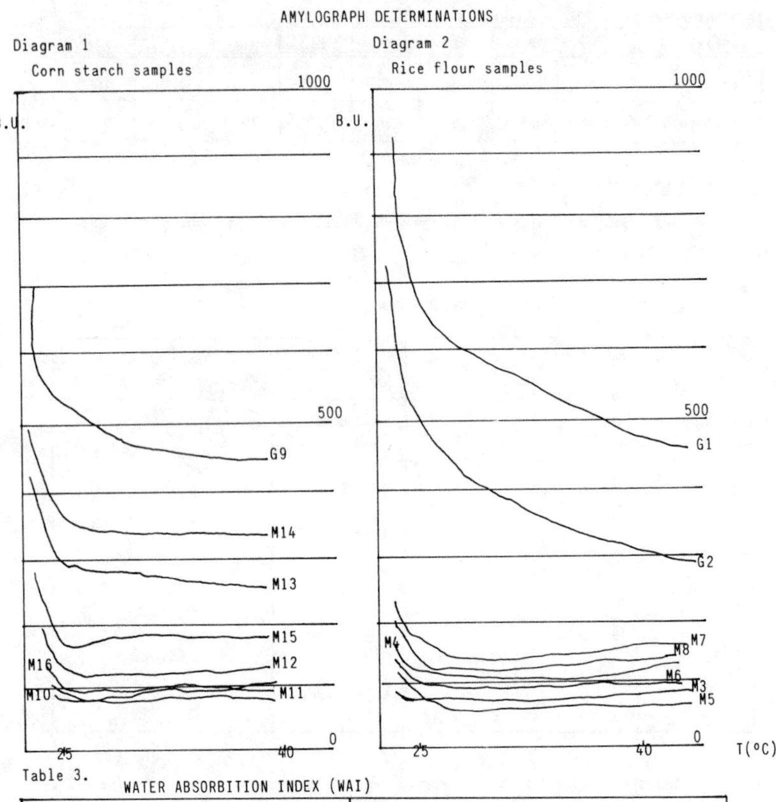

Table 3. WATER ABSORBITION INDEX (WAI)

Sample (Rice flour)	WAI	Sample (Corn starch)	WAI
G1	10.8	G9	7.4
G2	6.9	M10	6.6
M3	9.1	M11	7.3
M4	8.8	M12	6.9
M5	9.1	M13	6.9
M6	8.4	M14	7.0
M7	8.8	M15	6.7
M8	8.7	M16	6.8

4. CONCLUSION

These results demonstrate that the degree of internal shear undergone by the starch samples during extrusion-cooking is of utmost importance in determining the functional characteristics of the final product, especially its thickening power and its resistance to syneresis.

This behaviour is probably related to the minor damage to the starchy structure taking place during the low shear extrusion process.

The model of extruder-cooker on which the experiment will be carried out is a very important variable to be taken into account for a better understanding of structural changes in starch during extrusion-cooking.

COOKING EXTRUSION OF HORSE BEAN AND A FEED DIET FORMULATION

D.J. van Zuilichem
Department of Food Process Engineering, Agricultural University, 6703 BC Wageningen, De Dreijen 12, Netherlands.

L. Mościcki
Institute of Food Engineering, Agricultural University, 20-612 Lublin, pkwn 28, Poland

Summary

The extrusion-cooking potential of horse beans has been investigated using a single screw extruder. Attention was focussed on process requirements as well as on nutritional effects of extrusion-cooked horse bean in a chicken feed. The process optimum thermal conditions required for good texturized products out of horse bean concentrate and horse bean flour are given. Feeding trials with diets that contain up to 80 % of extrusion-cooked horse bean are reported to be successful. It can be concluded that nowadays conventional protein sources such as soya, present in chicken feed can be replaced by them out of extrusion-cooked horse bean.

1. INTRODUCTION

The serious global food protein shortage has led to intensive efforts towards exploring novel and under utilized indigenous protein resources.

Horse bean /vicia faba/, a legume with high content of both protein and lysine, cultivated in many countries, is used to different extents as a food as well as an animal feed. In general, the utilization of this crop is less than its potential offers, particularly in human consumption.

The process for preparation of protein isolates and concentrates from horse bean has been described in the literature /3/. The results of horse bean application in human and animal diets show, however, that the authors are divided in their conclusions.

One way in which horse bean can be utilized is to process them by extrusion-cooking, which makes their proteins available and leads to a new group of attractive enginered products. With regard to their protein properties such extruded proteins have to compete with soya, a conventional, widely used component in food and feed formulations.

The extrusion-cooking of horse beans has been investigated as a route to an available and acceptable protein supplement in feed products. Attention was directed to process requirements and to the nutritional effects of extruded horse bean in chicken feed rations.

2. MATERIALS AND METHODS

During this investigation two differet types of raw materials were used, horse bean flour from Polish crops /Bobik Nadwiślański/ and horse bean concentrate, produced by IRSPAP-Wageningen, Netherlands /see Table I/.

TABLE I Composition of materials used

Material	Protein Nx6,25 %	Carbohydrates %	Fat %	Ash %	PDI %	Particle size m x10^{-3}	Fibre %
Horse b. flour	26,6	48,36	1,85	3,8	30,0	± 50	5,06
Horse b. concentr.	62,0	15-28	2,00	6,5	90,0	± 20	2,50

The horse bean flour used was prepared by conventional cleaning and milling after a first step in which the hulls were seperated /6/.

Several samples were processed under different working conditions in a single screw extruder,2" Battenfeld Fleisner design,to produce different product properties.The investigated process variables were: screw r.p.m., moisture content of the feed,compression ratio of the screw,die diameter , process temperature profiles along the barrel. The samples were produced in replicates for statistical validaty.Temperatures were measured with a thermocouple lance supported inside the tip of hollow screw and at the rear of the extruder.At corresponding locations in the barrel thermocouples were also mounted.At the same points the corresponding pressures were measured with Kistler-piezo-quarts crystal pressure probes.

The production capacity,the expansion ratio,and the mechanical and structural proporties of the product were determined after each run.

The quolity of an extruded product can be estimated in several ways depending on the purpose and application of the product in the particular sector of food industry.In this case we considered mainly chemical changes, bulk density,PDI number,water solubility and water absorption,texture and colour as important data for extruded horse bean products /1,2,7,8/.

The nutritional effects of extruded horse bean flour were assessed by feeding trials on 3 groups of 48 broiler chicken /Cornish x White Rock/, aged from 5 to 8 weeks by supplementing the normal soya protein in their daily feed ration,following the international standardized rules for feeding trials /5/.The comperable feed material conteined:maize-55 %,soya-19 %, wheat-10 %,oats-6 %,other macro and micro components -10 %.

3. RESULTS AND DISCUSSION

During this investigations results were obtained,which enabled the optimum process conditions for the extrusion-cooking of horse bean with a single screw extruder to be determined.The results given in Table II are those very near to the optimum.For these extrusion conditions it proved necessery to define also an optimum temperature profile,in order to achieve the results of Table II.This temperature profile is given in Fig1.

In literature investigations about the application of horse bean in human food products like e.g. bakery products only deal with the use of horse bean isolates /3,4/.It is also known that trials in Canada with extrusion-cooked horse bean in receoees for e.g. sausages were unsuccessful.

On the contrary animal feed application of horse bean are known to be more popular. The investigations done by Ryś /5/ in this field show the positive effects in pig and in chicken feeding.The horse bean contains 6 g/16g N of lysine which is slightly higher than the values measured in toasted soya grits.A methionine deficiency,0,8 g/16 g N, can be easily supplimented in the feed formulation.It is interesting that extrusion-cooking of horse bean reduces the trypsin inhibitor activity to a sufficiently low level.

TABLE II Chosen extruder conditions corresponding to optimum product

Material	Moisture %	Compress. ratio -	Die diameter mx10^{-3}	Press. MPa	Bulk dens. kg/m^3	Water absorp. %	Output kg/h	r.p.m.
Horse bean flour	28	3,0	6-8	-	419	143	12-13	80
Horse bean concentr.	28-30	2,4	8	2,0	490	130	11-12	80

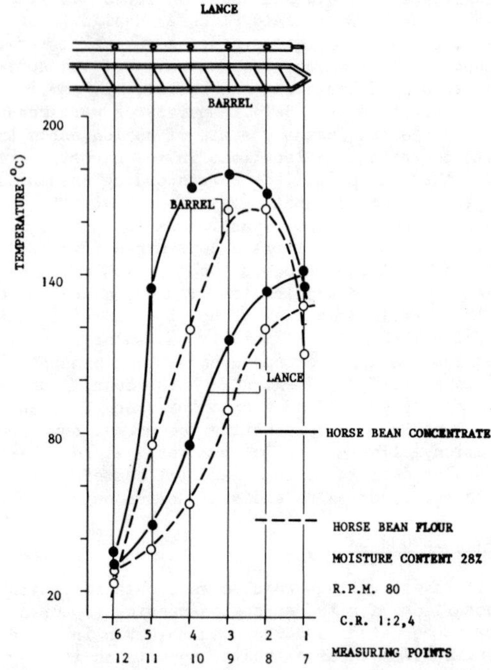

Fig.1. Temperature profile during extrusion-cooking of horse bean

In this investigation a much better chicken growth could be obtained with extruded horse bean than with the same, but unprocessed, raw material /see Table III /. The increase in growth was about 20 % at a feed conversion decrease of about 15 %, which is noticed to be very favourable.

Although not reported here in data the feed digestion was observed to be better for the diets conteining extrusion-cooked horse bean, in particular with respect to organic substance.

Further observations were made on the best way to replace soya grits wit the extruded horse bean products in the chicken daily ration.

TABLE III Effect of horse bean diet of chicken

Material	Growth g	Feed conversion kg/kg [x]
Unprocessed horse bean	510	3,8
Horse bean extruded	611	3,25

[x] kg of feed per 1kg of growth

Results showed that particularly a partial replacement of soya by the horse bean product gave the best results.The results of a product,consisting for 80 % out of the horse bean product resulted however in an unattractive and high feed conversion factor /see Table IV /.A good optimum was determined near a 50 % replacement.This investigation has established the extrusion-cooked horse bean in the feed industry.Depending on availability it can replace at least 50 % of soya in chicken ration.At that level a good effect of chicken growth and a good feed conversion factor is noticed.Extrusion-cooking of horse bean gives the possibility to penetrate into the local feed markets with a new category of feed products.Such products score high in proteins and can easily be compounded with other raw materials which score poor in proteins and amino acids patterns,such as many local crops will do.

TABLE IV Effect of different proportions of extruded horse bean in feed dose

Feed used	Growth of chicken g	Feed conversion kg/kg [x]
Control sample	761	2,90
Soya grits - 25 % Extruded horse bean - 40 % Methionine - 0,2 %	820	3,08
Extruded horse bean - 80 % Methionine - 0,2 %	768	3,51

[x] kg of feed per 1 kg of growth

REFERENCES

1. A.O.C.S. /1965/.Tentative Method Ba-11-65
2. Jowitt,R. /1974/. Journal of Texture Studies, 5,351-358
3. Patel,K.M. /1975/. Horsebean as Protein Supplementin Bread Making II, Cereal Chem.,52,794-800
4. Patel,K.M. and Johnson,J.A. /1975/.Horsebean as Supplement in Bread Making I, Cereal Chem.,51,693-700
5. Ryś ,R. /1974/.Nasiona roślin strączkowych w świetle badań Zakładu Żywienia Zwierząt IZ,Nowe Rolnictwo,6,9,/in Polish/
6. Watson,J.W.,Mc Ewen,T.J.,Bushuk,W. /1975/. Note on Dry Milling of Faba beans,Cereal Chem.,52,2,272
7. Zuilichem,D.J. van,Lamers,G.H.J.,Stolp,W./1975/.Proceedings of the EFChE Conf. Eng. and Food Quolity,6th Europ. Symp.,Cambridge
8. Zuilichem,D.J. van,Stolp,W.,Witham,I./1977/.Texturization of soya with a single screw extruder,Proceedings CPCIA,Extrusion-cooking Conf.,Paris

CHEMICAL MODIFICATION OF STARCH BY HEAT TREATMENT AND FURTHER REACTIONS OF THE PRODUCTS FORMED

O. THEANDER and E. WESTERLUND
Department of Chemistry and Molecular Biology, Swedish University of Agricultural Sciences, S-750 07 Uppsala, Sweden

Summary

The analytical dietary fibre value of various extruded starch-rich products or such products treated with various lengths of time at 180 °C was shown to increase with the extent of heat treatment. This increase was found to be mainly caused by modification of starch to enzyme-resistant glucan structures and to some extent also to the Maillard reaction. Levoglucosan (1,6-anhydro-β-D-glucopyranose) and oligosaccharides with such end-units have been shown to be present in 80 % aqueous extracts from heat-treated potato or wheat starch and indications have been obtained that such 1,6-anhydroglucose units react further by external transglycosidation reactions to enzyme-resistant, branched structures. HPLC and ^{13}C-NMR have been shown to be valuable tools for fractionation and identification of 1,6-anhydrosaccharides.

1. INTRODUCTION

After participation in an international collaborative study, organized by the Department of Food Chemistry and Technology, University of Helsinki (1), on the dietary fibre analysis of extrusion-cooked and in other ways heat-processed cereal and potato products, we became interested in investigating further the chemical reactions of starch under various food technology conditions. Several studies have been made on various processing conditions by extrusion cooking on the gelatinization of starch, leading to an expanded texture. The physical and chemical changes of starch after different extrusion cooking conditions have been extensively studied by Mercier's group (2 and previous publications). The more profound modification and degradation of starch leading both to complex carbohydrate products, which occur under the more extreme conditions during the production of technical dextrins and pure pyrolysis has been reviewed by Horton (3) and Greenwood (4). The extent of chemical changes of the starch components amylose and amylopectin is of course mainly dependent on the temperature and time involved, but the pattern of products is also expected to vary with moisture content, pH and other factors.

2. RESULTS AND DISCUSSION

2.1. The effect of thermal modification of starch on the dietary fibre (DF) analysis of foods

The Helsinki study (1) involved wheat flour, whole wheat meal - untreated and extrusion cooked (twin screw extruder Creusot Loire BC 45) at different temperatures - as well as potato - untreated, boiled and fried. On these samples we applied a method for chemical characterization and analysis of water-soluble and insoluble DF, which previously had been developed and applied on various types of foods (5). The removal of starch in that procedure and/or our quantitative analysis of starch is based on a combined gelatinization/hydrolysis by a thermostable amylase, Termamyl (at 96 °C), followed by incubation with an amyloglucosidase (at 60 °C). The neutral sugar and uronic acid fibre constituents are determined by gas-liquid chromatography (GLC) and decarboxylation respectively and the Klason (sulfuric acid) lignin gravimetrically. We found an increase in the amount of DF for all heat-treated samples compared to untreated samples and that this increase involved both a glucan and Klason lignin increase.

The results given in Table I on extruded wheat flour have been taken from later studies. The samples were prepared at SIK - The Swedish Food Institute - in the same type of extruder as the Finnish samples. Among the extrusion conditions used, the temperatures measured in the material flow are given in the Table.

Table I. DF-analysis of extruded wheat flour

DF-components	Extrusion temperature		
	105°C	150°C	200°C
Neutral water insoluble polysaccharide components:			
Glucans (non-starch)	0.8	1.1	2.0
Arabinoxylans	0.9	0.9	0.3
Others	0.1	0.1	0.1
Neutral water-soluble polysaccharide components:			
Glucans (non-starch)	0.1	0.2	0.6
Arabinoxylans	0.6	0.7	1.2
Others	0.2	0.2	0.2
Uronide components:	0.2	0.2	0.2
Klason lignin:	0.1	0.4	0.4
Total DF:	3.0	3.8	5.0

All figures are given as percentages of the dry matter of the treated material.

A significant increase of the DF-value with the extent of temperature is noted even under the short treatment prevailing under extrusion cooking. It is also obvious that modified starch causes an increase of both water insoluble and soluble glucans, thus giving the main DF-contribution for this product together with some increase of the Klason lignin. To some extent that glucan which is not removed (or hydrolysed) by the enzyme system used might be such retrograded, enzyme-resistant starch found by many workers in processed foods. According to a recent procedure by Englyst et al. (6) such starch can be estimated after alkaline solubilisation. Most likely that fraction is also resistant to the human enzyme system and can therefore be considered as a DF-constituent. The main part of the glucan increase, however, is probably caused by chemical modifications, such as those discussed below. The increase of the Klason lignin value by heat treatment is probably mainly an effect of the Maillard reaction and we have also found how the nitrogen content of that residue increases with the extent of heat treatment by extrusion (7).

In order to study the thermal starch modification further and get enough products for an organo-chemical identification we are also studying the changes of wheat flour and starches that occur during longer treatment times. An example of this is shown in Table II for wheat flour treated at 180 °C (7).

Table II. Changes in DF-compositions and enzymatically determined starch contents in wheat flour treated at 180 °C

Component	Treatment time (h)			
	0	0.5	2	4
Starch	77.2	78.3	72.8	64.4
Glucans (non-starch)	0.9	1.1	1.6	2.3
Arabinoxylans	2.0	1.9	1.7	1.7
Other neutral polysaccharide components	0.5	0.3	0.4	0.3
Uronide components	0.2	0.2	0.2	0.2
Klason lignin	0.3	0.4	1.2	2.0
Total DF	3.9	3.9	5.1	6.5

All figures are given as percentages of the dry matter of the treated material.

The amount of starch available with the Termamyl/amyloglucosidase system first shows a slight increase and then a steady decrease in time, while the DF-value increases with time. It is previously known that dextrinization products from starch and amylose have a high resistance to starch-digesting enzymes, a resistance which increases with the extent of the heat treatment (3, 4).

2.2. Identification and analysis of saccharides containing 1,6-anhydro-β-D-glucopyranose (levoglucosan) end-units

In a heat-treated starch product one would expect the presence of fragments ranging from high degrees of polymerisation (DP) down to oligosaccharides and monomeric products. As the more low-molecular ones are easier to isolate and characterise we have worked with the products soluble in 80 % aqueous ethanol. Mercier's group has previously fractionated oligosaccharides, soluble in aqueous ethanol, from potato starch extruded at 190 °C, on Biogel P_2 (2). Based on that method and an enzymatic study the saccharides were reported to be glucose, maltose, maltotriose and higher linear oligomers. The individual components were not, however, isolated for further chemical identification. Notably, however, this group found no such oligosaccharides when extrusion cooking cereal products and cereal starches.

Before going into more detailed investigations of extracts from extruded and technically processed food, we are developing a methodology on pure starches given a more extended heat treatment and without extra addition of water. Fractionation by high-pressure liquid chromatography (HPLC) on Dextropak has been found to give a much sharper separation than Biogel. Figure 1 shows a fractionation of the extract from potato starch, which has been treated at 180 °C for 4 h. Pure wheat starch treated in a similar way shows a very similar pattern. The extract from potato starch amounted to 12 % of the heat-treated sample. Almost exclusive presence of anhydro-saccharides such as maltosan (4-O-α-D-glucopyranosyl-1,6-anhydro-β-D-glucopyranose) and the higher homolog maltotriosan was found, but only traces of maltose and maltotriose, after isolation by preparative HPLC and comparison with authentic, synthetic samples. Also the presence and identification of free levoglucosan (1,6-anhydro-β-D-glucopyranose) was proved after isolation.

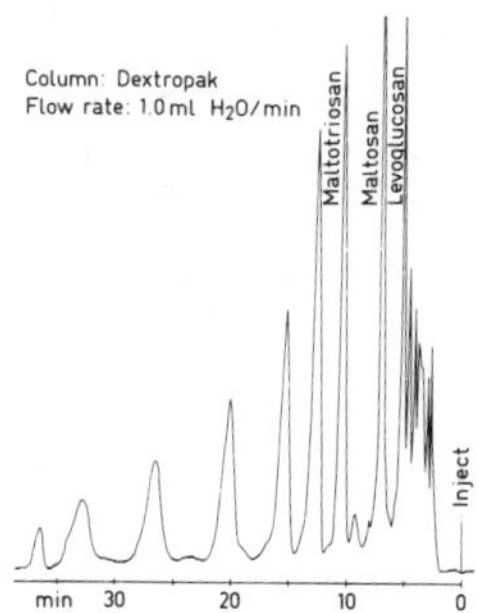

Figure 1. HPLC of 80 % aqueous ethanol extract of potato starch, treated at 180 °C for 4 h.

A very useful tool for identification of these 1,6-anhydro-compounds and also for a rapid characterization of a whole extract is ^{13}C-NMR. The anomeric carbon (C-1) as well as the CH_2-group carbon (C-6) all give rise to signals at characteristic chemical shifts in the anhydro-glucose units (see Figure 2 and 3 for pure maltotriosan and the extract from heat-treated potato starch respectively).

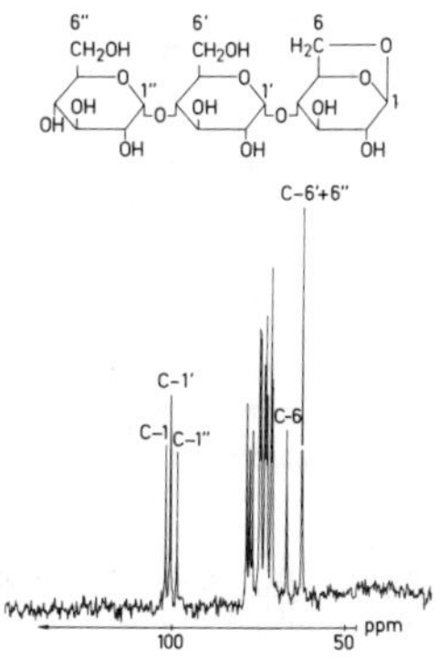

Figure 2. ^{13}C-NMR of maltotriosan.

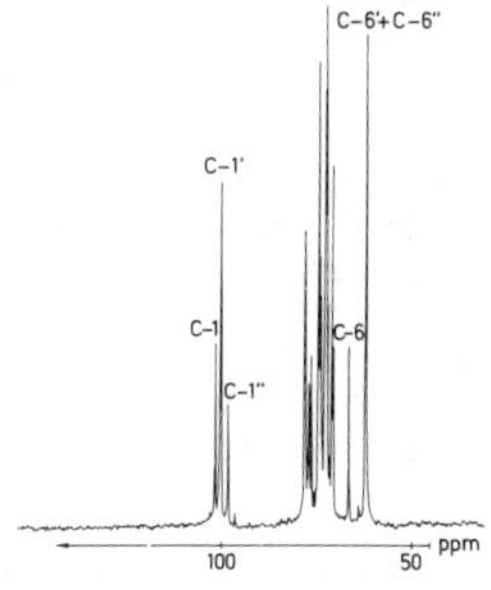

Figure 3. ^{13}C-NMR of 80 % aqueous ethanol extract of potato starch, treated at 180 °C for 4 h.

By treating the extracts with the Termamyl/amyloglucosidase system it was found (Figure 4) that the α-1,4-bond in maltosan was not cleaved and that all higher homologs were cleaved to maltosan and glucose. This observation is the basis for a GLC-method which we are at present developing for a quantitative determination of the total amount of 1,6-anhydro-β-D-glucopyranose units in heat-treated starch products. In that method we treat the whole product and not only the aqueous extract enzymatically and if necessary we enrich the monomeric and dimeric components by gel filtration on Sephadex G-15. By this procedure we have found 0.9 % maltosan units after enzymatic hydrolysis in wheat starch extruded at 150 °C and 1.5 % in wheat flour treated at 180 °C, for 1 h, as calculated on the 80 % ethanol extract. To our knowledge this is the first time the presence of 1,6-anhydro-β-D-glucopyranose-containing saccharides have been proved to be present in extruded products. Our present studies indicate that also "normal" maltosaccharides may be formed and that the ratio between the two types of saccharides is very dependent on the original moisture content and other factors.

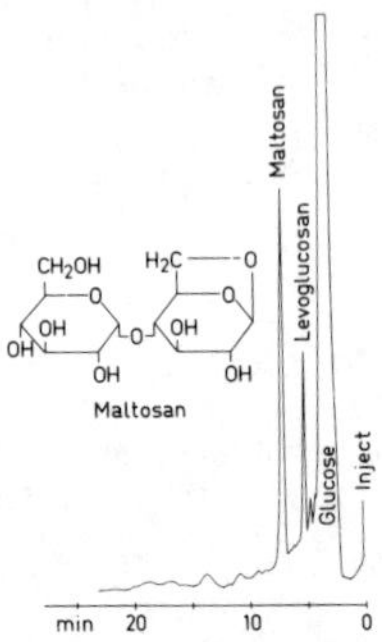

Figure 4. HPLC after enzymatic hydrolysis of 80 % aqueous ethanol extract of potato starch, treated at 180 °C for 4 h.

The present findings are in agreement with the extensive investigations of Wolfrom et al. on so-called pyrodextrins, which are formed by roasting starch under more extended conditions (see in ref. 3 and 4). They thus isolated levoglucosan and various oligosaccharides - including such with 1,6-anhydro-glucose end-units - with different types of glycosidic bonds. The widely accepted mechanism (3, 4) is that under the conditions for the dextrinization process, with only smaller amounts of water present, the 1,6-anhydro-compounds are formed via internal transglycosidation and chain rupture. The 1,6-anhydro end-units are then supposed to react subsequently with other available hydroxyl groups to form various branched structures (external transglycosidation).

At present we are studying these further reactions under the conditions of extrusion cookings and related thermal conditions of the probably rather short-lived 1,6-anhydro-saccharides. Using methyl β-xylopyranoside as a model for a xylan we have thus observed how the xylose unit reacts with low-molecular 1,6-anhydro compounds forming new branched structures. An indication that most of the increase in the glucan content of the DF-values, which we noticed in starch products after heat-treatment, corresponds to the formation of new branched structures is given in the following finding; namely that 18 times more glucose was released by hydrolysis of an 80 % ethanol extract from wheat starch treated at 180 °C for 2 h with mineral acid than with our enzyme system.

REFERENCES

1. VARO, P., LAINE, R. and KOIVISTOINEN, P. The effect of heat treatment on dietary fiber: A collaborative study. J. Assoc. Off. Anal. Chem., in press.
2. MERCIER, C., CHARBONNIERE, R., GALLANT, D. and GUILBOT, A. (1978). Structural modification of various starches by extrusion cooking with a twin-screw French extruder. In: Polysaccharides in Food, Eds. Blanshard, J.M.V. and Mitchell, J.R., Butterworths, London, 153-170.
3. HORTON, D. (1965). Pyrolysis of starch. In: Starch - Chemistry and Technology, Eds. Whistler, R.L. and Pashall, E.F., Academic Press, New York and London, Vol. I, 421-437.
4. GREENWOOD, C.T. (1967). The thermal degradation of starch. Advan. Carbohydrate Chem., 22, 483-515.
5. THEANDER, O. and ÅMAN, P. (1979). Studies on Dietary Fibres. 1. Analysis and chemical characterization of water-soluble and water insoluble dietary fibres. Swedish J. agric. Res., 9, 97-106.
6. ENGLYST, H., WIGGINS, H.S. and CUMMINGS, J.H. (1982). Determination of the non-starch polysaccharides in plant foods by gas-liquid chromatography of constituent sugars as alditol acetates. Analyst, 107, 307-318.
7. THEANDER, O. (1983). Advances in the chemical characterisation and analytical determination of dietary fibre components. In: Dietary Fibre, Eds. Birch, G.G. and Parker, K.J., Applied Science Publishers, London and New York, 77-93.

INFLUENCE OF DIFFERENT EXTRUSION PARAMETERS ON THE VITAMIN STABILITY

C. MILLAUER
W. M. WIEDMANN, Werner & Pfleiderer, Stuttgart
U. KILLEIT, Hoffmann-La Roche, Grenzach-Wyhlen

1. INTRODUCTION/PROBLEM

There is up to now only little information on the influence of extrusion on vitamin stability, and even that is contradictory. The present investigations were mostly carried out on laboratory extruders with low throughputs (60-110 kg/h), the stability of vitamin B_1 (thiamin) being described the most extensively.

The purpose of the present investigation is to show the influence of throughput and moisture on the stability of vitamins B_1, B_6, B_{12} and folic acid when extruding flat bread. To check the results gained on a pilot plant, a trial was carried out under production conditions, other B-vitamins being included.

2. METHOD/TEST MATERIAL

The flour pre-blend of 95% wheat flour T 550, 1% salt, 1% sugar 1% full milk powder and 2% vitamin blend is fed to the extruder with 13% moisture and 3 to 11% additonal water so that the total water content is 16-24%. Over 96% of the vitamin content is reached by means of this addition (Table I).

The laboratory tests were made on a Continua 58 with 60-110 kg/h throughput, 16 to 24% total water content, 300 rpm, 150°C temperature, about 178 maximum product temperature and 30 to 60 s residence time.

The production tests were made on a flat bread plant based on the Continua 120 twin-screw co-rotating and self-cleaning mixing and cooking extruder with a throughput of 500 kg/h roasted and packed flat bread. Thiamin and riboflavin were determined fluorimetrically and the other vitamins of the B complex by microbiological methods. The results of the vitamin analyses are arithmetic mean values from 2-4 readings.

3. TESTS RESULTS

Vitamin stability improves with increasing throughput and increased water content (Fig. 1).

Significant negative correlations result between the specific energy and the degree of vitamin stability (Fig. 2), with the exception of vitamin B_6. According to test results, the order of stability is vitamin B_6, vitamin B_{12}, thiamin, folic acid.

Vitamin stability in production conditions was at the top limit of the stability range for the 4 sensorially best products of the pilot plant. For the 4 other B vitamins, there do not seem to be any serious losses (Table II). After-drying should not cause any particular stability problems either (Table III).

Vitamin levels of flat bread mix

Vitamin	Content, mg/100 g dry weight		
	Wheat flour 550	Addition	Flat bread mix
Vitamin B1	0,13	3,57	3,70
Vitamin B6	0,11	4,09	4,20
Vitamin B12	–	$11,5 \times 10^{-3}$	$11,5 \times 10^{-3}$
Folic acid	$18,0 \times 10^{-3}$	0,90	0,92

Vitamin	Percent retention	
	Pilot plant (approx. 100 kg/h)	Production plant (approx. 500 kg/h)
Vitamin B1	38 – 65	62
Vitamin B6	71 – 83	90
Vitamin B12	65 – 96	99
Folic acid	35 – 45	56
Vitamin B2	85	
Niacinamide	80	
Calcium-pantothenate		91
Biotin		74

Effect of drying

Plant	Vitamin	Percent retention
Pilot plant	Vitamin B1	90
	Vitamin B6	100
	Vitamin B12	93
	Folic acid	95
	Vitamin B2	92
	Niacinamide	93
Production plant	Calcium-pantothenate	100
	Biotin	100

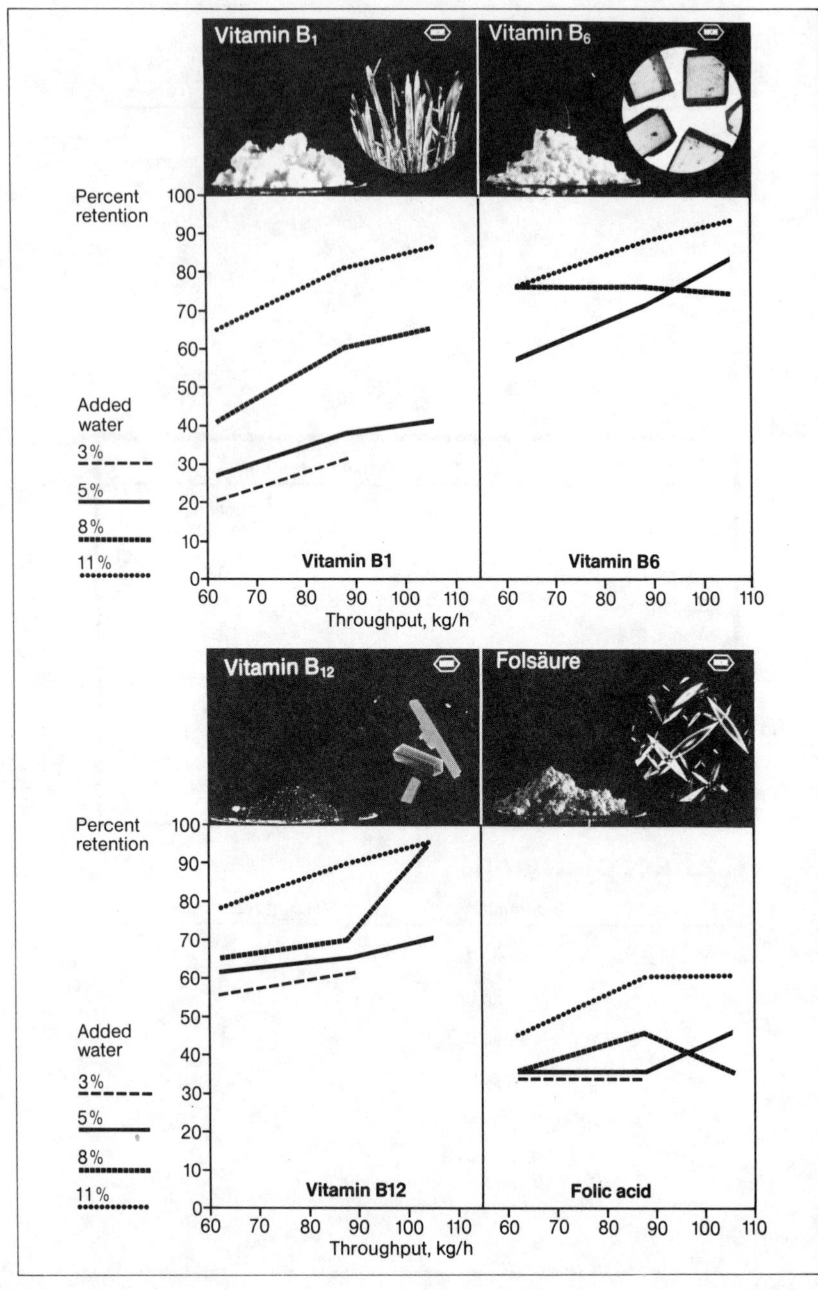
Vitamin B_1
Vitamin B_6
Percent retention
Added water
3%
5%
8%
11%
Vitamin B1
Vitamin B6
Throughput, kg/h
Vitamin B_{12}
Folsäure
Percent retention
Added water
3%
5%
8%
11%
Vitamin B12
Folic acid
Throughput, kg/h

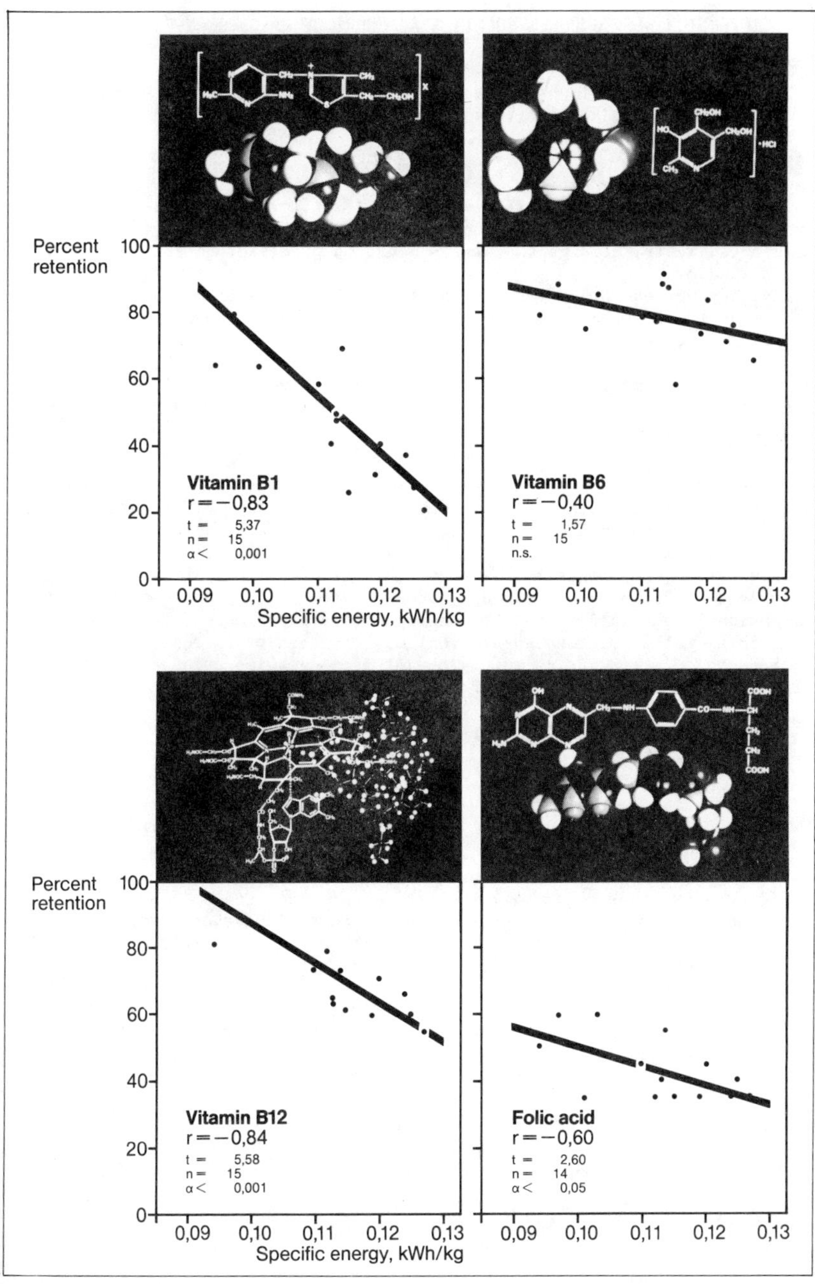
Percent retention
100
80
60
40
20
0
0,09
0,10
0,11
0,12
0,13
Specific energy, kWh/kg
Vitamin B1
r = −0,83
t = 5,37
n = 15
α < 0,001
Vitamin B6
r = −0,40
t = 1,57
n = 15
n.s.
Vitamin B12
r = −0,84
t = 5,58
n = 15
α < 0,001
Folic acid
r = −0,60
t = 2,60
n = 14
α < 0,05

RETENTION DES COMPOSES VOLATILS AU COURS DE LA CUISSON-EXTRUSION

J. CROUZET, A. SADAFIAN, B. DOKO et H. CHOUVEL
Laboratoire de Biochimie Appliquée et Laboratoire de Biochimie et Technologie Alimentaires. U.S.T.L. MONTPELLIER - France

Résumé

La rétention de composés volatils modèles : linalol, acétate de terpenyle, β-ionone, p-hydroxyphényl butanone-2 par différentes matrices : amidon, amidon-caséinate, mélange biscuit au cours du processus de cuisson extrusion a été étudié.
L'addition de ces composés en émulsion dans l'eau entraîne une élimination très importante, probablement par suite de leur entraînement par la vapeur d'eau. Une rétention plus importante est obtenue en assurant la protection de ces composés : présence de parois naturelles ou artificielles (encapsulation) ou formation de complexes d'inclusion avec la β-cyclodextrine. Une rétention quantitative a été obtenue pour la β-ionone en utilisant une double encapsulation.

Summary

Retention of model volatile components : linalool,terpenyl acetate, β-ionone, 4-hydroxyphenyl-2-butanone in starch, starch-caseinate, biscuit mixture during extrusion-cooking was studied. Avery important loss was observed when volatile components are added to the mixture as emulsion in water, this loss occurs probably through water evaporation. A more important retention is obtained if volatile components are protected : natural or artificial walls (microencapsulation) or formation of inclusion complexes with β-cyclodextrin. Using multiple wall microcapsules a quantitative retention was obtained for β-ionone.

1. INTRODUCTION

La rétention des arômes au cours du traitement des produits alimentaires a donné lieu à de nombreuses études, plusieurs mécanismes impliqués dans ce phénomène lors de la concentration de produits liquides (1) ou différents procédés de séchage, ont été developpés (2-4).
En ce qui concerne la cuisson-extrusion on admet généralement que ce processus, par suite des températures élevées mises en jeu, conduit à une perte importante des composés volatils de l'arôme (5,6). Cependant les données chiffrées sont relativement peu nombreuses, Palkert et Fagerson (7) signalent que lors de l'extrusion de farine de soja le pourcentage de récupération de composés modèles représentatifs de l'arôme de viande, introduits avant le processus varie de 4 % pour les sulfures de propyle et de butyle à 22 % par le diméthyl-2,4 thiazole. Le pourcentage moyen de rétention pour les 10 composés étudiés par ces auteurs est de 9,5 %.

Nous nous sommes proposés au cours du présent travail, d'une part d'étudier la rétention de composés volatils caractéristiques d'arômes de fruits ou présents dans certaines huiles essentielles, d'autre part d'étudier les modes et formes d'introduction favorisant la rétention des arômes : microencapsulation ou complexes d'inclusion.

2. MATERIELS ET METHODES

La rétention des composés volatils a été étudiée au cours de la cuisson-extrusion soit de différents mélanges : caséinate 80 % - amidon 20 % ; caséinate 6 %, isolat protéique de soja 7,4 %, amidon de maïs 23 %, farine de blé 42 %, saccharose 15 %, chlorure de sodium 1 % (mélange biscuit) soit d'amidon pur. Les composés volatils étudiés ont été additionnés au mélange solide soit en émulsion dans l'eau : linalol, acétate de terpényle, β-ionone, huile essentielle de cardamone, soit sous forme solide : p-hydroxyphényl butanone-2, arôme de framboise et β-ionone encapsulés, β-ionone et acétate de terpényle sous forme de complexe d'inclusion dans la β-cyclodextrine. Les quantités de composés volatils utilisées sont de 0,5 g pour 100 g de produit extrudé. Les produits solides ont été mélangés pendant 30 minutes dans un appareil Lödige M 20 G. Nous avons utilisé un extrudeur bi-vis Creusot-Loire BC 45, après différents essais préliminaires nous avons opéré à 160 °C et 150 bars.

L'extraction des composés volatils présents dans les mélanges avant extrusion et dans les produits extrudés a été réalisée en utilisant un extracteur solide-liquide (Soxhlet), le solvant étant soit le dichlorométhane, soit un mélange éther-pentane.

Les extraits organiques séchés ont été concentrés par distillation jusqu'à 0,8-1 ml (8).

Les composés volatils ont été déterminés par CPV : colonne capillaire de verre de 60 m x 0,4 mm, garnie de Carbowax 20 M (50 à 180 °C à 4°C/minute) selon le composé volatil étudié nous avons utilisé l'heptanol ou l'allyl-2 phénol comme étalon interne.

3. RESULTATS ET DISCUSSION

Les pourcentages de récupération des différents composés volatils utilisés au cours de cette étude sont donnés tableau 1., dans le cas le plus défavorable (linalol) l'efficacité de l'extraction est d'environ 75 %, pour tous les autres composés la récupération est quantitative.

TABLEAU 1 : Récupération des composés volatils avant extrusion.

Composés	Solvant d'extraction	% de récupération
Linalol	dichlorométhane	74 ± 2
Acétate de terpényle	dichlorométhane	95 ± 3
β-ionone	éther-pentane	98 ± 2
p-hydroxyphényl butanone-2	éther-pentane	98 ± 2

Compte tenu des concentrations relativement élevées en composés volatils utilisés au cours de la présente étude nous n'avons pas observé d'interférences avec les composés formés par dégradation thermique des substrats au cours du processus d'extrusion (fig. 1).

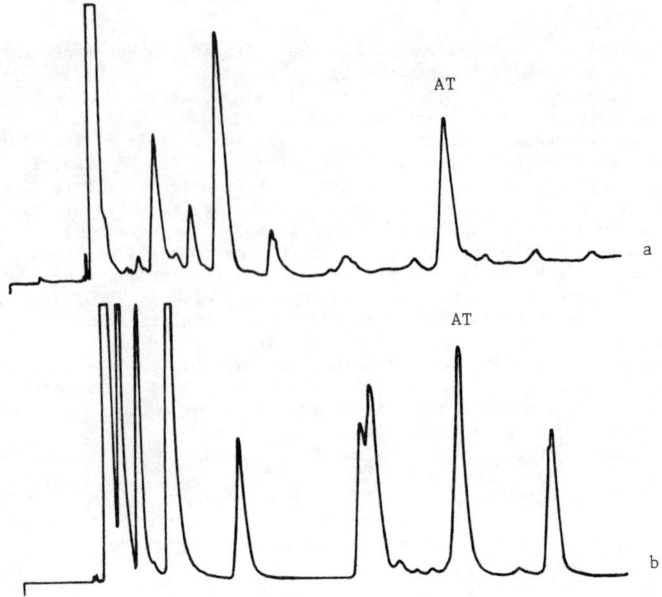

Fig. 1 - Chromatogramme de l'huile essentielle de cardamone a) après extrusion b) avant extrusion. AT = acétate de terpényle.

Au cours de la cuisson-extrusion à 160 °C on observe une élimination quasi-totale des différents composés volatils étudiés (tableau 2).

TABLEAU 2 : Récupération des composés volatils introduits sous forme libre après cuisson-extrusion.

Composés	Mélange	% de récupération
Linalol	caséinate-amidon	4
Acétate de terpényle	biscuit	
- produit pur		1
- huile essentielle de cardamone		1,5
β-ionone	amidon	< 1
p-hydroxyphényl butanone-2	amidon	< 1

Les pourcentages de rétention observés sont très inférieurs à ceux obtenus par Palkert et Fagerson (7). Il ne semble pas que l'on puisse évoquer pour rendre compte de ces pertes en composés volatils la formation de complexes avec les composés solides, et en particulier de complexes d'inclusion avec l'amidon (9-12) puisque la méthode d'extraction utilisée permet de récupérer quantitativement la β-ionone du complexe d'inclusion β-ionone - β-cyclodextrine. Par contre on peut penser que le faible taux de rétention des composés volatils est imputable à leur entraînement par la vapeur d'eau au cours du traitement (5).

Dans ces conditions nous avons été conduit à envisager l'utilisation de différentes techniques d'incorporation des composés volatils permettant de limiter ce phénomène d'entraînement (13-14).

Deux composés ont été plus particulièrement étudiés : l'acétate de terpényle qui est un constituant de l'huile essentielle de cardamone, dont la graine est utilisée avec succès pour aromatiser certaines patisseries orientales et la β-ionone qui est un composé connu comme intervenant dans l'arôme de framboise.

TABLEAU 3 : Influence du mode d'incorporation sur la rétention de la β-ionone et de l'acétate de terpényle.

Composés	% de rétention				
	1	2	3	4	5
Acétate de terpényle	1,5*	7,5	-	4	-
β-ionone	< 1	13	2,5	-	99

* huile essentielle de cardamone.
1 = composés volatils libres, 2 = complexe d'inclusion dans la β-cyclodextrine, 3 = simple encapsulation, 4 = broyat de graines de cardamone, 5 = double encapsulation.

Les résultats rassemblés dans le tableau 3 indiquent que la rétention de l'acétate de terpényle passe de 1,5 à 4 % lorsqu'on substitue à l'huile essentielle de cardamone un broyat de graines. On peut penser que dans le second cas une partie de l'huile essentielle se trouve protégée à l'intérieur des cellules qui ont résisté au broyage.

Un effet protecteur du même ordre de grandeur est observé lorsqu'on procède à l'encapsulation de la β-ionone dans la gomme arabique.

La formation de complexes d'inclusion composés volatils - β-cyclodextrine permet d'augmenter le taux de rétention de la β-ionone et de l'acétate de terpényle jusqu'à des niveaux acceptables, respectivement 13 et 7,5 %. L'utilisation de tels complexes faciles à préparer peut constituer une solution aux problèmes posés par l'aromatisation des produits avant cuisson-extrusion.

Cependant une solution plus efficace réside très certainement au niveau de l'utilisation d'arômes protégés par un double encapsulage. En effet on note une rétention totale de la β-ionone présente dans un arôme de framboise doublement encapsulé.

Les auteurs remercient les Etablissements LAURENT, GIVAUDAN FRANCE et ROQUETTE FRERES qui leur ont fourni respectivement les arômes de framboise, l'acétate de terpényle et la β-cyclodextrine.

REFERENCES

1. BOMBEN, J.L., BRUINS, S., THIJSSEN, H.A.C. et MERSON, R.L. (1973). Aroma recovery and retention. Advances in food research vol. 2C 1-104 Acad. Press. N.Y. London.
2. FLINK, J. et KAREL, M. (1972). Mechanisms of retention of organic volatiles in freeze dried systems. J. Fd. Tech. 7, 199-211.
3. THIJSSEN, H.A.C. et RULKENS, W.H. (1968). Retention of aromas in drying food liquids. D. Ingénieur., 80, ch. 45.
4. KERKHOF, P.J.A.M. et THIJSSEN, H.A.C. (1979). The effect of process conditions on aroma retention in drying liquid foods - Aroma research. 167-197 Pudoc Wageningen.
5. KRUKAR, R.J. (1971). Flavor stability in extruded snack foods Symposium on extrusion process and product development. Am. Assoc. Cereal Chem. St Paul.
6. LYON, L. (1980). Popularity of extrusion processing places added demands on flavorings. Food Product. Dev. 58-61.
7. PALKERT, P.E. et FAGERSON, I.S. (1980). Détermination of flavor retention in pre-extrusion flavour textured soy protein. J. Food Sci., 45, 526-528
8. BEMILMANS, J.M.H. (1979). Review of isolation and concentration techniques - Progress in flavour Research 79-98. Applied Science Publishers Ltd London.
9. OSMAN-ISMAIL, F. et SOLMS, J. (1972). Interaction of potato starch with different ligands. Die Stärke, 24, 213-216.
10. OSMAN-ISMAIL, F. et SOLMS, J. (1973). The formation of inclusion compounds of starches with flavor substances. Lebensm. Wissen. U. Technol., 6, 147-150.
11. SOLMS, J., OSMAN-ISMAIL, F. et BEYELER, H. (1973). The interaction of volatiles with food components Can. Inst. Food Sci. Technol. J., 6, 10-16.
12. WYLER, R. et SOLMS, J. (1981). Inclusion complexes of potato starch with flavor compounds. Flavour' 81, 693-699 Walter de Gruyter and Co Berlin. New-York.
13. BALASSA, L.L. et FANGER, G.O. (1971). Microencapsulation in the food industry. CRC Critical Reviews in food technology, 245-265.
14. LINDNER, K., SZENTE, L. and SZEJTLI, J. (1981). Food Flavoring with β-cyclodextrin complexed flavour substances. Acta Alimentaria, 10, 175-186.

LE CONCHAGE DU CHOCOLAT EFFECTUE DANS UN CUISEUR-EXTRUDEUR BI-VIS

(CHOCOLATE CONCHING PERFORMED BY A DOUBLE SCREW EXTRUDER)

H.CHAVERON,H.ADENIER,A.KAMOUN
Laboratoire de biophysicochimie et technologie alimentaires.
Université de Compiegne.BP233 60206 Compiegne cedex (France)
M.BILLON
Département Presses,Extrudeurs,Pompes.
CREUSOT-LOIRE,usine de l'Ondaine,42701 Firminy (France)
J.PONTILLON
Laboratoires de recherches.CACAO BARRY BP8 78250 MEULAN (France)

summary
The purpose of this work is to replace conching,wich is an expensive and discontinuous phase of the processing of chocolate,by a continuous process in wich a double screw extruder is brought into play. Conching is,with roasting,grinding and tempering one of the most important phases in the processing of chocolate.
Extrusion finds use in the various fields of the food industry but has not yet been studied thoroughly in order to totally replace conching.The assays show,concerning Casson plastic viscosity,Casson yeld value and granulometry,that the values are very near those of the traditionally conched chocolate.The flavour,analysed by gas chromatography of dynamic head-space and measured by Differential Olfactory Stimulator (STOD) show that loss of aromatic compounds is less important than when traditional conching is applied.
This study has allowed us to introduce a prototype of a double screw extruder specially designed for the chocolate conching.

Résumé
L'objectif des travaux était de substituer,dans le procédé de fabrication du chocolat,à une phase discontinue (le conchage),longue, empirique et couteuse en énergie,une phase continue mettant en oeuvre un cuiseur-extrudeur bi-vis.
L'opération de conchage est,avec la torréfaction,le broyage et le tempérage l'une des étapes-clés de la fabrication du chocolat.La cuisson-extrusion n'a jusqu'a présent fait l'objet d'aucune étude en chocolaterie en vue de se substituer totalement au conchage. Les essais montrent que,pour la viscosité plastique selon Casson, la limite d'écoulement selon Casson et la granulométrie,les caractéristiques des produits obtenus sont proches de celles du chocolat conché traditionnellement.Pour l'arôme,la chromatographie en phase gazeuse de l'espace de tête dynamique ainsi que la mesure de l'intensité aromatique globale,estimée au stimulateur olfactif différentiel (STOD) montrent que la perte de composés aromatiques au cours du procédé est moins importante que lors du conchage traditionnel.Les travaux effectués permettent de réaliser un prototype d'extrudeur bi-vis spécialement adapté au conchage du chocolat.

1. INTRODUCTION

Situé entre le broyage fin de la pâte de chocolat (raffinage) et la cristallisation de la phase grasse (tempérage),le conchage,longue opération de malaxage,modifie favorablement le profil aromatique obtenu par torréfaction,principalement par départ de substances volatiles.(MANIERE et DIMICK 1979).

En marge de la flaveur mais toujours au plan organoleptique,le conchage à été de tout temps l'opération permettant de rendre le chocolat"onctueux". Ce caractère organoleptique assez mal défini semble être lié aux modifications physiques de la pâte de chcolat,modifications qui permettent d'obtenir une structure dans laquelle les particules non grasses constituent une dispersion dans une phase lipidique continue.

Au plan rhéologique,l'élimination d'eau,consécutive à l'échauffement de la pâte de chcolat au cours du travail mécanique permet d'abaisser la viscosité et la limite d'écoulement de cette dernière.

Un progrès important dans le procédé de conchage est apparu en différenciant les deux finalités de cette opération:

1)élimination d'eau et de substances volatiles

2)obtention de la phase lipidique continue.

De cette différenciation est né le conchage réalisé en deux phases:

La première,dite " à sec",au cours de laquelle le travail mécanique est appliqué à la pâte de chocolat avec une teneur en matiere grasse juste suffisante pour mener à bien l'opération précédente de broyage (raffinage). Cette basse teneur en matieres grasses favorise l'élimination de l'eau et des substances volatiles.

La seconde phase du conchage consiste à poursuivre le malaxage en incorporant le complément de matières grasses et plus tardivement les tensio-actifs(lécithine).C'est au cours de cette phase que se réalise la dispersion des particules non grasses dans la phase grasse (plastification).

Les nouvelles technologies de conchage adoptent des matériels différents pour réaliser les deux phases,(SCHMITT,1974),(TSCHEUSCHNER,SCHEBIELLA et FORSTER,1981),(KLEINERT,1974),(ROBINSON et ZIEGLEDER,1980).Dans ces deux derniers procédés,la première phase est réalisée sur le cacao seul.

Les matériels nouveaux de conchage concernent généralement la première phase,la plastification continuant à être réalisée dans des matériels classiques.Dans le procédé LINDT (KLEINERT,1974),la dispersion des particules non grasses dans la phase lipidique continue est partiellement réalisée dans un matériel bi-vis.

Il n'existe aucune publication concernant l'extrusion en remplacement total du conchage.Un brevet(OCKER,1972) décrit un système pour "produire une masse de chocolat en continu".Ce système s'éloigne sensiblement de celui dont il est question ici.Il ne porte pas spécifiquement sur le conchage.Aucune évaluation des critères de qualité des produits obtenus n'a été faite.

2.PARTIE EXPERIMENTALE.

Les essais ont été réalisés sur une formulation de chocolat noir riche en matières grasses,correspondant à une couverture destinée à la confiserie de chocolat,à l'aide d'une extrudeuse bi-vis CREUSOT-LOIRE BC-45.

Dans un premier temps,les essais ont porté sur l'extrusion "à sec".Les produits extrudés ont subi un traitement dans un malaxeur GUITTARD pour ajuster la teneur en matières grasses du chocolat à la formule finale.

Dans un deuxieme temps,le travail a porté sur la plastification.

Au cours des différents essais,plusieurs paramètres ont été étudiés: profil de vis,température du fourreau,vitesse de rotation des vis,débit d'alimentation.

Les échantillons obtenus ont été comparés au chocolat conché traditionnellement.L'évaluation des produits finis a été faite tant au plan physicochimique que psychophysique.

2.1 Rhéologie:

Mesure de la viscosité et de la limite d'écoulement selon Casson,sur la pâte de chocolat fondue à 40°C,à l'aide d'un Rhéomat 15.

2.2 Granulométrie:

La dispersion des particules a été évaluée par la jauge de NORTH.Avec cet appareil,il est possible de quantifier les particules d'une taille donnée pour chaque étalement d'un même volume de pâte de chocolat.Cette estimation doit être comparée à celle d'un échantillon témoin.

2.3 Analyse physicochimique de l'espace de tête dynamique:

Chromatographe en phase gazeuse (CPG) équipé d'un systeme de piègeage des arômes sur Tenax.

50mg d'échantillon constitué de 40% de pâte de chocolat et de 60% de chlorure de lithium subissent un balayage par un courant d'hélium à 37°C pendant 15mn.Les arômes sont adsorbés sur le Tenax refroidi à -10°C par un courant d'azote liquide.
L'introduction des composés volatils en tête de colonne se fait par chauffage rapide du piège à 250°C.Les conditions de la CPG sont les suivantes:
.colonne capillaire en verre FFAP 50m
.température du four:de 50°C à 200°C avec un taux de programmation de 2°C par mn.

2.4 Examen psychophysique de l'intensité aromatique globale par le stimulateur olfactif différentiel (STOD):

Cet appareil est basé sur le principe neurophysiologique de l'inhibition réciproque des deux bulbes olfactifs.
Si l'on soumet simultanément chacune des deux voies olfactives d'un même individu à deux impulsions d'odeur d'intensité différente,seul le signal correspondant à la plus forte intensité est perçu par le cerveau.(GUILLET, GOMES et MAC LEOD,1981).

Cette mesure ne donne pas d'information sur la qualité de l'arôme.
100g d'échantillon broyé sont placés dans des sacs de Tedlar thermosoudés pendant 48h à 37°C avant examen.
.durée de stimulation:200 millisecondes
.référence:Butanol+air à concentration constante.

3.RESULTATS

Comparaison des caractéristiques de deux chocolats noirs de formulation identique;l'un est conché par extrusion,l'autre traditionnellement (témoin)

3.1 Rhéologie:

	Témoin	Extrudé
Viscosité η_{ca} (poises)	10 ± 1	7 ± 1
Limite d'écoulement τ_{ca} (dynes/cm^2)	50 ± 5	60 ± 5

Les caractéristiques du produit extrudé sont tres proches de celles du produit conché traditionnellement.

3.2 Granulométrie:

Nombre de particules d'une dimension donnée:

Dimension (μ)	Témoin	Extrudé
90	0	0
80	1	1
70	2	2
60	2	6
50	4	13
40	11	20
30	30	plus de 30

Dans le domaine de 70 à 90 microns,le nombre de particules est identique pour les deux types de procédés.

3.3 Analyse physicochimique de l'espace de tête dynamique:

Les chromatogrammes de l'arôme du chocolat extrudé et de celui conché traditionnellement (témoin) sont représentés cidessous:

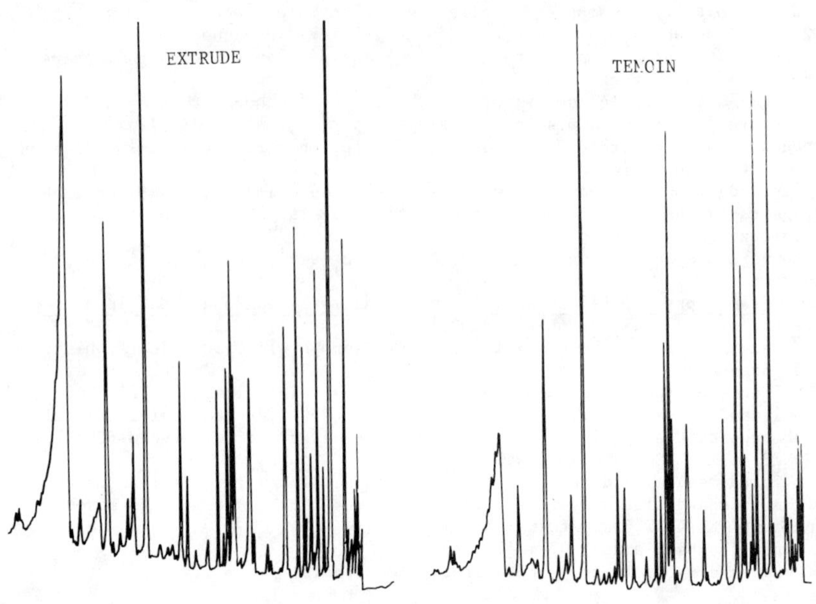

Les chromatogrammes"Extrudé"et"Témoin" apparaissent très proches l'un de l'autre sur le plan qualitatif.Ils se distinguent sur le plan quantitatif:la surface totale des pics du chocolat extrudé est supérieure à celle du témoin.

3.4 Analyse psychophysique de l'arôme:

Mesure au stimulateur olfactif différentiel (STOD).
La figure suivante regroupe,pour un individu,les résultats obtenus en examinant comparativement trois types de chocolat :non conché,extrudé et conché traditionnellement (témoin).

Chaque courbe relie les points qui correspondent à la quantité d'échantillon (en ordonnée) nécéssaire pour égaliser l'intensité arômatique d'une quantité connue de butanol(en abscisse).

Par exemple,pour égaliser 5,8ml de butanol (log 5,8=0,76),il faut 4,8ml (log 4,8 = 0,68) de chocolat non conché,ou 6,2ml (log 6,2 = 0,79) de chocolat conché traditionnellement ou 5,3ml (log 5,3 = 0,72) de chocolat extrudé.

L'analyse de l'intensité globale de l'arôme par le STOD permet de mettre en évidence pour le conchage traditionnel une perte de l'ordre de 20% de l'intensité globale de l'arôme par rapport au chcolat non conché,alors qu'avec l'extrusion,la perte n'est que de 10% (KANE,1983).

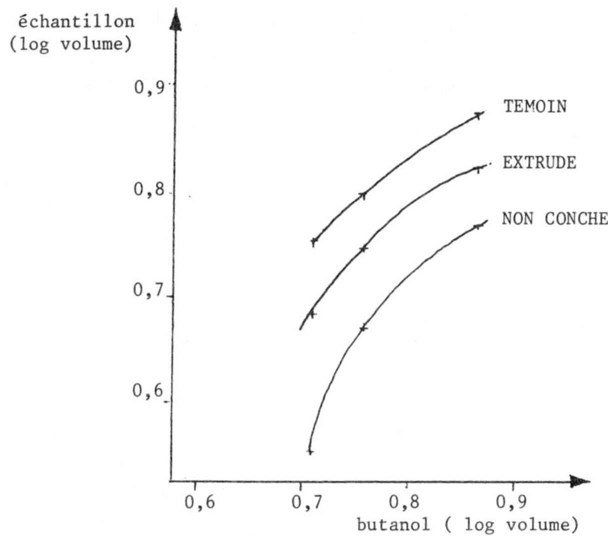

4.CONCLUSION

Les travaux effectués,qui se poursuivent actuellement sur le chocolat au lait permettent de remplacer le conchage par un procédé faisant intervenir un extrudeur bi-vis.

L'utilisation de ce dernier permet une réduction très importante du temps de conchage qui ne représente plus que quelques minutes.

Les caractéristiques du chocolat extrudé diffèrent peu de celles du chocolat conché traditionnellement.

Les travaux sont en cours pour la réalisation d'un prototype d'extrudeur bi-vis spécialement adapté au conchage du chocolat.

5.BIBLIOGRAPHIE

GUILLET PH.,GOMES A.,MAC LEOD P., 1981 .Mesures des odeurs.Le stimulateur olfactif différentiel (STOD). Industries Alim.Agric.,98(3),191-193.

KANE A., 1983 .Optimisation de l'utilisation du stimulateur olfactif différentiel en industrie alimentaire.Effet du conchage et de la cuisson-extrusion en chocolaterie.These de docteur ingénieur,Université de Compiegne.

KLEINERT J., 1974 .Von der kakaopulvertechnologie zur schokoladenherstellung ohne conchierung.in"1 internationnalerKongress über kakao und schokoladeforschung"Dissertations u fotodruck Franck editeur,München.

MANIERE F.Y.,DIMICK P.S., 1979 .Effects of conching on the flavour and volatile components of dark semi-sweet chocolate.Lebensmittel.Wissenschaft u Technol.12(2),102-107.

ROBINSON L.,ZIEGLEDER G., 1980 .Raffinage de la pâte de cacao in"publication 202 des Ets Bauermeister ,Hamburg.

OCKER H., 1972 .Device for producing a mass of chocolate ,US patent n° 3 682 086.

SCHMITT A., 1972.Concerning conching:considerations of external and internal causative factors,Candy and Snack Industry,137(9),44-48.

TSCHEUSCHNER H.D,SCHEBIELLA G.,FORSTER H., 1981. Continuerlisches intensiv conchieren von schokoladenmassen. Kakao Zucker,33(5),122-126.

PHYSICAL AND RHEOLOGICAL ASSESSMENT OF EXTRUSION COOKED MAIZE

S.I.FLETCHER, T.J.MCMASTER, P.RICHMOND AND A.C.SMITH
ARC Food Research Institute
Colney Lane
Norwich

Summary

The use of an intermeshing, co-rotating twin screw extruder for the cooking of maize grits is discussed. Cooked and semi-cooked samples were obtained from the extruder by a dead-stop technique for different operating conditions. Scanning electron microscopy reveals the effect of the extrusion process on the maize microstructure. The dilute solution rheology of the maize samples is similarly sensitive to progress down the extruder and varies with extrusion conditions. The effect of post-extrusion cooking treatments on the dilute solution viscosity has also been explored.

1. INTRODUCTION

The recent use of twin screw cooking extruders has ranged from the modification of starches (1) to the pre-treatment of biopolymers prior to biotechnological processes such as syrup manufacture and fermentation (2). The extrusion cooking of cereals has received considerable attention in relation to the development of snack foods, breakfast cereals and health products. A number of parameters have been used to characterise the extrudate, including colour, water content, density, expansion ratio, breaking force and dilute solution viscosity (3 - 5). The viscosity of solutions of ground extruded maize grits has been studied by Seiler et al (6) using a Brabender viscometer. The implications of solution and paste viscosities in estimating the degree of cooking of extruded products have been discussed by Paton and Spratt (7). Mercier and co-workers extended the study of cooking extruder operation by examining material extracted from the extruder following a dead-stop procedure (8). They examined the properties of maize starch using optical and electron microscopy and intrinsic viscosity measurement. The viscosity of maize starch solutions as a function of heat treatment has also been studied by Bagley and co-workers (9).

The present work used a twin screw cooker extruder to obtain material from the barrel processed under different conditions after a dead-stop procedure.

The dilute solution viscosity of extruded and partially extruded material was obtained and compared with that of raw maize. The effect of post-extrusion heat treatments on the dilute solution viscosity was also investigated. Scanning electron microscopy (SEM) was used to explore the microstructure of maize samples removed from the extruder.

The results indicate the complexity of the viscosity changes that occur as a result of extrusion and post-extrusion heating. Some aspects of the data may be inferred from changes in the starch fraction of the maize although the swelling and molecular configuration changes may not be fully identified at present.

2. EXPERIMENTAL APPROACH

Commercial maize grits of 600 μm nominal size were extruded using a Baker Perkins MPF 50 intermeshing, co-rotating twin screw extruder. The barrel could be split longitudinally and the screws built up from various conveying and paddle elements. Two screw geometries were used in these experiments as shown in fig 1. The extruder was instrumented to measure the temperatures and pressures along the barrel together with the torque, screw speed and feed rate. The extrusion conditions used in these experiments are given in table I. Under steady conditions of pressure, torque and temperature (after 20-30 minutes) a sample of the extrudate was collected. The pressure as a function of barrel length at equilibrium is shown in fig 2. The extruder was then stopped by cutting the feed supply, the electrical heating and the screw rotation. The barrel was then cooled by the refrigeration system to ambient temperature over a period of approximately 20 minutes and dismantled. Samples were taken from points along the barrel as shown in fig 1.

Table I. Extrusion Conditions

Experiment	Heating Zone Temperature (^{o}C) Feed -------------- Die	Screw Config. (Fig 1.)	Screw Speed (rpm)	Maize Feed Rate (Kg/hr)	Water Feed Rate (Kg/hr)
1.	27, 52, 93, 120, 120	(i)	390	60	3.3
2.	27, 60, 110, 140, 140	(i)	390	60	2.5
3.	25, 40, 80, 100, 100	(i)	405	50	0
4.	27, 52, 93, 120, 120	(ii)	390	60	1.6

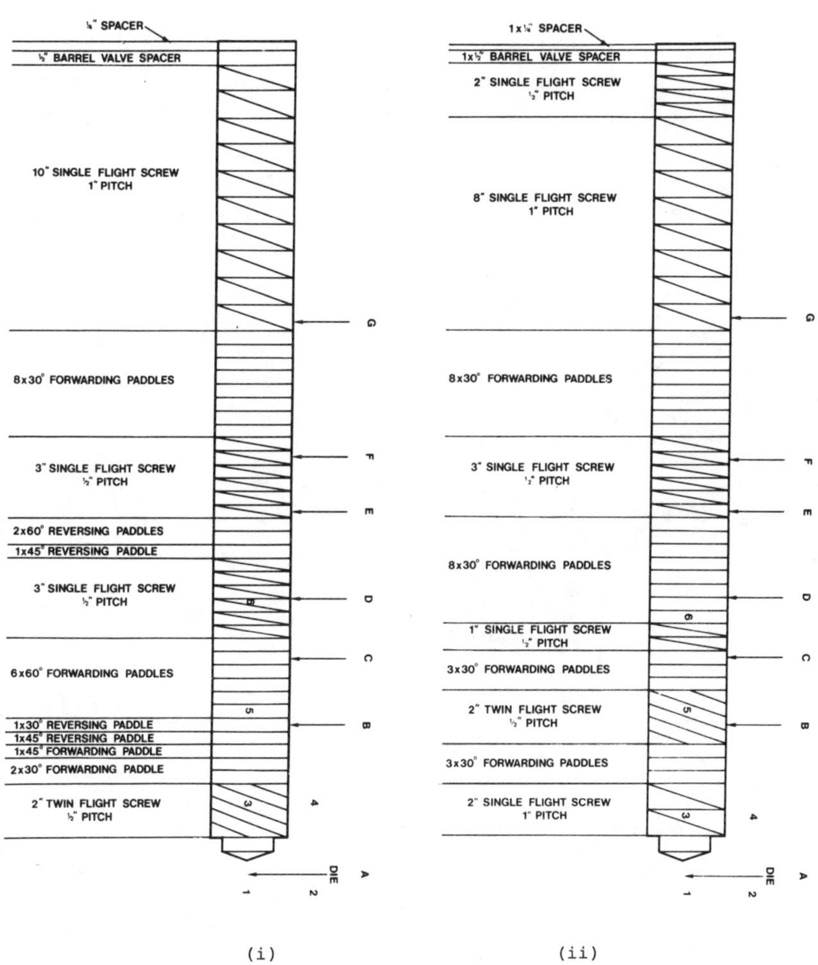

Fig. 1. Screw geometries used in the extrusion experiments. Pressure transducer points are lettered. Sampling points are numbered.

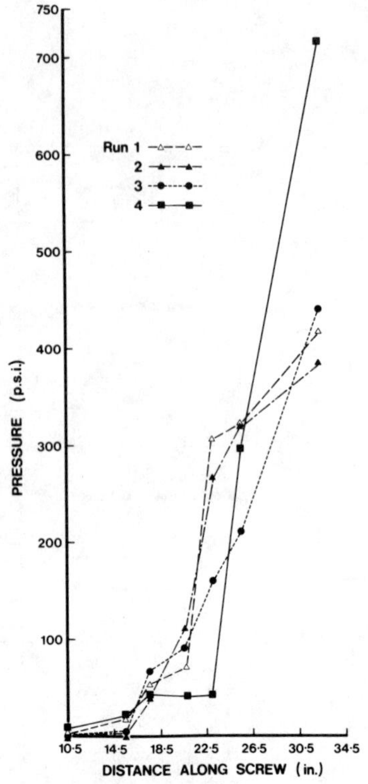

Fig. 2. Pressure in the extruder as a function of extruder length. Extrusion conditions are given in Table I.

The raw maize, barrel samples and the extrudate were milled in a centrifuge mill to particle sizes in the range 180-250 μ m. Rheological studies were carried out with a Couette viscometer (Contraves Rheomat 115). The milled material was added to distilled water to give a 9.1% by weight solution which was stirred continuously for 30 minutes. The solution was then sheared in the viscometer at ambient temperature, according to the following shear history: i) the rate was increased linearly to $148s^{-1}$ in 431 sec., ii) the rate was maintained constant at $148s^{-1}$ for 120 sec., iii) the rate was decreased linearly to zero in 431 sec. The solution was then heated and stirred for 15 minutes at 40^oC followed by repetition of the above shear procedure at ambient temperature. This was repeated for heat treatment at 60, 70 and 80^oC.

The viscosity as a function of shear rate for different barrel samples, from one extrusion condition, is shown in figs 3-6. The viscosity at the maximum shear rate ($148s^{-1}$) at ambient and 80^oC as a function of sampling point is shown in fig.7 for another extruder experiment. The extrudate viscosity from each extrusion condition is shown as a function of post-extrusion heating treatments in fig.8.

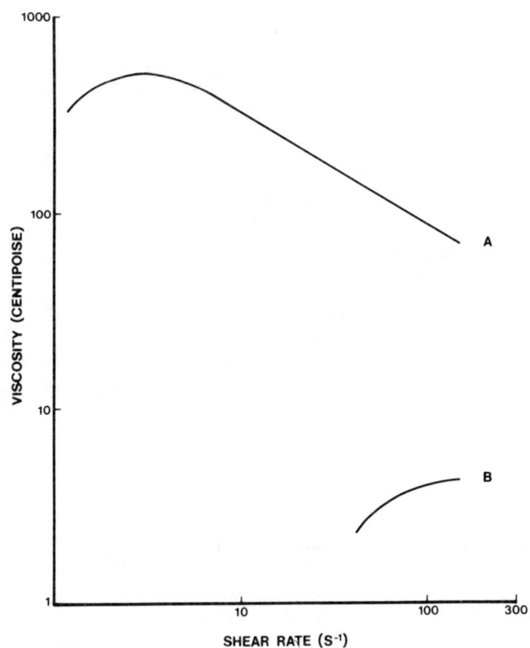

Fig. 3. Solution viscosity as a function of shear rate for raw maize at ambient temperature. A. Heated to 80oC; B. As made.

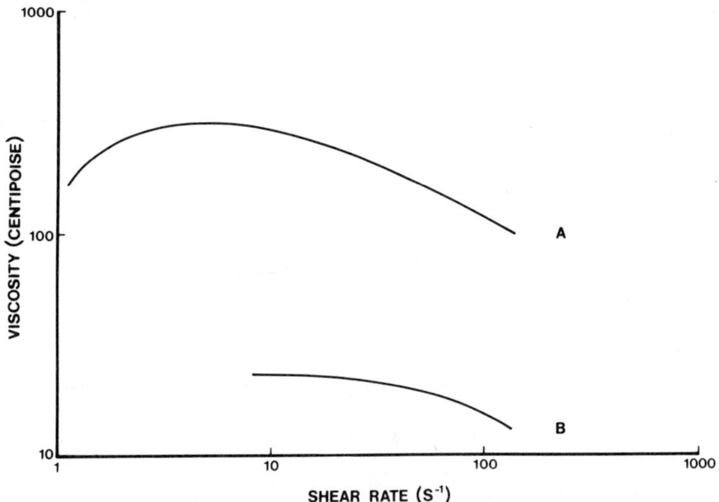

Fig. 4. As for Fig. 3 but maize from sample point 6 in experiment 3 (Fig. 1). A. Heated to 80°C; B. As made.

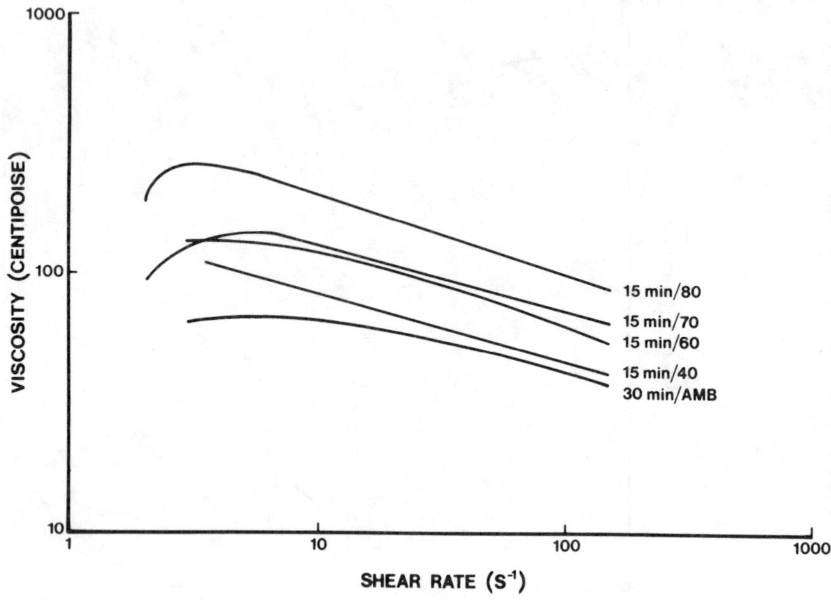

Fig. 5. As for Fig. 4 but maize from sample point 5 in experiment 3. Post-extrusion heating times and temperatures (°C) indicated. AMB = ambient.

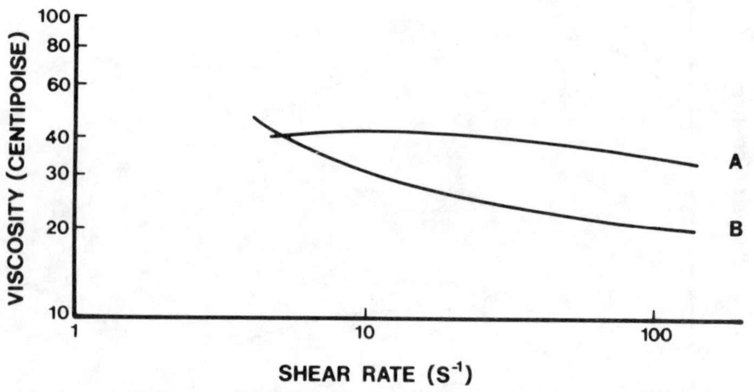

Fig. 6. As for Fig. 3 but maize extrudate in experiment 3. A. Heated to 80°C. B. As made.

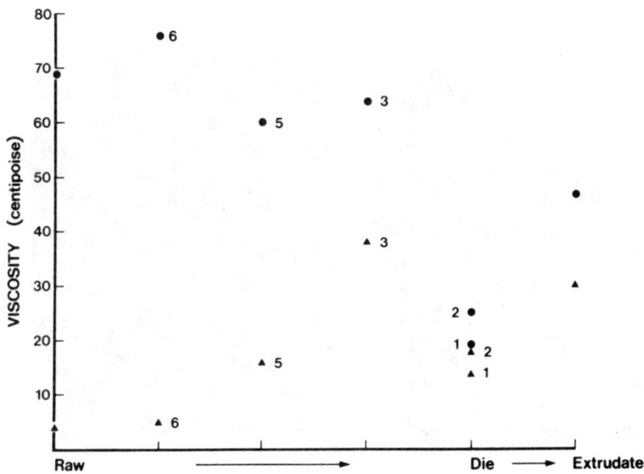

Fig. 7. Solution viscosity as a function of sampling point (Fig. 1) at ambient temperature from experiment 2 (Table I). Heated to 80°C, ●. As made, ▲.

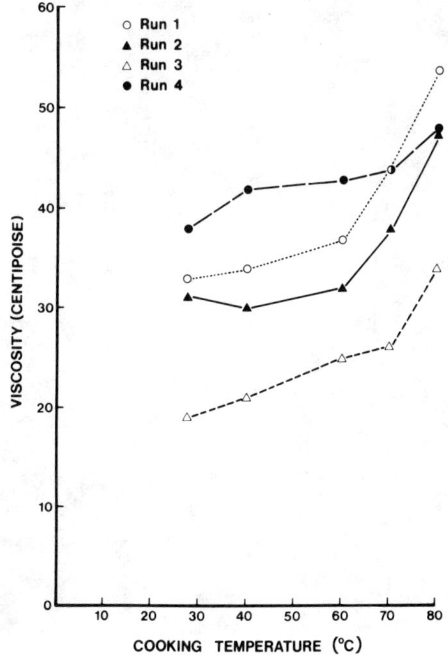

Fig. 8. Extrudate solution viscosity as a function of cooking temperature sequence (see Section 2) for each extrusion condition (Table I).

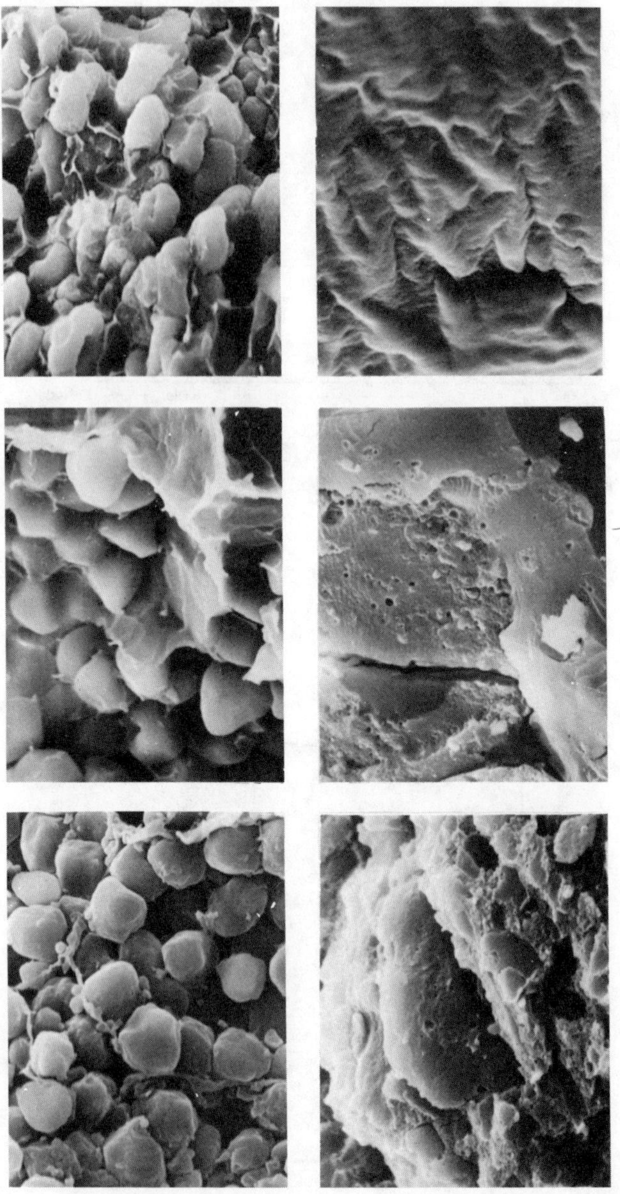

Fig. 9. SEM sequence for samples from experiment 2.

Samples from the extruder barrel were mounted with silver paint and coated with gold before examination by SEM using a Philips 501B at 30KV. A sequence of micrographs corresponding to position in the extruder is given in fig.9.

3. DISCUSSION

The dependence of the viscosity on the shear rate for raw maize solutions shows deviation from Newtonian behaviour. The as-made solution exhibits dilatant behaviour in common with the solution heated up to 80°C at shear rates less than $6s^{-1}$. The cooked solution is however shear thinning for shear rates greater than $6s^{-1}$ (fig. 3). The form of these results may be compared with data for maize starch solutions. The latter exhibit dilatant behaviour after negligible heat treatment and shear thin after heat treatment at 80°C (9).

The rheology of samples extracted from the barrel also shows a change from dilatant to shear thinning behaviour with increasing shear rate. The maximum in the viscosity shifts to lower shear rates with increasing heat treatment (figs 4-5). Extruded samples in solution are almost Newtonian, becoming slightly shear thinning with heat treatment (fig. 6). These data have been fitted to the equation:

$$\eta = K\dot{\gamma}^{\,n-1} \qquad . \ (1)$$

where applicable. The viscosity is η, the shear rate is $\dot{\gamma}$ and K, n are constants. The values of the power law index, n are given in table II. These data indicate a tendency to Newtonian viscosity with progress down the extruder and a wider applicability of eq (1) after post extrusion heating.

Table II Values of the power law index, n, for maize solutions.

Sample point (fig.1)	Highest cooking temperature (°C)	n	Shear rate range (s^{-1})
Raw Maize	80	0.43	> 5
6	Ambient	0.53	> 50
6	80	0.58	> 10
5	Ambient	0.75	> 50
5	80	0.69	> 6
Extruded	Ambient	0.87	> 50
Extruded	80	0.91	> 9

The transition from dilatant to shear thinning behaviour in the rheology of starch solutions was attributed to the rigid starch granule becoming swollen and elastically deformable (9). These arguments may be applied to maize solutions although sedimentation effects contribute in the uncooked raw maize solution. The shear thinning behaviour suggests that macromolecular entanglement and orientation occurs in raw and partially extruded maize solutions, compounded by starch granule swelling after post-extrusion heating. Fully extruded maize solutions are rheologically simpler indicating that molecular degradation (e.g. dextrinisation) may be occuring as suggested in the extrusion cooking of maize starch (8). (Heat treatment of the extrudate solution results in shear thinning behaviour which suggests that there may be some swelling of residual ungelatinised starch).

The SEM evidence to some extent supports these observations. The sequence of micrographs (fig.9) shows the progressive distortion of the starch granules, their eventual disappearance and the formation of a textured structure at the die. They resemble those of gelatinised and dextrinised maize as reported by Gomez and Aguilera (10). The break-up of the starch granules along the extruder is consistent with a reduction in the degree of starch swelling in the dilute solutions of this study. The contribution of the swollen granule deformation to the shear thinning behaviour of the dilute solutions is therefore reduced. The viscosity of solutions containing starch granules increases markedly on heating to 80°C corresponding to gelatinisation. The viscosity of the extrudate solution also increases on heating although this may not be uniquely attributed to the swelling of starch particularly since discrete granules are no longer evident in the micrographs of the extruder die material. The heating of dilute solutions constitutes a very different environment to that in the extruder and therefore reaction and molecular association involving the cereal components may occur when the extruded material is heated in solution.

The dilute solution viscosity at a given shear rate varies non - monotonically with distance along the extruder (fig 7). The variation in the unheated viscosity is complicated by sedimentation effects at the first sampling points. Thereafter the viscosity reaches a maximum and falls with further extrusion distance. The cooked solution viscosity shows a maximum after some compaction of the maize in the extruder. The viscosity then falls with increasing distance as a result of decreasing starch gelatinisation and increasing dextrinisation by the extruder. There are evidently further contributions to the observed viscosity behaviour. Colonna et al (8) reported intrinsic viscosity measurements for maize starch which varied non-linearly along the extruder barrel. They also mentioned that the intrinsic viscosity varied radially with sampling point. The present experiments also reveal different viscosity values for material extracted near the extruder barrel and that from the root of the screw in the die and metering zone regions. (figs 1, 7).

The viscosity of the extrudate solutions and its variation with heating differs for each of the extrusion conditions (fig 8). In particular the screw configuration affects the product viscosity and its variation with post-extrusion heating. The influence of screw geometry on the Brabender viscosity of maize was also emphasised by Seiler et al (6).

Paton and Spratt (7) suggested that viscosity profiles of cooked cereal products could be used to indicate their "degree of cooking". These experiments clearly demonstrate that the dilute solution viscosity is a sensitive indicator of the effect of different extrusion conditions on the product properties.

REFERENCES

1. COLONNA, P. and MERCIER, C. (1983) Macromolecular Modifications of Manioc Starch Components by Extrusion - Cooking with and without Lipids. Carbohydrate Polymers 3:87
2. LINKO, P. (1982) HTST-Extruder as a Tool for Bioconversions. Koch und Extrudier Techniken, ZDS, Solingen Oct.11-13.
3. ANTILA, J., SEILER, K., LINKO, P. and SEIBEL, W. (1983) Production of Flat Bread by Extrusion Cooking using Different Wheat/Rye Ratios, Protein Enrichment and Grain with Poor Baking Ability. J. Food. Eng. 2:189.
4. VAN ZUILICHEM, D.J. and STOLP, W. (1976) Theoretical Aspects of the Extrusion of Starch Based Products in Direct Extrusion Cooking Process. International Snack Seminar. ZDS, Solingen. Oct. 18-21
5. FLETCHER, S.I., MCMASTER, T.J., RICHMOND, P. and SMITH, A.C. (1983) Physics and Extrusion of Soft Solid Foodstuffs Proceedings of the Third International Congress on Engineering and Food, Dublin, Ireland. To be published by Applied Science, London.
6. SEILER, K., WEIPERT, D. and SEIBEL, W. (1980) Viscosity Behaviour of Ground Extrusion Products in Relation to Different Parameters. Food Process Engineering Vol. 1. Food Processing Systems ed. Linko, P., Malkki, Y., Olkku, J. and Larinkari, J. Applied Science, London.
7. PATON, D. and SPRATT, W.A. (1981) Simulated Approach to the Estimation of Degree of Cooking of an Extruded Cereal Product Cereal Chemistry 58:216
8. COLONNA, P., MELCION, J-P., VERGNES, B. and MERCIER, C. (1983) Flow, Mixing and Residence Time Distribution of Maize Starch within a Twin-Screw Extruder with a Longitudinally-Split Barrel. J. Cereal Science 1:115
9. CHRISTIANSON, D.D. and BAGLEY, E.B. (1983) Apparent Viscosity of Dispersions of Swollen Cornstarch Granules. Cereal Chemistry 60:116
10. GOMEZ, M.H. and AGUILERA, J.M. (1983) Changes in the Starch Fraction During Extrusion-Cooking of Corn. J. Food Science 48:378

ACKNOWLEDGEMENTS

The authors would like to thank G. Gooday and T. Hurn for their technical assistance. They are grateful to R. Turner for the scanning electron micrographs.

A COMPARISON OF PAH LEVELS IN EXTRUDED FOODS WITH THOSE IN CONVENTIONALLY COOKED FOODS

G C CRIPPS

Polytechnic of the South Bank, Department of Applied Biology and Food Science, Borough Road, London, SE1 0AA, England

and

M J DENNIS

MAFF Food Science Laboratory, Haldin House, Queen Street, Norwich, England

INTRODUCTION

Polycyclic aromatic hydrocarbons (PAH) are a group of contaminants which occur widely but at very low levels in foods. Methods of processing and cooking can influence the level of PAH's in the foodstuff. The chemical changes occurring in foods during extrusion cooking and which affect PAH concentration are not well understood and in this preliminary study we have investigated PAH levels in commercially available extruded foods, conventionally cooked foods and laboratory products.

THE PAH LEVELS OF SOME COMMERCIAL CEREAL PRODUCTS

Figure 1 indicates the generally low levels of PAH in these products. The extruded foods appear to have higher levels of PAH's than the conventionally baked bread. However, the levels of PAH's present are broadly similar to those found in a total diet survey of British foods. (M J Dennis, R C Massey, D J McWeeny, M E Knowles and D. Watson. Fd. Chem.Toxic. (1984) in press). eg. Benz(a)pyrene averaged 0.32 μg/kg with a range of 0.12 - 0.79 μg/kg for a cereals composite and averaged 0.17 μg/kg for the total diet.

THE PAH LEVELS OF A DOUGH SAMPLE BEFORE COOKING AND AFTER EXTRUSION AND CONVENTIONAL BAKING

The data in Figure 2 shows that the cooking procedure produces only marginal increases in the PAH levels of the final product. Here, the extruded product has lower PAH levels than the baked bread.

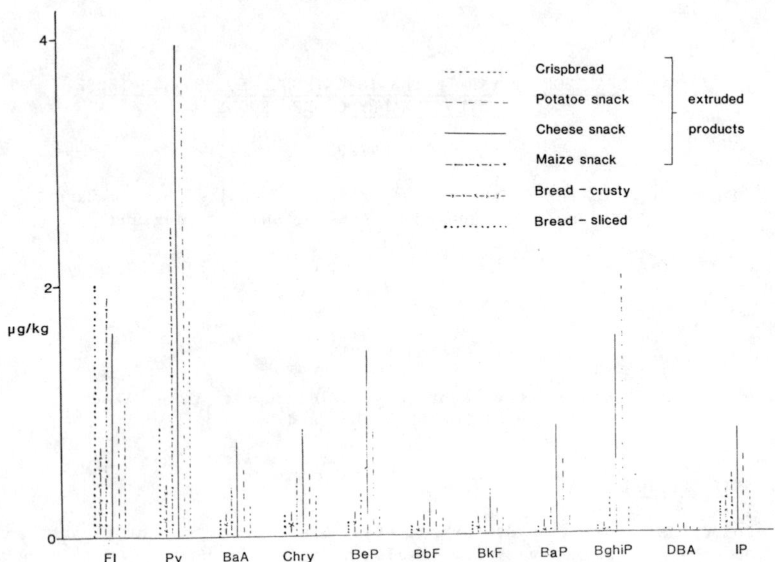

Figure 1. The PAH Levels of Some Commercial Cereal Products

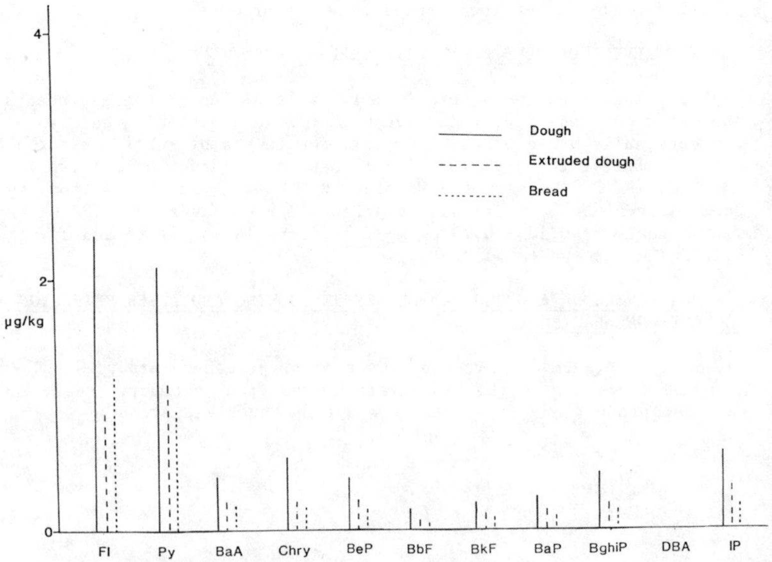

Figure 2. The PAH Levels of a Dough Sample Before Cooking and After Extrusion and Conventional Baking

CONCLUSION

The data on PAH levels in bread and in extrusion-cooked dough indicates that any increase during processing is generally smaller with extrusion cooking than with conventional baking. The slightly higher levels found in commercial extruded products may be due either to higher levels in the starting ingredients or to PAH production during post-extrusion cooking; further studies to elucidate this point are in progress.

ABBREVIATIONS USED IN FIGURES

PAH	Polycyclic Aromatic Hydrocarbons	BbF	Benzo(b)fluoranthene
Fl	Fluoranthene	BkF	Benzo(k)fluoranthene
Py	Pyrene	BaP	Benzo(a)pyrene
BaA	Benz(a)anthracene	BghiP	Benzo(g,h,i)perylene
Chry	Chrysene	DBA	Dibenz(a,h)anthracene
Bep	Benzo(e)pyrene	IP	Indeno (1,2,3-c,d)pyrene

ENERGY ASPECTS IN EXTRUSION COOKING OF STARCHES AND FLOURS

I. BEN-GERA, O.B. SMITH and G.J. ROKEY
Wenger International, Inc. and Wenger Manufacturing, Inc.

Summary

The extrusion cookers discussed here convert electrical energy supplied to their main drive motor into sensible heat. This is achieved through controlled friction in the extruder barrel and is influenced by the design of the barrel components as well as the characteristics and composition of the processed materials. Cooking starches and cereal or tuber flours in this way is efficient and energy saving. Scaling up of grain or flour extrusion cooking operations results in a marked decrease in electrical power requirements. A system with production capacity of 250 kg/h requires 0.114 kW/kg while a system with 1250 kg/h production rate requires only 0.058 kW/kg. A further capacity increase, up to 9000 kg/h results in electrical power requirement of only 0.028 kW/kg. Depending on the desired properties of the finished product, steam may be added to the product at a level of up to 10 percent. Steam requirements for heating of the extruder barrel are negligible.

1. INTRODUCTION

Starches of a wide range of different properties are available today to the food, feed, paper, textile, oil, pharmaceutical and other industries. Most of these products are produced by conventional starch gelatinization and modification technology. This conventional technology, with drum dryers and reaction vessels is being challenged now by modern extrusion cooking systems. Applying the technology of extrusion cooking to the needs of the different industries which use gelatinized and modified starches has not been an easy task. Although certainly an area of great interest, motivated by several reasons, one of which is the desire for lower production costs, considerable research and development work had to be done, and was done, over a period of some 30 years now. This is not surprising, in view of the fundamental differences between extrusion cooking and the conventional and older technology, and in view of the need if not always than at least in many cases, to duplicate existing products, with which the end users are already familiar, and do accept or even require their exact functional properties. Cereal flours are precooked, as single ingredients or as components in formulated food and feed products.

When starch and water are worked together in the extruder to form starch doughs, these doughs are extensible and expand easily. The extrudate of such cooked dough sets up quickly after extrusion and shows a characteristic cellular structure.

Starch or starch containing flours can be gelatinized during an extrusion cooking process. They can be extruded as thick boiling cooked at high moisture contents and at low extrusion temperatures, or as a thin boiling at low moisture contents and at relatively high extrusion temperatures.

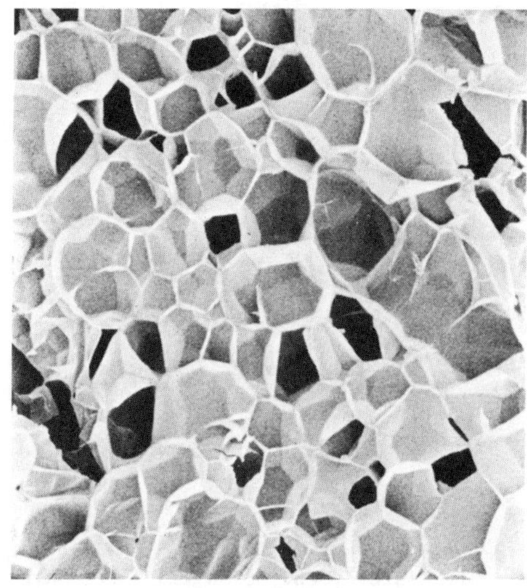

Figure 1 - Cellular Structure of Extruded Starch

2. THIN AND THICK BOILING STARCHES

Raw native starch does not dissolve in cold water and takes up water very slowly. Pregelatinized starch does dissolve in cold water. It has attained its water absorption, water holding capacity, viscosity characteristics in cold water due to the fact that it had been pregelatinized.

It is possible to produce gelatinized starches through extrusion cooking. These products can be tailor made and once rehydrated, to exhibit their maximum viscosity in different temperatures as required. The difference between the processing conditions of these different starches relate mainly to their moisture content during extrusion cooking (Figure 2).

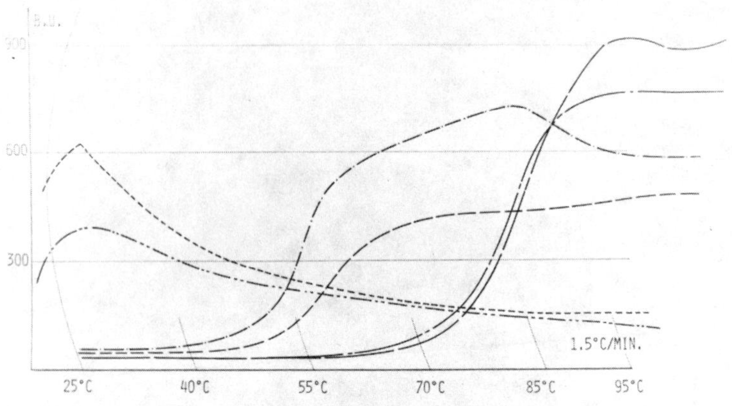

Figure 2 - Effect of Moisture Content on Viscosity Characteristics of Corn Starch

Figure 2 shows the pasting characteristics of corn starch, pregelatinized through extrusion cooking, on a medium shear extruder configuration.

It is important to note that from an economial point of view, it would be not only advantageous, but probably crucial to obtain higher levels of viscosity throughout the temperature range shown in this slide. It is also important to bring moisture levels down from 50 percent for the thick boiling starch. Such reduction will make economics of the extrusion process more attractive.

3. EXTRUSION COOKING OF HIGH VISCOSITY, THICK AND THIN BOILING STARCHES

The new process for the gelatinization of starches or cereal and tuber flours which is discussed here, has a unique combination of advantages over other methods of pregelatinizing starchy raw materials. This new process is a double extrusion process, in which the starch is moistened and gelatinized in an extrusion cooker, after which it is conveyed into a secondary extruder. This second extruder is a cooling and forming extruder (Figures 3 and 4).

Figure 3 - Photograph of Wenger F-25/20 Starch Cooker Including Cooking Extruder and Cooling and Forming Extruder

Figure 4 - Photograph of X-255 Extrusion Cooker/Cooling and Forming Extruder - Capacity to 2,200 Kilograms per Hour

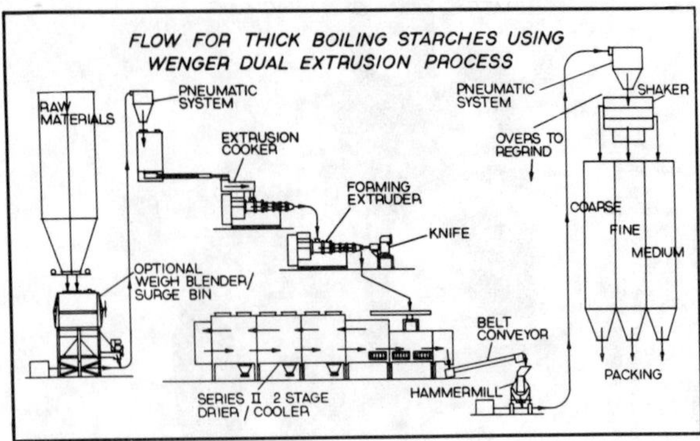

Figure 5 - Flow Sheet for the Extrusion Cooking of Thick Boiling Starches

The extrudate is cut at the die of the second unit into thin sheets. Thereafter the extrudate is dried and cooled. After drying and cooling, the product is put through suitable size reduction equipment to reduce particle size as desired (Figure 5). By the proper choice of processing temperatures, moisture contents, pressures, feed rates, barrel assembly length and choice of components like flow constrictors, screws and sleeves, die configurations and other factors, pregelatinized starches can be produced in which viscosity characteristics, solubility rate of rehydration, thickening power, gel strength, and gel setting times are tailored.

The new process has a unique combination of advantages over other methods of pregelatinizing starches which will be discussed elsewhere.

Figure 6 compares a rate of amylose leaching which is an indicator of shear level during processing, in drum dried drilling mud starch, drum dried food thickening agent, and extrusion cooked food thickening agent.

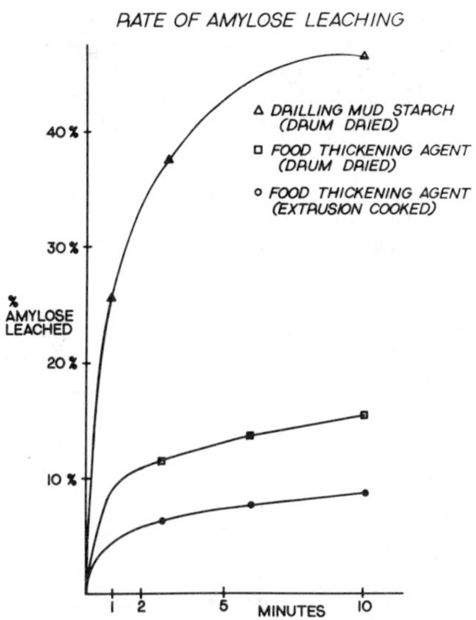

Figure 6 - Rate of Amylose Leaching, Drum Dried and Extrusion Cooked Starches

4. DRUM DRIED PREGELATINIZED STARCHES AS COMPARED TO EXTRUSION COOKED PREGELATINIZED STARCHES

A drum dryer is the most widely used equipment to manufacture pregelatinized starches and flours, especially those processed for high paste viscosity. A drum dryer operates at low speeds and imparts low shearing forces to a starch slurry during cooking. Upon rehydration, the starch granules imbile water and swell and produce high viscosities. However, cooking starch on a drum dryer has many limitations. The production rate is low, maintenance costs and operating costs are high, and the process is difficult to control. The drum dryer is inefficient in transferring heat to cook and dry the starch. Drum dryers require much higher moisture contents (60-70% moisture) than do extrusion cookers (15-30%). Energy costs will continue to be of a major concern and an important factor in the overall processing costs of pregelatinized starch produced on a drum dryer.

Extrusion processing is advantageous over drum drying because extruders require less energy and lower equipment and lower labor costs per unit of dry product; the process is easier to control; it produces at much higher production rates and is much less costly to maintain than are drum dryers.

Total energy used per kilo of product in the process which is presented here is much lower than the energy used per kilogram of same product when produced by a drum dryer.

Tables I, II and III, show the energy requirements of drum dried starch as contrasted to energy requirements for the new Wenger process.

TABLE I
EXTRUSION SYSTEM AND DRUM DRYER
ENERGY CONSUMPTION PER KG OF FINISHED PRODUCT

PROCESS	CAPACITY at 12% MCWB KG/H	ELECTRICITY KWH/KG	GAS M3/KG	STEAM KG/KG
EXTRUSION	180	0.378	0.02	0.05
EXTRUSION	230	0.304	0.02	0.05
EXTRUSION	350	0.267	0.02	0.05
DRUM DRYER	350	0.051	-	1.33
EXTRUSION	2200	0.088	0.02	0.05

As one can see from Table I, increasing the throughput or utilizing extrusion systems of higher capacity will result in lower energy requirements per kilogram of throughput.

The energy consumption values for extruder throughputs of 180, 230, 250, 2200 kilograms per hour are based on Wenger model numbers as shown in this table, on the right.

The 350 kilograms per hour rate makes a practical direct comparison of utility requirements for Wenger systems and drum dryers of 350 kilograms per hour capacity.

Table II converts energy requirements to actual costs of electricity, gas and steam. Utility costs will of course vary from one location to another but since the utility costs are shown at the bottom of this table, it will be a simple matter to calculate utility costs, based on your local costs for these utilities.

TABLE II
EXTRUSION SYSTEM AND DRUM DRYER
ENERGY COST PER KG OF FINISHED PRODUCT

PROCESS	CAPACITY KG/H	ELECTRICITY $/KG	GAS $/KG	STEAM[1] $/KG	TOTAL WITHOUT STEAM $/KG	TOTAL WITH STEAM $/KG
EXTRUSION	180	0.026	0.003	0.001	0.029	0.030
EXTRUSION	230	0.021	0.003	0.001	0.024	0.025
EXTRUSION	350	0.019	0.003	0.001	0.022	0.023
DRUM DRYER	350	0.004	-	0.027	0.031	0.031
EXTRUSION	2200	0.006	0.003	0.001	0.009	0.010

NOTE [1]: MANY EXTRUDED STARCH PRODUCTS DO NOT REQUIRE STEAM.
ASSUMPTIONS: ELECTRICITY = $0.07/KWH
GAS = $0.15/M3 (9000 KCAL/M3)
STEAM = $0.02/KG (550 KCAL/KG)

Table III shows the total energy consumption per kilogram of finished product on extrusion cooking system of various capacities as well as for the drum dryer at capacity of 350 kilograms per hour.

TABLE III
EXTRUSION SYSTEM AND DRUM DRYER
TOTAL ENERGY CONSUMPTION (1) PER KG OF FINISHED PRODUCT

PROCESS	CAPACITY KG/H	TOTAL ENERGY WITHOUT STEAM KCAL/KG	TOTAL ENERGY WITH STEAM KCAL/KG
EXTRUSION	180	500	531
EXTRUSION	350	406	435
DRUM DRYER	350	781	781
EXTRUSION	2200	254	282

NOTE: (1) ACCUMULATED TOTAL ENERGY ADDITIONS MADE FROM TABLE I

TUNING CONTROL OF FOOD-EXTRUSION BY MANIPULATION WITHIN THE DIE-HEAD DURING PROCESSING

J. McN. DALGLEISH
Applied Biology & Food Science Dept.,
Polytechnic of the South Bank, London

Summary

A comparison procedure is outlined for checking that the configuration within the die-head is uniformly suitable for conveying a range of molten foodstuffs having formulae commonly used in the process of extrusion-cooking. The method is extended to a tunable device.

1. INTRODUCTION

In an extrusion-cooker, operational settings in the barrel/screw part can be manipulated for a variety of mixes to give a fluid compound of desired characteristics. The die(s) can be chosen to give the product characteristics required. The part in between, the die-head, is normally of fixed dimensions, and has to suit whatever is happening before and after it.

Preferably this part should not upset what is intended, but the user, as a rule, is presented with 'fait accompli',since the designer of the die-head has had to generalise this 'joining part' to match a future selection of 'end' particular to an unknown mix or product. It has had to be treated as a rectilinear pipe with no wall rugosity, for a supposedly incompressible fluid on permanent flow.

On this basis, for all shapes of pipe-like section, from Bernouilli and Poiseuille, we would obtain $\left(\frac{P_1 - P_2}{\rho}\right) = \frac{N \mu \bar{v} \ell}{\rho m^2}$ with laminar flow, where the symbols have their usual meaning; m being the hydraulic mean depth = a/w = sectional area / wetted perimeter.

From $\tau w \ell = (P_1 - P_2) a$ or viscous stress at boundaries $\tau = (P_1 - P_2)\frac{m}{\ell}$ ......(1)

we may then write, non-dimensionally

$$\frac{\tau}{\rho \bar{v}^2} \cdot \frac{\rho \bar{v} m}{\mu} = N \quad \text{......... a constant..................(2)}$$

m is particular to the configuration of the channel, so we may rearrange to obtain

$$\frac{\tau}{\bar{v}\mu} = \frac{N}{m} \quad \text{for any given part of the 'pipe' section.}$$

The left-hand-side of this is based on the material flowing, while the right-hand-side is based on the construction of the channel. The pragmatic step is to assume this applies to the contours of the die-head.

If we want to avoid any integral changes in the shape of the die-head from influencing unduly the flow characteristics of the various materials which may be selected to pass through the die-head, then N/m should remain substantially the same for each successive portion of the particular die-head.

Normally, but not necessarily, this contrived sequence of the design, once constructed, will remain fixed.

2. APPLICATION

For most doughs which exhibit a pseudoplastic behaviour, the decreasing viscosity with increasing shear rate will then produce a mean velocity at each section which will have reacted to compensate, as in equation (2). The duration of the effect will be noted in equation (1). Between the end of the screw(s) and the actual die(s) this simplistic approach may then form a guide to design of the various pathways within the die-head.

Transitions between the prescribed shapes should of course be gradual.

Boussinesq (1868) reported by Davies & White[1] was the first to find values of 2N for various Depth/Breadth ratios of channel thus:-

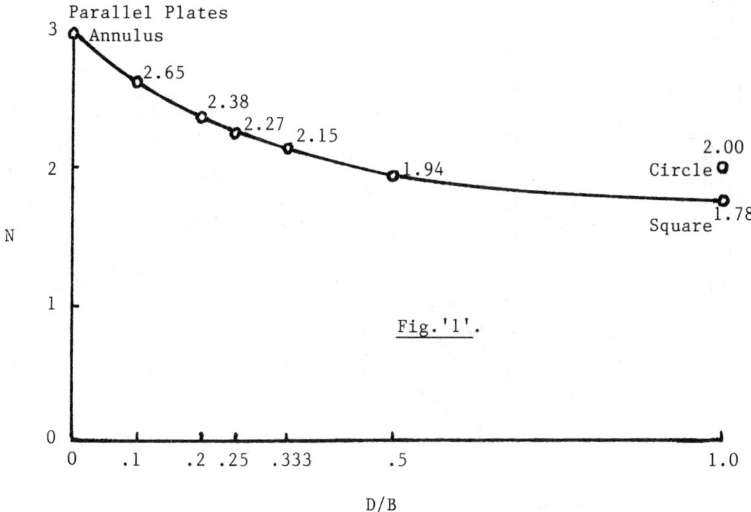

Fig.'1'.

Testing this in a particular die-head which changes shape from a circle at the end of the screw to three "flat-bread" die-slots, Fig.'2', produced the relationships shown in Fig.'3':-

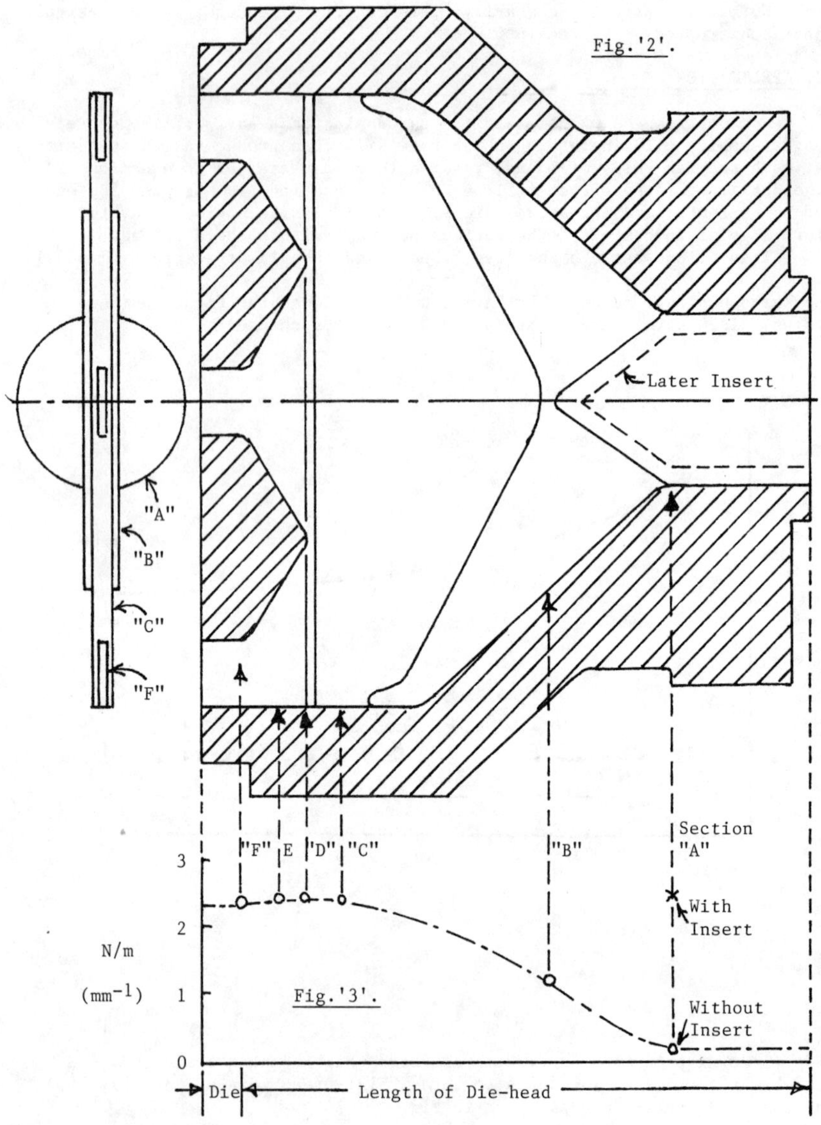

Provision of a tapered-nose 'piece', inserted to form an annulus as indicated, brought the relationship at "A" approximately to the values obtained for sections "C" through "F".

It is not uncommon practice in plastics extruders to keep them filled, moving out the old plastic material only by pumping in the new. This is not hygienically acceptable with foodstuffs. Barrels, screws, die-head and die(s) must be easily and quickly dismountable for thorough cleaning.

For the same reason there should be no "dead-spots" in the design which permit foodstuffs to lurk, nor become subject to overprocessing for long periods. Grooves in the barrel are disadvantageous in this respect, unless perhaps rifled contrary to the screw spiral. The D/B ratio of the screw spiral groove should not be too low and this can be improved with a two-start thread.

For a single screw with single die it is usual to taper the screw-shaft end and match the die-head, to form an annulus leading to the die. For multi-dies on a pitch circle radius around the screw axis, dead-spots may be avoided by the die-head construction illustrated in Fig.'4'.

Should the number of dies requiring to be accommodated be more than can be placed on the line of the screw diameter, then the filler pieces should be tapered in the appropriate direction.

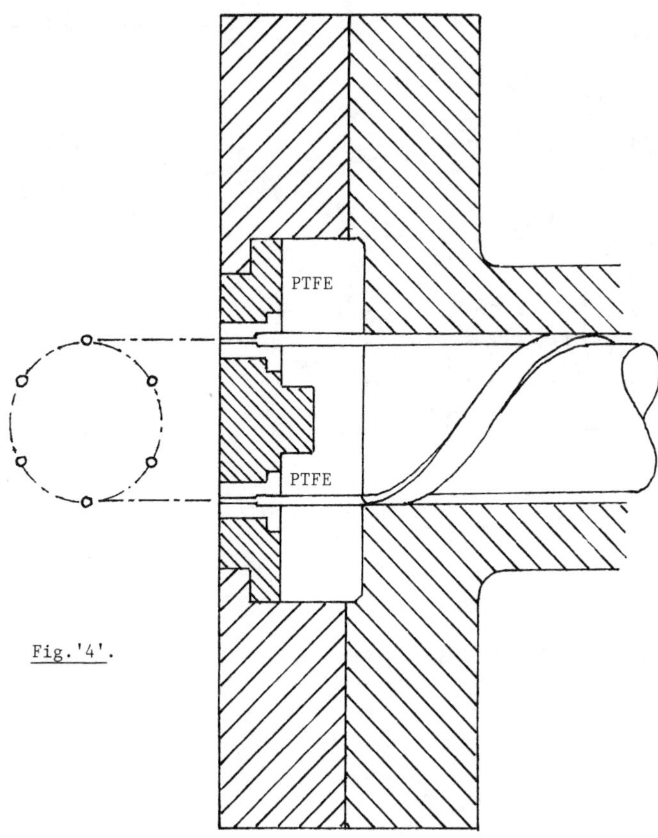

Fig.'4'.

It is not always practicable to cater for extreme variations, in the nature of the "food-melts" which may be formulated, by means of a particular fixed-section die-head, although in other respects most machines may be looked on as being "universal". The route through the die-head has to be suitable for the particular "melt" and extrusion.

A right-angle die-head, originally used for the plastics coating of metal wires, has been adapted in an experimental food-processing rig, both for examining the condition of the "melt" and for adjusting the configuration of the die-head internal section during processing.

The adaptation is fashioned in the style of Fig.'5'.

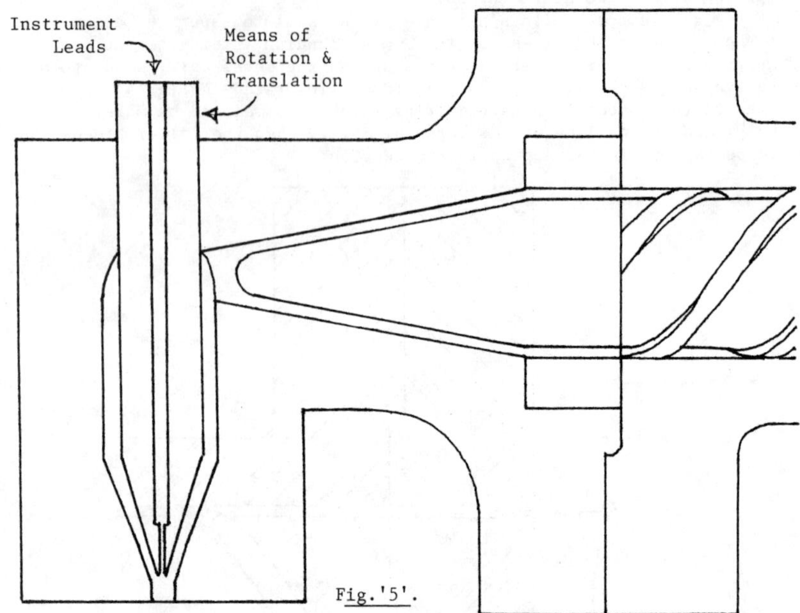

Fig.'5'.

This permits fine adjustment on feed-back of information to control the characteristics of a particular melt in its approach to the die.

REFERENCE

1. DAVIES, S.J. and WHITE, C.M. (1929). Review of Flow in Pipes & Channels. Engineering. 19th & 26th July, and 2nd August.

EFFECT OF SOME EXTRUSION PARAMETERS ON THE SOLUBILITY AND VISCOGRAMS OF EXTRUDED WHEAT FLOUR

J.C. Kim
Institute for Cereals, Flour and Bread TNO
Wageningen, The Netherlands.

1. INTRODUCTION

The high temperature/short time extruders (HT/ST extruders) are used for the production of snacks, breakfast cereals, flat breads and binders for food and other industries. It has also been found that by extruding wheat or rye flour with a HT/ST extruder one can obtain products with which production process of bakery products and other foodstufs can be simplified or economized. The extrusion modified wheat flour improves also quality of sponge cake [1]) and retarts the staling of bread.

The quality of the final product and the suitability of the extruded flour for the above mentioned applications depend on the kind and the extend of changes in the flour, especially in the starch. The conversion of raw starch to a cooked and digestable material by the application of heat and moisture is called gelatinization. In the conventional way of gelatinization, water is absorbed and bound to the starch molecules with a resulting change in structure. Under extrusion conditions (with HT/ST extruder) which provide high temperature, shear stress, and relatively low amount of moisture available for the starch, one can also expect a breakdown of the starch molecules to dextrins. A recent publication of Paton[2]), in which he mentioned differences in pasting charcarteristics of wheat starch pregelatinized by drum-drying and by cooking extrusion, confirms the idea on the effect of extrusion on starch.

There are several ways to assess the effect of cooking-extrusion on wheat and other starch rich materials. It is found, however, that the solubility in water and the pasting characteristics by making viscograms, give usefull information on the structure and application potentials of extruded wheat flour[3]).

In this paper the relationship between the solubility and the pasting characteristics (viscogram) of extruded wheat flour prepared by the Creusot-Loire twin screw extruder BC-45, and the influence of the moisture content of raw material and the screw configuration on this relationship is discussed.

2. METHODS

A Creusot-Loire co-rotating twin screw extruder type BC-45 was used, the screws having a constant root diameter and a decreasing pitch.

2.1 Extrusion conditions

Length of screw : 500 mm
Distance, between the axes of the two screws: 45 mm
Composition of screw elements : two different shear elements were used.

	direct pitch				reverse pith (shear element)
length (mm)	200	100	50	100	50
pitch (mm)					12
number of passes		–			3 x 6 mm (width)
in shear elements		–			3 x10 mm (width)

The shear element with 3 passes of 6 mm produces a higher shear than the other shear element.

Screw speed : 150 r.p.m
Number of dies : 2 circular dies
Diameter of die : 2 mm
Moisture content of raw material (wheat flour) (%) : 20, 25, 30 en 35 %
Extrusion temperature : ± 125 °C

2.2 Properties of wheat flour (raw material)

For the extrusion tests a commercial wheat flour for bakery purpose was used.

Analysis of the flour:

Total protein (%) (N x 5.7) : 12.80
Ash (%) : 0.53
Soluble carbohydrates (%, as glucose) : 4.80
Soluble proteins (%, N x 5.7) : 3.13

2.3 Assessment of extruded products

The following properties of extruded products are determined:

- content of water soluble carbohydrates
- viscosity characteristics by making viscogram of Brabender

2.3.1 Preparation of samples

The extruded products are dried at 50 °C and then ground, using a pin-mill and sieved. The flour fractions with particle size smaller than 150 μm were used for the assessment. For simplicity's sake, this flour is termed "extruded wheat flour".

2.3.2 Determination of water soluble carbohydrates

The method of determination is based on the A.O.C.S. Official Method Ba-11-65 for the determination of Nitrogen Solubility Index (NSI). The principle of our method is as follows:

Water soluble components of the sample were extracted at 30 °C. The extract was hydrolyzed by boiling in a reflux cooler for 1½ hour with 3 % hydrochloric acid. The reaction product was neutralized and filtered. The content of reducing sugars in the filtrate was determined. The content of soluble carbohydrates is expressed as per cent glucose.

2.3.3 Assessment of viscosity characteristics of extruded wheat flours

The viscosity characteristics of the extrudates are studied by using the Brabender viscograph. The detailed description of the method of assessment is given in an other publication of the author[4]).

3. RESULTS AND DISCUSSION

3.1 Influence of moisture content of the raw material on the content of soluble carbohydrates

Table 1 shows the effect of moisture content of the wheat flour (raw material) on the content of soluble carbohydrates of extruded wheat flour. In the extrusion process with the Creusot-Loire twin screw extruder the content of soluble carbohydrates of extrudates decreases with increasing moisture content of the raw material. (see table 1).

Table 1: Influence of moisture content of the wheat flour (raw material) on the content of soluble carbohydrates of extrudates.
Extrusion temperature: 125 °C
Die : 2 circular dies with diameter of 2 mm.

Moisture content of wheat flour (raw material) (%)	Soluble carbohydrates (%)	
	Shear element with 3 passes of 10 mm (los shear)	Shear element with 3 passes of 6 mm (high shear)
20	38.9	46.4
25	24.6	23.8
30	15.5	21.3
35	21.2	18.6

Mercier and Feillet[5]) performed extrusion studies on wheat starch and found also that soluble starch increases with decreasing moisture in raw material. When patato starch was extruded, however, the effect of moisture in the raw material on the content of soluble starch in extrudates was just the other way round[6]).

The relationship between moisture content and soluble carbohydrates depends on the level of shear in the barrel as shown in table 1. With the combination of 20 % moisture and the high shear screw element the highest soluble carbohydrates content was obtained. The reduction of this content by increase of the moisture content from 20 to 25 % is higher with the high shear screw element than with the low shear element; from 46.4 % to 23.8 % with the high shear and from 38.9 to 24.6 % with the low shear. When the low shear element was used, we found the lowest content of soluble carbohydrates at 30 % moisture (15.5 %), while in the extrusion with the high shear element the content of soluble carbohydrates decreases gradually with increasing moisture through the wole range.

I believe that the intensity of shear between starch granules during the extrusion process is responsible for the relationship between the moisture content and the content of soluble carbohydrates. The difference in the results of extrusion with the low and high shear elements indicates that there is more than one mechanism for the formation of soluble carbohydrates during extrusion process.

3.2 Influence of moisture content of the raw material on the pasting characteristics of extruded wheat flour

The influence of moisture content of the raw material (wheat flour) on the viscogram of extruded wheat is clealy seen in fig. 1 and 2. The analysis of the both figures shows also the effect of the level of shear in the barrel. Independent to the shear level, cold water viscosities (initial viscosity of viscogram) increase gradually with increasing moisture content of the raw material. The extruded wheat flour obtained from the raw material with 20 % moisture has not only the lowest cold paste viscosity, but also the viscogram curve differs from that of the other extruded wheat flours. The difference becomes stronger when wheat flour is extruded with the high shear element (fig. 2).

These results indicate that in cooking-extrusion of starches and starch rich materials, the starch is not only gelatinized by heat and moisture but also degradated to soluble dextrins by friction between starch granules. The relative importance of the two reactions depends on the extrusion parameters, especially shear in the barrel and moisture content of the feed product.

REFERENCES

1. J.C. Kim and W. Rottier, Modification of aestivum wheat semolina by extrusion. Cereal Foods World, 25 (2), 62, 1980.
2. D. Paton and W.A. Spatt, Simulated approach to the estimation of degree of cooking of an extruded cereal product. Cereal Chem. 58 (3), May/June 1981.
3. J.C. Kim, Veränderung von Weizen durch Extrusion. Int. Zeitschrift für Lebensmitteltechnologie und -Verfahrenstechnik, 33 (5), 334-343, (Aug.-Sep. 1982).
4. J.C. Kim, Influence of the type of extruder on the viscosity characteristics of extruded wheat flour. Proceedings of the Symposium "Koch- und Extrudiertechniken" organized by the Internationales Süsswaren Institut in Solingen, W.-Germany, 11-13 October 1982.
5. C. Mercier and P. Feillet, Modification of carbohydrates components by extrusion-cooking of cereal products. Cereal Chem., 52 (3), 283, 1975.
6. C. Mercier, Effect of extrusion-cooking on potato starch using a twin screw French extruder. Stärke, 29 (2), 48, 1977.

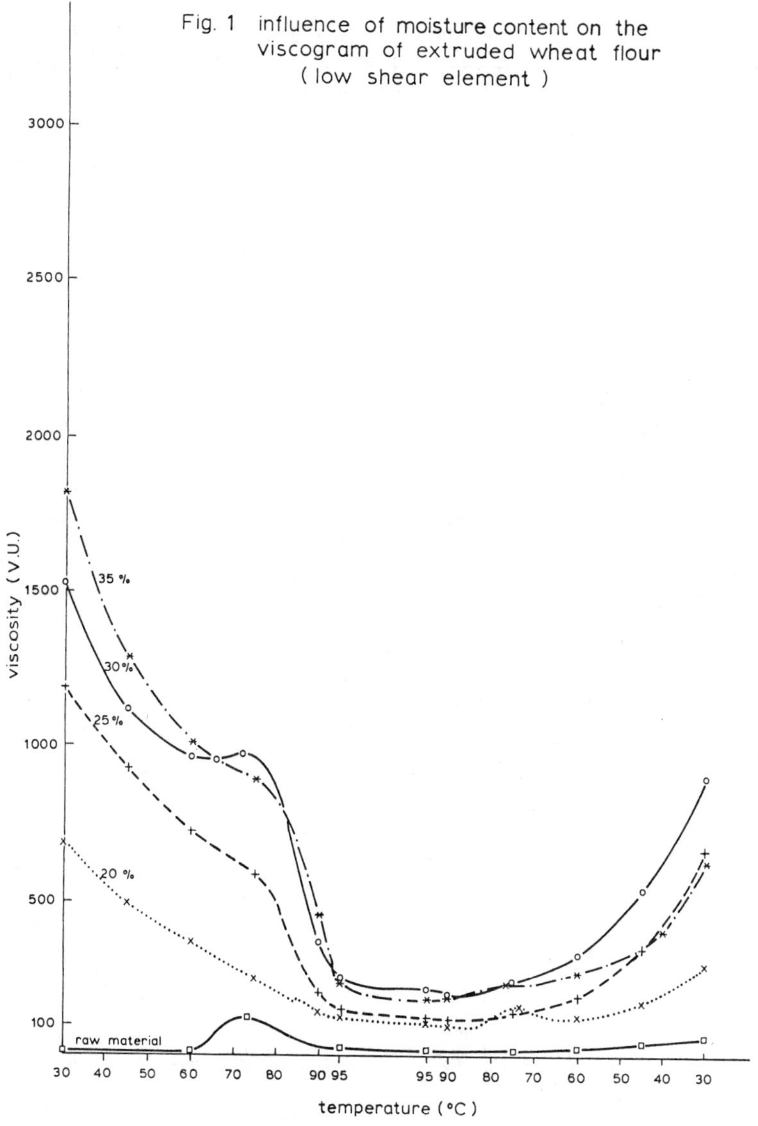
Fig. 1 influence of moisture content on the viscogram of extruded wheat flour (low shear element)
3000
2500
2000
1500
1000
500
100
viscosity (V.U.)
35 %
30 %
25 %
20 %
raw material
30 40 50 60 70 80 90 95 95 90 80 70 60 50 40 30
temperature (°C)

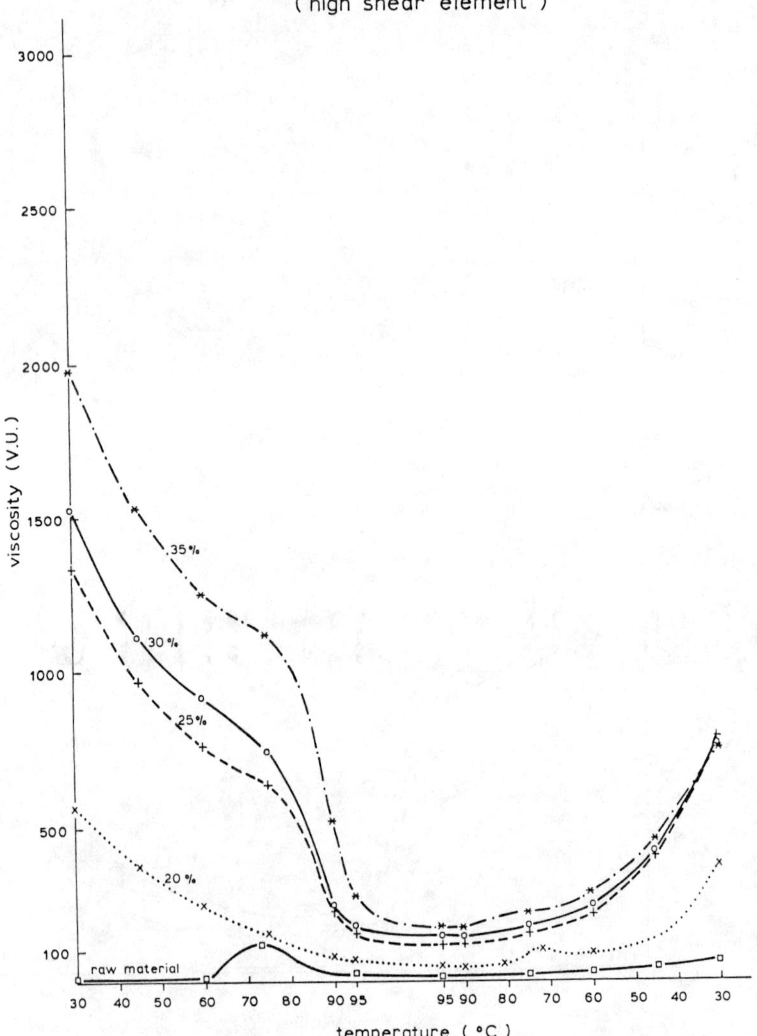
Fig. 2 influence of moisture content on the viscogram of extruded wheat flour (high shear element)
viscosity (V.U.)
3000
2500
2000
1500
1000
500
100
35 %
30 %
25 %
20 %
raw material
30 40 50 60 70 80 90 95 95 90 80 70 60 50 40 30
temperature (°C)

WHEAT STARCH EXTRUSION COOKING WITH ACID OR ALKALI

R. KERVINEN and P. LINKO,
Helsinki University of Technology, Department of Chemistry, Laboratory of Biochemistry and Food Technology, SF-02150 Espoo 15, Finland

T. SUORTTI and J. OLKKU*
Technical Research Centre of Finland,
Food Research Laboratory, SF-02150 Espoo 15, Finland

Summary

Wheat starch was extruded by a twin-screw Creusot-Loire BC 45 extrusion cooker, varying feed moisture, mass temperature, and phosphoric acid or sodium hydroxide contents. The effects of extrusion process parameters on product characteristics were determined by measuring water absorption index, water solubility index, consistency development of water suspended milled product, molecular weight distribution by HPLC and X-ray diffractograms. The results showed that an increase in phosphoric acid content and a decrease in feed moisture remarkably increased starch breakdown. The effect of alkali was normally less marked. Although the average molecular weight of starch decreased, only little small molecular breakdown products were formed.

1. INTRODUCTION

Extrusion cooker as a reactor for physical, chemical or enzymatic conversions of biopolymers (starch, cellulose, protein) offers numerous possibilities. Linko et al. (1) have described extrusion cooking comprehensively and Olkku (2) has discussed the basic phenomena involved. Linko (3) has investigated the enzymatic hydrolysis of cereal based material during and after twin-screw extrusion cooking, and Faubion et al. (4) the effect of pH on the extrusion of wheat starch. Lai and Sarkanen (5) have shown that in 0.01-0.1 N sodium hydroxide solutions at 100 °C amylose becomes totally degraded. Bazua et al. (6) have investigated extrusion cooking in the presence of 0,2 % calcium hydroxide as an alternative method of producing corn dough for tortillas. The aim of this study was to investigate the effects of acid or alkali treatment of wheat starch under various extrusion conditions on the structure and functional properties of the product.

2. MATERIALS AND METHODS

Wheat starch (Raisastarch, Raision Tehtaat Oy, Finland) was used as cereal raw material. The pH was regulated by orthophosphoric acid or sodium hydroxide.

The experiments were performed by a Creusot-Loire BC 45 twin-screw extruder varying feed moisture content (15-60 % of total feed), mass temperature (130-190°C) and phosphoric acid (0-1,5 % w/w of total liquid

* Present address: Orion Co Ltd, PL 8, SF-02101 Espoo 10, Finland

content) or sodium hydroxide (0-4 % w/w of total liquid content) contents, and keeping feed rate (300 g d.m./min), screw rotation speed (150 rev/min), and die diameter (5 mm) and length (27 mm) constant. The 500 mm screws were composed of four compression screw elements, and one reverse screw element before the die. The process conditions were chosen by the aid of statistical experimental design (7). The second-order equations were fitted by least squares to the primary results obtained, and regression equations were visualized by contour surfaces.

Water absorption index (WAI) and water solubility index (WSI) were determined by the method of Anderson et al. (8). The pasting properties of the milled products were investigated by the Brabender Viscograph. Before the measurement the pH of the starch-water suspension was adjusted to 6 either by phosphoric acid or sodium hydroxide. The suspension was heated to 95 °C at the rate of 1,5 °C/min, held constant for 5 minutes, and cooled to 25 °C at the same rate. Starch breakdown was followed by high performance gel permeation chromatography (HPLC) using M-6000A pump, M-710B automatic injector, μ-Bondagel E-1000, E-500 and E-125 columns in series and M-401 refractive index detector (Waters Associates Inc., Ma., USA). Mobile phase was dimethylsulfoxide, the flow rate 0,3 ml/min and the sensitivity of the refractive index detector 8x. The samples were dissolved in DMSO as 1 % (w/v) solutions, of which 20 μl injections were made. Dextran standards T 2000 ($\overline{M}_w$ 2000000), T 500 ($\overline{M}_w$ 500000), T 70 ($\overline{M}_w$ 70000), T 40 ($\overline{M}_w$ 40000) and T 10 ($\overline{M}_w$ 10000) from Pharmacia Fine Chemicals were employed in calibration. The effect of extrusion cooking on the organized structure of starch was investigated by Phillips PW 1051 X-ray diffractometer (CuK radiation with Ni filters, 36 kV, 16 mA, slits 1/0,2/1°, scanning speed 1° 2θ/min, time constant 3s, and range 3 x 10^2).

3. RESULTS AND DISCUSSION

3.1 Functional properties

The addition of acid or alkali markedly affected the performance of the extruder. An increase in feed moisture, acid content and mass temperature decreased the electrical current, indicating partial breakdown of starch structure and reduced mass viscosity. The addition of phosphoric acid decreased the consistency of water suspended milled product (Fig.1a).

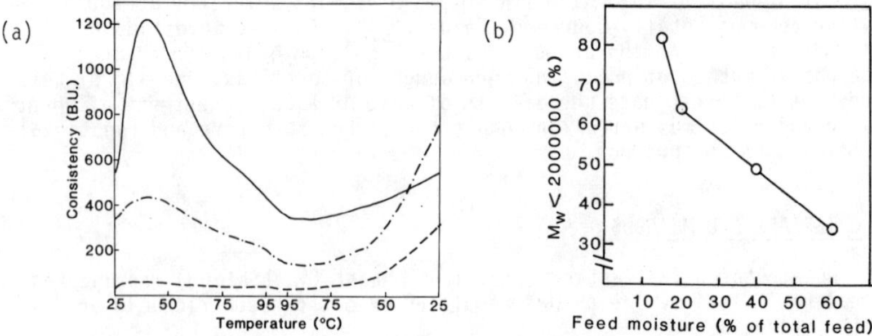

Fig. 1 (a) The influence of the addition of phosphoric acid during extrusion cooking on consistency development of water suspended milled extrudate. Extrusion conditions: feed moisture 31 %, mass temperature 178 °C, phosphoric acid content 0 % (——), 0,29 % w/w (-.-) or 1,21 % (---). (b) The effect of feed moisture on the quantity of starch molecules of M_w<2000000.

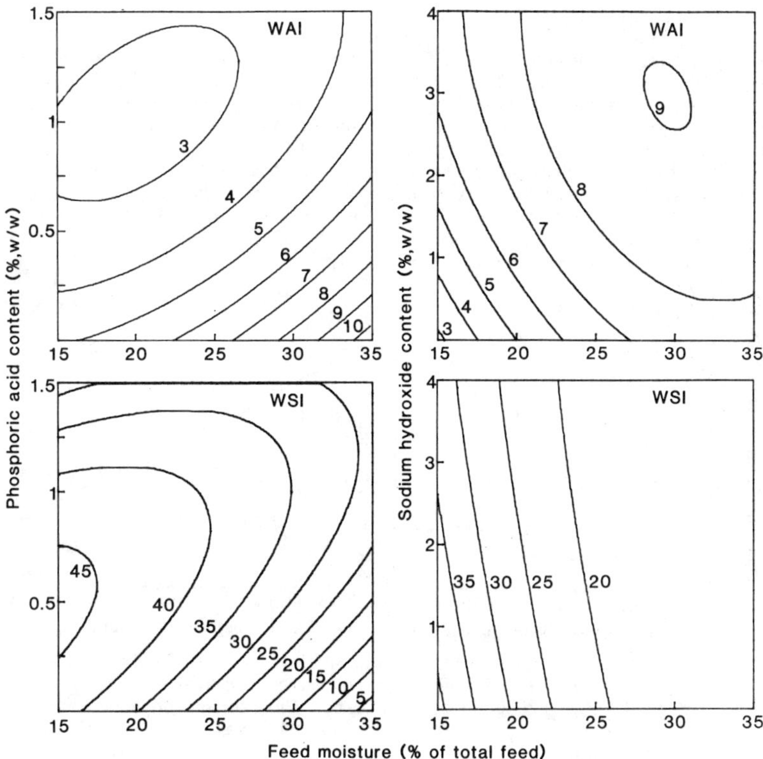

Fig. 2 The effect of feed moisture and acid (left) or alkali (right) content at mass temperature of 160 °C on WAI and WSI of extrudate. The regression equations for WAI were significant at 0.1 % level and for WSI at 1 % level, respectively. The multiple correlation coefficient r = 0.98 (upper left), r = 0.93 (upper right), r = 0.92 (lower left) and r = 0.91 (lower right).

Fig. 2 illustrates the effect of feed moisture and acid or alkali at constant mass temperature (160 °C) on water absorption index and water solubility index of the extrudate. Increasing feed moisture generally increased WAI and decreased WSI, except at high levels of phosphoric acid. An increase in phosphoric acid content generally decreased WAI and increased WSI, whereas an increase in sodium hydroxide content increased WAI and decreased WSI sligthly. The effect of mass temperature on WAI and WSI was less significant. The highest WSI of 54 % was obtained after extrusion at 15 % feed moisture, 190 °C mass temperature and 1,5 % phosphoric acid content. Also according to Faubion <u>et al.</u> (4) the solubility of wheat starch increased sharply at extrusion pH below 4.

3.2 Structural changes

Both acid and alkali treatment remarkably decreased the average molecular weight of starch. Fig. 3 illustrates the effect of feed moisture and acid or alkali content at constant mass temperature (190 °C) on molecular weight ratio M_w<2000000/M_w>2000000 which indicates the degree of starch breakdown. An increase in acid content and simultaneous decrease in feed moisture increased starch breakdown. At low moisture levels an increase in alkali decreased the molecular weight ratio, until at a certain alkali content a minimum was reached. Under the experimental conditions with alkali the lowest average molecular weight of starch was obtained after extrusion treatment under high moisture and alkali contents.

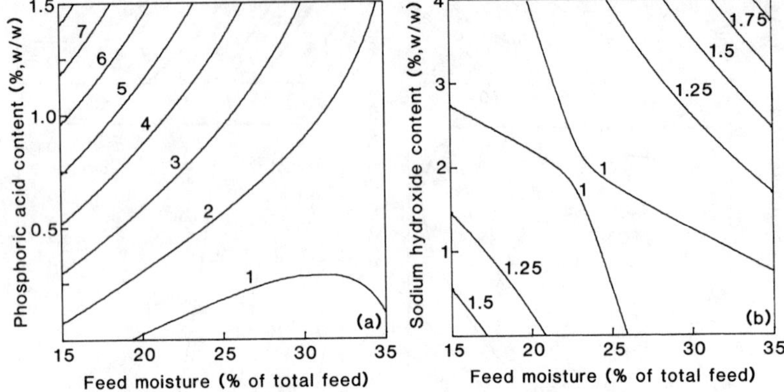

Fig. 3 The effect of feed moisture and acid (a) or alkali (b) content at mass temperature of 190 °C on the molecular weight ratio M_w<2000000/M_w>2000000 of extrudate. The regression equations were significant at 0,1 % level (a) and 1 % level (b). The multiple correlation coefficient r = 0.96 (a) and r = 0.92 (b).

Fig. 4a illustrates the molecular weight distributions of extruded starch, and starch extruded under similar conditions with 0,29 and 1,21 % w/w phosphoric acid. The quantity of starch breakdown products of 2000 <M_w <2000000 increased from 32 to 79 % when the acid content was increased from 0 to 1,21 % w/w. Increasing feed moisture from 15 to 60 % (mass temperature 160 °C, phosphoric acid content 1 % w/w) decreased the amount of molecules of M_w<2000000 from 82 to 34 % (Fig. 1b). Apparently there is a combined effect of pH and shear on starch breakdown. Decreasing feed moisture at constant mass temperature increases frictional heat generation, which may contribute to the breakdown of starch. The X-ray diffractograms showed that the organized structure of starch was destroyed in extruder to a great extent. However, the small amount of starch molecules of M_w<2000 (at most 6 %), and low DE-values obtained showed, that only little small molecular breakdown products were formed. Mercier and Feillet (9) observed no formation of maltodextrins during extrusion of cereal starches. Also Linko et al. (10) observed low DE-values after barley starch extrusion. The degree of hydrolysis with alkali was smaller than with acid. An increase in sodium hydroxide content from 0 to 3,2 % w/w under similar process conditions increased the quantity of molecules of 2000<M_w<200000 from 27 to 59 % (Fig. 4b).

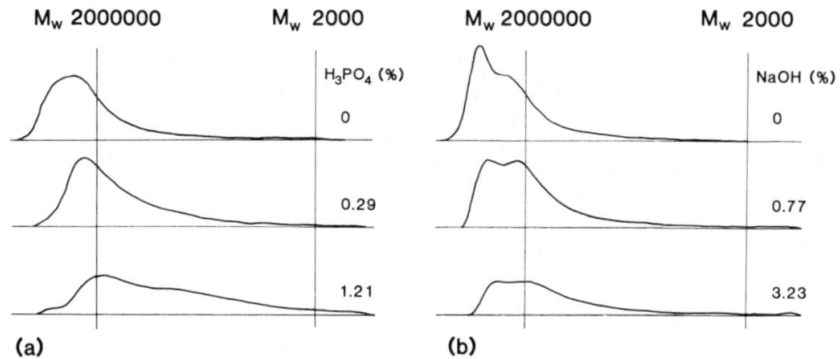

Fig. 4 The molecular weight distributions of starch extruded without or with added acid (a) or alkali (b). Extrusion conditions: mass temperature 178 °C and feed moisture 19 % (a) or 31 % (b).

REFERENCES

1. LINKO, P., COLONNA, P. and MERCIER, C. (1981). High-temperature, short-time extrusion cooking. In Advances in Cereal Science and Technology, Vol. IV (Y. Pomeranz, ed.), A.A.C.C., St Paul, Minnesota, 145 - 235.
2. OLKKU, J. (1981). Extrusion processing - A study in basic phenomena and application of systems analysis. In Developments in Food Preservation - 1 (S. Thorne, ed.), Applied Science Publ., London, 177 - 214.
3. LINKO, P. (1982). HTST-(High-Temperature-Short-Time-)Extruder als biochemischer Reaktor. Getreide, Mehl Brot. 36, 326 - 332.
4. FAUBION, J.M., HOSENEY, R.C. and SELB, P.A. (1982). Functionality of Grain. Cereal Foods World 27(5), 212 - 216.
5. LAI, Y.-Z. and SARKANEN, K.V. (1969). Kinetic study on the alkaline degradation of amylose. J. Polymer Sci.: Part C, (28), 15 - 26.
6. BAZUA, C.D., GUERRA, R. and STERNER, H. (1979). Extruded corn flour as an alternative to lime-heated corn flour for tortilla preparation. J. Food Sci, 44, 940 - 941.
7. COCHRAN, W.G. and COX, G.M. (1957). Experimental Designs. 2nd ed., John Wiley & Sons, Inc., New York, 335 - 369.
8. ANDERSON, R.A., CONWAY, H .F., PFEIFER, V.F. and GRIFFIN, Jr., E.L. (1969). Gelatinization of corn grits by roll- and extrusion-cooking. Cereal Sci. Today. 14(1), 4-12.
9. MERCIER, C. and FEILLET, P. (1975). Modification of carbohydrate components by extrusion-cooking of cereal products. Cereal Chem. 52, 283 - 297.
10. LINKO, Y.-Y., VUORINEN, H., OLKKU, J. and LINKO, P. (1980). Effect of HTST-extrusion on retention of cereal α-amylase activity and on enzymatic hydrolysis of barley starch. In Food Process Engineering (P. Linko and J. Larinkari, eds.), Vol. 2, Enzyme Engineering in Food Processing, Applied Science Publishers, London, 210 - 216.

EXTRUSION COOKING OF HIGH α-AMYLASE FLOUR FOR BAKING

C. MATTSON, J. ANTILA,* Y.-Y. LINKO and P. LINKO
Helsinki University of Technology, Department of Chemistry,
Laboratory of Biochemistry and Food Technology,
SF-02150 Espoo 15, Finland
*Technical Research Centre of Finland, Food Research Laboratory,
SF-02150 Espoo 15, Finland

Summary

A blend of wheat and barley malt flours with high amylolytic activity (falling number 62) was extrusion cooked using a Creusot-Loire BC 45 twin-screw extruder with 500mm screws having either reverse- or co-pitched screw elemnts next to extrusion die. Total feed rate varied from 200 to 500 g/min. Feed moisture was held constant at 18%, set barrel temperature varied between 82° and 87°C, and the mass temperature at the die from 120° to 128°C. In all cases α-amylase was sufficiently inactivated. Ten to 50% (w/w) of milled extrudate was substituded for unprocessed wheat flour in bread baking, and in some experiments vital wheat gluten was added to supplement proteins denatured during extrusion cooking. Satisfactory loaves with some decrease in volume were obtained with 20% of extrudate and 2% of gluten. Bread containing extruded flour was softer and more elastic, and remained longer fresh during storage.

1. INTRODUCTION

Cereal grain of excessively high amylolytic activity has only limited, if any, use in bread baking. This can be a serious economic problem in cold and rainy climates. Starch breakes down during the early stages of baking, resulting in a darker crust color, an open crumb pore structure, and at higher levels of α-amylase activity in moist and sticky crumb (1). It has been shown that cereal grain α-amylase can be effectively inactivated by extrusion cooking (2-5), and the baking potential of pregelatinized starch has been investigated (6). We have also shown that high α-amylase flour can be used in direct extrusion baking of crispbread of excellent quality (7). In the present study the effects of partial substituting of wheat flour with milled extrudate processed from flour of high amylolytic activity were investigated.

2. EXPERIMENTAL

Commercial baker's wheat (moisture 12.8%, falling number 354) and barley malt (moisture 9.8%, °L 118, maltose equivalent 470 mg maltose/100 g flour)were obtained from Raision Tehtaat, Raisio, Finland as a gift. To simulate high α-amylase activity flour, 6.3% (w/w) of malt flour was added to wheat flour.

Creusot-Loire BC 45 twin-screw extruder with 500 mm screws having either reverse- or co-pitched 50 mm screw elements next to extrusion die was used for processing wheat/malt flour blend under various extrusion conditions. Total feed rate was varied from 200 to 500 g/min, set barrel temperature varied between 82° and 87°C, and the mass temperature at the die from 120° to 128°. Feed moisture was held constant at 18%. Circular dies of 5 mm diameter were used. All extrudates were kept for 18 h in open polyethylene bags, after which they were milled with a Wiley laboratory mill through 0.8 mm sieve. Milled samples were stored at room temperature in sealed jars.

The extrudate processed at total feed rate of 500 g/min and mass temperature of 120°C using only co-pitched screw elements was employed in test baking experiments. A series of breads was baked by replacing 10, 20, 30, 40, and 50% (w/w) of flour with the extruded and milled wheat/malt flour blend. In addition, 10, 15, and 20% of vital wheat gluten (extrudate basis) was added. Dough water quantity was varied to obtain a final consistency of 475, 450 or 425 B.U. The following basic recipie was used:

flour	100 g
water	55-65 g
yeast	5.6 g
fat	4.2 g
sugar	4.2 g
salt	2.0 g

Doughs were mixed in a Hobart A200 mixer for 8 min. Floor time was 20 min at room temperature, after which 180 g of dough was weighed in small metal pans (12 × 7.5 × 7 cm), proofed for 50 min at high humidity, 37°C, and baked for 20 - 25 min at 210°C.

In addition to conventional physical dough testing, amylolytic rest-activity, quantity of damaged starch as expressed either as mg maltose/g (d.m.) or in % of maximum damage obtained, and the viscosity behaviour both with Brabender Anylograph and Ottawa Starch Viscometer were determined.

3. RESULTS AND DISCUSSION

3.1. Extrusion Cooking of Wheat/Malt Flour Blends

The moisture content of milled extrudates varied from 11.3 to 12.0% with co-screw, and from 9.1 to 9.9% with reverse screw combination. Starch damage was highest with reverse screw combination, varying from 387 - 391 mg maltose/g. With co-screw combination an increase in total feed rate (200, 300, 400 and 500 g/min) decreased starch damage significantly (336, 296, 252 and 220 mg maltose/g, respectively). Amylolytic enzymes were effectively inactivated in all cases.

Microscopic examination revealed that with co-screw combination at the lowest feed rate about 50% of starch granules had lost their birefrigence, whereas at the highest feed rate of 500 g/min only about 30% of granules were affected in this manner. Typical consistency curves are shown in Fig. 1 clearly illustrating the milder extrusion treatment of sample chosen for baking tests (A) as compared with a sample extruded using a reverse screw element (B). Water absorption of extrudates was considerably higher than that of the native unprocessed flour. Consequently, water absorption increased with an increase in the relative amount of extrudate in the dough.

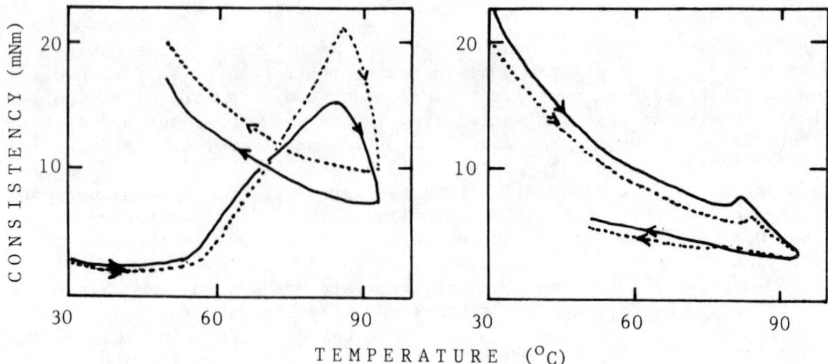

Fig. 1 Typical viscograms (Ottawa Starch Viscometer) of milled extrudate suspensions (15% d.m.) in water (———) and in 0.1% $HgCl_2$ (·····); A, extruded at 120°C mass temperature, 500 g/min total feed, co-screw combination; B, extruded at 123°C, 300 g/min, reverse screw next to die.

3.2. Baking Experiments

Bread volume decreased markedly with an increase in the relative quantity of milled extrudate in dough as can be seen from Fig. 2. The effect could be in part reversed by the addition of vital wheat gluten (Fig. 2). The effect of gluten was more drastic at higher levels of extrudate.

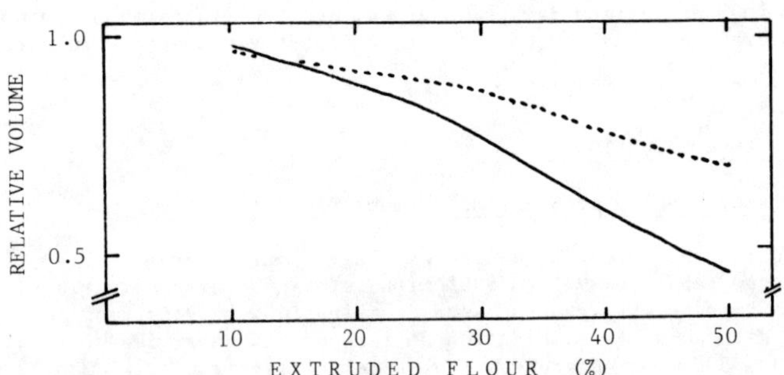

Fig. 2 The effect of the addition of milled extrudate on relative bread volume without (———) and with (·····) 10% (w/w) vital wheat gluten (extrudate basis).

Fig. 3 illustrates the effect of vital wheat gluten on the volume of bread baked with 20% (w/w, flour basis)(———) and 50% (·····) of milled extrudate. Satisfactory bread quality was obtained with 20% extrudate and 20% vital wheat gluten (extrudate basis). Loaves obtained with extrudate possessed a darkened, shiny crust color.

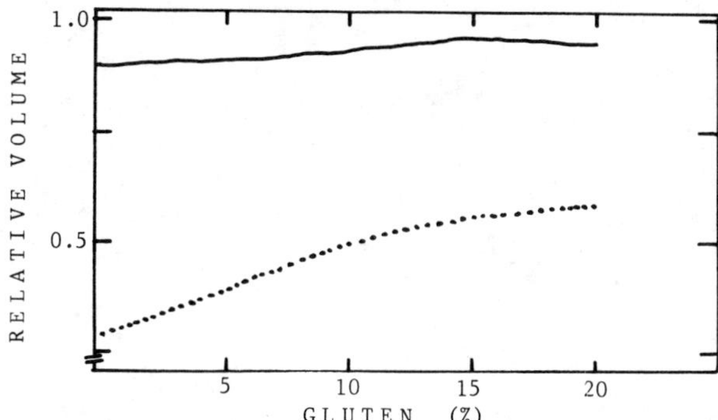

Fig. 3 The effect of vital wheat gluten addition on the relative volume of bread baked with 20% (———)and 50% (·····)(w/w, flour basis) of milled extrudate.

The effect of the addition of extrudate on staling and softness of bread was followed by Instron Universal Testing Machine by determining the Young's module (N/m^2), elasticity, and maximum power (N) after 1 and 3 days of storage at room temperature. Bread containing 20% of extrudate remained soft and elastic even after 3 days. Crumb pore structure of such loaf was equal to that of the reference bread (Fig. 4). The addition of gluten also appeared to have a crumb softening effect. If the relative quantity of the extrudate was increased to 50%, the crumb was moist and sticky.

4. CONCLUSIONS

In conclusion it can be said that it is possible to inactivate cereal α-amylase in high amylase (falling number about 60) flour sufficiently for bread baking purposes by extrusion cooking under mild enough conditions to avoid excessive starch damage. Up to 20% (w/w) of wheat flour could be replaced by milled wheat flour extrudate without other significant adverse effects but a slight decrease in loaf volume, which could in part be compensated by the addition of vital wheat gluten. The additions of extrudate improved the keeping quality of bread, which remained soft even after 3 days of storage at room temperature.

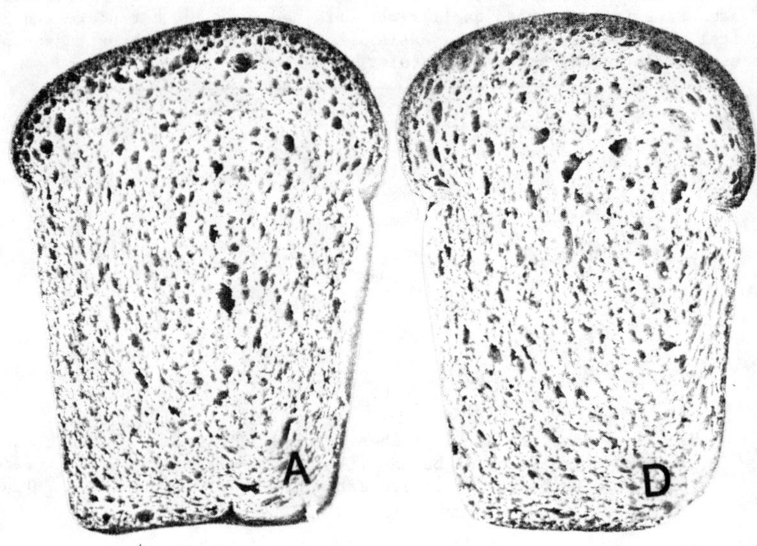

Fig. 4 Crumb structure of wheat bread; A, reference loaf; B, bread baked with 20% (w/w) extrudate (flour basis) and 20% of vital wheat gluten (extrudate basis).

5. REFERENCES

1. CHAMBERLAIN, N., COLLINS, T. H. and McDERMOTT, E. E. (1981). Alfa-amylase and bread properties. J. Food Technol. 16, 127-152.
2. LINKO, P., OLKKU, J., ANTILA, J. and ROSENBERG, K. (1980). Reduktion der Enzymaktivität während der Hochtemperatur-Kurzzeiterhitzung beim Extrudieren. Getreide, Mehl u. Brot 34, 78-81.
3. LINKO, Y.-Y., VUORINEN, H. and LINKO, P. (1980). The effect of HTST-extrusion on retention of cereal α-amylase and on enzymatic hydrolysis of barley starch. In "Food Process Engineering," Vol. 2, "Enzyme Engineering in Food Processing" (P. Linko and J. Larinkari, eds), Applied Science Publishers, London, pp. 210-223.
4. LINKO, P., COLONNA, P. and MERCIER, C. (1982). High-temperature, short-time extrusion cooking. In "Advances in Cereal Science and Technology," Vol. 4 (Y. Pomeranz, ed.), A.A.C.C., St Paul, USA, pp. 145-235.
5. NIERLE, E., El BAYA, A. E., SEILER, K., FRETZDORFF, B. and WOLFF, J. (1980). Veränderungen der Gegreideinhaltstoffe während der Extrusion mi einem Doppelschneckenextruder. Getreide, Mehl u. Brot 34, 73-78.
6. LORENZ, K. and KULP, K. (1981). Heat-moisture treatment of starches. II Functional properties and baking potential. Cereal Chem. 58, 49-52.
7. LINKO, P., MATTSON, C., LINKO, Y.-Y. and ANTILA, J. (1983). Production of flat bread by continuous extrusion cooking from high α-amylase flour J. Cereal Sci. 1 (in press).

SUGAR AND SALT IN EXTRUSION COOKING FOR INTERMEDIATE MOISTURE PRODUCTS

J. VAINIONPÄÄ, E.-K.RAUTALINNA and R. KARPPINEN
Technical Research Centre of Finland
Food Research Laboratory
SF-02150 ESPOO 15, FINLAND

R. KERVINEN and P. LINKO
Helsinki University of Technology, Department of Chemistry
Laboratory of Biochemistry and Food Technology
SF-02150 ESPOO 15, FINLAND

Summary

A Creusot-Loire BC 45 twin-screw extrusion cooker was employed for production of intermediate moisture foods (IMF) by two methods. In one method, mixtures of wheat flour, sugar and salt were extruded varying sugar content, salt content, screw rotation speed, feed moisture and mass temperature. A finished product of desired a_w was obtained. In another method, wheat flour was extruded alone at 20 % moisture varying the screw rotation speed and mass temperature in the extrusion barrel. Sugar and salt were mixed with extruded wheat flour, and sufficient water was added to reach the final product moisture content of 25 %. The result showed, that 23 % sugar and 1.5 % salt in wheat flour-sugar-salt blend resulted in a water activity of 0.85 for 25 % moisture extrudate. When sugar and salt were mixed afterwards with the milled extrudate 30 % sugar and 1.5 % salt were needed to reach the same a_w.

1. INTRODUCTION

The principle in the production of intermediate moisture foods (IMF) is to adjust the product water activity (a_w) to the level where the growth of micro-organisms is prevented or greatly reduced (a_w from 0.65 to 0.9) but the water content remained high enough for palatability (from 10 to 40 %). The a_w is lowered by humectants, such as sugars, salts and/or polyhydric alcohols. Consideration should also be given to retardation of microbial growth by antimycotic agents and inhibition of chemical deterioration reactions (1).

The effects of sugar (0 - 30 % of d.m.) and salt (0 - 4 % of d.m.) on wheat flour extrusion cooking under moisture range from 10 to 20 % (of total feed) have been previously investigated (2). Linko *et al.* (3) and Vainionpää *et al.* (4) have recently described the employment of extrusion cooker for cereal based IMF. The aim of this study was to investigate the employment of extrusion cooker in the production of IMF using sugar and salt as humectants.

2. MATERIALS AND METHODS

2.1 Materials

Commercial wheat flour (Vaasa Mills Ltd, Finland) was used as cereal raw material. Other materials added were crystal sugar and sodium chloride containing potassium iodide 25 mg/kg.

2.2 Extrusion Cooking

The experiments were performed by a Creusot-Loire BC 45 twin-screw extruder with a screw composition CRCCC, where C is a compression screw element and R a counter pitched screw element located 50 mm before the die. Die diameter was 5 mm and lenght 27 mm. Two main methods were used in preparation of the IM product. In one method mixtures of wheat flour, sugar and salt were extruded varying the sugar content (0 - 60 % of total d.m.), salt content (0 - 8 % of total d.m.), screw rotation speed (100 - 200 rev/min), feed moisture (20 - 40 % of total feed) and mass temperature (120 - 180°C), and keeping the feed rate constant (300 g d.m./min). A finished product of desired a_w was obtained. In another method, wheat flour was extruded alone at 20 % moisture varying the screw rotation speed (100 - 200 rev/min) and mass temperature in the extruder barrel (130 - 190°C), and keeping the feed rate constant (400 g total feed/min). Sugar (0 -30 % of total d.m.) and salt (1.5 % of total d.m.) were mixed after extrusion with extruded wheat flour, and the mixture was moistened to obtain an IM product of 25 % moisture content. Second order regression equations were computed for a_w and product moisture.

2.3 Analytical Methods

Sample moisture content was determined by drying for one hour at 130°C. For the measurement of water activity the samples were allowed to reach equilibrium overnight at 20°C. Water activity was measured by Humicap HM 14 tester (Vaisala Ltd, Finland) calibrated with LiCl (12.4 % RH) and K_2SO_4 (97.2 % RH).

3. RESULTS AND DISCUSSION

3.1 Extrusion of wheat flour, sugar and salt blends

Various ingredients, such as sugar and salt, have been observed to have a profound effect on extrusion cooking of wheat flour. According to the experiments of Kervinen (2) with a Creusot-Loire BC 45 twin-screw extruder (screw rotation speed 100 - 200 rev/min, feed moisture 10 - 20 % of total feed, mass temperature 130 - 190°C, sugar content 0 - 30 % of d.m., salt content 0 - 4 % of d.m., feed rate 400 g d.m./min and screw composition CRCCC) an increase in sugar content in wheat flour-sugar-salt blend increased starch gelatinization and certain product functional properties such as water absorption ability and cold water consistency of a water suspended milled product, until at a certain sugar content a maximum was reached. With a further increase of sugar content, these product properties began to decrease. Fig. 1 shows the effects of sugar content and mass temperature during extrusion cooking on starch gelatinization and cold paste consistency (5,p.177; 6,p.117). Even small salt quantities at the order of 1 % added to wheat flour clearly increased starch gelatinization and water absorption ability under the experimental conditions described above (2).

Fig. 2a illustrates the effects of sugar and salt on a_w and product moisture in the intermediate moisture range. An increase in sugar and salt content decreased a_w as expected. However, the effect of small salt contents normally used in foods was less marked. The quantity of humectants needed for IMF was affected to some extent by the process conditions. Under low screw speed (100 rev/min), low mass temperature (120°C) and salt content of 1.5 %, the sugar content of 23 % resulted in a_w 0.85 for 25 % product moisture (Fig. 2b).

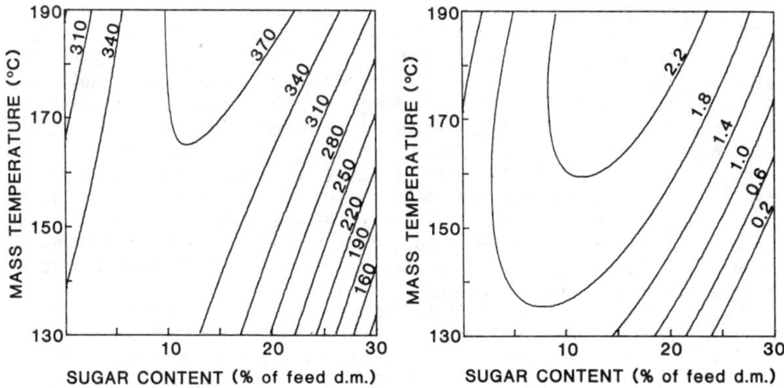

Fig. 1 The effects of sugar content and mass temperature (a) on the degree of starch gelatinization (maltose, mg/g, d.m. of extruded wheat flour), and (b) on the cold paste consistency at 30°C (mNm, Haake Rotovisco). Conditions: screw speed 150 rev/min, feed moisture 15 % of total feed and salt content 2 % of feed d.m. The graphs are based on regression equations, which were significant at 0.1 % level, and the lack of fit terms at 5 % level (a) and not even at 5 % level (b). The multiple correlation coefficient r = 0.98 for both equations.

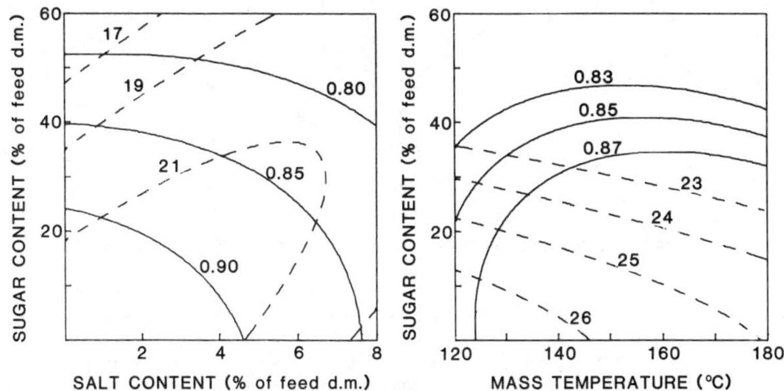

Fig. 2 The effects of salt and sugar contents (a), and mass temperature and sugar content (b) on a_w (—) and product moisture (---) of extruded product. Conditions: (a) screw speed 150 rev/min, mass temperature 150°C and feed moisture 30 % of total feed, (b) screw rate 100 rev/min, feed moisture 35 % of total feed and salt content 1,5 %. The graphs are based on regression equations, which were significant at 0,1 % (a_w) and 1 % (moisture) levels, and the lack of fit term at 5 % (a_w) and 1 % (moisture) levels. The multiple correlation coefficient r = 0.99 (a_w) and r = 0.98 (moisture).

3.2 Addition of humectants in wheat flour extrudate

In addition to the method, where removal of water by extrusion cooking resulted in final IMF, another method was investigated, where the addition of water to the blend consisting of extruded wheat flour, sugar and salt resulted in the final a_w. In this method, about 30 % sugar (instead of 23 % in the method described above) was needed to be blended with extruded wheat flour and salt to obtain an IM food at a_w 0.85 and moisture content of 25 % (Fig. 3a). The final product a_w could be minimized by extrusion of the flour component either under low screw speed and high mass temperature (Fig. 3a) or under high screw speed and low mass temperature (Fig. 3b). Under high screw speed more frictional heat is generated by the rotation of the screws, allowing low temperature, whereas under low screw speed and low shear more additional heat is needed.

The amount of sugar and salt in IMF may be reduced by adding other humectants. It has been shown that a mixture containing 17 % sugar, 1.5 % salt, 1.4 % glycerol and extruded wheat flour resulted in an a_w of 0.85 at 25 % product moisture. The addition of propylene glycol allowed a further reduction of sugar quantity. A mixture consisting of 9 % sugar, 1.5 % salt, 1.6 % glycerol, 1 % propylene glycol and extruded wheat flour resulted in the same a_w as above (3,4).

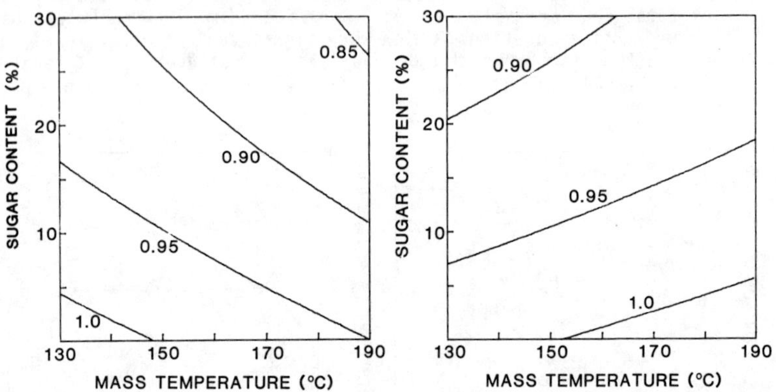

Fig. 3 The effects of mass temperature during wheat flour extrusion cooking and sugar content (% in d.m. of extruded wheat flour-sugar-salt blend) on the a_w of final product containing 25 % moisture. Conditions: screw speed 100 rev/min (a), and 200 rev/min (b). Salt content was 1.5 %. The graphs are based on the regression equation, which was significant at about 10 % level and the lack of fit not even at 5 % level. The multiple correlation coefficient r = 0.85.

REFERENCES

1. KAREL, M. (1976). Technology and application of new intermediate moisture foods. In Intermediate Moisture Foods (R. Davies, G. G. Birch and K. J. Parker, eds.), Applied Science Publishers Ltd, London, 4.
2. KERVINEN, R., LINKO, P. and OLKKU, J. (1981). Extrusion cooking induced changes in functional properties of wheat flour. In Proc. Nordic Cereal Chemists Assoc. 21. Congr., 9. - 11. June 1981, Espoo, Finland, Kirjapaino Painorengas Oy, Helsinki, 219.
3. LINKO, P., KERVINEN, R., KARPPINEN, R., RAUTALINNA, E.-K. and VAINIONPÄÄ, J. (1983). Extrusion cooking for cereal-based intermediate-moisture products. In Proc. Int. Symp. on the Properties of Water, 11. - 16. September 1983, Beaune, France. (in press).
4. VAINIONPÄÄ, J., KARPPINEN, R., RAUTALINNA, E., KERVINEN, R. and LINKO, P. (1983). Adsorptive and desorptive modes in extrusion of cereal based intermediate moisture foods. In 3rd International Congress on Engineering and Food, 26. - 28. September, 1983, Dublin, Ireland.
5. LINKO, P., COLONNA, P. and MERCIER, C. (1981). High-temperature, short-time extrusion cooking. In Advances in Cereal Science and Technology, Vol. IV (Y. Pomeranz, ed.), A.A.C.C., St Paul, Minnesota, 145.
6. OLKKU, J., HAGQVIST, A. and LINKO, P. (1983). Steady-state modelling of extrusion cooking employing response surface methodology. J. Food Engineering 2, 105.

THE MEASUREMENTS OF THE RHEOLOGICAL PROPERTIES OF STARCH-RICH MATERIALS

J.P. Sęk[*], L.P.B.M. Janssen
Laboratory of Physical Technology
Delft University of Technology
The Netherlands

D.J. van Zuilichem
Department of Food Processing
Agricultural University Wageningen
The Netherlands

(* - on leave from Institute of Chemical Engineering, Lódź Technical University, Poland)

1. INTRODUCTION

Extensive literature exists describing the technological aspects of food extrusion but the design of extruders for food materials is impeded by the lack of data on physical properties of processing media. Therefore a new version of the annular shear cell is developed in which the friction between granular material and solid surfaces as well as the internal friction could be measured. These data are indispensable in modelling the feed zone of a single screw extruder.

2. APPARATUS

Annular shear cells are commonly used in determining the frictional properties of granular materials. Their basic principles have been described by Carr and Walker (1). However in their traditional form these apparatus allow for measurements at room temperature and at relatively low values of normal and shear stresses. This does not correspond with high temperatures and high stresses during extrusion.
Figure 1 shows the construction of the new shear cell. The inner wall of the shearing area has a diameter of 4 cm. The outer diameter is 8 cm. These dimensions permit the use of small samples. The cell can be used for two different types of measurements. By inserting the rotating cup (Fig. 1a) it is possible to measure the internal friction of the material. This cup can be replaced by a rotating disc (Fig. 1b) in order to measure the friction between the material and solid surfaces. The body of the cell can be heated by electrical elements for measurements at elevated temperatures. The sealing system ensures a constant moisture content of the food material tested. The shear cell is connected to the drive by a torque meter, which allows the measurements of changes in the friction with time - Fig. 2.

3. EXPERIMENTAL RESULTS

The cell was used to determine the wall friction and the internal friction of La Plata maize grits with a starch content of about 80%. The moisture content of the samples was varied up to 35% while the temperature was kept at 25°C. Figure 3 shows the internal friction as a function of normal stresses for different moisture contents. Analogously figure 4 shows the wall friction of maize grits on ordinary steel. From these data the friction coefficient for internal and for wall friction can be calculated as shown in figure 5.

4. DISCUSSION

The shear cell used allows the study of frictional phenomena in food material at conditions similar to those during extrusion. The coefficients of dynamic friction for maize grits at a steal wall as well as the coefficients of internal friction increase with moisture content up to 30%. Above this moisture content the friction decreases because of the lubricating action of the free water between the particles. At a fixed moisture content the coefficient of internal friction is always larger than the wall friction coefficient. The data obtained give some indication of the influence of the moisture content of the material during extrusion and may be applied for the design of the feed zone of the extruder.

5. REFERENCES

1. CARR, J.F., WALKER, D.M. Powder Technology 1 (1967/68), 369.

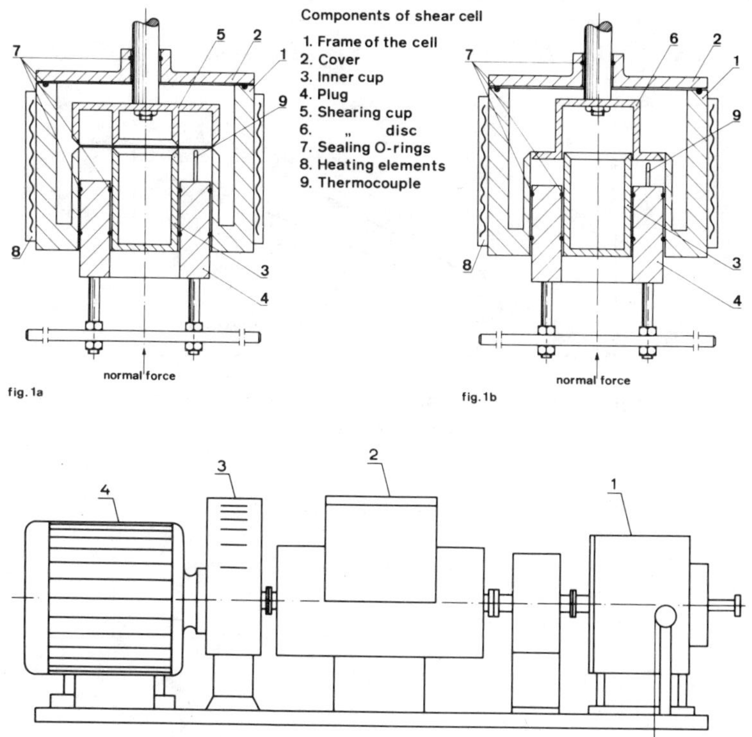

1 - shear cell, 2 - torquemeter, 3 - gear box, 4 - engine

Fig. 2 Experimental set - up.

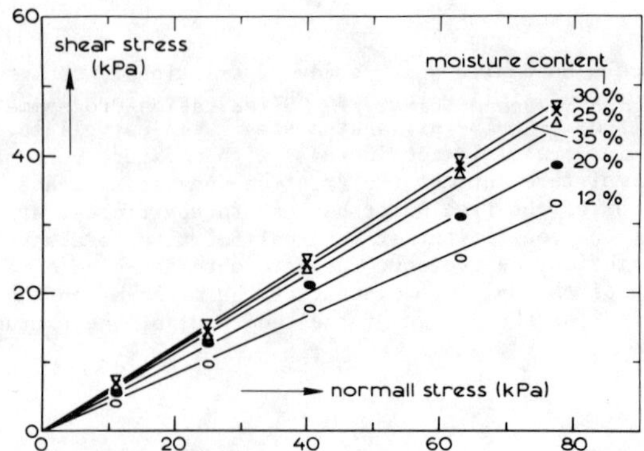

Fig.3 Effective yield loci for maize grits at different moisture content.

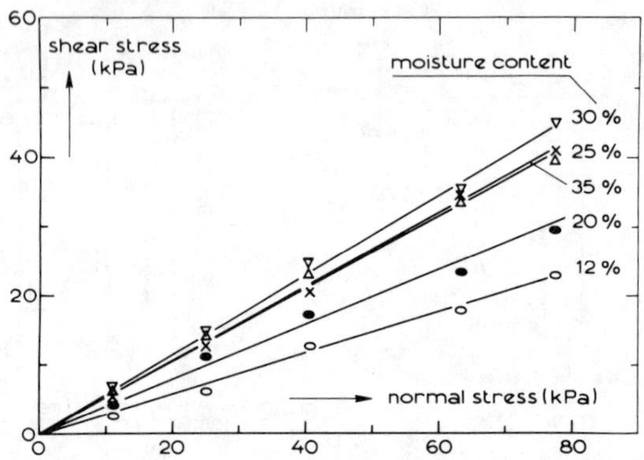

Fig.4 Wall friction measurements for maize grits at different moisture content.

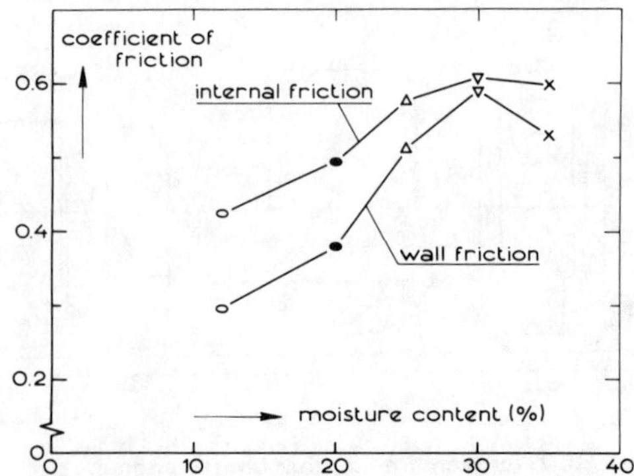

Fig.5 Dynamic coefficients of wall and internal friction for maize grits at different moisture content.

EXTRUSION COOKING: GUIDELINES FOR PUBLICATIONS

Worked out by Subgroup 1 (Extrusion Cooking)
of the European Scientific Collaboration Programme
COST 91 (Thermal Processing and Quality of Foods)

1. OBJECTIVE AND SCOPE

During the course of a 4 year collaborative research programme on extrusion cooking, it quickly became apparent that detailed information on the extruder and raw materials used, and on the processing conditions of extrusion cooking, was necessary for a proper interpretation of experimental results and to allow meaningful comparisons between the results of different teams. It also appeared from the scientific literature that adequate reporting was an exeception rather than a rule.

The Subgroup has therefore prepared a list of recommendations for reporting equipment used and experimental conditions in a manner as complete and as informative as possible. The list given below is intended to help research scientists and technicians, not to deter them from publishing! Hopefully it can be of use also for planning extrusion experiments and for performing measurements.

It is well understood that these recommendations are not compulsory and should not be considered, for example, as criteria for acceptance of papers. Not all of the listed guidelines are important, or even applicable in all cases. The Subgroup is fully aware of the fact, for example, that the precise measurement of temperature, pressure, viscosity and residence times of food mixes within different zones of an operating extruder is a difficult task. Such measurements require sophisticated and costly equipment, and only a few laboratories can perform them at the present time.

2. GUIDELINES FOR PUBLICATIONS

2.1 Extruder, Instrumentation and Other Equipment

The make, type, model, size and number of the extruder should be reported for exact identification. Non-standard individual modifications should be described.

The following indications are necessary concerning the extruder barrel: length and diameter (section in the case of a twin screw extruder); presence of grooves; location and type of heating and cooling systems surrounding the barrel; identification of zones heated or cooled; position of feed opening and of other eventual openings; provision for rapid opening of barrel.

Because of the importance of screw profile, detailed indications or preferably dimensional drawings should be given concerning the screw(s): number of screws; whether co- or counter-rotative, self-wiping; length, diameter, dimension of flights and number and angle of pitch for each successive screw segment (such as compression element, reverse flow element, mixing disc, steam lock, etc). The clearance between screws, between screw(s) and barrel, between screw(s) and die should be reported. The overall length/diameter ratio of the screw(s) should be indicated, together with the compression ratio, as usually defined by the ratio of height between barrel and screw axis at the beginning and end of compression zone. It is of interest to report if wear (of screws and of barrel grooves) has been checked recently. Wear of screw elements can be determined by making plastic impressions at different times.

Detailed information is also desirable concerning die plate and inserts: dimensional drawings should indicate dimensions of dies and give die internal volume. The number and position of die inserts should also be given.

Temperature probe(s) are necessary for adequate process control. The make, type and location of temperature probes should be reported. It is important to indicate if these probes are inserted in the barrel or screw metal, or if they are in close contact with the food mix being extruded. In the latter case, indicate how well they are insulated from the barrel heat. Check that temperature probes have been calibrated recently.

Similar indications should be given for pressure transducer(s). Report where pressure is being measured (e.g. just before the die, or on a thrust bearing.) State whether the transducers are of diaphragm, strain gauge, piezoelectric or other type. Check that they are free from temperature effects and have been calibrated recently. For simple calibration, it is possible to use a pipe attached to a hand-operated hydraulic pump and connected to both the pressure transducer and (through a different hole) an accurately calibrated standard gauge.

Other instrumentation being eventually used should also be indicated, e.g. device(s) for measuring energy input (mechanical and/or thermal energy in kW.h absorbed per kg of dry product); device(s) for measuring torque (correction for drive train and motor inefficiency should be made with a no-load run); device for measuring viscosity of the food dough in the die (viscometer/rheometer die; the viscosity can be measured even with a short die, provided end correction factors are taken into account); on-line radioactivity detectors for residence time determination; regulation and automation devices (e.g. pressure regulation through feed rate).

Ancillary equipment (upstream and downstream of the extruder) should also be reported.

2.2 Feed Materials and Extruded Product

2.2.1 Mixing and conditioning operations before extrusion

It is often important to report equipment used for mixing, mixing time and conditions, and granulometry of ingredients. The following questions should be answered: has the homogeneity of the powder mix been checked? Was the mix preconditioned for moisture and/or temperature? What was the holding holding time before extrusion? It is of interest that pre-heating the food mix just before extrusion can result in increased capacity.

2.2.2 Food mix during extrusion

The full specification of each component/ingredient (including any minor additive(s) used) should be given.

It is necessary to indicate the proportion/percentages (state basis) of each ingredient in each food mix including water added as liquid or steam. These values may have to be calculated from the input feed rates of powder mix, water and oil introduced into the extruder.

If not clear from the foregoing, it is important to state, for each food mix, the total water content (% wet weight basis) and the aggregate proportion of other components (protein, lipids, and possibly starches, fibre, ash, low molecular weight sugars)(% dry weight basis). Give method of determination in sufficient detail where relevant to interpretation of results.

The type of water used; hardness or other detailed data can be given if relevant and available.

Where appropriate, a_w and pH of food mix should be indicated (the pH can be determined by progressive rehydration and extrapolation to initial moisture content).

2.2.3 Extruded product

The following indications are useful:

Moisture content after cooling (% wet weight), and variations of this moisture content with time (either for just-extruded products or for products cooled for different times).

Moisture content after drying, and how product was dried.

Other significant constituents, including methods of determination (e.g. lipid content, solvents used, prior acid hydrolysis. The lipid content of the extruded product may be lower than that of the food mix, due to melting and exudation of oil at the die exit).

Dimensions, bulk density, expansion coefficient (ratio of product dimensions to die dimensions, or preferably ratio of product section area to die section area), size and distribution of voids. The methods and number of determinations should be indicated.

Product colour and relationship to feed colour.

Functional properties of ground extruded product (water absorption rate and capacity, viscosity characteristics, solubility of carbohydrate and protein constituents, etc.).

Any relevant data on texture (tensile strength, impact test, high rate shearing, stickiness, crispiness, etc.).

Any relevant chemical determination (extent of vitamin or available lysine loss, degree of starch gelatinization, etc).

Any relevant indication on taste and flavour.

2.3 Extrusion Conditions

2.3.1 Process parameters

Sufficient time (at least 10 minutes) of normal operation should be allowed for equilibration of the extruder (before starting an experiment or between experiments). It is necessary to indicate whether steady state operation has been attained. Checking should be carried out perferably by determination of constant torque or power, or constant pressure before the die. Constant product temperature is less indicative of equilibrium. Equilibrium cannot be reached when input dry solids feed rate is different from output dry solids flow rate. It may be necessary, for proper equilibration, to approach gradually towards the final temperature and feed rate conditions.

It should be indicated whether the extruder is starve-fed (with a screw feeder for instance) (as is most frequently done with twin-screw extruders) or fed "at will" by overflow (as usually carried out with single screw extruders).

For reporting feed rates, determinations should be repeated several times with sample collection during appropriate time intervals. The input feed rate (kg/h) should refer preferably to the total food mix (including added liquid water and/or oil). If not, it should be indicated whether the input feed rate refers to dry solids or to the powder food mix. Specific input feed rates should be given for each separate feeding or pumping device, listing equipment, material being fed or pumped, and fluctuations in each specific feed rate. The output flow rate (kg total weight/h, or kg dry weight/h) should be determined and reported, together with an assessment of its fluctuations. It is to be compared to the input feed rate, and indicates extruder capacity.

The true measured rate of rotation of screw(s) should be reported, in rpm, together with the method of measurement and the possible fluctuations.

The ratio of input feed rate to rate of rotation of screw(s) may be a relevant parameter indicating the level of filling of screw flights or chambers.

The temperature of barrel at different locations should be reported.

It is of interest to mention any problems encountered during extrusion (surging, equilibrium difficult to reach, blocking, etc).

2.3.2 System parameters

The following parameters should be reported, whenever available:

Temperature of food mix during extrusion and where it is measured. State if temperature fluctuations were observed.

Pressure within the extruder, at different locations. State if pressure fluctuations were observed.

Complete temperature and pressure profiles along the extruder, if measured.

Mechanical and/or thermal energy inputs (kW.h/kg dry weight of product).

Torque values: indicate method of measurement and variations.

Residence times: indicate method of measurement (pulse method with dye or with radioactive material. Indicate which dye or material, the amounts added, the duration of addition. The details of sample collection and of dye extraction should be given). Results should be expressed by the residence time distribution curve, and the values for minimum, median, mean, maximum residence times, and dispersion. The % recovery of dye or radioactivity should be given.

Viscosity of extrudate in the die channel.

2.3.3 Collection of extruded product, and post-extrusion processing

Indications should be given on the following operations:

Procedure for collecting and sampling of product (frequency, duration, size of samples, randomisation).

Packaging and storage arrangements.

Post-extrusion processing: cutting, stretching, flattening between rollers, drying, toasting, flavouring, packaging.

2.3.4 Experimental design

State whether duplicate experiments were carried out and checked for constant conditions of product temperature, residence time, etc.

When comparative experiments are carried out, it should be stated which independent and dependent variables were changed, and how. Caution: changes in water or lipid content may extensively modify friction effects and therefore product temperature, even if the barrel temperature is kept constant.

If multifactorial statistical experimental design and response surface analysis or any other appropriate experimental arrangement is used, indicate details of experimental design, give regression equations with coefficients (values and significance), result of validity test, coordinates of optimum points and boundaries of regions of interest.

3. RECOMMENDATIONS: LIST OF DESIRABLE MINIMUM INFORMATION FOR REPORTING EXTRUSION COOKING EXPERIMENTS

3.1 Extruder and instrumentation

Make, type, model of extruder.

Description (dimensional drawings) of extruder barrel, screw(s) and dies. The screw profile (or configuration) is of great importance (with location of transport, compression, reverse flow, etc. screw elements).

Equipment for measuring temperature and pressure, and its location. It is necessary to state whether the temperature of the food mix or the temperature of the barrel metal was measured.

3.2 Feed materials and extruded product

The exact composition of the food mix during extrusion should be reported: percentage of each food ingredient (% of total weight). The total and added water contents (% of total weight) are also of great importance.

In some cases it is necessary to ensure that the feed materials are well mixed before extrusion. Pretreatment and/or preconditioning of ingredients should be reported.

Moisture content and expansion coefficient (or bulk density) of the extruded product are useful indications.

3.3 Extrusion conditions

3.3.1 Process parameters

True rate of rotation of screw(s) (rpm).

Input feed rate (kg/h) for the total food mix (including added liquid water), or feed rates of solids and of added water. Feeding device(s) and fluctuations in feed rate(s). Whether the extruder is starve-fed, or fed "at will".

Time allowed for equilibrium before starting an experiment, and between subsequent experiments.

Temperature of barrel (in different zones, if available).

Problems during extrusion.

3.3.2 System parameters

Temperature of food mix during extrusion. Indicate where measured and whether it remains constant.

Pressure within the extruder. Indicate where measured and whether it remains constant.

Residence time distribution, if available. Indicate method of measurement.

3.3.3 Procedure for collecting and sampling extruded product and Post-extrusion processing.

3.3.4 Experimental design

Duplicate experiments. Comparative experiments (state which independent and dependent variables were changed). Statistical experimental design.

4. CONTRIBUTORS

COST 91/1 Subgroup members, especially:

Prof. Cheftel, Laboratoire de Biochimie et Technologie Alimentaires, Université des Sciences, 34060 Montpellier, France, Chairman of the Subgroup.

Prof. Linko, Helsinki University of Technology, Finland, Chairman of COST 91.

Prof. Asp, University of Lund, Sweden.

Dr. Frazier, Dalgety Spillers Group Research, United Kingdom.

Dr. Jeunink, Créalis-BSN, France.

Dr. Seiler, Cereal Institute, Detmold, Germany.

Mr. Van Zuilichem, Agricultural University, Wageningen, The Netherlands.

The suggestions of Prof. J.M. Harper and Mr. R. Jowitt are gratefully acknowledged.

INDUSTRIAL APPLICATIONS OF EXTRUSION COOKING
(present or potential)

1. HUMAN FOODS

1.1 Mixing + shaping ± moderate heat treatment

Doughs (possibly precooked):

- Pasta products
- Bread, tortillas
- Frozen pastry } amylose-lipid complex: no starch retrogradation
- Frozen pizza }

Indermediate moisture foods:

- Food bars
- Dry sausage (beefsticks) } good compaction at low moisture easier drying; blend various flesh; mix cereal + meat
- Mechanically deboned flesh }
- Various fillings

1.2 Mixing + decrystallisation + shaping

Confectioneries:

- Sucrose + glucose syrups } anhydrous melting of sucrose
- Sweets, liquorice, caramel }
- Some chocolate products
- Insoluble starch (glassy state) (moulding)

1.3 Crushing + mixing + cooking ± sterilization ± shaping

Purées: fruit, vegetables (potatoes), meats (or blends); possibly followed by aseptic filling (baby foods)

1.4 Grinding + pressing + moderate heat treatment

Facilitate subsequent oil extraction from oilseeds

1.5 Thermal treatment + texturization + shaping ± fibre or nutrient enrichment

Cereal products:

- Snacks (direct or indirect expansion; crispiness)
- Breakfast cereals + flakes (low rehydration; crispiness)

Flat breads, biscuits, cookies, crackers, wafers, croutons (crispiness, low stickiness)

Bread crumbs (using old bread etc.)

Quick-cooking pasta products, for soups, etc. (quick rehydration, low stickiness)

Reconstituted rice grain (from broken grains or flours)

Various dietetic foods (gluten-free biscuits, etc)

Potato products:

Analogues of fries, etc.

Vegetables and pulses:

Textured vegetable proteins, meat analogues (with or without incorporation of fish mince, mechanically-deboned meat or casein) (quick rehydration + mechanical and thermal resistance)

Others:

Production of sterile process cheese

Co-extrusion of raviolis, of snacks with cheese filling, of pastry with fruit-gel fillings, etc.

1.6 Thermal treatment + partial drying + obtention of functional properties ± nutrient enrichment

(e.g. instant solubility or dispersibility, high viscosity, etc. after grinding and rehydration)

Pregelatinized starches

Precooked flours or grits for use in the baking industry (bread, biscuits, cakes...)

Inactivation of enzymes in high-enzyme flours

Precooked couscous

Treatment of malt for the brewing industry

Precooked (instant) powder gruels or instant foods: baby foods; weaning foods

Corn-Soy-Milk; Whey-Soy-Blend; other blends (potato, rapeseed, fish mince); full fat soy flour (with destruction of enzymes, antinutritional factors, microorganisms; better shelf life)

Roasting of coffee grains: coffee powder

Compaction of coffee fines

Cocoa processing

Instant powder drinks: coffee or cocoa-based

Instant rice pudding

Instant dry soup mixes: legumes treated individually or blended before extrusion

Dairy, blood or other proteins: e.g. production of caseinates (improvement of solubility, emulsifying, thickening properties; pH modifications at low water content, sterilization)

Maltodextrin powders with encapsulated flavours

Starch, flour, sugar, spices, with low microbial charge

2. ANIMAL FEEDS + WASTE TREATMENTS

Mix, Blend Grind; Press, Agglomerate, Shape;

Cook (pregelatinize starch, predigest cellulose, destroy anti-nutritional factors, enzymes, pasteurize); Dry (partly);

Puff.

Oil-seed meals

Cattle feed with pregelatinized starch + urea

Pet foods; some with intermediate moisture. Coextrusion of foods with multicolour layers.

Fish food
(fast hydration; should not disperse, nor float, nor sink too fast)

Agglomeration of cereal dust

Compaction of rice bran, inactivation of lipases, facilitate oil extraction

Use of brewers yeast and single cell protein
(kill microorganisms, increase digestibility by cell wall disruption and protein denaturation)

Agglomeration of fodder

Treatment of cellulosic wastes, eventually with alkalis (increase digestibility for ruminants; facilitate hydrolysis and conversion by fermentation)

Treatment of grass and leaves, e.g. lucerne
(remove excess water and some soluble compounds, destroy lipoxygenase, improve digestibility, improve storage)

Destruction of aflatoxins by alkaline treatment of oilseed meals

Destruction of gossypol, glucosinolates, phytates, oligosides

Treat wet grains with organic acids

Continuous insolubilization-gelification of proteins from blood, offals, with encapsulation of free amino acids

Treatment of horn, hooves, feathers

Mixing waste blood with flour
(effluent treatment: feeds)

Mixing wastes and chemicals to make fertilizers
(with ideal composition, convenient handling - no powder, good hydration properties)

Sawdust agglomeration

Preparation of paper pulp

Tobacco puffing

3. THE EXTRUDER AS A CHEMICAL AND BIO-REACTOR

HTST, pressure, shear, low hydration, + reagents ± gas,
+ sequential reactions (long barrels)

Hydrolysis:

Cellulosic waste hydrolysis, with HCl, H_2SO_4
(for conversion into glucose and later ethanol)

Starch hydrolysis, with HCl or thermostable amylases

Protein hydrolysis, with HCl or proteases

Other reactions:

For modifications of functional properties such as solubility, emulsification, resistance to heat treatments

Protein acetylation, succinylation, etc.

Covalent binding of various derivatives, amino acids, etc.

Starch cross-linking, stabilization, fixation of emulsifying agents, etc.

Flavour generation, using maillard precursors, meat by-products, etc.

Other recations of synthetic and biopolymers.

INDUSTRIAL COOKING

INTRODUCTION AND CONCLUSION

C E ERIKSSON
Deputy director of SIK-the Swedish Food Institute, PO Box 5401, S-402 29 Göteborg, Sweden

Industrial cooking is a very broad subject and it was at an early stage decided to divide it into three process groups (chairman):

A. Boiling, frying, roasting and similar processes (C. Skjöldebrand, S)
B. Bread baking incl. its fermentation (O. Tolboe, DK)
C. Further treatment for catering (G. Glew, UK).

Data of some fifty European projects belonging to the three process groups were collected from all participating countries and widely circulated as a project catalog.

There is a growing tendency in industrial cooking to change the production from upscaling of ordinary home kitchen procedures towards the application of process engineering, sometimes in fully continuous processes. This change, however, leads to increasing demands on the implementation into the processes of proper control and measurements to enable the production of high quality, nutritious and safe foods. Much effort is made towards a better understanding of food processes in food laboratories in Europe and elsewhere. To reach a necessary and good cooperation between these laboratories it is essential they talk the same "cooking language", plan, perform, and report their experiments and results in a reproducible and comparable way, and accept some common basic principles of food processing. Therefore, it was decided already at an early stage of this subgroup's work to appoint three working parties with responsibility to study:

1. Terminology in industrial cooking (L.E. George, UK *et al.*)
2. Quality assessment comparision (Y. Mälkki, SF, *et al.*)
3. Principles of food process modelling (K. Paulus, FRG, *et al.*)

both for the present cooperation within subgroup 2 and a wider future use.

About once a year workshops were arranged at different food research institutions, jointly for the three process groups and working parties. The former also met separately in the meantime to discuss common matters. The annual workshops consisted primarily of group labor for each separate process group and working party with intermediate reporting and discussion sessions when all participants were together. The final effect of this concerted action, as reported here, is to a large extent a cumulative effect reached during these workshops.

The three working parties have elaborated a terminology list, guidelines for planning and reporting industrial cooking investigations and principles of process modelling respectively. The terminology and guidelines are briefly presented in these proceedings. The terminology, in English, French and German, and the guidelines will, however, form a later separate publication. Dr. Paulus presents in an individual chapter a categorization of the most common cooking processes. This will be of great importance both for the systematization of the field as such and also for describing quality and nutritional changes during cooking as well as for the optimization of the processes with regard to a number of quality, nutritional and other parameters.

Process group A concentrated its efforts on three areas selected according to the already mentioned process model principles (see introduction to process group A). These areas were cooking (heating and cooling) in scraped surface heat exchangers, oven baking of meat products and boiling of potatoes in catering establishments. The prime outcome of the work is modelling of the first two processes and a new boiling equipment primarily for potatoes. Process group B also concentrated its work into three areas, namely, process conditions vs bread quality, the importance of salt and fiber in bread baking from both a nutritional and a process point of view, and finally optimization of the baking process with regard to bread quality. Progress is reported on the sponge process, instrumental and sensory measurements of bread aroma and texture, the technical need of salt in baking and energy issues using both conventional and unconventional heating. Process group C has worked on four catering subjects: meals prepared from fresh and pre-prepared components respectively, warmholding, and reheating of food. This subgroup's work can lead to altered routines and better equipment for the preparation of high quality and nutritious catered food.

At this concluding seminar reports are presented, both orally and as posters from the three working parties and the three process groups. In addition, there are contributions regarding the state of the art of industrial cooking in the United States (D. Lund, USA), recommendations for future research needs for the baking industry (W. Seibel, FRG), consumer aspects (P. Koivistoinen, SF, and R. Remy, B.) and nutritional aspects (A. Bender, UK) on industrial cooking.

Finally, I express my sincere thanks to the invited speakers and to the participants of subgroup 2 for their excellent work and enthusiasm which contributed significantly to a better understanding of a number of important industrial cooking processes. It also brought many European research groups in this field together into a fruitful cooperation which, hopefully, will continue even now after the COST 91 exercise is finished.

IMPACT OF INDUSTRIAL COOKING OF FOODS ON ITS NUTRITIONAL AND QUALITY CHARACTERISTICS

D. B. LUND
Professor of Food Process Engineering
University of Wisconsin-Madison

Summary

Industrial cooking systems in the U.S. have undergone relatively modest changes recently. Most improvements are in production capacity and energy efficiency. Aseptic processing of liquid foods and retort pouch processing are gradually assuming a market share although many predict that there will be widespread and rapid adoption by the food industry in the next five years. Microwave applications in the food industry have not developed as rapidly as originally anticipated. As long as relatively modest geometries are used for food products and the economics of energy favor steam or natural gas for heating, application of microwave heating may continue to be adopted at a low rate.

The impact of industrial cooking on quality characteristics of food has been well-documented especially for nutrients and colors. Additional research is needed on the effect of processing on texture and starch gelatinization.

1. INTRODUCTION

The term "cooking" of foods conjures up the image of a chef (professional) or amateur preparing a food item which will be served hot. The term "industrial cooking" of foods, on the other hand, suggests preparation of foods in an industrial setting which requires the application of heat in the process. The resulting "processed" food product may or may not be heated immediately prior to consumption. The application of heat to achieve temperatures above ambient for prescribed times is usually done to extend the shelf-life of the product but there are concommitant changes in the food besides the reduction in microorganisms and enzymes. Thus industrial cooking may result in changes in texture, color, and nutritional quality. Changes in these quality characteristics frequently dictate the application of the process in industry. The purpose of this paper is to review the state-of-the-art of industrial cooking in the U.S. and the impact of heating on quality characteristics of foods.

2. STATE-OF-THE-ART OF INDUSTRIAL COOKING IN THE U.S.

Industrial cooking can conveniently be categorized into thermal processes for extension of shelf-life and baking. Blanching, pasteurization and retorting are all applied to reduce microbial or enzymic activity to extend the shelf-life of the product. Improvements in systems that are used for these processes have come primarily in production capacities and energy efficiency. In blanching, closed steam blanchers have been adopted for frozen vegetables to improve efficiency. Energy efficiency is limited, however, by the necessity of venting the blancher to remove non condensible gases released by the tissue upon heating.

Improvement in conventional retorting has also been primarily in the area of energy efficiency. For example, California Canners and Growers (1) have modified atmospheric can cookers with a vapor recycle steam jet which reduced steam usage. This is now standard equipment on all FMC belt-type steam sterilizers, pasteurizers and blanchers and on screw-type cookers and steamers.

In 1978, the development of the retort pouch by U.S. Army Natick R and D Command, Continental Flexible Packaging and the Flexible Packaging Division of Reynolds Metal Company received the IFT Food Technology Industrial Achievement Award (2). Many technologists expected the pouch to achieve commercialization to a much larger degree than it has. Currently Europe and Japan have sizeable markets for pouch processed foods and the market is growing. Problems in the U.S. market centered around approval of adhesives and verification of constituent migration rates as well as poor line rates and marketing approaches to introducing the pouch. The major consumer of retort pouch products remains the U.S. military with their Meals-Ready-to-Eat (MRE) concept. Undoubtedly, the institutional market for products in retort pouches is sizeable and sales will continue to grow in that area. Pouch packaging line speeds continue to make steady improvements with current line speeds in the area of 300 pouches per minute.

Rigid plastic ovenable up to 220°C has been developed by Eastman Chemical Products and will appear soon in the marketplace (3). It can be heated in conventional or microwave ovens and has the advantage (over ovenable paperboard and aluminum foil containers) of looking like china. The containers are made from polyethylene terephthalate (PET) and cost less then thermoset polyester trays. With approximately 20% of all American households currently equipped with microwave ovens and preditions of 40% by 1985, it is anticipated that processing of products in ovenable containers will be steadily increasing.

Aseptic packaging and processing in the U.S. is expected to exceed 500 million units in 1983 (4) and the food industry is moving rapidly based on the following advantages of aseptic processing: (1) ability to use thermoplastic and paper-based materials in applications previously confined to metal or glass, (2) inexpensive packaging creating a single service market, (3) in many cases, superior flavor, texture and nutrient in the product and (4) decreased energy costs in processing. All current aseptic applications in the U.S. are liquid products which are normally pasteurized or acid products (fruits). In the near future aseptically processed dog food (currently marketed in Europe) will appear in the U.S. market. This is a non-acid product consisting of separately processed chuncks of meat and gravy. Adoption of aseptic processing appears to be currently limited by mechanical design, not process design.

Microwave applications in food processing and industrial cooking has followed the same tortuous path to commercialization as aseptic processing. Heralded as a major innovation in the forties and fifties, microwave heating has made relatively minor impact on commercial food processing. Recently Ohlsson (5) summarized food industry applications of microwave and high frequency heating in the following table.

Table I. Present application of dielectric heating in the food field on an industrial (I) or pilot-plant (P) scale for high frequency (HF) and microwave heating (MW).

Process	Products	Application Scale HF	MW
Thawing-tempering	Meat, fish, berries	I	I
Cooking	Chicken, bacon, meat patties, fish	-	I
	Sausage	-	P
Rendering	Lard and tallow	P	I
Blanching	Corn on the cob, potato, fruits	P	P
Proofing	Doughnuts, bread	-	I
Baking	Doughnuts	-	I
	Bread	P	P
Roasting	Peanuts, coffee, cacao	I	P
Pasteurization	Yoghurt, bread, potatoes	I	I
	Meat products	-	P
Sterilization	Pouched foods	-	P
Finish drying	Pasta, onions, chips	-	I
	Biscuits, cakes, crisp bread	I	-
Vacuum drying	Fruit juice concentrates	-	I
	Cereals, beans, nuts	-	P

From: Ohlsson (5)

The potential for microwave application is tremendous but there are few installations for industrial cooking in the U.S.

Industrial baking processes in the U.S. are undergoing improvement for energy efficiency primarily through better process instrumentation and control. Microprocessor monitor and control systems have the potential of revolutionizing food process systems not only in baking but in other unit operations also. Energy efficiency improvements also maintain a high priority in institutional reheating of foods. This is an area which has never received the attention from equipment manufacturers that it deserves. For manufacturers, energy is finally becoming a basis for development and adoption of technology.

3. IMPACT OF INDUSTRIAL COOKING OF FOODS ON NUTRITIONAL AND QUALITY CHARACTERISTICS

Improvements in thermal processes applied to foods is identified through two mechanisms: experimentation and prediction. The method of choice is prediction because it is less costly and much less time consuming. However, ability to predict consequences of a thermal process on a chemical or physical property of a food constitutent implies *a priori* knowledge of a model for the effect of process conditions on rates of change. Recently there have been significant quantities of data generated on the effect of process parameters on food constituents. The effects of temperature on rate constants for various physical and chemical processes and the relative rate of the process at 121°C (D-value) are given in Table II.

Table II. Order-of-magnitude of D_{121} and E_a values for physical and chemical processes

Process	D^1 121 (min)	E_a^2 (KJ/mol)
Physical Properties	widely	2-40
Water Vapour Pressure		40
Water Diffusion Coefficient		8-40
Heat Transfer Coefficient	varying	2-30
Physical Rate Process (drying, crystallization, etc.)	varying	17-60
Chemical Reactions		
Vitamins	100-1000	80-120
Color, Texture, Flavor	5-500	40-120
Enzyme Inactivation	1-10	50-400
Vegetable Cells	0.002-0.02	400-500
Spores	0.1-5.0	200-300

[1] D_{121}-time at 121°C for a first order response to decay by 90%.
[2] E_a-Arrhenius activation energy.

Generally if a process has a small activation energy, factors other than temperature are used to improve the process (e.g. increased agitation, decreased particle size, etc.). From Table II, it is observed that physical properties are less temperature dependent than chemical reactions and thus if a physical process governs the rate of the process, temperature will not necessarily be a primary consideration.

3.1 Effect of Heat on Nutrients in Foods.

There are many current reviews in the literature on the effects of heat processing on nutrients in foods (e.g. (6), (7)) and they will not be reviewed here. Many nutrients are susceptible to loss of biological activity upon the application of high temperatures for short periods of time and their relative rates of destruction and dependence on temperature are compared to vegetative cells and spores in Table II.

Improvements in canning with respect of nutrient retention has not been a result of systematic investigation. Only recently have "optimization" procedures been applied to this problem. Previously the driving forces for process change were either process throughput or product quality. Interestingly, any change in the process resulting in increased throughput generally resulted in improved nutrient retention. Basically the industry moved toward higher temperature-shorter time (HTST) processes to maximize throughput. This concurrently maximized product quality and nutrient retention for most products. The fact that the reaction resulting in inactivation of micro-organisms and/or spores is much more temperature dependent than those for quality factors and nutrients accounts for the simultaneous maximization (Table II). However, increasing retort temperature was basically limited by safety consideration in maintaining a large pressurized vessel. Furthermore, at very high temperatures (e.g. 140°C), enzymes were active following the heat treatment and these limited the shelf life of the product. For conduction heating foods, there was also a loss of color or excessive softening of product at the periphery of the container resulting in loss of product quality.

To overcome this problem, aseptic canning was developed particularly for viscous liquid foods. The limit on maximum temperature of processing for particulate foods, however, is still around 132°C as pointed out by Hersom and Shore (8). They describe an aseptic processing system for sauce and solids and concluded that for vegetables such as carrots, the maximum cross-section dimension is about 2 cm(3/4 in) with a practical upper limit on temperature of about 132°C (270°F). These maxima are based on exessive softening and subsequently possible sloughing and damage on the outside of the vegetable piece and enzyme survival and regeneration on the inside. Thus for all practical purposes the maximum working temperature for retaining product quality and increasing product throughput for heat processing natural tissue systems which heat rapidly is about 130-140°C.

The problem of maximizing nutrient retention for products which heat by conduction is more complicated than just described. Several investigators (9-13) have considered maximizing nutrient retention in constant and variable retort temperature profiles. Results clearly show that unique control strategies exist for each process.

A word of caution should be mentioned here with regard to application of kinetic parameters for process improvements. Thompson (6) concluded that there is evidence that kinetic models developed under isothermal conditions may not be integratable and thus applicable to nonisothermal conditions. Furthermore, for microbial inactivation, there may be some history effect which precludes applying isothermally derived models to nonisothermal conditions. Lund (21) also cautioned on the interpretation of models applied to foods. Thus, although existing data can be used to suggest strategies for process improvement, the final proof must be a relatively large scale industrial experiment.

3.2 Effect of Heat on Quality Characteristics of Foods.

Texture, flavor and appearance are perhaps the most important characteristics of foods because they are attributes the consumer can readily assess. Although the consumer is becoming increasingly aware of attributes such as nutritional content (including presence of toxic factors) and microbiological quality, these attributes are more difficult to assess, quantify and use as a basis upon which to distinguish between competing products. Frequently, however, development of operations (machine and process design) is a trial and error procedure because there is a lack of quantitative data on the effects of process parameters on quality attributes. Recently Lund (14) reviewed the effects of heat on reactions influencing color, flavor and texture.

It was concluded that temperature dependence of color and flavor change should be similar to that of vitamins because similar chemical mechanisms are involved. For texture changes, however, there have been relatively few attempts to develop parameters which can be used to quantify the effect of heating. The "kinetics" of change in texture as a function of process have generally been reported as first order models. Lund summarized existing data in Table III.

Several observations can be made upon examination of the data. First, most of the studies have been done on high moisture samples. The effect of water activity during heating on texture has not received much attention.

Table III. E_a for texture changes in food systems.

Component	pH	Temp. (°C)	E_a(kcal/mol)
Beef semitendinosus muscle	natural	55-59	140
Kangaroo tail tendon	1.8	50-70	67
	4.2		145
	7.2		152
	12.5		77
Kangaroo tail tendon	6.0		141
Kangaroo tail tendon with			
0.1 mol/l NaCl	6.2		95
1.0 mol/l NaCl	6.3		93
2.0 mol/l NaCl	6.2		83
4.0 mol/l NaCl	6.1		87
sat. with NaCl	6.0		75
External portion aponeurctic sheet	natural	60-70	106
Ligamentum muscle	natural	60-70	76
Deep pectoral muscle	natural	50-60	44
Semimembranous muscle	natural	60-100	23
Peas	natural	77-93	20
Cucumbers	natural	71-93	22

From: Lund (14)

The second observation is that there are no studies in which moisture is migrating into the sample resulting in modification of texture. Although cooking of dry legumes and cereal grains are examples of processes involving simultaneous moisture migration and softening, there has been no attempt to develop a generalized model to describe the processes.

Finally, generally temperature dependence of textural characteristics of meat (muscle) is greater than that for plant tissue. The apparent activation energy for muscle is in the range associated with structural changes induced by proteins, whereas for plant tissue it is in the range associated with chemical reaction. For muscle tissue, the temperature dependence of change in texture is extremely complex because of the collagen shrinkage reaction and the collagen-gelation transformation. In addition there are changes in the structure of the muscle protein itself. For plant tissue, hydrolysis of cell wall constitutents, swelling due to expansion of gases and heat induced changes in water holding capacity (e.g. gelatinization of starches), can all affect texture.

Starch gelatinization is widely used in food processing to provide unique textural and structural characteristics to products. Studies on the kinetics of starch gelatinization are very limited, even though the phenomenon of starch gelatinization has long been recognized. Currently there is no definitive kinetic model for starch gelatinization. Moreover, the phenomenon of gelatinization itself is still not fully understood. However, several investigators have used a first-order model for starch gelatinization.

A summary of the reported studies on kinetics of starch gelatinization together with the values of activation energy is given in Table IV. Suzuki et al (17), Bakshi and Singh (16) and Wirakartakusumah (18) (DSC Isothermal Method) reported similar activation energies (19-25 kcal/mol, respectively), while Kubota et al (15) reported 14 kcal/mol. The difference among these four values was probably due to the difference in the nature of raw material studied and the method of choosing a zero heating time reference.

Table IV. Summary of reported studies on kinetics of starch gelatinization.

Type of Study	Method	Temp. Range (°C)	E_a (kcal/mol)
Gelatinization of rice starch (15)	Flow consistency with viscometer	70-85	14
In situ gelatinization of brown rice (16)	Amlose/iodine blue value	50-85	25
Cooking of rice (17)	Compressibility with plastometer	75-100	19
Gelatinization of rice starch (18)	DSC-Isothermal	65-75	25
	DSC Non-Isothermal	70-75	44-73[d]

[d]The activation energy increased with temperature of gelatinization.

The E_a's from the Non-Isothermal Method of Wirakartakusumah (18) are much greater and appear to be dependent on temperature, E_a increasing with increasing temperature. This can be rationalized by considering starch gelatinization as a semi-cooperative melting process. It has been well-documented that a starch granule is semicrystalline and consists of amorphous and crystalline regions. Due to its lower order of crystallinity, the amorphous region hydrates initially and is more labile to heat treatment with water present than the crystalline region. Since the amorphous region is an integral part of the starch granule, the destabilization and swelling of the amorphous region during heating at low temperature facilitates further destabilization of the crystalline region by tearing molecules from crystallites as the heating temperature is increased. At low temperatures of heating near the onset temperature, gelatinization occurs randomly in each granule primarily in the amorphous regions of the granule. Upon continued heating at the same temperature, eventually all of the amorphous regions are destabilized and crystalline regions begin to "gelatinize." However, the extent of gelatinization is dependent on temperature and as the heating temperature increases, the extent of crytallites that are gelatinized also increases. When the temperature is sufficiently high, both the amorphous and crystalline regions are gelatinized. Thus, the amorphous regions may be interpreted as a "promoter" for further gelatinization of more crystalline regions of the granule. During the destabilization process, a more open structure of the starch develops. If this is the case, then the isothermal data should represent gelatinization of the more crystalline regions. Generally, the more resistant granules would exhibit a higher E_a, presumably because of the large number of hydrogen bonds. The high order of crystallinity produces a sharper peak on X-ray diffraction (8) and higher gelatinization temperatures (9). Donovan and Mapes (90) separated the amorphous regions from the crystalline regions of potato starch using acid hydrolysis. In the present study, the E_a from the isothermal procedure is 25 kcal/mol which is lower than the E_a from the non-isothermal procedure. This may be due to the role that amorphous regions play in destabilizing the crystalline regions during gelatinization, a role which results in lowered activation.

In summary, the process of starch gelatinization does not appear to be a first order process although it can be modeled as a first order process over limited extents of gelatinization. Rate constants are highly dependent on the starch source and its history.

4. SUMMARY

At the opening COST meeting in Dublin in 1977, Burton (19) identified several areas requiring further research in thermal processes. Some of those recommendations are still valid today including: (1) thermal death data for microorganisms need to be extended to higher temperatures; (2) discrepancies between laboratory thermal death data (generally generated isothermally) and results obtained for practical plant at high processing temperatures (UHT) need investigation; (3) vitamin losses during processing and storage, and the interaction between different food constituents, need definition; (4) methods of container sterilization during aseptic filling should be investigated and given a scientific basis; and (5) techniques for ultrahigh temperature processing of liquids containing particles should be studied. To this list could be added: (1) starch gelatinization phenomena under extrusion and non-extrusion conditions needs to be investigated; (2) effect of process variables on textural changes in food requires research which includes the correlation of mechanical properties measured objectively to sensory analysis; (3) increased application of on-line monitor and control of physical and chemical properties of foods should be expanded and will lead to improved products; and (4) development of improved packaging materials is necessary for continued development of retort pouch and aseptic processing.

Finally, a recent poll (20) of the top 100 food companies in the U.S. offered some interesting insights into what they perceive as significant developments in the last five years (Table V) and the biggest problems facing the food industry that research can help solve (Table VI). As there is a shift in the U.S. from publicly funded research to industrially funded, undoubtedly the areas of highest interest to industry will receive the attention (Table VI).

Table V. What would you consider the five most significant products/processes/technological breakthroughs within the food industry during the past five years?

Aseptic packaging (58%)
Biotechnology/genetics/enzyme technology (42%)
Retort pouches/products (36%)
High fructose syrups (13%)
Microwaveable packaging (13%)
Energy/efficiency (13%)
Reformed/restructured foods (10%)
Irradiation (10%)
Freeze-Flo technology (10%)
Tamper-resistant/evident packaging (6%)
Membrane technology (6%)

From: (20)

Table VI: What do you consider the biggest problems facing the food industry that research can help solve?

(In order of most mentions)
Energy-cost/efficiency
Nutrition/health
Processing costs
Packaging costs
Labeling
Commitment to research
Wastewater/wastes
Consumer perception/misconceptions about ingredients/foods
Regulatory activity

From: (20)

Acknowledgment: This is an contribution from the College of Agricultural and Life Sciences, University of Wisconsin-Madison.

REFERENCES

1. _______. (1983). Vapor-recycling retrofit for atmospheric cookers saves $4,000-$8,000/unit/yr. Food Processing 44(7):82.
2. Mermelstein, N.H. (1978). Retort pouch earns 1978 IFT Food Technology Industrial Achievement Award. Food Technol. 32(6):22.
3. _____. (1983). PET challenges ovenable board. Food Engineering 55(1):91.
4. _____. (1983). The aseptic report. Food Engineering 55(7):65.
5. OHLSSON, T. (1983). Food industry applications of microwave and high frequency heating. Food Engineering News May.
6. THOMPSON, D.R. (1982). The challenge in predicting nutrient changes during food processing. Food Technol. 36(2):97.
7. LUND, D.B. (1982). Influence of processing on nutrients in foods. J. Food Protection 45:367.
8. HERSOM, A.C. and SHORE, D.T. (1981). Aseptic processing of foods comprising sauce and solids. Food Technol. 35(5):53.
9. LUND, D.B. (1977). Design of thermal processes for maximizing nutrient retention. Food Technol. 31(2)71.

10. SAGUY, I. and KAREL, M. 1979. Optimal retort temperature profile in optimizing thiamin retention in conduction-type heating of canned foods. J. Food Sci. 44:1485.
11. TEIXEIRA, A.A., DIXON, J.R., ZAHRADNIK, J.W. and ZINMEISTER, G.E. (1969). Computer optimization of nutrient retention in the thermal processing of conduction-heated foods. Food Technol. 23(6):137.
12. TEIXEIRA, A.A., ZINMEISTER, G.E. and ZAHRADNIK, J.W. (1975). Computer simulation of variable retort control and container geometry as a possible means of improving thiamin retention in thermally processed foods. J. Food Sci. 40:656.
13. THIJEESEN, H.A.C., KERKHOF, P.J. A.M. and LIEFKENS, A.A.A. (1978). Short-cut method of the calculation of sterilization conditions yielding optimum quality retention for conduction-type heating of packaged foods. J. Food Sci. 43:1096.
14. LUND, D.B. (1982). Quantifying reactions influencing quality of foods: Texture, flavor and color. J. Food Proc. Pres. 6:133.
15. KUBOTA, K., HOSOKAWA, Y., SUZUKI, K., and HOSAKA, H. (1979). Studies on the gelatinization rate of rice and potato starches. J. Food Sci. 44:1394.
16. BAKSHI, A.S. and SINGH, R.P. (1980). Kinetics of water diffusion and starch gelatinization during rice parboiling. J. Food Sci. 45:1387.
17. SUZUKI, K., KUBOTA, K., DOMICHI, M., and HOSAKA, H. (1976). Kinetic studies of cooking of rice. J. Food Sci. 41:1180.
18. WIRAKARTAKUSUMAH, M.A. (1981). Kinetics of starch gelatinization and water absorption in rice. PhD Thesis. Department of Food Science, University of Wisconsin-Madison.
19. BURTON, H. (1977). Quality aspects of thermal sterilization processes. In "Food Quality and Nutrition." W.K. Downey (ed.) Applied Science Publishers. London.
20. ________. (1983). 1983 R and D survey results: Top 100 Food Companies. 44(7):42.
21. LUND, D.B. (1983). Considerations in modeling food processes. Food Technol: 37(1):92.

TERMINOLOGY OF COOKING

L.E. GEORGE

Ministry of Agriculture, Fisheries and Food, Great Westminster House, Horseferry Road, London SWIP, England

B. DRAKE

and

C.E. ERIKSSON

SIK-The Swedish Food Institute, P.O. Box 5401, S-402 29 Göteborg, Sweden.

A working party, developed a special publication with the aim that it be of assistance for food scientists and technologists in selecting appropriate cooking terms. It is intended to be used together with the Guidelines for Quality Assessment Comparision (developed by the working party of (Mälkki _et al_.), for the planning of experiments and the presentation of experimental data involving industrial cooking and catering (foodservice). The number of terms used in gastronomical cooking is very large and sometimes inconsistent. This publication is limited to the most common words and expressions used in the home and in industrial cooking (using heat) as well as in catering (foodservice). A few ancillary terms are also included.

Since most scientific and technical reports for international readers appear in one of the languages of English, French or German, this publication has been made multilingual. It consists of two parts. One basic part, in English, French and German, contains (a) definitions of physical principles and mathematical expressions concerning cooking of food, and (b) definitions of cooking terms (nouns and verbs) in alphabetical order. The other part consists of a number of supplements, one for each language other than English. Each supplement contains a plain alphabetical list of the English terms translated into the other language and vice versa. With the basic volume and the supplement in for instance Finnish, it is possible to select the right term or definition in either English, French or German starting from the correct word or expression in Finnish. The publications can of course also be used in the reversed way. Presently, supplements are available in Danish, Dutch, Finnish, French, German, Greek, Italian, Serbo-Croatian, Spanish and Swedish. Other languages are planned to be added later. The basic part, in English, French and German, which also contains the English word list, as well as the above guidelines for quality assessment comparision, will be issued as a separate publication from the COST 91 concluding seminar. Inquiries about this publication can be made to C.E. Eriksson.

(L.E. George served as chairman of this working party until his retirement September 1982 and was succeeded by C.E. Eriksson.)

QUALITY ASSESSMENT COMPARISON

Y. MÄLKKI[1], J.-M.BRÜMMER[2], A. BOGNAR[3] and R. GOUTEFONGEA[4]

[1]Technical Research Centre of Finland,
Food Research Laboratory
SF-02150 Espoo 15, Finland

[2]Bundesforschungsanstalt für Getreide- und Kartoffelverarbeitung,
D-4930 Detmold, Germany (F.R.)

[3]Bundesforschungsanstalt für Ernährung,
D-7000 Stuttgart 70, Germany (F.R.)

and

[4]Laboratoire des aliments d'origine animale I.N.R.A.
F-44072 Nantes, France

In the framework of COST 91 Subgroup 2 (Industrial Cooking), the above mentioned working group has studied guidelines for defining process conditions and quality evaluation for heat processed foods, in order to facilitate the comparison of results from different studies. Because the critical quality criteria to be monitored vary according to the commodity, country and the process, no uniform procedure is recommended. However, as far as possible, the following points should be considered in planning the experiments and presenting the data:

1. PROCESSING CONDITIONS

1.1 Time-temperature programmes of the treatment, both of the heating and cooling medium and of the product surface and centre. Reference should be given to typical time-temperature programmes of the commodity and process.

1.2 When the heating or cooling medium is air, its velocity and absolute or relative humidity or data enabling estimaton of their order of magnitude.

1.3. If heating medium is a liquid, data enabling the estimation of possible leaching effects.

1.4 If water is present in a limited amount, its amount in relation to the food or to some specific component of it.

1.5 Water activity of the product surface and product centre at the beginning and at the end of the treatment.

1.6 C-values of the treatment, stating for which property the determination has been made.

2. PROPERTIES OF THE MATERIAL TO BE HEATED

2.1 When using vegetable material, a statement on the plant variety, maturity and any possible pretreatment.

2.2 When using meat or fish material, a statement on the species and on the part of the carcass, content of fat and connective tissue and data on any ante-and post-mortem treatments and quality indices determined. For cut and comminuted products dimensions of particles and direction of fibres should be included.

2.3 Presence or absence of ingredients known to have an influence on thermally induced changes(e.g. those participating in Maillard-reactions, acids, neutralizing, chelating or complexing agents etc.)

2.4 Information concerning components which are thermally unstable, the activation energies and reaction velocities of their decompositions and other reactions within relevant temperature areas.

2.5 z - values of the thermally induced changes, stating within which temperature range the value has been determined.

3. CHEMICAL, PHYSICAL AND SENSORY STUDIES

3.1 Analytical data on heat-sensitive substances used as criteria for the effect of the treatment (e.g. enzymes, vitamines, hydroxymethylfurfural).

3.2 For determining colour changes, determining the colour, if possible, in the tristimulus system or using Colour Atlas comparison.

3.3 For texture properties, any data making possible the calculation of forces of deformation in physical units and giving the exact conditions of deformation, such as size and dimensions of the sample, measuring device, speed of deformation and depth of penetration.

3.4 For flavour changes, any supporting instrumental or analytical data on the content of volatile or taste-giving soluble components, threshold determinations, comparisons with chemically defined reference substances, or use of standardized vocabulary if available. Any known information about the relation between stimulus intensity and response should be indicated.

3.5 As far as possible, trained panels and procedures standardized either generally (e.g. 1,2,3) or for the commodity studied (e.g. 4) should be used. For defining methods and conditions of the sensory evaluation, the guidelines of the IFT Sensory Evaluation Division (5) should be used.

REFERENCES

1. Sensory analysis of Food, General Guide to Methodology. ISO/TC34, Subcommitte 12 "Sensory Analysis". AFNOR, Paris 1979.
2. DIN-Normen für Sensorik, Beuthverlag, Berlin, 1981.
3. ASTM Special Technical Publications Nos. 433, 434, 682, 758.
4. SEIBEL, W., BRÜMMER, J.-M. and HAUPTMANN, S., Sensorik-Ausbildung für Sachverständige der DLG-Qualitätsprüfung für Backwaren. Getreide, Mehl und Brot 36 (1982) 54-56, 77-84, 104-112, 136-140, 162-167.
5. ANON., Guidelines for the preparation and review of papers reporting sensory evaluation data. Food Technology 35 (1981):4, 16-17

MODELLING IN INDUSTRIAL COOKING

K. O. PAULUS
Federal Research Centre for Nutrition, D-75oo Karlsruhe
(Report of the Process Model Working Party)

Summary

This report summarizes considerations and discussions within COST 91/ Subgroup 2 on process models. As industrial cooking comprises many different cooking processes it is necessary to establish categories for the most important processes and effects in the products, in order to systematize the whole field. Cooking is realized primarily by heat and mass transport, mechanisms which are responsible for quality changes. Cooking processes for solid foods are divided into 3 subgroups with regard to process conditions, and corresponding models for transport mechanisms can be established. Another model category comprises kinetic models to describe quality changes; examples are given. A third model category for process optimization is mentioned but not discussed in detail.

1. INTRODUCTION

According to the definition of the term cooking as interpreted by the Working Party Terminology, cooking refers to a thermal process leading to positive and/or negative changes in the quality of a food product. Thus cooking is an essential food engineering process within COST 91 and indeed the most important process in food engineering, as the majority of foods is consumed only in cooked form.

No fundamental differences are found if one compares conventional industrial food processing to industrial cooking. The generalized flow diagram in figure 1 shows that processing in large kitchens also comprises a different number of unit operations. They are supplemented by the decisive element of preparation as the actual goal of the entire process (1). There are still some more points in which industrial processing and industrial cooking differ. There is, first, a variety of different cooking processes characterized by highly different processing conditions. Food intended for out-door feeding, secondly, is mainly produced by smaller companies which usually do not produce the meals on a large scale, but have to adapt their production to requirements changing from day to day. Also for this reason it seems advisable to systematize cooking processes with regard to the relevant process parameters, with the aim of obtaining information what quality changes are supposed to take place in what product under the given processing conditions of a certain cooking process.

2. SYSTEMATIZATION OF THE PROBLEM

During cooking chemical, physical and also microbiological changes are taking place. To ensure the desirable changes, and to minimize undesirable ones it is necessary to quantify these changes, at least for the essential product parameters, as a function of the process conditions. Data obtained from systematic studies in this respect are rare (2).

Figure 1

Generalized flow sheet of industrial cooking

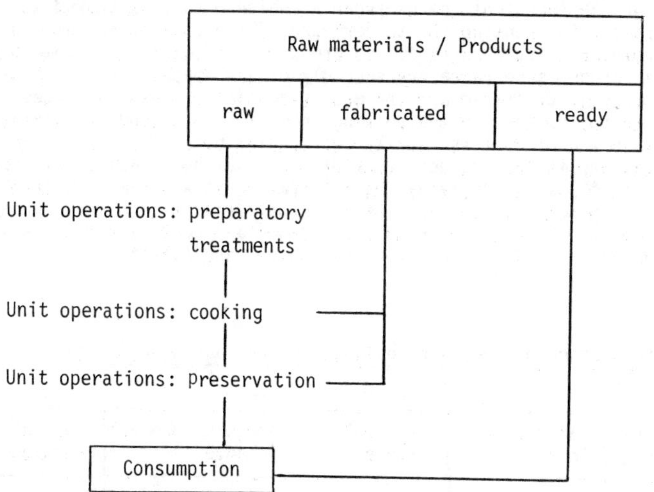

The cooking process is of primary interest. The majority of cooking processes may be classified into two categories of processes, namely moist and dry cooking. The most important processes are characterized in table 1.

Table 1

Characterization of some important cooking processes

Category	Moist cooking		Dry cooking		
Process	boiling	steaming	baking	frying	grilling
Heating medium	water $\vartheta \geq 100^{o}C$ $a_w = 1$ $p \geq 1$ bar	steam $\vartheta \geq 100^{o}C$ $a_w = 1$ $p \geq 1$ bar	air $\vartheta \gg 100^{o}C$ $a_w < 1$ $p = 1$ bar	oil/fat $\vartheta > 100^{o}C$ $a_w < 1$ $p = 1$ bar	IR $\vartheta \gg 100^{o}C$ $(a_w < 1)$ $p = 1$ bar

There are still much more variants in each category than the processes listed in the table; it is also possible to combine processes of either category or processes not directly classifiable with one of the cate-

gories, such as cooking by using microwaves, for instance. The table shows that the essential process parameters are:
- temperature of the cooking medium
- water activity of the cooking medium, and
- pressure at which the food is cooked.

In view of the fact that the pressure at which a food is cooked is primarily a factor influencing the temperature of the cooking medium, and that, furthermore, quality changes due to pressure fluctuations of some bar in the relevant pressure range are scarcely of any importance, temperature and water activity indeed are the most essential process parameters. Depending on the purpose of the cooking process, the intended temperature increase in the product is usually accompanied by a reduction of the water activity in the product. Cooking processes hence may be defined, in a strict sense, as insteady thermal treatments which may be accompanied by a drying process in the case of a process falling into category dry cooking.The main categories of possible changes and general effects are presented in table 2. Their influence on sensory and nutritive quality, the

Table 2

Characterization of important quality changes during cooking

Category	Material loss	Physical change	Physicochemical change	Chemical reaction
Distinct changes	dissolved solids volatiles	structure colour permeability	turgor solids	protein aminoacids lipids carbohydrates vitamins minerals flavour
Effect	loss	softening discolouration increase	loss loss	denaturation destruction oxidation hydrolysis loss loss change

two essential quality criteria, is predominantly negative. However, as far as the different mechanisms such as the cooking process itself, for instance, are concerned, positive effects appear first, until, from a certain intensity of the treatment on, negative effects prevail. For these processes which are typical for food processing, certain model concepts and pertinent date exist (3).

One recognizes from this brief survey that a subdivision of [illegible] problems into transport mechanisms and mechanisms effecting qua[illegible]ges is advisable (see also figure 2). Consequently also two ty[illegible]ls, namely transport models and kinetic models, are required, as [illegible] figure 3. First, a description of the engineering process is [illegible] as

Figure 2

Industrial cooking and relevant mechanisms with regard to process models

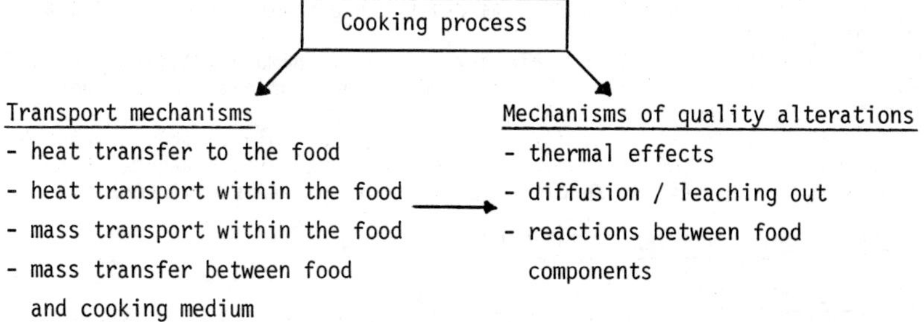

Figure 3

Model categories for industrial cooking

Category I: Transport models

Models to describe within the product:

- o $\vartheta = f(t, x)$
- o $a_w = f(t, x)$
- o $c = f(t, x)$

Category II: Kinetic models

Stochastic models	Interpretation models
o empirical	o scientific
o black box models	o kinetics explainable

Category III: Optimization models

To establish the best process conditions

a basis for construction of the transport models. A description of temperature, moisture or concentration of a certain component in the product as a function of time and place, although governed by the classical laws, is in practice frequently possible only by means of several simplifying assumptions.

During processing, many constituents in and properties of the product may be influenced (see table 2). Resulting quality changes may in principle be described in two ways, namely by stochastic models and by interpretation models. Both are suitable to describe the cooking kinetics; the decisive difference is that in the case of stochastic models the data obtained are described in the best possible way by an empirical approach, whereas in the case of interpretation models the pertinent chemical or physical laws or reaction mechanisms are known and may serve as the basis for interpretation.

Finally there is a third category of models, namely optimization models which help in selecting the best processing conditions with regard to the preservation of essential quality attributes.

This model structure relates in principle not only to industrial cooking, but to all forms of food processing. What matters, however, is to adapt this concept to the specific requirements of industrial cooking.

3. EXAMPLES TO CHARACTERIZE THE MODELS

3.1 Transport models

If not all, so the majority of cooking processes may be allocated to one of the model situations shown in figure 4; one must bear in mind, how-

Figure 4

Transport model situations for cooking solid foods

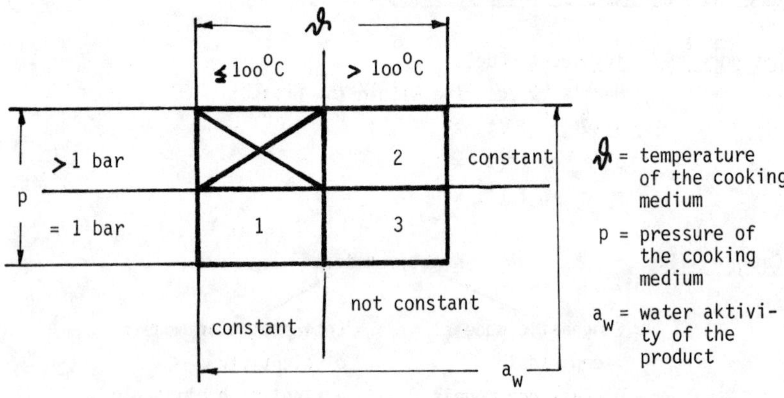

ever, that these apply only to solid foods. In the case of liquid and pumpable products, heat transport is of utmost importance; the description of heat transport generally presents not too many difficulties, however.

The important cooking processes listed in table 1 may be allocated as follows:
Situation 1: boiling, steaming
Situation 2: pressure cooking, steaming
Situation 3: baking, frying, grilling.

Situation 3 undoubtedly is the most difficult one since it includes many complex processes concerning the entire range of heat and mass transport. For the case of baking of meat loafs in a convection oven these transport processes were described in detail; resulting quality changes were also described and interpreted (4 - 6).

3.2 Kinetic models

Stochastic or blackbox models must be used in all cases where the change to be described refers to a complex property of the product. Sensory attributes, for instance, are in the majority of cases caused by changes in different components, and it is therefor scarcely possible to explain changes in these attributes by one simple reaction. In these cases the changes resulting from precisely defined processing conditions have to be measured and described mathematically.

So a special system of categories was developed to control the sensorial quality during boiling of potatoes, for instance (7). This procedure allowed to describe the changes in taste and texture of potatoes during boiling at constant temperature by simple equations of 0-order as a function of time. The temperature dependence of the rate constant could be described by using the Arrhenius equation.

The effect of a cooking process may be indicated by the cooking degree which contains an important coefficient to determine the temperature dependence of a certain change, namely the z-value (8). For taste and texture of potatoes the z-value is about 18°C, depending on the variety (9). The importance of knowing the kinetics was demonstrated in experiments in 21 potato varietes in which the rate constants varied considerably and in which optimal cooking times differed by more than factor 3 (1o).

Whereas the empirical blackbox models reflect only one special situation, interpretation models supply mathematical descriptions of a certain reaction in a generally valid form and can be applied to different products and different reactions/changes; only the constants or coefficients contained in the general mathematical description are subject to change.

As mentioned before, interpretation models concern certain defined ingredients; such models are applicable only when the reaction mechanisms producing a certain change are known. An important substance group are vitamins. Kinetic date were elaborated for different vitamins in different products under different processing conditions; it is remarkable that for a change in vitamins the same interpretation model may frequently be used. The majority of reactions may be described by a zero- or 1st-order mechanism, a general form of which is shown in figure 5 (11). The temperature dependence of the rate constant is demonstrated by means of the second equation, the Arrhenius equation in fact.

In some cases the decrease in vitamins may be described by a first-order mechanism; the decrease in vitamins as a function of temperature corresponds in practically all cases to an Arrhenius equation. From the activation energy the before mentioned z-value may be calculated which is essential for the determination of the decrease in vitamins during the entire process. Some date concerning vitamin changes during cooking processes are compiled in table 3 (12 - 17).

Figure 5

Arrhenius kinetics to describe quality changes in foods

$$\frac{dc}{dt} = k \cdot [c]^n$$

c = concentration of the component
k = rate constant
n = order of the expression

$$k = k_o \exp\left[-\frac{E_a}{RT}\right]$$

E_a = activation energy
R = gas constant
T = absolute temperature

Table 3

Vitamin degradation during cooking

Vitamin	Cooking process	Product	Temperature °C	z-value	Ref.
C	boiling and pressure cooking	spinach	7o - 13o	65	(12)
	pressure cooking	peas	11o - 125	51	(13)
	extrusion cooking	modelfood	6o - 9o		(14)
		a_w=o.69		36	
		a_w=o.8o		34	
		a_w=o.9o		42	
B_1	boiling and pressure cooking	spinach	7o - 13o	6o	(12)
	baking	minced meat	7o - 1oo	22	(15)
	retorting	milk	12o - 15o	31	(16)
A	pressure cooking	beef liver puree	1oo - 125	25	(17)

Such data, supplemented by the time-temperature-history, and if necessary, also by the a_w-time-history, allow to determine the effect of cooking processes on different quality attributes of a certain product.

3.3 Optimization models

All descriptions actually have the aim of determining the optimal processing conditions, provided that all parameters are known which may influence the optimal result. This optimization problem is a problem in the whole field of food engineering which need not be discussed here in greater detail. In recent years, however, this problem was subject of increasing activities in research and development.

4. CRITICAL EVALUATION OF THE ACTIVITIES OF THE WORKING PARTY

The conference held in Dublin in 1977 which lastly initiated the COST 91 activities, made also recommendations in this respect. One of these recommendations said: "A limited amount of information is available about nutritional changes in manufactured food as purchased. Foods can be cooked well or badly and be kept hot for varying periods before being eaten in both institutional and domestic situations." These sentences indicate the entire field of problems to be dealt with by Subgroup "Industrial Cooking". The activities concerning "Modelling in Industrial Cooking" should indeed not have run parallelly to the remaining activities, but should have preceded the latter. For these considerations were supposed to provide the basis for concrete projects, ie. provide a structure for concrete research projects. Even if this was not possible for reasons of organization, continuous occupation with this problem as well as reports submitted and pertinent discussions provided valuable ideas for the work in Subgroup 2. The report now submitted however, should help all future working groups concerned with problems of industrial cooking in planning such activities and defining a systematic approach. Whereas in the past studies concentrated primarily on individual problems out of the entire field, it will be necessary in the future to describe the processes comprehensively. To solve this task, modelling seems inevitable.

REFERENCES

1. PAULUS, K. (1983). Verfahrenstechnik der Speisenproduktion. In: PAULUS, K., WENGER, R. and BRANDSTETTER (Ed.). Gemeinschaftsverpflegung: Erfahrungen und Anwendungen in der Praxis. Wissenschaftliche Verlagsgesellschaft mbH, Stuttgart, 93-118
2. LUND, D.B. (1982). Quantifying reactions influencing quality of foods: texture, flavor and appearance. J. Food Proc. Pres. 6: 133-153
3. LUND, D.B. (1982). Influence of processing on nutrients in foods. J. Food Protect. 45: 367-373
4. SKJÖLDEBRAND, C. (1979). Frying and reheating in a forced convection oven. Thesis, Lund/Sweden
5. SKJÖLDEBRAND, C. (198o). Convection oven frying: Heat and mass transfer between air and product. J. Food Sci. 45: 1354-1358
6. SKJÖLDEBRAND, C. (198o). Convection oven frying: Heat and mass transfer in the product. J. Food Sci. 45: 1347-1353
7. PAULUS, K. and TIRTOHUSODO, H. (1982). Untersuchungen zur Garungskinetik von Kartoffeln. Ernährungs-Umschau 29: 255-259
8. Report of the working party "Terminology", COST 91/Subgroup Industrial Cooking.
9. HARADA, T., TIRTOHUSODO, H. and PAULUS, K. (1984). Cooking kinetics of potatoes. II. Influence of temperature and time of the heat treatment on cooking. Will be printed.
1o. HARADA, T., TIRTOHUSODO, H. and PAULUS, K. (1984). Cooking kinetics of potatoes. III. Influence of composition of the potatoes on cooking. Will be printed.

11. LABUZA, T.P. and RIBOH, D. (1982). Theory and application of Arrhenius kinetics to the prediction of nutrient losses in food. Food Technol. 36: (1o) 66-74
12. PAULUS, K. (1979). Nomographs to determine alterations of essential components in leafy products during thermal treatment in water. J. Food Sci. 44: 1169-1172
13. RAO, M.A., LEE, C.Y., KATZ, J. and COOLEY, H.J. (1981). A kinetic study of the loss of vitamin C, color, and firmness during thermal processing of canned peas. J. Food Sci. 46: 636-637
14. LAING, B.M., SCHLUETER, D.L. and LABUZA, T.P. (1978). Degradation kinetics of ascorbic acid at high temperature and water activity. J. Food Sci. 43: 144o-1443
15. SKJÖLDEBRAND, C., ANÄS, A., ÖSTE, R. and SJÖDIN, P. (1983). Prediction of thiamine content in convective heated meat products. J. Fd. Technol. 18: 61-73
16. HORAK, F.P. and KESSLER, H.G. (1981). Thermische Thiaminschädigung - Eine Reaktion 2. Ordnung. Z. Lebensm. Unters. Forsch. 173: 1-6
17. WILKINSON, S.A., EARLE, M.D. and CLELAND, A.C. (1981). Kinetics of vitamin A degradation in beef liver puree on heat processing. J. Food Sci.: 32-33, 4o

INTRODUCTION TO PROCESS GROUP A (FRYING, GRILLING, BOILING)

C. SKJÖLDEBRAND

Division of Food Engineering, Lund University, P.O. Box 50, Alnarp, Sweden.

This introduction will give a summary of the work that has been done within process group A in the COST 91 Subgroup 2. It will also make a presentation of the research cooperation in the process group as well of the papers that are representing these projects.

The work of the process group A started in Gothenburg 1980. Subgroup 2 was divided into three process groups one of which is process group A called "Frying, grilling, boiling and similar processes". The intentions were to try to follow the recommendations made at the Dublin seminar 1977 for the dependence of these processes on the quality and the nutrition in the collaboration work.

DESCRIPTION OF THE PROCESSES

Boiling, roasting, frying and grilling of foods are processes wellknown since the origion of mankind. Heat is supplied, in different ways to the foodstuff to increase the flavour and the taste and increase storage time. Furthermore due to protein denaturation and other changes in the product the digestability is increased. In recent years the usage of industrial cooking was incresed tremendously and the demand for nutritional, welltasting "convenience" foods forced the food industry and the food science research departments to start developing processes and to start basic research work within these operations.

When heat is supplied to a foodstuff chemical and physical reactions and microbiological destruction take place. These reactions may be desired or not desired. The factors initiating and influencing them must be known to be able to optimize the different operations. The reactions involved are often interrelated depending on time/temperature treatment and water activity.

What differ these three processes? It is very difficult to clearly separate them from each other. The terminology is also different in different countries and languages. In the following frying,and grilling differ from boiling in such a way that the ready made product will have a brown dry crust in the first two processes. The difference is caused by the origin and the temperature of the heating media. In frying roasting and grilling the heating media are liquid fat, heated surfaces, air (moist or dry) or radiators. Boiling is very often carried out in water or steam and at temperatures below or around 100°C throughout the process. Thus the operation is also boiling when sauces and soups (also consisting small pieces) are cooked. In the heating media first mentioned the temperatures often are very high and the surface is easily dried out causing the crust formation.

Many of the chemical changes when frying or grilling differ from when boiling, giving different types of ready made products.

In the following frying and grilling will be separated from boiling. The processes will be described more in detail. What happen in the different operations will be discussed.

Frying, roasting and grilling

Heat is transferred in three different ways to the product when frying, roasting or grilling that is via conduction, via convection and via radiation. In panfrying a heated surface is in direct contact with the product surface and thus heat is transported due to conduction. In deep fat frying and in an ordinary oven or in a forced convection oven, heat will be transported due to convection via fat or air. Finally in some ovens radiation from infrared radiators will be the heating media (microwaves will be excluded from this discussion).

The principial difference between the heating processes in heat transport are the heat transfer coefficients which are measurement of the resistance of the heat supplied to the surface. Thus heating by means of conduction gives the fastest response in the surface temperature and consequently the highest heat transfer coefficient and futhermore the shortest frying time comparing an operation of a product with equal geometry. The slowest method is oven roasting.

As the heat is transported through the product changes will occur. These changes are not only depending on the temperature raise but also on the time the different parts will be held at different temperature levels. Water and fat will be transported out of the product. This will cause weight losses. The mechanism for this is due to different phenomena. In the beginning of the operation free water occurs on the surface. This water is evaporated and is transferred to the surroundings. When the temperature in the meat reaches 40°C proteindenaturation will start and change the structure of the product. At the same time water and perhaps some fat will be pressed out from the protein network. Also due to concentration differences in the product water will diffuse from the inner part towards the surface. The amount of water and fat leaving the product i.e. the weight losses depend on the recipe.

When the surface temperature has reached 100°C the evaporation zone starts to move towards the centre of the product. The surface and parts outside the evaporation zone gets drier and the temperature increases towards the temperature of the heating media. A crust starts to form. Earlier the Maillard-reaction is initiated and when the crust gets drier and the temperature raises this reaction will proceed giving a brown colour. Also the fat will oxidize at higher surface temperature. Caramellization will take place. Due to these different reactions the aroma of a fried or grilled product will develop and thus are very important for the taste and smell of the ready made product.

Due to all the reactions mentioned and the lack of water in the pores the crust will be porous which will give a certain crispiness. This porosity will also cause a decrease in heat conductivity and the crust will act as an insulation. The crust is also less hygroscopic than the inner part of the product.

In meat products the most heat labile vitamin is thiamine. This means that when the nutrition value of meat is to be optimized the kinetics of this vitamin should be known for each product and then the reduction may be calculated if the heat treatment is well defined. It is found that the thiamine loss is great in the crust but very little in the inner parts.

As for the sensoric behaviour there is research work going on at different places, among other places in Gothenburg, Sweden. The sensorial properties are very important factors when optimizing the quality.

Boiling

According to the definition made boiling means cooking at temperatures below or around 100 ^{o}C. The products boiled are mostly potatoes and vegetables but of course meatproducts are also boiled. Including in the boiling operations should also be blanching of vegetables. This is a stage before freezing pasteurization and sterilization. Earlier, the aim was to destroy enzymes that would do harm and to destroy toxical substances when the product is further treated. Today its is found that the blanching process may be used for other purposes.

In the boiling, heat is transferred to the product due to convection. When the heating media is water the heat transfer coefficient is higher than for air (at the same temperature). Also when boiling (including blanching) there are chemical and physical changes and microbiological destruction. The water losses are not as high but at the same time the water may leach the product, that is salts, proteins and starch may be transferred to the water. In vegetables there are a high content of carbon-hydrates which are hydrolyzed, or gelatinized during the boiling process. It has been found that when a vegetable for instance carrots or peas are heated at 60 ^{o}C before sterilization the texture will be better after the total process. This is due to an enzyme pection-metyleseras that influence the pectine acid giving a solid texture where calcium is included. If the temperature is low this enzyme will be able to act (about 60 ^{o}C).

There is a tedency to re-use the water when boiling in the industry. That means that the water is circulated which give a water with higher dry-substance, containing for example sugar. This gives a decrease of losses and better quality, colour and nutrition.

In low temperature boiling it would be able to boil bigger pieces and still get a good quality as the temperature profile in the product will not be steep. Research work on the low temperature boiling of meat is carried out at SIK in Gothenburg. Results have been that this operation has given product with better aroma. The collagen in the meat is changed due to enzyme influences.

To be able to evaluate the influence on quality of different cooking methods calculating the so called cook value could be very convenient. In this calculation the combined influence of time and temperature treatment on different quality factors will be taken into account. The cook value is defined as

$$C = \int \frac{T(t) - 100}{10^{z}} \, dt$$

The cook value means the equivalent number of minutes at 100 ^{o}C compared to the treatment at the temperature. T.

Z_c is the necessary temperature to increase the reaction rate 10 fold. The Z_c value differs (from product to product) and from one reaction to the other.

By use of computer programes of the temperature distribution and integration of the C-values the quality changes in the food during processing can be estamited and optimized.

PROCESS GROUP WORK

After the Versailles workshop in 1981 it was proposed that the processes that fit within this group be divided and syste mized according to three principles as follows.

A1: Processes at media temperatures above 100°C and a constant water activity in the product i.e. sterilization and pressure cooking.

A2: Processes at media temperature above 100°C and a changed water activity in the product i.e. frying, meatbaking, grilling and roasting. Some of these processes are comparable with bread baking process in process group B.

A3: Processing at temperatures less than or equal to 100°C and constant product water activity i.e. boiling, steaming, pasteurization, blanching and low temperature boiling.

The definitions of all these operations are in the COST 91 terminology list.

The gaps and the knowledge about these processes were tried to be surveyed, when different researches, who were interested in the field, met in Versailles. Many of the gathered researchers were interested in "doneness" and quality. The interaction of time/temperature and doneness required more work to be carried out. What new kinds of products would be required in the future and how would one set out to produce them. It was also found that the group covered a huge area.

One thing that could be a collaboration work was to do process modelling on a similar basis.

It was suggested in a discussion that the work should be concentrated on needs for selected models set up for the processes. For example kinetic values and cook values would be some examples for tools to use for the data from the investigation covered with this area. If individual research projects supported the group with data a suggestion for a model which could be published as a collaboration work could be done.

The basis for these models have been presented earlier . The researchers should use the terminology and quality assessments decided by the subgroup. The process parameters may be described in this way as well as the effects on quality and nutrition.

24 different research projects have been participating in the cooperation within process group A. Data from these are collected in the project catalogue.

Six of these projects may be classified within the above mentioned process group A1, eight within A2 and ten within group A3. The projects come from seven different countries in Europe. A wide range of equipments and foodstuffs have been and are investigated. Furthermore quality and nutritional effects are covered from many aspects.

In Zeist, the Netherlands, the process group met again and several researchers made a report from their own investigations. After this it was decided that this process group should concentrate on technological problems because it was difficult to get proper quality recommendations for the processes. However regarding the nutritional values it was possible to tell something concrete.

In Zagreb, Jugoslavia an outline for what should be presented in this final seminar from process A and in a last meeting in Lund, Sweden it was decided that the 24 different projects could be represented by the three contributions written by together 10 different researchers. They all are about process modelling covering the three different categories of projects mentioned earlier.

They also cover three types of foodstuffs one of which is liquid and two are solid. Different quality and nutritional aspects of the special process are reported.

The first paper is about boiling in scraped surface heat exchangers. A liquid foodstuffs is investigated. This process belongs to the first or the third process category i.e. the water activity is constant and tempreatures cover the whole possible range. A model for optimizing the process is presented. This is collaboration between projects carried out in France, Finland, the Netherlands and Sweden.

The second paper concerns boiling of potatoes i.e. a process from the third category where temperatures are less or equal to 100°C and water activity is constant in the product. This paper is a collaboration between researchers from two different countries i.e. West Germany and Sweden. These two countries are involved in the third paper which concerns baking of meat loaves in convection ovens. This process belongs to the category A2 where the temperature is above 100°C and the water activity will change.

Before these papers are presented I will thank all the researchers and project leaders of the 24 different projects participating in the collaboration work in process group A for their cooperation and wish them all good luck for the future.

PERFORMANCE OF SCRAPED SURFACE HEAT EXCHANGERS

M. Härröd
SIK - The Swedish Food Institute
J.F. Maingonnat
INRA-Laboratoire de Génie Industriel Alimentaire

Summary

It is easier to control time, temperature and pressure in a scraped surface heat exchanger than in a kettle. This makes it possible to achieve more even and improved quality in cooking of high viscous products. There are also advantages in cooling such products in a scraped surface heat exchanger.

Knowledge of the flow pattern is fundamental to the understanding of heat transfer and residence time distribution. The flow can be regarded as the sum of one axial and one rotating flow. In industrial applications the axial flow is always laminar and the rotating flow either laminar or turbulent.

If the rotating flow is turbulent, changes in heat transfer and temperature profile due to changes in operating conditions can be predicted with a relative standard deviation of less than 10%. If the rotating flow is laminar, channeling may occur and the deviations can increase tremendously.

Equations for heat transfer are based on experiments with one specific type of scraped surface heat exchanger. However, if these equations are applied to other scraped surface heat exchangers the relative standard deviation is sometimes more than 50%.

The residence time distribution is important in some processes, e.g. sterilization but there is a lack of knowledge how the residence time distribution is related to the operating conditions in a scraped surface heat exchanger.

1. BACKGROUND

Cooking of high viscous products in the food industry mainly takes place in kettles, where the possibilities of controlling and optimizing heat treatment are generally very limited. Increased demand for efficient and labour-saving processes in the food industry makes the use of continuous cooking processes important. A scraped surface heat exchanger (SSHE) is well suited to heat or cool high viscous foods. In Scandinavia SSHE have, to a great extent, replaced kettles for cooking of fruit sauces, jams and marmalades. In these processes, heating is followed by cooling. There are also industrial applications that include some cooking time before cooling, e.g. for rice porridge. Figure 1 shows a general flow sheet for a SSHE application.

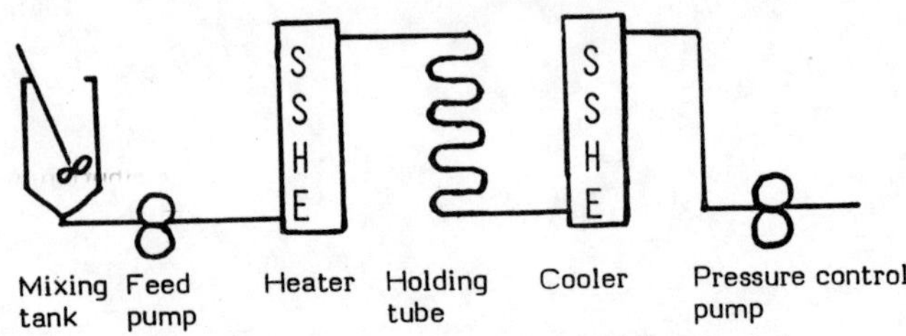

Figure 1. A general flow-sheet for a SSHE application.

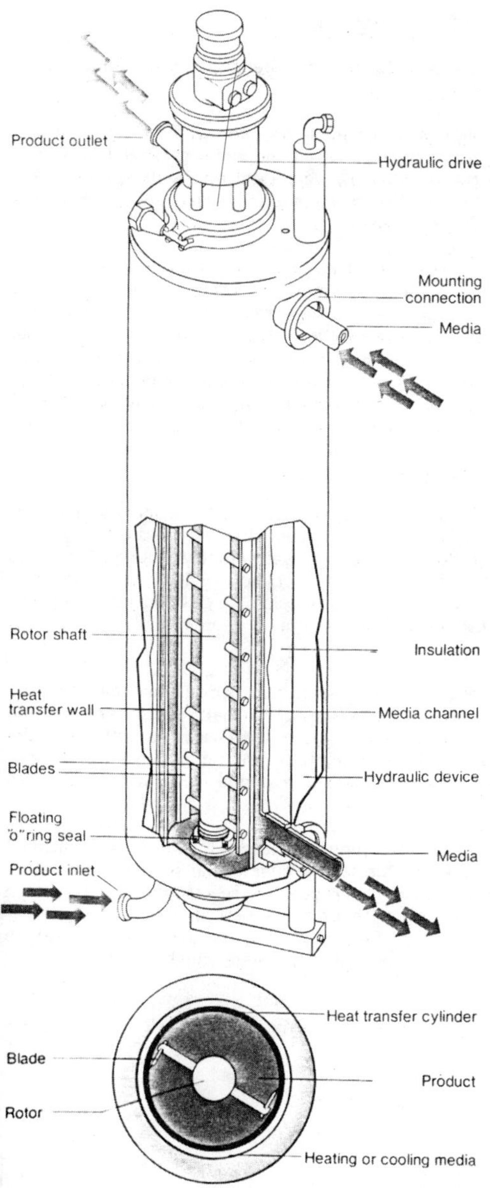

Figure 2. A scraped surface heat exchanger

1.1 Function

An SSHE, see figure 2, consists primarily of a pipe. The outer surface of the pipe is surrounded by a cooling or a heating medium. The product is pumped through the pipe whilst the inner surface of the pipe is kept clean by rotating blades. With this design it is possible to achieve efficient heat transfer even with high viscous liquids without fouling on the heat transfer surface.

1.2 Problems and limitations

Products containing solid particles up to a diameter of 20 mm can be treated in an SSHE on one condition - the product must be pumpable.

If the product contains solid particles of a density that differs from the liquid a large number of practical tests with the mixing and pumping equipment must generally be performed, before a constant ratio between particles and liquid can be achieved.

In some cases channeling occurs, which means that some of the product passes through the SSHE without being properly heated or cooled.

For some processes, e.g. HTST sterilization, it would be possible to obtain a better quality product if the residence time distribution and the temperature profile were known in the SSHE. The reason for this is that safety calculations are based on time and temperature relations in the holding tube and the heat treatment in the SSHE is added just as a safety factor.

One other area where improved equations for heat transfer, temperature profile and residence time distribution would be valuable is on scaling from pilot plant to industrial scale.

2. FLOW PATTERN AND TEMPERATURE PROFILES

Knowledge of flow pattern is fundamental to the understanding of heat transfer and residence time distribution.

Plug flow with perfect radial mixing is the ideal flow for an SSHE. Under these theoretical conditions, all volume elements are treated the same time and along the same axial temperature profile in the SSHE. However, axial mixing is not desirable because it causes an expansion of the residence time distribution and reduces the available temperature difference for heat transfer.

The real flow can be regarded as the sum of two components, one axial and one rotating. In industrial applications the axial flow is always laminar and the rotational flow is either laminar or turbulent. Experiments have verified that the transition from laminar to turbulent rotational flow takes place in an SSHE when the rotational Reynold number is approximately equal to the critical rotational Reynold number, which can be calculated only from the diameter of the tube and the diameter of the rotor. The rotational Reynold number can be calculated from the rotational speed of the rotor, the diameter of the tube and the kinematic viscosity.

2.1 Laminar rotational flow - Couette flow

The flow pattern in an SSHE, when the rotational flow is laminar, is schematically illustrated in figure 3a-c. The radial mixing, between the rotor and the blades, is very poor, see figure 3a. The axial flow causes an axial velocity distribution. Each volume element follows a helical flow line. They rise through the SSHE with different rate at different radii, see figure 3b-c. The axial velocity distribution along the radius causes a residence time distribution which may be amplified by viscosity profiles due to heating or cooling.

In the heating process the heat is concentrated to a small mass close to the heating surface, due to poor radial mixing. The viscosity normaly decreases when the temperature rises and the axial velocity distribution increases. In this case the maximal flow rate occurs close to the heating surface. In the cooling process the influence on viscosity is the other way round, and the maximal flow rate occurs close to the rotor.

A general temperature profile along the length of an SSHE, when the rotational flow is laminar, is illustrated in figure 3d. Due to the poor radial mixing tremendous temperature differences may occur along the radius. A result of this is that in measuring the average outlet temperature without previous radial mixing the error can be large. In some experiments with good radial mixing the average temperature scattered, due to a random variation of the ratio between much cooled and less cooled product in the outlet.

There are no equations for heat transfer or residence time distribution available when the rotational flow is laminar.

FLOW PATTERNS and TEMPERATURE PROFILES in SCRAPED SURFACE HEAT EXCHANGERS

Rotational flow laminar - Couette flow
Axial flow laminar

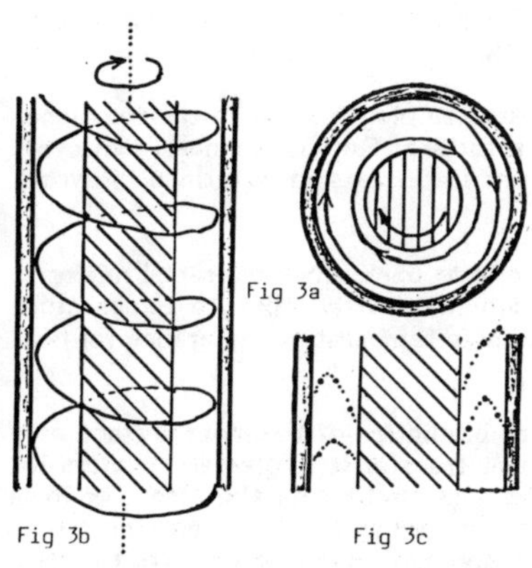

Figure 3a-c. Flow pattern

The blades are omitted in the drawings

Rotational flow turbulent - Taylor vortices
Axial flow laminar

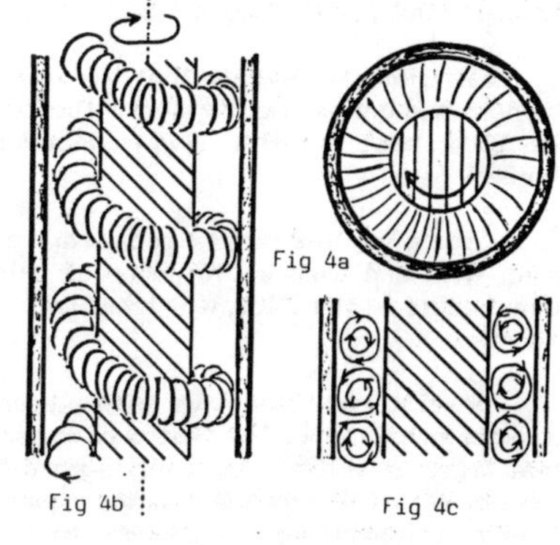

Figure 4a-c. Flow pattern

The blades are omitted in the drawings

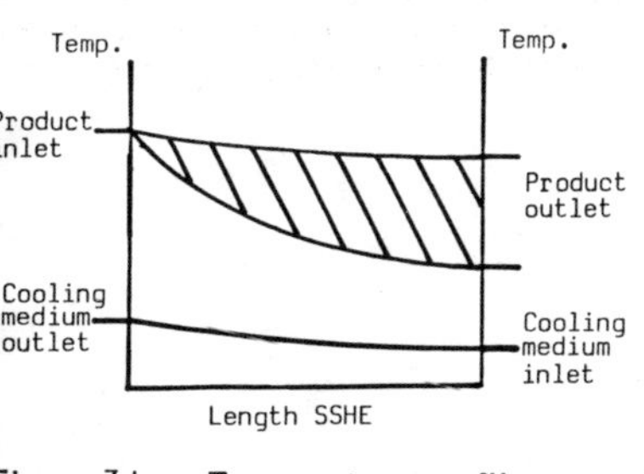

Figure 3d. Temperature profile

Temperature range along the radius

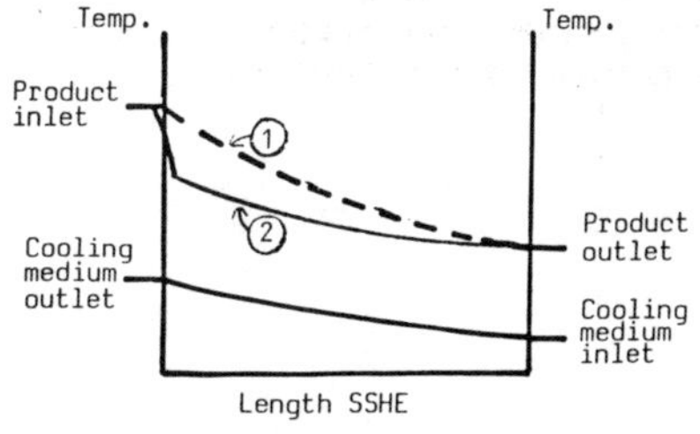

Figure 4d. Temperature profile

(1) good radial mixing and no influence of axial mixing
(2) good radial mixing and high influence of axial mixing

2.2 Turbulent rotational flow - Taylor vortices

The flow pattern in an SSHE with Taylor vortices is schematically illustrated in figure 4a-c. The vortices can be illustrated with circles in the annular space, see figure 4c. They move slowly along the SSHE due to the axial flow, see figure 4b. The higher the rotational Reynold number the more irregular the vortices will be. The vortices cause both radial and axial mixing, see figure 4a and figure 4c. The influence of the axial mixing depends on the ratio between axial mixing and axial flow rate. Generally, on scaling from pilot plant to industrial scale, the axial mixing is kept constant but the axial flow rate is increased. Thus the influence of axial mixing is larger in pilot plants than in industrial applications.

Experiments indicate that the flow is closest to plug flow when the rotational Reynold number is slightly above the critical rotational Reynold number. However, the total heat transfer always increases with an increasing rotational Reynold number.

Figure 4d illustrates a temperature profile in an SSHE when the radial mixing is good, with and without influence of axial mixing. With axial mixing a temperature jump occurs at the inlet, which reduces the available temperature difference for heat transfer.

Several equations have been presented for prediction of the inner surface heat transfer coefficient. The best way to predict heat transfer is to give an equation for heat transfer without axial mixing and to make a correction for the effect of axial mixing. With this method existing equations predict changes in heat transfer due to changes in operating conditions with a relative standard deviation of 10%. Unfortunately the equations are valid only for one type of SSHE. If the equations are used on another type of SSHE the relative standard deviation may be considerable, sometimes more than 50%.

3. RESEARCH WITHIN COST 91

At VTT, in Finland, an SSHE has been used as a chemical reactor for wheat starch gelatinization and enzymatic hydrolyzation (4). At TNO, in Holland, the heat transfer in an SSHE with water and high viscous foods has been studied (5). At INRA, in France, a basic investigation on heat transfer, temperature profile and pressure drop in an SSHE is being made (6). At SIK, in Sweden, a basic investigation on heat transfer, temperature profile and residence time distribution in SSHE during heating and cooling is in progress (7).

REFERENCES

1. Trommelen, A.M. and Beek, W.J. (1971)
Flow phenomena in a scraped surface heat exchanger.
Chem.Eng.Sci. 26(1971):11, p. 1933-42

2. Trommelen, A.M., Beek, W.J. and van de Westerlaken, H.C. (1971)
A mechanism for heat transfer in a Votator-type scraped surface heat exchanger.
Chem.Eng.Sci. 26(1971):12, p. 1987-2001

3. Weisser, H. (1972)
Untersuchungen zum Wärmeübergang in Kratzkühler.
Thesis at Karlsruhe Universität (T.H.) 1972

4. Hakkarainen, L., Linko, P. and Olkku, J. (1983)
State vector model for Contherm scraped surface heat exchanger, used as an enzyme reactor in wheat starch conversions.
(Submitted for publication)

5. van Boxtel, L.B.S. and de Fielliettaz Goethart, R.L. (1983)
Heat transfer to water and some highly viscous food systems in a water-cooled scraped surface heat exchanger.
(J. Food Process Eng. in press)

6. Maingonnat, J.F. and Corrieu, G. (1983)
A new approach to model in the thermal behaviour of a scraped surface heat exchanger. (Submitted for publication in the proceedings of Third International Congress on Engineering and Food, Dublin 1983)

7. Härröd, M. (1982)
Scraped surface heat exchangers. Literature and market survey.
(To be published)

BOILING PROCESSES AND EATING QUALITY OF POTATOES

Yngve Andersson
SIK-The Swedish Food Institute
Box 5401
S-402 29 Göteborg
Sweden

SUMMARY

The deleterious effect on potato quality during handling in catering units is well documented. In this paper, the effects of various heat treatments on the nutritional and sensory quality of potatoes are summarized. The importance of a thorough knowledge of the time-temperature development in the potato tissues is discussed and in this connection, the use of Z-and C-values to determine the "cooking degree" of potatoes is described. Finally, equipments used for potato cooking are discussed.

INTRODUCTION

In the Scandinavian countries as well as in Germany, potatoes are one of the most important staple foods. For example, in Sweden the consumption of fresh potatoes (excl. industrially treated) was 1981 539 million kg, a figure that has decreased since 1960 when it was 638 million kg (Anon. 1982). One of the reasons for this decreasing trend may be that the boiled potatoes served in different catering units do not have a sufficiently high quality to compete neither with industrially refined potato products like french fries nor with other products like rice or pasta.

The deleterious effect on potato quality during handling in catering units is well documented. The negative effects of long warm holding times on sensory quality and C-vitamine content were described by Hansson et. al., 1972, Jonsson, 1980, and Karlström, 1982. In a study performed in twelve different catering units Österström and Gustavsson, 1982, showed that the potatoes often were over-cooked, in one case up to 700 %. The over-cooking was caused by difference in potato size within the same batch, uneven temperature distribution within the cooking cavity, too long cooking times, warm-holding caused by holding periods between the time when the potatoes were finished cooked and when they were served, and warm-holding caused by too large batches. In many cases it was also shown that the potato cooking equipments used gave a slow cooking rate caused by too low input of electrical power in comparison to the amount of water and potato that was heated.

MAIN EFFECTS OF COOKING

What criteria determine when potatoes are finished cooked? Of course, the eating quality is strongly associated with taste and texture. Partly as a consequence of dissolvation of the pectic substance, which glues the cells together, the texture will gradually change from "raw" to "boiled". When potatoes are cooked under home preparation conditions, i.e. in small batches with an effective energy supply, they are "finished cooked" after 20-30 minutes. If the cooking procedure is more prolonged, as often in catering, an overcooked taste and texture results.

During the heat treatment, also nutritional changes occur. As an effect of the gelatinization, the starch in the starch granules will be available for human digestion. Heat-labile vitamines are destroyed and minerals are lost through the heating medium.

COOKING VALUE

When potatoes are cooked, the heat penetration and the cooking time depend on the potato size, the temperature and the heat transfer properties of the cooking medium, and the cooking properties of the tissues. Paulus and Tirtohusodo (1982) showed that the logarithm of the hardness measured instrumentelly or sensory was directly proportional to the heating time at specific temperatures. They also calculated Z-values of ca 23°C for the potato tissues (from Desiree potatoes). The Z-value is the temperature change needed to reach a tenfold increase in tissue-softening rate. One should observe that the results of calculation of Z-values may depend of the techniques used for the instrumental and the sensory analyses. Paulus and Tirtohusodo, 1982, used a universal testing machine in which a plunger penetrated the potato tissue with a constant speed. Dagerskog and Österström, 1978, had another technique for measuring "instrumental hardness". They used a SuR-penetrometer in which a needle penetrated the potato tissue with a constant force. With this technique they reported Z-values of ca 17°C (for potato species: Bintje, Magnum Bonum and King Edward). When determining Z-values a number of experiments must be performed at specified temperatures. However, in practice the potato tissue is subjected to variable temperatures during the heating. In order to take this into consideration, Mansfield, 1962, suggested the use of a cooking value ("C-value") determined by:

$$C = \int_0^t 10^{\frac{T(t)-100}{Z}} dt$$

According to this definition the C-value corresponds to the time during which the tissue should be heat treated at 100°C in order to reach the same degree of tissue softening as in the case in question. (Observe the analogy with the F-value used for microorganisms.) The C-value was determined by e.g. Dagerskog and Österström, 1978, to 8-12 min. depending on potato species.

NUTRITIONAL QUALITY

Potatoes have a high nutrient density, which means a high content of nutrients per unit of energy. However, the nutritional value of potatoes may be considerably decreased during preparation as can be illustrated by the following table:

Table 1 Comparisons between water and pressure (steam) cooking regarding some important nutrients.

	Water cooking 98°C	Pressure Cooking (steam) 110-120°C
C-vitamine	14-20%[a)] 23-28%[b)]	9-12%[a)] 11-14[b)]
Thiamin	14-24%[a)] 20-24%[b)]	8-14%[a)] 8-14%[b)]
Riboflavin	10%[b)]	1-2%[b)]
Minerals	19-30%[a)] 22-29%[b)]	7-16%[a)] 12-15%

a) According to Zacharias and Bognar (1982)
b) According to Bognar (1983)

As can be seen in Table 1 pressure (steam) cooking of potatoes gives lower losses of vitamins and minerals. In accordance with these results, Bognar (1983) showed a higher occurrence of vitamins and minerals in the water used, when water-cooking potatoes, than in the condensed water from the pressure cooker. This difference could explain the differences between cooking methods in Table 1, and Bognar concluded that the decrease in vitamine content due to heat degradation was not significant.

SENSORY QUALITY

The sensory quality of potatoes served in catering establishments can vary within broad limits. To explain why it is necessary to establish basic knowledge about the relationship between on one hand the art and the degree of heat treatment and on the other hand the sensory quality. In a comparative study between potatoes cooked according to two different time-temperature courses (one fast and one slow) Österström et.al., 1982, showed that potatoes cooked by the fast method (=potatoes put into cooking water and than cooked to specified C-values) were judged by a trained panel as more mealy, less hard/firm and less sticky/gluey than those cooked by the slow method (=potatoes put into cold water and than cooked to specified C-values). On the latter, a "skin" was formed during cooking and an off flavor was developed. With a similar approach, Zacharias and Bognar, 1982, also compared potatoes which was put into hot water and then cooked with potatoes put into cold water and then cooked. They concluded, contrary to Österström et.al., 1982, that no difference could be observed in potato quality between the two methods. This different result may be explained by the fact that Österström et.al. compared potatoes that had received the same total amount of heat (i.e. cooked to the same C-values), while Zacharias and Bognar, 1982, did not specify the heat dose.

EQUIPMENT FOR POTATO COOKING

As a consequence of the correlation between time-temperature development in the potato and the potato quality, it is obvious that the characteristics of the cooking equipment is important for the final result. In both steam cookers and water cookers, the heating capacity chosen is often far too low for optimal cooking rate, while the small table top high-pressure steam cookers usually have quite adequate heat generation capacity for full load conditions. In water cookers the time for cooking potatoes may thus be 50-60 min., while optimal heating conditions would result in a finish-cook in 20-25 min. with superior food quality (Bengtsson 1979).

Österström and Andersson, 1979, studied the temperature distribution in a water cooker and two different pressure cookers (working at 107 and 111°C, respectively). In the non-optimized pressure-steam cookers it was shown that large temperature differences were obtained between the coldest and warmest points. The consequence of this is that a potato located in the coldest point will be undercooked in comparison to a potato located at the warmest point. In order to get more even temperatures within pressure cookers, the equipment could be optimized simply by a prolongation of the time for venting the air from the cooking cavity. This is a very simple operation which will lead to better control of the cooked potato quality. It was observed that there was no difference in sensory quality between potatoes cooked in the two pressure steam-cookers, while the degrees of sloughing and mealiness were considered higher for steam-cooked potatoes and the wateriness were higher in water-cooked potatoes.

An alternative to the steam and water cookers is the convection steamer. In this equipment potatoes are cooked with steam at 100°C. The cooking time to get a finished cooked potato (to a C-value of 6-8 minutes) is about 30 minutes. In a convection steamer the possibilites for a high vitamine retention is great. In comparison with water the vitamine extraction by steam is low and the low steam pressure gives a low thermic degradation. (Österström, 1980).

Two of the main problems when serving cooked potatoes in catering establishments are the long warm holding times and the risk that the potatoes will become overcooked. In an attempt to avoid these problems a special semi-continuous potato cooker was constructed and tested at SIK-The Swedish Food Institute (Andersson and Gustafsson, 1983). The cooker can be loaded in advance and automatically started at a pre-determined time. It contains a number of separately controlled cooking spaces, which make it possible to boil smaller amounts of potatoes at each time in relation to the serving frequency and thus avoid long warm holding times. The boiling is stopped automatically just before the potato tissue is completely cooked and the potatoes are then chilled with cold water to a proper serving temperature.

FINAL WORDS

Much work has been devoted to studies of potato cooking, but still there is much more that should be done. The relations between potato quality and thermal processing should be further elaborated. The equipment used for potato cooking should be improved by i.e. getting lower temperature spread within the cooking cavity and by increasing the rated power input. Finally, also the kitchens must devote much more interest in the handling and preparation techniques in order to minimize the over-cooking and to shorten the warm-holding.

REFERENCES

Anon., 1982
Statistical Abstract of Sweden.
Stockholm (Liber Förlag/Allmänna Förlaget), 1982, p. 230.

Andersson, Y., Gustafsson, U.
Cooking of potatoes in a semi-continuous equipment.
SIK-Rapport No 520 (1983).

Bengtsson, N., 1979
Catering equipment design and food quality.
In Advances in catering technology (Ed. G. Glew), London,
(Applied Science Publishers Ltd) p. 121.

Bognar, A., 1983
Unpublished results.

Dagerskog, M., Österström, L., 1978
Boiling of potatoes. An introductory study of time-temperature
relations for texture changes.
SIK:s Service-Serie No 584 (1978).

Jonsson, L., 1980
Nutritional changes in food handled in catering establishments.
Doctoral thesis from Department of Clinical Nutrition,
University of Göteborg, 1980.

Karlström, B., 1982
Quality changes during warm-holding of foods in catering establishments.
Doctoral thesis from Chalmers University of Technology,
Göteborg, 1982.

Mansfield, T., 1962
High temperature - short time sterilization.
Proc. 1st Int. Congress.
Food Sci. and Techn., London, Vol. 4, p. 311.

Paulus, K, Tirtohusodo, H., 1982
Untersuchungen zur Garungskinetik von Kartoffeln.
Ernährungs-Umschau (1982):8, p. 255.

Zacharias, R., Bognar, A., 1982
Schulverpflegung mit Speisen aus eigener Zubereitung - Mischküche Stufe I.
Bundesministerium für Ernährung, Landwirtschaft und Forschung und der Bundesforschungsanstalt für Ernährung, Stuttgart (1982) p. 91.

Österström, L., Andersson, Y., 1979.
Cooking of potatoes by steam and water.
In Advances in Catering technology.
(Ed. G. Glew) London (Applied Science Publishers Ltd) p. 247.

Österström, L. 1980
"Convection steamer": A new heat treatment method" for catering and food industry.
SIK-Rapport No 467 (1980).

Österström, L., Gustafsson, U., 1982
Calculation of the cooking value of potatoes cooked in catering establishments.
SIK-Rapport No 499, 1982.

Österström, L. et.al., 1982
Boiling of potatoes: Effects of time and temperature on sensory attributes.
SIK-Rapport No 501 (1982).

MODELING THE BAKING PROCESS OF MEAT PRODUCTS USING CONVECTION OVENS

E. HOLTZ and C. SKJÖLDEBRAND
Lund University, Div. of Food Engineering, P.O. Box 50, S-230 53 Alnarp, Sweden

A. BOGNAR and J. PIEKARSKI
Federal Research Centre for Nutrition, Garbenstrasse 13, 7000 Stuttgart 70, West Germany

Summary

This paper describes the basis for a mathematical model, regarding heat and mass transfer, for baking of meat loaves in convection ovens. Some experiments have been carried out where quality and nutritional aspects have been studied. The crust formation and the time temperature relationships have also been measured. The work has been done in two research institutes (Federal Research Centre for Nutrition in West Germany and Div. of Food Engineering in Sweden) as a collaboration within the EEC project COST 91.

Among other things the crust thickness was found to be dependent on the amount of weight loss and sensorial aspects have been correlated to the oven conditions. These results will later be used in a computer programme containing the mathematical model that is under construction.

INTRODUCTION

Research concerning convection ovens has been carried out for a long period of time at both Lund University, Sweden and the Federal Research Centre for Nutrition in Stuttgart, West Germany. The COST 91 collaboration has given a good opportunity to do comparable work at the two departments eventhough the special interests are a little bit different.

Baking is defined according to the COST 91 terminology as "cooking of food in an oven in air to which water vapour may or may not be added". The convection ovens are used a lot for baking and reheating of foodstuffs in both the food and the catering industries. Parameters as temperature, air velocity and humidity of the air may be varied in the oven which will influence the product. It is of great importance to achieve proper conditions in the oven in order to get a product with a desired crust thickness, crust colour and that is ready in the centre at the same time.The meat baking process has been investigated earlier and reported in the literature (1, 2, 3, 4, 11). Attempts to make models have been done, but only parts of the total aspects have been considered.

A mathematical model is needed in order to optimize the final quality in the meat products during the baking process, with regard to technical, physical, chemical, nutritional and sensorial aspects.

This paper describes the basis for such a model, and some experiments of more practical character for baking of meat loaves in convection ovens. Experiments have been carried out in Stuttgart in order to examine how different parameters in the oven influenced the changes in vitamin content and the sensorial quality. In Lund time-temperature relationships and crust formation have been studied. Work is also proceeding in Lund to make a model in order to optimize the process.

BASIS FOR THE MODEL

Heat is transferred from the air to the surface by convection and further towards the centre of the product by conduction during the heat treatment. The rate of the heat which is transported in the crumb may be calculated with help of the thermal diffusivity or the conductivity.

The temperature of the surface may reach a high level (> 120°C) during the process. Together with a low water activity are these conditions favourable for the Maillard reaction and the crust formation. With these reactions follow same positive aspects such the formation of a brown colour and aroma in the crust and on the negative side formation of mutagenic substances. During the heat treatment the properties of the product will change due to e.g. fat melting, protein denaturation and water evaporation. Furthermore microorganisms are killed and reduction of vitamin content occur. Figure 1 shows a schematic picture of some process parameters and some quality aspects that should be included in the model.

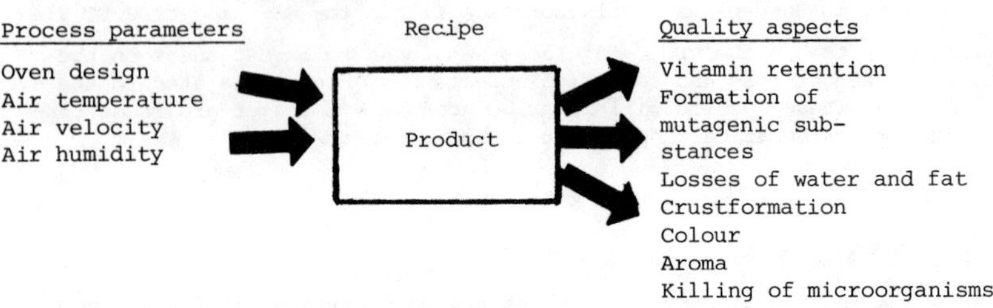

Fig. 1 Schematic picture over the process.

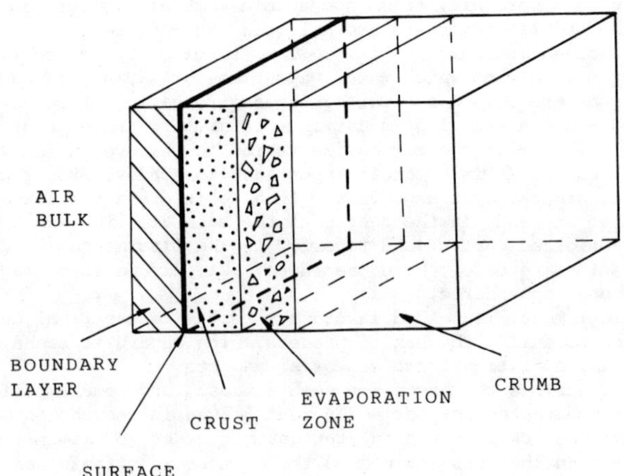

Fig. 2. Picture of the different zones in the product and its surroundings (5).

The temperature of the air bulk is considered constant. The boundary layer (see fig. 2) is a thin layer of stagnant air around the product which is of great importance for heat and mass transfer (5, 6). The velocity, temperature and humidity of the air influence the heat transfer through the boundary layer to the surface, e.g. with higher velocity and higher humidity of the air follows an increase of the heat transfer coefficient.

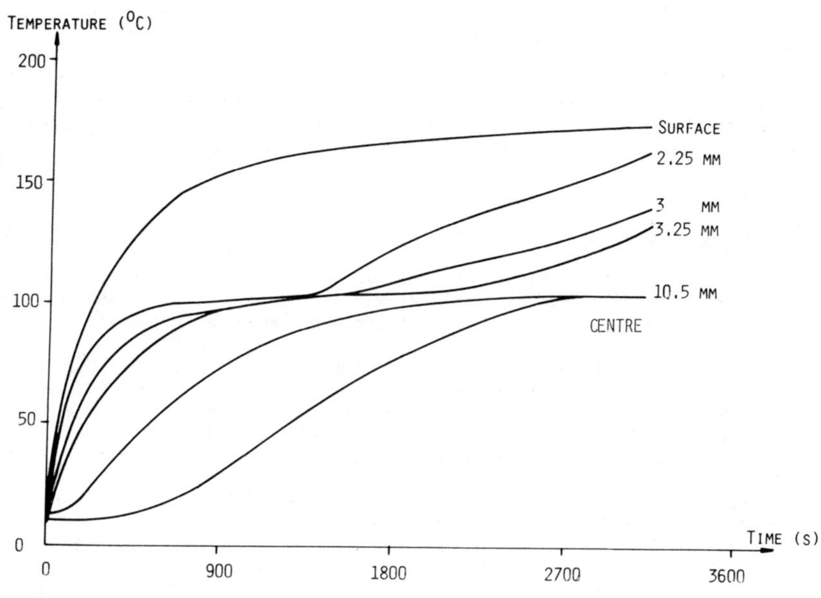

Fig. 3. The temperature profile in a meat product. (7).

The temperature of the surface increases rapidly up to 100°C (see fig. 3), stays there for a while before it leaves the 100°C - isotherm and increases towards the temperature of the air bulk. The 100°C - isotherm or the evaporation zone moves towards the centre versus the time (7). With this model system the crust is defined as those parts of the product that have had a temperature over 100°C. (4). This definition correlates well with differences in colour between the crust and the crumb. (9).

Many physical properties in these types of product will change during the heat treatment, for instance: (7).

- The product will shrink due to protein denaturation and water evaporation
- Due to protein denaturation the water holding capacity will decrease
- During the heat treatment the pores will be emptied by water. The porosity of the crust is different compared with the crumb.
- The a_w is equal to 1 in the crumb but less than 1 in the crust.

MATERIAL

A minced meat loaf was used in all the experiments. Minced meat products are common in many countries and it is also considered as a rather homogeneous material to be used in the research work. The product mixture consisted of minced beef (M. bicepsfemoris), lard, water, potato starch and spices. 3% starch was added in order to increase the water holding capacity. The raw meat mixture contained 65.5 g water, 18.0 g protein and 11.4 g fat per 100 g product.

The oven used in Stuttgart was a commercial convection oven (Juno), which had been changed for research purposes. The air temperature range is 50-250°C and the air velocity 0-5 m/s. The air can be humidified by a boiler in the oven. The humidity of the air may be measured using a special equipment. It contains 10 shelves (of the dimensions 1/1 Gastronorm) and the distance between them are 60 mm. The oven used in Lund is a research convection oven which has been described earlier in the literature (4).

METHODS

Experimental procedure

About 600 gram of the minced meat mixture was made into loaves of the dimensions 200*50*50 mm. The meat loaf was placed in a teflon coated net cage of stainless steel in order to maintain the shape and dimensions.

In Stuttgart the baking procedure was carried out at 150, 175 and 200°C and the air velocity was held constant at 4-5 m/s. The air humidity varied only due to evaporation from the product. The temperatures of the air and in the product were measured using thermocouples (Fe-Co). The accuracy of these was found to be ± 0.5 K. All data were collected in a computer system.

In Lund all the temperatures were measured using thermocouples (Cr-Al) except for the surface temperature for which an IR-pyrometer was used (4). The air velocity was varied from 3-9 m/s and the temperatures were 175 and 200°C. The crust thickness was examined with the help of a sliding caliper.

The product was weighed before and after the process in all experiments.

The meat loaves were analysed after baking according to the following procedures: In the sensorial analysis five different criteria were used to judge the properties of the product; colour of the crumb and the crust, odour, taste and consitency. The panel consisted of seven to nine well trained persons, whom in a first test judged the optimal baking conditions using a scheme with five different categories for each criteria (See table 1).

In the second test the optimal (for each air temperature) baked meat loaves were further investigated using a specific 9 score scale (10). These notes were allocated according to a specific evaluation scheme containing a product specific description of the characteristics over the entire scale range.

Water, protein and fat were analysed according to the usual methods. Thiamine and vitamin B_6 were analysed by HPLC (12, 13). Arithmetic mean values and standard deviations were calculated according to usual statistical calculation and regression.

Table 1. Specific sensorial test scheme for the judgement of the quality of the meat product. (transleted from German)

Category	Readiness	Colour of the surface	Colour in the crumb	Odour-Taste	Consistency
A	still raw	redish beige-grey	all through red	as raw meat no noticeable roasting aroma	very soft
B	not quite well done	pinkish, middle-brown, redish too thin crust	the outer parts of the crumb grey-beige, the centre part still redish	a little flat, a little bit roasting aroma, still some raw meat odour and taste	soft, lika paste
C	well done	brownish-redish, shiny, as uniform, as possible due to product components crust thickness 1-2 mm	all through beige-pink-brownish-grey, shiny	a clear odour and taste of roasting aroma no off-flavour	all through juicy, tendery bound hard together
D	to well done	dark brown, a little bit uniform some brown-black parts	-	a clear burnt taste and odour a little bit bitter	noticeable dry, tough and dry crust
E	overdone	burnt, too dark brown towards the black-brown colour	-	a clear strong burnt, taste, bitter	dry, too firm too crispy and cracked crust

Mathematical model

As the heating process proceeds the product will be divided into different zones due to different transport mechanisms and different physical and thermal properties. Table 2 shows the transport mechanisms for the product and its surrounding (8).

When making a mathematical model, one way is to start from the outer part i.e. the surrounding air. Calculation of the amount of energy that is supplied to the product from the hot surrounding air is a first step. The surface temperature is calculated from the supplied energy and a heat transfer coefficient (α). The thermal properties are different in the different zones and if these values are known together with the physical properties the temperature profile in the crust and the crumb can be determined.

Water is transported from the inner parts towards the surface. In the evaporation zone there is a change of phase. Energy is needed for the evaporation and this amount of energy has to be included when calculating the heat transfer coefficient. As the weight losses are relatively high in these types of products this energy is of great importance.

	Boundary layer	Crust	Evaporation zone	Crumb
Mass transter	Vapour diffusion	Vapour diffusion	Vapour diffusion Surface diffusion Capillary flow	Capillary flow
Heat transfer	Conduction Convection	Conduction Flow of vapour	Conduction Flow of vapour and liquid water	Conduction

Table 2. Overview for heat and mass transfer in different zones (8).

First it is necessary to have a mathematical model and from that simulate temperature and water content profile in the product.
Second the kinetics of the change in quality properties of interest should be known and together with the simulated temperature and water content profile estimate losses or changes.

To calculate the combined effect of time, temperature and quality different theories can be used. The most common used theoretical equations is the Arrhenius relation combined with the cooking value (15). There are othe theories, mostly made for microorganisms. One of them is the one Casolari has reported (16).

RESULTS

Figure 4 shows the weight losses found in the experiments as a function of time at different air temperatures.

The crust thickness was found to be highly correlated to the weight loss which is shown in figure 5. Furthermore the crust thickness increased with increasing air temperature and air velocity (See fig. 6).

In figure 7 is shown the found optimal baked meat loaves regarding the colour on the surface and in the crumb. The baking time as a function of air temperature is shown in the diagram as well as the centre temperature for each meat loaf.

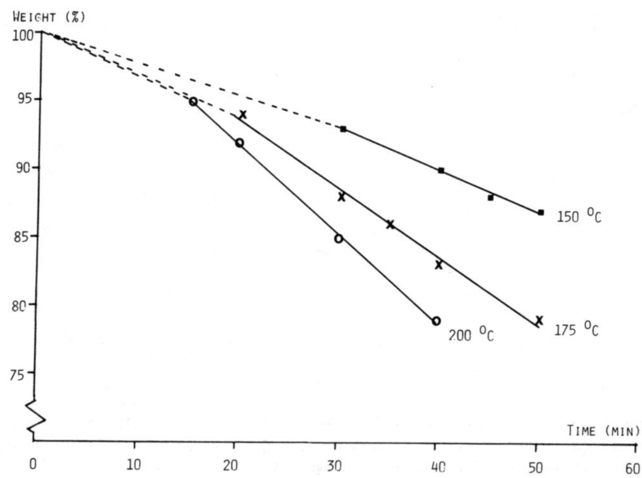

Fig. 4. The weight as a function of time at different air temperatures. The linear regression lines are drawn in the diagram with the following correlation coefficients:
$r_{150°C} = 0.99$, $r_{175°C} = 0.99$, $r_{200°C} = 0.99$ (Stuttgart).

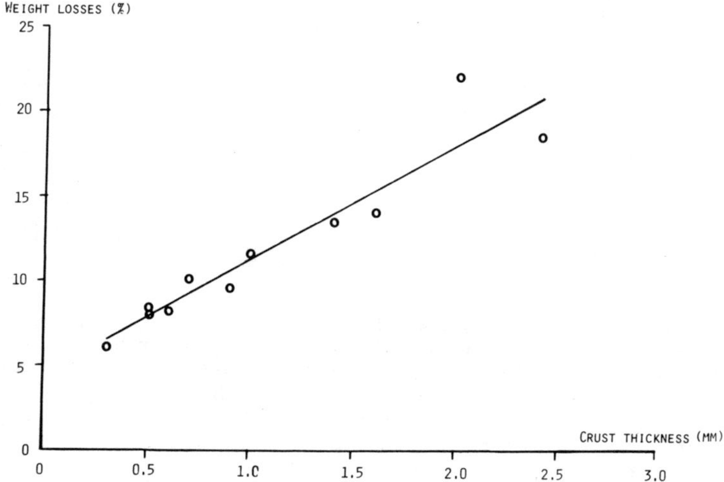

ig. 5. Weight losses as a function of crust thickness. A linear regression ave a correlation coefficient of 0.95. (Lund).

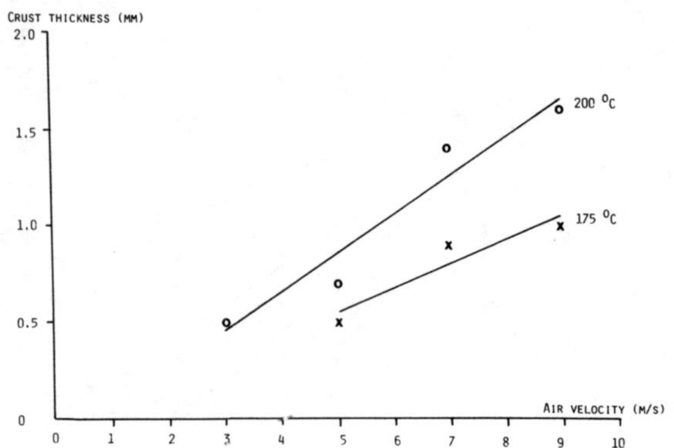

Fig. 6. The crust thickness as a function of air velocity at different air temperatures. All meat loaves were baked for 30 minutes. A linear regression gave the following correlation coefficients:

$r_{175^{\circ}C} = 0.94$, $r_{200^{\circ}C} = 0.97$ (Lund).

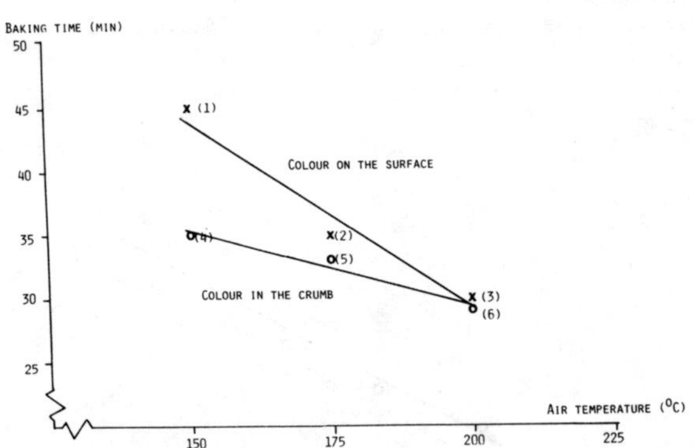

Fig. 7. Baking time as a function of air temperature for optimal baked meat loaves. The sensorial criteria were colour on the surface respectively in the crumb. A log linear regression was used to draw the lines with correlation coefficients as follows:

$r_{surface} = 0.99$ $r_{crust} = 0.99$

t_c = reached temperature in the centre

$t_{c1} = 92^{\circ}C$, $t_{c2} = 87^{\circ}C$, $t_{c3} = 81^{\circ}C$, $t_{c4} = 83^{\circ}C$, $t_{c5} = 85^{\circ}C$, $t_{c6} = 79^{\circ}C$

(Stuttgart).

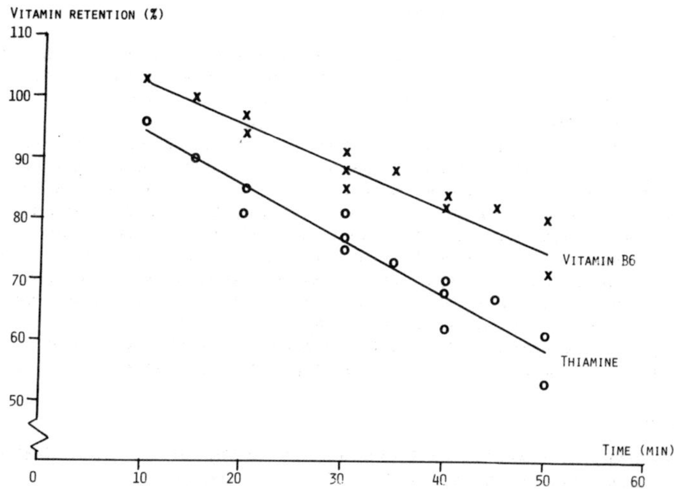

Fig. 8. Vitamin B_6 and thiamine retentions as functions of heating time. A linear regression gave a correlation coefficient of 0.90 for vitamin B_6 and for 0.96 for thiamine. (Stuttgart).

A diagram of the nutritional value represented by thiamine and vitamin B_6 of the different meat loaves are drawn in figure 8. The vitamin retention as a function of baking time is shown.
In all diagrams each point is a mean value of several experiments. The regression coefficients are calculated from the mean values.

DISCUSSION

Figure 4 shows that the weight losses were dependent on both the convection oven air temperature and the baking time.
The weight losses consisted mainly of water and the fat losses were of minor importance. The losses of protein and minerals were almost zero.
The crust formation is a consequence of the drying of the outer parts of the product due to the conditions in the oven. As shown in figure 5 the crust thickness was highly correlated with the weight losses which is in accordance with the above mentioned drying process. The crust thickness and the weight losses did not only depend on the air temperature. The air velocity was of a great importance which is shown in figure 6.
As said earlier an optimal ready baked meat loaf should have a desirable crust colour and thickness and should be ready made in the crumb at the same baking time. This is illustrated in figure 7 where sensorial analyses have been used in order to find the optimal air temperature at a constant air velocity. The results showed that a temperature of 200°C is needed to obtain these criteria at the investigated air velocity (4-5 m/s). At lower air temperature the crumb was ready baked at a lower baking time than the crust colour, which would result in an overdone crumb when the crust colour is acceptable. This could be prevented by an increased air velocity as is shown in figure 6 e.g. using a higher air velocity the air temperature

may be decreased still obtaining an acceptable crust colour together with a ready made and not overdone crumb.
The thiamine and the vitamin B_6 content decreased linearly with heating time independent on the investigated air temperature. This is due to the fact that the heat treatment in the crumb is rather equal independent of the air temperature during the actual time range (< 50 min).

Making a model for simulation of time/temperature relationships the described types of technical results are needed. Also the basic knowledge of the mechanisms of heat and mass transfer is necessary. These simulations together with the nutritional and sensorial aspects will form the optimization of the process. Further work has to be made to find the time-temperature dependence on these aspects i.e. some kind of kinetics of the reactions are needed. An example of the thiamine retention with the same meat system has been studied in an earlier investigation (14). These results obtained here fit well with the kinetic model.

This type of research work will be carried out in the near future to get an optimization model which is as complete as possible.

REFERENCES

1. DAGERSKOG, M. Stekning av livsmedel, Ph.D. Thesis, Chalmers University of Technology, Göteborg, Sweden, 1978, (in Swedish).
2. GODSALVE, E.W., Heat and mass transfer in cooking meat, Ph.D. Thesis, University of Minnesota, Minnesota, USA, 1978.
3. HUNG, C-C. Water migration and structural transformation of oven cooked meat. Ph.D. Thesis, University of Minnesota, Minnesota, USA, 1980.
4. SKJÖLDEBRAND, C. Frying and reheating in a forced convection oven. Ph.D. Thesis, University of Lund, Lund, Sweden, 1979.
5. CHRISTENSEN, A., SKJÖLDEBRAND, C., HALLSTRÖM, B. Surface drying in heat processing of solid foods, in Drying '82. A.S. Mujumdar. Ed., Hemisphere publishing Corporation, Washington, USA, 1982, p. 6-11.
6. WELTY, J.R., WICKS, C.E., WILSON, R.E. Fundamentals of momentum, heat and mass transfer, John Wiley & Sons, New York, USA, 1969.
7. SKJÖLDEBRAND, C., HALLSTRÖM, B. Convection oven frying: Heat and mass transport in the product. J. Food Sci., 1980, 45, 1347-1353.
8. HALLSTRÖM, B., SKJÖLDEBRAND, C. Heat and mass transfer in solid foods, in Developments in Food Preservation -2, S. Thorne Ed., Applied Science Publishers Ltd, Essex, England, 1983, chapter 3.
9. SKJÖLDEBRAND, C., OLSSON, H. Crust formation during frying of minced meat products, Lebensm.-Wiss. u. -Technol. 1980, 13, 148-151.
10. DIN-Norm 10952, Sensorishe Prüfverfahren; Bewertende Prüfung mit Skale, Beuth-Verlag, Berlin, West Germany, 1982.
11. BOGNAR, A., PIEKARSKI, J. Langzeitgarverfahren in feuchter und trockener Hitze, 5. Wissenschaftlich-Technisches Ernährungsforum (WTE) Herborn '81, Garverfahren - Technologie und Andwendung, p. 38-57.
12. BOGNAR, A. Bestimmung von Riboflavin und Thiamin in Lebensmitteln mit Hilfe des Hochdrückflüssigkeitschromatographie (HPLC), Deutsche Lebensmittel Rundschau 1981, 77, 431.
13. BOGNAR, A., not published.
14. SKJÖLDEBRAND, C., ÅNÄS, A., ÖSTE, R., SJÖDIN, P. Prediction of thiamine content in convective heated meat products, J. Food Technol. 1983, 18, 61-73.
15. JEN, Y., MANSON, J.E., STUMBO, C.R., ZAHRADNIK, J.W. A procedure for estimating sterilization of and quality factor degradation in thermally processed foods, J. Food Sci., 1971, 36, 692-698.
16. CASOLARI, A. A model describing microbial inactivation and growth kinetics. J. Theor. Biol. 1981, 88, 1-34.

INTRODUCTION TO PROCESS GROUP B (BAKING)

O. Tolboe

Jutland Technological Institute, Chemistry and Food Technology Department, Aarhus, Denmark.

Summary

As an introduction to the presentation of the projects of Process Group B (Baking) a survey is given of the consumption of cereals in some European countries and the role of bread in the diet. The recommendations from the Dublin COST Seminar as well as the themes of the projects related to these recommendations are mentioned.

The role of cereals in man's diet is well known and well documented (table I), and so is the overconsumption in the Western World of fat and sugar.

Table I. Consumption of cereals in 1978 in some OECD-countries.

Rice is not included.

	per capita consumption kgs/year			
	wheat[1)]	rye[1)]	total cereals[1)]	cereals as flour[2)]
Belgium - Luxembourg	68.8	0.6	71.2	70
Denmark	44.3	17.2	68.4	63
Finland	47.1	20.8	73.0	73
France	73.7	0.4	75.8	74
German Federal Rep.	50.7	14.0	68.6	68
Greece				111
Ireland	80.0	0.3	84.6	87
Italy	122.4	0.2	126.5	131
Netherlands	54.3	4.2	60.8	64
Spain	70.8	1.6	72.4	73
Sweden	43.2	12.6	59.2	61
Switzerland	65.5	4.2	76.0	76
United Kingdom	65.5	0.3	72.2	69
Yugoslavia	145.4	-	167.3	-

1) Food Consumption Statistics 1964-68. OECD, Paris 1981.

2) Statistical Information to the Community. Eurostat, Luxembourg, 1982.

The recommendations from nutrition scientists to cut down our consumption of fat and sugar and to increase our intake of dietary fibre can be met in a simple and inexpensive way by increasing the consumption of bread, one of the world's oldest processed foods. During the latest decades we have seen a decrease in consumption of bread and other processed cereals - except alcoholic beverages - in the western countries. Now a slight increase in consumption is registered in some countries and we must as food technologists support this changing trend as far as possible to increase consumption of cereal-based products.

Bread is one of our most traditional foods with deep roots in our past and with local varieties in huge numbers. Therefore, we should not only think of developing "a nutribread", which will solve the nutrition problems in the western world, but we must base our development on the local bread varieties. Since people are not buying nutrition, but food, we must supply them with attractive food, attractive bread. This is first of all done by producing optimum quality of the traditional varieties, but also by adding new types of tasty and preferably fibrerich bread based on whole meal and/or with added fibre.

The participants of Process Group B are consequently convinced that our area of research is an essential one and we hope that the results of our research and our cooperation will be of value to our community.

The Proceedings of the COST Seminar in Dublin 1977 where the COST-91 project was prepared contain among other things recommendations regarding research on thermally processed foods, including thermally processed cereals (Annex 1).

We have unfortunately - but not unexpectedly - not been able to run projects on all of the recommended topics and not been able to solve all problems mentioned in the recommendations, but we hope we have made a move in the right direction in cooperation as well as in results.

Table II shows in a very simplified form the subjects of the projects in our group and the main parameters studied:

- influence of raw material characteristics and product composition (extraction rate, fibre, salt)
- process conditions (conventional and unconventional baking, process optimization, fermentation conditions)
- quality evaluation of baked products (sensory, physical, chemical, nutritional)

The group had no projects on influence of variety on functional properties as recommended.

For reporting to this Symposium we have divided our 11 projects into three groups, which cover the main themes:

- process conditions and quality
- salt and fibre
- baking process optimization and quality

Three of my colleagues will now present reports from the three project groups before Professor W. Seibel will give us recommendations for future research.

May I finally thank the leaders of the 11 projects for their excellent cooperation.

Table II

Parameter	CH-1	D-1	D-2	D-3	NL-1	S-1	S-2	S-3	S-4	SF-1	SF-2
	PROJECT NO.										
Composition											
flour extraction	x	x									
recipe	x		x	x			x	x		x	
salt				x						x	
Process											
fermentation	x		x	x	x			x			
conventional baking	x	x	x	x	x	x	x	x		x	x
IR-baking							(x)		x		
storage			x					x			
Quality evaluations											
sensory		x	x	x	x	x		x	x	x	x
physical, chem.		x	x	x				x		x	x
nutritional	x						x				

Annex 1

RESEARCH AND DEVELOPMENT REQUIREMENTS ON SOME ASPECTS OF THE QUALITY AND NUTRITIVE VALUE OF THERMALLY PROCESSED CEREALS

Christiane Mercier and J. Delort-Laval
Institut National de la Recherche Agronomique
Centre de Nantes
44072 - Nantes, France

ABSTRACT

Variation in raw material quality and development of new industrial processes, consideration of hygienic and nutritive value of cereal products, lead to the proposal of the following topics for R and D in Western Europe:

- *Genetic research to breed varieties of wheat better adapted to industrial requirements which involves development of adequate European tests for the machinability of the dough and for the breadmaking.*
- *Evaluation of the technological aptitude of cereals taking into account all the characteristics of the raw material (genetic origin, growing and harvesting conditions, storage) and their properties for a definite process (bread, biscuits, pastries, etc).*
- *Technical aspects of new treatments of non baking cereals and their effects on the organoleptic and nutritive properties of the end-products.*
- *Contamination of cereal raw materials and end-products by mycotoxins during storage. Behaviour of additives (antimicrobial, texturising agents) during technological processing.*
- *Development of methodology to evaluate sensory properties and nutritive value of cereal products containing unusual components such as bran, protein, modified starch.*

Other topics are submitted for consideration to the following panels:

- *Chilling and freezing* : *chilling during rest after dough fermentation in French breadmaking.*
- *Cooking* : *evaluation of cooking quality of pasta with relation to raw material and industrial treatment.*
- *Dehydration* : *effect of storage and drying treatment of cereals on the technological properties of cereals for starch extraction and breadmaking.*

BREAD QUALITY AND BREAD MAKING PROCESSES

P. Sluimer
Institute for Cereals, Flour and Bread TNO
Wageningen, The Netherlands,

J.M. Brümmer
Bundesforschungsanstalt für Getreide-und Kartoffelverarbeitung,
Detmold, Bundesrepublic Deutschland,

U. Stöllman
The Swedish Food Institute,
Göteborg, Sweden.

INTRODUCTION

In countries all over the world wheat bread is produced in a great variety of types and shapes with different characteristics. The extraction rate of the flour e.g. has a great influence on bread characteristics; whole meal-, brown or white bread are made. The recipe is another important factor influencing the characteristics of the bread. A lean recipe, with only the basic raw materials, wheat flour, yeast, salt and water, is often used in the production of French bread, but also of Arabic flat bread. The opposite is a recipe for e.g. German Weihnachtstollen, or Italian Panetone with a large proportion of butter, fat, eggs, milkpowder and dried fruits and nuts. A third factor that influences the characteristics of bread is the way it is baked. It can be baked in open or closed tins, on baking sheets or directly on the hearth of the oven, as loaves or as buns or rolls.

These and other factors affect the characteristics of the bread to a large extent. However, all wheat breads, if properly prepared, have also a few characteristics in common, irrespective of the way of production and consumption habits. This is what is named quality from technological view point.

BREAD QUALITY

Objective and measurable quality characteristics for bread are hardly available. The assessment of bread quality is usually made sensorially. Research on the determination of quality characteristics by objective methods is important for industry since it would allow a check of the compliance with contracts. One of the projects of this presentation deals with the objective characterisation of bread quality. (ref. 1).

The three aspects of quality of wheat bread we want to deal with are the lightness of its texture, its eating characteristics and its taste and flavour.

- An important characteristic of wheat bread, either brown or white, is its lightness. It is the only product parameter which can be easily measured. It can be expressed as a specific volume. The specific volume of white bread is in the order of 5.10^{-3} m^3/kg of bread (5 l/kg), whole meal bread in the order of 2.5 to 4.0 10^{-3} m^3/kg, while the specific volume of rye bread is in the order of 10^{-3} m^3/kg. Wheat bread is one of the lightest of our basic foods

- The second factor in bread quality we mentioned, are the eating characteristics. They are far more difficult to describe in objective terms than the lightness of the texture. Good bread had a sufficient coherance, is not crumbly or dry. The crumb should have some resistance to chewing and it must easily regain its original form after compression. If this is not the case we call the bread sticky. One of the problems is that the eating characteristics of bread depend on storage time and storage conditions. Figures 1, 2 and 3 shows the changes in bread characteristics during storage, which are usually called staling. Figure 1 shows the decrease in softness of the crumb with increasing storage time. On the other hand, dryness of the crumb, or crumbliness increased during storage (fig. 2). Crispness of the crust disappeared within one day after baking (fig. 3).
- A third quality aspect of bread is its taste and flavour. Wheat bread has a very specific taste. Important factors affecting the flavour of bread are the fermentation and the baking process. Immediately after baking the bread flavour is very strong. During storage the intensity of the flavour diminishes and also its character changes. To describe the flavour of bread or the change in it in objective terms is even more difficult than the eating characteristics. Assessment of flavour is only done sensorially.

The consequence of this staling process during storage is that bread is preferrably sold and consumed fresh. Bread that is to be sold in the morning will generally have to be baked at night. Therefore, bakers' labour is often associated with night labour. The wish to reduce this night work, either by shortening the length of the bread making process or reducing the labour requirement, has led to various new processes in bread making.

BREAD MAKING PROCESSES

The following bread making processes will be reviewed.

straight dough process with bulk fermentation
green dough method
mechanical dough development
activated dough development
sponge and dough processes
dough retarding and dough freezing.

Straight dough process with bulk fermentation

This process is the established way of bread making in large parts of Europe. After mixing all ingredients, a bulk fermentation is applied ranging from less than one hour to several hours, most often interrupted by a knock back. Dividing and rounding are followed by an intermediate proof of 15 to 45 minutes. The total time for bread making ranges from 3 h to 5.45 h. An advantage of this process is that it is flexible. Variations in the duration of bulk fermentation time affect bread quality only slightly. The process is less suitable for mechanized bakeries. Bowl handling is labour consuming and gives rise to errors; in addition the quality of fermented, aereated doughs is imparted by the mechanical action of the divider. Furthermore dividing fermented doughs results in a wider range of dough pieces than does dividing immediately after mixing. These are the reasons that a process with a long bulk fermentation is only rarely found in larger mechanized bakeries.

Green dough method

One of the developments in larger mechanized bakeries was the development of the so-called green dough method. The dough is divided directly after mixing and two intermediate proof times are applied both of approximately 40 minutes. In this process, the disadvantages of the bulk fermentation system have been eliminated. Proper conditions in the overhead provers are indispensable for an uninterumpted process and a good bread quality. The total length of the process is approximately 3 hrs.

Activated and mechanical dough development

These two methods have been developed between 1955 and 1965. In both processes only one short intermediate proof and no bulk fermentation is given, before the final proof and baking. In activated dough development (ADD), this is achieved by adding cysteine to the dough, in mechanical dough development (MDD) mixing is carried out in a so called high-speed mixer. In both processes the use of a large quantity of a flour improver and of an emulsifier or fat is required. To accelerate dough development, a high dough temperature is applied.

The advantages of these processes is that they save time and labour. The total process time is in the order of two hours. A disadvantage is that the bread quality differs in some respects from that of bread made by the straight dough or green dough method. The crumb is shorter and the taste and flavour are less pronounced. Lightness of the crumb is not affected.

Sponge and dough processes

In particular in the USA, sponge and dough processes are often used. The total amount of flour and water, or part of them, is mixed together with the yeast and this sponge is fermented for one to several hours.

There are three reasons to apply a sponge. The yeast is activated: the rate of the carbon dioxide production is faster if a sponge is used. Secondly, the dough is developed better. The machinability of doughs from sponges is better than that of straight doughs with the same formula. Thirdly, during the sponge fermentation, products are developed that contribute to the formation of bread flavour.

A sponge may have different forms. It may have the consistency of a dough, and it may also be slightly slacker by changing the flour/water ratio. Continuous bread making systems often use liquid sponges in which the ratio flour/water is smaller than 1.

One of the projects of this presentation deals with the improvement of bread quality made with short bread making systems like mechanical and activated dough development, using sponges (ref. 2).

Dough retarding and dough freezing

The need for work at night can be considerably reduced by splitting up the process of bread making into two parts, one from mixing upto and including moulding, and a second part including baking. This can be done by cooling the moulded doughs to temperatures below 0 °C and keeping them at this low temperature. If the storage temperature is in the order of -5 to 0 °C, the procedure is called dough retarding. It is often used in smaller bakeries to avoid night work. The maximum allowed storage time is in the order of 48 hrs. When the storage temperature is around -15 °C it is called freezing. This is often practized in larger, wholesale bakeries. They supply retail bakeries with frozen dough pieces. After warming, the dough is further processed in the reatil shop. This way of handling dough is called bake-off.

Bread from these systems of bread making sometimes has typical defects. The crust is discoloured in a reddish-brown hue, sometimes accompanied with blisters, the crumb is slightly irregular and coarse and its volume is low. This may lead to unsatisfactory eating characteristics.

One of the project which is carried out in the frame work of the COST-91 programme was aimed at improving the bread quality from retarded and frozen doughs (ref. 3). The bread quality can be improved by using a short bread making process, thus with extra flour improver and the addition of fat and emulsifier, and a fast cooling and a stepwise warming of the dough before the final proof.

Straight dough process with bulk fermentation	
mixing	0.20 h
bulk fermentation with knock back	0.45 - 3.0 h
dividing and rounding	
intermediate proof	0.10 - 0.40 h
moulding	
final proof	1.0 h
baking	0.30 h
total duration of process	2.45 - 5.30 h

Green dough method	
mixing	0.20 h
dividing and rounding	
first intermediate proof	0.40 h
knock bak	
second intermediate proof	0.40 h
moulding	
final proof	1.0 h
baking	0.30 h
total duration of process	3.10 h

Activated and mechanical dough development	ADD	MDD	
mixing	0.20	0.05	h
dividing and rounding			
intermediate proof	0.10 - 0.30	0.10 - 0.30	h
moulding			
final proof	1.0	1.0	h
baking	0.30	0.30	h
total duration of process	2.0 - 2.20	1.45 - 2.05	h

Sponge and dough process	
sponge fermentation	2 - 5 h
mixing	0.15 h
bulk fermentation	0.15 h
dividing and rounding	
intermediate proof	0.10 - 0.30 h
moulding	
final proof	1.00 h
baking	0.30 h
total duration of process	4.10 - 7.30 h

REFERENCES

1. U. Stöllman, Sensory and instrumental analyses of bread. COST-91-project B-S-3.
2. P. Sluimer, The use of sponges in short bread making processes. COST-91-project B-NL-1.
3. J.M. Brümmer, Special dough processing at lowered temperatures. COST-91-project B-D-2.

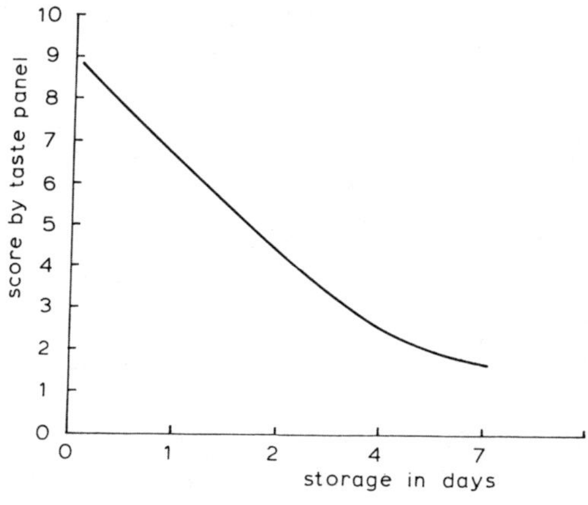

Fig. 2; changes in crumbliness of crumb during storage

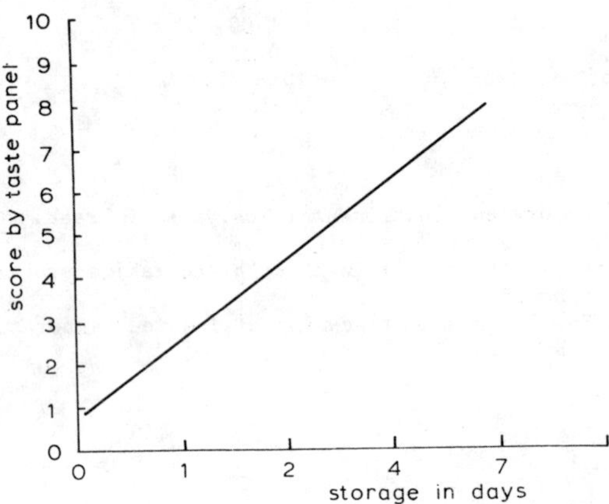

Fig. 3 ; changes in crispness of the crust during storage

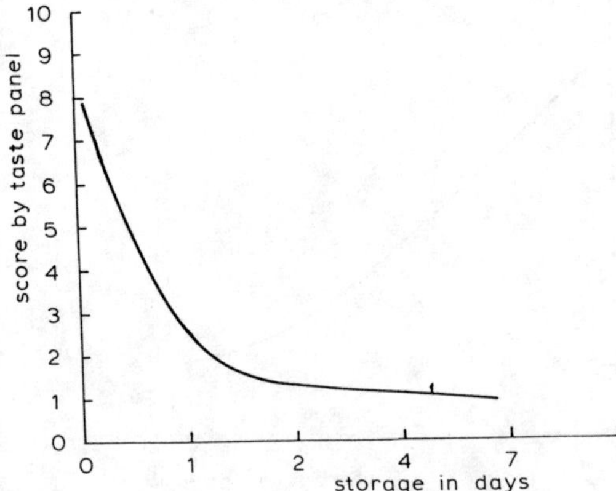

THE ROLE OF SALT AND DIETARY FIBRE IN BREAD BAKING AND NUTRITION, TECHNOLOGICAL AND NUTRITIONAL ASPECTS

J.-M. BRÜMMER[1], N.-G. ASP[2], P. LINKO[3] and P. SCHEFFELDT[4]

[1]Federal Research Centre of Grain and Potato Processing, Institute of Baking Technology D-4930 Detmold, Federal Republic of Germany

[2] Chemical Center, University of Lund, Food Chemistry S-22007 Lund, Sweden

[3]Helsinki University of Technology, Department of Chemistry, Laboratory of Biochemistry and Food Technology SF-02150 Espoo 15, Finland

[4] Institute for Nutrition Research Foundation "Im Gruene" CH-8803 Rueschlikon, Switzerland

Summary

Sodium and dietary fibre are today of more importance in food-evaluation than in the past. For both substances bread is a remarkable source. The effects of these ingredients in baking technology and the possibilities of salt reduction or fibre addition are discussed. Analytical aspects are pointed out. This paper is a conclusion of older experiences and newer results of investigations of COST 91, Sub-Group 2, Process Group Baking.

1. INTRODUCTION

Sodium chloride (common salt) has a number of functions in bread making. It effects rheological properties and biochemical reactions of dough, gas production and retention, and finally the quality of baked goods, mainly crust colour, volume and flavour. These technological effects of salt are wellknown but not looked for with special importance during the last years. Today there is a trend towards low salt diets, new attention is paid to this object.

Bread is an important source for higher intake of dietary fibre. Rye or wheat wholemeal breads have high fibre contents, and can be enriched by the addition of bran. Other fibre sources and problems of fibre determination are given in this paper.

2. SALT IN BREAD TECHNOLOGY

The technological role and importance of salt has mainly been investigated in wheat bread making. Less information on salt is available in rye processing and for bread baked from blends of barley etc. (7,11,18,19).

Numerous interactions between salt and wheat protein have been studied. The effects are : reduction in waterbinding capacity of wheat flour, increase in development time, instability of wheat doughs during mixing (Fig. 1 - 3, Tables I and II), decrease in solubility of gluten proteins.

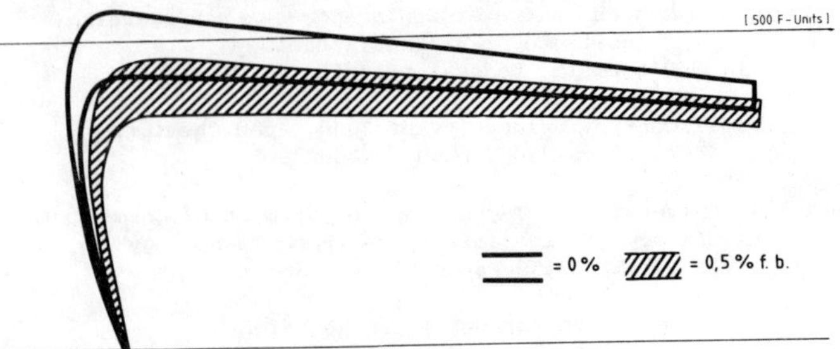

Fig. 1 : Farinograph-curves of doughs from commercial german wheat flour, same water adsorption (59 %) but different sodium chloride concentration on flour basis (f.b.)

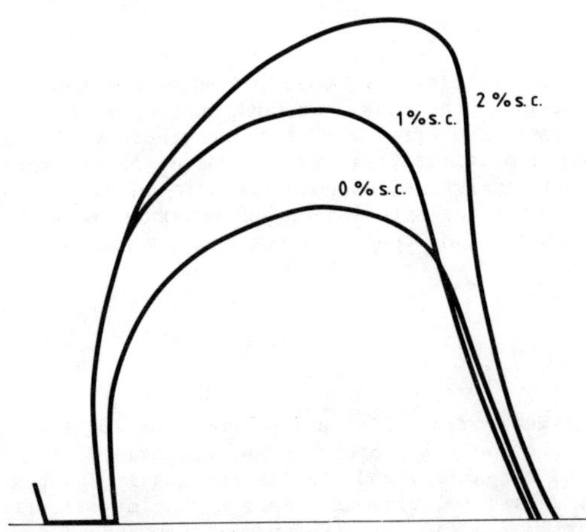

Fig. 2 : Short-time extensograph-curves of doughs from commercial german wheat flour, same consistency (600 FU) but different sodium chloride (s.c.) concentrations on flour basis

Table I : Dates of short-time-extensograph-curves (Fig. 2)(23)

wheat flour with	0 %	0.5 %	2.0 % sodium chloride
water adsorption for 600 F-Units	55.1	53.7	53.0
SE-number (KEZ)	66	73	96
Max-height (MH)	500	650	780
Relation-number (RZ)	7	12	12

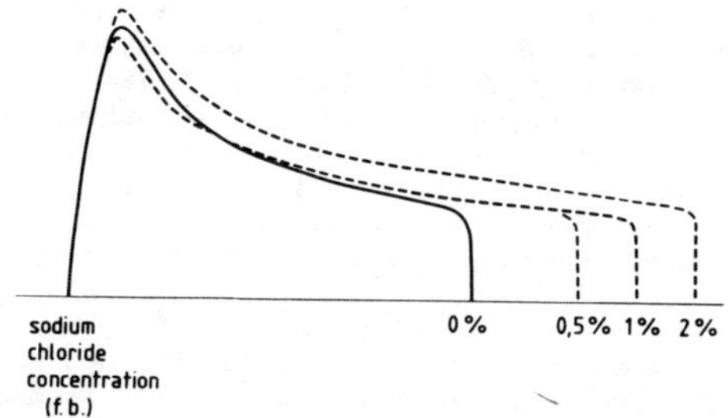

Fig. 3 : Alveograph-curves of doughs from commercial german wheat flour, same consistency but different sodium chloride concentrations on flour basis (f.b.)

Table II : Dates of alveograph-curves (Fig. 3)

sodium chloride	W $\times 10^{-4}$J	P	G	L	$\frac{P}{L}$
0	140	55	19	75	0.25
0.25	160	52	21	92	0.23
0.50	165	52	21	93	0.23
1.00	176	51	23	103	0.22
1.50	185	51	24	114	0.21
2.00	195	57	24	114	0.21

Salt also effects starch gelatinization temperature and peak viscosity. This is shown for two rye flours with different enzymatic activities in Fig. 4.

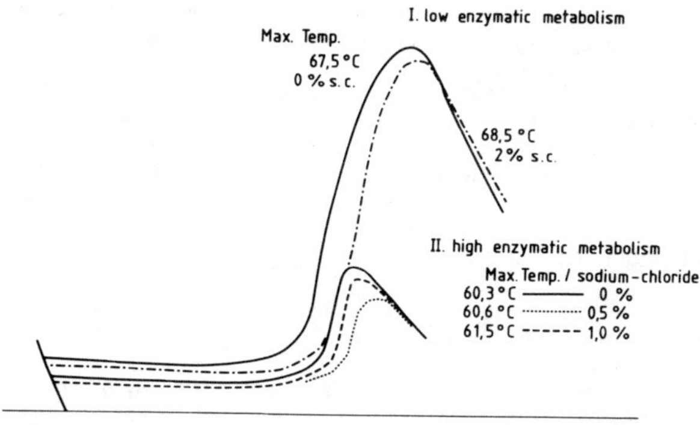

Fig. 4 : Amylograph-curves of slurries from two commercial german rye flours, with different enzymatic properties, same flour and water ratios but different sodium chloride (s.c.) concentrations on flour basis

It is also well known that not only the presence of salt or acid itself causes the best rye bread quality. Both, optimum salt and acid content, are necessary for a good dough and bread quality. Salt together with an optimum pH value improves dough consistency, gas retention, and loaf volumes. Salt reduces the solubility of pentosans of rye proteins, and so increases viscosity, so that rye doughs are more resistant against enzymatic attack. Salt is further known to inhibit the growth of lactic acid bacteria, and tends to shift the dough microflora towards more homofermentative lactobacilli (11).

These effects of salt are normally intensified by higher salt contents. The optimum levels are about 0.5 % for wheat flour dough properties, 1.0 to 1.5 % for the characteristics of rye flour doughs. From a sensoric point of view salt contents from 1.5 to 1.8 % (calculated on flour weight) are most effective. This addition is close to the average quantity of salt added for flours in Europe (about 1.5 to 2.0 %). Sodium and sodium chloride in milling products and cereals foods can be determined by an ion sensitive electrode with good reproducibility (17).

It may also be of interest to examine very high and very low salt levels (7,11,18,19) in bread processing. Too much salt will result in very sticky doughs, low loaf volume, a round shape, intensive crust colour and with very close but uniform crumb porosity. On the other hand very little salt shows stiff doughs of short softening times. The bread has a large volume, pale crust, poor crumb coherence and spreadability and poor flavour.

3. NUTRITIVE ASPECTS OF SALT IN CEREAL PRODUCTS

Nowadays with a trend towards low salt diets, new attention is paid to the salt content in food. The salt addition in bread production can be decreased by 10 - 20 %. The flavour of well fermented rye bread will not suffer by this reduction.

The same reduction in wheat bread causes less flavour. Sodium chloride replacement by other anorganic salts will not give the same flavour as the original quantity of sodium chloride. The costs are higher (7,18,19).

High sodium intake together with smoking, high body weight and lack of exercise may cause high blood pressure. Salt is one but not the only factor of this risk. In earlier discussions on the effect of sodium intake very often only the sodium added by the food producers was calculated. It has been shown, however, that the total sodium intake depends on the sodium content of the raw material and the sodium additions during commercial food processing, cooking and eating. So it is necessary to inform the consumers of the role of the salt shaker in the kitchen or during a meal.

4. DIETARY FIBRE IN CEREALS AND CEREAL PRODUCTS

The recommendation of a higher daily intake of dietary fibre has resulted in special developments. Fibre enriched breads were developed and are on the market (1, 21).

Systematic investigations were started to study fibre contents in raw materials. There are still difficulties in the interpretations about the characteristic of dietary fibre. Different methods give different results depending on different degrees of soluble or insoluble fibre. Consequently, information on fibre contents in food should be given only with the analytical method (5,6,8,9,12-16,20).

For grain and some fruits and vegetables a ranking list of fibre contents has been published. Rye and barley have a higher fibre level than oat, rice, corn or wheat; rye, barley and oat have the highest content of soluble fibre (5, 16). In the case of cereal products the extraction rate of the milling process has to be taken into consideration.

Bread is a good source for a higher supply of dietary fibre. This can be further improved. Wheat and soybean bran have been mainly investigated as a fibre source. In baking technology an addition of 10 % wheat bran with low starch content is recommended. This results in a higher dough yield,but still gives normal dough properties. The tolerance during the final proof decreases, the crust colour of the bread is darker. Crumb colour depends on the particle size of the fibre source. Coarse bran is more favourable in comparison to finer bran. Fine ground products give non-bread crumb colour in flour breads (1).

The crumb texture itself is shorter and less elastic. The flavour of wheat and wheat-rye mixed bread with 10 % wheat bran is enhanced. The same addition to rye flour or rye-mixed bread gives a poorer flavour. In wholemeal rye bread 10 % wheat bran does not change the taste, colour etc. Such a bread has one of the highest fibre contents.

In the Federal Republic of Germany some ready mixes with excellent high fibre addition are already available. So it is possible to offer a wide range of fibre enriched bread and roll varieties (1,21).

5. DIETARY FIBRE IN NUTRITION

Too low fibre levels in food may result in constipation and are also considered as a risk factor for the development of several deseases including diabetes mellitus. Fibre from cereals have the best functional properties because these fibre types influence intestinal transit time and fecal weight, and they are one possibility to lower calory content in bread (2,4, 10,22).

Today 40-50 % of dietary fibre intake in Germany, 45 % in Sweden (16) and 21 % in Switzerland (6) originate from cereals. Even wheat bread baked from low extraction rate flours has an important dietary fibre content. The intake of dietary fibre from white bread is usually higher than thought, and is higher than in the same amount of most fruits and vegetables. An agreeable fibre intake is guaranteed by 6 to 8 slices of wholemeal-rye bread; the same quantity of fibre is in 3 to 4 cabbage heads, or 13 green peppers, or 30 tomatos (22).

It is of great importance for the human diet to increase the daily fibre intake by an increased consumption of fibre containing foods. There is no doubt that the easiest way to get sufficient fibre is a higher bread consumption, especially with bread produced from flours with a high extraction rate or wholemeals or with an addition of bran.

REFERENCES

1. ALTROGGE, L., SEIBEL, W. and STEPHAN, H. (1980). Möglichkeiten der Ballaststoffanreicherung bei Brot und Kleingebäck. Getreide Mehl und Brot 34. 9, 243 - 247
2. ARNBJÖRNSSON, E., ASP, N.-G. and WESTIN, S.I. (1982). Decreasing incidence of acute appendicitis, with special reference to the consumption of dietary fiber. Acta Chir. Scand. 148. 461 - 464
3. ASP, N.-G. et al. (1981). Dietary fibre in type II diabetes. Acta Med. Scand., Suppl. 656. 47 - 50
4. ASP, N.-G. et al. (1981). Wheat bran increases high-density-lipoprotein cholesterol in the rat. Br. J. Nutr. 46. 385 - 393

5. ASP, N.-G. et al. (1983). Rapid enzymatic method for assay of insoluble and soluble dietary fiber. J. Agric. Food Chem. 31. 3, 476 - 482
6. BLUMENTHANL, A., SCHEFFELDT, P. and SCHÖNHAUSER, R. (1983). Zum Nährstoffgehalt schweizerischer Brote und deren Beitrag zur Bedarfsdeckung der Bevölkerung (Nutrient contents of different types of Swiss bread and their contribution towards covering requirements of the population). Mitt. Gebiete Lebensm. Hyg. 74. 80 - 92
7. BRÜMMER, J.-M. and BAUMANN, K. (1980). Möglichkeiten zur Verminderung des Kochsalzgehaltes in Brot und Kleingebäck. Unpublished
8. ELCHAZLY, M. and THOMAS, B. (1976). Die Verteilung der Ballaststoffe in dem Gewebe des Weizenkornes. Getreide Mehl und Brot 30. 265 - 269
9. FRØLICH, W. and ASP, N.G. (1981). Dietary fiber content in cereals in norway. Cereal Chem. 58. 1, 524 - 527
10. HARMUTH-HOENE, A.E., MEIER-PLOEGER, A. and LEITZMANN, C. (1982). Der Einfluß von Johannisbrotkernmehl auf die Resorption von Mineralstoffen und Spurenelementen beim Menschen. Z. Ernährungswiss. 21. 3, 202 - 213
11. LINKO, P., HÄRKÖNEN, H. and LINKO, V.V. (1983). Process technological effects of sodium chloride in bread baking. Journal of Cereal Sci., in press
12. MENGER, A. (1980). Zum Problem des Erfassens von "Dextrinen" und löslichen Ballaststoffen in Backwaren. Getreide Mehl und Brot 34. 6, 143 - 146
13. MENGER, A. (1980). Die Problematik der Ballaststoff-Analytik bei Lebensmitteln auf Getreidebasis. Mühle und Mischfuttertechn. 117. 48, 655 - 659
14. MENGER, A. (1981). Die Bedeutung von Ballaststoffen aus Getreide für die menschliche Ernährung. Brot in unserer Ernährung. Eine wissenschaftliche Bestandsaufnahme. Hrsg. GMF, Bonn, 18 - 22
15. MEUSER, F., SUCKOW, P. and KULIKOWSKI, W. (1983). Analytische Bestimmung von Ballaststoffen in Broten, Obst und Gemüse. Getreide Mehl und Brot 37. in press
16. NYMAN, M. et al. (1983). Dietary fiber content and composition in six cereals at different extraction rates. Cereal Chem. in press
17. RABE, E. (1983). Zur Natrium- und Kaliumbestimmung mit ionensensitiven Elektroden. Z. Lebensm. Unters. Forsch. 176. 4, 270 - 274
18. SALOVAARA, H. (1982). Kalium- und Magnesiumchlorid als Substitut des Kochsalzes im Roggenbrot - eine sensorische Untersuchung. Getreide Mehl und Brot 36. 1, 20 - 22
19. SALOVAARA, H. (1982, 1983). Studien über den Ersatz von Kochsalz durch andere Salze bei der Brotherstellung. 1. Teil : Einfluß auf die Rheologie und die Backeigenschaften von Roggenteigen. Getreide Mehl und Brot 36. 11, 295-298. 2. Teil : Einfluß auf das Gasbildungsvermögen im Weizenteig und im gesäuerten Roggenteig. Getreide Mehl und Brot 37. 8, 240 - 243
20. SOUTHGATE, D.A.T. (1969). Determination of carbohydrates in foods. J. Sci. Fd. Agric. 20. 331 - 335
21. TRÜMPER, M. (1982). Erprobte Rezepte für Brot und Kleingebäck mit Ballaststoffen. Deutsche Bäcker-Ztg. 69. 29, 904
22. WISKER, E., JESSEN, H. and FELDHEIM, W. (1982). Einfluß verschiedener Brotsorten mit unterschiedlichem Ballaststoffgehalt auf das Stuhlgewicht. - Akt. Ernährungsmed. 7. 4, 161 - 163
23. BRUMMER, J.M. (1978) Getreide Mehl und Brot 32.8, 220-224

OPTIMIZATION OF THE BAKING PROCESS AND ITS INFLUENCE ON BREAD QUALITY

Y. MÄLKKI[1], W. SEIBEL[2], C. SKJÖLDEBRAND[3] and Ö. RASK[4]

[1]Technical Research Centre of Finland,
Food Research Laboratory
SF-02150 Espoo 15, Finland,

[2]Bundesforschungsanstalt für Getreide- und Kartoffelverarbeitung,
D-4930 Detmold, Germany (F.R.),

[3]University of Lund,
S-23053 Alnarp, Sweden

and

[4]SIK-Swedish Food Institute,
S-402 29 Gothenburg, Sweden

Summary

The results of four coordinated research projects are reviewed. As criteria for adequate baking, several physical, chemical and sensory criteria were studied. For crumb ripening, elevation of the crumb centre temperature and resiliency of the crumb were found to be the most reproducable. At this point rupturing of starch granules as observed by electron micrography and gelatinization, as revealed by hydrolysability with ß-amylase, has occurred. In developing crust colour, the temperature is the most important determining factor. Penetration of heat into the crumb can be enhanced by elevated temperatures, air velocity, short-wave infrared radiation, and by using heat conduction from the hearth, as well as their combination. Equations for estimating effects of the process parameters on baking time, weight loss and energy consumption were developed. The energy consumption was found to be mainly determined by the amount of water evaporated.

1. INTRODUCTION

Baking is one of the most critical steps of the process for the quality of bread and other bakery products. It is, however, also the major energy consuming step in the bakery industry. Current interest in studying this process has arisen for several reasons: for achieving savings in the cost of energy, for increasing the speed of production, for selecting the best alternative for each purpose from the available types of baking ovens, for evaluating possibilities for using new types of heating, such as infrared or micro-wave heating in this process, and for maintaining or even improving quality when introducing alterations in the process or equipment, or increasing the scale, mechanization and automation of production.

The aim of these coordinated research projects has been to obtain fundamental data for selecting optimal conditions in baking, both with respect to economy and quality. The research in Detmold, Germany, project B-D-1 (1, 2, 3, 4,), has concentrated on the penetration of heat in a conventional baking oven at three initial oven temperatures and using three types of wheat flour with differing extraction rates. Influence on quality has been tested using both sensory evaluation and physical measurements. The changes in the starch were also followed by electron micrography.

The reseach group in Lund, Sweden, project B-S-1 (5, 6) has studied heat penetration in a convection oven as a function of temperature, air velocity and application of moisture. Along with sensory evaluations, texture of the crumb and colour of the crust were measured physically.

The research group in Gothenburg, project B-S-4 (7, 8) has concentrated on the effects of infra-red heating on heat penetration and bread quality.

The Finnish group, project B-SF-2 (9, 10), has studied the effects of combinations of convection, radiation and conduction on heat transfer and quality parameters. In the experiments of the three latter groups, the same dough formulation and pre-baking conditions were used.

2. ADEQUACY OF BAKING

To be able to determine the minimum time required for the process, objective criteria for the adequacy of heat treatment are needed. The criteria used should have sufficiently high correlations with the subjective evaluations of the majority of customers. Thus for testing the usefulness of various physical or chemical criteria, reference was made to the evaluations of sensory panels or expert judgements. To support the panels, guidelines for the optimal properties of both crust and crumb were prepared (7, 11), including colour tables for evaluating the crust colour (7).

In regard to the crust development, both in commercial practice and in research the most common criterion is the crust colour, which can be checked by comparison with colour tables and by light absorption measurements. For crust thickness, standards and opinions vary from country to country and depend on the type of bread.

Towards the end of baking, the temperature of the crumb in the centre of the loaf approaches asymptotically the boiling point of water. In earlier studies the crumb centre temperature has often been suggested as the most reliable criterion for the adequacy of baking. In the present research programme, Vassileva, Seibel and Stephan (2) have compared the results of sensory evaluations with the temperature history of loaves. The samples studies represented loaves from three different types of wheat flour, and for each type of bread, three different temperature programmes were applied. An optimal sensory quality was achieved when the baking was continued for 5 to 10 minutes after reaching 98°C in the centre of the loaf, the total baking time needed for optimal sensory quality being 125 to 130 per cent of the time required to reach 98°C in the centre. When the same temperature programme and loaf size was used, the optimal time was independent of the flour type. In the Finnish studies (9, 10) performed with pan breads of smaller size, and at a lower temperature range, the panel already found the loaves sufficiently baked after the crumb temperature had reached 90°C.

Resiliency of bread crumb has been recently suggested as a criterion for the adequacy of baking by Marston and Wannan (12),who found it had a good correlation with the judgements of the majority of consumers. This property was studied also by our German group (2). The inflection point in the resilience pattern also coincided in this study for each flour type and for each temperature programme with the optimal sensory quality of the crumb. Our Swedish group in Lund (5, 6) used the inflection point of resiliency as a criterion for the adequacy of baking. It must be pointed out, however, that according to Marston and Wannan, this criterion is not applicable in cases using rapid heating methods such as micro-wave heating.

Related to the same properties might also be the stability of volume during cooling and storing. Rask and Gustafsson (7) observed that the stability of volume when cooling reached its maximum after an adequate heat treatment according to sensory evaluation. Penetrometric measurements (2, 9) did not give any clear inflection point or other limit which could be useful for determining the adequacy of baking.

Changes in starch granules during the baking were followed by scanning electron micrography by our German group (3, 4,). In the crumb, the deformation and folding of starch granules was substantially increased when the temperature reached 97 to 98°, and further deformation and loss of the crystalline structure was observed when the heating after this temperature was prolonged. This was evidently caused not only by the elevated temperature itself, but also in part by the increased amount of free water liberated in the coagulation of proteins and transferred to the starch. The Finnish group (9, 10) studied the changes in starch by light microscopy, and by studying its hydrolysability by β-amylase. The majority of the starch granules had already lost their birefringency after passing 70°C. Hydrolysability by β-amylase increased rapidly until the crumb temperature reached 98°C, after which the increase was slight.

A further method to check the adequacy of baking is to follow the loss of weight. Vassileva, Seibel and Stephan (2) found that the loss of weight varied from 9 to 11 %, depending on the flour type, when the bread was adequately baked. Correspondingly, Marston and Wannan (12) reported losses of 10 to 12 per cent.

The criteria presented refer specifically to the experimental conditions of each study. Deviations are to be expected, for reasons such as varying loaf size, and when the dough formulation deviates from that of standard baking recipes in a way influencing the amount or availability of water or increasing starch gelatinization and/or protein denaturation temperatures.

Furthermore, the temperature recordings bear often some uncertainties: the slowest heating point is not in the geometric centre of the loaf, some types of sensors have excessive dimensions for measuring only the point of slowest heating, or are influenced by the outer part of the crumb or by oven temperatures. For some rheological measurements, samples greater than just the point of slowest heating are needed. Finally, the criteria of the sensory panels for adequate baking can differ depending on national taste and texture expectations.

3. BAKING TIME AND ITS INFLUENCE ON QUALITY

3.1. Conventional oven (1-4)

The oven temperature programmes of the German studies started with higher initial temperatures, which were continued until the crumb temperature reached 60°C (end of oven spring), after which the oven temperature was allowed gradually to decrease. For loaves of 1000 g, the following rates of heating and baking times were found:

Initial oven temperature °C	Time until 98°C in the centre min	Time until acceptable crumb quality min	Time until optimal crumb quality min
280	24	25	30
260	28	30	36
240	33	36	43

As expected, both minimal and optimal baking times are highly dependent on the baking temperature. Provided an optimal baking time was used, the temperature programme had no influence on the volume yield, porosity pattern or other quality criteria of the crumb. The length of baking time had a more pronounced influence on the sensory quality scores at higher baking temperatures than at lower. The level of the temperature had the most marked effect on the crust colour, whereas prolonged baking times increased the thickness of crust and consequently the loss of weight. The baking time did not have an influence on the speed of becoming stale for 24 and 36 hours after baking.

For the temperature of the oven hearth, an optimum from 250 to 260°C was found.

3.2 Convection oven (5, 6)

In an experimental oven designed and constructed by the institute the temperature range studied was from 170 to 260°C, air speeds between 0.5 and 5 m/s, while air humidities corresponded to dew points between 80 and 90°C The sizes of the loaves were 100 and 300 grams, the latter being baked in moulds of black aluminium foil.

Under these conditions, limiting for the adequacy of baking was in some cases the colour of the crust, in others the heating of the crumb, and thus for the end point of the baking the time was taken when the crust had obtained the desired colour range and the crumb resiliency was above 0.55. As in the conventional oven, the colour development was highly dependent on the temperature. In addition, it also depended on air velocity. To obtain the same level of browness the time required at 170°C was 29.3 min, whereas it was at 260°C only 7.2 min. using the same air velocity of 2 m/s. However, according to the crumb resiliency the minimum time at 260°C was 10.6 min. At 220°C, the corresponding times for colour development varied between 18.1 and 13.7 min, when the air velocity varied between 0.5 and 2 m/s, but the time decreased only to 12.3 min when the air velocity was further increased to 5 m/s. Browning time was further influenced by the application of moisture. Without steam injection, the baking time according to colour development was at 220°C and with 2 m/s air velocity ca. 5 min longer than when steam was applied.

3.3 Infra-red heating (7, 8)

In an experimental oven constructed by the institute, short-wave infra-red power was applied in order to obtain better heat penetration than in the conventional medium wave-length range. The energy was applied constantly, periodically or by pulsing. The surface effect was limited so as not to exceed 2 W/cm^2 to avoid overheating. Control bakings were made using a conventional oven and a convection oven at 200°C.

Compared with the conventional oven control, a reduction in the baking time of about 25 % was achieved, but the elevation of the centre temperature was about the same as that with the convection oven. It may be noted however, that these results were obtained for pan bread. For free-standing bread the infra-red heating gave more rapid heat penetration. Of the three means of IR applications, periodical was found to be the most efficient.

Losses of weight were comparable to that of a conventional oven when the baking times were comparable in respect to the crumb temperature development. Within the limits of experimental error, the volume yields were comparable to the controls. The density of the crumb in cylindrical samples from the centre was of the same level as that in the control, but for samples of greater size it was higher, showing evidently a different moisture distribution than in the control. The colour of the crust could be adjusted to be comparable to the control by regulating the infra-red effect. The crust showed a tendency to light spots or bubbles, this being evidently due to drying effects. No steam was applied during the baking.

3.4 Combined heating (9, 10)

In an experimental oven constructed by the institute, conduction, convection and radiant heat and their combinations were applied for baking. Furthermore, an equipment was constructed to control the humidity in the oven. Baking tests were made for loaves of 300 g in weight in black aluminium foil moulds as in the both Swedish projects. Oven temperatures were 200, 220 and 240°, air velocities 0 to 2 m/s, air humidities from 10 to 310 g/m^3. For conduction heating, a removable hearth was used, with the same temperature as the oven temperature to enable good control of the oven temperature. Times required for adequate baking (crust colour, centre temperature 98°, sensory evaluation) varied between 12 and 20 minutes.

The regression equations calculated on the basis of experimental results allow the calculation of the magnitude of the effect of the variables, singly or in combination, on the baking time, weight losses, crust colour, volume yield and energy consumption. As in the other coordinated studies, oven temperature was the most important determining factor on the baking time as judged on the basis of crust colour and crumb centre temperature. Use of conductive heating could cut the baking time at each oven temperature by 2 to 2.4 minutes, and use of air circulation by additional 1.2 minutes. Losses of weight were most dependent on the baking time, the other influencing factor being the air velocity. Both high levels of oven temperature and use of air circulation slightly reduced the volume yield, although only the effect of air velocity was found to be significant.

It is evident that the levels of the variables chosen influenced the results, and that for instance, at higher oven temperatures the effect of radiant heat would be more marked. The effect of conduction heating, which was already great using the same hearth temperatures as that of the oven, could be accentuated by using hearth temperatures higher than the oven temperature. The effect of air moisture on crumb temperature elevations remained in these experiment negligible, as it did in the experiments of our group in Lund. Although the steam condensed leaves a considerable amount of energy on the surface, the condensation is limited only to the first one to three minutes of baking, when the surface temperature is sufficiently low to allow condensation. The effect of condensation is counteracted by the absorption of infra-red heat by the water vapour, this diminishing the effect of radiant heating .

4. ENERGY CONSUMPTION

Several studies on the total energy consumption in bakery plants have recently been performed in various countries. A part of the Finnish project (9, 10) was also devoted to the total energy economy of bakeries. In the plants studied, the energy consumption of the ovens was 30 to 46 per cent of the total energy consumption of the bakery plants. The total costs of energy were at the time of the study on average 2.2 per cent of the turnover and are today about 3 %. A theoretical study on the heat and mass transfer in baking has been recently made by Nebelung and published in a dissertation in Dresden (13).

Energy consumption at the baking step has been studied in the framework of this project by the two Swedish groups (5-8) and the Finnish group (9-10). In the convection oven (5, 6) the energy consumption was found to be highly dependent on the amount of water evaporated. Thus all the factors having an influence on the loss of weight and crust thickness can also be directly regarded as increasing the energy consumption. The energy consumptions observed varied thus between 518 and 925 kJ/kg, the theoretical consumption being between 400 and 500 kJ/kg. The energy efficiency in the infra-red oven (7, 8) used was relatively low, ca 11 % due to its construction, but previous experience of the institute on this type of ovens indicates that efficiencies up to 25 to 30 % can be achieved. The application of infra-red radiation can thus not be based on energy economy as compared to other means of heat transfer. In the studies on combined heating, (9, 10), the dominating factor in energy consumption was the baking time. The effect of air circulation was statistically significant, but of another order of magnitude. In earlier studies it has been observed that the amount of steam applied to moisten the oven atmosphere varies widely in practice, and it was found in a recent Danish study (14) to form 4 to 29 per cent of the total energy consumption of the ovens studied.

5. CONCLUSIONS

From the experimental results of this coordinated study and from the previous studies it can be concluded that in enhancing the speed of heat penetration and in reducing the baking time the oven temperature is the most important factor. Its effect can be accentuated by increasing the air velocity and by using conductive heating. These means are also advantageous for reducing weight losses and for cutting the costs of energy. A limiting factor for the use of high oven temperatures and for the hearth temperature is the crust colour development. At higher temperatures the heating of crumb is slower than the crust colour development and thus full advantage cannot be obtained from too high an oven or hearth temperature. When using infra-red heating, the most effective heat penetration is achieved using short wave length energy and periodical application. The baking times are comparable to those obtained in other types of ovens using high oven temperatures or air velocities. A similar final quality of product can be achieved using each type of heating by regulating the levels of the variables or their combinations.

REFERENCES

1. VASSILEVA, R. and SEIBEL W., Backparameter und Brotqualität, 2. Temperaturmessungen an freigeschobenen Weizen(mehl)broten. Getreide Mehl und Brot 35 (1981): 3, 238 - 239, COST-Project 91 B-D-1
2. VASSILEVA, R., SEIBEL, W. and STEPHAN, H. Backparameter und Brotqualität. 3. Einfluss von Backzeit und Backtemperatur auf die Weizen(mehl)-brotqualität in Abhängigkeit von Mehlausbeute. Ibid. 259 -263, COST-Project 91 B-D-1
3. VASSILEVA, R, SEIBEL, W. and MEYER, D., Backparameter und Brotqualität 4. Strukturveränderungen der Stärke in der Brotkrume beim Backen. Ibid. 303 - 305. COST-Project 91 B-D-1
4. VASSILEVA, R., ANTILA, J., SEILER, K., SEIBEL, W., STEPHAN, H. and MEYER, D., Backparameter und Brotqualität. Bundesforschungsantalt für Getreide- und Kartoffelverarbeitung, Jahresbericht 1981, D 17. COST-Project 91 B-D-1
5. CHRISTENSEN, A., BLOMQVIST, I. and SKJÖLDEBRAND, C., Optimering av bakningsprocessen med avseende på energiminimering (konvektionsugnar) Univ. Lund, Dept. Food Engineering, Report 83-01/9. COST-Project 91 B-S-1
6. CHRISTENSEN, A., BLOMQVIST, I. and SKJÖLDEBRAND, C., Optimization of the baking process (Convection ovens). In: Zeuthen & al. (Eds.) Thermal Processing and Quality of Foods. Applied Science, London 1984. COST-Project 91 B-S-1
7. RASK, Ö. and GUSTAFSSON, U. Optimering av bakningsprocessen vid tillverkning av matbröd med avseende på energiminimering (IR-värmning). SIK- The Swedish Food Institute Up 2128. COST-Project 91 B-S-4
8. RASK, O., Optimization of the baking process (Infra-red radiation). In: Zeuthen & al., Loc.cit. COST-project 91 B-S-4
9. PALOHEIMO, M., MÄLKKI, Y. and KAIJALUOTO, S., Optimering av bakningsprocessen med avseende på energiminimering. Technical Research Centre of Finland, Report, 1983-01-26. COST-Project 91 B-SF-2
10. PALOHEIMO, M. and MÄLKKI, Y. and KAIJALUOTO, S., Optimization of baking process: Combined heat transfer. In: Zeuthen & al., Loc.cit. COST-Project 91 B-SF-2
11. BRÜMMER, J-M., Zur Beurteilung der sensorischen Qualität von Backwaren-Anleitungs-und Schulungsbeispiele. Getreide Mehl Brot 36 (1982) 77 - 84. COST-Project 91, B-D-1.
12. MARSTON, P. E. and WANNAN, T. L., Bread baking - the transformation from dough to bread. Bakers Digest 50 (1976):8, 24 - 28, 49
13. NEBELUNG, M. (1978) Dissertation. Dresden
14. THØGERSEN, L. K., Brancheenergiundersøgelse for rugbrødsindustrien. Jydsk Teknologisk Institut, JTI-Sag-Nr 62-32288-7, Århus, 1978.57 pp.

RESEARCH NEEDS OF THE EUROPEAN BREAD INDUSTRY, STATE OF THE ART OF BAKING AND RECOMMENDATION FOR FURTHER WORK

W. Seibel
Federal Research Centre for Grain and Potato Processing
Institute of Baking Technology
Detmold, Federal Republic of Germany

Summary

The nutrition of the world's people is mainly based on grain and bread. More than 50 % of the food energy comes from grain products. An improvement of the nutritive situation can therefore only be reached with grain and bread.
Further research is necessary in many fields of baking technology. Since wheat breeding was very successful we should concentrate now on rye. With rye we get more bread varieties and therefore a higher bread consumption.
Within the baking procedure the effect of sponge doughs and sour doughs should be investigated. Extrusion cooking will be a new parameter in bread production. We have to improve the sensoric evaluation and we should know more about the nutritive value of different breads. Grain research will be also responsible to translate the scientific results, which are of interest to the consumers, to the public.

1. Introduction

It is the goal of grain processing and especially of baking technology, to produce different types of bread with good flavour and texture and sufficient shelflife. The original energy of the grain should be kept as far as possible.

More than 50 % of the energy for the nutrition of the people originates from grain and therefore the nutrition problems of the world can be only improved or partly solved by producing more grain and more bread.

Practical trials have demonstrated that each region has its own types of bread and it is almost impossible to exchange bread varieties from one region to the other. Some countries in Central Europe use also rye as bread grain in order to produce rye- or rye-mixed breads from ryeflour and/or rye meal. All efforts to produce and sell those breads outside these countries were - until now - not very successful. Therefore research needs in the field of baking and recommendations for further work can be only discussed on a regional basis.

2. Bread production in Europe

In Europe bread is produced either by the industry or by handcraft bakeries. The relationship differs from country to country. But no significant differences in bread quality have been observed between the two types of production. In some countries the production in small bakeries is mostly done by hand, in other countries, there is already a high degree of mechanisation in the production of bread. Therefore further research should take specially in consideration the situation of small bakeries.

3. Baking process

The baking process is the most important factor within the production of bread. Baking conditions can influence the sensoric quality (crust-/ crumb characteristic / flavour), the shelflife and the nutritive value (5,12,13). Experiments have shown, that the reduction of the baking time by introducing other heat transformation systems mostly causes a decrease in the over all bread quality.

It is now possible to measure the changes in temperature in the bread crumb and crust exactly. Figure 1 shows a typical temperature profile during the baking process (12).

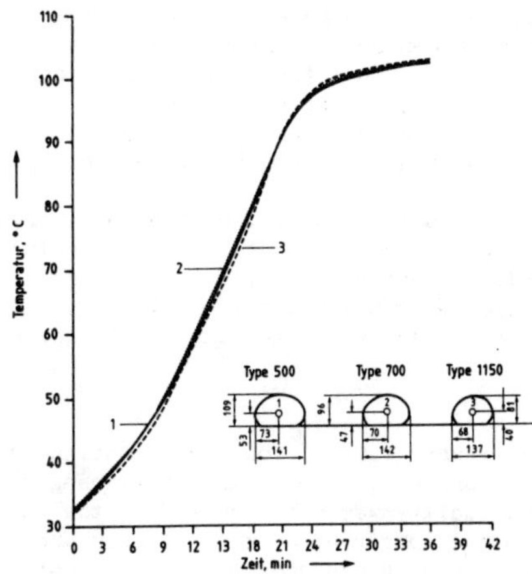

Figure 1 Temperatures in wheat breads during baking (baking temperature 240° C)

For a bread with good quality the final temperature in the centre of the crumb should reach 102-105° C. Under these conditions bread is produced with excellent flavour and a very good shelflife. Baking time and temperature have to be adjusted to the type of bread. Table I shows baking conditions, which are now practised in the Federal Republic of Germany.

Table I Bread varieties and Baking conditions

Bread variety	baking temperature ° C	baking time min
wheat bread (750 g)	240	30 - 35
toast bread (500 g)	220	35 - 40
wheat mixed bread (1250 g)	260 ⟶ 220	50
rye mixed bread (1500 g)	260 ⟶ 220	60
rye bread (1500 g)	300 ⟶ 220	60
rye meal bread (1250 g)	190	70

We also practise a so called "pre-bake-system" with a "pre-bake-oven" followed by a normal baking oven. In the pre-bake-oven we use a temperature between 330-430° C for 2 - 5 min with or without steam injection in order to close the surface of the bread. Under these conditions less bread flavour can evaporate.

4. Further research

The over all bread quality depends on different factors: starting from the quality of the raw material and the recipe until the wrapping and storage of bread (8).

4.1 Raw material and recipe

Grain breeders were very successfull during the last 20 years. The yield and baking quality, especially with wheat, were improved extraordinary. There is practically no sprout problem with wheat.

Because bread consumption can be increased by a large number of bread varieties, it is now necessary to concentrate breeding research on rye. With rye we can increase the number of bread specialties, and rye bread has more intensive flavour and a longer shelflife than wheat bread (10). The main research objects are: high yielding varieties and sprout resistance.

In all countries are discussions about food and chemistry. We should therefore also try to bake good bread without or little baking improvers. Also the salt problem needs further research.

There is no real pesticide or heavy metal problem with grain. Nevertheless we should control the positive status and try to decrease these substances (4).

4.2 Process engineering

Years ago a so called modern bakery should use continuous mixing and a very short production process. Sour dough was exchanged by acidification agents. But the bread quality was not accepted by the customers. Now we observe more and more a longer production process and sour doughs for the rye bread production and sponge douhgs for the wheat bread production are discovered again (11). Customers prefer breads rich in flavour, and therefore fundamental research is necessary why sponge doughs respectively sour doughs and longer production times give a better bread.

The classical system for the production of bread (dough mixing, fermentation, baking) has been partly changed for some breads, because crisp bread can be produced now by HTST-Extrusion cooking, directly from meal and/or flour. Energy requirement compared with the classical system of crisp bread production is decreased remarkable. Therefore we should investigate how this modern process can be used also for the production of other baked goods (1).

Especially in summer time a considerable part of the wrapped sliced bread is spoiled by mould. Of course there are possibilities to protect bread against these mircroorganism (for example: preservatives, pasteurisation, micro waves, atmospheric exchange in the package). But we should find better production conditions to prevent mould infestation after baking (2).

4.3 Sensoric evaluation

Numerous grain institutes around the world tryed to determine bread flavour by gaschromatography. More than 100 flavour compounds were detected, but by chemical means it is still not possible to decide whether a bread has a good or poor flavour.

But intensive studies in sensoric evaluation have demonstrated (7,14) that trained experts can describe very reproducible the specific aroma patterns of different breads. It is even possible to decide by sensoric evaluation whether rye bread was baked with acidification agents or with sour dough.

Table II Sensoric evaluation of rye- and rye-mixed bread, baked with sour dough or acidification agents

Addition of	rye mixed bread	points*)	wheat mixed bread	points*)
3/3 sour dough	fully aromatic	5	aromatic	5
2/3 sour dough) 1/3 acidif.agents)	mild aromatic	5	aromatic	5
1/2 sour dough) 1/2 acidif.agents)	little aromatic	4	little aromatic	4
3/3 acidif.agents	tasteless	3 - 2	little aromatic/ tasteless	4 - 3

*) 5 = excellent; 1 = not satisfactory

Further research is necessary to investigate the influence of each part of the baking process on the sensoric value of bread.

Bread has to have a good quality, not only the time when it is bought but also when it is consumed by the customer. Therefore we should know more how the customers keep their bread in the household and which factors can influence the bread quality (especially shelflife and flavour) under the special storage conditions in the household. Based on such investigations, the producer should give proposals to the customer regarding optimum storage conditions and maximum shelflife. According to the new labelling law in the Common Market area, there has to be a declaration about shelflife on every wrapped bread, starting January 1rst, 1984.

4.4 Nutritive evaluation

Bread plays a very important part in human nutrition. In the Federal Republic of Germany 20 % of the energy and protein intake and 15 % from iron within the daily nutrition comes from bread.

But there is a remarkable difference in the nutritive value between the grain and most of the bread varieties, because we use mostly low extracted wheat flour and not whole wheat meal. Low extracted wheat flour has of course less minerals, vitamines, and dietary fibre (6). The difference in mouth feeling, texture and appearance is the reason that most people prefer white bread instead of meal bread. On the other hand there is a remarkable deficiency in fibre in the daily nutrition.

Therefore we should try to develop bread specialties with the mouth feeling of white bread and a composition of the important compounds of whole meal bread. It is a real challenge to us, to produce those grain products, but the flavour of these breads should be at least satisfatory.

In some countries there are deficiencies of certain nutrients in the daily food (for example: Ca, Fe). Then the flour is enriched with the missing components. This enrichment of flour is an excellent system because bread is eaten every day and the people get a more balanced diet. All missing nutrients can be added via flour and bread, the homogenous distribution even in small amounts in flour is no technical problem.

For special diseases people need special health food, including bread. We should make more effort to develop bread specialties for diabetes or "Cöliakie" with good quality, because these groups should enjoy their daily bread too.

We do not know very much how the value and digestability of certain bread components are influenced during baking or extrusion cooking. The biological value of protein is diminished with about 30 % in the bread crust during baking, but the energy seems not to be influenced by these heat treatments (3). Future test trials should give an answer about the over all nutritive value from normal bread and also the numerous bread specialties, depending on the different production conditions.

4.5 Nutritive information for the customer

In former times research projects were finished when we received the final results. But now it is necessary to inform all interested groups so far nutrition problems are involved. This information is very difficult because we have to translate our scientific figures in a very simple language, so it will be understandable to the public.

Grain and bread research should find methods, how results can be transfered to all interested groups as fast as possible. In many countries exists a movement of young people in the direction of an alternative way of life including alternative nutrition f.e. alternative bread (2). One reason for this development may be an insufficient information about our normal bread varieties, their recipes, production conditions, special nutritive values, and shelflifes.

It has been demonstrated that a continuous information to the public and specially the physicians (which are the most important food advisors in every nation) about facts and not stories, can change the image of certain foodstuffs. It will be very important for further research to investigate this particular field. Since nutrition experts all over the world advise a higher bread consumption to balance the inbalance between nutrients intake and nutrients requirements, grain research should support all efforts in this field.

REFERENCES

1. Antila, J., Seiler, K., Seibel, W. and Linko, P. (1983). Production of flat bread by extrusion cooking using different wheat/rye ratios, protein enrichment and grain with poor baking ability. Journ.Food.Eng. 2, 189-210
2. Brümmer, J.-M., Seibel, W. and Spicher, G. (1981). Schimmelbefall bei Brot. Forschung 1, 25-27
3. Menden, E. and Horchler, V. (1978). Über den Einfluß der Krustenbildung bei Brot auf Proteinqualität und kalorische Ausnutzung. Getreide Mehl und Brot 7, 184-188
4. Ocker, H.-D. (1983) Schadstoffe in Getreide. - Kenntnisstand und Lükkenkatalog. Getreide Mehl und Brot 1, 3-7
5. Pomeranz, Y. (1982). Mehlbestandteile und Backverhalten. Getreide Mehl und Brot 10, 264-272
6. Rabe, E. and Seibel, W. (1981) Nährwert von Brot und Spezialbroten. Getreide Mehl und Brot 5, 129-135
7. Seibel, W. (1982). Sensorik-Ausbildung für Sachverständige der DLG-Qualitätsprüfungen für Backwaren (1.Mitteil.). Getreide Mehl und Brot 2. 54-56
8. Seibel, W. (1982) Optimierung der Backwarenqualität. Getreide Mehl und Brot 8, 207-211
9. Seibel, W. (1983). Alternative Backwaren - Herstellung und Qualität. Getreide Mehl und Brot 1, 7-13
10. Seibel, W., Brümmer, J.-M. and Stephan, H. (1978). Produktion und Qualitätseigenschaften von deutschen Spezialbroten. Getreide Mehl und Brot 11, 301-310

11. Spicher, G. and Stephan, H. (1982). Handbuch Sauerteig. Behr's Verlag, Hamburg
12. Vassileva, Radka and Seiler, K. (1981). Backparameter und Brotqualität (1.Mitteil.). Getreide Mehl und Brot 7,184-187
13. Vassileva, Radka, Seibel, W. and Meyer, D. (1981). Backparameter und Brotqualität (4.Mitteil.). Getreide Mehl und Brot 11, 303-305
14. Weber, Angelika and Seibel, W. (1980). Das neue DLG-Prüfschema für die Qualitätsbeurteilung von Backwaren. Getreide Mehl und Brot 3, 71-73

SUB-GROUP 2 INDUSTRIAL COOKING
INTRODUCTION - PREPARATION, WARMHOLDING AND REHEATING OF FOODS IN CATERING

G. GLEW, Director, Hotel and Catering Research Centre, Catering Studies Department, The Polytechnic, Queensgate, Huddersfield, U.K.

Summary

The catering industry is of growing importance in Europe and eating out will increasingly substitute for food consumed in the home. The preparation, warmholding and reheating of food by caterers affect the quality of the end-products. Warmholding and reheating of food are heating processes not used by food manufacturers but are very important in the catering industry. The practices of food preparation for catering consumers from both raw or pre-prepared products often differs from domestic or food manufacturing practice. In this section emphasis will be placed on the quality changes observed by these different practices.

1.0 The Catering Industry

Food consumption outside the home constitutes a significant proportion of total food consumed in all European countries. In the 1970's the percentage of food expenditure devoted to eating out was about 20% in France, 14% in Germany, 15% in Sweden and 12% in Britain (1). In no West European country has the level of eating outside the home reached the scale found in N. America where one in three meals is eaten away from the home (2). Eating out is likely to continue to increase in all developed countries and the proportion of the total diet consumed at home will therefore fall.

It is clear that more attention must be paid in the future to the growing importance of the catering industry (the North American term is "foodservice industry") by governments, economists, food manufacturers, sociologists and nutritionists. The associated industries of tourism and the provision of all types of accommodation are part of this growth area and their importance to western economies can no longer be ignored. For example, in the United Kingdom the workforce in the hotel and catering industry numbers 2.14 million people representing just less than 10% of the total workforce (3).

A further development which, in Europe, is beginning to accelerate is the so-called "fast-food revolution". Fast foods are not new but the ownership pattern and marketing strategy employed is new. The new fast food catering operations are based on food products that are supplied almost exclusively by food manufacturers. Indeed many of the fast food chains are owned by food manufacturers (e.g. General Foods, Pillsbury, United Biscuits) so there is a major international business element moving rapidly into an industry which in many European countries still consists of single-unit family operations.

In public sector catering many countries operate school lunch systems, hospital catering, and operations for feeding the elderly, the defence forces and people in prisons. In addition, feeding factory workers is an important element of social welfare. The catering in these operations tends to be on a larger scale and is often wholly or partly subsidised by the state. It is in this area that "industrial cooking" becomes most significant.

2.0 Industrial Cooking in Catering

One important activity within the area of "industrial cooking" is cooking of food immediately before consumption in the catering industry. Food manufacturers process food by boiling, steaming, frying, roasting and baking. Caterers use the same processes to cook food for the table. These processes are dealt with elsewhere in this volume and will not be considered further in this section. The physical, chemical and biological processes taking place in the food should be the same whether the process takes place on a large food manufacturing scale or on the smaller catering scale, although scale-up problems exist and are yet to be overcome by food manufacturers (4). However, caterers cook food in equipment of much smaller scale and often of different design to food manufacturers. Furthermore, the cooking process is usually controlled by a chef, not by a food technologist and the process is regarded as an art not a science. Hence, the chef's control over the cooking conditions depends on the degree of development of his skill and not on the use of scientific instruments. Canned baked beans from one manufacturer look the same and taste the same all over the world. Coq au vin is different in different countries, in different styles of restaurant and may even be slightly different each time it is prepared by the same chef. Within this section it was therefore felt appropriate to consider the quality of meals prepared from fresh components and also from preprocessed components. The use of preprocessed foods by caterers is increasing rapidly particularly in public sector catering.

In addition to heat processing of fresh and prepared meal components many caterers use two other processes which are unique to catering and not used by food manufacturers. These two processes are warmholding and reheating of food.

Warmholding is the process of keeping food hot and ready to eat for periods of a few minutes to many hours. The ideal of serving freshly cooked food to consumers is often impossible to attain in practice for economic reasons. It is clearly more cost effective to prepare food on a large scale and distribute to satellite service points. The traditional method of attaining this is to cook the food centrally and distribute the food hot to the service points. There is always some delay between kitchen and table but in many catering operations the delay is lengthy. The food is transported in heated or insulated containers and often kept hot at the service point before consumption. Warmholding can take place for as long as 5 to 6 hours in certain circumstances. Temperatures at which food is warmheld vary widely. The minimum temperature (for bacteriological reasons) allowable by law in most countries varies between 60 and 70 deg.C. However, no maximum temperature has been recommended. It is clear that during warmholding the cooking process continues and when temperatures and times of storage are unknown and variable then the quality of the end product is also likely to vary.

On the other hand reheating is a process which is usually carried out quickly and often under controlled conditions. Prediction of the quality effects of heat processing during reheating should be easier when the geometry and processing parameters are known and this is more likely than in warmholding. Reheating does result in quality changes which are discussed in this section.

In general, caterers have great concern for the quality of their products. However, the high organoleptic quality of a dish produced by a skilled chef can be easily lost by subsequent warmholding and/or reheating (after possibly chilling or freezing). Not enough is known about the effect of this practice on organoleptic quality.

Acknowledgements

Authors of the papers in this section would like to acknowledge the help given by their colleagues in many countries and for their permission to use their work. Particular thanks are due to Drs. Paulus (Karlsruhe), Zacharias and Bognar (Stuttgart), Fielliettaz Goethart (Zeist), O'Sullivan (Huddersfield), Ohlsson (Goteborg), Jonsson (Goteborg), and Piekarski (Stuttgart).

References

1. Glew, G. (1980). Background and trends in catering to 1990 in Western Europe. In Glew, G.(Ed.) Advances in Catering Technology. 3-15. Applied Science Publishers Ltd., London.

2. Livingstone, G.E. & Chang C.M. (1980). Background and trends in catering to 1990 in the U.S.A. In Glew, G. (Ed.) Advances in Catering Technology. 16-30. Applied Science Publishers Ltd., London.

3. Glew, G. (1983). The Future. Hospitality No.43, September. Published by Hotel, Catering and Institutional Management Assc., London.

4. Anon. (1982). Report of Workshop III. Catering/Foodservice Industry. In Koivistoinen, P., Hall, R.L., & Malkki, Y. (Eds.) The role and application of food science and technology in indusrialised countries. Proceedings IUFOST/OECD Symposium, Helsinki. VTT Symposium 18, Espoo, Finland.

SENSORY QUALITY AND NUTRITIVE VALUE OF MEALS PREPARED FROM FRESH AND PREPROCESSED COMPONENTS

M. TURNER, J. MOTTISHAW,
The Hotel and Catering Research Centre, Catering Studies Department, The Polytechnic, Queensgate, Huddersfield, U.K.

R. ZACHARIAS, A. BOGNAR.
Federal Research Centre for Nutrition, Karlsruhe/Stuttgart, W. Germany.

Summary

This paper presents the results of 2 investigations into the quality of preprocessed foods. At the Hotel and Catering Research Centre in England, food samples from 5 hospitals and 1 school were examined to compare the quality of food prepared and served using conventional, cook-chill, cook-freeze and cook-freeze-thaw catering systems. Quality measurements included determinations of ascorbic acid, thiobarbituric acid numbers, total viable numbers of bacteria and the temperature monitoring of food and equipment. Adherence to recommended time/temperature conditions is necessary to conserve the expected quality of preprocessed foods. At the Federal Research Centre in West Germany, nutritional and organoleptic quality measurements were performed on 5 carbohydrate dishes prepared from fresh, dehydrated, chilled, frozen and sterilised components. The dishes chosen were boiled potato, potato salad, creamed potato, pasta and rice. Nutritional measurements included determinations of water, protein, fat, carbohydrate, minerals, ascorbic acid, thiamin and riboflavin. Taste panellists evaluated the products for organoleptic quality. There was little or no difference in the protein, fat and energy content of products prepared from different components. Overall, the organoleptic quality of freshly prepared products was superior. Vitamin fortification of preprocessed components was necessary to provide similar or higher levels of ascorbic acid and riboflavin to those found in freshly prepared products.

1. Introduction

The caterer may choose whether he purchases fresh or preprocessed food with which to provide his customers with meals. To help him choose from fresh, chilled, frozen, dehydrated or sterilised foods he needs information about the quality of these foods which are now available. Most preprocessed foods require additional preparation and heat treatment by the caterer and he should be aware how these procedures will affect the organoleptic, microbiological and nutritional quality of the foods.

2. Background

In England 4 types of hospital catering systems were examined by the Hotel and Catering Research Centre in an attempt to evaluate the systems in terms of cost, labour requirements, system operation and food quality. Hospitals were chosen which used conventional, cook-chill and cook-freeze catering systems although a mixed system was also utilised by each hospital. In addition, an experimental cook-freeze-thaw catering system was implemented in a hospital and this new system and the original conventional catering system was evaluated. Cook-chill, cook-freeze and cook-freeze thaw catering systems all serve meals prepared from preprocessed components.

For the purposes of this research, the 4 catering systems can be defined as follows.

A conventional catering system is one in which food is prepared and cooked on the same day as its consumption. The described conventional system in hospital 1 used a centralised trayed meal service.

A cook-chill system is one in which food is cooked, then rapidly chilled, stored in controlled low temperature conditions (0 - 3 deg.C) and reheated just before consumption. In this study, food was prepared and chilled in hospital 2(a) and transported to hospital 2(b) where it was stored in chilled conditions before being reheated in bulk centrally and then transported hot in bulk containers to the wards for distribution.

A cook-freeze system is one in which food is prepared and cooked, then rapidly frozen, stored in controlled low temperature conditions (-18 deg.C) and reheated just before consumption. Cook-freeze systems often incorporate freshly prepared food items into the system and for this reason 2 hospitals had to be studied to obtain the different quality measurements required. At Hospital 3 food was prepared, cooked, blast frozen, stored and when required this food was transported to a ward kitchen where it was reheated in bulk in forced air convection ovens and served immediately to the patients. At Hospital 5 the food was also prepared, cooked, blast frozen, stored and was later reheated centrally in bulk in forced air convection ovens. A central trayed meal distribution service was used.

A cook-freeze-thaw system is based on a cook-freeze system which has an additional thawing stage before the reheating procedure. This cook-freeze-thaw system used frozen food which was processed in a production unit at hospital 4(a). Frozen food was transported to hospital 4(b) where multiportion packs of food were thawed centrally in a rapid thaw cabinet. Thawing was carried out at 13 - 20 deg.C for approximately 4 hours and then the food was rapidly chilled to temperatures below 5 deg.C. Chilled food was then plated centrally, chilled meals were transported to the wards and these meals were reheated in infra red ovens close to the patients. Before the implementation of the trial of the cook-freeze-thaw food system, hospital 4(b) used a conventional catering system.

3. Food Quality Measurements

Food quality measurements included assays of ascorbic acid in selected vegetable items, determination of TBA numbers in selected meat items, total viable numbers of bacteria in potentially hazardous food items and temperature monitoring of food and equipment at critical control points in the food system.

Losses of ascorbic acid can occur at all stages of food processing, as this nutrient is susceptible to oxidation, is water soluble and is heat labile. It is one of the most readily destroyed of the vitamins and loss of the vitamin may therefore be used as an indicator of the severity of food processing which the food has undergone.

Ascorbic acid was assayed using the dye reduction method (reduction of 2,6 dichlorophenol indophenol).

The unsaturated fatty acids present in food oxidise to produce malonaldehyde which reacts with thiobarbituric acid to give a red pigment which can be measured. The thiobarbituric acid (TBA) number is the concentration of malonaldehyde expressed as milligrams per kilogram of food and this number is used as an indicator of rancidity in foods. During storage in air, oxidation of meat occurs, the TBA number increases and the food becomes increasinly unacceptable (1).

A distillation method was used to determine the TBA number of meat samples.

Microorganisms in food which are of public health concern may proliferate during food processing. During initial cooking vegetative bacterial cells are killed but contamination of the food may occur due to the presence of airborne bacteria or due to handling. The subsequent time/temperature conditions of processing will affect the growth of these microorganisms and also the growth of any bacterial spores which survive the initial cooking procedures.

The total viable mesophilic organism count was estimated using the droplette technique (cook-chill system) and pour plate technique (conventional and cook-freeze system). Plate count agar (tryptone glucose yeast agar) was used for both methods.

Time temperature parameters of food items are critical control points in the Hazard Analysis Critical Control Point (HACCP) models (2) for catering systems. These time temperature standards for different systems are designed to minimise the time that the temperature of foods are in a zone of growth for microorganisms. Adherence to these standards during food processing should not only control the microbiological quality but in many cases should also conserve the nutritional and sensory quality of the food items.

The temperature history of food items and the air temperatures of process equipment was monitored in different catering systems.

4. Sampling.

4.1 Ascorbic Acid Samples

For each system samples of different batches of 4 vegetables were collected. Cabbage and peas were available from all the hospitals and 2 types of potato were selected, namely creamed potato (reconstituted dehydrated) or steamed new potato and roast or chipped potato (both incorporated fat during cooking). Vegetables were examined at different stages of processing but 2 common points were established in each system. The reference point was considered to be when the vegetables had been just cooked as subsequent handling or processing took place in all of the systems. Sampling at the service of the vegetables occurred after the final heat treatment in each of the 4 systems. Food consumption took place immediately in hospital 4(b), after plating in hospital 3, after hot transport in hospitals 1 and 5 and after hot transport and plating in hospital 2(b).

4.2 TBA Number Samples

Samples of different batches of beef and pork were collected at different stages of processing in the conventional, cook-chill and cook-freeze catering systems. The 2 points which were common to all systems and considered relevant were the just cooked and at service stages.

4.3 Bacteriology Samples

Food items were selected which could be potentially hazardous and these fell into the following categories - minced meat, cottage pies, meat pies with pastry, chicken dishes and sweet dishes. It was not possible however to sample the same dishes from all catering systems. Meat pies with pastry were not sampled in the conventional catering system, likewise a chicken dish was not sampled in a cook-chill system. Samples were taken from different batches of each menu item in the conventional and cook-chill system, but this was not possible in the cook-freeze system since 1 batch of a food item in frozen storage supplied the hospital for several weeks or months, and to sample 3 or more batches could have taken many months.

Numbers of bacteria present were estimated at the "at service" stage to give a direct comparison of bacteriological food quality.

Conventional system samples were taken from hospital 1, cook-freeze samples from hospital 5, and cook-chill samples from a school meals service.

4.4 Temperature Monitoring Procedures

The temperature of samples of different food items were monitored at different stages during handling after the initial cooking process. Special packs of mashed potato were monitored during the cook-chill and cook-freeze system procedures. The operating temperatures of refrigeration and freezing equipment was measured. Spot checks were performed using a hand held digital thermometer with a Ni-Cr/Ni-Al needle probe. Continuous temperature measurement was carried out using a data logger with Cu/Con fine wire thermocouples.

5. Results and Discussion

5.1 Ascorbic Acid Content of Vegetables

Table I shows the comparative values for the ascorbic acid content of the vegetables when sampled at the service point of the 4 catering systems. The content of peas and cabbage were highest in the conventional system and levels of ascorbic acid in creamed potato and roast potato were comparatively high in this system, but other systems also provided similar levels in these potatoes. The retentions of the just cooked vegetable ascorbic acid were highest in the conventional system and apart from one batch of creamed potato and one batch of peas, ranged from 69 - 93%. These relatively high levels are attributable to the use of a batch system of pressure steaming the vegetables which ensured that cooking times were controlled and warmholding before service was kept to a minimum. Only the creamed potato had long periods of warmholding before service (75 - 100 minutes) as it was made up in large batches.

Cook-chill processing and reheating reduced the levels of ascorbic acid significantly in 7 out of the 11 batches of vegetables which were assayed. The system examined only stored the food for approximately 24 hours and apart from 1 batch of creamed potato and 1 batch of peas, retention was between 39 and 78%. Further daily losses of 6% could be

Table I

Ascorbic Acid Content of Vegetables (mg/100g) Sampled At The Service Point From 4 Different Catering Systems.

Vegetable	Conventional Hospital 1				Cook-Chill Hospital 2(b)				Cook-freeze Hospital 3				Cook-Freeze-Thaw Hospital 4(b)			
	$\bar{x}$	s	(n)	% Ret.	$\bar{x}$	s	(n)	% Ret.	$\bar{x}$	s	(n)	% Ret.	$\bar{x}$	s	(n)	% Ret.
Cabbage	37	4.0	(12)	77	38	4.2	(3)	73	11	1.8	(12)	51	17	2.9	(6)	46
	34	4.0	(12)	77	14	2.6	(3)	73	5	1.4	(6)	28	6	1.3	(6)	54
	35	3.3	(12)	78					1	0.4	(6)	22	4	1.7	(6)	27
									1	0.4	(6)	9				
Peas	16	2.7	(12)	85	2	0.2	(3)	33	5	0.8	(6)	45	6	0.6	(6)	68
	11	2.2	(12)	69	5	1.3	(3)	39	4	1.0	(6)	44	5	0.8	(6)	54
	8	2.3	(8)	50	7	0.6	(3)	71	3	0.5	(6)	37	3	0.7	(6)	43
Creamed Potato	4	0.3	(12)	11	19	1.3	(3)	101					21	1.5	(6)	83
	22	1.3	(12)	70	17	1.8	(3)	78					26	1.5	(6)	84
	21	0.7	(12)	69	16	0.3	(3)	71					18	0.6	(6)	80
New Potato									10	2.2	(12)	73				
									9	2.2	(10)	80				
									12	2.6	(10)	70				
Roast Potatoes	10	1.5	(12)	75	5	7.6	(3)	46								
	13	1.5	(12)	93	12	2.7	(3)	62								
	11	1.0	(12)	80	6	1.4	(3)	44								
Chipped Potato									14	2.6	(10)	76	2	0.8	(6)	22
									13	2.7	(10)	68	4	0.3	(6)	98
									7	1.7	(10)	46	2	0.8	(6)	49

$\bar{x}$ = mean ascorbic acid value (mg/100g). s = standard deviation. (n) = number of samples
% Ret. = content of ascorbic acid expressed as a percentage of the content of ascorbic acid in the vegetable when just cooked.

expected during extended storage of vegetables in the system (3). Samples of the same types of vegetables taken from a cook-chill system supplying school meals showed ascorbic acid retention of 12 - 69% after 4 or 5 days chilled storage (4).

The cook-freeze results show relatively low levels of ascorbic acid present in cabbage and peas and the retention of the vitamin was between 22 and 51% (excepting one batch of cabbage). Significant losses were incurred, between 19 and 42%, because of delay in the blast freezing process after cooking (20 - 60 minutes) and the addition of water to the product at this stage to aid the final reheating. Although initial delays also occurred with potato products, there was no significant decline in the ascorbic acid content. Final values of the vitamin in potato products was relatively high and retention ranged from 69 - 90% (apart from one batch of chipped potato).

The cook-freeze-thaw cabbage and peas had levels of ascorbic acid similar to those in the cook-freeze system but such large losses did not occur between initial cooking and blast freezing. Losses ranging from 3 - 23% occurred before freezing and frozen storage. Thawing, and chilled storage (approximately 15 hours) was responsible for further losses of 0 - 25% of the vitamin. Reheating in infra red ovens incurred further losses and the final retention of the vitamin was 27 - 68%. Levels in chipped potato were low and reheating was responsible for the greatest losses during processing. Creamed potato however, retained high levels, even after reheating.

The final levels of ascorbic acid in the processed vegetables from all the systems were dependent upon the levels which were present at the reference point, namely when the vegetables were just cooked. Table II shows these levels for the 4 systems. It can be seen that the conventional system started off with the most consistent and highest levels of ascorbic acid in the cooked peas and cabbage. This was the only system which cooked small batches of vegetables in pressure steamers, the other systems all used large boiling pans which resulted in poor control and variability between batches. Creamed potato which was fortified with ascorbic acid had a standard production method in all systems and had fairly consistent levels of the vitamin. Deep fried roast potato (hospital 1), oven cooked roast potato (hospital 2(a)) and deep fried chipped potato (hospital 3) were consistent in the initial levels of the vitamin but lower levels were found in cook-freeze-thaw chipped potato which used frozen "oven" chipped potato.

The consistent and relatively high levels of ascorbic acid found in cabbage and peas just after cooking in hospital 1 should not be expected in all conventional catering systems. When results from hospital 1 were compared to results from 2 conventional systems which were also operated at hospitals 2 and 4 (Table III), these hospitals provided lower and variable levels of the vitamin. Once again creamed potato provided consistent levels of the vitamin in all three hospitals. 2 out of 3 of the batches of deep fried chipped potato from hospital 4 were low in ascorbic acid levels.

5.2 TBA Number of Meats

Table IV shows the TBA numbers of cooked beef and pork at the service point from conventional, cook-chill and cook-freeze catering systems. The conventional system had relatively low numbers ranging from 0.1 to 1.3. Turner et al (5) found that rancidity could be detected by taste panellists in samples of ground pork with TBA numbers

Table II.
Ascorbic Acid Content of Vegetables (mg/100g) Sampled When Just Cooked From 4 Different Catering Systems

Vegetable	Conventional Hospital 1			Cook-Chill Hospital 2(a)			Cook-Freeze Hospital 3			Cook-Freeze-Thaw Hospital 4(b)		
	$\bar{x}$	s	(n)	$\bar{x}$	s	(n)	$\bar{x}$	s	(n)	$\bar{x}$	s	(n)
Cabbage	48	4.3	(12)	52	4.3	(3)	23	1.5	(12)	37	4.9	(6)
	45	1.7	(11)	19	1.8	(3)	17	1.9	(10)	11	0.8	(6)
	44	3.2	(11)				5	0.9	(10)	16	2.1	(6)
							10	1.4	(10)			
Peas	18	1.0	(11)	7	0.8	(3)	10	0.6	(12)	10	0.8	(6)
	15	1.2	(11)	12	0.3	(3)	9	0.3	(9)	9	1.1	(6)
	17	2.3	(8)	9	0.1	(3)	9	0.7	(10)	8	0.9	(6)
Creamed Potato	28	0.9	(11)	19	1.5	(3)				25	0.5	(6)
	31	1.2	(12)	22	0.5	(3)				31	0.8	(6)
	31	2.1	(11)	22	1.8	(3)				23	0.4	(6)
New Potato							12	1.6	(12)			
							11	1.5	(9)			
							17	3.5	(10)			
Roast Potato	13	2.9	(12)	16	7.8	(3)						
	14	1.4	(12)	19	2.9	(3)						
	14	1.1	(11)	14	3.4	(3)						
Chipped Potato							18	1.9	(18)	11	1.5	(6)
							19	1.5	(9)	4	0.3	(6)
							15	2.5	(9)	5	0.5	(6)

$\bar{x}$ = mean ascorbic acid value (mg/100g). s = standard deviation. (n) = number of samples.

Table III

Ascorbic Acid Content of Vegetables (mg/100g) Sampled When Just Cooked from 3 Different Hospitals Using Conventional Catering Systems.

Vegetable	Hospital 1			Hospital 2(a)			Hospital 4(b)		
	$\bar{x}$	s	(n)	$\bar{x}$	s	(n)	$\bar{x}$	s	(n)
Cabbage	48	4.3	(12)	52	4.3	(3)	10	1.2	(6)
	45	1.7	(11)	19	1.8	(3)	8	0.9	(6)
	44	3.2	(11)				14	2.2	(6)
	30	1.2	(13)				10	0.8	(6)
	29	2.0	(12)						
Peas	18	1.0	(11)	7	0.8	(3)	7	1.5	(6)
	15	1.2	(11)	12	0.3	(3)	5	0.7	(6)
	17	2.3	(8)	9	0.1	(3)	8	0.5	(6)
	15	1.6	(12)						
	12	0.4	(11)						
Creamed Potato	28	0.9	(11)	19	1.5	(3)	24	1.6	(6)
	31	1.2	(12)	22	0.5	(3)	26	1.1	(6)
	31	2.1	(11)	22	0.4	(3)	22	1.5	(6)
	28	0.6	(12)						
	28	0.6	(12)						
Roast Potato	13	2.9	(12)	16	7.8	(3)			
	14	1.4	(12)	19	2.9	(3)			
	14	1.1	(11)	14	3.4	(3)			
	18	2.2	(12)						
	13	5.4	(12)						
Chipped Potato							14	1.1	(6)
							4	0.8	(6)
							3	3.8	(6)

$\bar{x}$ = mean ascorbic acid value (mg/100g). s = standard deviation. n = number of samples.

Table IV.
TBA Number of Meats Sampled at the Service Point From 3 Different Catering Systems.

Meat	Conventional Hospital 1			Cook-Chill Hospital 2			Cook-Freeze Hospital 3		
	$\bar{x}$	s	(n)	$\bar{x}$	s	(n)	$\bar{x}$	s	(n)
Roast Beef	0.3	0.2	(12)	4.7	1.6	(6)	0.6	0.1	(12)
	0.2	0.1	(12)	10.1	2.0	(6)	0.4	0.1	(6)
							0.9	0.2	(6)
							1.3	0.6	(6)
Roast Pork				11.7	0.7	(6)	2.7	0.7	(12)
				13.9	3.3	(6)	3.5	1.8	(6)
				10.3	3.2	(4)	3.3	1.3	(6)
Pork Chops	0.8	0.2	(10)						
	1.3	0.2	(11)						
	1.0	0.4	(12)						

$\bar{x}$ = mean TBA number (mg malonaldehyde/kg food). s = standard deviation. (n) = number of samples.

Table V.
TBA Number of Meats Sampled When First Cooked From 3 Different Catering Systems.

Meat	Conventional Hospital 1			Cook-Chill Hospital 2			Cook-Freeze Hospital 3		
	$\bar{x}$	s	(n)	$\bar{x}$	s	(n)	$\bar{x}$	s	(n)
Roast Beef	0.2	0.2	(11)	0.4	0.2	(12)	0.4	0.01	(12)
	0.2	0.1	(12)	0.9	0.5	(12)	0.3	0.1	(6)
	0.1	0.1	(12)				0.5	0.1	(6)
							0.5	0.1	(6)
Roast Pork				0.6	0.1	(12)	0.4	0.1	(12)
				0.6	0.2	(12)	0.4	0.2	(6)
				0.6	0.3	(12)	0.4	0.2	(6)
Pork Chops	0.6	0.4	(12)						
	0.9	0.3	(10)						
	0.5	0.1	(10)						

$\bar{x}$ = mean TBA number (mg malonaldehyde/kg food). s = standard deviation. (n) = number of samples.

above 1.2, although the products with such low levels would not necessarily be unacceptable, especially when the food is included in a complete meal. Low TBA numbers are attributable to the fact that after cooking, the meat was served as quickly as possible, warmholding times ranging from 6 to 75 minutes.

There were significant increases in TBA numbers of meat samples during processing, storage and reheating in the cook-chill system. These TBA numbers which ranged from 4.7 to 13.9 are found to be detectable by taste panellists (5). At this hospital after the meat was cooked, it was cooled for a short time in ambient conditions and then stored overnight in a refrigerator before being blast chilled and stored. The overall storage time for the meat was 2 days and storage of up to 5 days could be expected in a cook-chill system during which time TBA numbers would further increase. Samples of roast beef, from a cook-chill catering system supplying school meals reached a TBA value of 4.1 and samples of roast pork reached a TBA number of 15.3 after 4 days chilled storage (6). The relatively high TBA numbers found in the hospital food samples after only 2 days storage can be attributed to the preliminary storage of the cooked meat before blast chilling, a practice which is not recommended (7).

In the cook-freeze system the cooked meat batches examined were stored for approximately 6 weeks in the deep freeze before being reheated for service. The TBA numbers for roast beef were relatively low, being only 0.4 - 1.3. Values for roast pork increased significantly during processing and storage and resulted in final values of 2.7 -3.5. Roast meats were cooked and then stored in the refrigerator overnight before being sliced, packaged and blast frozen and the TBA numbers of roast pork rose during this preliminary storage.

Table V illustrates that the TBA numbers for roast meats when just cooked in all 3 systems were low and similar, as would be expected.

5.3 Total Viable Counts of Bacteria

Bacteria are reduced in numbers, but not eliminated by cooking. Bacterial spores are able to survive cooking, and cooked foods may be contaminated by bacteria from the surrounding environment. It can be seen from Table VI that low numbers of bacteria were present in most conventionally prepared foods, despite cooking.

In cook-chill systems the food is cooked, then chilled. Some bacteria are able to reproduce slowly at chill temperatures, so numbers will increase slightly. This is shown in Table VI for cottage pie in particular, and to a lesser extent for minced beef and pork pie. Up to 4.5×10^3 bacteria per gram survived reheating.

The lowest numbers of bacteria were found in the cook-freeze samples. In this system foods are cooked, which destroys many bacteria, frozen (numbers of viable bacteria are reduced during freezing) and the foods to be served hot are reheated from frozen, which destroys a very large proportion of the already depleted bacterial population. That bacteria do survive freezing is demonstrated in Table VI by the numbers found in cold sweets (1.6×10^2 - 1.1×10^3/g). Cold sweets were left to thaw at kitchen temperatures before service in hospital 5.

Samples from all catering systems were satisfactory, although 1 batch of conventionally prepared cottage pie had 10^4 bacteria per gram of food - rather a high figure for a freshly cooked menu item. Numbers of bacteria increased slowly during storage in the cook-chill catering

Table VI.

Number of Mesophilic Bacteria Per Gram Of Food Sampled At the Service Point From 3 Different Catering Systems..

	Conventional Hospital 1.			Cook-Chill School.			Cook-Freeze Hospital 5.		
Category	Food Item	$\bar{x}$	n	Food Item	$\bar{x}$	n	Food Item	$\bar{x}$	n
Minced Meat Dishes	Beef	6.3	8	Beef *	9.8×10^{2} (1)	6	Lamb	1.5×10^{1}	10
		5.6×10^{1}	8		1.2×10^{3} (2)	6			
		1.3×10^{1}	8		1.5×10^{3} (3)	6			
Cottage Pie	Cottage Pie	1×10^{4}	4	Cottage Pie	0 (2)	3	Cottage Pie	3.6	10
		6.6×10^{2}	4		7.5×10^{1} (3)	3			
		2.5×10^{1}	8		4.5×10^{3} (5)	3			
Meat Pies with pastry				Pork	6.6×10^{2} (2)	3	Beef & Mushroom	1.0	10
					2.4×10^{2} (3)	3			
					9.5×10^{2} (4)	3	Steak & Onion	1.6×10^{1}	10
					1.3×10^{3} (5)	3			
							Chicken & Vegetable	1.0	10
Chicken Dishes	Fricasse	1.3×10^{1}	8				Curry	1.0	10
		1.9×10^{1}	8				Supreme	5.5	10
		0	8						
Sweet Dishes	Egg Custard	1.3×10^{2}	8				Cheesecake	6.6×10^{2}	10
		3.1×10^{1}	8					1.1×10^{3}	10
		1.3×10^{1}	8				Gateau	1.6×10^{2}	10

$\bar{x}$ = mean no. of bacteria/g.   n = no. of samples.
* = Sampled before reheating. (1) = Day 1 of chilled storage (day of preparation) (2) = Day 2 of chilled (3)storage. (3) = Day 3 of chilled storage. (4) = Day 4 of chilled storage. (5) = Day 5 of chilled storage.

system. Very low numbers of bacteria were present in reheated dishes in the cook-freeze system, with higher numbers present in cold desserts. Numbers of bacteria found in cold desserts were well below the 10^5 /g and 5×10^5 /g given as maximum permissible counts in frozen desserts in 2 American states (8).

During thawing in the cook-freeze-thaw system air at temperatures of 13 - 20 deg.C is circulated around the frozen food. Small lightweight items thaw quickly (sandwiches thaw within an hour) and so it is possible that numbers of bacteria increase on thawing. The potential for bacterial growth was tested by inoculating trial packs and by determining the numbers of bacteria present in selected food items. It was found that:

i. In large (1Kg) packs of food there was little evidence of bacterial growth during thawing.

ii. In small packs, e.g. sandwiches, bacterial growth was found in inoculated packs - a threefold rise in bacterial numbers was found. No evidence of proliferation during thawing was found in sandwiches not inoculated.

iii. In thawed sandwiches, numbers of mesophilic bacteria ranged from 1.6×10^2 to 9.8×10^3. (This compared very well with bacterial numbers found in retailed sandwiches by Christiansen and King (9) who found 1.3×10^3 to 5.3×10^7 bacteria per gram in roast beef, egg salad, turkey and various sausage and burger sandwiches, and who found no sandwiches with less than 10^3 bacteria per gram).

Because of the danger of bacterial growth during thawing, and the subsequent growth of bacteria during chilling it was decided that all food should be consumed within 24 hours of the start of thawing or discarded.

5.4 Temperature Monitoring of Food and Equipment

In the conventional system, observations showed that food was cooked on the day of consumption and warmholding was kept to a minimum. The temperatures of food items recorded at plating during service were variable and many food items were not at 70 - 80 deg.C which was the food temperature recommended by the manufacturers of the heated pellet system, which was required to ensure the food remained hot during delivery for up to 45 minutes. Satisfactory temperatures were recorded for those food items which were batch cooked as required. Unsatisfactory temperatures were recorded for items such as liquids, creamed potato, meat dishes and food with a low thermal mass e.g. thin slices of bacon or fish.

The running temperature of the deep freeze was satisfactory but the cold room air temperature was unsatisfactory, being 6 - 9 deg. C during the study period.

Guidelines for precooked chilled food (7) include the temperature/time recommendations for each stage during processing. At the cook-chill hospital 2(a), the blast chiller operated efficiently and the chilled storage cabinet was capable of running at less than 3 deg.C. In spite of this, spot checks revealed that some food items in the cabinet were at 6 or 7 deg.C due to the cabinet door being kept open which allowed the air temperature to rise to 13 deg. C and stay above 3 deg.C for approximately 1 hour. During the transport of the chilled food to hospital 2(b), the food temperatures could rise and subsequent chilled storage of the food at the hospital was higher than recommended, food temperatures rising to 10 - 13 deg.C during the 24 hour storage of food. Chilled food which reaches temperatures of above 10 deg.C in a cook-chill system should be discarded.

These unacceptably high storage temperatures of chilled foods are held partly responsible for the relatively low retention levels of ascorbic acid which were found in vegetables sampled from the system. Relatively high TBA numbers found in roast beef and roast pork samples can also be partly attributed to these storage conditions. Extended storage of foods in this cook-chill system to the recommended 5 days could have resulted in meals which were unacceptable both microbiologically and organoleptically.

Guidelines for precooked frozen foods (10) include time/temperature recommendations for the handling of foods during cook-freeze processing. At the cook-freeze hospital 3, the blast freezer was capable of lowering the temperature of hot cooked foods to -5 deg.C within the recommended time of 90 minutes. However, food did not always achieve this air temperature drop within the time as it was not always portioned and packaged quickly enough. The air temperature of the deep freeze was satisfactory for food storage. The refrigerators used for the storage of raw foods were not maintained at the recommended temperatures as doors were difficult to close and on occasions were left open.

The cook-freeze-thaw time/temperature guidelines were based on existing recommendations for precooked frozen foods (10). The additional time/temperature parameters specified for the rapid thaw stage, subsequent chilled storage, portioning, chilled transport and reheating were developed for the system implemented. Food was thawed at 13 - 20 deg.C for approximately 4 hours, blast chilled for 1 hour to reduce food temperatures to less than 5 deg.C. The subsequent chilled storage, portioning, transport and storage of food should be carried out below 5 deg.C and within 19 hours (allowing a total period of time of 24 hours for the thawing and storage of thawed food). Food should be adequately reheated before immediate consumption.

The rapid thaw cabinet was capable of thawing and chilling food as specified. During the plating of chilled food the temperatures of items rose and ranged from 4 - 18 deg.C. Rise in food temperatures was unavoidable as the air temperature at plating was not cold enough to maintain adequately low temperatures. Although refrigerated trolleys for the plated meals operated at temperatures of less than 5 deg.C they were not capable of reducing the temperature of relatively hot food to 5 deg.C during the storage and transport of food to the wards. Generally food temperatures evened out, but a range of 7 to 18 deg.C was recorded. Ridley and Matthews (11) found that central plating and transport of chilled food to hospital wards resulted in food items with temperatures ranging from 1 to 23 deg.C. During reheating of the meals some items did not reach adequate temperatures as they had not been processed to the required specification by the cook-freeze production unit e.g. the pieces of liver were too large in dimensions to reheat adequately during the time allowed.

Results of temperature monitoring of food and equipment in different hospitals has shown that in all catering systems it is necessary to regularly monitor the temperature of food during processing and the equipment used to prepare, process and store food.

6. Background

The comparison of the quality of food prepared using 4 systems must be qualified as food samples collected at hospitals may reflect the initial quality of the raw food and the intermediate processing which the food undergoes. A project at the Federal Research Centre for Nutrition in West Germany has evaluated how specific constraints affect the organoleptic and nutritional quality of carbohydrate dishes. One of the areas investigated was to compare the nutritional and organoleptic quality of dishes prepared from fully prepared foodstuffs and dishes prepared from commercially prepared or preprocessed components. (12)

6 carbohydrate dishes were evaluated and this paper discusses the results of 5 of the dishes : boiled potato, potato salad, creamed potato, pasta and rice. Dishes were prepared using ingredients which were initially processed in different ways, namely fresh, dehydrated and frozen components. Additional data for frozen, sterilised and chilled food was used from previous experiments.

7. Methodology

7.1 Nutritive Value.

Water, protein, fat, raw fibre, ash and sodium chloride were determined according to the usual analytical methods.

Ascorbic acid was assayed photometrically.

Thiamine and riboflavin were determined spectrofluorometrically using high pressure liquid chromatography.

The carbohydrate content was determined by the difference method and the calorific value was calculated using the following formulae : 1g fat = 38.9 kJ. 1g protein = 17.1 kJ. 1g carbohydrate = 19.15 kJ.

7.2 Organoleptic Quality

Five trained tasters evaluated the products and scored them on a 9 point scale for colour, appearance, odour, taste and texture. This score took into account both general and specific product aspects. The score range from 9.0 to 7.0 (quality class A) corresponded to very good or good quality, all requirements fulfilled, the range from 6.9 to 5.5 (quality class B) to a satisfactory quality, requirements almost fulfilled, the range from 5.4 - 4.0 (quality class C) to an unsatisfactory quality, requirements not completely fulfilled, and the range from 3.9 to 1.0 (quality class D) to an unacceptable quality, requirements not fulfilled.

7.3 Treatment of Results

Mean values and standard deviations were calculated for the nutritional and organoleptic results. The nutritional values were tested using the following statistical procedure: normal distribution tests, F and t-Tests. Differences in organoleptic scores were tested using the t-Test. The overall organoleptic quality of each product was graded A,B,C or D, but if the ratings for different sensory attributes showed large differences so that this overall grading was not clear cut, an assessment was made. Nutritional and organoleptic results were examined to find out which method of preparation gave the highest quality product.

8. Results and Discussion

8.1 Boiled Potato.

Up to 9 varieties of potato were boiled under domestic and large scale catering conditions. The varieties selected were Clivia, Calpa, Desiree, Granole, Hansa, Jetta, Maya, Montana and Nicola, which were mainly firm textured and medium early crop potatoes. Culpa and Jetta potatoes were also steam cooked.

Existing results for chilled, frozen and sterilised foods were used for comparative purpses.

8.1.1 Organoleptic Quality

The overall results for the quality of boiled potatoes are shown in Table VII. Different varieties of potatoes were awarded different sensory ratings, and differences were most marked for the attributes, colour, appearance and texture. The majority of ratings were greater than 7.0, and 11 out of 14 of the boiled potato products were awarded Class A overall quality. The remaining 3 products were in Class B due to a heavy rather bitter taste, a floury dry texture or a rather cracked surface.

It can be seen that the frozen and sterilised potato products were inferior in quality to the freshly prepared ones. The quality of deep frozen potatoes was dependent upon the quality of the potato crop used which varied from year to year because of the growing conditions. Sterilised products from cans or pouches were classified in category D due to poor scores for odour and taste. Improvements in flavour leading to classification B were obtained when pouch packaged potatoes were heated in a water bath.

Chilled Potatoes were almost as highly scored as freshly prepared potatoes.

Table VII

Organoleptic Quality Classification of Boiled Potato Products Made From Different Starting Materials

Initial State (Potato)	No. Varieties Or Processors	Preparation Process	No. Samples Per Quality Class A	B	C	D
Raw,	9	Boiled in 4 litre pan	9	2	-	-
Quartered	3	Boiled in pressure cooker	2	1	-	-
	2	Steamed in 4 litre pan	2	1	-	-
	1	Steamed in pressure cooker	1	-	-	-
Chilled	4	Air circulated at 130 deg.C	1	3	-	-
Frozen	5	Air circulated at 150 deg.C	-	3	4	1
Sterilised	2	Air circulated at 120 deg.C	-	-	-	4
	2 (1)	Water bath at 95 deg.C	-	2	-	-

(1) flexible pouch processed.

8.1.2 Nutritive Value

Some nutrients, namely the carbohydrate, ascorbic acid and thiamin contents of boiled potatoes were also dependent upon the variety and origin of the raw potato. Freshly cooked potato provided from 12.5 -

20.2g of carbohydrate, 6.4 - 12.1 mg ascorbic acid, and 0.049 - 0.102 mg thiamin per 100g of edible portion. Dry matter was almost unchanged by either boiling or steaming being 96 - 100% of the original value. Mineral content was reduced during cooking with losses of between 5 and 30%, the lowest losses being incurred by pressure steaming. The ascorbic acid content was reduced by 14 - 20% by boiling and 9% by steaming. At least 14% thiamin was lost during boiling and 8 - 14% by steaming. Losses of 1 - 17% riboflavin were incurred.

Table VIII shows the protein, fat, carbohydrate and energy values for freshly prepared, chilled, frozen and sterilised boiled potatoes and it can be seen that they are similar for all products. The differences in fat, carbohydrate and energy are attributable to the differences in the variety of the potatoes and to the addition of fat in some commercial products.

Table VIII
Protein, Fat, Carbohydrate and Energy Content Of Boiled Potato Products Made From Different Starting Materials.

Initial State (Potato)		Nutrient Content Per 100g Edible Portion							
		Protein (g)		Fat (g)		Carbohydrate (g)		Energy (kJ)	
	n	$\bar{x}$	s	$\bar{x}$	s	$\bar{x}$	s	$\bar{x}$	s
Raw	4	2.1	0.3	0.2	0.1	15.4	3.2	294	44
Chilled	3	2.2	0.1	0.1	-	18.1	1.7	352	30
Frozen	7	1.9	0.1	0.7	0.7	16.4	1.5	343	50
Sterilised	6	1.7	0.4	0.1	0.1	13.6	1.3	262	22

n = no. of varieties or processors.

The differences in vitamin content of the different products can be seen in Table IX. Ascorbic acid and thiamin levels in freshly prepared products were on average 35 - 47% and 4 - 9% higher respectively, than in chilled or frozen products. Sterilised products had highest levels of ascorbic acid, presumably this was added during manufacture. Riboflavin levels were highest in chilled and frozen products and lowest in sterilised products.

8.2 Potato Salad

This product, normally prepared from potatoes cooked in their skins, was also prepared using freshly boiled, dehydrated and sterilised potatoes. 4 different varieties of boiled potato and 5 varieties of potatoes cooked in their skins were tested. For all samples the organoleptic and nutritional evaluation took place 1 hour after mixing the product with the marinade. Results obtained earlier for chilled potato salad were also used.

8.2.1. Organoleptic Quality

The classification of the quality of the different products shown in Table X indicates that potatoes cooked in their skins were of consistently high quality being graded A. Within this category the varieties Sieglinde and Nicola were superior in texture and taste. Greater differences in quality were noticed between varieties of boiled potatoes, Sieglinde and Culpa being awarded the highest scores for

Table IX.
Ascorbic Acid, Thiamin and Riboflavin Content of Boiled Potato Products Made From Different Starting Materials.

Initial State (Potato)		Vitamin Content Per 100g				Vitamin Content per 1000kJ			
		mg		%		mg		%	
	n	$\bar{x}$	s	$\bar{x}$	s	$\bar{x}$	s	$\bar{x}$	s
Ascorbic Acid									
Raw, peeled, quartered	10	8.4	1.8	100	21	30.0	4.0	100	13
Chilled	3	4.5	2.5	53	19	12.7	6.6	42	22
Frozen	7	5.5	1.4	65	16	16.0	3.9	53	13
Sterilised	3	61.8	7.5	735	90	241.7	44.6	805	148
	3	10.1	2.9	119	34	38.4	12.9	128	43
Thiamin									
Raw, peeled, quartered	10	0.070	0.015	100	22	0.253	0.056	100	22
Chilled	3	0.064	0.032	91	46	0.178	0.078	70	31
Frozen	7	0.067	0.021	96	30	0.207	0.070	82	28
Sterilised	6	0.019	0.015	27	22	0.073	0.061	29	24
Riboflavin									
Raw, peeled, quartered	10	0.025	0.001	100	6	0.093	-	100	6
Chilled	3	0.073	0.071	290	285	0.199	-	214	200
Frozen	7	0.034	0.020	134	82	0.098	-	106	63
Sterilised	6	0.019	0.008	73	46	0.070	0.031	75	33

n = no. varieties of processors

texture and taste. Overall, boiled potatoes were graded A and B. Further hedonic taste panel tests did not reveal a definite preference for the salad prepared from potatoes cooked in their skins over that prepared from boiled potato. When dried potato slices were tested almost all the sensory quality scores were reduced and in particular the taste was found to be insipid, stale or burnt and the texture was found to be rubbery or pulpy, thus a classification of B and C was given. Salad prepared from sterilised pouch potatoes had a bitter taste and was a brownish colour, although the texture was satisfactory and so was classified B. In 3 out of 4 products, chilled potato salad was graded A quality.

Table X.
Organoleptic Quality Classification of Potato Salad Products Made From Different Starting Materials.

Initial State (Potato Slices)	No. Varieties or Processors	Preparation Process	No. Samples Per Quality Class			
			A	B	D	D
Potato in skins.	5	Mixed with marinade	6	-	-	-
Boiled Potato	4	Mixed with marinade	2	3	-	-
Dehydrated	2	Cooked and mixed with marinade	-	1	1	-
Sterilised	1	Mixed with marinade	-	1	-	-
Chilled	4	-	3	1	-	-

Table XI.
Protein, Fat, Carbohydrate and Energy Content Of Potato Salad Products Made From Different Starting Materials.

Initial State (Potato slices)	n	Protein (g) $\bar{x}$	Protein (g) s	Fat (g) $\bar{x}$	Fat (g) s	Carbohydrate (g) $\bar{x}$	Carbohydrate (g) s	Energy (kJ) $\bar{x}$	Energy (kJ) s
Potato in skins/ boiled potato	4	1.6	0.3	3.0	1.0	12.1	1.2	353	29
Dehydrated	2	1.2	0.1	4.2	0.6	13.3	1.0	412	2
Sterilised	1	1.7	-	3.5	-	13.5	-	397	-
Chilled	3	2.0	0.2	2.7	2.2	12.3	0.2	350	93

Nutrient content per 100g Edible Portion

n = no. varieties or processors.

Table XII.

Ascorbic Acid, Thiamin and Riboflavin Content of Potato Salad Products Made From Different Starting Materials.

Initial State (Potato)	n	Vitamin Content per 100g mg $\bar{x}$	s	% $\bar{x}$	s	Vitamin Content per 1000kJ mg $\bar{x}$	s	% $\bar{x}$	s
Ascorbic Acid									
Potato in skins	2	8.0	0.5	100	6	21.1	0.5	100	2
Boiled potato	2	6.3	0.3	79	5	17.6	0.1	83	1
Dehydrated	2	3.8	0.6	48	8	9.2	1.4	44	7
Sterilised	1	13.4	-	168	-	33.8	-	160	-
Chilled	3	2.9	2.4	36	30	8.4	7.6	40	36
Thiamin									
Potato in skins	2	0.070	0.003	100	4	0.187	0.018	100	10
Boiled potato	2	0.056	0.005	80	7	0.156	0.020	83	10
Dehydrated	2	0.021	0.004	30	6	0.051	0.010	27	5
Sterilised	1	0.027	-	39	-	0.068	-	36	-
Chilled	3	0.064	0.035	91	50	0.193	0.130	103	69
Riboflavin									
Potato in skins	2	0.027	0.001	100	3	0.071	0.004	100	6
Boiled potato	2	0.023	0.004	85	14	0.064	0.010	90	14
Dehydrated	1	0.020	-	74	-	0.048	-	68	-
	1	0.245	-	907	-	0.592	-	834	-
Sterilised	1	0.064	-	237	-	0.161	-	226	-
Chilled	2	0.040	0.027	148	100	0.113	0.096	159	135
	1	0.109	-	400	-	0.403	-	567	-

n = no. varieties or processors

8.2.2. Nutritive Value.

Table XI shows that potato salad prepared from all types of potato had similar levels of protein, fat, carbohydrate and energy value, any differences in fat and energy content were attributable to the marinade used.

The ascorbic acid, thiamin and riboflavin content of potato salad depended upon the variety of potato and the initial cooking procedure used on the potato. Salad prepared from potatoes cooked in their skins had 15 - 20 % more vitamin than salad prepared from boiled potatoes. These differences were much smaller than differences due to the use of different varieties of potato, for example, the ascorbic acid content of salad made from Sieglinde potatoes was 100% higher than that of salad made from Culpa potatoes. Table XII shows that the use of dehydrated potatoes resulted in lower levels of ascorbic acid and thiamin than when freshly boiled potatoes were used. Sterilised potato slices also yielded lower levels of thiamin than freshly boiled potatoes. High levels of ascorbic acid and riboflavin in dehydrated and sterilised products indicate the fortification of these products during manufacture. Chilled storage of 1 day incurred losses of ascorbic acid in potato salad.

8.3 Creamed Potato.

Freshly prepared and dehydrated products were tested. Results for chilled and frozen products were examined from work carried out previously.

8.3.1. Organoleptic Quality

All freshly prepared creamed potato, irrespective of the 4 different varieties of potato used were awarded classification A (Table XIII). The different varieties have only insignificant differences in their scores for colour, taste and texture.

Table XIII.
Organoleptic Quality Classification of Creamed Potato Products Made From Different Starting Materials.

Initial State (Potato)	No. Varieties or Processors	Preparation Process	No. Samples Per Quality Class A	B	C	D
Raw	4	Boiled	4	-	-	-
Dehydrated						
Granules	4	Mixed with water/milk	-	3	1	-
Flakes	2	Mixed with water/milk	2	-	-	-
Granules	4	Mixed with water	1	2	3	-
Flakes	2	Mixed with water	1	-	1	-
Chilled	4	Air circulated at 130 deg.C.	2	2	-	-
Frozen	3	Air circulated at 150 deg.C.	2	3	-	-

Creamed potato products prepared from dehydrated potato flakes were considered superior in quality to products prepared from granules, mainly because of better odour and taste. Creamed potato prepared from both flakes and granules was light yellow to yellow in colour and any addition of riboflavin introduced a greenish tone to the yellow colour. Products made from flakes had a light smooth texture and were classified overall as A whereas those made from granules had a thin gluey texture and were classified B and C, the C classification being due to a pungent odour and an insipid, sour or burnt taste.

Dehydrated potato products which did not contain dried milk and had milk added during preparation were generally scored higher than products which already contained dried milk, as some of these varieties developed an "off" or "old" milk flavour. It was also found desirable to add butter or margarine to those products which recommended it as this improved the organoleptic quality.

Chilled creamed potato made from boiled potatoes was classified A and when made from dehydrated potato was classified B for quality.

Taste and odour of batches of frozen mashed potato were scored differently when tests were performed in 1973 and 1981, overall classification A and B being awarded. Another product was awarded a B for quality because of a slightly "old" odour and an insipid taste.

8.3.2. Nutritive Value

There were no significant differences in protein, fat, carbohydrate and energy contents of the different creamed potato products (Table XIV).

Table XIV.
Protein, Fat, Carbohydrate and Energy Content of Creamed Potato Products Made From Different Starting Materials.

Initial State (Potato)	n	Nutrient Content Per 100g Edible Portion							
		Protein (g)		Fat (g)		Carbohydrate (g)		Energy (kJ)	
		$\bar{x}$	s	$\bar{x}$	s	$\bar{x}$	s	$\bar{x}$	s
Raw (1)	1	1.7	-	1.5	-	12.4	-	304	-
Dehydrated (2)	7	1.9	0.2	1.7	0.3	9.9	1.2	267	25
Dehydrated (3)	5	1.6	0.2	0.8	0.3	10.6	1.2	240	16
Chilled	3	2.8	0.6	1.9	0.7	11.8	1.1	323	52
Frozen	5	1.9	0.2	2.6	1.0	11.1	0.5	327	51

n = no. varieties of processors. (1) = mixed with milk and fat.
(2) = mixed with water/milk and fat. (3) = mixed with water.

There were differences in vitamin levels of mashed potato products made from different potato varieties and from different types of

Table XV.

Ascorbic Acid, Thiamin and Riboflavin Content Of Creamed Potato Products Made From Different Starting Materials.

Initial State (Potato)	n	Vitamin Content per 100g mg $\bar{x}$	s	% $\bar{x}$	s	Vitamin Content per 1000kJ mg $\bar{x}$	s	% $\bar{x}$	s
Ascorbic Acid									
Raw (1)	2	8.1	2.7	100	33	26.6	8.7	100	33
Dehydrated (2)	6	2.8	0.6	35	7	10.6	2.0	40	8
Dehydrated (3)	7	4.7	1.6	54	20	17.5	6.4	66	24
Chilled	3	2.7	1.3	33	16	7.6	3.6	29	14
Frozen	3	2.7	0.9	33	11	8.2	3.9	31	15
Thiamin									
Raw (1)	2	0.068	0.012	100	18	0.224	0.042	100	19
Dehydrated (2)	6	0.041	0.011	60	16	0.155	0.039	69	17
Dehydrated (3)	7	0.037	0.015	54	22	0.143	0.040	64	18
Chilled	3	0.049	0.027	72	40	0.146	0.102	65	45
Frozen	3	0.048	0.013	70	19	0.135	0.024	60	11
Riboflavin									
Raw	2	0.094	0.031	100	32	0.309	0.103	100	-
Dehydrated	2	0.084	0.005	89	5	0.328	0.003	106	1
Dehydrated	4	0.493	0.191	524	203	1.897	0.812	614	362
Dehydrated	1	0.040	-	42	-	0.162	-	52	-
Dehydrated	6	0.440	0.105	468	112	1.763	0.508	570	188
Chilled	2	0.084	0.013	89	14	0.250	0.078	81	25
	1	0.327	-	349	-	0.850	-	275	-
Frozen	3	0.824	0.524	876	557	2.599	2.120	841	686

n = no. varieties or processors. (1) mixed with milk and fat. (2) mixed with water/milk and fat. (3) mixed with water processed potatoes. The ascorbic acid and thiamin content in mashed potato prepared from raw potatoes was on average 46 - 65% and 40 - 46% higher respectively than in mashed potato prepared from dehydrated products.

The high levels of riboflavin in samples prepared from dehydrated potato (see Table XV) indicate the addition of riboflavin during manufacture and the level of additives varied in products from different firms. Products from different manufacturers provided different levels of all 3 vitamins and so some mashed potato made from dehydrated potato had levels of ascorbic acid, thiamin and riboflavin similar to those in mashed potato made from freshly boiled potatoes.

Chilled and frozen mashed potato had on average similar levels of the vitamins as the freshly prepared dehydrated product. High riboflavin levels indicate that dehydrated potato was used in the preparation of chilled and frozen products.

8.4 Pasta

The pasta dishes examined were noodles, macaroni and spaghetti. Results for previous samples of chilled, frozen, and sterilised pasta dishes were considered.

8.4.1. Organoleptic Quality

Table XVI shows that the boiling of 1 part pasta to 5 parts water resulted in a cooked product of quality class A. Decreasing the food:water ratio to 1:10 did not significantly affect the overall quality but increasing the food:water ratio to 1:2.5 resulted in an unsatisfactory product as the viscous cooking liquor was very slow to drain off the pasta. The cooking time required and quality of the products were dependent upon the proportion of egg in the pasta recipes. Overcooking resulted in a doughy and soft texture and these unfavourable changes were more readily noticeable in the egg free pastas. To prevent cooked pastas sticking together 10 - 15g oil should be added to each kilogram of cooked food.

Table XVI.

Organoleptic Quality Classification of Pasta Made From Different Starting Materials.

Initial State (Pasta)	No. Varieties or Processors	Preparation Process	No. Samples Per Quality Class A	B	C	D
Raw	4 (1)	Cooked in 4 litre pan	8	-	-	-
	4 (1)	Cooked in pressure cooker	8	1	-	-
Chilled	4	Air circulated at 130 deg.C.	4	-	-	-
Frozen	5	Air circulated at 150 deg.C.	3	5	1	-
Sterilised	1	Air circulated at 120 deg.C.	-	1	-	2

(1) pasta:water = 1:10 and 1:5

Chilled pasta foods were judged to be of very good quality and classified A. 3 deep frozen products from 1 manufacturer were also very good (class A) but those from 2 other manufacturers were classed B and C. Sterilised pasta products showed deterioration in all the organoleptic qualities measured and 2 samples were unacceptable (class D).

8.4.2 Nutrititive Value

There was little difference in the main nutrient contents and energy values of the 4 pastas. (See Table XVII). The slight variations in protein, fat, carbohydrate and energy content were attributable to the quality of the raw food, the different volumes of water added during cooking and the different weights of fat added. The yield of cooked pastas amounted to 232 - 282g per 100g dry food, the relative proportion of raw food to cooking water does not significantly influence the weight of water absorbed. Cooking of pastas did not cause any significant loss of protein or fat, slightly higher losses of carbohydrate occurred when the quantity of cooking water was increased. Mineral content was reduced by 13 - 90% (relative to the dry weight) and the highest losses were induced when large volumes of water were used during cooking.

Table XVII.
Protein, Fat, Carbohydrate and Energy Content of Pasta Made From Different Starting Materials.

Initial State (Pasta)		Nutrient Content Per 100g Edible Portion							
		Protein (g)		Fat (g)		Carbohydrate (g)		Energy (kJ)	
	n	$\bar{x}$	s	$\bar{x}$	s	$\bar{x}$	s	$\bar{x}$	s
Raw	4(1)	5.4	0.6	1.9	0.5	25.4	1.0	597	35
Chilled	3	4.9	0.6	2.0	0.6	25.5	3.8	600	56
Frozen	9	5.1	0.2	4.8	2.4	24.0	2.2	689	78
Sterilised	3	5.1	0.3	5.8	0.9	25.0	1.4	738	55

n = no. varieties or processors.
(1) = pasta:water = 1:5, oil added after cooking.

It was also established that cooking in different proportions of water resulted in different losses of thiamin and riboflavin but results were inconsistent. Cooking in 5 times its weight of water incurred thiamin losses in pasta of 7 - 22% lower and riboflavin losses of 6 - 7% lower than cooking in 10 times its weight of water.

Freshly prepared, chilled, frozen and sterilised pastas all had similar levels of protein and carbohydrate (Table XVII) and differences in energy content were due to higher levels of fat in frozen and sterilised products.

Table XVIII shows that freshly prepared pastas provided the highest levels of thiamin, being 39 - 64% higher than chilled, frozen and sterilised pasta products. Riboflavin levels were 27% lower in chilled products than in freshly prepared pasta, some frozen and sterilised products had similar levels and some had 10 times higher levels of this vitamin.

Table XVIII.

Thiamin and Riboflavin Content of Pasta Made From Different Starting Materials

Initial State (Pasta)	n	Vitamin Content per 100g mg $\bar{x}$	s	% $\bar{x}$	s	Vitamin Content per 1000kJ mg $\bar{x}$	s	% $\bar{x}$	s
Thiamin									
Raw									
- pasta:water = 1:5	7	0.044	0.010	100	23	0.074	0.015	100	20
- pasta:water = 1:10	8	0.036	0.009	82	20	0.061	0.013	82	18
Chilled	3	0.020	0.005	45	11	0.033	0.011	45	15
Frozen	5	0.027	0.008	61	18	0.040	0.010	54	14
Sterilised	3	0.016	0.006	36	14	0.021	0.007	28	9
Riboflavin									
Raw									
- pasta:water = 1:5	2	0.026	-	100	-	0.045	-	100	-
- pasta:water = 1:10	2	0.023	-	88	-	0.039	-	87	-
Chilled	3	0.019	0.005	73	19	0.031	0.008	69	18
Frozen	4	0.027	0.006	104	23	0.043	0.007	96	16
	1	0.282	-	1085	-	0.347	-	771	-
Sterilised	3	0.038	0.015	146	57	0.052	0.023	116	53

n = no. types of processors

8.5 Rice

Cooked rice was prepared from parboiled, high grade patna, and standard grade rice. The effects of different proportions of rice and water during cooking and the addition of fat during cooking were examined.

8.5.1 Organoleptic Quality

When the parboiled and high grade rice were boiled in different proportions of water, 1:10, 1:5 and 1:2.5, there was no significant difference in their qualities and they were classified grade A. (Table XIX). Standard rice boiled in 10 times its weight of water was almost as good as high grade patna rice, but the samples of standard rice cooked in rice:water ratios of 1:5 and 1:2.5 were not so fluffy and softer and were classed quality B.

Table XIX.
Organoleptic Quality Classification of Rice Made From Different Starting Materials.

Initial State (Rice)	No. Varieties or Processors	Preparation Process	No. Samples Per Quality Class A	B	C	D
Raw	3 (1)	Boiled in 4 litre pan	8	2	-	-
	1 (1)	Boiled in pressure cooker	3	-	-	-
	1 (2)	Steamed in 4 litre pan	1	-	-	-
	1 (2)	Steamed in pressure cooker	1	-	-	-
Chilled	3	Air circulated at 130 deg.C.	3	-	-	-
Frozen	5 (1)	Air circulated at 150 deg.C.	2	5	-	-
Sterilised	2 (1)	Air circulated at 120 deg.C.	-	1	-	2

(1) rice:water = 1:10, 1:5 and 1:2.5
(2) rice:water = 1:2.5, seasoned, fat/onions added.

Samples of patna and parboiled rice cooked in both domestic and large scale catering conditions were all classified as quality A. Steaming of parboiled rice with fat and onions also resulted in a high quality product.

The quality of rice after 1 days chilled storage was grade A. It is possible for frozen rice to be of high quality but products were variable, having A or B quality. Sterilised rice samples were unacceptable in quality being awarded low scores for colour, odour, taste and texture.

8.5.2 Nutritive Value

The effect of different cooking conditions on the nutrients nutritive value of cooked rice can be seen in Table XX.

Table XX.
Protein, Fat, Carbohydrate and Energy Content of Rice Made From Different Starting Materials.

Initial State (Rice)	n	Protein (g) $\bar{x}$	s	Fat (g) $\bar{x}$	s	Carbohydrate (g) $\bar{x}$	s	Energy (kJ) $\bar{x}$	s
Raw									
parboiled	1(1)	2.4	-	0.5	-	26.9	-	519	-
parboiled	1(2)	2.7	-	1.9	-	26.6	-	577	-
patna	1(1)	2.1	-	0.2	-	21.9	-	419	-
long grain	1(3)	2.2	-	2.0	-	20.2	-	462	-
Chilled	3	2.8	0.5	1.1	0.7	32.6	2.3	651	17
Frozen	6	2.9	0.3	2.9	2.0	28.9	3.7	658	77
Sterilised	3	3.3	0.5	3.6	1.4	33.3	6.1	768	63

Nutrient Content Per 100g Edible Portion

n = no. varieties or processors.
(1) = rice:water = 1:2.5.
(2) = rice:water = 1:2.5 and fat added during cooking.
(3) = rice:water = 1:5 and margarine added after cooking.

Variations in protein, carbohydrate, mineral and energy values are attributable to the different water intakes during cooking. The water content of parboiled rice was always significantly lower and thus the protein and carbohydrate value was always higher than in patna rice. The yield was dependent upon the cooking process used, as boiling in 2.5 times its weight of water showed 1 - 6% higher yield than boiling in 5 and 10 times its weight of water. There was almost no loss of protein, fat or carbohydrate during cooking. Mineral losses were dependent upon the raw material and cooking process, and the cooking of rice in 2.5 times its weight of water resulted in little or no loss of these nutrients.

As can be seen in Table XXI, losses of thiamin and riboflavin were also affected by the ratio of food to water, greater losses being incurred by higher proportions of water. Steaming of rice retained more thiamin than boiling in water and the addition of onions and stock to steamed rice increased its level of riboflavin.

The fat, carbohydrate and energy content of commercially produced chilled, frozen and sterilised rice was on average higher than freshly boiled rice.

The thiamin content of chilled, frozen and sterilised rice was approximately the same as parboiled rice cooked in 5 times its weight of water. (Table XXI). This content was relatively high compared to the patna rice, so it was assumed that commercial products used parboiled rice or vitamin enriched rice. Riboflavin levels followed a similar trend as thiamin.

Table XXI.

Thiamin and Riboflavin Content of Rice Made From Different Starting Materials.

Initial State (Rice)	n	Vitamin Content per 100g mg $\bar{x}$	s	% $\bar{x}$	s	Vitamin Content per 1000kJ mg $\bar{x}$	s	% $\bar{x}$	s
Thiamin									
Raw - Parboiled									
- rice:water = 1:2.5	2	0.071	-	100	-	0.135	-	100	-
- rice:water = 1:2.5	2 (1)	0.067	-	94	-	0.117	-	87	-
- rice:water = 1:5	2	0.047	-	66	-	0.089	-	66	-
- rice:water = 1:10	2	0.040	-	56	-	0.075	-	56	-
Raw - Patna									
- rice:water = 1:2.5	1	0.013	-	18	-	0.030	-	22	-
- rice:water = 1:5	1	0.010	-	14	-	0.020	-	15	-
- rice:water = 1:10	1	0.006	-	8	-	0.016	-	12	-
Chilled	3	0.059	0.021	83	30	0.091	0.031	67	23
Frozen	3	0.043	0.015	61	21	0.054	0.023	40	17
Sterilised	3	0.046	0.015	65	21	0.061	0.025	45	19
Riboflavin									
Raw - Parboiled									
- rice:water = 1:2.5	2	0.011	-	100	-	0.021	-	100	-
- rice:water = 1:2.5	2 (1)	0.022	-	200	-	0.038	-	181	-
- rice:water = 1:5	2	0.009	-	81	-	0.016	-	76	-
- rice:water = 1:10	2	0.004	-	36	-	0.008	-	38	-
Raw - Patna									
- rice:water = 1:2.5	1	0.006	-	54	-	0.014	-	67	-
- rice:water = 1:5	1	0.005	-	45	-	0.013	-	62	-
- rice:water = 1:10	1	0.002	-	18	-	0.005	-	24	-
Chilled	3	0.025	0.025	227	227	0.038	0.040	181	190
Frozen	3	0.023	0.006	209	54	0.030	0.014	143	67
Sterilised	3	0.029	0.009	264	81	0.038	0.011	181	52

n = no. varieties or processors. (1) = steamed with added fat.

9. Conclusion

The comparison of the quality of food menu items sampled from 4 different catering systems used by hospitals and a school has shown that the final quality of the ready to serve foods is dependent upon the time/temperature conditions that the foods undergo during preparation and processing.

The actual contents and retention levels of ascorbic acid were generally highest in freshly cooked vegetables served in a conventional system as the high pressure steaming of small batches of vegetables not only retained high levels of the vitamin but enabled warmholding before and during service to be minimised. Vitamin losses incurred during cook-chill processing and 1 days chilled storage were higher than expected due to the unacceptably high storage temperatures during chilled storage. Cook-freeze processing of vegetables at a hospital incurred excessive losses due to delays and the addition of water during processing. Pre-freezing losses were smaller in the cook-freeze-thaw vegetables but the overall losses due to processing and final reheating in infra red ovens were similar to those in cook-freeze vegetables.

As expected TBA numbers were low in the foods sampled from the conventional food system. Relatively high TBA values for roast pork in the cook-freeze system and for roast pork and roast beef in the cook-chill system were attributed to the overnight storage of the cooked joints before blast freezing and blast chilling. Increases in TBA numbers are expected in certain foods stored in cook-chill conditions but the high numbers measured after only 1 days storage were partly attributed to the high temperatures of storage. Prolonged storage of cooked meats would have rendered them organoleptically unacceptable due to rancidity.

The microbiological quality of foods sampled from all systems was satisfactory. The reheated samples of foods taken from the cook-freeze system had the lowest numbers of bacteria present. Controlled rapid thawing of foods did not appear to increase the numbers of bacteria present in frozen foods. Generally numbers were also low in foods from the conventional system. As expected numbers of bacteria increased during chilled storage of foods.

It was found that regular temperature monitoring of food items and equipment during the preparation, processing and storage of food is necessary. It is also recommended that all refrigeration, chilling and freezing equipment is fitted with temperature recorders and alarms.

The comparison of nutritional and organoleptic quality of carbohydrate foods prepared from differently processed starting components showed general trends which can be applied to all 5 dishes.

The organoleptic quality of freshly prepared carbohydrate dishes was generally found to be superior to chilled, deep frozen and sterilised products. This superior rating applied especially to the taste attribute of the food, but also to a lesser degree to the colour, appearance and texture attributes.

The established shortcomings of colour, appearance and texture of some commercial products could be avoidedby a different manufacturing process. The quality of products depended upon the raw material used. The variety, crop growing and storage conditions of potatoes can all affect the final organoleptic quality of potato products.

The protein, fat and carbohydrate content of products prepared freshly from raw materials did not differ significantly from those prepared from chilled, frozen or sterilised materials.

The levels of ascorbic acid, thiamin and riboflavin were usually highest in freshly prepared products, provided that the commercial products were not fortified with vitamins. Vitamins and mineral contents of dishes were affected by the choice of raw material and the cooking process.

Deterioration in nutritional, organoleptic and microbiological quality can occur during the preprocessing of food components. Some of these losses could be minimised by the control and use of optimal time/temperature conditions during processing. The choice of raw materials and the method of cooking affect the level of some nutritional losses. Fortification of some nutrients are necessary in certain preprocessed foods to maintain expected levels of these nutrients.

Information concerning the quality of preprocessed food components should be available to the caterer so that he may choose the type of product suited to his particular system. Having purchased the food item, any further handling and processing by the caterer should maintain the highest possible level of product quality. Personnel should be trained to understand catering operations and emphasis should be placed on adherence to the time/temperature conditions recommended during food handling.

Acknowledgements

M. Turner wishes to acknowledge the help of the staff from the hospitals and schools involved. Thanks are due to the Department of Health and Social Security for financial support during the research. Thanks are also due to other members of the research staff in the Hotel and Catering Research Centre who worked on this project under the direction of the head of the Catering Studies Department, Mr. G. Glew.

References

1. Witte, V.C., Krause, G.F. and Bailey, M.E. A New Extraction Method for Determining 2 - Thiobarbituric Acid Values of Pork and Beef During Storage. J. Food Science. Vol.35, p.582.

2. Bobeng, B.J. and David, B.D. HACCP Models for Quality Control of Entree Productions in Foodservice Systems. J. Food Protection. Vol.40, No.9, p.632-638. 1977.

3. Bognar, A. Nutritive Value of Chilled Meals. Advances in Catering Technology. Ed. G. Glew. Applied Science Publishers, London. p.387-408. 1980.

4. Catering Research Unit. The Determination of Ascorbic Acid in Food Samples Collected at a School Meals Production Unit. Lab. Report No.140. Catering Studies Dept., The Polytechnic, Huddersfield, England.

5. Turner, E.W., Paynter, W.D. Montie, E.J., Bessert, M.W., Struck, G.M. and Olson, F.C. Use Of The 2 - Thiobarbituric Acid Reagent To Measure Rancidity in Frozen Pork. Food Technology, Vol.8, p.326-330. 1954.

6. Catering Research Unit. Pilot Scheme for the Thiobarbituric Acid Testing of Precooked Chilled Food. Lab. Report 139. Catering Studies Dept., The Polytechnic, Huddersfield, England.

7. Department of Health and Social Security. Guidelines on Precooked Chilled Foods. Her Majesty's Stationery Office. 1980.

8. Jay, J.M. Microbiological Standards for Various Food Products. Modern Food Microbiology. 2nd edition. p.313. 1970.

9. Christiansen, L.N. and King, N.S. The Microbial Content of Some Salads and Sandwiches at Retain Outlets. J. Milk Food Technol. Vol.34, No.6. 1971.

10. Department of Health and Social Security. Precooked Frozen Foods. D.H.S.S. Medicines and Foods Division. 1970.

11. Ridley, S.J., Matthews, M.E. Temperature Histories of Menu Items During Meal Assembly. Distribution and Service in a Hospital Food Service. Journal of Food Protection. Vol.46, No.2. p.100-104. February 1983.

12. Zacharias, R., Bognar, A. Quality of Carbohydrate Containing Side Dishes. In: Schulverpflegung mit Speisen aus eigener Zubereitung und industrieller Herstellung. Mischkuche - Stufe I. Publ. BMELF, Research. Centre for Nutrition. Stuttgart, (W. Germany). 1982, p. 91-140.

A REVIEW OF THE LITERATURE CONCERNING WARMHOLDING OF FOODS IN CATERING

E.W. HOLYNSKI, J.N. AUCKLAND AND G. GLEW.
Department of Catering Studies, The Polytechnic, Queensgate, Huddersfield, HD1 3DH, England.

Summary:

Warmholding of food in catering can affect various quality factors before food is presented to the customer. Surveys of actual warmholding practices in the catering industry found wide variation in methods being used and a large range of times and temperatures. Holding times ranged from 0-7.1/2 hours but most were not greater than 4 hours, and food temperatures at service could be between 29 and 93°C. Provided that temperatures are above those required in law, warmholding time does not influence the microbiologic quality of food that is being warmheld. In fact longer times at temperatures above 60 °C are better from a microbiological viewpoint. This would not be the case though if a high standard of nutritional quality is the objective. In general, vitamin C was less stable than thiamine during warmholding with greater losses occurring at the higher temperatures. Small losses of 2-5% per hour have been found for riboflavin, B-carotene, niacin, pyridoxine and retinol and a few studies indicate that changes in protein quality and essential fatty acids on warmholding are also small. It is the organoleptic quality which has been found to be the most limiting factor in warmheld food. Rate of organoleptic deterioration depends on the food item. Fresh potatoes are particularly sensitive, and in some cases a maximum warmhold time of only one hour has been recommended. Future work is needed in determining maximum warmholding times for individual food items and developments in warmholding equipment/distribution systems could enable caterers to accurately warmhold food to achieve a high quality end product.

1. The Practice.

Warmholding (or hot-holding), is the process of keeping fully cooked food hot until it is to be consumed. This is usually the final stage in a conventional catering system and in theory should not be required in cook-freeze or cook-chill systems although in practice it has been found to occur (1).

Warmholding is carried out in various catering situations for a number of reasons, and in certain cases is unavoidable. In the industrial and welfare sectors of the industry, large numbers of people are served in a short space of time which can result in large amounts of food being produced in the kitchen before the service period to cope with the predicted demand. Thus, the need for warmholding arises when the rate of food production is higher than the rate of consumption. If production and consumption take place in separate sitings, then warmholding will occur during transport. Examples of catering situations where food has to be transported include "meals-on-wheels", hospitals, school meals and certain large industrial plants. In commercial operations such as hotels and restaurants, cooking of food may be spaced out over many hours and in this case warmholding could be used for convenience or to make it possible for a small number of staff to cook the amount of food required.

2. Surveys.

Warmholding surveys are summarized in Table 1. Little is known of the actual methods of warmholding employed by the catering industry. This is partly due to the vast range of methods and equipment available to caterers for adaption to their own ´unique´ catering operation. There are various designs of bain-maries and hot cupboards for holding food hot in the kitchen and also in areas where food is actually served and consumed. Furthermore, there is a large number of distibution systems utilizing mobile trolleys - heated or unheated, transportable hot locks and the type of container/plate used for the food all add to the variation in warmholding techniques.

Temperatures and times used are the most important factors to consider in any warmholding system and in practice both vary widely. In a survey of hospitals in the United Kingdom (2) it was found that the time elapsed after completion of cooking before service of vegetables to patients was less than 7 minutes in only 19% of small hospitals (less than 100 beds). In the larger hospitals (greater than 100 beds) warmholding times varied from 7 minutes to 1 hour or more. A more recent survey was carried out by the U.K. Catering Research Unit (3) which concerned a hospital tray distribution system incorporating the use of heated pellets to maintain the temperature of the food on the plates. After 35 minutes holding in unheated trolleys food temperatures ranged from 43 to 61°C. Most items were between 50 and 55° C. Temperature at plating was found to be important in controlling final temperature of food, as was the "bain-marie" containers used for holding food during plating. A further hospital survey (4) of a heated pellet system recorded temperatures of 31 to 91°C at plating of food. After 30 minutes certain meal components were 41 - 73 °C, which is within the manufacturers recommended time of 45 minutes for the system to maintain an adequate serving temperature. In Sweden, a comparison of hospital tray distribution systems (5) found that warmholding times ranged from 15 minutes to 3.1/2 hours. These systems resulted in a maximum temperature of 60°C on delivery of food to patients, and the last patients to be served received food at a temperature nearer to 30° C. Another Swedish study (6) monitored warmholding times and temperatures in hospitals, nursing homes, schools and restaurants. Times involved varied from 5 minutes to 7.1/2 hours but generally they were less than 3 hours 50 minutes for potatoes, fried and baked fish, and hamburgers. The longest times were found in cafeterias with schools next. Serving temperatures ranged from 29 to 93°C and a trayed meals system in hospitals was able to retain heat during service better than a bulk system. When a centralized tray system was compared to a decentralized tray system in one hospital (7) temperatures on service ranged from 51 to 66°C.

Studies in other catering systems have also revealed variation in warmholding practices. In the United States of America (8) a study of two college and three healthcare practices indicated that whipped potatoes were held for up to 3 hours 50 minutes between production and delivery of food to service or tray areas. Another survey (9) discovered that service temperatures in four cafeteria style units ranged from 33°C to 92°C over a 2.1/2 hour period for a number of food items. Although items were replenished during the service period, holding times took at least 15 minutes except for chopped steak with pineapple. Holding equipment utilizing water or steam heat was found to be more efficient than dry heat equipment in maintaining temperature.

Meals for the elderly have been investigated in the United States of America. One study involved a system transporting meat loaf from a production site to two service points (10). Holding times were found to vary from 30 minutes to 2 hours 57 minutes. Although food was assembled at 84 to 88°C, temperatures on service at site 1 (shortest distance) were 53 to 74°C and at site 2, 49 to 60°C. The food was kept in insulated trays during transport. In a meals-on-wheels service in the U.K. (11) it was found that meals were stored from 3 minutes to 1.1/2 hours in a hot lock before the first delivery took place. The time from the first to the last delivery was 35 minutes to 2 hours 40 minutes and therefore the meals could be in the hot locks from 50 minutes to 3 hours 20 minutes. At the first delivery no food item was maintained at 65 °C and the average serving temperature for a main course was only 41°C and for a dessert 38°C, when the last meals were delivered. Another meals-on-wheels survey in 1979 (1) showed that similar times were involved, with meals remaining in the hot locks for up to 3 hours 50 minutes. Temperatures at service of the first lot of meals were between 44 and 70°C, whereas final meals ranged from 30°C to 56°C.

3. The Bacteriological Consequences of Warmholding.

A very high bacteriological safety standard is required for all foods, and for this reason the Food Hygiene Regulations in the U.K. state that a minimum temperature of 62.7°C must be maintained in cooked food that is being held or transported. The Department of Health and Social Security recommend a minimum temperature of 65°C.

When beef loaves were heated and held under controlled conditions (12) baking to at least 60°C was sufficient to destroy most of the bacteria present. Bacterial growth did not occur on holding at 71 °C for 60 to 90 minutes and the final product on service had an excellent microbiologic quality. It was reported (10) that a 26 to 40 fold reduction in the total microbial population of meat loaf occurred during cooking and that during transport in insulated containers a further 2 fold reduction in total counts had occurred. The longer transportation time (to site 2) appeared to reduce the total plate count compared to the shorter time taken to transport the food (to site 1). Final numbers of surviving organisms at service was also partly due to the position of the meat in the stacks of trays, where the temperature was found to vary (see table). It was also demonstrated (13) that holding of roast beef up to 18 hours with a maximum temperature of 67°C and a minimum tempeature of 56°C indicated that no food poisoning hazard existed. The time/temperature relationships involved apparently destroyed the _Salmonella_ _Seftenberg_ organisms under investigation.

It would appear that if the temperatures required in law are maintained, then bacteriological quality of warmheld food will be at a

Table I
Surveys of Warmholding Practices

Country. Year. Ref.No.	Type of Catering Operation	Method	Time (Range)	Food Temp. (^{o}C) at Service.	
				Mean	Range
U.K. 1963 (2)	Hospitals (152)	Bulk heated food trolleys or food boxes, except - very small hospitals-food direct to patient. Larger hospitals - motor transport.	$<$ 100 beds = 0-55 mins $>$ 100 beds = 7-55 mins	Not Recorded	
U.K. 1982 (3)	Hospital (1)	Conventional system, trayed meals. Heated pellets in base under plate	35 mins on average	50-55	43-61
U.K. 1983 (4)	Hospital (1)	Cook freeze system, trayed meals. Heated pellets in base under plate	30 mins on average	40-56	40-73
Sweden 1978 (5)	Hospitals	Tray distribution systems and few portion pack	15 mins - 3½ hours	-	Max. 60 Last meals 30
Sweden. 1977 (6)	Hospitals (2) Schools (2) Nursing homes Restaurants	Bulk and tray systems. 1 central kitchen, 1 serving kitchen.	5 mins - 7½ hours	-	29-93
U.S.A. (1960) (7)	Hospital (1)	Decentralized tray - bulk food conveyors and centralised tray - hot food trolleys.	Not recorded	54-60	51-66
U.S.A. (1961) 9)	Cafeterias (4)	3 wet steam tables, 1 electric dry table	2½ hour service period Individual items$>$15 mins (except steak)	-	33-92

Contd.

Table I. (continued)

Surveys of Warmholding Practices

Country. Year. Ref.No.	Type of Catering Operation	Method	Time (Range)	Food Temp. (^{o}C) at Service.	
				Mean	Range
U.S.A. 1979 (10)	Transported meals for the elderly.	Transported in insulated containers from 1 production site to 2 service sites.	30 mins - 2Hrs. 57 mins.	site 1 64 site 2 52	53-74 49-60
U.K. (1982) (11)	Transported meals for the elderly.	Held in hot locks, transported to homes.	50 mins - 3 Hrs. 20 mins	M.C. * 1st meal 47 last meal 41 Dessert 1st meal 46 last meal 38	M.C. * 30-62 Dessert 26-60
U.K. (1979) (1)	Transported meals for the elderly.	Held in hot locks (charcoal heated) transported to homes.	51 mins - 3 Hrs. 50 mins	-	30-70

* = Main Course

high enough level no matter how long the warmholding period may be. Temperatures suitable for growth of food poisoning bacteria must be avoided (i.e. optimum growth occurring around temperatures of 30 - 45° C. However, it is known that such bacteria can often grow at temperatures of 40 - 50°C and the surveys to date indicate that such temperatures unfortunately are often encountered in practice (See Table I).

Table II

Microbial Evaluation of Meat Loaf as Influenced by Transportation to Two Different Sites

(from Cremer and Chipley 1979 (10))

Sample Point	Total Plate Count (count/g of product)	Time Mins (mean)	Internal Temp °C (mean)
Raw Meat	91,000 - 118,000	-	2.5
Before heating for assembly	2,300 - 3,700	-	7.3
Assembly	205 - 400	6.6	84.1-87.5
Site 1 - position 1	180	21.2	61.0
Site 1 - position 2	70	21.2	67.0
Site 1 - position 4	155	21.2	62.0
Site 2 - position 1	105	72.7	52.0
Site 2 - position 2	165	72.7	51.0
Site 2 - position 4	120	72.7	52.0

Site 1 is shortest transportation distance
Site 2 is longest transportation distance
position 1 is at top of stacked containers
position 2 is in the middle of stacked containers
position 4 is at the bottom of stacked containers

4. The Nutritional Consequences of Warmholding.

Food service practices and their effect on nutrient retention continues to contribute an increasingly important input to the nutritional well being of individuals, especially as the trend towards eating outside the home increases (14).

Other authors (15,16) have drawn attention to the lack of information on nutrient retention during warmholding and meal distribution in practical catering situations. It is these situations only which are dealt with in the following discussion. Evaluation under controlled conditions in the laboratory are shown separately in Table III. Much of the difficulty in collecting the information relates to the fact that techniques of warmholding and food production are themselves continually changing. A review of work from 1940-1960 (15) clearly indicates that content of certain nutrients in cooked food is greatly reduced during warmholding. Later studies of institutional food handling highlight the variation in susceptibility of nutrients to warmholding conditions.

The survey of British hospitals (2) found 36 - 90% losses of ascorbic acid in vegetables due to prolonged cooking and delays in service, i.e. warmholding. When ascorbic acid was analysed in meals for the elderly (11) average losses between end of cooking and the first meal served ranged from 31 - 54%. Additional losses occurred during delivery but were relatively small because of the low temperatures involved. This resulted in total losses in vegetables of 46 - 58%. In another meals-on-wheels study (1) small losses of ascorbic acid were found in vegetables during delivery due to low storage temperatures. The ascorbic acid intake of 100 elderly subjects during a survey in Portsmouth, England was indeed found to be lower on those days when meals-on-wheels were supplied compared to the intake for the rest of the week. (17).

In a study of a school "commissary" service (18) food was transported in insulated carts from one central kitchen to a number of schools. Samples of various foods were measured for nutrients at the kitchen and at six of the schools. Mean losses of 19% ascorbic acid and 20% iron were found at the schools. There was no detectable loss of riboflavin, less than 2% thiamine, and less than 10% calcium lost during transport. In eight of the schools food was sampled at various stages and losses of ascorbic acid, thiamine, iron, calcium and protein were found to be small from just after cooking to the moment of service.

Problems of accounting for such a large variation have stimulated more controlled studies. These are summarized in Table III. From this it can be seen that methods used vary greatly and conditions of warmholding reflect practices of different countries. For example, studies in Sweden and Germany have used longer holding times than most American studies. The study of beef roasts (19) however, did reflect the delayed service method for meat cookery used in the United States of America. In this, the cooking of meat roasts actually takes place during long warmholding times with the consequence that relatively low temperatures are required for cooking.

A recent review of nutrient losses showed that an inverse relationship existed between warmholding temperatures and the retention of vitamin C and thiamine (20). Time of warmholding was also shown to offset the retention of vitamin C and thiamine although the effect varied with temperature. It was concluded that thiamine was more stable than vitamin C during simulated warmholding conditions. The literature also shows small decrease in riboflavin and vitamin B6, and a limited amount of work also indicates negligible losses of nicotinic acid, vitamin A, carotene, essential amino acids and essential fatty acids (20).

For beef warmheld under controlled conditions, (12,19,21,22) thiamine retention was found to relate to method of warmholding. (i.e. sliced or unsliced) and time/temperature of holding. 79.2% retention was obtained in the beef roasts held unsliced for 1.1/2 hours, and a greater retention of 83.5 - 92.4% occurred when roasts were held for 16 and 24 hours respectively. In beef stew after 3 hours warmholding at 82 °C thiamine retention was 63%, but the same percent retention was found in beef loaf after only 1.1/2 hours hot storage at 71 °C. This would indicate that the ´form´ of the food can play an important part in nutrient retention.

Studies in Sweden have been carried out on boiled or steamed potatoes (5,6,23,30,35). Using temperatures of 50 - 90 °C and warmholding times of up to 5 hours, ascorbic acid retention was 8 - 75%, with the greatest loss occurring within the first hour. Thiamine retention in the potatoes was 65 - 90%, and losses of vitamin B6 were small resulting in 84 - 100% retention. Vitamin loss in mashed potato (fresh and dehydrated) under controlled conditions has been studied in

Table III
Nutritional Evaluations of Warmholding under Controlled Conditions in the Laboratory.

Country Year Ref.No.	Menu Items	Method	Holding Times	Nutrients Studied	% Retention
U.S.A. 1966 (19)	Beef Roasts	Compared 'just cooked' roasts with roasts browned and then held in electric drying oven. Internal temp. of food 60°C and 70°C	16 Hours 24 Hours	Thiamine Riboflavin	Retention on raw. Roast 16 Hr. Roast 24 Hr. 71.2 83.5 72.3 92.4 98.8 88.6 93.2 87.0
U.S.A. 1970 (21)	Beef Stew Chicken a la King Shrimp Newburg Peas in Cream Sauce	Compared different heating methods. Food held in pouches by immersion in a water bath at 82°C	1 Hour 2 Hours 3 Hours	Thiamine	Average Retention on Cooked. (All Products) 1 Hour 2 Hours 3 Hours 78.2 73.9 67.4
U.S.A. 1972 (22)	Beef Roasts	Held before and after slicing. Used dry heat. Internal temp. of food 60°C	1½ Hours	Thiamine	Retention on raw. Unsliced 79.2 Sliced 76.5 Refrigerated, sliced & reheat 67.8 Sliced and served 78.8
Sweden 1972 (23)	Potatoes Peas	Held in dry heat at different temps. 50°C, 70°C, 80°C, 90°C	30 mins 1 Hour 2 Hours 3 Hours 5 Hours	Ascorbic Acid Thiamine	Retention on Cooked. Potatoes Peas 75-17 59-2 Retention on raw. Potatoes Peas 90-65 100-64
U.S.A. 1973 (24)	Chicken Pot Pie	Reheated by different methods and held on simulated steam table. Internal temp. of food 82°C	30 mins	Total Vitamin C Ascorbic Acid Thiamine	Retention on raw. Losses on heating only 82 (mean)
U.S.A. 1975 (25)	Mashed Potato (dehydrated) Pot roast with gravy (fresh) Peas with onions (frozen) Beans & Frankfurters (Canned frozen) Diced Carrots (frozen) Fried fish portions (fresh)	Held at 93 - 99°C, dry heat. Internal temp. of food 82°C.	30 mins 1½ Hours 3 Hours	Riboflavin Ascorbic Acid B-Carotene Thiamine	Retention on Cooked and Frozen. 30 mins. 1½ Hours 3 Hours 98-89 98-91 97-83 66 51 40 92 88 92 100-93 92-86 88-77
U.K. 1975 (26)	Peas	Evaluated an insulated food service tray. Food temp. at start of holding 56°C	15 mins 30 mins 45 mins 60 mins 75 mins	Ascorbic Acid	Retention on cooked. 89-50
U.S.A. 1975 (27)	Turkey breast quarters	Held in a warming tray covered and uncovered. Internal temp. of food 80°C.	15 mins 1 Hour	Vitamin B_6	Retention on cooked. 84 Moisture and fat free basis 74

Contd.

Table III (continued)

Nutritional Evaluations of Warmholding under Controlled Conditions in the Laboratory.

Country Year Ref.No.	Menu Items	Method	Holding Times	Nutrients Studied	% Retention
Germany 1976 (28)	Food Groups : Vegetables Meat/Fish/Egg Potato/Pastry/Rice Stews	Held at 70 - 80°C	1 Hour 2 Hours 3 Hours	Ascorbic Acid Thiamine Riboflavin Niacin Pyridoxine B-carotene Retinol	Average All Foods 1 Hour / 2 Hours / 3 Hours 92 / 85 / 78 95 / 91 / 86 97 / 95 / 92 98 / 96 / 94 97 / 94 / 91 95 / 90 / 85 97 / 94 / 91
U.K. 1977 (29)	Potatoes (Fresh, dehydrated) Peas (frozen) Cauliflower (Fresh, frozen) Cabbage (fresh)	Reheated food and held in hot storage cabinet at 70°C, 80°C.	15 mins 30 mins 45 mins 60 mins 75 mins 90 mins	Ascorbic Acid	Retention on cooked. 30 mins / 60 mins / 90 mins All foods 80°C: 24-99 / 24-88 / 24-73
Sweden 1977 (6)	Potatoes Cod (Fresh,frozen) Herring (fresh)	Held at 60°C, 75°C, 90°C. Internal temp. of food 65°C.	1 Hour 2 Hours 3 Hours 4 Hours	Vitamin B6 Amino Acids Fatty Acids Ascorbic Acid	Retention on Cooked. Potatoes / Cod / Herring 100-84 / 89-70 / - - / 96-89 / - - / - / 100 26-8 / - / -
Sweden 1977 (30)	Potatoes	Held to maintain centre temps. of 60°C, 75°C, 80°C.	≤ 4 Hours	Vitamin B6 Ascorbic Acid	100 50 after ½ Hour
U.S.A. 1978 (12)	Beef Loaf	Held at 83°C in chamber of 50% relative humidity. Food Temp. 71°C.	1 Hour 1½ Hours	Thiamine	62.5 (mean)
Sweden 1977 (5)	Potatoes	Compared potatoes held in nitrogen gas, CO2 atmosphere at 75°C with those held in air	3 Hours	Ascorbic Acid	After 3 Hours: Air 28.0 CO2 85 N_2 100
U.S.A. 1978 (31)	Fried Chicken Beef Patties	Compared different heating methods. Held in dry heat. Internal food temp. 71-82°C	½ Hour 1½ Hours 3 Hours	Thiamine Riboflavin	Retention on Frozen, Raw & Char-Broiled 95 - 74 99 - 80
Germany 1978 (32)	24 meal items. Meat, Fish, Vegetables One-dish meals, Potatoes, Rice, Cereal Products.	Reheated and held at centre temp. 60°C, 70°C, 80°C	≤5 Hours	Amino Acids Vitamin C Vitamin B_1 Vitamin B_2 Retinol	100 62 - 75 71 - 95 100 100

Contd.

Table III (Continued

Nutritional Evaluations of Warmholding under Controlled Conditions in the Laboratory.

Country Year Ref.No.	Menu Items	Method	Holding Times	Nutrients Studied	% Retention
U.S.A. 1981 (33)	Mashed potato (dehydrated)	Held on steam table at approx $93^{o}C$. Internal temp. of food 77 - $82^{o}C$.	15 mins 30 mins 1 Hour	 Ascorbic Acid Riboflavin Niacin Vitamin B_6 Folic Acid	Retention on Uncooked. (Dry Weight) Granules Flakes 95 - 66 98 - 63 95 - 89 103 - 101 98 - 94 103 - 93 112 - 107 102 - 98 102 - 87 81 - 56
Sweden 1981 (34)	Peas (frozen) * Cod au Gratin +	Compared different systems, also effect of the atmosphere. Internal temp. of food $75^{o}C$.	1 Hour 2 Hours 3 Hours	 * Ascorbic Acid + Vitamin B_6	Retention on Cooked. 1 Hour 2 Hours 3 Hours 37 7 1 90 80 60
Sweden 1981 (35)	Potatoes	Compared effect of different conditions. Held at $75^{o}C$ different atmospheres $60^{o}C$, $75^{o}C$, $90^{o}C$ in air.	1 Hour 2 Hours 3 Hours 4 Hours	 Ascorbic Acid Vitamin B_6	Retention on Cooked. (after 4 Hours) Air Nitrogen 19 88 89-94 -
Sweden 1981 (36)	Hamburgers	Influence of degree of cooking to $70^{o}C$, $90^{o}C$. Held in warming cupboard at $75^{o}C$.	3 Hours	 Protein Quality Linoleic Acid	$70^{o}C$ $90^{o}C$ ← No Difference → 99 95
U.S.A. 1981 (37)	Chicken	Held on steam table at $66^{o}C$ Internal temp. of food $89^{o}C$.	1½ Hours	Thiamine	No significant losses during holding. Significant loss on cooking.
U.S.A. 1982 (38)	Italian Spaghetti	Compared freshly cooked, warmheld, and reheated on steam table at approx. $66^{o}C$.	1½ hours	Thiamine	Retention on Cooked 99.2 - 90.8 (means)
U.S.A. (8)	Potatoes (dehydrated)	Held at $82^{o}C$. 50% relative humidity. Internal temp. of food 71 - $74^{o}C$.	1 Hour	Vitamin C	64 (mean)

the United States of America and the United Kingdom (8,25,29,33). At temperatures of 70 - 82°C and warmholding times of 15 minutes to 3 hours, retention of ascorbic acid was 24 - 98%, with dehydrated mashed potato samples having a greater retention than the fresh product. Retention of riboflavin, niacin and vitamin B6 were all greater than 89%. After 3 hours thiamine retention was found to be 82% which is greater than that obtained in boiled/steamed potatoes. For mashed potatoes made from dehydrated flakes and granules, the folic acid content was found to drop to 56% retention after 1 hour in flakes as compared to 87% in granules.

Peas have been shown to lose considerable amounts of ascorbic acid on warmholding (23,25,26,29,34). At temperatures of 50 - 90°C for 15 minutes to 5 hours, ascorbic acid retention was 2 - 94%. Reported losses in thiamine appear to be lower (21,23,25). At temperatures of 50 - 90°C with warmholding times of 15 minutes to 5 hours retention was 64 - 100%. Experiments on other food items can be seen in Table III.

In the German study investigating a number of different food groups held at 70 - 80°C (28) losses of ascorbic acid, thiamine, riboflavin, B-carotene, niacin, pyridoxin, and retinol were calculated. These amount to 8%, 5%, 3%, 5%, 2% and 3% loss per hour respectively. This indicates that ascorbic acid is the most sensitive vitamin to warmholding with thiamine next for any food group. The results obtained for this study under controlled conditions compared favourably with experiments on the same food items using simulated mass catering conditions.

5. The Organoleptic Consequences of Warmholding.

Although the nutritional quality of warmheld food may be of great importance in certain catering situations, the view has been expressed that nutritional value alone is not limiting (32). Organoleptic consequences of warmholding also contribute greatly to the overall quality of food that is to be served to the consumer. - No firm conclusions can yet be made on how this effects consumer acceptance; but attempts have been made to compare results obtained from a trained taste panel with that of preferences for untrained consumers. One such study (39) used 17 trained panellists to determine maximum storage times, (at each temperature) at which potatoes could be warmheld to still be considered 'acceptable'. A preference study using 55 untrained consumers was also carried out and the results of the two experiments compared. Consumer preferences were found to fall within the limits of 'acceptability' determined by the trained panel. The methodology used here would therefore seem to be of practical importance to consumer preferences.

Organoleptic changes during warmholding have been found to vary according to the conditions used, and between food items. Under controlled conditions the organoleptic effect of warmholding has been measured in the following types of food:

"Meat"

For beef loaf the score for overall acceptability was greater in simulated cook-chill and cook-freeze systems than a conventional system including warmholding for 1 to 1.1/2 hours at 71°C (12). Different methods of holding were found to effect the scores given for beef roasts (40). Meat held unsliced for 1.1/2 hours and roasts that were sliced and served immediately, were given higher mean scores for all attributes than meat which had been sliced/held or refrigerated/sliced/reheated. Aroma was highly correlated to flavour of the lean meat and other good correlations were found between colour and flavour of the lean, and between juiciness and tenderness. In beef roasts prepared under conventional conditions and by the delayed service method held for 6 hours and 18 hours, conventionally cooked roasts scored higher for all attributes except aroma (13). Roasts held for 18 hours were given the same quality score for tenderness as conventional roasts, but were more tender than those held for 6 hours. When beefburgers were held for up to 5 hours at 75°C, no significant change could be detected in appearance, texture, taste, and colour (23). However, when hamburgers were fried to a temperature of 70°C and 90 °C the initial degree of ´cooking´ or ´overfrying´ had a negative influence on sensory quality during warmholding for 3 hours (36). Overfrying produces hamburgers that are firmer and drier (6).

A comparison of sliced beef and hamburgers subjected to three different pretreatments before holding for 2 hours (41) concluded that warmholding after the reheat should be as short as possible so that the increase gained in sensory quality by proper pretreatment is not lost. The presence or absence of gravy was found to be the most important factor influencing the sensory quality of meatballs. (42). Warmholding with gravy appears to result in a higher sensory quality, which also holds for other types of meat (28).

"Fish"

Cod was held for up to 5 hours under wet and dry conditions at 75° C. A warmholding time of 1 hour resulted in a significant difference in taste and texture (23). It took 4 hours warmholding for the untrained panel to detect a significant difference in texture and taste, with a difference in appearance being detected at 5 hours warmholding. In another study (6) cod was held at 60°C, 75°C and 90°C for up to 4 hours. Temperature had no effect on flavour, and time of warmholding was found to significantly affect total flavour although in practice it was unimportant on its influence upon consumer acceptability. Industrially prepared cod au gratin and breaded cod were held under the same conditions (6). These were judged to be better than the cod on its own when newly cooked and total flavour decreased with warmholding time at 75°C. However, after 4 hours both items had acceptable scores. Breaded herring in the same study decreased in total flavour and other properties after 4 hours at 75 °C. The affect of temperature was greater at 60 - 75°C than 75 - 90°C, and greatest loss in quality occurred at 60°C.

"Potatoes"

In most studies the sensory quality of potatoes has been found to be relatively susceptible to warmholding. Studies on boiled/steamed potatoes holding in laminated cardboard cartons at 75 °C showed a significant decrease in quality of appearance after 1 hour, with texture and taste taking 3 hours for the decrease to be significant (23). When potatoes were held in aluminium packs with closed lids, scores for total impression, odour, appearance, flavour and texture decreased rapidly with warmhold time (6). Greatest deterioration occurred in the first few hours. When temperature of warmholding increased, 60 °C to 75°C to 90 °C, most properties significantly decreased, such as total flavour. A slight increase in firmness and wateriness occurred but mealiness decreased. Off flavours produced during warmholding contributed to the total flavour decrease and discolouring contributed to the decrease in total appearance.

Significant interactions between time and temperature were reported on all properties except firmness and mealiness (30). Similar results were obtained in a Dutch study (42). Here, acceptable warmholding times at 60°C and 80°C were taken when less than 15 - 20% of the panel assessed samples to be bad or very bad. (5 point scale equivalent). For potatoes it was concluded that warmholding should not be longer than half an hour and definitely less than 1 hour. Further investigations in Sweden (30) revealed that it took 3 hours for flavour to become unacceptable at 60°C compared with 1.1/2 hours at 90°C. The method of cooking may contribute to these varying results, as seen when potatoes were cooked by boiling and steaming (43). After warmholding for 1 hour at 70°C the total texture of the pressure cooked potatoes was better, but sloughing and mealiness were higher in the boiled potatoes.

Determination of preferences of warmheld potatoes for appearance, odour, flavour and texture, indicated that changes in flavour were the most important factor in determining preference (39). Recent studies compared different potato mixes with freshly prepared potato and warmheld these for 4 hours (44). It was concluded that first flavour, then texture are related to overall preference. Overall low scores were obtained for the warmheld samples with the fresh potato being particularly sensitive to warmholding. The results indicated that except for one potato product, warmholding was the least preferred treatment. A further breakdown of textural components revealed that stickiness and coarseness explained 91% of the preference for texture in all the samples (45). Another comparison of fresh creamed potato and creamed potato using dehydrated mix (29) showed that the fresh potato changed in flavour after 15 minutes storage at 80 °C, but the dehydrated potato did not show a significant difference in flavour even after 60 minutes storage at 80°C. Although the dehydrated potato mix was shown to be less sensitive to warmholding, the panellists preferred the fresh potato and this was significant at zero storage time.

"Other Vegetables"

When frozen peas were held for up to 5 hours at 75°C a significant difference in taste and colour could be detected after 1 hour warmholding. It took 3 hours for the difference in texture to decrease significantly (23). When the change in colour on warmholding for 2 hours at 80 °C was measured by the conversion of chlorophyll to phaeophytin (46), chlorophyll retention in peas and sprouts after warmholding was markedly reduced. The panellists showed a greater preference for products higher in chlorophyll. The chlorophyll content of spinach on warmholding has been found to relate to the time of warmholding at each temperature (28). Regression equations revealed an increasing loss at 60°C, 70°C and 80°C. A strong positive correlation was also obtained between chlorophyll content and colour rating determined by panellists, both decreasing with time.

Warmholding of spinach, peas and cabbage at 80°C resulted in spinach and cabbage being unacceptable after 3 hours (42). Under these conditions a sauce was also found to improve the quality of vegetables after warmholding. A maximum warmholding time of 1 hour was recommended for a savoy cabbage.

When a comparison was carried out on the effects of warmholding on frozen peas, fresh and frozen cauliflower and cabbage (29), the result was that peas were the only vegetable where a flavour difference could be detected. This occurred after 30 minutes hot storage. The panellists did not show any significant preference for peas and cauliflower after 0, 30 and 60 minutes storage at 80°C.

From these studies it would appear that the rate of organoleptic deterioration therefore depends on the food item that is being warmheld. A German study (28) collected standard values for warmholding at 70 - 80°C and determined tolerance quality levels for each food type. (See Table IV). It is interesting to note that mashed potatoes could be held for 4 hours compared with the recommendations of less than 1 hour at not too high temperatures by Swedish, Dutch and British workers.

Again, the results on organoleptic quality reflect the different methods used in large scale catering from country to country. Actual assessments of the organoleptic quality of food produced in catering establishments are very few. An evaluation of meat loaf served in a commissary system found that the practices of reheating and transport were acceptable with regard to sensory quality (10). However, texture and flavour were given lower overall scores compared to appearance and colour. A survey of cafeteria style food establishments (9) noted that during extended warmholding all vegetables tended to become overcooked, soft and mushy. Conditions were then simulated in the laboratory because in practice many food items did not remain on the steam table for long enough to determine food ´quality´ accurately. Visible changes of food quality over a 2.1/2 hour holding period were recorded and these can be seen in Table V.

Table IV
Standard Quality Values for Warmholding at 70 - 80°C

Food Group	Food types which have reached the lower quality limit of class B (i.e. taste value: satisfactory).				
	2 Hours	3 Hours	4 Hours	5 Hours	Not Reached after 5 Hrs
Meat/fish/ egg	Baked fish fillet	Roast Pork. Pork cutlet breaded. Rissoles. Scrambled egg with Ham.		Chicken Fricasse* Fish fillet in sauce. Egg in mustard sauce*	Beef Goulash
Vegetables		Green Beans. Peas. Carrots. Spinach.	Cauliflower. Sprouts	Red Cabbage.	Sauerkraut.
Potato/ Pastry/ Rice.	Salt Potatoes.	Mashed potatoes (dehydrated) Potato pieces. Rice pudding.	Mashed potatoes. Egg Pie. Jacket potatoes (without skin.) X	Rice *	Potato salad.
Stews		Green Bean Casserole.		Pea Casserole	

*Improved Preparation. X Held in salt water.
Note: Quality grades: A= 9.0-7.0; B= 6.9-5.5; C= 5.4-4.0; D= 3.9-1.0.

6. Practical Questions Concerning Warmholding.

It is evident that conditions of warmholding vary throughout the catering industry but an important question to be considered is "how do these conditions effect the overall quality of the food and its perception by the consumer ?" This is a difficult question to answer as there are many influences that can play an important part in any assessment of food quality, for example social values and expectations. Grater, (47) put forward a caterers definition of food quality as, "to match or exceed customer expectation in terms of appearance, taste, aroma, nutrition and value at the moment of consumption". As we have seen warmholding can affect the organoleptic and nutritional qualities of food according to time/temperature relationships, but in relation to bacteriological quality, high standards can be maintained provided food is kept at 60°C or above. If these relationships are important, do caterers really know at what temperatures to serve food; and is it possible to maintain these temperatures during the warmholding period ?

In answer to the first question, relatively few studies have been carried out on the effects of serving temperature on consumption and acceptability of foods. Two experiments investigated temperatures within the range of normal serving limits. It was found (48) that

Table V

Visible Food Quality Deterioration Under a Simulated Cafeteria System.

Food Item	Food Temp. (^{o}C)	Viable signs of quality at relevant holding times
Mashed Potato	>65.6	>20 mins 'waxy! Stiff, crusted, discoloured + From the top under wet conditions, from the bottom in dry conditions.
Potatoes (dehydrated powder)	<60.0	>30 mins - deteriorates rapidly
Baked Potatoes	71	1 Hour = soggy, shrivelled
Cereal Products	>71	30 - 45 mins. Drying and discolouration
Sliced meats - no gravy - with gravy	>71 >71	15 mins. Dehydrated - Surface drying even with stock $2\frac{1}{2}$ hours. Surface drying and scum formation
Cream Soups	77 - 60.0	>1 hour. Darker. No evidence of curdling
Portioned Meats	38 - 89.0	30 - 45 mins. Drying
Ham and Creamed Ham	51 - 78.0	>$1\frac{1}{4}$ hours. Creamier in colour, surface drying and scum formed. $2\frac{1}{2}$ hour. Curdling.
Green Vegetable	 88 61 Higher temps.	30 mins on steam table - slightly discoloured. 15 mins hot cabinet + 15 mins steam table - markedly discoloured 15 - 30 mins - discolouration >30 mins - discolouration 45 mins - discoloured most
Yellow Vegetable	>65.6	1 hour - Drying - no discolouration

preferred eating temperatures for soups and beverages were 63 - 66 °C and for vegetables, potatoes and entrees 60 - 63 °C. For surgical patients it was found that the preferred eating temperature of meat was cooler than for vegetables and potatoes i.e. 66 - 71°C, compared to 71 - 77 °C (49). It is interesting to note that both these ranges were higher than the first study (48). The difference in these preferred eating temperatures could reflect the contribution of learned cultural factors. This has been shown to be important by other workers (50). They investigated the acceptability of beverages and foods over a wide range of serving temperatures. The results showed that the effect of serving temperature on the items studied was a function of the temperature at which that beverage or food is normally served. For example, apple pie, ham and biscuits were more acceptable than other foods (e.g. beef stew, scrambled eggs) at ambient temperature because they were the most likely of all the tested food items to be served at ambient temperatures or below in practice. It may be noted though that similar preferences may not be observed in poor rural areas or underdeveloped countries where different heating and cooling conditions for food may be minimal or absent.

Various systems of temperature maintenance have been studied, both in practice and under controlled conditions, producing some interesting results. One study evaluated the heat insulating properties of different tray distribution systems (51). Time taken to reach 65°C and 60 °C was studied. These were regarded as optimum and maximum warmholding times. For mashed potato, a starting temperature of 80°C resulted in a maximum warmhold time of 35 minutes, however tomato soup had a maximum warmhold time of 63 minutes starting at 95°C. If plates were preheated, warmholding times increased. The soup bowls had better heat retention than plates and a double packeted plastic lid and underplate was found to have the best insulating properties. In another study (52) 3 different tray distribution systems were compared over a period of 2 hours. Times for the temperature of the food to fall from 70°C to 60°C were 14.5 minutes, 17.5 minutes and 45 minutes. After 15 minutes the double packeted plastic tray with compartments maintained a significantly higher temperature than the other two that utilised china plates with steel underplates. A plastic lid with a hole lost heat quicker than a steel lid with no hole. Under practical conditions there was no significant difference in initial temperatures recorded but the plastic tray maintained a significantly higher temperature than the other two after 5 - 15 minutes and during transport the same system proved to be significantly better. Using meat loaf, mashed potato and diced carrots the heat retaining capabilities of delivery systems were studied (53). Relatively short warmholding times were recorded for items to reach their acceptable temperatures. In the case of carrots, cooking rates were extremely rapid giving a time of 11 minutes as the maximum for maintaining a temperature of 71°C. Knowing the difficulties that hospital caterers face in getting food to patients, this system would mean a very tight schedule to deliver food at preferred eating temperatures. This type of delivery system does require initial high temperatures to be achieved. By increasing the portioning efficiency for an insulated tray service system, 400g of canned rice pudding had a temperature of 71°C after 2 hours (26). This involved an initial temperature of 90°C. A further series of experiments led to the identification of factors which affect the ´quality´ of the food served on insulated trays.

These were:
1. ambient temperature,
2. initial temperature of each food,
3. mass of the food portions,
4. location of the tray within the stack.

In meals-on-wheels heat retention is also important because of the hot storage periods involved. Heat retention of various containers (using rice pudding) which could be used in a meals-on-wheels situation has been investigated (54). Wooden ´hot-lock´ boxes with polyurethane insulation maintained temperatures of greater than 65°C for up to 3 hours 50 minutes (12 meal size) and 4 hours 13 minutes (6 meals size). In a simulated run removing meals at 10 minute intervals the last meals to be served were greater than 65°C after 2 hours 20 minutes. For polystyrene bulk containers which hold 9 main meals and 18 sweet containers, on a simulated run the containers kept the meals at 65 °C and above for up to 1.1/2 hours (55). Temperature at loading and speed of removal from the box are crucial. Temperatures in a Temprite ´citizen´ stack during a simulated run were 58.0 - 62.0°C after 1 hour 10 minutes (56).

In commercial situations display of hot food means that warmholding does not take place in enclosed containers. A study of temperature retention in a bain marie for tuna and noodles, spaghetti and meat suace and yankee pot roast (57) found that a 5lb aluminium foil container was less efficient at heat retention than the same size container made from steel. Also, temperatures could be maintained at a higher level in a 10lb stainless steel container. Tuna and noodles had a temperature loss (all containers) after 1 hour of 20 - 34 °C. Similarly the temperature loss of spaghetti was 21 - 30°C and pot roast 19 - 47°C. This meant that with starting temperatures of 77 - 88 °C after 1 hour the temperature would only be over 60°C if the 10lb stainless steel container is used.

In the surveys of actual warmholding practices great temperature variation was found (see Table I). Perhaps with further developments in warmholding equipment, caterers could warmhold foods more accurately at controlled temperatures. This temperature control could then be related to information about warmholding behaviour of foods in order to achieve a high ´quality´ product. Finally, once this has been achieved caterers could adapt methods to their own situation. Perhaps a new breed of food and nutrition experts will emerge who are capable of utilising ´computerised´ catering systems to their full potentials (58).

References

1. TURNER, M., RYLEY, J., KERWIN, M. (1979). The Nutrient Content of Meals on Wheels in Leeds. In Glew, G. (Ed.) Advances in Catering Technology. Applied Science Publishers Ltd., London.
2. PLATT, B.S., EDDY, J.P., PELLET, P.L. (1963). Food in Hospitals. Oxford University Press, London.
3. CATERING RESEARCH UNIT (1982). A Study of a Conventional Hospital Catering System. Lab. Report No. 157. The Catering Research Unit, The Polytechnic, Queensgate, Huddersfield, W. Yorkshire, England.
4. CATERING RESEARCH UNIT (1983). A Study of a Cook-Freeze Hospital Catering System.

5. JONSSON, L. (1977). Hospital Food - Effect of Hot Holding on Nutritional Value. Naringsforskning 22 (4), 344-350.
6. KARLSTROM, B., JONSSON, L. (1977). Quality Changes During Warmholding of Foods. In Glew, G. (Ed.) Catering Equipment and Systems Design. Applied Science Publishers Ltd., London.
7. THOMPSON, J.D., HARTMAN, J., PELLETIER, R.J. (1960). Two Types of Tray Service Studied Side by Side. Hospitals. 34:82 (Feb)
8. SNYDER, P.O., MATTHEWS, M.E. UNPUBLISHED. Conventional Food Service Systems - Percent Retention of Vitamin C in Whipped Potatoes After Pre-Service Holding.
9. BLAKER, G.G., RAMSEY, E. (1961). Holding Temperatures and Food Quality. J. Am. Diet. Assc. 38:450.
10. CREMER, M.L., CHIPLEY, J.R. (1979). Time and Temperature, Microbiological and Sensory Quality of Meat Loaf in a Commissary Food Service System Transporting Heated Food. J. Food Sci. 44,2.
11. TURNER, M. GLEW, G. (1982). Home Delivered Meals for the Elderly - A Nutritional Study. Food Technol. July p.46.
12. BOBENG, B.J., DAVID, B.D., (1978). HACCP Models for Quality Control of Entree Production in Hospital Foodservice Systems - II. J. Am. Diet. Assc. 73, 530-535.
13. FUNK, K., ALDRICH, P.J., IRMITER, T.F., (1966). Delayed Service Cookery of Loin Cuts of Beef. Comparison Methods for Quantity Foodservice. J. Am. Diet. Assc. 48:210.
14. GLEW, G. (1979). Advances in Catering Technology. Applied Science Publishers, London.
15. LACHANCE, P.A. (1975). Effects of Food Preparation Procedures on Nutrient Retention with Emphasis upon Food Service Practices. In Harris, and Karmas, Nutritional Evaluation of Food Processing. p.463 2nd Ed. AVI Publishing Co.
16. HENDRICKS, T.S. (1981). Effects of Current Foodservice Practices on the Nutritional Quality of Foods. In Hospital Patient Feeding Systems - Proceedings of Symposium, Radisson South Hotel, Minneapolis, Minnesota, Oct. 19-21. National Academy Press, U.S.A.
17. DAVIES, L., HARSTRUP, K., BENDER, A.E. (1973). Ascorbic Acid in Meals on Wheels. Modern Geriatrics. July/Aug.
18. HEAD, M.K. (1974). Nutrient Losses in Institutional Food Handling. J. Am. Diet. Assc. 65:October p.423:427.
19. GAINES, M.K., PERRY, A.K. AND VAN DUYRIE, F.O. (1966). Preparing Top Round Beef Roasts. J. Am. Diet. Assc. 48 : 204.
20. LAWSON, J. (1982). Nutrient Losses During the Warmholding of Food. Paper presented at COST Symposium on Nutrition in Catering 7-8th January.
21. KAHN, L.N., LIVINGSTON, G.E. (1976). Effect of Heating Methods on Thiamine Retention in Fresh or Frozen Prepared Foods. J. FoodSci. 35 p.349.
22. BOYLE, M.A. (1972). Thiamine in Roast Beef Held by Three Methods. J. Am. Diet. Assc. 60 : 5 May, p.398.
23. HANSSON, E., OLSSON, H., BOSUND, I., RASMUSSEN, I. (1972). Changes During Warmholding of Food Products. Naringsforskning 16 p.106.

24. LACHANCE, P.A., ARVIND, S., RANADIVE, A.S., MATAS, J. (1973). Effects of Reheating Convenience Foods. Food Tech. 27:36 (Jan).
25. ANG, C.Y.W., CHANG, C.M., FREY, A.E., LIVINGSTON, G.E. (1975). Effects of Heating Methods on Vitamin Retention in Six Fresh or Frozen Prepared Food Products. J. of Food Sci. 40. p.997.
26. BISHOP, K., ARMSTRONG, J.F., GLEW, G. (1975). Some Preliminary Investigations on an Insulated Food Service Tray. Lab. Report No. 52. Catering Research Unit, The Polytechnic, Huddersfield.
27. ENGLER, P.P., BOWERS, J.A. (1975). Vitamin B6 In Reheated, Held and Freshly Cooked turkey Breast. J. Am. Diet. Assc. 67 p.43.
28. BOGNAR, A., ZACHARIAS, R. (1976). Service with Warmheld Foods. Herausgegeben vom Bundesministerium fur Ernahrung. Landwirtschaft und Forsten und der Bundesforschungsanstalt fur Ernahrung Bearbeitet vom Institut fur Hauswirtschaft, Stuttgart.
29. HILL, M.A., BARON, M., KENT, J.S., GLEW, G. (1977). The Effect of Hot Storage After Reheating on the Flavour and Ascorbic Acid Retention of Pre-Cooked Frozen Vegetables. In Glew, G. Catering Equipment and Systems Design. Applied Science Publishers, London.
30. KARLSTROM, B., LUNDREN, B. AND JONSSON, L. (1977). Warmholding of Potatoes. The Influence of Time and Temperature on Sensory Quality, Vitamin B6 and Ascorbic Acid. S.I.K. Rapport No. 417.
31. ANG, C.Y., BASILLO, A.L., CATO, B.A., LIVINGSTON, G.E. (1978). Riboflavin and Thiamin Retention in Frozen Beef - Soy Patties and Frozen Fried Chicken Heated by Methods used in Foodservice Operations. J. Food Sci. 43:1024.
32. PAULUS, K., NOWARK, J., ZACHARIAS, R., BOGNAR, A. (1978). Influence of Heating and Keeping Warm on the Quality of Meals. Amm. Nutr. Alin. 32. 447-458.
33. AUGUSTIN, J., MAROUSEK, G.A., ARTZ, W.E., SWANSON, B.G. (1982). Vitamin Retention During Preparation and Holding of Mashed Potatoes Made from Commercially Dehydrated Flakes and Granules. J. Food Sci 47:274.
34. JONSSON, L., DANIELSSON, K. (1981). Vitamin Retention in Foods Handled in Foodservice Systems. Lebensmittel Wissenschaft + Technologie 14:(2) p.94-96.
35. JONSSON, L. (1981). Studies on Vitamin Retention in Steamed Potato During Warmholding in Air and in a Nitrogen Atmosphere. Lebersmittel Wissenschaft + Technologie 14 p.43-46.
36. JONSSON, L., KARLSTROM, B. (1981). Effect of Frying and Warmholding on Protein Quality, Linoleic Acid Content and Sensory Quality of Hamburgers. Lebensmittel Wissenschaft + Technologie 14 (3) p.127-130.
37. LEE, F.V., KHAN, M.A., KLEIN, B.P. (1981). Effect of Preparation and Service on the Thiamin Content of Oven Baked Chicken. J. Food Sci. 46 :5 p.1560-1562.
38. KAHN, M.A., KLEIN, B.P., LEE, F.V. (1982). Thiamin Content of Freshly Prepared and Left-Over Italian Spaghetti Served in a University Cafeteria Foodservice. J. Food Sci. 47:2093.

39. LUNDGREN, B., KARLSTROM, B., LJUNGQUISV, A.C. (1979). Effect of Time and Temperature of Storage After Cooking on the Sensory Quality of Binkje Potatoes. J. of the Sci. of Food and Agriculture. 30 (3) 305-308.

40. BOYLE, M.A., FUNK, K. (1970). Holding Roast Beef by Three Methods. J. Am. Diet. Assc. 56:1 p.34-38.

41. DAGERSKOG, M., KARLSTROM, B., BENGTSSON, N. (1976). Influence of Degree of Precooking on Quality of Frozen Sliced Beef and Patties. Paper presented at 22nd European Meeting of Meat Research Workers.

42. DE FIELLIETTAZ GOETHART, R.L., LASSCHE, J.B., VAN GEMERT, L.J. Unpublished. The Influence of Warmholding on the Organoleptic Quality of Potatoes, Meatballs and some Vegetables.

43. OSTERSTROM, L., ANDERSSON, Y. (1979). Cooking of Potatoes by Steam and Water - A Comparison of Eating Quality. In Glew, G. Advances in Catering Technology, Applied Science Publishers, London.

44. NOVAIS, A., HANSON, S.W., RYLEY, J. (1982). The Texture of Mashed Potato in Catering. I.- The Background. The Contribution of Hedonic Texture to Overall Preference. Lebensm. Wiss + Technol. 15, 295-302.

45. NOVAIS, A., HANSON, S.W., RYLEY, J. (1982). The Texture of Mashed Potato in Catering. II - The Breakdown of Hedonic Texture into Non-Hedonic Textural Components. Lebensm. - Wis. + Technol. 15, 343-347.

46. RYLEY, J., BROOKES, G.C., PAUL, M. (1977). The Effect of End Cooking Methods on the Colour of Frozen Vegetables. In Glew, G. Catering Equipment and Systems Design. Applied Science Publishers, London.

47. GRATER, C. (1977). Constraints on Food Quality. In Glew, G. Advances in Catering Technology. Applied Science Publishers, London.

48. BLAKER, G.G., NEWCOMER, J.L., RAMSEY, E. (1961). Holding Temperatures Needed to Serve Hot Foods Hot. J. Am. Diet. Assc. 38 p.455.

49. THOMPSON, J.D., JOHNSON, D. (1963). How Hot Is Hot Enough? Hospitals 37:61 (Sept 1).

50. CARDELLO, A.V., MALLER, O. (1982). Acceptability of Water, Selected Beverages and Foods as a Function of Serving Temperature. J. Food Sci. 47 p.1549.

51. DE FIELLIETTAZ GOETHART, R.L. (1979). Comparison of Heat Insulating Properties of Different Food Service Trays. In Glew, G. Advances in Catering Technology. Applied Science Publishers, London.

52. JONSSON, L., OHLSSON, L., LINDHOLM, B. (1977). Comparison of the Temperature Holding Capacities of 3 Different Tray Distribution Systems. In Glew, G. Catering Equipment and Systems Design. Applied Science Publishers, London.

53. SHEA, L.A. (1974). Heat Retaining Capabilities of Selected Delivery Systems. J. Am. Diet. Assc. 65 p.430.

54. CATERING RESEARCH UNIT. (1977). Temperature Testing of Wooden Hot-Lock Boxes for North Yorkshire Social Services. Lab. Report No. 100. The Polytechnic, Huddersfield.

55. CATERING RESEARCH UNIT. (1977). Temperature Testing of Polystyrene 'Bulk' Food Containers. Lab. Report No. 101 - The Polytechnic, Huddersfield.
56. CATERING RESEARCH UNIT. (1977). Simulation of a Meals-on-Wheels Round to Test the Temperature Holding Performance of an Insulated Temprite 'Citizen' Stack. Lab. Report No. 93. The Polytechnic, Huddersfield.
57. HANCOCK, D.A. (1973). Planning the Change to Convenience Foods for a Hospital. J. Am. Diet. Assc. 63 (4) 418-422.
58. BALINTFY, J.L. (1979). Catering to Consumers Food Preferences. In Glew, G. Advances in Catering Technology. Applied Science Publishers, London.

REHEATING OF FOOD IN CATERING

C. SKJÖLDEBRAND[1)], Th. OHLSSON[2)], K. O'SULLIVAN[3)] and M. TURNER[3)].

1) Department of Food Engineering, Lund University, Sweden
2) SIK - Swedish·Food Institute, Gothenburg, Sweden
3) Catering Research Unit, Huddersfield, Great Britain

Summary

This paper presents three different investigations on reheating in Catering. At SIK - Swedish Food Institute the microwave oven has been examined and some recommendations are reported on how to use these type of ovens for reheating of some items. The convection ovens have been studied at the department of Food Engineering, Lund University, Sweden. Some results and recommendations are presented in the paper.
The third investigation concerns reheating in Steam a Multibeam Cooker. Dry reheating and moisture reheating are compared from quality and cost point of view. This study was done at the Catering Research Unit Huddersfield, Great Britain.
The paper is a presentation of investigations collaborated in COST 91 subgroup 2, processgroup Catering (C).

INTRODUCTION

One alternative system of handling foods in catering is to use prepared foods that are reheated shortly prior to eating. The production of the food will take place either in the food industry or in special designed production kitchens related to the catering. The prepared food is either frozen or chilled before the reheating procedure.

This total procedure may have its advantages as the main production may be spread more evenly throughout the day and the week than in a conventional system. There is a need for reheating of these products in connection to the serving. The problem is not however as easy as that. One has to take into consideration the total procedure i.e. time/temperature process to be able to govern quality when the food is reaching the consumer. The storage after preparation and chilling or freezing procedure does also effect the food quality. This means that it is important to control the reheating procedure carefully as well as to know the time/temperature history and quality status of the product before reheating. A lot of studies have earlier been carried out to compare a system containing a reheating step to a system including warmholding for long times. (1, 2, 3).

The reheating procedure as such has been studied in the literature (4, 5, 6). The most common reheating equipment are convection ovens and microwave ovens. Reheating in steam is also used.

Within COST 91 three different participating projects have been dealing with reheating

- Microwave heating of ready to serve foods, T. Ohlsson and U. Thorsell SIK - Swedish Food Institute, Sweden.
- Frying and reheating equipment in catering establishments, especially convection ovens. C. Skjöldebrand, Lund University, Sweden.
- The implementation of cook - freeze - thaw catering in hospitals and cook-chill catering in the health and social services, K. O'Sullivan and M. Turner, Department of Catering Studies, Huddersfield, Great Britain.

A summary of the results obtained in these studies will be reported here.

PROBLEMS IN MICROWAVE REHEATING OF CHILLED FOODS (7, 8).

At SIK - The Swedish Food Institute a field study was done to penetrate problems in microwave reheating of prepared foods. Instructions to overcome these problems have also been worked out.

It was found from the field study that the personnel appreciated the convenience of the microwave ovens but worried about the risks of leakage of microwave energy from the ovens. The other great problem was that heating result was not always acceptable. The following comments were recieved.

* Too low average temperature
* Uneven heat distribution between different components and within large components.
* Dehydration of thin whole meat and fish components
* Skin formation on white boiled potatoes
* Boiling-over of soups
* Deformation of plastic cups
* Rapid cooling, after microwave reheating
* Burnt edges on mashed potatoes and casserols
* Watery texture of vegetables
* Cracks in breaded products

With these aspects in mind some laboratory experiments were carried out. In these, three basic approaches were used to solve the problems:

* Microwave acceptable plastic cups and materials
* Proper control of geometry of food and packages
* Proper selection of rate of heating (power level).

Commercial microwave ovens with variable power dials but with different power ratings were used in the experiments.

The following results and recommendations to reheating using microwave oven were obtained from the experiments.

- Gravy and Soups in plastic cups.

It was found that cups of polypropylen or expanded polystyren performed well in microwave heating. The simple polystyren cups (the most common disposable cup) were easily deformed and should not be used. It was also found that a cup with straight walls gave less overheating than with tapered wall Cylindrical cups with straight walls should preferably not have too wide diameter. In addition the cups should not be filled more than 2/3 of the total.

Following these recommendations it was possible to heat 100 g of soup or gravy at full power level (1300 W). For 250 g, the power level should be reduced to about 900 W. It is recommended that stews, that easily flow out on the plate during normal reheating, are reheated in cups. 150 g portion of stew can be reheated at 900 W to good cooking results.

- Heating uniformity of mashed potatoes, purees etc.

If mashed potatoes is plated without any consideration on the microwave heating thin edges will usually be present. They will become brown during microwave reheating. It is important to have proper control of the geometry of the mashed potatoes. A good method for this is to use an ice-cream spoon. Using such spoons a good reheating result was obtained for 250 g mashed potatoes at 600 W and at 1300 W for 150 g mashed potatoes.

It should be pointed out that the ice-cream spoon should have a diameter of about 55 mm. With this there will be a partial centre heating effect and this together with a normal microwave heating of the outer parts will give a very good uniformity.

- Reheating of boiled potatoes.

It was found that the problems of skin formation of boiled potatoes were not caused by the microwave heating itself but by poor cooking and cooling of the potatoes. It is important that the cooking time is properly controlled. If possible, the potatoes should also be cooled in water and not in a refrigerator, where skin formation will be promoted.

In the reheating it is recommended that potatoes of uniform size are used. Size of about 75 g is optimal. Larger potatoes should preferably be divided and heated with the cut surface facing the plate. Using these recommendations there was not any problems to reheat potatoes even at 1300 W. Also during the reheating the potatoes should be covered with a plate.

- Dehydration of meat slices.

The drying out of thin meat slices can be prevented if the slices are placed on top of each other in a stack or as roof tiles. The idea is to have control of the geometry so that the thickness is between 10 and 20 mm and as uniform as possible. Thicker slices, i.e. of meat loaves should be placed single but close to each other. The power level have to be reduced in the reheating of thin meat slices. It was found that 450 W is a suitable power level.

- Heating uniformity in for example gratins.

Large food portions of above 25-30 mm thickness can be difficult to heat uniformily as the thickness approaches the penetration ability of the microwaves. Such portion should never be placed on top of each other but close side by side. It is an advantage if the thickness is uniform and not above 25 mm. It is recommended a power level of 450 W.

So called differential heating containers were studied. These equipment come with a sleeve where aluminium foil covers the edges. Microwaves were allowed to enter the container only in the middle of the portion. It was found that for mashed potatoes 1300 W could be used with good heating results using these containers whereas without them severe drying out and burned edges occured. In these cases only about 450 W could be used.

- Heating of complete meals containing several components.

This part has only partly been investigated. It was found that the recommendations given above for the individual components, will also improve the heating uniformity substantially over "normal plating" techniques. But the temperatures in the different components will differ even if these recommendations are used i.e. further developmentwork is needed.

REHEATING USING CONVECTION OVENS

As said in the introduction the convection oven is besides the microwave oven the most common reheating equipment. This method has been thouroughly investigated at Lund University in Sweden.

Reheating of meat loaves were examined in a special designed oven (9). This has been reported earlier in the literature (10). Heating of some different prefabricated meals from an industry were also investigated from the reheating point of view.

A meat product with a topping and fish product were examined. Five different commercial ovens were used and these were all of the same size.

The problems when using a convection oven for reheating are

* If the air velocity is too high a dry and burnt product will be obtained.
* If a too high temperature is used the product will be over heated. As the product is prepared the heating procedure has to be adjusted to this. The aim is to achieve eating temperature with minimum changes of the product.
* It was difficult to understand the proper use of the air humidity.
* In a commercial oven the air velocity is uneven causing drying out and controling problems.
* Proper process time for reheating is a problem in the aspect that it is difficult to measure quality and that the oven is uneven.

The results from the investigations may be summarized as follows:

- The air temperature in the oven should be between 125-175°C during most reheating procedures. It is important that this interval is not exceeded. Otherwise a dry and burnt product will be a result of the process.

- The convection oven may be used for reheating directly from frozen stage if proper air conditions are used. These have to be examined for each product.
- The products have to be distributed equally in the oven. A space between the product and the shelf above this should be more than 20 mm.
- The air velocity should be less than 5 m/s above all the products and controlled carefully.
- The vitamin losses were moderate at the actual temperatures in the investigated meat loaves.
- Heating of prefried flat ground meat products should be placed as roof tiles, not too compact. Some water should be placed in the bottom of the tray.
- It is important to test the heating time for each used product.

- The steam or a tray with water should be used during the whole process.
- Fish may be heated in a package without lid and with steam injected in the oven.
- It was found that a prefried meat loaf, that had been prepared at optimal conditions could be reheated giving a good quality. The oven temperature should be between 125 °C and 150 °C and the air velocity about 5 m/s and a high humidity would minimize the weight and nutritional losses. The reheating time is important to control and should not exceed 25 minutes.

REHEATING OF MEALS IN A SUNBEAM MULTICOOKER AND IN WATER (11).

At the Catering Research Unit in Huddersfield, Great Britain some research on reheating has been done in a project for feeding housebound elderly. Steam and water were the heating methods.

The Sunbeam Multicooker (GMC 11) is a piece of domestic equipment. This is an aluminium deep sided pan with an electrical element in the base, and fitted with a high dome lid with vents. The removable temperature control fits into a socket on the side of the pan and the control is calibrated in °F. A 13 amp power outlet can be used, the electricity consumption being 1.24 kilowatts. There are two handles for lifting and four plastic feet raise the pan of working surfaces. The Multicooker can be used to roast, stew, braise, poach, boil and bake.

Experiments were carried out to evaluate the performance of the Multicooker as a reheating device for individual frozen main courses and sweets. It was assessed with regard to:

- Quality and appearance of food when reheated.
- Safety and ease of handling of food and equipment during process.
- Time required to reheat food to overall adequate temperature.
- Cost of reheating one meal.

Individual frozen meals packed in foil trays with foil lined cardboard lids used and they were obtained from an industry. A range of six main courses of average weight 300 grammes and six sweets of average weight 145 grammes were used. Average meal temperature in frozen state: -15 °C to -20 °C.

It was decided to test the Multicooker as a dry reheating device and as a steam reheating device. To prevent burning a metal grid was used to raise the meal of the base of the Multicooker.

The frozen main course was placed on the grid in the cooker and the lid fitted in position.

The temperature in the cooker was about 230 °C and the reheating time for the main course was 1 hour.

In the other reheating method the frozen main course was placed in the cooker. Cold water was added. The temperature was 190 °C and reheating time was 40 minutes.

The results from these investigations were

- Quality and appearance of food reheating in dry heat could well be compared with reheating in a convection oven. However potato products got soft in the steaming method.

- Safety and ease of handling of food and equipment during process. The dry heat process was easier to use than an oven. Steam heating could be more dangerous.
- Time required to reheat the food. 60 minutes were required to get a food temperature of 70°C in the dry heat method. In steam the time was less than in dry heat only 40 minutes. This means that dry heat required about the same time as the oven and steam somewhat less.

The overall conclusion from this investigation was that this cooker used provides a satisfactory, safe and economical method of reheating frozen meals in the homes of the elderly. Either dry heat or steam heat can be used but dry heat is recommended.

DISCUSSION AND CONCLUSIONS

A lot of work has been done to examine the reheating procedure. These three investigations are some examples.

It is important that they fit into the total handling system, which has to be examine in total. Investigations have shown that not all foodstuffs in a meal may be prepared and reheated (2). This shows that the questions to organize a catering system is difficult when it comes to quality and nutrition of a food. Firstly, anyhow, a well trained staff is of necessity to take can of all the research results that have been reported on for example reheating. This means that the investigations here reported have to be translated to a programme for training and quality the kitchen staff.

They are also a tool to the kitchen manufacture on how to construct equipment which is the second important point. Organizing and knowing what foodstuffs may be reheated and training how to reheat are also important.

In such a system where this criteria are fulfilled these advises reported here are valuable but only if they are known to the people using them.

The paper here written is one example where three different institutions have been doing research in the same area. One of them is valuable but together they cover a great deal of important results.

To conclude

* The microwave oven, the convection oven and the multibeam cooker may be used for reheating
* The three equipments may be used for different items depending on the properties of the food.
* The advises here written have to be followed getting a god quality.
* All food items can not be reheated, for example some vegetables as potatoes have to be fresh cooked.
* The staff have to be trained to get a system that works.

REFERENCES

1. KARLSTRÖM, B. (1982). Kvalitetsförändringar vid varmhållning av livsmedel i storhushåll. Thesis Gothenburg, Sweden. (In Swedish).

2. Anon. Schulverpflegung, mit speisen aus eigener Zubereitung und industrieller hesternung Mischküche - Stufe I. Report from Bundesforschungsanstalt für Ernährung. 1982.
3. JONSSON, L. (1980). Näringsvärdesförändringar i livsmedel hanterade i storhushåll. Thesis Gothenburg, Sweden. (In Swedish).
4. WALKER, R.B. and GLEW, G. (1974). The Selection forced convection ovens for reheating pre-cooked frozen food. HCIMA Review Issue No 1.
5. WALKER, R.B and GLEW, G. Forced convection ovens for reheating pre-cooked frozen food. J. Fd. Technol. (1969), 4, 307-318.
6. BOGNAR, A. (1980). Nutritive Value of Chilled meals in Advances in Catering Technology ed G. Glew. Applied Science Publishers.
7. OHLSSON, Th. and THORSELL, U. Improved microwave reheating of ready-meal in food service. STU-report No 9-6355 (In Swedish) 1982.
8. OHLSSON, Th. (1982). Fundamentals of Microwave Heating. SIK service series.
9. SKJÖLDEBRAND, C. (1979). Frying and reheating in a forced Convection oven. Thesis, Lund University.
10. SKJÖLDEBRAND, C. and ÖSTE, R. (1980). Reheating of Minced meat Products in a Convection oven. In Advances in Catering Technology ed. G. Glew Applied Science Publishers.
11. ARMSTRONG, J, O'SULLIVAN, K and TURNER, M. (1980). The housbound Elderly Technical Innovations in food Service. Report Catering Research Unit, Huddersfield.

WHAT SHOULD THE CONSUMER DEMAND FROM INDUSTRIALLY PREPARED FOODS

P. KOIVISTOINEN
Department of Food Chemistry and Technology
University of Helsinki, Finland

Summary

It seems to me that the most important decisions about the whole food economy are taken by consumers, who are tending to form various groups with different attitudes, values and dietary habits. Each group represents a potentiality for the food markets. The old traditions are breaking down faster than expected. New means of communication, information and policy are needed between consumers and producers. People need information and advice, but they themselves decide what food is best for them. The producers have to know what the people really want to have. We must learn to listen more to the consumers.

1. INTRODUCTION

I find my topic "What should the consumer demand from industrially prepared foods" a confusing one and it does not fit into my philosophy about an average consumer in relation to food.

If I should accept the topic as such I had to be a God Almighty knowing what is the best food for the people. Even if I had that knowledge I had to assume that those foods and diets were made available by the producers and the consumers were taking benefits of that best food. This would mean that I, an outsider, with my experience or authority would tell to the two partners, the producer and the consumer, the rules of behavior without knowing their real possibilities and willingness to act so. For this reason I limit myself to the description of the consumer's behavior, as I see it, in relation to food. The demands are results of that behavior to be detected by food industry and respected by food scientists and technologists when seeking applications acceptable to consumers.

My lecture - far from a scientific paper - is based on my own experiences and observations made from different positions in academic and business life as well as the papers of three IUFoST symposia held in Einsiedeln, Switzerland 1959 (1), in Helsinki-Espoo, Finland 1981 (2) and Ottawa, Canada 1982 (3). Further, my thoughts are modified by two excellent papers given by Professor Schutz at the IUFoST Congress in Dublin this year (4) and by Professor Hawthorn at the opening session of this seminar (5).

2. CONSUMER, THE DECISION MAKER

In our developed countries the food markets have been saturated since the World War II. The traditional forms of agriculture have been converted into an agro-food business. In the past when the food supply was limited the availability and price of food were the primary determinants of dietary habits. Now with food markets of everything and with high income level the consumers themselves are determining the dietary patterns. They have become, in addition, the real decision makers of the entire agro-food business. A food product for having a chance to contribute to people's nutrition and the producer's account has to pass the acceptance by the consumers in a free market situation. For traditional foods it is easier than for the novel foods, but it is not any more a selfevident privilege. Consumer behavior is changing often more rapidly than expected and new competing products are developed and offered in the food markets.

3. DEMANDING CONSUMERS?

Do we have a demanding consumer? How and by whom the demands of the consumer are formulated? To my understanding there is a majority of consumers, quite silent and happy, living regular life and eating traditional food, listening to the advice of experts, and agreeing with the good aims, but usually ignoring them without making any action. These individuals, most of us, seem to use the power of the consumers only through purchasing quietly. Their purchases form the food markets and their decisions to buy represent a quiet, but very true specification of demand to the producers and of the foods. Most of their purchase decisions are traditional, but they might be ready to try something new or even accept it into regular use. This readiness represents the chance of the new products and the decline of traditional products. These are the things the food producers have to know to stay in business.

Lately, it has become popular to establish organizations specified to represent the interest of consumers and to define and express the demands, often in a very loud voice generally in a society and especially to the administrators and producers. Dr Aebi in the book "Criteria of Food Acceptance" states: "The consumer of today is like a sphinx; God-like and powerful on one side, unpredictable in his intentions and reactions on the other. According to the declarations of their tough and popular representations, everything that is produced by food manufacturers should be absolutely safe, cheap, taste good and be healthy as well" (6). In recent years the implementation of these demands in specified form into food legislation has required an enormous effort from scientists and administrators. As a result food industry has to observe a great number of specific regulations intended to serve the purchasing process and health of the consumers, but often for extra price. The legal regulations cover only a fraction of the product characteristics which might be important to the quiet consumer when determining the food markets. There are many other challenges to be met by food industry.

4. NEW VALUES, NEW CONSUMER GROUPINGS

In the present food system the distance from the raw material producer or food processor to the consumer has lengthened. By distance here I mean space, time, knowledge of production and psychological coherence. As a result, the consumers are inclined to dream about illusions of green, pure and healthy. In days gone by when families lived on farms and each member of the family worked to produce food, there was no place for the issues that are common in consumerism today. In society as we know it today, when we are not directly involved in food production, and obtaining food is no longer main purpose of our activities, food as a commercial item has to be readily identifiable, convenient to obtain and prepare, tasty, nutritious, safe, suitable for specific health or disease conditions, consistent with individual ethical concepts, etc.

The consumers of to day have become - thanks to the mass information - increasingly aware of food, they are asking in their minds questions, and they are becoming more and more difficult to convince. Much of this, as I said before, is due to the distance from the raw product producer and the food processor to the consumer. Consumers ask questions which present no problems to those familiar with agriculture and the food industry. Many are answered through food leagislation and package labeling, but many, even some that are vital, have no good answer. These concern the safety, nutrional value, and suitability of goods for specific health, disease and ethical situations. These are aspects which are difficult to analyse and define, even by experts, and they are even more difficult for the consumers to grasp and understand. What is more, consumers are exposed to a mass of highly inconsistent and thus confusing information. How can the public be expected to sift the really important from the meaningless?

5. FEAR OF CHEMICALS

In past years there has been much discussion on the chemicals which are used in food production: fertilizers, pesticides, and additives. This discussion has aroused fears in consumers despite the strict regulations that are applied. A considerable sector of the population appears to react negatively to the use of chemicals and prefers products claimed to be free of them; these are believed to be safer than products produced by conventional technology. To what extent this attitude will spread among consumers is difficult to predict, but it may present a permanent opening for food markets. Be that as it may, this antichemical stand seems to be so entrenched that the use of a number of chemicals is to be limited to a practical minimum. Chemical contaminants in the environment may also play an important role in food markets.

6. FOOD - HEALTH - DISEASE

It is generally accepted that the right food promotes good health and that an inadequate diet may increase the risks of certain diseases. For consumers, food, health and disease constitute a very confusing and serious problem. Nutritional awareness is increasing, and clear trends can be seen in food consumption as a result of changes in consumer behaviour. Some of the changes are much more rapid and extensive than was previously believed and are reflected both in conventional and health food markets.

There seem to be several different groups of people attempting to improve their diets in different ways. There are those who listen to nutrition experts and modify their diets by cutting down on calory intake, fats, salt, sugar, by increasing their use of vegetables, by taking note of cholesterol, etc. This attitude has already had a marked impact on food markets.

Then there are the groups of people suffering from or trying to prevent a disease who change their diets for ones with a specific therapeutic effect. This behaviour opens the market for foods with therapeutic value.

Another is composed of health food store customers. These tend to differ from the above groups. It seems to me that many of them go to health food stores because they believe that normal foods are inferior to particular health foods or health foods containing specific qualities good for health. These people have lost their faith in conventional food markets. This is a serious problem for the responsible businesses and for the administrators of conventional food systems.

Further, I should like to mention three groups of people who currently use or may require specific food markets. Babies need particular foods, and baby foods have already established extensive food markets. Then there is the ever-growing sector of old people, who for many reasons need certain foods and diets. Specific food markets, I believe, have great potential for them. The third group of people is made up of those devoting much time and effort to maintaining or improving physical fitness by exercise, proper diets and an active lifestyle. They believe that they need specific, very effective food items for their physical performance and muscle condition. The great sport stars serve as exciting idols for this sector of the population, and new markets are opened for specific food items.

7. ATTITUDE PATTERNS

In general, there seem to be three major food qualities - pleasure, nutritional value and safety - that according to a study by Solveig Wikström (7), are considered to be of prime importance by modern consumers:

- Those who think that nutritional qualities are important also care for safe food.
- Those who think that food as a pleasure is very important pay little attention to nutritional and safety aspects.
- Those who think that nutrition and safety are very important pay small attention to high food prices.

Future food marketing will certainly have to put increasing emphasis on the selection of target groups and on associated qualities of food like pleasure, nutrition and safety. Products will have to be compatible with the attitudes of consumers, and effective means of communication will have to be developed to inform the consumers about the qualities of the products.

8. ETHICAL ASPECTS

There is nothing new about ethical aspects modifying dietary habits. Many religions have ethical rules about food. One wonders if new situations may arise that may call for yet more ethical considerations determining the acceptance of foods. When I say this I have in mind animal lovers and movements for protecting animals. It is easy to see that people dedicated to the protection of animals refuse to eat animal products and consider that the mass production of animals for consumption is unethical and also unacceptable, regardless of its former or current value as a source of food. Ethical aspects of this type may be enhanced as the distance from producer to consumer increases. In the old days the farmers' children were used to and accepted animal husbandry, hunting, fishing and so on as natural means of obtaining food for the family; after all, alternatives were few. Nowadays, when most people do not have to worry about the availability of food and food varieties abound, the ethical aspects may start to affect food acceptance and even the acceptance of the technologies commonly involved in agriculture and the food industry. I think that in future there are going to be many more questions and even problems of this type to be faced by the conventional food industry. This can be a barrier of practical applications of scientifically sound technologies which are rejected by the consumers for one reason or another.

REFERENCES

1. SOLMS, J. and HALL, R. L. (Eds.) (1981). Criteria of Food Acceptance. LWT-Edition 6. Forster Verlag AG/Forster Publishing Ltd, Zurich.
2. KOIVISTOINEN, P., HALL, R. L. and MÄLKKI, Y. (Eds.) (1982). The Role of Application of Food Science and Technology in Industrialized Countries. VTT Symposium 18. Technical Research Centre of Finland, Espoo.
3. ANON. (1983). Research management for food industries. IDRC-MR75eR. Int. Devel. Res. Centre, Ottawa.
4. SCHUTZ, H.G. (1983). Consumer perceptions of food quality (manu-scirpt). 6th World Congress of Food Science and Technology, Dublin.
5. HAWTHORN, J, (1983). Quality of processed foods. COST '91 Concluding Seminar, Athens.
6. AEBI, H. E. (1981). Food acceptance and nutrition: Experiences, intentions and responsibilities. An introductory paper. In: SOLMS, J. and HALL, R. L. (Eds.), Criteria of Food Acceptance. LWT-Edition 6. Forster Verlag AG/Forster Publishing Ltd, Zurich, pp. 3-11.
7. WIKSTRÖM, S. R. (1982). Food consumption and consumer behaviour in the future. In: KOIVISTOINEN, P., HALL, R. L. and MÄLKKI, Y. (Eds.). The Role and Application of Food Science and Technology in Industrialized Countries. VTT Symposium 18. Technical Research Centre of Finland, Espoo, pp. 45-58.

NUTRITIONAL ASPECTS OF INDUSTRIALLY PREPARED FOOD

A.E. Bender
Department of Nutrition, Queen Elizabeth College, University of London.

Summary

Caterers (Food Service Managers) have, or will have, an obligation to follow national nutritional goals by; (a) supplying meals that follow such policies; (b) encouraging the consumer to follow the guide-lines; (c) making available dishes such as sugar-free, low-sodium, fibre-rich, low-fat; and (d) providing nutritional information about the dishes served.

It is not widely appreciated, particularly by those who allocate research funds, what a large segment of the food industry is involved in food service (catering) and that this segment, in contrast with the manufacturing industry, is in direct contact with the consumers.

A number of recent developments will have an increasing influence on the catering industry, and if the industry is to develop along acceptable lines all sectors will have to take note. These include; (1) A growing awareness by the public(i.e. consumers, i.e. customers) of the influence of diet on their health; (2) the growth of food regulations; (3) the promotion of national nutrition policies.

The last, which will have the greatest impact on the catering, even more than the manufacturing industry, is illustrated by the large-scale health campaign currently in operation with considerable success in Norway. Such nutrition policies are based on nutritional 'goals' which include reduction of surplus body weight, reduction of fat consumption with a change towards a higher proportion of polyunsaturated fatty acids, reductions in salt and sugar, and increases in the consumption of whole grain cereals, fruits and vegetables.

How can the catering industry reconcile its meals and menus with such public campaigns? Is it feasible to consider offering dishes which run counter to such advice?

Industrial feeding consists of many sectors ranging from institutional feeding where the consumers are provided with all their food to commercial meals such as in restuarants which may make only an occasional contribution to the diet. It includes school meals intended not only to supply the required nutrients and to compensate for any possible deficiencies in the home, but also intended to introduce children to a wider variety of foods, to inculcate good eating habits (meaning good food selection) and to introduce or re-inforce good social behaviour. Meals in factories are certainly not intended to achieve the last three lessons but merely to satisfy the wants of the consumer bearing in mind that they may be eaten five days a week and may be the main meal of the day.

Despite this varying importance of industrial feeding in the diet of the consumer what will be the position of the caterer when the time comes when national bodies are attempting to achieve nutritional goals through public education and his meals do not satisfy these criteria? Two examples where catering flies in the face of nutritional recommendations are establishments that proudly boast 'we serve only the best butter', and those where only the finest white bread and rolls are served.

There are two tasks for the caterer which, unless accepted and undertaken, will put industrial feeding in conflict with public opinion; one is to provide information about the food served and the second is to make available such foods as are demanded.

The advent of nutritional labelling, whether voluntary or by regulation, varies in type and detail but is a response to public demand. Consumers may require to know a diversity of facts - energy content of the dish provided just as they do on a packet of food, total fat or even saturated fat content, possibly the cholesterol, sodium, sugar and fibre content. It may at present appear far-fetched to expect the caterer to provide such information on a restuarant menu but if there is a national campaign regarding such ingredients will the caterer not eventually be forced to do so? There have been occasional attempts in the past by some restaurants to include figures of energy content of dishes but, again in the past, there has been so little interest that most of these attempts have been abandoned. The climate of opinion is now changing. Why should the consumer not know exactly what he is eating not merely in gourmet terms but in nutritional terms?

The second task is to make available the variants of the standard foods required by consumers. There appear to be few catering establishments which make available diabetic jam, low-sodium dishes or even brown or wholemeal bread as alternatives. It is possible to do so, as exemplified by the American drink available in six varieties, namely with and without sugar, normal or low in sodium, and with and without caffeine. This must provide great difficulties to the manufacturer and no less the retailer but should be not quite so difficult to the caterer.

It is often a cause of considerable irritation to the customer - who finally votes with his pocket-book rather than makes a complaint - when fresh fruit has unrequested sugar sprinkled on top, when the waiter does not know (or care) whether the fruit juice has been sweetened, does not know the difference between brown and wholemeal bread, or whether the soup or sauce has been thickened with wheat flour.

This topic is well illustrated by the recent finding from the Department of Catering at the Polytechnic, Huddersfield, England, that meals from restaurants and public houses contained 57% fat - at a time when almost every organisation concerned with dietary health education is proposing a reduction of fat intake from 40% to 30-35%.

LES CONSOMMATEURS ET L'ALIMENTATION EN 1983

R. REMY
Responsable des études alimentaires
Association de Consommateurs.
Bruxelles

Résumé

Depuis leur création, les organisations de consommateurs ont toujours été soucieuses de la qualité des aliments. Leur vision des choses reposent avant tout sur les enseignements de l'histoire des aliments au travers des âges. Actuellement, les soucis essentiels de leurs membres ont trait à la protection de la santé, à la qualité nutritionnelle des aliments et au coût réaliste des denrées. Ce sont ces objectifs qu'en matière alimentaires poursuivent les organisations de consommateurs.

Summary
From their creation, consumers'association attach a real importance at the quality of foods. Their objectives are, at first, derivated from the food story throughout all ages. Actually, their members are essentially senzibilized to the problems of security with foods, nutritional quality and realistic cost of food products. These are some of the tasks for consumers'associations.

Au cours des âges, l'alimentation de nos populations ainsi que nos habitudes alimentaires ont subi de profondes modifications ; modifications dues à divers facteurs qui ont pour nom la situation économique dans nos divers pays (évolution du budget des foyers - coût des aliments...), l'évolution des idées, l'éducation alimentaire et la publicité. Tous ces éléments font que les aliments que nous consommons aujourd'hui sont bien différents de ceux que nos aïeux consommaient il y a quelques décennies ou plus. L'on peut raisonnablement admettre que dans nos pays occidentaux la fin de la 2ème guerre mondiale constitue une date importante. C'est en effet de cette époque que datent les premières réelles innovations technologiques importantes ayant conditionné "l'industrialisation" de notre alimentation. Et puis, bénéficiant de progrès scientifiques évidents, d'une meilleure perception des phénomènes physiques, chimiques, biologiques, les aliments se sont davantage industrialisés. Et, pour nous, organisation de consommateurs, ce concept d'industrialisation n'est pas forcément synonyme de moins bonne qualité, bien loin de là. Mais ce phénomène nous oblige cependant à préciser certaines de nos positions.

Et tout d'abord : pourquoi et comment les organisations de consommateurs se préoccupent-elles de l'alimentation ?

Une telle question peut, à notre avis, être abordée de deux façons différentes : d'abord, en examinant l'histoire de l'alimentation humaine et ensuite en débouchant sur les préoccupations essentielles et actuelles des organisations de consommateurs.

Alimentation et Histoire

Lorsque l'on survole l'histoire de l'alimentation, une première conclusion générale se dégage : les hommes on toujours du peiner pour se nourrir. Ce qui nous permet d'en tirer une première leçon : les revendications des consommateurs n'ont rien de "réactionnaires". Il ne s'agit pas d'être des nostalgiques du bon vieux temps, d'un âge d'or que les conservateurs de tous les temps ont toujours défendu par fonction. Pas de refus systématique donc des évolutions technologiques... mais un réalisme prudent.

Autre enseignement de l'Histoire : se nourrir est aussi un acte social. Pour certains, il sera considéré comme un élément sécurisant, comme un moyen d'extérioriser sa personnalité et son humeur - pour d'autres, il servira de compensation à un échec ou un moyen de démontrer son appartenance. Raison pour laquelle les organisations de consommateurs attachent non seulement de l'importance à ce que sont les aliments du point de vue nutritionnel mais aussi à leurs propriétés objectives.

Autre enseignement : l' alimentation est un problème politique, ce qui est d'autant plus vrai depuis que les hommes se sont groupés en agglomérations, que les aliments ne sont plus produits sur leurs lieux de consommation et qu'il a donc fallu structurer la production, le stockage et la distribution.

Nul n'est besoin de s'étendre longuement sur l'usage de l'arme alimentaire dans les rapports de force entre pays.

C'est donc un élément dont nous devons, de force ou de gré, tenir compte.

Le besoin de s'organiser pour se défendre...

Ces quelques préliminaires expliquent partiellement le besoin qu'ont ressenti les consommateurs de se grouper d'abord essentiellement pour maîtriser le prix et la qualité des aliments. Par après, se développèrent les organisations de consommateurs telles que celles dont je fais partie. Elles furent très souvent l'émanation de personnes qui, pour des raisons diverses (professionnelles, sociales,...) se rendirent compte très tôt qu'ils étaient manipulés, tenus dans l'ignorance des diverses innovations, démunis de pouvoir exercer un choix en connaissance de cause.

Se fixant trois buts bien précis, à savoir, former, informer et défendre le consommateur individuel, ils offraient, de la sorte, une possibilité de rééquilibrer le marché, de faire entendre une voix autre que celle des producteurs et distributeurs.

Et aujourd'hui, pourquoi se préoccuper d'alimentation ?

Parmi les nombreux contacts et dialogues que nous entretenons avec nos membres, une préoccupations essentielle (sans être la seule) se dégage : que faut-il manger pour protéger sa santé ?

Ce désir de sécurité est donc l'un de nos soucis permanents. Il n'est pas question pour nous de nier ou d'ignorer les progrès technologiques. Les fraudes grossières (mouillage du lait...), les épidémies collectives à l'échelle mondiale sont certes quasi disparues. Ce qui ne veut pas dire pour autant que tout est pour le mieux dans le meilleur des mondes. Au cours de ces dernières années, bien des problèmes importants ont retenu notre attention : hormones et antibiotiques dans les viandes, aflatoxines, additifs, qualité microbiologique,...

Plus récemment encore, la commercialisation en Espagne d'une huile inapte à la consommation humaine nous a rappelé le problème posé par la dénaturation des denrées alimentaires et l'urgence de mesures réglementaires visant à réduire au strict minimum les risques de fraude ou d'erreur humaine. Nos législations nationales ou internationales font, en effet, trop souvent penser à un mur dressé morceau par morceau et où les pierres de taille voisinnent avec les briques, les dalles de béton avec les planches chancelantes. Ce qui forcément provoque des lézardes ou des trous...

Sécurité mais aussi qualité nutritionnelle...

Autre souci des organisations de consommateurs : la qualité nutritionnelle. Les divers procédés technologiques nouveaux ne peuvent en aucun cas, à nos yeux, permettre la commercialisation d'aliments déséquilibrés du point de vue diététique : la floraison du "fast-food" qui permet la vente de viandes hachées ou "hamburgers" à 30 % de matière grasse, la distribution de fromages blancs à 90 % d'eau, la mise au point de fruits à haut rendement mais à faible teneur en vitamines en sont quelques exemples parmi d'autres...

L'abus de sel dans les conserves alimentaires constitue un autre aspect très important à une époque où hygiénistes et diététiciens portent tous leurs efforts pour en réduire la consommation.

En fait, ce que nous réclamons, c'est davantage de recherches et de connaissances scientifiques, une place plus importante de la nutrition dans l'enseignement et le développement de recherches en nutrition humaine.

Enfin nous estimons également que la sécurité de notre alimentation passe non seulement par des législations précises mais aussi par la mise en place de services de contrôles rigoureux.

Le souci de la qualité...

Définir le mot "qualité" pourrait assurément faire l'objet d'un séminaire complet. Ce n'est pas notre but mais nous dirons cependant que la baisse de la qualité gustative est exprimée par de nombreux consommateurs. Ce n'est nullement le signe d'une nostalgie mais une réalité qui, bien que difficile à mesurer, se vérifie dans les faits. Plutôt que guérir, mieux vaut prévenir. A voir certaines évolutions - la qualité des jambons, la valeur boulangère de certains blés,... nous pensons qu'il est temps de réagir.

Et le prix ?

A nos yeux, une meilleure qualité n'est pas impérativement liée à un prix plus élevé. D'abord parce qu'une mauvaise qualité se paie également : pertes plus élevées, gestion des excédents, perte de marchés. Ensuite parce que certaines améliorations ne sont pas synonymes de coûts supplémentaires.

Des études plus nombreuses et de toute façon accessibles aux oragnisations de consommateurs sur le coût de la qualité sont indispensables.

Autre risque de l'innovation en agro-alimentaire : la tendance à développer des produits à forte valeur ajoutée au détriment des aliments simples. Notre position est claire à ce propos : chacun doit pouvoir choisir son type d'alimentation en fonction de ses goûts, de ses besoins et de ses possibilités financières.

Pour terminer et respecter le temps d'audience qui nous est imparti, nous évoquerons brièvement et superficiellement quelques unes parmi nos autres revendications en matière d'alimentation.

1. Quelles que soient les techniques nouvelles mises au point et développées, elles ne peuvent avoir d'effets défavorables sur notre environnement ni sur la santé des travailleus appelés à les manipuler.

2. Le consommateur doit être traité en adulte et non infantilisé. Il a droit à toute l'information dont il a besoin pour exercer un choix judicieux et sage.

3. Les organisations de consommateurs doivent pouvoir accéder facilement à l'information. Tout comme les producteurs, elles désirent mener une lutte farouche contre les informations fausses ou tronquées. Que ne voit-on d'évaluations nutritionnelles fausses par le seul fait que les techniques actuelles mettent au point des aliments dont la composition s'éloigne fortement des tables alimentaires existantes, tout simplement par faute d'informations.

Conclusions

Mon exposé s'est voulu aussi général et aussi concis que possible. Je n'oserais pas prétendre y être parvenu mais parallèlement aux communications hautement scientifiques et techniques que nous avons entendues et entendrons durant cette semaine, j'ai cru bon tenter de donner la vision des organisations de consommateurs sur les problèmes actuels de l'alimentation. Ce n'est qu'au travers d'un échange de vues sain, honnête et objectif entre les diverses parties concernées que nous pourrons atteindre les buts que nous nous sommes fixés.

Références

1. J. Trémolières - Y. Serville - R. Jacquot - H. Dupin. Les bases de notre alimentation (1980).
2. J. Lederer - Hygiène alimentaire
3. Bulletin d'Information du Laboratoire Coopératif - n° 146 septembre 82
4. Test-Achats - Mensuel de l'Association des Consommateurs - Bruxelles

UNE NOUVELLE MODELISATION DE LA DIFFUSION AXIALE DE CHALEUR DANS UN ECHANGEUR DE CHALEUR A SURFACE RACLEE

J.F. MAINGONNAT et G. CORRIEU

Chargé et Maître de Recherches à l'I.N.R.A.
Laboratoire de Génie Industriel Alimentaire

Résumé

La diffusion axiale de chaleur dans un échangeur de chaleur à surface raclée a été étudiée en mesurant la température du produit au sein de l'appareil. Les profils longitudinaux de température du produit montrent qu'il existe une brusque augmentation de température à l'entrée de l'échangeur. Ce saut de température augmente avec la vitesse de rotation des lames, avec une diminution du débit de produit et du diamètre du rotor. La diffusion axiale de chaleur est classiquement modélisée à l'aide du piston à diffusion axiale. Les auteurs comparent les valeurs du paramètre de ce modèle qu'ils obtiennent à celles fournies par la littérature. Les corrélations entre ce paramètre et les conditions opératoires n'étant pas satisfaisantes, les auteurs proposent une relation entre le taux d'augmentation de température à l'entrée de l'échangeur et le rapport entre le nombre de Taylor et le nombre de Reynolds axial. Ce rapport permet de prendre en compte la vitesse de rotation de l'arbre, le débit de produit et le diamètre du rotor.

Abstract

The axial heat diffusion in a scraped surface heat exchanger has been studied by measuring the temperature in situ on the product itself. The longitudinal temperature profiles showed a jump just after the stream enters the exchanger. This jump increased when increasing the rotational speed and when decreasing the product flow rate and the shaft diameter. The axial heat diffusion is usually modelized by the dispersion model. The authors compared their experimental parameters values of this model vith those given by litterature. No satisfactory correlation between this parameter and operating conditions were obtained. From their experimental results, the authors correlate the temperature jump ratio at the inlet with the ratio between the Taylor number and the axial Reynolds number. This simple and original relation take into account the variations of the rotational speed, the flow rate and the rotor diameter.

1. INTRODUCTION

Les échangeurs de chaleur à surface raclée sont fréquemment utilisés dans les industries agro-alimentaires pour traiter des produits fortement visqueux ou chargés en particules. Dans ces appareils, la rotation d'un arbre muni de lames qui raclent la paroi d'échange crée des phénomènes de convection favorisant le transfert thermique au sein du produit.Cette fonction de brassage provoque également une diffusion axiale de chaleur dans l'appareil qui pénalise ses performances thermiques en diminuant la différence moyenne de température entre les fluides. Il est donc important de caractériser ce phénomène et de le corréler aux conditions opératoires afin de l'estimer aussi précisément que possible lors du dimensionnement ou de la détermination du régime de fonctionnement d'appareils industriels.

La diffusion axiale de chaleur dans un échangeur de chaleur à surface raclée a été mise en évidence de manière indirecte par la détermination de la dispersion des temps de séjour (2, 3) et de manière directe par des mesures de températures du produit dans l'appareil (4, 5, 6). Les études thermiques et en particulier les profils longitudinaux de température du produit dans l'échangeur montrent que la diffusion axiale de chaleur se traduit par une brusque augmentation de la température juste après l'entrée du produit dans l'appareil.

Les travaux cités s'accordent pour montrer que qualitativement la diffusion axiale de chaleur augmente avec la vitesse de rotation de l'arbre et avec une diminution du débit de produit. Afin d'étudier quantitativement l'influence des conditions opératoires sur la diffusion axiale de chaleur, celle-ci a été modélisée à l'aide de l'écoulement piston à diffusion axiale. Ce modèle, traduit par l'équation différentielle 1, est à un seul paramètre, D_E (la diffusivité thermique équivalente) qui est mis sous forme adimensionnelle par l'intermédiaire du nombre de Bodenstein global ($v.L/D_E$).

$$\frac{d^2\theta_{zp}}{dz^2} - Bo \cdot \frac{d\theta_{zp}}{dz} + Bo.\beta.(\theta_{mi} - \theta_{zp}) = 0 \qquad (1)$$

La résolution de cette équation permet de déterminer la température en un point en fonction de son abscisse z, du nombre d'unités de transfert, β et du nombre de Bodenstein. Inversement, la mesure de la température du produit juste après son entrée dans l'échangeur permet de déterminer le nombre de Bodenstein. Cette méthode d'identification du paramètre du modèle est utilisée par PENNEY et BELL (5) et WEISSER (8).

Dans cet article, nous utilisons également cette méthode et comparons les résultats ainsi obtenus à ceux de ces auteurs. Nous cherchons également une méthode plus simple pour corréler le saut de température à l'entrée de l'appareil aux conditions opératoires.

2. MATERIEL ET METHODE

Les essais thermiques ont été effectués sur un échangeur à surface raclée TR 13 x 30 DUPRAT disposé horizontalement, dans lequel du sirop de sucre à 68 brix a été chauffé à l'aide de vapeur. La longueur et le diamètre de la paroi d'échange cylindrique sont 0,37 m et 0,13 m. Nous avons utilisé deux rotors de diamètre différent (0,079 m et 0,106 m) équipés de deux ou quatre lames. Les débits de fluides et leurs températures à l'entrée et à la sortie de l'échangeur sont mesurés afin d'établir les bilans thermiques. L'originalité de notre dispositif expérimental réside dans la mise en oeuvre sur chaque rotor d'un système de capteurs permettant de mesurer la température du produit en cinq points le long de l'échangeur avec une précision de ± 0,2°C. Une description plus détaillée de ces dispositifs expérimentaux et des conditions opératoires est fournie par ailleurs (6). Nous avons étudié les variations de la diffusion axiale de chaleur en fonction des paramètres suivants :

. diamètre du rotor (0,106 m et 0,079 m)
. débit de produit ($3,6.10^{-2}$ kg/s à $11,3.10^{-2}$ kg/s)
. vitesse de rotation de l'arbre ($1,67\ s^{-1}$ à $6,67\ s^{-1}$)

. nombre de lames (2 ou 4).

3. RESULTATS ET DISCUSSION

3.1. Profils longitudinaux de température

Les programmes thermiques des essais n'étant pas rigoureusement identiques, nous présentons ces profils sous la forme du taux d'augmentation de température en un point rapporté à l'augmentation totale, en fonction de l'abscisse adimensionnelle de ce point (Fig. 1). Les points de coordonnées (0,0) et (1,1) traduisent respectivement, les températures d'entrée et de sortie du produit dans l'échangeur. Les extrémités de la surface d'échange proprement dite sont figurées par les repères A et B. Nous avons représenté en traits pleins, numérotés 1 et 2, les profils longitudinaux théoriques correspondant aux écoulements de type piston dans le cas des vitesses de rotation extrêmes.

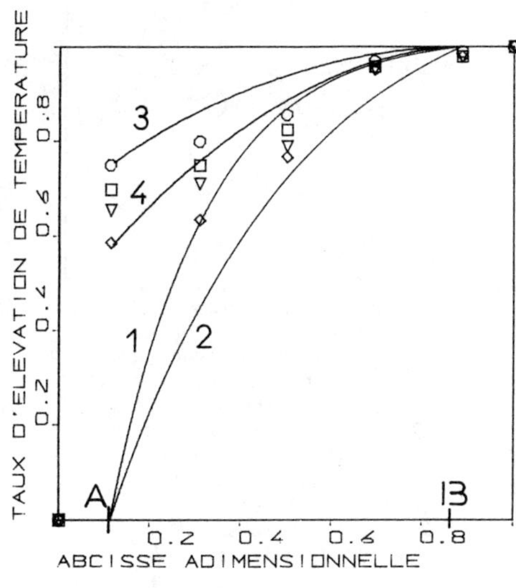

Figure 1 :

Profils longitudinaux de température pour différentes vitesses de rotation.

○ : 5,8 s^{-1} ; □ : 4,17 s^{-1} ; ▽ : 3,33 s^{-1} ; ◇ : 2,5 s^{-1} ; ($\dot{m}p$ = 5,5.10^{-2} kg/s ; d_r = 0,079 m).

Courbes 1 et 2 : écoulement piston

Courbes 3 et 4 : écoulement piston à diffusion axiale.

La principale différence entre profils expérimentaux et théoriques réside dans la brusque augmentation de température du produit au début de la paroi d'échange. Après cette brusque augmentation, la courbure des profils expérimentaux et théoriques diminue avec le nombre d'unités de transfert. Nous considérons, par conséquent, que le saut de température constitue une bonne caractérisation de la diffusion axiale de chaleur dans notre appareil. Conformément aux résultats fournis par la littérature, le saut de température à l'entrée de l'échangeur augmente avec la vitesse de rotation des lames et une diminution du débit de produit. Nous avons également montré qu'une augmentation de la taille de l'entrefer entraînait une augmentation du saut de température.

3.2. Modélisation à l'aide du piston à diffusion axiale

PENNEY et BELL (5) et WEISSER (8) comparent leurs résultats à ceux que CROOCKEWITT, HONIG et KRAMERS (9) ont obtenu sans lame. Dans cette présentation des résultats (Fig. 2 et 3), la diffusivité équivalente est mise sous forme adimensionnelle en la rapportant au produit d'une vitesse et d'une longueur caractéristiques de l'écoulement rotationnel, en l'occurence, la vitesse en bout de lame (v_L) et le diamètre hydraulique de l'espace annulaire ($d_{mi} - d_r$). Ils comparent l'évolution de ce nombre sans dimension à un nombre de Reynolds faisant intervenir les vitesse et longueur caractéristiques précédemment définies. Ces figures montrent que nos résultats sont cohérents avec ceux obtenus par WEISSER mais sont très dispersés, le coefficient de corrélation r^2 n'excédant pas 0,3 quelle que soit la configuration de l'échangeur. La dispersion de nos points et la disparité des résultats obtenus par les différents auteurs, rendent aléatoire l'utilisation de ce type de représentation pour estimer la diffusivité thermique équivalente, et par conséquent le saut de température, à partir des conditions opératoires.

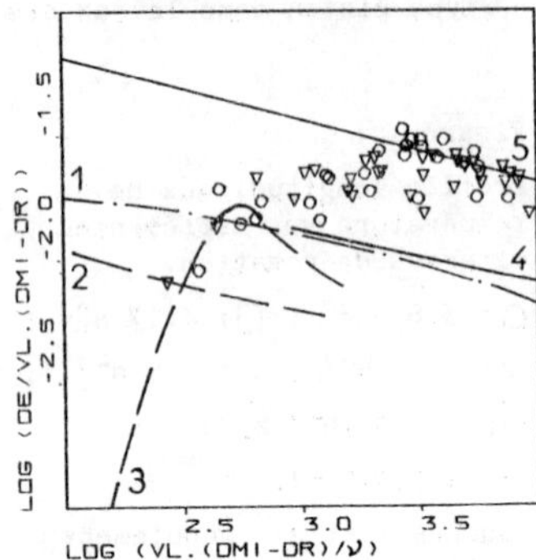

Figure 2 : Influence du nombre de Reynolds modifié sur le rapport des diffusivités $D_E/v_l.(d_{mi}-d_r)$.
○ : 4 lames ; ▽ : 2 lames ;
d_r = 0,079 m.
1 CROOCKEWITT et al (9) ;
2 et 3 PENNEY et BELL (5) ;
4 WEISSER, 2 lames (8) ;
5 WEISSER, 4 lames (8).

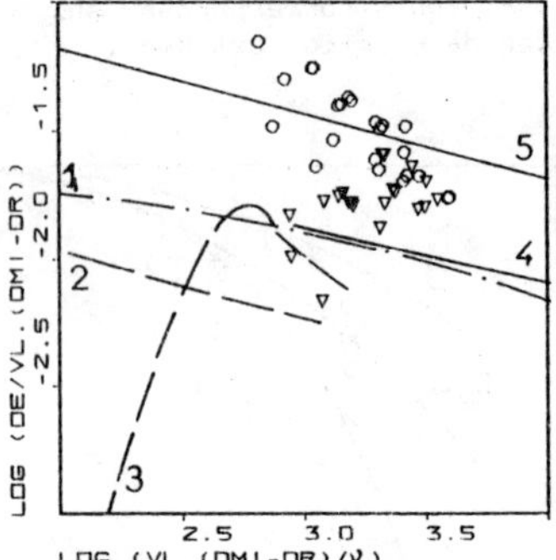

Figure 3 : Influence du nombre de Reynolds modifié sur le rapport des diffusivités $D_E/v_l.(d_{mi}-d_r)$.
○ : 4 lames ; ▽ : 2 lames ;
d_r = 0,106 m.
1 CROOCKEWITT et al (9) ;
2 et 3 PENNEY et BELL (5) ;
4 WEISSER, 2 lames (8) ;
5 WEISSER, 4 lames (8).

Par ailleurs, nous avons vu que la solution analytique de l'équation différentielle 1 permettait de tracer des profils longitudinaux de température correspondant au modèle d'écoulement piston à diffusion axiale. Après avoir identifié le paramètre de ce modèle par la méthode décrite précédemment, nous avons tracé ces profils longitudinaux. Ils sont présentés sur la figure 1, numérotés 3 et 4, pour les vitesses de rotation extrêmes. Il apparaît que les températures ainsi calculées sont systématiquement supérieures aux températures mesurées. Ceci provient probablement du fait que les hypothèses restrictives nécessaires à la résolution analytique de l'équation 1 ne sont pas vérifiées. En particulier, ni les caractéristiques physiques du produit variant le long de l'échangeur, ni les phénomènes de convection, ni le coefficient de transfert de chaleur ne sont constants le long de l'appareil.

3.3. Modélisation du saut de température à l'entrée de l'échangeur

Nous avons comparé qualitativement nos résultats aux études sur l'écoulement d'un fluide dans un espace annulaire lorsque le cylindre intérieur est en rotation (10, 11). Il semble exister un lien entre l'influence des paramètres étudiés sur la diffusion axiale de chaleur et sur la formation et l'amplitude des tourbillons toriques de Taylor qui caractérisent ce type d'écoulement. Aussi, avons nous cherché une corrélation entre le taux d'augmentation de température (relation 2) à l'entrée de l'échangeur et les nombres sans dimensions utilisés pour caractériser la structure de l'écoulement.

$$\gamma = (\theta_{op} - \theta_{ep})/(\theta_{sp} - \theta_{ep}) \qquad (2)$$

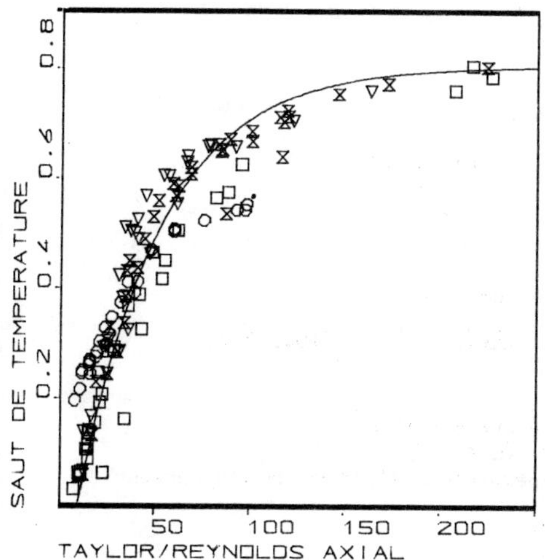

Figure 4 :

Influence du rapport Ta/Re_a sur le saut de température à l'entrée de l'échangeur.

▽ : d_r = 0,079 m et 4 lames.

⧖ : d_r = 0,079 m et 2 lames.

○ : d_r = 0,106 m et 4 lames.

□ : dr = 0,106 m et 2 lames.

La figure 4 montre la variation de γ en fonction du rapport entre le nombre de Taylor et le nombre de Reynolds axial. Ce rapport est une fonction de trois facteurs : la vitesse de rotation, la vitesse axiale moyenne du produit et un terme prenant en compte la variation du diamètre du rotor. Le coefficient de corrélation r^2 entre la relation 2 et les points expérimentaux est égal à 0,85.

$$\gamma = 0{,}8 - e^{-2{,}33\ 10^{-2}.Ta/Re_a} \qquad (2)$$

Par ailleurs, la figure 4 semble indiquer que le nombre de lames (2 ou 4) a peu d'effet sur la diffusion axiale de chaleur.

4. CONCLUSION

La diffusion axiale de chaleur dans un échangeur de chaleur à surface raclée a été mise en évidence à l'aide de mesures de température du produit au sein de l'appareil. Cette diffusion caractérisée par une brusque augmentation de la température du produit à l'entrée de l'échangeur, augmente avec la vitesse de rotation des lames, avec une diminution du débit de produit et une réduction du rapport entre le diamètre du rotor et celui de la paroi d'échange. La diffusion axiale de chaleur est classiquement modélisée à l'aide du piston avec diffusion axiale. Nos résultats montrent que les corrélations fournies par la littérature entre le paramètre de ce modèle, calculé à partir du saut de température à l'entrée et les conditions opératoires sont inapplicables à notre échangeur. Par ailleurs, les profils longitudinaux de température prévus par ce modèle ne correspondent pas aux profils expérimentaux.

Les auteurs proposent une corrélation simple et originale entre le taux d'augmentation de température à l'entrée de l'appareil et le rapport entre le nombre de Taylor et le nombre de Reynolds axial. Cette corrélation permet de caractériser la diffusion axiale de chaleur en fonction du régime de fonctionnement de l'échangeur (vitesse de rotation du rotor et débit de produit) et du diamètre du rotor.

LISTE DES SYMBOLES UTILISES

A : Surface (m^2)
Bo : Nombre de Bodenstein ($v.L/D_E$)
Cp : Chaleur spécifique (J/kg.K)
D_E : Diffusivité thermique équivalente (m^2/s)
d : Diamètre (m)
L : Longueur (m)
$\dot{m}$: Débit massique (kg/s)
N : Vitesse de rotation (1/s)
Re_a : Nombre de Reynolds axial ($v.(d_{mi} - d_r)/\nu$)
Ta : Nombre de Taylor ($2.\pi.(\frac{d_{mi} - d_r}{d_r})^{0,5} . \frac{N.d_r.(d_{mi} - d_r)}{4.\nu}$)
z : Abscisse adimensionnelle
α : Coefficient de transfert de chaleur ($W/m^2.K$)
β : Nombre d'unité de transfert ($\alpha.A/\dot{m}.Cp$)
γ : Taux d'augmentation de température à l'entrée de l'échangeur
θ : Température (°C)
ν : Viscosité cinématique (m^2/s)

Indices

e : Entrée
i : Interne
m : Paroi d'échange
p : Produit
s : Sortie
o : Point d'abscisse 0 (juste après l'entrée)

BIBLIOGRAPHIE

(1) TAYLOR, G.I. (1923). Stability of a viscous fluid contained between two rotating cylinders. Phil. Trans., A, 223, 289-343
(2) CHEN, C.Y.A. et ZAHRADNIK, J.W. (1967). Residence time distribution in a swept surface heat exchanger. Trans. ASAE, 10, n° 4, 508-511
(3) MILTON, J.L. et ZAHRADNIK, J.W. (1973). Residence time distribution in a votator plant using a non newtonian fluid. Trans. ASAE, 16, n° 6, 1186-1189
(4) BLAISDELL, J.L. et ZAHRADNIK, J.W. (1959). Longitudinal temperature distribution in a scraped surface heat exchanger. Food Technol., 13, n° 11, 659-662
(5) PENNEY, W.R. et BELL, K.J. (1969). The effect of back mixing on the mean temperature difference in a agitated heat exchanger. Chem. Eng. Progr. Symp. Ser., 65, n° 92, 21-33

(6) MAINGONNAT, J.F. et CORRIEU, G. (1983). Etude des performances thermiques d'un échangeur de chaleur à surface raclée. 2ème partie. Influence de la diffusion axiale de chaleur sur les performances thermiques. Entropie, n° 111, 37-48

(7) DANCKERTS, P.V. (1963). Continuous flow systems distribution and residence time. Chem. Engng. Sci., 2, n° 1, 1-13

(8) WEISSER, H. (1972). Untersuchungen zum Wärmeübergang im Kratzkünhler. Thèse Université de Karlsruhe.

(9) CROOCKEWITT, P., HONIG, C.C. et KRAMERS, H. (1955). Longitudinal diffusion in a liquid flow through an annulus between a stationary outer cylinder and a rotating inner cylinder. Chem. Engng. Sci., 4, N°1, 111-118

(10) KATAOKA, K., DOI, H. et KOMAI, T. (1977). Heat/Mass transfer in Taylor vortex flow with constant axial flow rates. Int. J. Heat Mass Transfer. 20, 57-63

(11) LEGRAND, J. et COEURET, F. (1982). Transfert de matière global liquide paroi pour des écoulements associant tourbillons de Taylor et circulation axiale forcée. Int. J. Heat Mass Transfer, 25, n°3, 345-351

HEAT TRANSFER IN A SCRAPED SURFACE HEAT EXCHANGER DURING COOLING OF SOME HIGHLY VISCOUS FOOD PRODUCTS

L.B.J. VAN BOXTEL and R.L. DE FIELLIETTAZ GOETHART
Institute CIVO-Technology TNO, Zeist, The Netherlands

Summary

Heat transfer in a water-cooled scraped surface heat exchanger (Crepaco model VT-422) has been investigated. The overall heat transfer coefficient in the heat exchanger is composed of three elements: heat transfer coefficient in the coolant jacket, resistance to heat flow in the separation wall and heat transfer coefficient inside the scraped cylinder. In contrast with studies published elsewhere, heat transfer was investigated with three regular non-newtonian food systems with a complicated and unknown flowing behaviour at higher shear rates. For the three starch-based food products studied (starch content 12-18 %) the heat transfer coefficients inside the scraped cylinder were measured for shaft speeds ranging from 1.67 to 10 revolutions/s. The experimental results were compared with heat transfer coefficients calculated with a model based on the penetration theory. For the starch-based products, in general, no consistent interactions between mass flow rates and internal heat transfer coefficients were observed. The results obtained showed that for roux correction factors of 0.24 and for ragout of 0.33 suffice to describe the internal heat transfer coefficients with the modified heat penetration model. For the less viscous velouté sauce the correction factor increased from 0.2 to 0.5 with increasing shaft speeds. For water the correction factor equals unity in the same test equipment (2).

1. INTRODUCTION

Heat transfer to flowing products is a frequently applied operation in the food industry. Particularly in this industry it is often not practical to heat or cool certain products in conventional plate heat exchangers, tubular heat exchangers or batch kettles on account of their viscous character. Heat transfer is often impeded by the formation of a stagnant layer at the heat transferring surface and can be improved by continuously scraping the surface. In this way formation of a stagnant film is obviated and the rotation results in the scraped off layer continuously mixing with the bulk, while fresh material is moved to the heated or cooled wall. This mechanically induced convective transport improves heat transfer.
Scraped surface heat exchangers (further referred to as SSHE) are finding increasing use for thermal purposes, especially for flowing food systems with suspended solids and/or high viscosity. Very few studies have been published dealing with heat transfer characteristics in SSHE for regular, highly viscous food products.

This paper describes cooling experiments carried out with three food products, i.e. velouté sauce, roux and ragout. These are highly viscous products of non-newtonian nature; ragout even contains small pieces of solid meat. The influence of the type of product and of operating conditions on heat transfer at the scraped wall side were items of interest in this study. Data concerning the heat transfer in the test equipment from the cooling water to the scraped wall have been published elsewhere (2).

2. HEAT TRANSFER CALCULATIONS

2.1 General

The physicotechnical aspects of heat transfer, design considerations and applications of SSHE's are described by several authors (1, 3,4, 6, 7, 8, 9, 11). A recent literature survey of the several equations for estimation of the heat transfer at product side is given by Härröd (5). The most elaborate investigations on heat transfer in SSHE's using newtonian flowing glycerol-water mixtures are those reported by Skelland et al. (8) and Trommelen (10). The basic principles of heat transfer can be applied to a SSHE just as well as to other forms of heat transfer equipment. Assuming that a plug flow consists of coolant and product, the heat flow from scraped side to cooling jacket is defined with the formula:

$$q = U.A.\Delta T_{ln} \qquad [1]$$

In this relation the heat transferring area A is based on the logarithmic mean radius R_{ln} of the cylindrical heat transferring wall.
As R_{ln}/R_i and R_{ln}/R_e approach one for the SSHE under study, it is justified to express the overall heat transfer coefficient U without significant loss in accuracy as:

$$\frac{1}{U} = \frac{1}{h_s} + \frac{x}{k_w} + \frac{1}{h_j} \qquad [2]$$

Using ΔT_{ln} in formula [1] implies that in the evaluation of the overall heat transfer coefficient U axial dispersion or backmixing effects are neglected.

2.2 Physical aspects of heat transfer at product side

Skelland et al. (8) performed an extensive experimental study on heat transfer in a Votator. Starting from a dimensional analysis they determined the dimensionless groups which should be included in a correlation for heat transfer at product side. Their cooling experiments with glycerol-water mixtures could be correlated by means of the dimensionless equation:

$$\frac{h_s \cdot d_t}{k_p} = \alpha \left(\frac{c_p \cdot \mu_p}{k_p}\right)^{\beta} \cdot \left(\frac{(d_t - d_s) \cdot v_p \cdot \rho_p}{\mu_p}\right)^{1.00} \cdot \left(\frac{d_t \cdot N}{v_p}\right)^{0.62} \cdot \left(\frac{d_s}{d_t}\right)^{0.55} \cdot n^{0.53} \qquad [3]$$

where for cooling viscous liquids α is 0.014 and β is 0.96 and for cooling thin mobile liquids α is 0.039 and β is 0.70. The formulated equation is not satisfactory in all respects; the choice of the dimensionless groups is rather arbitrary and the determination of the exponents is debatable. Moreover, the influence of the heat conductivity on the heat transfer coefficient is of minor significance in the relationship ($h_s \sim k_p$-0.04), although the mechanism of heat transfer for viscous fluids will be largely based on conduction (9, 10). A more fundamental approach is given by Latinen (7) and Harriott (4). Their model is based on the penetration theory from which they derived a theoretical relationship for the heat transfer coefficient:

$$h_s = 1.13\ (k_p.\rho_p.c_p.N.n)^{0.5} \qquad [4]$$

Application of this correlation in experiments with viscous fluids showed deviations between theoretical and experimental results (10). This inevitably leads to the mechanism being more complicated than suggested by the penetration model. For water the theoretical and experimental values did match, while for more viscous fluids the experimental values were lower than expected. Trommelen (10) suggested describing the heat transfer by an equation that results from the penetration theory modified by an empirically determined correction factor ϕ:

$$h_s = 1.13\ (k_p.\rho_p.c_p.N.n)^{0.5}.\phi \qquad [5]$$

2.3 Heat transfer coefficient in the water-cooled jacket

The heat transfer coefficient at coolant side in SSHE's was calculated with the Dittus Boelter formula by other authors (6, 8, 10):

$$Nu = 0.0225\ Re^{0.8}\ .\ Pr^{0.4} \qquad [6]$$

in which the characteristic length in Nu and Re is the hydraulic diameter. Originally this formula was developed for calculating film heat transfer coefficients in tubular heat exchangers under turbulent flow conditions. The above mentioned authors used this relationship also for estimating heat transfer coefficients for the helical path in the cooling jacket. Studies of heat transfer with water as coolant in the jacket of the SSHE used in our experiments will be published elsewhere (2). A better fit for heat transfer in the watercooled jacket was obtained with the empirical equation [7], which is used for estimating the h_j values in the reported experiments:

$$Nu = 0.0158\ Re^{0.8}.Pr^{0.4} + 18.2 \qquad [7]$$

3. MATERIALS AND METHODS

3.1 Equipment

The heat transfer measurements were carried out with a vertical type scraped surface heat exchanger (CREPACO VT-422; Crepaco Inc., Chicago, USA). Cooling water and product were fed into the SSHE at the bottom section and flowed co-currently to the topsection. Two rows of scraper blades were mounted on the shaft, which was driven by a hydraulic motor with an infinitely variable speed. Details of the SSHE-equipment are listed in Table I.

Table I Specifications of the water-cooled tube of the Crepaco VT-422 scraped surface heat exchanger

shaft diameter (d_s)	:	38.0×10^{-3} m
internal diameter scraped wall ($d_t=2R_i$)	:	98.0×10^{-3} m
external diameter scraped wall ($2R_e$)	:	101.6×10^{-3} m
thickness heat transferring scraped wall (x)	:	1.8×10^{-3} m
axial length scraped wall (L)	:	510.0×10^{-3} m
logarithmic mean area heat transferring wall (A)	:	0.16 m^2
rows of mounted scraper blades	:	2
jacket cooling water helix: depth	:	4.2×10^{-3} m
width	:	80.0×10^{-3} m
hydraulic diameter jacket cooling helix (d_h)	:	8.0×10^{-3} m
thermal conductivity wall material	:	16 W/mK

The number of revolutions of the shaft could be varied between 0 and 10 rev/s and was measured with a flashlight stroboscope. Tap water was used as coolant. The product for the cooling experiments was prepared in a cooking vessel. From there it was pumped by means of a variable hydraulically driven lobe pump through an automatically controlled, steam-heated SSHE-cylinder connected in series to heat the product to about 70-95 °C before entering the water-cooled cylinder under investigation. Temperatures of cooling water and product inlet and outlet were measured in line with copper-constantan thermocouples and registered on a temperature-recorder (Philips PM8235-multipoint recorder). Mass flow rate of cooling water and product were measured by weighing a suitable amount having passed through the SSHE for a certain time. When the steady state for an experimental run had been established, mass flow rates, shaft speed, inlet and outlet temperatures were recorded. Heat flow was calculated from enthalpy changes in coolant. The overall heat transfer coefficient was determined with formula [1].

3.2 Product formulations and physical properties

For the experimental work the following products were used: velouté sauce, roux and ragout. The composition and the relevant physical properties of the three food systems are given in Table II. The density and the specific heat for the starch products were calculated on the basis of literature values found for the recipe components water, flour, fat and beef. The thermal conductivities were obtained from heat penetration tests (temperature range 20-80°C), carried out by the authors. The starch-based products were prepared as follows: fat was melted and heated to 80°C in a steam-heated, mixing cooking vessel. Flour was added and mixed to a homogeneous mass with the melted fat; hot water and in the case of ragout the meat cut in small pieces were added; the mass was heated to 95°C and kept at this temperature for at least 4 minutes before starting the cooling experiments.

After cooking the food systems have a firm consistency due to gelatination of the starch. Velouté sauce, the most mobile of the three, showed a pronounced pseudoplastic flowing behaviour. The apparent viscosity measured in a rotoviscosimeter (Contraves Rheomat 15 T) at

Table II Composition and physical properties of the food products

	composition (%)				physical properties		
	water	flour	fat	meat*	ρ_p(kg/m^3)	c_p(J/kg K)	k_p(W/mK)
velouté sauce	82.0	12.0	6.0	-	1000	3760	0.56
roux	73.0	18.0	9.0	-	1000	3550	0.50
ragout	66.4	16.3	8.2	9.1	1000	3520	0.50

* cooked beef cut in small pieces.

60°C ranged from 18.0 Ns/m^2 at a shear rate of 2.927 s^{-1} to 2.6 Ns/m^2 at a shear rate of 59.22 s^{-1}. Although for roux and ragout no reliable viscosities could be measured in the rotoviscometer, they are in any case considerably higher than for velouté sauce.

3.3 Levels of experimental conditions

The ranges of the operating conditions and some calculated axial and rotational Reynolds numbers are summarized in Table III. The ranges of the product inlet and outlet temperatures are also included in order to give an idea of the performance of the cooling section of the SSHE. Due to lack of clearness with respect to the viscosity behaviour inside the SSHE the axial and rotational Reynolds numbers for the starch-based food products are very uncertain. For velouté sauce these values are estimated on the basis of a measured apparent viscosity at 60°C and a shear rate of 59.22 s^{-1}. For roux and ragout they are unknown, but will undoubtedly be lower than for velouté sauce.

Table III Ranges of experimental conditions

experimental condition	product/food system		
	velouté sauce	roux	ragout
mass flow rate product (kg/s)	0.032-0.114	0.017- 0.112	0.035- 0.103
shaft speed (s^{-1})	1.67 -9.92	1.67 - 8.83	1.67 - 8.33
product inlet temperature (°C)	95	70 -80	70 -82
product outlet temperature (°C)	29.3 -83.4	19.0 -61.0	29.2 -62.4
axial Reynolds number Re_a	0.1 - 0.4	-	-
rotational Reynolds number Re_r	6 -37	-	-

4. RESULTS AND DISCUSSION

4.1 Scraped-side heat transfer coefficients

During the experimental work with the food products some difficulties arose. Due to the heavy consistency, the product was not completely mixed when leaving the SSHE, which made it impossible to accurately measure the average outlet temperature. In some experiments also a part of the fat-fraction of the product crystallized when the temperature sank to below 40 °C. For reasons already mentioned, the overall heat transfer coefficients were obtained from the quantity of heat taken up by the coolant. Knowing q, the overall heat transfer coefficients were calculated with formula [1]. A moderate cooling water flow rate of about 0.2 kg/s was used to allow the temperature difference to be measured with sufficient accuracy. In the experiments the heat transfer coefficients at the scraped surface were calculated, using formula [2]. The values for the heat transfer coefficient in the jacket -h_j- were obtained from equation [7]. The values for the empirical correction factor ϕ in formula [5] followed from the ratio experimental h_s-value and those expected on account of the penetration theory.

4.2 Velouté sauce

The heat transfer coefficients at the scraped surface increased for this product at the different mass flow rates from 500 W/m^{2}K at a shaft speed of 1.67 rev/s to 3150 W/m^{2}K at 10 rev/s. The empirical correction factor ϕ as plotted in Figure 1 was not constant in the explored range of operational conditions. It increased from 0.2 at 1.67 rev/s to 0.5 at 10 rev/s. This implies that the influence of shaft speed on heat transfer is stronger than to be expected from the penetration theory. This phenomenon may be accounted for by a gradual transition of the flow type in the SSHE. At low shaft speeds a Couette flow type is likely to occur. This is a simple shear flow in which the radial velocity component is zero outside the area of the scraper blades. If shaft speed increases Taylor vortices may gradually start to develop, introducing an extra radial transport (5, 10, 12). No clear interaction effect of mass flow rate and shaft speed could be observed (Figure 1).

4.3 Roux

Roux has a firmer consistency than velouté sauce. This results in lower heat transfer coefficients, which increased from 670 to 1330 W/m^{2}K with shaft speeds increasing from 1.67 to 8.33 rev/s. A decrease in h_s with increasing mass flow rates is noticed at higher shaft speeds. The reason for this strange and inconsistent phenomenon is not clear. With ragout, a quite similar product, this effect did not occur. The empirical correction factor ϕ fluctuated between 0.2 and 0.3 (see Figure 2).The average ϕ-value amounted to 0.24. The factor $N^{0.5}$ in the penetration model is in conformity with the results of the experiments. A Couette-like flow type is most likely to occur for this product. Radial convective heat transport is probably limited, due to suppression of vortex formation. This may explain the low values for ϕ.

4.4 Ragout

The composition of ragout is almost identical to that of roux. It differs in that it contains 10 % meat, cut in small pieces. The heat transfer coefficients showed a more pronounced scattering between the different experiments. Values ranged from about 780 W/m^{2}K at 1.67 rev/s to 1900 W/m^{2}K at 8.33 rev/s. Mass flow rate had no consistent effect on the heat transfer. The average ϕ-value is 0.33 and is independent of shaft speed (Figure 3). The heat transfer coefficients, being circa 20% higher than in the experiments with roux, must be caused by the presence of the meat, which to some extent presumably disturbs the laminar Couette-flow. However, the depth of the disturbances will be limited in consequence of the small dimensions of the solid pieces.

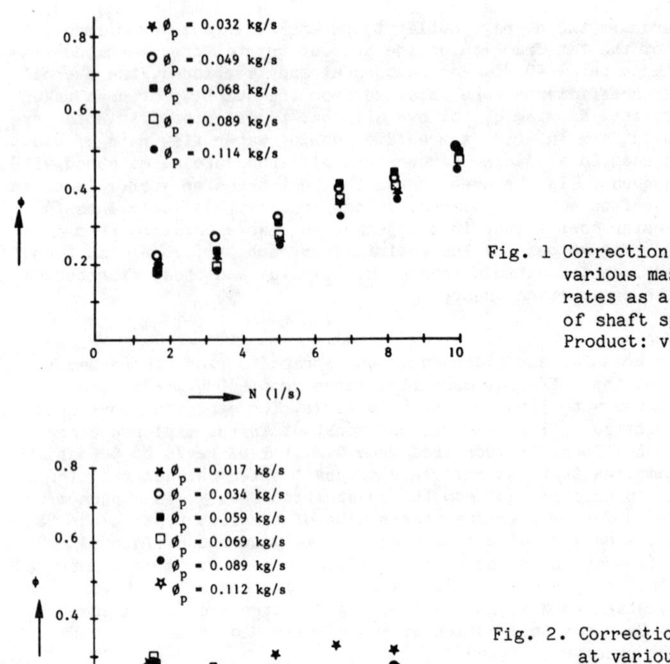

Fig. 1 Correction factor ϕ at various mass flow rates as a function of shaft speed
Product: velouté sauce

Fig. 2. Correction factor ϕ at various mass flow rates as a function of shaft speed
Product: roux

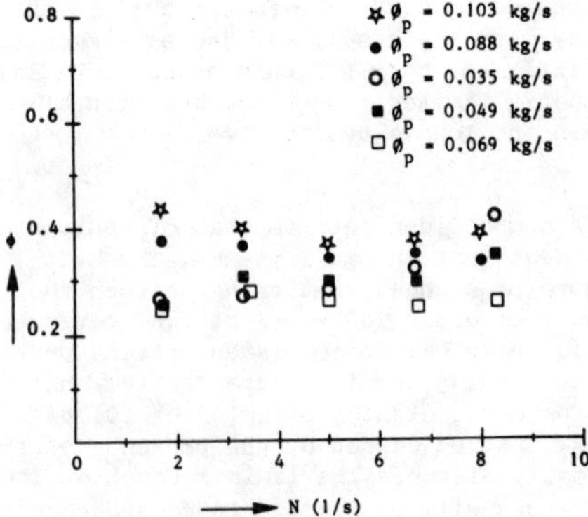

Fig. 3. Correction factor ϕ at various mass flow rates as a function of shaft speed.
Product: ragout.

5. CONCLUSIONS

The flowing behaviour of the food products used in the experiments is very complicated. This, in combination with the uncertainty with respect to local temperature gradients and shear rates inside the SSHE, made it impossible or at least doubtful to correlate the heat transfer observed to dimensionless numbers like Re and Pr as is done by several other authors mostly experimenting with newtonian model fluids. For water in the same SSHE heat transfer coefficients at scraped surface were obtained which were in conformity with the ones predicted by the penetration theory (2). For the three food products studied, the penetration theory model requires addition of an empirical correction factor, which is in all cases well below unity. For roux the experimental heat transfer coefficients increased from 670 to 1330 W/m^2K with shaft speeds increasing from 1.67 to 8.33 rev/s. With an empirical correction factor of 0.24 the data fitted to the values expected on account of the penetration theory. Heat transfer coefficients for ragout ranged from 780 to 1900 W/m^2K in the explored shaft speed range. The correction factor was independent of shaft speed and mass flow. With ϕ is 0.33 the results fitted to the penetration model. The presence of the meat had a slightly favourable effect on heat transfer. With respect to the velouté sauce, which has a weaker consistency, the situation was different. Heat transfer coefficients varied from 500 W/m^2K at 1.67 rev/s to 3150 W/m^2K at 10 rev/s. The correction factor increased in this shaft speed range from 0.2 to 0.5. Presumably, a gradual transition of the flow type had taken place (Couette flow to Taylor vortices).

In the literature various data are to be found concerning heat transfer to newtonian fluids in a SSHE. However, there is little information available on heat transfer to highly viscous non-newtonian fluid food products. This study has supplied information about heat transfer to some food products with a rather firm consistency, which will prove to be suitable for designing and engineering purposes.

NOMENCLATURE

A	logarithmic mean area heat transferring wall, m^2
c_p	specific heat product, J/kg K
d_t	inside diameter product cylinder, m
d_s	shaft diameter, m
d_h	hydraulic diameter, m
h_s	heat transfer coefficient at scraped side, W/m^2K
h_j	heat transfer coefficient in cooling jacket, W/m^2K
k_c	thermal conductivity coolant, W/mK
k_p	thermal conductivity product, W/mK
k_w	thermal conductivity wall material, W/mK
N	shaft speed, 1/s
n	number of rows of scraper blades
Nu	Nusselt-number
Pr	Prandtl-number
q	heat flow, W
Re	$vd_h\rho/\mu$ Reynolds number
Re_a	$v(d_t-d_s)\rho/\mu$ axial Reynolds number
Re_r	$Nd_t^2\rho/\mu$ rotational Reynolds number
R_e	external radius heat transferring wall, m

R_i	internal radius heat transferring wall, m
R_{ln}	logarithmic mean radius heat transferring wall, m
ΔT_c	temperature difference inlet and outlet coolant, ^oC
ΔT_p	temperature difference inlet and outlet product, ^oC
ΔT_{ln}	logarithmic mean temperature difference, ^oC
U	overall heat transfer coefficient, W/m^2K
v	average flow velocity, m/s
x	thickness heat transferring wall, m
ρ_p	density product, kg/m^3
ρ_c	density coolant, kg/m^3
ϕ	correction factor
μ	viscosity, Ns/m^2
$\emptyset_c$	mass flow rate coolant, kg/s
$\emptyset_p$	mass flow rate product, kg/s

REFERENCES

1. ANTON, J.D. (1977). The Contherm scraped-surface heat exchange. I.F.S.T. Proceedings 10 (3), 137-142.
2. BOXTEL, L.B.J. van, and FIELLIETTAZ GOETHART, R.L. de (1983). Heat transfer to water and some highly viscous food systems in a water-cooled scraped surface heat exchanger. Submitted for publication to J. Food Process Eng. (1983).
3. CUEVAS, R. and CHERYAN, M. (1980). Heat transfer to a liquid food system in a scraped-surface heat exchanger. In: Foods Processing Engineering, Vol 1 (P. Linko, Y. Mälkki, J. Olkku and J. Larinkari, eds) 506-510, Applied Sciences Publishers, Ltd., London.
4. HARRIOTT, P. (1959). Heat transfer in scraped-surface exchangers. Chem. Eng. Prog. Symp. Ser 55 (29), 137-139.
5. HÄRRÖD, M. (1982). Scraped surface heat exchangers. Literature and market survey. SIK:s Service Serie no. 718, Sweden.
6. HOULTON, H.G. (1944). Heat transfer in the Votator. Ind. Eng. Chem. 36, 522-528.
7. LATINEN, G.A. (1959). Discussion of the paper "Correlation of scraped film heat transfer in the Votator". Chem. Eng. Sci. 9, 263-266.
8. SKELLAND, A.H.P., OLIVER, D.R. and TOOKE, S. (1962). Heat transfer in a water-cooled scraped-surface heat exchanger. Br. Chem. Eng. 7, 346-353.
9. TROMMELEN, A.M. (1967). Heat transfer in a scraped-surface heat exchanger. Trans. Inst. Chem. Eng. (London) 45, T176-T178.
10. TROMMELEN, A.M. (1970). Physical aspects of scraped-surface heat exchangers. Thesis, University Delft, The Netherlands.
11. TROMMELEN, A.M., BEEK, W.J. and WESTELAKEN, H.C. van de (1971). A mechanism for heat transfer in a Votator-type scraped surface heat exchanger. Chem. Eng. Sci. 26, 1987-2001.
12. TROMMELEN, A.M. and BEEK, W.J. (1971). Flow phenomena in a scraped-surface heat exchanger (Votator-type) Chem. Eng. Sci. 26, 1933-1942.

CRUST FORMATION DURING BAKING OF MEAT PRODUCTS IN A CONVECTION OVEN AND ITS EFFECT ON THE FORMATION OF MUTAGENIC SUBSTANCES

HOLTZ, E., Lund University, Div. of Food Engineering, P.O.B. 50, S-230 53 Alnarp, Sweden.
NILSSON, U., Lund University, Div. of Applied Nutrition, P.O.B. 740, S-220 07 Lund, Sweden.

INTRODUCTION

Baking in a convection oven is defined according to the COST 91 terminology as "cooking of food in an oven in air to which water vapour may or may not be added". During the baking process of meat products a crust is formed, which includes e.g. the browning reaction with the aroma as positive aspect. With the formation of the crust follows also negative aspects as formation of mutagenic substances. The formation of mutagenic substances in meat products during the frying process has been studied a lot in the literature (1).

This collaborative work was supposed to make a link from the technical to the nutritional points of view, due to the different knowledge of the two institutes - Div. of Food Engineering and Div. of Applied Nutrition at Lund University, Sweden. In a pilot-plant convection oven (2) it was possible to bake the products under well controled circumstances. We wanted to survey in more detail how the parameters in the heating medium influence the formation of mutagenic substances and also try to explain the different mechanisms that are important. In order to optimize a process both technical and nutritional aspects have to be included. Work is proceeding at the Div. of Food Engineering in order to make a mathematical model regarding heat and mass transfer in a minced meat mixture system. Using this model it is possible to simulate the temperature profile in the product. Quality aspects have to be included. One way is to make kinetic studies on the quality changes. In this work the Arrhenius equation has been used for kinetic studies of the formation of mutagenic substances. The amount of mutagenic substances was analysed using Ames test at the Div. of Applied Nutrition (4, 5), where work concerning formation of mutagenic substances in food have been done for a long period of time.

In an earlier work at the two departments the prediction of thiamine retention in convective heated meat products was studied with help of the Arrhenius equation (6). Good correlation between the simulated and analysed values were obtained.

MATERIAL AND METHODS

In this investigation it was supposed that the formation of mutagenic substances in the meat product follows a first order reaction. Equations explained in the literature (3, 6) may be used to get the reaction rate constant and a value for the activation energy. The cook value (C_0 value) is defined as the equivalent time at a reference temperature of 100°C (compare the F-value). A Z-value can be determined from its temperature dependence which may be expressed by Equation 1 (6).

$$\log \frac{D_2}{D_1} = - \frac{T_2 - T_1}{Z_c} \qquad (1)$$

The meat loaf used, may be divided into a number of sections for which a temperature may be simulated. In this investigation we have only used the outer parts of the products due to that the mutagenic substances are mainly formed in these parts. In order to get the amount of mutagenic substances for the whole crust an integration has to be made according to Equation 2 .

$$1 \times 10^{-C/D} = \int_0^1 \frac{b}{b_0} \, dv \qquad (2)$$

From these equations the concentration of mutagenic substances may be determined.

The experiments were done in two steps, in both these a minced meat mixture was used consisting of beef (M. bicepsfemoris),lard, water, spices and potato starch. In the kinetic model evaluation,the mixture was freeze dried until a water content of about 15% was reached (a_w=0.65).

This was done in order to get a sample which would be like a crust. The samples were placed in sealed pounches and heated in an oil bath for various lengths of time at different temperatures. Kinetic data was calculated from the received results.

In the second step experiments were carried out in the described convection oven. The same meat mixture was formed into meat loaves and were baked at different temperatures and velocities of the air. The temperatures were measured with help of thermocouples (Cr-Al) except for the surface where an IR-pyrometer was used (2). The crust was peeled off and further analyzed with the Ames test to get the amount of mutagenic substances (4, 5).

RESULT AND DISCUSSSION

In figure 1 the analysed versus simulated content of mutagenic substances at different air conditions and baking time is shown.

These preliminary investigation shows good coorelation between the simulated content of mutagenic substances and the analysed. However the factor between analysed and simulated values is a little bit too high. The model has to be further developed and more accurate data have to be calculated.

The work will proceed in the near future with the model and the kinetic evaluation in order to predict the amount of mutagenic substances in baked meat products.

REFERENCES

1. HATCH, F.T., FELTON, J.S., BJELDANES, L.F. Mutagens from the cooking of food: Thermic mutagens in beef, in Carcinogens and Mutagens in the Environment, Stitch, H.F. Ed., CRC Press, Inc., Florida, USA, 1982, vol. 1. Food Products, chapter 12.
2. SKJÖLDEBRAND, C. Frying and reheating in a forced convection oven. Ph.D. Thesis, University of Lund, Lund, Sweden, 1979.
3. YEN, Y., MANSSON, J.E., STUMBO, C.R., ZAHRADNIK, J.W. A procedure for estimating sterilization of and quality factor degradation in thermally processed foods. J. Food Sci. 1971, 36, 692-698.

4. AMES, B.N., MCCANN, J., YAMASAKI, E. Methods for detecting carcinogens and mutagens with the Salmonella/Mammalianmicrosome mutagenicity test. Mutat. res. 1975, 31, 347-363.
5. JÄGERSTAD, M., LASER REUTERSWÄRD, A., OLSSON, R., GRIVAS, S., NYHAMMAR, T., OLSSON, K., DAHLQVIST, A. Creatin(in)e and Maillard reaction products as precursors of mutagenic compounds: Effect of various amino acids. Food Chemistry, 1983, 12.
6. SKJÖLDEBRAND, C., ÅNÄS, A., ÖSTE, R., SJÖDIN, P. Prediction of thiamine content in convective heated meat products. J. Food Technol. 1983, 18, 61-73.

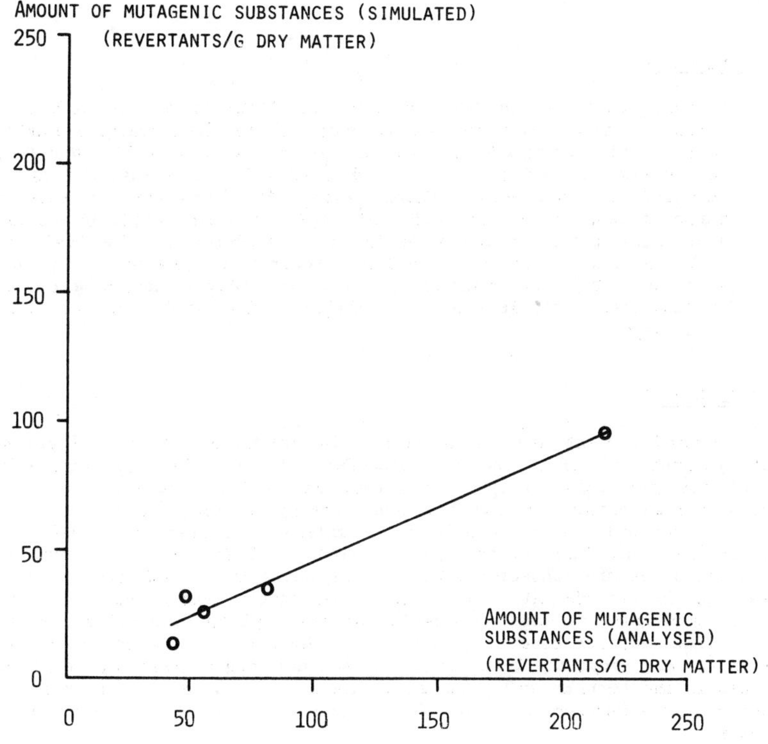

Fig 1. The simulated amount of mutagenic substances versus analysed. A linear regression line gave a correlation coefficient of 0.98.

MODELLING OF CONTHERM SCRAPED SURFACE HEAT EXCHANGER IN STARCH PROCESSING

L. HAKKARAINEN, O. MYLLYMÄKI*, J. OLKKU*† and P. LINKO
Helsinki University of Technology, Department of Chemistry,
Laboratory of Biochemistry and Food Technology,
SF-02150 Espoo 15, Finland
*Technical Research Centre of Finland, Food Research Laboratory,
SF-02150 Espoo 15, Finland

Summary

Starch gelatinization and enzymatic hydrolyzation were studied in a scraped surface heat exchanger. Wheat starch was hydrolyzed with thermostable α-amylase at a temperature range of 90°-120°C and feed solids contents of 30, 40 and 50% (w/w). The limiting factors of gelatinization appeared to change between 100°C and 105°C. The importance of temperature range of 100°-110°C for starch liquefaction was also illustrated. A state vector was developed on the basis of complete mixing and first order irreversible kinetics. A steady state at 108°C was obtained from the model developed for a chosen product feed rate. Asymptotic stability of the steady state could be calculated.

1. INTRODUCTION

Scraped surface heat exchangers (SSHE) are units fitted with blades that continuously scrape the heat transfer wall, thus being particularly useful for processing of viscous foods. Mathematical description of a reactor allows process optimization and studying of reactor behaviour. In a state vector model a state vector consisting of linear and non-linear differential equations is employed to define all states of the system. State variables are connected with mass and energy flows and their properties (1). In the present work starch gelatinization and enzymatic liquefaction were used as model processes for the SSHE in developing a state vector from reaction kinetic and other experimental data. Starch gelatinization (2) and enzymatic liquefaction were considered as reaction analogous changes in the tertiary molecular structure of starch, and the irreversible reactions were assumed to follow first order kinetics in an ideally mixed reactor.

---

†Present address: Orion Corporation Ltd, SF-02101 Espoo 10, Finland

2. MATERIALS AND METHODS

2.1. Experimental design

Wheat starch (Raisio A-starch) was hydrolyzed with Termamyl 60L α-amylase (Novo Industri A/S) using a vertically mounted Contherm model 6x3 scraped surface heat exchanger (Alfa-Laval, Sweden) fitted with four scraping blades of floating type. The material was pumped by a Mohno Ne 30 feeding pump through Contherm to a 42m long reaction tube (Ø 38/35.5 mm), with outlets at 0m, 9m and 15m points.

A set of small scale batch experiments was performed in order to determine reaction rate constants and activation energies. 60g Samples of starch or starch-enzyme suspensions (30, 40 or 50% w/w) were heated in glass tubes (37x150mm) at different temperatures (90°, 100°, 105°, 108°, 110° and 120°C) and at two enzyme levels (2ml/kg d.m., 3ml/kg d.m.) for 2 min to 2 hours. Immediately after heating 0.1% $HgCl_2$ was added to terminate the enzymatic reaction and the samples were cooled rapidly. The fitness of the kinetic constants was studied with pilot scale experiments in the Contherm.

2.2. Analytical

The moisture content of starch and freeze-dried samples was determined by drying at 130°C in an atmospheric oven for one hour, and that of the hydrolyzates by drying at 60°C in a vacuum oven for 7 hours. The degree of starch gelatinization was followed by the quantity of reducing sugars (maltose) in a freeze-dryed sample. The degree of liquefaction was measured as dextrose equivalent value (DE).

3. RESULTS AND DISCUSSION

3.1. Development of the state vector

SSHE was treated as an ideally mixed reactor where irreversible, endothermic reactions take place. If starch gelatinization is assumed as the limiting factor in hydrolysis, also enzymatic liquefaction may be regarded as a first order reaction. Mass and heat balance equations (1) and (2) were derived from those developed for continuous stirred tank reactors (3):

$$\text{Mass balance: } V\frac{dc_t}{dt} = Qc_{to} - Qc_t - kVc_t \qquad (1)$$

$$\text{Heat balance: } V\rho c_p\frac{dT}{dt} = -\Delta HkVc_t - Q\rho c_p(T-T_o) - hA(T-T_a) \qquad (2)$$

Temperature dependence of the reactions was described by Arrhenius equation. Heat transfer coefficient depends both on fluid properties and rotational speed of blades. In fluids of high viscosity a boundary layer builds up on the blade during the time between two scrapings. Because the temperature equalization is not complete, an average heat transfer coefficient was used (4). Thermal conductivity, density and specific heat were assumed to be independent of temperature. Steady state solution of the system described by eqns (1) and (2) was obtained by setting the derivatives to zero. For endothermic reactions heat generation (P_{rs}) and heat removed from the system (P_{as}) could be described by eqns (3) and (4). The intersection of the two curves plotted against temperature represents a steady state for the reactor.

$$P_{rs} = -hA(T-T_a) \tag{3}$$

$$P_{as} = Q\rho c_p(T-T_o)+\Delta HkVc_{to}/(1+k\tau) \tag{4}$$

The mass and heat balance equations were linearized and a state vector shown in eqns (5) to (7) was formed. Output and input concentrations of unreacted starch, temperature of suspension and volumetric product feed rate were chosen for variables.

$$\underline{x}(t)=\underline{A}\underline{x}(t)+\underline{B}\underline{u}(t),\ \underline{A}=\begin{bmatrix} a_{11} & a_{12} \\ a_{21} & a_{22} \end{bmatrix},\ \underline{B}=\begin{bmatrix} b_{11} & b_{12} \\ b_{21} & b_{22} \end{bmatrix} \tag{5}$$

$$\underline{x}=\begin{bmatrix} c_t \\ T \end{bmatrix} \tag{6}$$

$$\underline{u}=\begin{bmatrix} c_{to} \\ Q \end{bmatrix} \tag{7}$$

3.2. Reaction kinetics

For the development of the state vector, starch gelatinization and liquefaction were assumed to follow first order kinetics. Reaction rate constants of 0.009-0.345 min^{-1} were obtained for gelatinization from batch kinetic measurements. When the logarithm of the reaction rate constant was plotted as a function of the reciprocal absolute temperature, a change of slope was observed between 100°C and 105°C as shown in Fig. 1. The limiting factors of gelatinization appear to change at this temperature range.

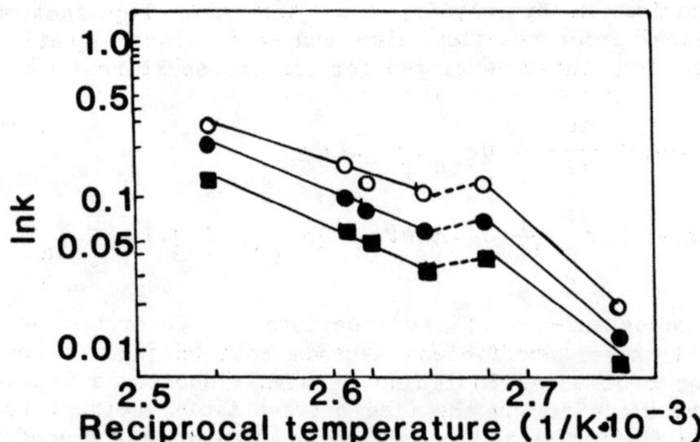

Figure 1. The logarithm of reaction rate constant for gelatinization as the function of reciprocal absolute temperature (o, 30% d.m.; •, 40% d.m.; ■, 50% d.m.).

The activation energies at 105°-$120^{o}C$ (90-100 $kJg^{-1}mol^{-1}$) were about half of the values obtained at the temperature range of 90°-$100^{o}C$. This phenomenon is typical to diffusion limited reactions (5). The reaction rate constants for liquefaction were considered valid only at process times less than 10 min. The residence time in the equipment used was shorter than 10 min at product feed rates higher than $4.6x10^{-3}m^{3}min^{-1}$. At temperature range of 100°-$108^{o}C$ the reaction rate constants were twice the value of the rate constants for gelatinization. At other temperatures the differences were even greater. Fig. 2. illustrates the importance of the temperature range of 100°-$110^{o}C$ for wheat starch liquefaction. An increase in temperature accelerates both the enzymatic degradation of gelatinized starch and the inactivation rate of the enzyme.

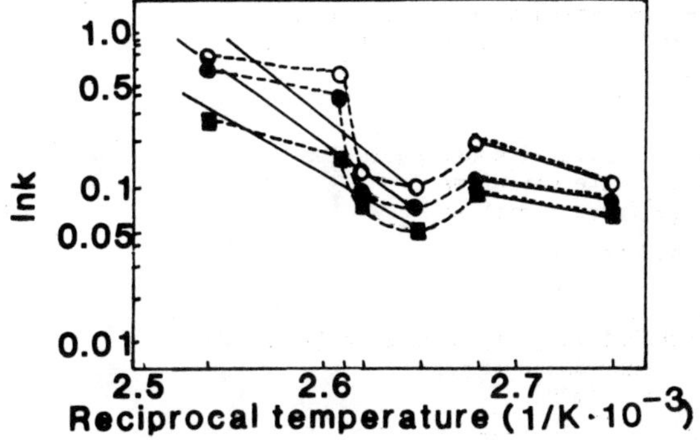

Figure 2. The logarithm of the reaction rate constant for liquefaction as the function of reciprocal absolute temperature (o, 30% d.m.; •, 40% d.m.; ■, 50% d.m.).
— assumed first order kinetics at temperature ranges of 90°-$100^{o}C$ and 105°-$120^{o}C$
--- experimental data

The activation energies obtained for 105°-$120^{o}C$ (110-160 $kJg^{-1}mol^{-1}$) were about 5 times greater than those obtained at the lower temperature range. The kinetic data for liquefaction was shown to be sufficiently close to that obtained from pilot scale SSHE runs, allowing the data to be used in the modelling. This cannot be said of the data for starch gelatinization under the experimental conditions employed.

3.3. Steady state and its stability

The steady state was solved from eqns (3) and (4) as described above. At the product feed rate of $6.9x10^{-3}m^{3}min^{-1}$ a steady state for liquefaction was obtained at $108^{o}C$. DE-values were between 16 and 27. Asymptotic stability of the steady state could be shown by calculations. The effect of changes in input and output variables on the reactor behaviour was studied by Laplace transfer functions. It could be shown that the reactor tends to stabilize relatively rapidly after temporary pertubations. As an example, a disturbance of $\Delta T=3^{o}C$ was calculated to return to the steady state in about 10 min at 30% (w/w) and in about 15 min at 50% (w/w).

4. NOMENCLATURE

A heat transfer area (m^2)
c_p specific heat of suspension (kJ/kg^oC)
c_t output concentration of unreacted starch (kg/m^3)
c_{to} input concentration of unreacted starch (kg/m^3)
h heat transfer coefficient ($kJ/min\ m^2\ ^oC$)
H heat of reaction (kJ/kg)
k reaction rate constant ($1/min$)
P_{as} heat removed from the system (kJ/min)
P_{rs} heat generation (kJ/min)
Q volumetric product feed rate (m^3/min)
T temperature of suspension (^oC)
T_o feed inlet temperature (^oC)
T_a temperature of heat transfer surface (^oC)
t time (min)
V reactor volume (m^3)

$\underline{x}$ matrix with derivated state variables
$\underline{x}$ matrix with state variables
$\underline{u}$ matrix with input variables
$\underline{A}$, $\underline{B}$ coefficient matrices

Greek letters
λ thermal conductivity of suspension (kJ/hm^oC)
ρ density of suspension (kg/m^3)
τ residence time (min)

5. REFERENCES

1. HARTMANN, K. (1978). Modellierung und Optimierung Verfahrenstechnischen Systeme. Akademie-Verlag, Berlin.
2. OLKKU, J., LINKO, P. (1977). Effect of thermal processing on cereal based food systems. In W.K. Downey ed. Food Quality and Nutrition. Applied Science Publishers Ltd, Barking.
3. COOPER, A.R., JEFFREYS, C.V. (1971). Chemical kinetics and reactor design. Oliver and Boyd, Edinburgh.
4. TROMMELEN, A.M., BEEK, W.J., VAN DE WESTELAKEN, H.C. (1971). A mechanism for heat transfer in Votator-type scraped-surface heat exchanger. Chem. Eng. Sci. 26:1987.
5. SUZUKI, K., KUBOTA, K., O'MICHI, M., HOSAKA, H. (1976). Kinetic studies on cooking of rice. J. Food. Sci. 45:1180.
6. BRUNS, D.D., BAILEY, J.E., LUSS, D. (1973). Steady state multiplicity and stability of enzymatic reaction systems. Biotechnol. Bioeng. 15:1131.

PROBLEMS IN MATHEMATICALLY MODELLING THE COOKING OF A JOINT OF MEAT

D. Burfoot and S.J. James
Agricultural Research Council, Meat Research Institute,
Langford, Bristol, BS18 7DY, U.K.

Summary

A simple mathematical model is used to demonstrate the need for reliable data when predicting the cooking times of meat. Currently, the accuracies of heat transfer coefficients and thermal properties of meat are not sufficient to enable a reliable prediction of the cooking time in a natural convection oven. Further results show that comparisons of predicted and actual cooking times require a good knowledge of the dimensions of, and fat distribution in, the meat.

1. INTRODUCTION

Relatively little meat and meat products are consumed uncooked. Cooking is the transfer of heat into the meat and the resulting temperature rise both destroys bacteria and changes the sensory properties. In the majority of investigations of cooking, heating times and weight losses have been reported for particular cuts or joints under specific conditions. It is difficult, if not impossible, to extrapolate such empirical results to other joints and/or cooking methods.

A more fundamental understanding of the cooking process in terms of both heat and mass flow must relate time and temperature to geometric and thermophysical properties of the meat and its environment. Previous studies (1,3) have revealed some of the problems of describing mathematically the complex phenomena that occur during cooking. Some difficulties are caused by the lack of data both on the thermal properties of meat, over the range of temperatures encountered during cooking, and accurate heat and mass transfer coefficients.

In this paper a very simple mathematical model is used to illustrate some of the errors that can occur in the prediction of cooking time when suitable data are either lacking or inaccurate. Although this work is related to dry air cooking in a natural convection oven, many of the conclusions are relevant to other modes of cooking.

2. MATHEMATICAL MODEL

Consider a homogeneous sphere of lean meat radius, R, initially at a constant temperature, Ti, placed in a natural convection oven operating at a constant temperature, Ta. Assuming that heat is transferred to the surface of the meat by convection from the air and by radiation from the oven walls, which behave as a black body, the heat fluxes by convection, qc, and radiation, qr, are given by:

$$qc = hc\ (Ta - Ts) \qquad [1]$$

and

$$qr = \varepsilon\ \sigma(Ta^4 - Ts^4) \qquad [2]$$

$$= hr\ (Ta - Ts) \qquad [3]$$

where hc and hr are the convective and radiative components of the surface heat transfer coefficient, Ts is the surface temperature, ε is the emissivity of the meat surface and σ is the Stefan-Boltzman constant.

The total heat flux, q, to the surface is given by

$$q = (hc + hr)\ (Ta - Ts) = h\ (Ta - Ts) \qquad [4]$$

where h is the effective surface heat transfer coefficient.

If the following assumptions are made

1. the sphere of meat has constant dimensions, properties and constituents,
2. initially, the temperature is uniform throughout the meat,
3. the surroundings are unaffected by the meat,

then:

$$\frac{\delta T}{\delta t} = \frac{k}{c\rho}\left(\frac{\delta^2 T}{\delta r^2} + \frac{2}{r}\frac{\delta T}{\delta r}\right) \qquad [5]$$

where r = radial coordinate, k = thermal conductivity, c = specific heat and ρ = density of the meat.

The boundary conditions are given by:

$$t = 0\ ;\quad r \leqslant R\ ;\quad T = Ti \qquad [6]$$

$$t \geqslant 0\ ;\quad r = 0\ ;\quad \frac{\delta T}{\delta r} = 0 \qquad [7]$$

$$t > 0\ ;\quad r = R\ ;\quad k\frac{\delta T}{\delta r} = h\ (Ta - Ts) \qquad [8]$$

Equations 5 to 8 were transposed into a finite difference scheme based on the method of Dusinberre (3). The resulting equations and their method of solution are presented by Bailey et al. (4).

The standard solution against which the effect of changes in properties, dimensions and new assumptions were considered was as follows. A 14 cm diameter sphere of lean meat with a uniform temperature of 5°C was instantaneously subjected to naturally convected air at 175°C and the change in centre temperature with time was calculated up to 74°C. The thermal properties of the meat (5) were calculated using:

$$k = 0.080 + 0.52x \quad W/m°C \qquad [9]$$

$$c = 1672 + 2508x \quad J/kg°C \qquad [10]$$

and

$$\rho = 890 + 110x \quad kg/m^3 \qquad [11]$$

where x = mass fraction of water = 0.75.

The coefficient, hc, was calculated using the correlation for laminar free convection around a smooth surface (6):

$$hc = 4.17\left(\frac{Ta - Ts}{L}\right)^{0.25} \quad W/m^2\,°C \qquad [12]$$

where L = R (cm) for a sphere (7).

An average hc of 8.4W/m²°C was calculated using Ts equal to 5 and 100°C in equation 12 and taking the mean of the two results. The same values of Ts were used in equation 2 with ε = 0.74 (5) to calculate an average hr of 10.1W/m²°C. Substituting both values into equation 4 produced a mean value for h of 18.5W/m²°C. Figure I shows the result of inserting these values into the computer program; the time needed to raise the centre from 5 to 74°C is 118 minutes. For interest a typical formula found in home economics books was 55 minutes per kg of meat + 25 minutes which results in a cooking time of 102 minutes.

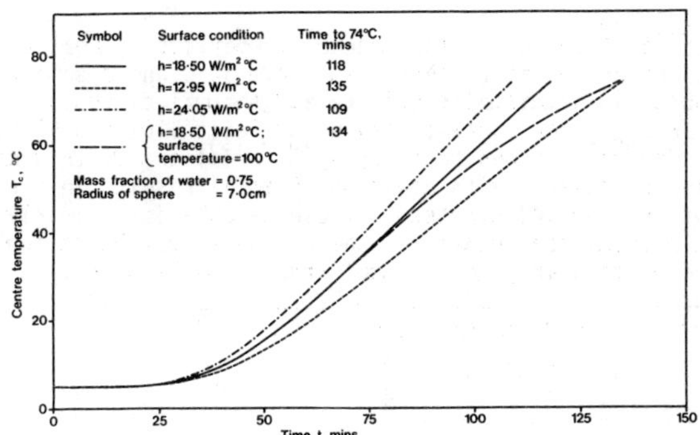

Figure I. The effect of surface boundary conditions on the predicted time-temperature histories within a sphere of meat roasted in a natural convection oven at 175°C

In the following sections the effect on the cooking time of variations and inaccuracies in the parameters present in equations 5 to 8 are examined. A number of features present in practice, but not accounted for in the model, are also discussed.

3. FACTORS WITHIN THE MODEL

3.1 Surface boundary conditions

The correlation used to estimate the convection heat transfer coefficient applies to bodies with simple geometries and smooth surfaces. These ideal conditions are rarely satisfied during meat cooking and further complications are caused by the transfer of heat along supporting structures. The emissivity of the meat's surface is not accurately known, the only values contained in the COST 90 compilation (5) are 0.74 and 0.9 for uncooked lean meat. No values were recorded for cooked beef. McAdams (6) states that convective surface heat transfer coefficients can only be predicted to an accuracy of ±20%. It is therefore unlikely that the effective coefficient during cooking can be predicted to within ±30%.

The effect on the time-temperature relationship of modifying the effective surface heat transfer coefficient by ±30% is shown in Figure I and results in cooking times to 74°C of 109 and 135 minutes.

Initially during roasting an imbalance exists between the flow of heat to the surface and from the surface to the centre of the meat: this is sustained by the evaporation of water. Some of this loss is replenished by the simultaneous diffusion of water outwards through the meat. However, for the conditions considered in this work, the rate of evaporation exceeds the diffusion rate and the boiling front slowly recedes into the meat, thereby forming a dry crust region at the surface. Although the rate of formation of this region has been investigated with meat loaves (2) and lean muscle (1), the thermal properties of such a crust on a meat joint are as yet unknown. While this precludes a detailed calculation of the effect of evaporation, a simple estimation was obtained by modifying the computer model such that the surface temperature was held at 100°C after it had reached that value. Figure I shows that this modification elongated the time-temperature curve with the result that the predicted cooking time increased from 118 minutes to 134 minutes.

3.2 Thermal properties

Morley's compilation (8) of the thermal properties of meat contains few data on the properties at temperatures above 30°C and a complete lack of data above 65°C. Thermal conductivity and thermal diffusivity for lean meat of near 75% water content vary by at least ±10% and ±20%, respectively. Figure II shows that applying variations of this magnitude to the properties calculated with equations 9 to 11 results in predicted cooking times of 102 and 142 minutes. These calculations clearly show that in using a mathematical model the thermal data must be reliable and pertinent to the particular piece of meat under consideration.

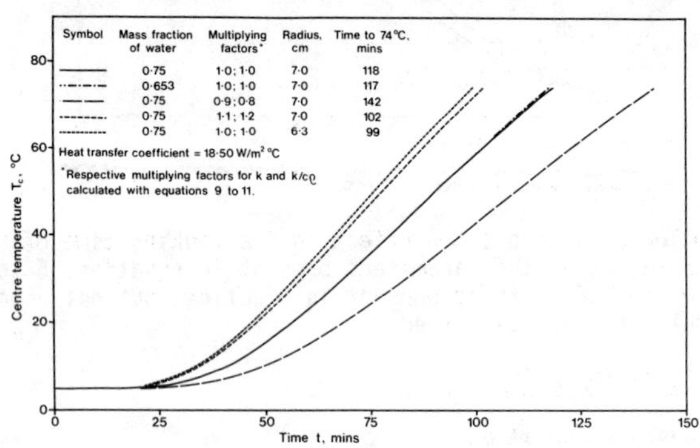

Figure II. The variations of predicted time-temperature histories caused by water losses, errors of the thermal data and dimension changes of a sphere of meat roasted in a natural convection oven at 175°C.

Figure II also demonstrates that the cooking time is affected only slightly by a 9.7% reduction of the water content of the sphere. This water loss was derived by assuming that all of the losses from the sphere were water and the total loss was 28%. This weight change was found when cooking a bovine M. semitendinosus as described in the following section.

3.3 Dimensional changes

Many papers report that meat shrinks during cooking, although dimensional changes have seldom been recorded. A simple experiment was undertaken to indicate the magnitude and regularity of dimensional changes during cooking. At various intervals along a bovine M. semitendinosus a small bore thin walled tube was pushed into the meat and lead spheres (1.1 mm diameter) were inserted so that a planar lattice of the spheres was produced. X-ray scans were taken before and after cooking the muscle at 175°C. The percentage movements of the spheres in each of the two directions of the plane are shown in Figure III. Shrinkage was greatest in the outer regions of the muscle with percentage changes varying from 4 to 46% throughout the muscle and the overall longitudinal movement was 20%. The rate of shrinkage was not measured but an estimate of the effect of shrinkage on cooking time was made by assuming that the radius of the sphere was immediately reduced by 10% at the start of cooking and remained at that value throughout the process. Figure II shows that the cooking time was reduced by 16% to 99 minutes.

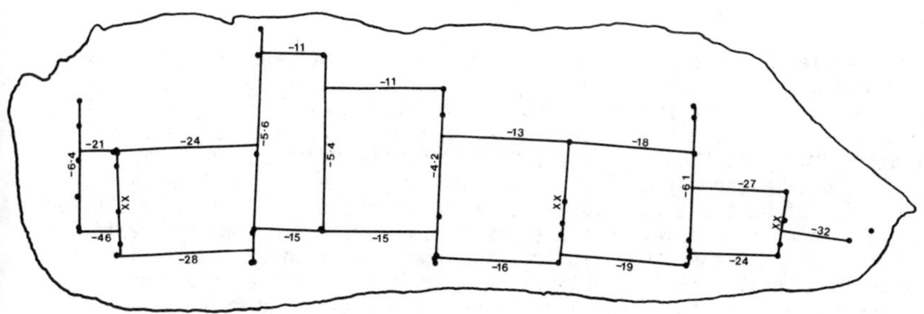

Figure III. Differential shrinkage within a bovine semitendinosus muscle cooked at 175°C. The data represent the percentage dimension changes along the indicated lines. The symbols, •, denote the final positions of the lead spheres. The symbols, xx, indicate that the dimension changes were insuffient for reliable determination.

4. FACTORS NOT CONSIDERED IN THE MODEL

A number of factors occur in practice that are not considered in the model, some of these are discussed briefly in the following sections.

4.1 Heats of reaction and melting

Although the measurement of the heats of reaction using differential scanning calorimetry has inherent problems (9), Karmas and Dimarco (10) used the method to obtain an estimate of 0.88kJ/kg for the enthalpy of reaction of lean beef muscle during cooking. This is less than 0.5% of the energy required to raise the centre temperature of the sphere from 5 to 74°C and therefore can be safely neglected in most models.

Lean muscle contains intermuscular fat and most commercial meat joints also contain extramuscular fat. The heat of liquefaction of fat has been found to be 112 kJ/kg (11), therefore the heat required to melt the fat in a joint with a 10% fat content represents approximately 5% of the total energy requirement. Since the thermal conductivity of solid fat is less than half of that of lean, the amount and position of fat could also affect the temperature distribution and/or cooking time.

4.2 Movement of water

The movements of water and heat were described earlier (Section 3.1) when considering the formation of a crust. A further consequence of the diffusion of water outwards from the cooler central regions is that this movement transfers heat and the resultant effect is comparable to increasing the thermal conductivity of the meat. This diffusion is further supplemented by the hydraulic force generated as each area of muscle contracts as it reaches the shrinkage temperature of collagen. Insufficient data is available to enable a thorough analysis of this phenomenon but it is expected that the low water flow rates and the small temperature difference, between the centre of the meat and the evaporation zone, will result in small transfers of heat relative to the overall heating requirement.

5. CONCLUSIONS

The use of a very simple mathematical model indicates that

deficiencies of data on many of the factors that influence the rates of heat and mass flow make it unlikely that cooking times can be predicted to an accuracy of better than ±20%. For simplicity, the combined effects of the errors have not been considered in this analysis, and the actual error in the predicted time may therefore be greater than 20%.

Complex mathematical models are being developed to simulate the diffusion of heat and mass during cooking and these will provide a clearer insight into the complex interactions that occur. Our simple investigation indicates that the practical applications of these models will be restricted until basic reliable data are available on heat and mass transfer coefficients, thermal properties and rates of diffusion and shrinkage.

REFERENCES

1. BENGTSSON, N.E., JAKOBSSON, B. and DAGERSKOG, M. (1976) J. Food Sci. 41, p1047.
2. SKJOLDEBRAND, C. (1979) "Frying and reheating in a forced convection oven", Ph.D. Thesis, University of Lund, Lund, Sweden.
3. DUSINBERRE, G.M. (1949) "Numerical analysis of heat flow", 1st Ed., McGraw-Hill, New York.
4. BAILEY, C., JAMES, S.J., KITCHELL, A.G. and HUDSON, W.R. (1974) J. Sci. Food Agric. 25, p81.
5. MILES, C.A., VAN BEEK, G. and VEERKAMP, C.H. (1981) "Calculations of the thermal properties of food", Paper presented at the final COST 90 seminar, Leuven, Belgium.
6. McADAMS, W.H. (1954) "Heat transmission", 3rd Ed., McGraw-Hill, New York.
7. KING, W.J. (1932) Mech. Eng. 54, p347.
8. MORLEY, M.J. (1972) "Thermal properties of meat: tabulated data", Special Report No. 1, Meat Research Institute, Langford, Bristol, UK.
9. WRIGHT, D.J., LEACH, I.B. and WILDING, P. (1977) J. Sci. Food Agric. 28, p557.
10. KARMAS, E. and DiMARCO, G.R. (1970) J. Food Sci. 35, p725.
11. POLLAK, G.A. and FOIN, L.C. (1960) Food Technol. 14, p454.

ON THE FORMATION OF "RESISTANT STARCH" DURING BAKING

C.-G. JOHANSSON, N.-G. ASP, I. BJOERCK, J. HOLM and M. SILJESTROEM
Department of Food Chemistry, Chemical Center, University of Lund
P.O.Box 740, S-220 07 LUND, Sweden

The formation of starch fractions that are more or less resistant to amylases during food processing is of interest from at least three different points of view:

1) Starch determination with enzymatic methods might underestimate the true amount of starch.
2) Dietary fibre might be overestimated due to contamination with starch. All methods for determination of dietary fibre depend on an efficient enzymatic starch hydrolysis.
3) Does "resistant starch" behave physiologically as starch (absorbed in the small intestine) or dietary fibre (not absorbed in the small intestine)?

Amylose-lipid complexes that occur naturally in cereals and are formed at processes such as extrusion cooking and baking, have been reported to resist amylase degradation *in vitro*. Holm et al. (1), however, showed that it was possible to hydrolyse the complexes slowly with excess of α-amylase. In addition, they demonstrated that amylose-lysolecithin complexes are completely absorbed in the rat small intestine, although at a slower rate than free amylose. Amylo-glucosidase at 60°C, often used for starch determination, hydrolysed the complexes only partially. Termamyl, a heat-stable microbial α-amylase, however, hydrolysed amylose-lipid complexes efficiently at 95°C. Inclusion of this enzyme in a gelatinization step is therefore suitable to hydrolyse amylose-lipid complexes for starch analysis.

An enzymatic gravimetric method for assay of total dietary fibre was recently developed in our laboratory (2). Starch degradation is carried out with Termamyl in the gelatinization step and pancreatin, containing porcine α-amylase. Amylose-lipid complexes do not contribute to dietary fibre values obtained with this method (1).

The formation of resistant starch during bread baking was studied by comparison of dietary fibre content in bread and the corresponding flour. The breads were made from wheat and rye flours of various extraction rates (3). The enzymatic gravimetric method including Termamyl was used (2). The dietary fibre residues thus obtained were analysed for "resistant starch" by glucoamylase after solubilization with 2M KOH for 30 min. at room temperature. Table I shows typical results.

Table I. Dietary fibre and "resistant starch" in bread made from low extraction wheat and rye flours (mg/100 mg dry matter in the flour and bread, respectively)

	Crumb	Crust	Flour
Dietary fibre	7.7	6.7	4.7
Resistant starch	1.6	1.2	Not detectable
Dietary fibre less resistant starch	6.1	5.5	4.7

The dietary fibre values in bread were always higher than expected from the flour analysis. The increase was more pronounced in the crumb than in the crust. Most of the increase could be accounted for as "resistant starch". GLC-analysis of the dietary fibre residues also verified an increase in glucan content. Drying and storage at room temperature for 4 weeks, or freezing at -20 C, did not alter the amount of resistant starch.

Extrusion cooking at severe conditions also increased the dietary fibre content. In wheat flour this was due to starch available to glucoamylase without KOH-solubilization, and no resistant starch was demonstrable (3,4).

REFERENCES

1. HOLM, J., BJOERCK, I., OSTROWSKA, S., ELIASSON, A.C., ASP, N.-G., LARSSON, K. and LUNDQUIST, I. (1983). Die Stärke (in press)
2. ASP, N.-G., JOHANSSON, C.-G., HALLMER, H. and SILJESTROEM, M. (1983) J. Agric. Food Chem. 31, 476-482.
3. JOHANSSON, C.-G., SILJESTROEM, M. and ASP, N.-G. (1983) Zeitschr. Lebensmitteluntersuch. u. Forschung (submitted)
4. BJOERCK, I., NYMAN, M. and ASP, N.-G. (1983). Cereal Chem. (submitted)

DIETARY FIBRE CONTENT AND COMPOSITION IN SIX CEREALS AT DIFFERENT EXTRACTION RATES

M. NYMAN [1], M. SILJESTRÖM [1], B. PEDERSEN [2], K.-E. BACH KNUDSEN[3]
N.-G. ASP [1], C.-G. JOHANSSON [1] and B.O. EGGUM [2]
[1] Dept. of Food Chemistry, University of Lund, Lund, Sweden
[2] Dept. of Animal Physiology and Chemistry,
[3] National Institute of Animal Science, Copenhagen, Denmark
Dept. of Biotechnology, Carlsberg Research Laboratory,
Copenhagen, Denmark

Summary

Insoluble and soluble dietary fibre was determined in flours with different extraction rates from wheat, rye, barley, rice, sorghum, and maize. An enzymatic, gravimetric method was used and compared with crude fibre, NDF and ADF determinations. The dietary fibre composition was determined at the lowest and the highest extraction rates.

1. INTRODUCTION

Cereal grains are important sources of digestible and undigestible carbohydrates. During milling to refined flours, the outer fibrous layers are removed and the dietary fibre content decreases with decreasing extraction rate.

The variation of insoluble and soluble dietary fibre with the extraction rate was investigated in flours of six cereals by using an enzymatic, gravimetric method. The composition of dietary fibre was determined by gas-liquid chromatography in flours having the lowest (approximately 65%) and the highest (100%) extraction rates. Crude fibre, neutral detergent fibre (NDF, with amylase) and acid detergent fibre (ADF) measurements were compared with determinations obtained by the enzymatic, gravimetric method (1).

2. MATERIALS AND METHODS

Hard winter wheat, rye and spring barley harvested in Denmark were used. Sorghum and rough rice from Italy were obtained from commercial sources. Different types of maize flours were produced from U.S. No. 3 maize. Total dietary fibre content was analysed by a gravimetric method, based on digestion of the samples with enzymes, as described by Asp et al. (2). Crude fibre was determined according to AOAC methods (3). ADF and NDF were analysed as described by Goering and Van Soest (4). The dietary fibre composition was assayed by gas-liquid chromatography as described by Theander and Åman (5).

3. RESULTS AND DISCUSSION

The cereals analysed showed increased dietary fibre content with increasing extraction rate. However, the slope, and start and end points of the curves, differed. In rye and sorghum, the dietary fibre content increased gradually with the extraction rate. Dietary fibre content in wheat was constant for extraction rates between 65 and 80%, but then increased rapidly with the extraction rate. In rice and to some extent in barley a similar relationship could be seen. This is partly due to differences in histological constitution between the cereals. Rice and barley contain, in contrast to the other cereals, husks that are rich in dietary fibre. This is also reflected in the dietary fibre content of whole grain, rice and barley flour which was 20%. Whole grain flour of wheat, sorghum and corn contained about 10% dietary fibre.

In all cereals, the content of soluble fibre was independent of extraction rate. This shows that the soluble polysaccharides arise essentially from the endosperm. The amount of soluble fibre was higher in rye, barley and wheat than in the other cereals investigated. This is of importance, especially in flours of low extraction rate, where almost half of the total dietary fibre was soluble. This also demonstrates that the soluble fraction should be included in a dietary fibre determination.

The main components of the dietary fibre were as expected arabinose, xylose and glucose. The relative monomeric composition of the fibre in wheat, rye and maize was the same in both extraction rates investigated. This indicates that the xylose-arabinose ratio is similar in both endosperm and outer layers of the kernel. In addition, the pentosan content in wheat, rye and maize was higher than in the other cereals investigated. The dietary fibre fraction in refined flour of barley, rice and sorghum was dominated by glucans. In whole grain flour of barley and rice, the xylose content increased with extraction rate. These cereals are surrounded by a husk rich in xylans. In sorghum, both arabinose and xylose content increased with increasing extraction rate.

The degree of lignification differed considerably with the rate of extraction. The lignin content was very low or not detectable in the refined flours. The highest values (about 4%)were found in whole grain of barley and rice, respectively.

Comparisons of different gravimetric methods show that crude fibre and acid detergent fibre measured only a fraction of insoluble and total dietary fibre. The slope of the regression equation between insoluble dietary fibre and neutral detergent-amylase determinations were 0.8-1.0 for wheat, rye, barley and rice and 1.2 for maize. Detergent methods applied to sorghum gave obscure results, probably due to proteins that are difficult to solubilize.

REFERENCES

1. NYMAN, M., SILJESTRÖM, M., PEDERSEN, B., BACH KNUDSEN, K-E., ASP. N.-G., JOHANSSON, C.-G. and EGGUM, B.O. 1983. Cereal Chem. (in press)
2. ASP, N.-G., JOHANSSON, C.-G., HALLMER, H. and SILJESTRÖM, M. 1983 J. Agric. Food Chem. 31:476
3. Association of Official Analytical Chemists. 1980. The Society, Washington, D.C.
4. GOERING, H.K. and VAN SOEST, P.J. 1970. Agricultural Handbook 379. Agric. Res. Serv. U.S. Dept. Agric., Washington. D.C.
5. THEANDER, O. and ÅMAN. P. 1981. Page 66 in: The Analysis of Dietary Fiber in Food. W.P.T. James and O. Theander, eds. Marcel Dekker, New York

EFFECTS OF SALT IN BAKING

H. HÄRKÖNEN and P. LINKO
Helsinki University of Technology, Department of Chemistry,
Laboratory of Biochemistry and Food Technology
SF-02150 Espoo 15, Finland

Summary

The effects of sodium chloride in baking of wheat, wheat/barley, and rye bread were investigated both from process technological and quality point of view. Farinograph absorption of wheat-containing flours increased and of rye flour decreased with an increase in salt content. Salt increased both mixing tolerance and extensibility of wheat dough. Salt had also a marked effect on starch gelatinization, increasing gelatinization temperature and hot paste viscosity. An increase in salt content decreased wheat bread loaf volume, but increased the volume of bread baked from wheat and barley blend. In most cases sponge baking resulted in best bread quality. Organoleptically best wheat bread was obtained with a minimum of 1.0% salt (flour weight basis), and rye or wheat/barley bread with 1.5% salt.

1. INTRODUCTION

Sodium chloride is believed to have invaluable functions in bread baking (1). Salt affects physical dough properties and loaf volume. It stabilizes yeast fermentation, and due to interactions with gluten proteins reduces water-binding ability of gluten (2). Furthermore, wheat dough development time and stability increase with an increase in salt content. Salt influences both wheat (3) and rye (4) starch gelatinization. In rye bread baking both optimal acidity and salt content are important from bread quality point of view (5). Salt controls lactic acid fermentation (6), in part by preventing excessive growth of hetero-fermentative lactic acid bacteria (7). Consequently, salt appears to be necessary in modern bread baking process. On the other hand, considerable criticism against high sodium intake in a typical diet in several countries has been raised recently (8). Thus a systematic study on the importance of salt in bread baking process seemed necessary.

2. EXPERIMENTAL

Commercial baker's wheat (moisture 14.5%, absorption 62.7%, falling number 273), rye (moisture 13.6%, absorption 84.0%, falling number 105), and barley (moisture 12.9%, absorption 86.0%) flours obtained as a gift from Vaasamill Ltd, Helsinki, Finland were used throughout this study. Sodium chloride was of analytical grade (Merck, Germany), and deionized water was employed in physical dough testing.

Farinogram values for wheat flour, a wheat and barley flour blend (60% wheat), and for wheat and rye flour blends (25, 40 and 50% rye) were determined by the AACC method. Salt was diluted in 135 ml of water, and all of the water needed was added during the first few seconds of mixing. The Bracht method (9) was used for rye and barley flour, and for wheat and rye flour blend (25% wheat) Farinograms. In this case salt was dissolved in 190 ml of water. Extensograms were determined by standard AACC method, and for the Amylograms 60 g of wheat and barley and 80 g of rye flour was used.

Panned and free-standing wheat and barley breads were baked using both straight- and sponge-dough methods. Rye breads were baked both by sour-sponge and by acidifying the dough to pH 4.4 by with lactic acid. The recipies used are given in Table I. Sponges were fermented for 13 hours. Doughs

Table I. Dough and sponge compositions

	Straight-dough		Sponge-dough			Sponge		
	Wheat (g)	Wheat/barley (g)	Wheat (g)	Wheat/barley (g)	Rye (g)	Wheat (g)	Wheat/barley (g)	Rye (g)
Wheat flour	1000	600	400	200	100	600	400	-
Barley flour	-	400	-	400	-	-	-	-
Rye flour	-	-	-	-	500	-	-	400
Water	630	700	-	150	100	600	500	600
Yeast	55	40	-	10	20	6	8	-
Sponge seed*	-	-	-	-	-	-	-	25
Sponge	-	-	1200	900	1000	-	-	-
Salt**	0-25	0-25	0-25	0-25	0-25	0-25	0-25	0-25
Sugar	40	20	40	20	-	-	-	-
Margarine	40	15	40	15	-	-	-	-
Milk powder	-	60	-	60	-	-	-	-

*A commercial mixed-culture of yeast and lactic acid bacteria

**In sponge-dough method salt was added either in sponge or during dough mixing.

were mixed in a Hobart A200 mixer for 6-8 min. Floor times for wheat and barley doughs varied from 20 to 30 min, depending on salt content. Scaling weights for wheat and barley doughs were 170g and 200g for pan baked and 450g and 500g for free-standing breads, respectively. Wheat doughs were proofed to optimum volume (25-90 min) at 30°C in a fermentation chamber at 70-80% relative humidity, and barley breads (30-70 min) at 34°C at 60-70% R.H. The floor time for rye dough was 60 min, scaling weight 500g and proof time 60 min at 34°C at 60-70% R.H. Wheat and barley breads were baked for 20 and 30 min, respectively, at 200°C, and rye breads for 50 min at 210°C. Bread volumes were determined by seed displacement, and organoleptic bread quality was judged by a professional test baker.

3. RESULTS AND DISCUSSION

3.1. Physical Dough Testing

Water absorptions as measured by Brabender Farinograph of wheat, barley and wheat/barley blend (60% wheat) flours decreased significantly with an increase in salt content (Fig. 1). An increase in the relative quantity of rye flour in wheat/rye flour blends increase Farinograph absorption. Even the lowest salt level of 0.5% (flour weight basis) used markedly decreased water absorption of wheat (from 62.7% to 61.6%) and barley (from 86.0% to 84.4%) flour doughs. Water absorption of rye flour dough increased with an increase in salt content (Fig. 1). A marked decrease in the absorption of rye flour dough was observed when the pH was adjusted to 4.4 with lactic acid, but this could be partly compensated by increasing salt level. Example Farinograms obtained with wheat (A,B) and with a blend of wheat and barley (E,F) flours both without (A,E) and with (B,F) 1.5% salt are shown in Fig. 2.

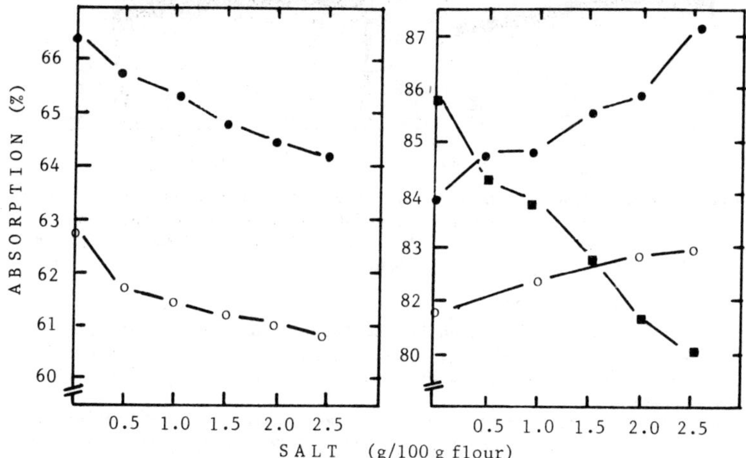

Fig. 1 The effect of salt on Farinograph water absorption at 500 B.U.; A, wheat flour (o) and a 60/40 blend of wheat and barley flours (•); B, barley flour (■) and rye flour (o, with, and •, without lactic acid).

Dough development time and stability of wheat flour dough increased and of barley flour dough decreased with increasing salt level. An addition of 2.5% salt nearly doubled the values for wheat, but decreased barley flour dough stability by about 40%. A blend of wheat and barley flours (40% barley) behaved essentially as wheat in this respect, although the stability was poorer than that of wheat flour dough. Development time and stability of sour rye flour dough increased slightly with an increase in salt content. The addition of 5.0% of dry salt (flour weight basis) to rye

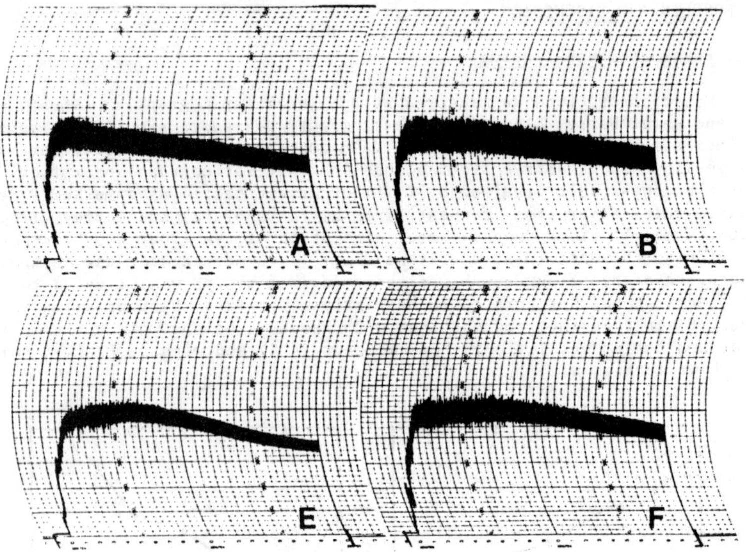

Fig. 2 Farinograms from A, wheat flour; B, wheat flour with salt (1.5 g/100g flour); E, a 60/40 (w/w) blend of wheat and barley flours; F, a 60/40 blend of wheat and barley flours with salt (1.5 g/100g flour).

flour containing doughs overmixed in the Farinograph could temporarily reverse the effect of overmixing, with a clear increase in dough consistency. A similar effect with a wheat flour dough mixed in a Mixograph has been reported (10).

An increase in salt level increased the resistance to extension and the extensibility of wheat flour dough after a rest period of 90 min, as measured by Brabender Extensograph. An increase in resting time increased resistance, but decreased extensibility. The high resistance to extension caused by sodium chloride has been explained by slight denaturation of gluten configuration (11). Finally, a rapid and marked increase in peak Amylograph consistency was observed both with wheat and barley flour suspensions, and an increase in initial and peak gelatinization temperatures with all flours studied, when salt content was increased.

3.2 Test Baking

Under our experimental conditions salt did not significantly affect the handling properties of wheat flour dough, whereas saltless dough made from wheat and barley flour blend was sticky and difficult to handle. Rye flour dough prepared without salt was slack and sticky, while doughs made both with acid and salt were easy to handle. For optimal proofing of wheat flour doughs a minimum salt content of 0.5% (flour weight basis) was necessary. A further increase in salt level did not affect proof-time in the straight-dough method. Sponge doughs in general required a longer proofing time, which was further extended by the addition of salt in the sponge. Proofing of rye flour doughs was delayed by salt, which is known to inhibit both the growth and the lactic acid formation of lactobacilli.

An increase in salt content drastically decreased wheat bread loaf volume. The effect was somewhat smaller with sponge-baked bread than with bread baked by the straight-dough method. The sponge method also improved crumb texture. Better quality bread was obtained when salt was added during dough mixing rather than to the sponge. In all cases, however, a salt content of 1.5% was judged organoleptically satisfactory. With bread baked from wheat/barley blend flour even the lowest (0.5%) salt level markedly increased loaf volume. Loaves of best structural characteristics and flavour were obtained by using 1.5% salt. The sponge-method improved crumb texture of unsalted bread, and the bread was softer and easier to cut.

The effect of salt in rye baking was quite different. Doughs prepared without salt and lactic acid spread during proofing, resulting in a flat loaf, often with the crust separated from the crumb. The loaf was difficult to cut, owing to a tough and rubbery crumb. Both acid and salt improved bread quality significantly, with best results obtained when both acid and salt were used. The synergistic effect of salt and acid has also been reported for wheat bread (12).

The results of this investigation will be published in detail elsewhere.

4. REFERENCES

1. NIMAN, C. (1981). Salt in bakery products. Cereal Foods World 26, 117.
2. GALAL, A. M., VARRIANO-MARSTON, E. and JOHNSON, J. A. (1978). Rheological dough properties as affected by organic acids and salt. Cereal Chem. 55, 683.
3. D'APPOLONIA, B. L. (1972). Effects of bread ingredients on starch-gelatinization properties as measured by the amylograph. Cereal Chem. 55, 532.
4. WASSERMANN, L. and DÖRFNER, H.-H. (1977). Die viskosen Eigenschaften von Roggenteigen unterschiedlicher Zusammensetzung. Getreide, Mehl u. Brot 31, 95.
5. DREWS, E. (1970). Studien über die Wirkung von Säure.und Salz bei der Herstellung von Roggenbrot. Brot u. Gebäck 24, 141.
6. SCHULZ, A. (1960). Untersuchungen über den Einfluss von Kochsalz auf die Mikroorganismen des Sauerteiges. Brot u. Gebäck 14, 199.
7. SPICHER, G. (1961). Die Erreger der Sauerteiggärung: Vergleichende Untersuchungen über den Einfluss von Kochsalz auf das Säuerungsvermögen der Milchsärebakterien des Sauerteiges. Brot u. Gebäck 15, 115.
8. ANONYMOUS (1980). Dietary salt. Food Technology 34, 85.
9. BRACHT, T. J. (1976). Rheologische Untersuchungen an Roggen mit dem Brabender Farinograph. Getreide, Mehl u. Brot 30, 5.
10. DANNO, G. and HOSENEY, R. C. (1982). Effect of sodium chloride and sodium dodecyl sulfate on mixograph properties. Cereal Chem. 59, 202.
11. TANAKA, K., FURUKAWA, K. and MATSUMOTO, H. (1967). The effect of acid and salt on the Farinogram and Extensigram of dough. Cereal Chem. 44,675.
12. BAKHOUM, M. T. and PONTE, J. G. (1982). Combined effects of sodium chloride and hydrochloric acid on wheat flour strength. Cereal Chem. 59, 37.

OPTIMIZATION OF THE BAKING PROCESS WITH RESPECT TO QUALITY AND ENERGY (CONVECTION OVENS)

CHRISTENSEN, A., BLOMQVIST, I., and SKJÖLDEBRAND, C.
Department of Food Engineering, Lund University, P.O.B. 50, S-230 53 Alnarp, Sweden.

INTRODUCTION

One result of the increasing interest in energy costs is an increasing interest on development of the industrial processes and apparatuses. The energy costs in for example the bakeries, counted as costs per kilogram bread, are still not very high but due to the tendency to increase the production volumes the real costs are high.

A few years ago a research project on baking and energy was started where three different laboratories cooperated. The aim was to find out the minimum energy needed to bake bread without getting a poor quality.

The laboratories involved were SIK - The Swedish Food Institute, Gothenburg and Department of Food Engineering, Lund both from Sweden and VTT - Technical Research Centre of Finland, Food Research Laboratory, Helsinki, Finland.

Measurements earlier carried out in some Swedish bakeries (1) showed that there is a saving potential with regard to energy if the processes were optimized in a better way than today.

The following questions started the project. Is it possible to decrease temperature in the oven and still get an acceptable quality within an acceptable time? What are the function of the steam injection in the oven? Is it possible to increase air velocity or the effect of radiation to get a more energy efficient process?

At the department of Food Engineering the experiments were carried out in a laboratory convection oven to investigate this type of heat transfer (2). Combining different air parameters such as temperature, velocity and relative humidity we wanted to find the minimum energy needed when baking bread, getting an acceptable quality or increasing the quality.

HEAT AND MASS TRANSFER DURING BREAD BAKING

To be able to find out the energy input to the bread the heat and mass transport from the oven to the bread have to be surveyed and calculated. A heat balance has to be made. The heat in the oven is transferred to the product in principle by the following mechanisms

* conduction via the plates
* convection via the air
* radiation via the heated walls in the oven compartment
* condensation due to steam injection during the first minutes of baking.

The driving force for the transport into the dough is the temperature gradient and coupled to this transport is the masstransport, which also contributes to the heat transport. Due to different mechanisms for the masstransport the total moisture content in the crumb does not change very much. The weight losses depend mainly on the mass transport from the crust formation.

The boundary between crust and crumb is in this investigation those parts where water is evaporated. The crust has had a temperature treatment above 100°C.

MATERIALS

A Swedish wheat bread was used in the investigation. Its ingredients in percent flour weight were: 55.6% water, 4.7% yeast, 1.7% sugar, 1.7% shortening and 1.7% salt.

The dough was formed to two different bread geometries, one weighing 100 grams formed like bun and one formed like a loaf weighing 300 grams.

The dough was proofed at 37-39°C for 45 minutes.

METHODS

To be able to compare the results a separate investigation was carried out to define when the bread was proper baked. As the bread consists of both a crumb and a crust the criteria has to include analysis of both parts. We chose to measure the colour and the thickness of the crust and the elasticity and the temperature of the crumb (3).

Two different references for the crust colour were used. The 100 g bread was compared to a similar bread baked in a small bakery. The bigger breads were compared to a colour scale decided by the three different laboratories together. The colour was measured by means of an Elrepho reflectance method.

The baking procedure in the oven was performed according to a scheme made up for the two different series. Different combinations of the air parameters were followed. The temperature was measured in the centre of the product during the total process. During the process the weight losses were measured by means of a balance. After processing the bread was weighed and the colour and the elasticity of the bread was measured.

The energy consumption was calculated by dividing the bread process into five different parts using the temperature and weight measurements (4).

The five energy consuming steps were:

- Heating the dough to 100°C
- Evaporation of water from the surface forming the crust
- Changes of the centre parts to a crumb by means of chemical reactions
- Superheating of steam transported though the crust
- Superheating of the dry crust to a mean temperature over the crust.

Using these an energy balance was made to calculated the energy need in the bread.

RESULTS AND DISCUSSION

Figure 1 shows one example of the surface temperature and centre temperature in the bread loaf baked at 210°C and an air velocity of 2 m/s.

Figure 2 shows the energy demand, weight losses and the crust colour as function of temperature and baking time.

The temperature measurements were used to calculated energy consumption. It was shown that the time to reach 98°C in the centre decreases when the air temperature increased.

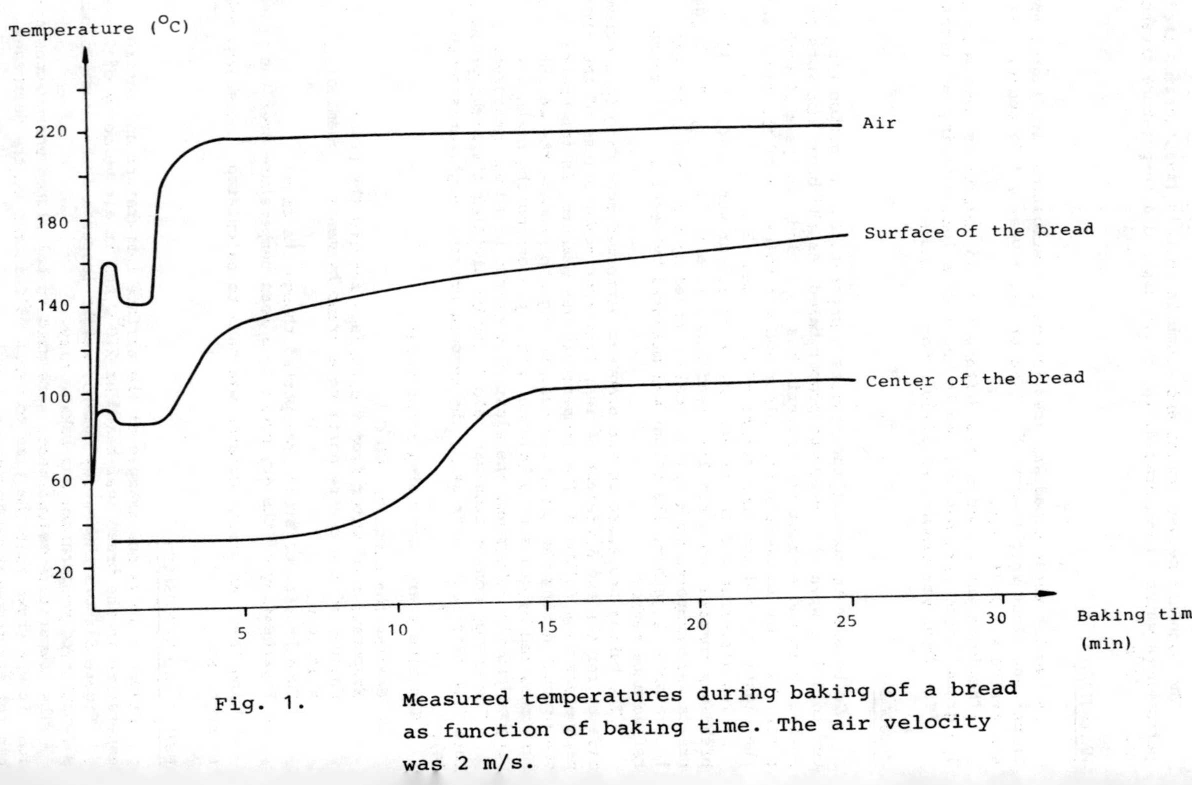

Fig. 1. Measured temperatures during baking of a bread as function of baking time. The air velocity was 2 m/s.

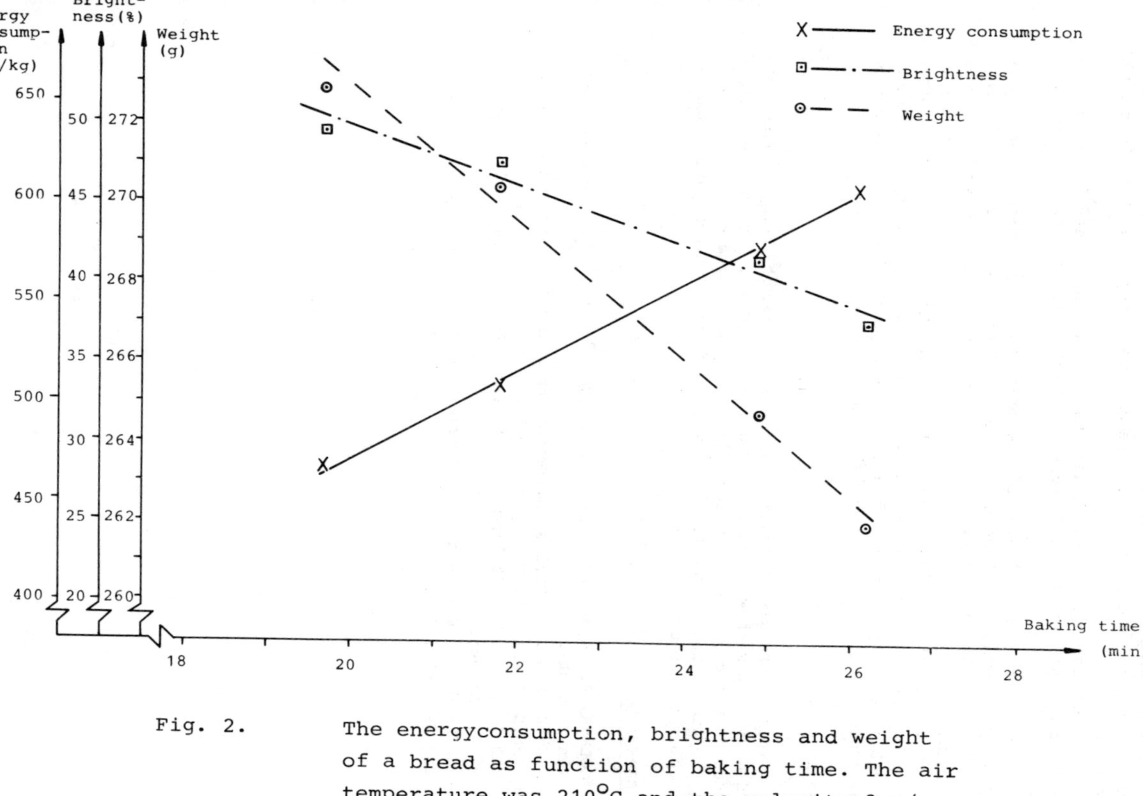

Fig. 2. The energyconsumption, brightness and weight of a bread as function of baking time. The air temperature was 210°C and the velocity 2 m/s.

The total energy was calculated from the five different steps after defining the proper baked bread. It was found for the 100 g bread that 220°C, 2 m/s and steam injection needed less energy than the other combinations.

The investigation also showed that the weight losses increase when temperature and air velocity increase. So did also the browning of the bread.

Concerning energy heated of the bread and evaporation of water are the biggest consumers. Heating of the bread is constant for all baking procedures. This means that to save energy the weight losses have to be decreased.

Measurements of elasticity in the crumb shows that this part often is ready before the crust colour is acceptable. This means that the crust formation is very important from the energy point of view. It is not only the crust colour but also crust thickness that is important.

CONCLUSIONS

This investigation concerns the consumption of energy in the bread and dough. This shows that it is the weight losses, i.e. evaporation of water in the crust, that consums the main part of the energy.

Optimization of the total process needs more investigation and calculation on the oven itself.

REFERENCES

1. SOLMAR, A. Bageriers energianvändning. Report 1979. Linköping Sweden. (in Swedish).
2. SKJÖLDEBRAND, C. Frying and reheating in a forced convection oven. Ph.D. Thesis. Lund, Sweden, 1979.
3. MARSTON, P.E. Breadbaking - the transformation from dough to bread. The Bakers Digest.
4. NEBELUNG, M. Heat- and mass transport during baking. Ph.D. Thesis. Dresden, DDR, 1978.

OPTIMIZATION OF BAKING PROCESS: COMBINED HEAT TRANSFER

M. PALOHEIMO, Y. MÄLKKI and S. KAIJALUOTO
Technical Research Centre of Finland
Food Research Laboratory

Summary

The effects of different heat transfer forms on the baking process were investigated. For this purpose an experimental baking oven, where different forms of heat transfer could be varied during experiments, was devised and constructed. In addition, a humidity meter for temperatures over 100°C was developed. A standard bread formula and constant baking procedures were used. Bread was taken from the oven when the crust had achieved normal browness and the crumb temperature had risen to 96...98°C. A trained panel furthermore confirmed using sensory evaluation that the loaves were thoroughly baked. Also enzymatic determinations of the starch gelatinization showed that in normal baking conditions the bread crumb was ready at this stage. Process parametres were temperature, air circulation in the baking chamber (convection), humidity during the initial stages of baking, the baking level in the oven and the possibility of heat conduction from the hearth. The most important factor was expectedly the temperature, although the hearth (conduction) and convection had a significant effect on the heat transfer. In the conditions used the effect of the hearth was more significant. The humidity and the baking level did not seem to be significant.

In the commercial baking ovens, heat is usually transferred simultaneously by convection, conduction, radiant heat, and at the beginning also by condensing steam. The present study was performed for elucidating the effects of various process parametres in combined heating, aiming at the same final quality of bread despite of variations in baking conditions.

1. EXPERIMENTAL ARRANGEMENT

1.1. Test baking oven

The oven construction allows variation in heating system and thus the different forms of heat transfer can be varied as desired. The oven can be heated by the heating jacket around the baking chamber, in which case the heat transfer is mainly radiation from the walls. The air in the baking chamber can be circulated, the rotary air can be subjected to separate heating when desired. Both heating forms are independent of each other. The products can be baked either on a grate or on a separate heatable hearth which can be placed into the oven when needed. The height of the grate/hearth in the oven can be changed, so that the distance from one radiation source (oven ceiling) can vary. The oven air can be humidified to a desired level by injecting steam (20 kPa) into the baking chamber from a separate boiler.

The oven is regulated and the baking programmes are followed by an BC-80 microcomputer. The humidity is measured by an infra red (IR) meter which was developed and produced in the framework of the study. The use of

IR-radiation in humidity measurements is based on the absorption properties of the water molecules. The IR transmitter and receiver are on opposite sides of the oven, absorption distance is about 1 m. The meter can monitor all changes in oven moisture.

1.2. Test baking

The experiments were carried out on a wheat pan bread (baking formula, % on flour basis: flour, 100 %, water, 58 %, yeast, 4.9 %, commercial baking margarine 1.7 %, sugar 1.7 %, salt 1.7 %). The formula and bread-making procedure were kept constant. Four loaves were baked at a time, 2 of them being taken for analysis.

The removal of loaves was to happen at the moment when they were baked through. By that time the crumb should be thoroughly baked and the crust should have a normal browness. Preliminary experiments were made to determine the correlation between crumb temperature and the moment when the crumb was thoroughly baked. The crumb temperature was compared with the starch gelatinization determined both microscopically and enzymatically after Sandstedt and Mattern (1).

The method chosen to determine the end point of baking was based partly on temperature measurements, and partly on a sensory evaluation of a panel. The baking time was mainly determined by crust colour, though still the crumb had to reach a temperatue of 96...98°C. After baking a sensory evaluation was made to confirm that the loaves were thoroughly baked and acceptable.

1.3. Experimental design

The experimental design was a half replicate of a 2^5 factorial design by Box and Wilson with four replications of the centre point to test the significance of the model. Variables chosen for the experiments and their levels are in Table I.

Table I. Variables and variable levels

Variable	Level		
	1	2	3
Temperature (°C)	200	220	240
Baking level (cm from the bottom)	7	17	27
Humidity after steaming (g/m^3)	10	160	310
Air circulation (m/s)	0	-	1...2
Hearth	no	-	yes

The level 2 was used only in the centre point replications. The humidity was adjusted to reach the desired level after 3 minutes measured from the beginning of the baking, whereafter it was allowed to escape naturally. The hearth either was not in the oven (level 1), in which case the pans were placed on the grate, or it was in the oven (level 3), when its temperature was adjusted to be the same as the particular temperature in the oven.

1.4.Determinations

The following determinations were performed:

- crumb temperature development during baking, determined from the temperature curve. The moment was estimated when the temperature reached or could have reached 98°C

- baking loss, 5 and 120 minutes after baking
- volume and specific volume
- thickness of the crust
- the colour of the crust and the bottom of the bread (compared with $BaSO_4$, 100 %)
- theoretical energy consumption

2. RESULTS

In preliminary experiments on starch gelatinization, microscopic examination showed that at a temperature of 70°C less than 10 % of starch granules were birefringent, while after the temperature had reached 80°C, almost all granules had gelatinized and lost their birefringence. At higher temperatures there was no difference compared with the 80°C situation. Similarly, longer treatment times at 98°C (5 min) also had no effect on the amount of birefringent granules. The method seemed to be insufficiently exact for the purpose. The enzymatical method to determine starch gelatinization (1) is based on the ability of β-amylase to hydrolyze gelatinized starch easily to maltose, as undamaged starch does not hydrolyze as easily. In Figures 1a and 1b the results of the enzymatical maltose determination of pan bread baked in constant conditions are shown. According to the results, the baking time does not seem to have any significant effect after the crumb centre had reached the temperature mentioned. After sensory evaluation the crumb was thorougly baked after the centre had reached 90°C.

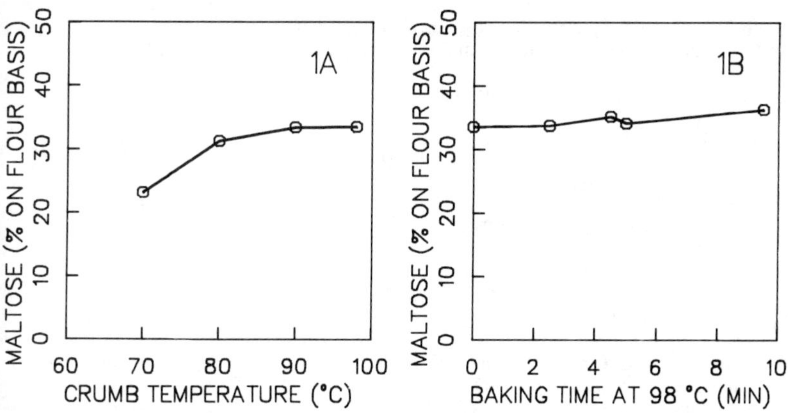

Fig. 1. Maltose formed by β-amylase from crumb centre samples, sample size 1.5 g (wheat pan bread, 280 g, baking temperature 220°C)
1a) as a function of the crumb temperature
1b) as a function of the baking time after the crumb temperature had reached 98°C

The experiments were planned so that in principle it would have been possible to estimate a model of the following type:

$$y = a_o + \sum_{i=1}^{5} a_i x_i + \sum_{i=1}^{4} \sum_{j=i+1}^{5} b_{ij} x_i x_j \qquad (1)$$

where $x_1,\ldots,x_5$ are process parametres and a_i and b_{ij} are parametres to be estimated. It proved, however, that coefficients of cross products were not worth estimating with so little experimental material because the accuracy of the measurements was not sufficient. The inaccuracy was most significant in determining the baking time. The end point of baking was established partly sensorially on the crust colour, and especially at low temperatures when crust colour was developing slowly, the reproducibility was poor. Therefore, the baking time had to be taken as an independent variable in the model.

Model parametres were estimated using multiple linear regression. Only those parametres which differed from zero with a 95 % probability were included. The models obtained are in Tables II and III. To facilitate comparison, Table II is normalized by dividing both sides of the equations by the arithmetic mean of the corresponding dependent variable. All the independent variables are normalized to the area 1...3 according to Table I. Baking time, which was taken into the model as an extra independent variable, is normalized by dividing by ten. R is a regression coefficient indicating the significance of the model.

The theoretical energy consumption is calculated from the equation (2)

$$q = (2.8\,(T_f - T_d) + 10.5 + \alpha\,(2260 + 3.370\,(T_y - T_f))/(1-\alpha) \text{ kJ/kg} \qquad (2)$$

where T_f, T_d and T_y are the temperatures of crumb, dough and surface (crust) respectively, and α baking loss after 5 minutes.

Table II Normalized model equations

$y_1 = 1,643 - 0,306x_2 - 0,0561x_4$	R = 0,924
$y_2 = 1,437 - 0,133x_2 - 0,0728x_4 - 0,0356x_5$	R = 0,945
$y_3 = 0,340 + 0,0449x_5 + 0,334x_6$	R = 0,874
$y_4 = 0,587 + 0,0229x_5 + 0,215x_6$	R = 0,902
$y_5 = 1,088 - 0,0298x_2 - 0,020x_5$	R = 0,681
$y_6 = 1,092 - 0,0523x_2$	R = 0,596
$y_7 = 1,274 + 0,0552x_1 - 0,163x_2 - 0,503x_3$	R = 0,911
$y_8 = 2,364 + 0,232x_2 - 0,169x_4 - 0,363x_6$	R = 0,847
$y_9 = 0,612 + 0,0261x_5 + 0,197x_6$	R = 0,865

y_1 = baking time (colour) min/17.5
y_2 = baking time (crumb) min/16.5n
y_3 = baking loss after 5 min %/7.85
y_4 = baking loss after 120 min %/10.43
y_5 = volume dm^3/0.917
y_6 = specific volume cm^3/g/3.29
y_7 = colour (surface) %/33.78
y_8 = colour (bottom) %/46.22
y_9 = theoretical energy consumption kJ/kg/428.3

x_1 = baking level
x_2 = temperature
x_3 = humidity
x_4 = hearth
x_5 = air circulation
x_6 = baking time

Table III. Model equations

baking time/min (colour) = 75.93 - 0.267 (temp/°C) - 1.96 *hearth (3)
baking time/min (crumb) = 41.64 - 0.110 (temp/°C) - 2.41 *hearth -1.18 *circ (4)
baking loss/% (5 min) = 3.025 + 0.705 *circ + 0.263 (time/min) (5)
baking loss/% (120 min) = 6.36 + 0.477 *circ + 0.224 (time/min) (6)
volume/dm^3 = 1.23 - 0.00137 (temp/°C) - 0.0358 *circ (7)
specific volume/cm^3/g = 5.14 - 0.0086 (temp/°C) (8)
colour/% = 90.96 + 1.86 (level/cm) - 0.276 (temp/°C) - 3.40 *hearth (surface) (9)
colour/% = 198.2 - 0.537 (temp/°C) - 15.6 *hearth - 1.68 (time/min) (bottom) (10)
energy/kJ/kg = 2.73 + 0.223 *circ + 0.845 (time/min) (11)

Discrete variables, hearth and air circulation receive values 1 or 0 depending on whether the variable is operating or not. Crust thickness could not be modelled, this probably being due to the inexact testing method. The baking time as an independent variable produced difficulties in the interpretations of results. It is not, after all, an independent variable like the others. Its appearance in the model means that the model indirectly includes the variables temperature and hearth, which are included in the model of the baking time.

Equation 3 for the baking time in Table III shows that by using, for example, a temperature of 220°C, the baking time according to the crust colour was 17.2 minutes without the hearth, while the use of the hearth decreased the time by about 2 minutes. The time needed to reach 98°C in the crumb represents the heat transfer in the bread. Equation 4 in Table III shows that e.g. at 220°C the baking time required was 17.4 minutes without the hearth and air circulation, while their use decreased the time by 2.4 and 1.2 minutes respectively. Two parametres, the baking level and the humidity, did not seem to have any significance. The steam condensed at the steaming stage on the bread surface was considered to transfer a significant heat quantity to the bread. The negative results of these experiments were probably due to the fact that by the steaming system used, the maximum humidity was attained within 3 minutes from the beginning, when the surface temperature was becoming too high for effective condensation.

The most important process variable in the baking was the temperature, as expected. Furthermore, the conducted heat from the hearth and the air circulation were of consequence in the heat transfer, in the conditions used the hearth more than the air circulation. It would be expected that the effect of using a hearth is still greater when the hearth temperatures used are higher. The relatively small effect of the air circulation is explained by the low air velocity used, 1...2 m/s. The significance of the air circulation in the baking loss (eq. 5 and 6, Table III) is increased probably as a result of the increased ventilation in the oven due to the leakages in the air channels. The model for the crust colour is included to give an idea of how great is the variation in baking time due to sensory evaluation. In the ideal case the bread crust colour should be independent of the process parametres. The model shows, however, that the colour correlates strongly with the temperature. This is caused by the rapid formation of crust colour at the highest temperatures used, so that the crumb did not have time to get thoroughly baked in those cases when the heat transfer was not intensified by the use of a hearth or air circulation.

REFERENCES

1. SANDSTEDT, R. M. and MATTERN, P. J., Damaged starch quantitative determination in flour. Cereal Chemistry 37 (1960): 3, 379 - 390.

THE INFLUENCE OF WARM-HOLDING ON THE ORGANOLEPTIC QUALITY OF POTATOES, MEATBALLS AND SOME VEGETABLES

R.L. DE FIELLIETTAZ GOETHART, J.B. LASSCHE AND L.J. VAN GEMERT
Institute CIVO-Technology TNO, Zeist, The Netherlands

Summary

Sometimes the organoleptic quality of institutional feeding is unacceptable.
Three components of hot meals, cooked potatoes (two varieties), roast meatballs and some vegetables were kept warm in order to establish the effects on the organoleptic properties and on the organoleptic quality.
One or two temperatures and three or four different times upto three hours were applied. The attributes were quantified by a selected and trained sensory-analytical panel (nine members), whereas the quality was determined by a consumer panel of about 50 members. The quantified attributes concern 13-18 aspects of appearance, odour, flavour and texture. The experiments resulted in recommendations for the practice of institutional feeding. Furthermore, the relationship between the organoleptic attributes and the quality was studied. It appeared that the decrease in quality due to warm-holding is in most cases a result of an increase in off-flavours and a decrease in positive flavour aspects.

1. INTRODUCTION

Nowadays it is quite common in institutions such as hospitals, schools and homes for elderly people, that meals are kept warm for some time before consumption (1).
Complaints have been made about the adverse effect of these distribution systems on the organoleptic quality of the meals.
Recently some studies in this field have been published (1-4). Our investigations comprised extensive experiments on the organoleptic quality and attributes of meal components during warm-holding. This paper summarizes the most important results.

2. MATERIALS AND METHODS

Two potato varieties, Bintje and Eigenheimer, were cooked in a vegetable/potato steamer and subsequently kept warm at 60 and 80 °C, for four different times, 0, 0.5, 1.5 and 2.5 hours.
Meatballs were roasted in a convection oven (5) and subsequently kept warm with and without gravy at 60 and 80 °C; warm holding times: 0, 1, 2 and 3 hours.
Three kinds of vegetables, namely frozen spinach, frozen peas and savoy cabbage, were cooked in water and kept warm at 80 °C for 0,1 and 3 hours. The spinach and the cabbage were prepared with and without a white sause.

A consumer panel consisting of about 50 untrained housewives assessed the organoleptic quality. The panel members were asked to score appearance, flavour, mouthfeel and overall quality on a category scale. For scoring the potatoes, the panel used a 5-point scale and for the meatballs and the vegetables a 7-point scale, both ranging from very good to very bad.
The sensory analytical panel which assessed the samples of cooked potatoes was composed of nine persons, selected from a group of 26. In a series of training sessions the panel generated a number of descriptive terms for the perceptible attributes of the odour and the flavour of the samples; the attributes were quantified with a graphic scale (6). A number of aspects, mostly concerning appearance and texture, derived from the P(otato) Q(uality) R(esearch) system (7) was quantified too in these training sessions. Thereafter, the actual study started.
The sensory-analytical panel which assessed the roast meat balls, consisted of nine persons selected from a technical panel, trained in quality evaluation of meat products. A list of descriptive terms was drawn up to cover the organoleptic attributes. Here, also the graphic scale was used.
The organoleptic attributes of the vegetables were not assessed.

4. RESULTS AND DISCUSSION

4.1. Potatoes

The assessments of the consumer panel have been presented in frequency diagrams with three categories. The evaluation of the flavour and overall quality is given in Figure 1. The influence of warm-holding on the appearance and the mouthfeel was similar.
From the evaluation by the sensory-analytical panel it appeared that the decrease in flavour quality during warm-holding is caused by the formation of off-flavours (Figure 2). Also a decreased perception of positive flavour aspects was observed. An important fact is that the results of the two types of panels were in agreement.
There were no apparent differences between the varieties Eigenheimer en Bintje.

4.2 Meatballs

As to the meatballs, we found that the presence or absence of gravy was the most important factor (Figure 3). Warm-holding with gravy is preferred. This is in agreement with other findings (1). Meatballs are less sensitive to warm-holding than potatoes. The assessment by the sensory-analytical panel revealed that warm-holding without gravy caused a dry and dark colour and also a less juicy product.

4.3 Vegetables

Warm-holding at 80 °C of all three vegetables results in a varying decrease in overall quality (Figure 4). Other aspects (colour, flavour, etc.) presented the same picture. After one hour of warm-holding slight differences with freshly prepared products were demonstrated. Spinach and cabbage had to be considered as unacceptable after three hours of warmholding. In general a sauce improved the quality after warm-holding.

FIG. 1 - FLAVOUR ASPECTS OF POTATOES DURING WARM-HOLDING QUANTIFIED BY SENSORY-ANALYTICAL PANEL

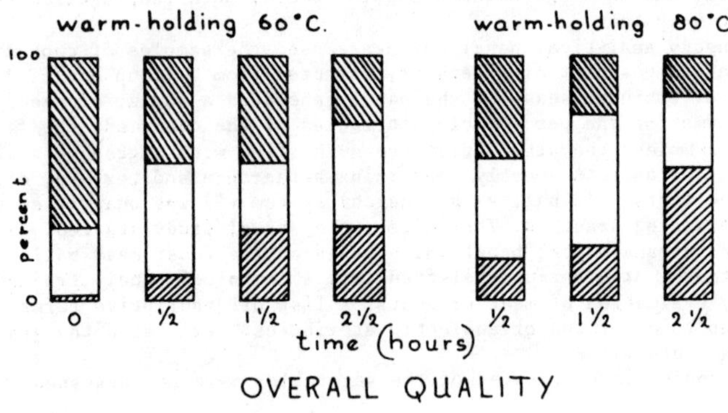

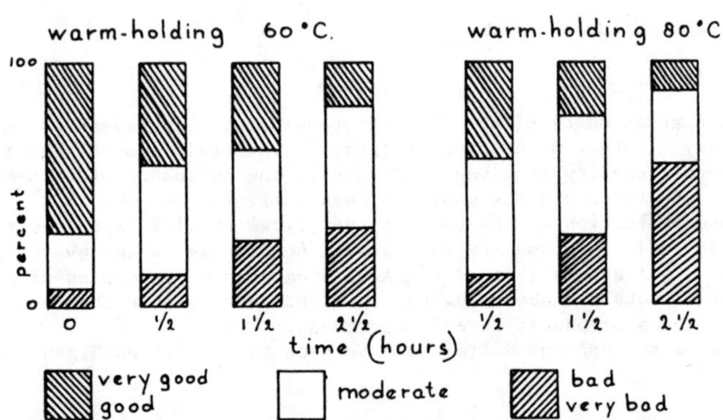

FIG. 2 - FLAVOUR ASPECTS OF POTATOES DURING WARM-HOLDING QUANTIFIED BY SENSORY-ANALYTICAL PANEL

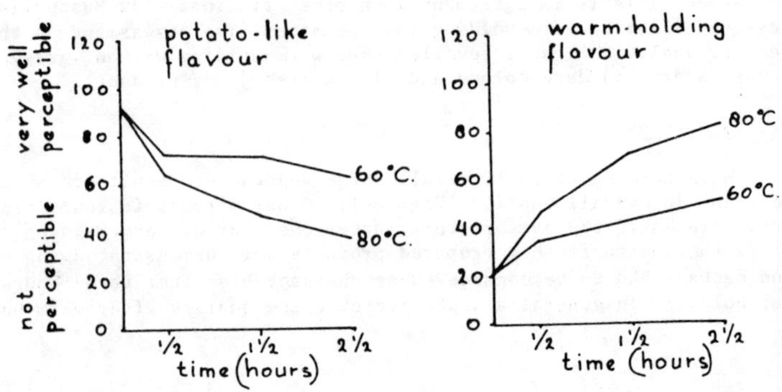

FIG. 3 - OVERALL QUALITY OF ROAST MEATBALLS

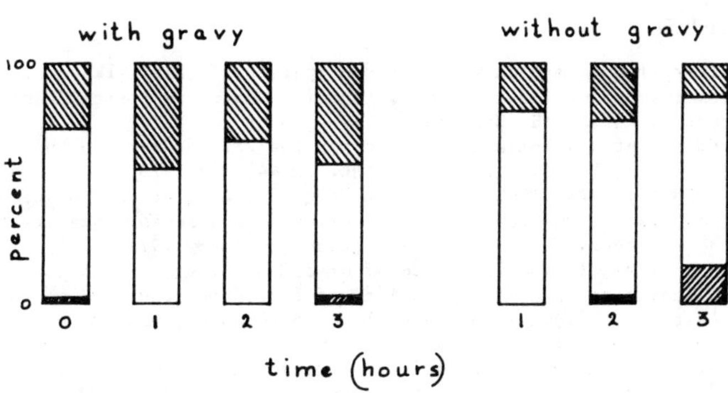

FIG. 4 - OVERALL QUALITY OF VEGETABLES

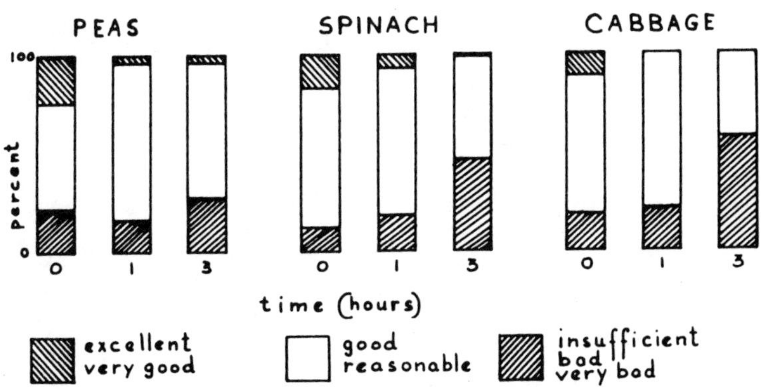

5. CONCLUSION

Potatoes appeared much more sensitive to warm-holding than meatballs and some vegetables. Cooked potatoes should preferably not be kept warm longer than half an hour and certainly not longer than one hour. The maximum warm-holding time of savoy cabbage is about one hour.

6. REFERENCES

1. Schulverpflegung mit warm gehaltenen Speisen aus Zentralküchen. Bundesministerium für Ernährung, Landwirtschaft und Forsten/Bundesforschungsanstalt für Ernährung, 1976
2. PAULUS, K. et al. Influence of heating and keeping warm on the quality of meals. Ann. Nutr. Aliment. 32 447-458,1978
3. LUNDGREN, B., KARLSTRÖM, B., and LJUNGQVIST, A.-C. Effect of time and temperature of storage after cooking on the sensory quality of bintje potatoes. J. Sci. Food Agric. 30: 305-318, 1979
4. GEMERT, L. et al. The influence of distribution systems on the organoleptic attributes and quality of cooked potatoes and meatballs In: Criteria of food acceptance, how man choosses, what he eats. Zürich: Foster Verlag, 403-413, 1981
5. Guidelines for cookery and sensory evaluation of meat (1978). American Meat Science Association, National Live Stock & Meat Board
6. STONE, H. et al. Sensory evaluation by quantitative descriptive analysis Food Technol. (Chicago), 24-34, Nov. 1974
7. WINIGER, F.A. and LUDWIG, J.W. Methoden der Qualitätsbeurteilung bei Kartoffeln für den menschlichen Konsum. Potato Res. 17 434-465, 1974.

FREEZING AND THAWING

INTRODUCTION AND CONCLUSION

M. JUL
Danish Meat Products Laboratory,
Ministry of Agriculture, Copenhagen, Denmark

1. INTRODUCTION

Within the area of research on the effect of treatment of foods with heat or cold, freezing and thawing was selected as one priority area for the following five reasons: **Firstly**, the sale of frozen foods is increasing in several countries, e.g. table 1. **Secondly**, food freezing is gaining in importance because of the increase in the use of home freezing and the use of frozen raw materials for further processing in industry, see table 2. **Thirdly**, freezing is becoming more and more common in institutional cooking and home food delivery systems, e.g. the so-called cook-freeze system. **Fourthly**, because after a considerable interest in research in that area of food science in the late 1950's in the USA and, as an upsurgeance of that, an activity in Denmark in the years 1965-78, research in that field has been grossly neglected. Data were needed because new packaging materials, new products and new distribution patterns have been introduced although often little was known about them. **Fifthly**, because many controversial views were held in this area, e.g. regarding the importance of freezing quickly, maintaining low temperatures during storage, the transportation of frozen foods, temperatures in sales cabinets, etc.

	1971	1981	1982
USA	34	40	
Sweden	11	20 (1979)	24.5
Denmark	11.5	22	23
UK	7*	15*	21
Switzerland	5	11	16
FRG	4	8	16
France	3	8	10
Japan		5	
Italy	1	3	4

Table 1. Changes per capita consumption in some countries. *) excluding poultry. (Jul, 1984).

70 kg per cap.	
1 kg	sold frozen
9 kg	has been frozen before sale
10 kg	frozen in homes

Table 2. Danish meat consumption, 1982, from Jul (1984).

The renewed interest in the area, which to a consideable extent was due to the COST 91 Project, has helped to throw new light on and get better knowledge of many important subjects in this area, including those just mentioned.

2. RATE OF FREEZING

It is still a prevailing view that the quality of frozen foods depends greatly on the rate with which they are frozen; very rapid freezing is considered to be very advantageous. This view has probably been promoted by equipment manufacturers and aggressive advertising regarding some frozen foods products.

In scientific circles, it was slowly being accepted that normally, the speed of freezing within the range generally occurring in commercial practice has little, if any, effect on quality. Exceptions existed. While for instance taste and texture, etc., of the thawed, cooked product was unaffected by freezing method, it is well-known that a rapid surface freezing of turkeys will give these a whitish appearance in the frozen state which is much preferred over an ordinarily frozen bird by the consumer. Therefore, in this case, a rapid crust freezing often precedes an ordinary, slower blast freezing.

Several COST 91 projects, among these were studies by Verbeke, van Hoof and Lauwers, and also by James, Nair and Bailey, indicate that no difference can be found between slowly and rapidly frozen meats. On the other hand, Canet and Espinosa established that for potatoes, carrots, and peas, a quick-frozen product showed some organoleptic advantage over a slowly frozen one. Apart from such cases, it is generally realized as pointed out by Veerkamp, that freezing rate is mainly a factor which can be left to technical considerations, e.g. energy consumption, optimum turn-over in freezing equipment, etc.

3. SHELF LIFE

3.1 TTT-PPP factors

In many circles, it has for long been a general conviction that the keeping time of a frozen food depended very much on the storage temperatures and that keeping times were much improved by lowering temperature. Originally, it was assumed that keeping qualities almost increase exponentionally with decreasing temperatures as suggested in curve D of fig. 1.

Later on, research demonstrated that the effect of decreasing temperatures below 20^{0}C often is not as large as originally assumed, and it was accepted that a curve more like the one indicated in curve A of fig. 1 represented a normal keeping quality diagram.

Later on, especially Danish experiments demonstrated that there are some cases, especially when cured meats are frozen and stored aerobically, where one actually arrives at so-called reverse stability, i.e. a decreased keeping time as a result of lowering temperatures, e.g. curve B in fig. 1.

Where cured meat products were packed anerobically, one often finds that the keeping time is completely temperature neutral, i.e. one finds the same shelf life whether storage at -10, -20 or -30^{0}C is used, e.g. curve E of fig. 1.

Also, more and more attention has lately been shown to the so-called PPP-factors, i.e. the influence that product, process and packaging have on shelf life.

Product. One may for instance for pig meat find a situation as indicated in curves A and B of fig. 2. Here, curve A may illustrate shelf life of meat from a pig fed a ration with an only moderately small content of fats containing unsaturated fatty acids, while curve B may show the keeping quality of meat from a pig fed a diet very low in unsaturated fats. In the fresh state, the taste difference between the two would hardly be noticed, in the frozen state one product keeps approximately twice as long as the other.

During the course of COST 91 Project, Crivelli, Bertolo, Maestrelli and Senesi demonstrated how harvest time has a considerable influence on the shelf life of frozen vegetables, while Ludovico-Pelayo, Hume and Love made a similar demonstration with regard to trout where shelf life of the frozen product was dependent on whether the raw material was obtained from a well-fed or a starved fish. Similarly,

Ristić demonstrated that in the frozen state, water chilled poultry had a somewhat longer shelf life than air chilled birds.

Process. Similarly, the process which the product undergoes prior to freezing may influence the shelf life of the frozen product demonstrably. Thus, curve C in fig. 2 may indicate how the keeping quality of pig meat derived from the same animal as that represented by curve B may be improved if the meat is cooked prior to freezing. During the COST 91 Project, Petrak, Roseg, Kosmerl, Hraste and Jelić showed how cutting-up of chicken will result in a shorter shelf life, probably due to more exposed cut surfaces than when freezing the whole bird.

Other process factors were examined by Jimenez-Colmenero, Garcia-Matamoros and Pelaez, by Fuster, Prestamo and Espinosa, by Moral, Tejada and Borderias, and by Gormley.

Packaging may also influence shelf life remarkably. Thus, curve D in fig. 2 may indicate the shelf life of a vacuum packaged piece of cooked pig meat similar to the one illustrated in curve C. Especially, good vacuum packing may have a pronounced effect on keeping quality, especially at warmer temperatures. During the COST 91 Project, the effect of packaging on fish was demonstrated by Christensen and Jessen. Similar data were obtained by Ahvenainen and Mälkki.

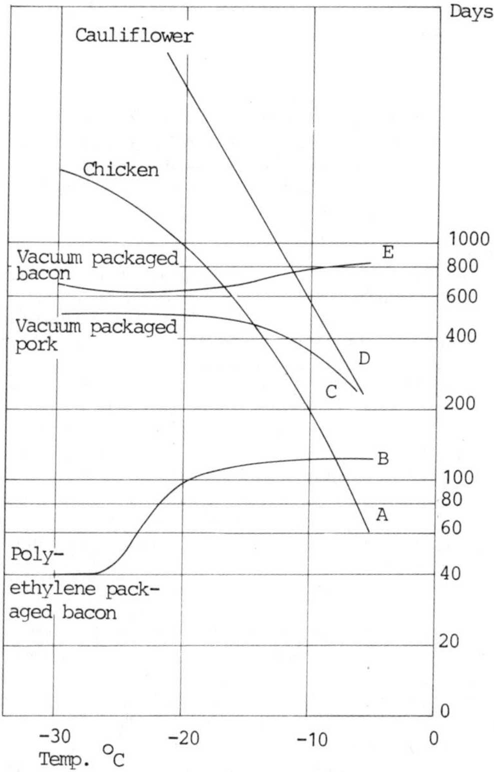

Figure 1. Estimated shapes of shelf life curves for various frozen foods.

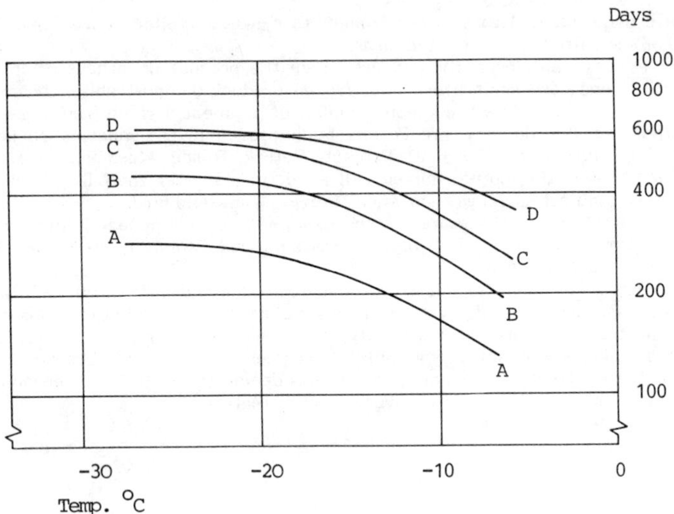

Figure 2. Examples of the influence of PPP factors on shelf lives of frozen foods. A. Pig meat with high iodine number in fat; B. Pig meat with low iodine number in fat; C. Same meat as B, but cooked; D. Same as C, but vacuum packaged.

3.2 Shelf life data

Prior to the COST 91 Project, our knowledge about the shelf life of frozen foods was not very satisfactory, and especially data on contemporary products, processes and packaging were lacking. Thus, fig. 3 from a literature study carried out by Spiess and quoted by Jul (1984) show that different authors have findings so different that the data are not usable in practice.

At the end of the COST 91 Project, Boegh-Soerensen summarized the data obtained for the shelf life of various frozen products during the Project, and has demonstrated that our knowledge about shelf life of frozen foods is now considerably improved. Firstly, Boegh-Soerensen has explained how it must be accepted that various groups of users in various geographic areas will have different requirements. Thus, shelf life depends not only on the condition of the product but also on the market and use for which the product is destined. This is a factor which need be kept in mind. It need also be stressed that it is a reality and that there is no singular definition of shelf life which will cover all cases.

During the Project, several attempts were made to find objective measurements which can be used in the determination of shelf life curves. Thus, Katsaboxakis and Papanicolaou showed how vitamin C analyses may be used. Others, like Bertelsen and Boegh-Soerensen, used a Hunter colorimeter to follow colour development in frozen beef, while Adams and Robertson compared objective and subjective colour measurements for vegetables. For minced fish, Lajolinna, Laine and Linko demonstrated the use of various chemical methods for shelf life determinations. From all these data, it must be concluded that such objective measurements can be useful tools but must correlate closely with the evaluation of quality as it accurs in the actual place for which the product is destined. By organoleptic methods, Ludwig found a shelf life of French fried potatoes of 18 months at -15°C.

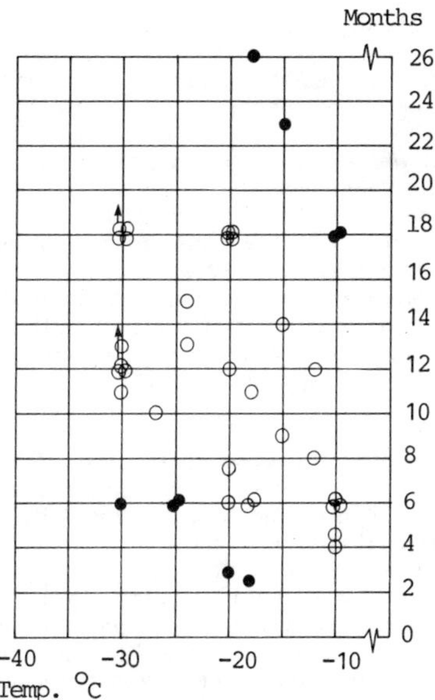

Figure 3. Shelf life (HQL) data for ready-to-eat meat dishes as reported by various authors, as collected by Spiess, and quoted by Jul (1984).

Thus, the COST 91 Project has provided good data with regard to shelf lives of various frozen products. Further, it has given good information about the shape of the curves to be expected from various products as exemplified in fig. 1. This means, as Winger has suggested, that we may use accelerated tests to obtain some idea of the shelf life of the frozen product. Thus, if we for a product know approximately what shape to expect of the shelf life curve, we may use tests at rather high temperatures, as warm as -5°C as suggested by Winger, to determine the level of the shelf life for the product in question. From this we may predict with a reasonably satisfactory degree of certainty the shelf life curve and use such predictions in our estimates with regard to the product's fate in the commercial freezer chain. The latter has been mapped out during the course of the project as it is described below.

Porsdal Poulsen has thrown additional light on our means of predicting the shape of the shelf life curve for various products.

3.3 Blanching

Above was mentioned that cooked products often keep better than uncooked products, a fact which normally is due to the destruction of enzymes in the product which otherwise might be active at freezer storage temperature.

The problem of pre-cooking, or blanching as it is called in this case, is particularly important for vegetables. The subject was under the auspices of the COST 91 discussed at a meeting in Paris in 1982, as described by Philippon, Ulrich and Zeuthen (1981). Where blanching of vegetables is used, the combination of time and temperature is very critical, e.g. as indicated by Katsaboxakis and also by

Katsaboxakis and Papanicolaou. As Steinbuch has suggested, one may obtain considerably better results by using a heat-shock method, i.e. very short time at high temperature exposure and immediate chilling. The blanching process may be monitored by determinations of enzyme destruction as suggested by Steinbuch and by McLellan and Robinson.

There may be cases where it is more advantageous not to blanch a vegetable. This is particularly true for vegetables which have very volatile flavour components, e.g. parsley. If they are not to be stored for too long, they may retain their flavour better if not heat treated at all. This will be particularly relevant where they can also be held at very low temperatures. This is hardly practical if they are to be sold through the normal commercial frozen food chain but where such products are held in the freezer storage rooms of one factory for use in further processing later, such freezing of non-blanched products may very well be practical and advantageous as it is described by Philippon, Rouet-Mayer and Abbas.

3.4 Calculating shelf life loss

Accepting that the losses in shelf life during freezer storage are both cumulative and commutative, it is simple to calculate the effect of a certain time and temperature treatment of a product once its shelf life curve is known, e.g. as described by Jul (1984). Thus, if one knew that a product would be exposed to a time-temperature experience as suggested in fig. 2, it would be easy to calculate if the product were likely to be of satisfactory quality when it reached the ultimate consumer. Of course, the time-temperature conditions to which various packages in a produced batch can be expected to be exposed will be very different. Fig. 4 indicates what might be expected for distribution of frozen foods in Denmark, fig. 5 what might be expected for frozen lamb exported from New Zealand to the UK.

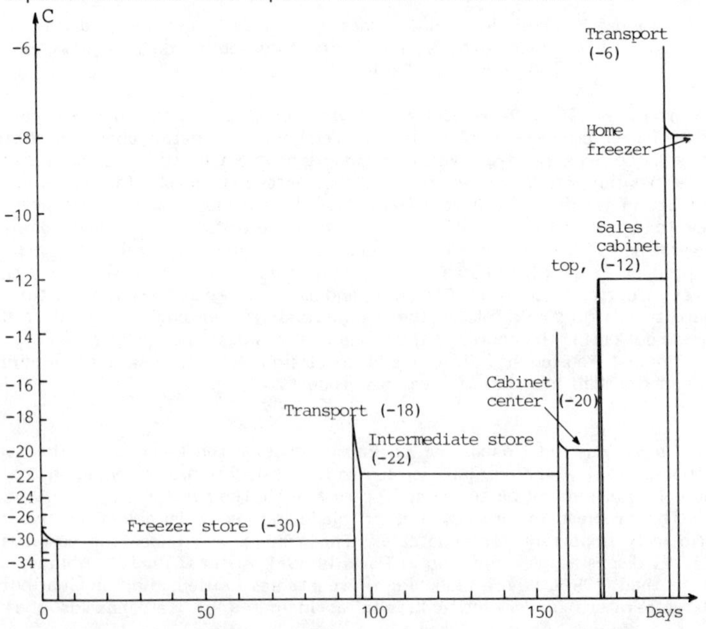

Figure 4. Times and temperatures in the distribution chain for frozen foods in Denmark, as collected by Kondrup and quoted by Andersen, Jul and Riemann (Jul, 1984).

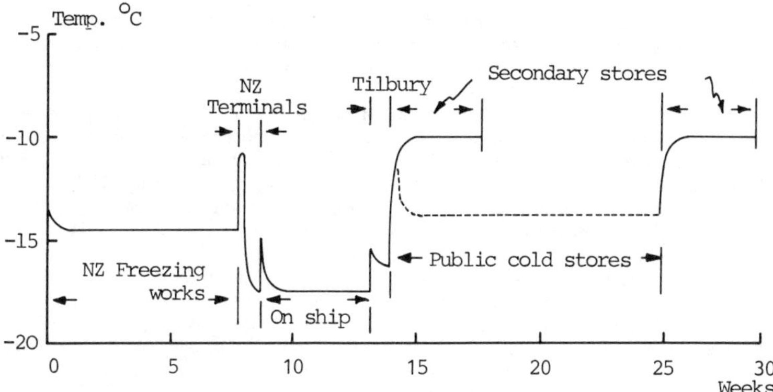

Figure 5. Times and temperatures for shipments of frozen lamb from New Zealand to UK, as collected by MIRINZ (Jul, 1984).

It is clear that such time-temperature curves can be designed to indicate ideal conditions, average condition, or even worst thinkable conditions considering that some products may stay in the factory or secondary freezer storage for long periods, others may move very quickly, some may stay in the top of a retail cabinet for long, while others may be sold almost immediately, etc.

3.5 Time-Temperature integrators

To help get some insight into this aspect of the frozen food chain, some frozen food integrators have been introduced, as described by Olsson. Such integrators could follow shipments of frozen foods. The idea is that they change colour at approximately the same rate as that with which shelf life is lost for a frozen product. Thus, a certain change in colour would indicate a certain loss of shelf life for the product. It is clear that integrators need be developed with many different time-temperature characteristics, cf. fig. 1. So far, such integrators have not been developed to the extent that they could be used for retail packages, but they are already in use for placement on pallets, etc. Here, allowance for differences between pallet surface temperature and product temperature need be made.

4. DATA FROM THE FROZEN FOOD CHAIN

4.1 Data collection

One large component of the COST 91 Programme has been studies of time and temperature conditions, etc., in the freezer chain. Very detailed surveys were made in the UK by Wignall, Potter and Storey, in Denmark by Boegh-Soerensen and Bech-Jacobsen, and for West Germany by Spiess and Folkers. With these studies we today have very good data for the distribution of time and temperatures in the various links of the frozen food chain. As indicated by Jul (1984), it may be practical to use in shelf life loss calculation the most unfavourable conditions of these applying at each stop for 75% of the products with regard to time and temperature. Since detailed data are now available on which to base such calculations we are now able with a considerably degree of accuracy to determine how various products will withstand the treatment to which they can be expected to be exposed under practical commercial conditions. It is particularly important that we now have these data also for various product groups. This is because there may be considerable difference, especially in storage times, between products which are seasonable like fruits and vegetables and those that are likely to be produced all year round, like meat and poultry.

Data for some products suggest that they stay only a very short time in the freezer chain; this applies for instances to cheeses as they are studied by Dalles, Kalatzopoulos and Kehagias and for bread by Gormley. For these, little need for concern seems to exist.

The time-temperature data obtained suggest that it may be reasonable and acceptable to consider some energy savings by increasing temperatures during freezer storage at the wholesale level. Such data are obtained by Houwing, Bon and Brünner. On the other hand, studies by Whittle, Young, Howgate, Craig and Smith suggest that at least for some fishery products it may well be necessary to advocate -30°C storage temperature. However, it does appear that if their quality criteria may be unnecessarily high.

The studies of the freezer chain also indicate that the transport conditions to which normally much attention is paid, is of very little importance, simply because the time during transport is generally very short. The only exception seems to be distant sea transports like transport of frozen lamb from New Zealand to Europe or the USA. However, here temperature managements seem to be adequate for the product transported, e.g. at an average of -17°C for lamb during a transport of approximately 6 weeks.

Studies by Spiess and Folkers suggest that it is advisable to sort products in a freezer storage room in such a way that delivery stops at retail outlets may be kept short. If the products are distributed unsorted, i.e. have to be sorted at the time of delivery from the van, considerable temperature abuse may ensue.

4.2 Freezer cabinets

Tests executed under the Project have thrown very important additional lights on conditions in freezer cabinets. It is generally stipulated that the temperature in these should be -18°C or lower with a tolerance of up to -15°C for short period under exceptional circumstances. Such requirements are in several countries legal instruments and they also form part of ISO/BSO standards. However, as Wignall, Potter and Storey have shown, the temperature in top layers of freezer cabinets in the UK may often be around -11 to -6°C, that means they differ quite markedly from the requirements. However, the same authors determined that the quality of the products as normally purchased in the retail trade in the UK is satisfactory. This suggests that frozen products are fully capable of maintaining adequate quality during the time and temperature conditions which are the prevailing ones in the existing frozen food distribution system. Thus, it seems that we need not enforce, let alone tighten, our present requirements with regard to freezer cabinets, as it is sometimes considered in several countries and also within the framework of the Commission of the European Communities. There may even be a case for making them less stringent and better in accordance with actual practice.

On the other hand, all the surveys clearly show that retail freezer cabinets are the weakest link in the distribution system for frozen foods. Thus, it may be useful as indicated by Philippon and described by Jul (1984) that results of time-temperature surveys in the retail chain periodically be made and the results communicated to the trade. Undoubtedly, such steps will serve to make the trade aware of the weak links and enable store suprvisors to maintain better cabinet management, etc.

Boegh-Soerensen and Bramsnaes have described that night covers might be used to considerably improve the condition in the cabinets and, as Sanderson-Walker suggests, Druelly glass in upright glass door cabinets may reduce heat radiation significantly.

As explained by Boegh-Soerensen and Bertelsen, exposure to light in freezer cabinets may very quickly be damaging to the appearance of frozen beef.

5. THAWING

In so far as usage in the homes is concerned, thawing frozen foods presents only minor problems.

However, as suggested in table 2, there is a fairly large and increasing use of raw material for further processing in the food industry. Here, special thawing installations are required. These are difficult to construct. One exacerbating fact is that the heat transmission rate is much lower in the thawed product than in the frozen product.

James, Nair and Bailey show that it is of some importance that thawing should be neither to slow or too fast. In general, it seems that programmed air thawing may be the method of preference, costs and benefits considered. Ohlsson studied microwave thawing systems but found that in general these are difficult to use. A combination microwave and vacuum thawing as suggested by James, Nair and Bailey may be a possibility, but appears expensive.

One process considered under this heading is tempering, i.e. heating the frozen product to just below the temperature where the main part of the ice would melt; this may be carried out advantageously by microwave or dielectric thawing. Such tempering is in widespread use, e.g. in the meat industry where the solidly frozen but not too cold products can easily be ground up and used in further processing. Since in tempering only little ice melting takes place, the cost of the energy input is less significant. Also, since the product is not actually thawed, run-away heating represents no problem.

6. NUTRITIONAL CONSEQUENCES OF FOOD FREEZING

It is generally recognized that freezing may lead to a considerable loss of vitamins, especially vitamin C for vegetables, as these are often blanched, i.e. a cooking process which involves a certain loss of vitamin C. Further, vitamin C loss is somewhat temperature dependent so that at cold storage temperature, i.e. $-20^{\circ}C$, the vitamin loss will be small. Somewhat surprisingly, Katsaboxakis showed that the loss in ascorbic acid in vegetables is not as high as normally expected at warm temperatures, e.g. $-12^{\circ}C$, if blanching has been adequate. Freezing and freezer storage also involves a certain loss in most vitamins of the vitamin B group as indicated by Mikkelsen, Rasmussen and Zinck. However, the loss is somewhat variable from 0 to about 40%. Interestingly, the loss seems to be practically temperature neutral, i.e. the same losses may be expected at $-30^{\circ}C$ as at $-8^{\circ}C$.

In all considerations regarding vitamin loss, etc., it is necessary to keep in mind as pointed out by Zinck that what matters is the viamin content in the final product as it is eaten. It very often turns out that the loss which was experienced during blanching and freezer storage to a consideably degree is compensated for by a lower loss during kitchen preparation. Therefore, in actual intake of nutrients it may be comparatively insignificant whether one component in the frozen food supply system has been frozen or not.

One somewhat critical factor may be pyridoxin. For the intake of this factor, the European margin is only slightly above what is recommended and for some groups actually under that level. In general, one may conclude as does Pietrzik that with the degree of freezing which is being used in a normal household today, one need not be overly concerned about losses in vitamin since intake seems to be fairly adequate. However, for special groups, or when new frozen foods are introduced, it seems prudent that considerations be given to this aspect.

In such considerations, however, it is generally overlooked that very considerable home freezing takes place. In any mapping of the nutritional consequences of food freezing, this aspect needs to be included.

More important is the question of the degree to which the introduction of food freezing may have altered our diets. Here, it seems obvious that intake of seasonal foods, i.e. many green vegetables and fruits and some fish, may have increased considerably because these foods may now be available in the frozen state all the year round. This suggests that the advent of food freezing actually may have meant a considerable improvement of the European diet.

Equally, important are considerations with regard to the use of freezing as a temporary preservation method for raw materials. As indicated in table 2, this practice is in quite widespread use, but few realize that many fresh, canned, frozen or cured products may at one stage during their manufacture have been frozen. It is necessary that further consideration be given to the nutritional consequences of this increasing practice.

A third important aspect is institutional feeding and food distribution systems. It appears as if cook-freeze systems and related practices are gaining in acceptance. Since special groups may be covered almost entirely by food supplied through such systems, these need to be monitored as regards the nutritional consequences.

7. PROPOSALS FOR FOOD FREEZING PRACTICES

It seems clear that the data collected in the course of the work of COST 91, Sub-Group on Freezing and Thawing, together with earlier data, when finally collected and analyzed, will provide a basis for drawing some important conclusions as regards these technologies. Some preliminary suggestions may already be mentioned:

7.1 Freezing

Generally, the method of freezing has little relevance for food quality except in a few cases where freezing rate may affect appearance of the frozen product. An example is poultry, where rapid crust freezing may result in a whitish appearance of the product, often preferred by customers. Another example is pig carcasses, where rapid blast freezing may result in an untidy appearance because of uncontrolled surface desiccation.

In general, freezing method may be selected in accordance with consideration for economy, including energy consumption, optimum utilization of freezer capacity and type of products frozen, etc. One ultimate demand is that all parts of the product should be below the freezing point before adverse microbiological activity can occur.

A product may be removed from the freezer, when sufficient heat has been removed from the product. This means that temperature may and will equilibrate in the freezer store or in the freezer transport vehicle.

7.2 Factory and wholesale storage

Especially for factory storage where often only one type of product, e.g. meat, fishery products or vegetables and fruits, are stored, temperatures may be selected in accordance with the shelf life characteristics of the product to be stored. It appears as if the following temperatures might be acceptable:

Beef and lamb carcass meat, poultry	$-15^{o}C$
Pork, processed meat except mechanically recovered meat	$-18^{o}C$
Vegetables (blanched where necessary) and fruits	$-18^{o}C$
Fish and fishery products, mechanically recovered meat and fish, the latter with limited storage time	$-24^{o}C$

7.3 Transport

Transport vehicles should maintain the temperatures indicated above but short time deviation up to $3^{o}C$ below the products freezing point, i.e. often up to $-4^{o}C$, may be tolerated. Retail shipments should be sorted prior to dispatch to various retail outlets.

7.4 Retail cabinets

Retail cabinets of contemporary good design are satisfactory. They should maintain a temperature of the most exposed product not warmer than -8°C. They need be so placed that they are protected from draft, excessive lighting, etc.

7.5 Retail management

The management of frozen foods at the retail level is in need of careful review. It is important to safeguard products against any exposure to temperatures above 3°C below their freezing point, and to ensure rapid turnover, i.e. a first in - first out type of management. A policy of sorting products periodically and arranging discount sales for products having stayed relatively long at the retail level might be encouraged.

7.6 Reverse stability

Since few products are actually harmed by colder temperatures within the temperature range normally found in the frozen food chain, no minimum coldest temperature need be observed.

7.7 Product, process and packaging factors

Each manufacturer may manipulate product, process and packaging factors which could result in products which would keep virtually unaltered even under abnormally adverse conditions in freezer storage and distribution. The above suggestions apply to products manufactured in accordance with what is presently considered good manufacturing practice.

7.8 Thawing

There is no hygiene or quality problem in using frozen raw materials for further processing into canned, cured or frozen products or for sale as fresh products. It is necessary, however, that thawing is carefully controlled and monitored.

7.9 Residence times

The above is suggested in the light of the data of residence times observed during the COST 91 Project in the various frozen food distribution chains, on the average less than 6 months for non-seasonal products and less than 12 months for seasonal products. In the above, a practical maximum time in the total chain of 18 months and a reasonable residence time in the retail sales cabinet is provided for.

8. LEGISLATION

The field of food freezing has caught the attention of many legislaters and food control authorities. Regulations exist as regards temperature, date marking, transport conditions, etc. It seems that many such regulations are based on insufficient data and an unjustified suspicion with regard to frozen food which in general are covered by more detailed regulations than many other foods. This is in spite of the fact that frozen foods are among those where the least health or quality concern exist.

One important result of the COST 91 Project in this regard is that many more data than were in existence hitherto have become available. In general, most of the information uncovered suggest that legislative steps are not required in this field. On the other hand, if such measures are found necessary as a result of public opinion and in spite of scientific advice, many data are now available on which such measures can be based. It seems that there is a potential for influencing legislation in such a way that the rules for food freezing and frozen food handling may be more relevant. Thus, Linko and Karhunen completed studies showing that bread, dough or similar products may be freezer stored at -10°C instead of the -20°C which are generally required in Finland, simply because the short times involved and the shelf life characteristics as determined by these authors have shown this temperature to be entirely satisfactory. Eventually, this led to modification of the rules regulating this practice.

9. CONCLUSION

It seems in general that the research which was a part of and sometimes even the result of the COST 91 Project in the feld of food freezing has resulted in a large number of data sufficient to justify stating that we today have a sound knowledge of the data on which both industry, trade and legislators can base whatever step they consider necessary in this field.

Further, the collaboration which was brought about through the existence of the COST 91 Project has proved most valuable. Contacts have been created between research establishments, between researchers and industry, and between researchers and users in such a way that these ties can be expected to continue. This is also important since new products, new processes, and new conditions will keep on appearing and be in need of further research.

10. REFERENCES

1. Jul, M. (1984). The Quality of frozen Foods. Academic Press, London.

2. Philippon, J; Ulrich, R; and Zeuthen, P. (1981). Compte Rendu de la Journée d'Etude Internationale sur le Blanchiment des Fruits et Légumes surgelés Etat des Connaissances Actuelles - Perspectves d'Avenir, CNRS, Laboratoire de Physiologie des Organes Végétaux aprés Recolte, Meudon.

THE TTT-PPP CONCEPT

LEIF BØGH-SØRENSEN
Department of Food Preservation,
The Royal Veterinary and Agricultural University

Summary

COST-91 projects on TTT (Time-Temperature Tolerance) and PPP (Product, Processing, Packaging) have greatly increased the existing knowledge of the storage life of frozen foods and of the factors (PPP) influencing storage life.
Product, i.e. nature and quality of raw materials, has been included in several studies: The seasonal variations in flavour change of fish, determination of optimum harvest time for vegetables (green beans), two different raw materials for manufacturing of cured herring, TTT-studies on cauliflower, French fries, minced beef, pork products, etc.
Processing. Blanching will be reviewed elsewhere, but it ought to be mentioned that several TTT-studies have been carried out on unblanched vegetables, in some cases vacuumpacked. Mechanically recovered meat and fish have been studied, as well as the influence of an ageing period prior to freezing of chicken parts.
Packaging. Besides the above mentioned, TTT-studies on different fish products have included determination of the influence of packaging.
PSL-data. Data on practical storage life (PSL) from different institutes are often inconsistent. Besides the PPP-factors, the different methods used for sensory evaluations and especially the different criteria used for assessment of PSL, are responsible for the conflicting data.
TTT-studies should preferably comprise a freezer chain simulation; especially, a certain residence time in display cabinets should be included.
Future. In spite of the comprehensive research carried out in the COST-91 project a number of important areas are in need of further study, e.g. the effect of using raw materials which have been frozen, and the storage life of prepared foods such as fish fingers, pizza, etc.

1. INTRODUCTION

Storage life experiments have been carried out since the beginning of the frozen food industry some 50 years ago, since it of course was necessary for the industry and the trade to know how long time the different frozen foods could be stored, still having an acceptable quality. The determination of storage life was in many cases carried out by experienced persons in the industry.

The TTT-concept (Time-Temperature Tolerance) was introduced in the "Albany" experiments, which were started in 1948. These experiments, and the resulting storage life data and the rule of additivity (cumulative quality loss) are very well known.

The PPP-concept (Product, Processing, Packaging) was introduced 20 years ago (Dalhoff & Jul, 1963). Since then, it has been widely recognized that the PPP-factors often have a remarkable effect on the storage life.

Now, storage life studies on frozen foods often are TTT-PPP studies, meaning that the storage life is determined at different storage temperatures, and the influence of an alteration in at least one of the PPP-factors is included. The impact of COST-91 on the present knowledge about the PPP-factors is dealt with in section 3.

2. DEFINITIONS OF STORAGE LIFE

A number of different definitions (types) of storage life is used, e.g. HQL, JND, stability time, shelf life, practical storage life (PSL), acceptability time.

2.1 HQL

The HQL concept was introduced in Albany. Rouet-Mayer et al (1983) proposes the following modification of the definition of HQL: "the time until a trained taste panel using triangle tests finds a statistically significant difference between the sample and a control kept at a very cold temperature". HQL is well suited for research work, but the HQL-concept has some drawbacks.

Control samples. For most frozen foods, the controls should be stored at -40°C or colder, preferably in special packaging, which will minimize quality changes. However, in recent years some frozen foods have been found with a very limited difference between storage life at -30°C and at -40°C, and even with storage life longer at -30°C than at -40°C, cf. Lindeløv (1978) Bøgh-Sørensen & Højmark Jensen (1981). In such cases, it would take a long time before a taste panel would be able to distinguish samples at -30°C from control samples at -40°C. Thus, the HQL found would be rather misleading, and in the case of reverse stability completely wrong - the test sample being better than the control.

When freezing products such as tomatoes, leafy vegetables, etc., HQL would be of very little value, as the controls as well as the samples would have a very low quality right after the freezing process.

Stability time is used in some Danish research institutes, where the samples are scored using a 10 (11) point scale. Stability time is defined as the time until the score has dropped one point from the initial value, i.e. the score right after freezing.

2.2 PSL (Practical Storage Life)

Whereas the types of storage life in 2.1 may be of value in research, they are of very limited interest for the frozen food industry, the trade and the authorities.

Here, the question is whether the consumers are likely to accept the products or not. PSL (Practical Storage Life) at a given temperature is the time the product can be stored and still be fully acceptable to the consumer; shelf life and acceptability time is about the same as PSL.

In practice, it is difficult to determine PSL at research institutes. Each institute has its own way of evaluating the quality and its own scale, system, etc., and it is not easy (perhaps not even worth-while) to harmonize the evaluation systems used. Also, each institute has its own way of defining what the consumer will accept and reject, which in any case is very difficult to decide, using - in most cases - a trained laboratory panel which tastes the samples alone without the normal ingredients used in the home. Such panel members detect off-flavours and other quality defects much quicker than do consumers, which makes it difficult to relate laboratory panel results to consumer reactions.

The definition of PSL, i.e. the limiting quality, varies between institutes. The Danish Meat Products Laboratory uses -1 in a +5/-5 hedonic scale, The Technological Laboratory, Ministry of Fisheries in Denmark uses 4 in a 0-10 scale, Karlsruhe uses 6 in a 0-10 scale, and the Instituto del Frio,

Madrid, 3 in a 1-5 scale; all these institutes use trained laboratory panels, e.g. food scientists, technicians, etc.

Obviously, the best way of determining PSL is using consumer tests, with a large number, e.g. several hundreds of tasters. This is time-consuming, impractical, and expensive, and it is not always possible to obtain that number of practically identical samples. Torry Research Station recently has used this method (Howgate, 1983), and SIK, Sweden, has sometimes used 200-300 housewives for establishing the storage life of a product (Olsson, 1981). Apart from these institutes there seems to be limited experience in using consumer tests for assessment of acceptability in TTT-experiments.

This leads to the conclusion that there is a need for exchanging information about the systems used for determining shelf life; this does not implicate the same evaluation system to be used everywhere, but simply that shelf life data from one institute should be understandable for another institute, e.g. in another country. This gives problems as consumers from different countries or from different regions in a country do not have the same likings. Steinbuch et al (1977) compared two varieties of green beans (Record from the Netherlands and Jutta in West-Germany). For the Jutta variety, the quality, especially the texture decreased much more rapidly than for the Record variety. The Record variety could be stored for more than two years at $-24^{o}C$, while the Jutta variety hardly tolerated one year. The authors also found that the taste panels in Karlsruhe preferred a dark green colour, while the panel in the Netherlands preferred a pale green colour, concluding that it was difficult to translate storage times found by subjective methods from one country to another.

Objective methods to determine storage life have been used extensively. In some Albany-experiments, the change of chlorophyl into pheophytin was measured. In many cases, the objective method resulted in the same storage life as did subjective (organoleptic) tests.

In a number of COST-91 projects objective methods have been used to follow the changes during freezer storage. Rouet-Mayer et al (1983) used chemical analysis for chlorophyl, the CENTURY instrument, and sensory evaluation. Until more knowledge is achieved, it must be concluded that objective methods should not be used unless followed with subjective methods.

A special case is analysis for vitamins. In Albany, the content of vitamin C (plus dehydroascorbic acid and diketogulonic acid) was analyzed in a number of fruits and vegetables, and in recent Danish experiments (Mikkelsen et al, 1983) the content of vitamin B_1, B_2 and B_6 has been determined in meat after freezer storage. It might be a good idea to lay down a certain loss of vitamins as a fixed storage life limit. For frozen fruits (e.g. strawberries) where the total of ascorbic acid, dehydroascorbic acid and diketogulonic acids remains constant, a certain loss of vitamin C could be a reasonable limiting quality (Guadagni & Kelly, 1958), but for a number of frozen vegetables this method would not be applicable because the total decreases during freezer storage (Dietrich et al, 1957). For meat and meat products, a certain loss in (one of the) B-vitamins could be used as a criterion, but due to highly fluctuating analytical values during freezer storage, and because the storage temperature seems to have little influence on the loss of B-vitamins, this approach is not applicable to-day.

2.3 TTT-curves

TTT-PPP experiments preferably should include several storage temperatures, at least three, e.g. $-12^{o}C$, $-18^{o}C$, and $-24^{o}C$. When the storage life has been determined at each storage temperature, the results are plotted in a diagram as shown in fig. 1, and TTT-curves are drawn.

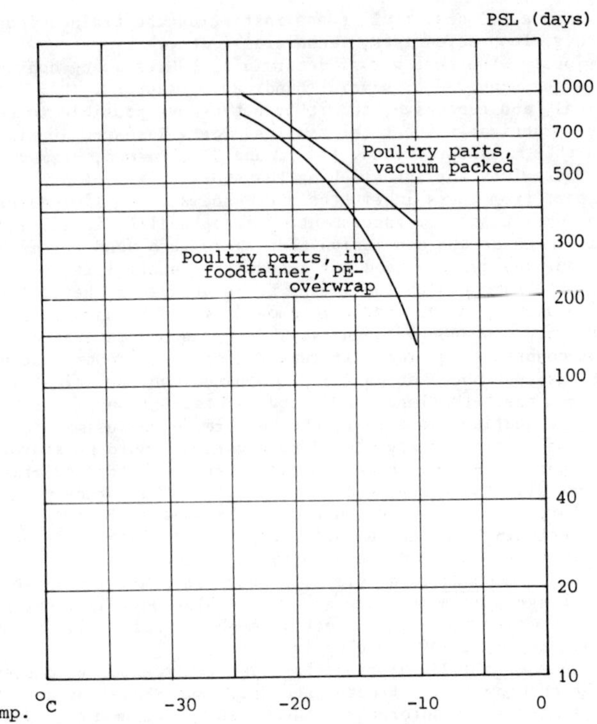

Fig. 1. PSL of chicken parts (drumsticks), packed in a foodtainer, overwrapped with PE, or vacuumpacked. After Bøgh-Sørensen et al (1978).

HQL-curves or PSL-curves can be drawn, and although in fig. 1 a semi-logarithmic ordinate axis has been used, a linear axis is often equally suited (Spiess, 1981).

TTT-curves are preferable to TTT-tables, as TTT-curves make it possibl to read the storage life (PSL) at other storage temperatures, which is of importance when calculating shelf life losses.

3. TTT-PPP in COST-91

Jul (1983) has reviewed existing data before the commencement of COST- regarding the knowledge about quality of frozen foods, including the role o the PPP-factors. In this section, the situation after COST-91 will be summa rized.

Most of the COST-91 research projects regarding TTT-PPP have dealt wit one product or one product group. Here, therefore, the situation within the main product groups studied will be described.

3.1. Fruits and vegetables

The "Albany" TTT-experiments were mainly on frozen fruits and vegetab since then, and until COST-91, comparatively few TTT-experiments have been carried out in this field.

Blanching is reviewed by Philippon (1983), and here it should only be mentioned that the possibilities of improving (reducing) or omitting the blanching process has drawn considerable interest in recent years. Baardseth (1978) stored blanched and unblanched vegetables at -20°C and -30°C, and concluded that for instance unblanched onions could be stored for at least 15 months, while carrot, cauliflower and French bean had to be blanched, a 5% residual activity of peroxidase being acceptable.

Variety. The importance of selecting the right variety of fruits and vegetables is widely recognized by research and industry, cf. Steinbuch et al (1977).

In COST-91, several projects have concerned frozen beans. Philippon et al (1983a) tested unblanched French cut beans for chlorophyl a/chlorophyl b in the raw beans, and for colour and flavour of cooked beans. The colour of the raw unblanched beans was fairly stable at temperatures at -20°C or colder, but the decrease in flavour is much more rapid than in blanched beans, see fig. 2. It must be stressed that the authors have drawn a straight line, and that the curve shown in fig. 2 is the responsibility of the author of this paper. However, it also indicates that there are quite large uncertainties in drawing TTT-curves, besides the well known large uncertainties when determining the storage life at one (constant) storage temperature. In fig. 2, HQL for blanched beans from the Albany-experiments is shown, too.

Katsaboxakis and Papanicolaou (1983) found unblanched green beans unacceptable after 1 month at -18°C.

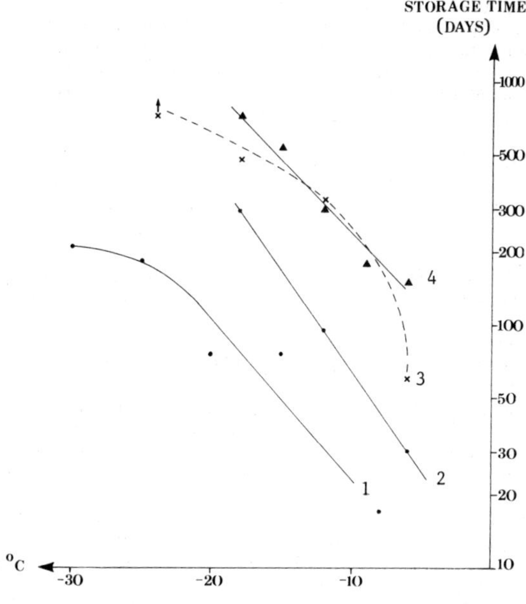

Fig. 2. TTT-curves for different frozen vegetables.
1: Unblanched beans, HQL. Adapted from Philippon et al (1983).
2: Blanched beans, HQL. After Guadagni (1968).
3: Blanched cauliflower, PSL. After Fuster et al (1983b).
4: French fries, PSL. After Ludwig (1983).

For green beans as for most other foods, an optimal quality of the raw material (Product) is essential. Crivelli et al (1983) determined the optimum harvest time for three varieties of green beans, using size distribution, seed/pod ratio and seed length as parameters. Adams and Robertson (1983) compared sensory evaluation and colour measurements of green beans blanched for 0-60 seconds; chlorophyl a correlated with good colour, and pheophytin a with poor colour.

Parsley. Philippon et al (1983b) found that for parsley the blanching process will cause serious aromatic losses, but also stabilize the aromatic quality during storage. Thus, the loss of aroma is due to enzymatic processes. Also, it was found that for small packages (500 g) of unblanched parsley the use of a plastic packaging with low permeability to oxygen and aroma substances will increase the storage life. The increase is approximately similar to the effect of lowering the storage temperature 8°C. Rouet-Mayer et al (1983) found for unblanched frozen parsley large differences from one year to another; in one year, HQL at -20°C was 106 days, the next year it was only 23 days. It was concluded that unblanched parsley should be stored at -25°C or colder. In the case of long storage times it is necessary to use blanching

Processing was studied by Fuster et al (1983a). For mushrooms, dipping in metabisulfite instead of blanching gave better quality; vacuum packaging was better than polyethylene (PE) bags. For apple slices, calcium and sulfite treatments improved quality. For four varieties of strawberries it was concluded that it is not beneficial to apply any treatment prior to freezing. Espinosa et al (1983) studied the influence of Process (blanching/not blanching) and Packaging (vacuumpackaging/normal packaging) for carrots. Preliminar results were that the blanched carrots can be stored at -12°C for about 280 days in a PE-pouch, and more than 310 days when vacuumpacked.

Fuster et al (1983b) stored blanched cauliflower, where colour (appearan is the critical parameter. At -6, -12, -18 and -24°C, the following PSL (mont were found: 2, 11, 16, more than 18, respectively. A PSL-curve is drawn in fig. 2.

Ludwig (1983) determined PSL for French fries at five temperatures from -6°C to -18°C, see fig. 2. It was concluded that if the maximum storage time required is 18 months, -15°C would be sufficient for French fries.

3.2 Meat and poultry

Bech-Jacobsen and Bøgh-Sørensen (1983) determined stability time and PSL of four retail packed pork products at -12°C, -18°C, -24°C, -30°C and -40°C. For most of the products the quality loss over time was linear at cold storage temperatures, and nonlinear at -12°C. The TTT-curves looked rather normal, except for vienna sausages where the max. storage life was at -30°C, -40°C giving about the same PSL as -12°C.

Bertelsen and Bøgh-Sørensen (1983) examined minced ground beef in trans parent PE where one of the main problems is the colour degradation during display. Three days exposure to the lighting in a display cabinet caused appreciable colour changes. The temperature conditions were of much less importance.

Mechanically recovered meat (MRM) could be a valuable contribution to the meat supply, but the storage life is rather limited, also in the frozen state. Jimenez-Colmenero et al (1983) have studied washing as a treatment prior to freezing. Washed, vacuum packed MRM could be stored at least 3 months at -12°C.

Ristic (1983) tested the influence of an ageing period of 1-8 days on the storage life of chicken parts. Freezer storage at -15°C and -21°C resulted in the same shelf life (PSL), which is given as 12 months for leg muscles and 16 months for breast muscles. Also, it was found that water chilling gave better quality than air chilling, and that ageing for 3 or 6 days improved the taste.

3.3 Bread

Storage life data for frozen bread are limited. Gormley (1983) tested sliced white bread, packed in PE-bags. After 4 weeks at -20°C, no difference was found between the frozen bread and fresh bread.

Linko and Karhunen (1983) found that quick freezing and cold temperatures had a negative effect on doughs, which should be stored for one week, at a temperature not colder than -10°C. The reason for this seemingly reverse stability is killing of yeast below -10°C.

3.4 Fish

Houwing (1983) stored mackerel and herring in blocks, matjesherring, cod and plaice fillets, boiled shrimps and cooked mussels at -19°C, -22°C and -25°C. After one year, the quality was significantly changed, but still acceptable. There was very little difference between -19°C and -25°C, indicating that it might be possible to store frozen fish at -20°C, thus saving energy, as a 1°C rise in temperature could save 6% energy. The influence of freezing method and packaging was less pronounced than predicted.

Christensen and Jessen (1983) stored cured fish products, i.e. matjesherring and "gravad" Greenland halibut at -3°C, -12°C, -24°C (and -30°C for gravad halibut). For matjesherrings two different raw materials were used; the use of frozen herrings resulted in better quality and better PSL than the use of fresh herrings. For "gravad" halibut vacuumpackaging resulted in considerably longer storage life. In all cases, a normal stability was found.

Ludovico-Pelayo et al (1983) studied the Product factor, i.e. the seasonal variations of rainbow trout. The degree of unsaturation of the fatty acids decreased from January to July, and less cold-store flavour was developed (during 6 weeks at -15°C). The authors concluded that, as in cod, there is a seasonal variation in the cold-store flavour of trout, which influences the storage life. But in other respects, frozen rainbow trout behave differently from frozen cod. One of the reasons is the relatively large quantity of triacylglycerols in trout, while in cod there is mainly phospholipids.

Tejada et al (1983) examined the stability of different parts of vacuum-packed rainbow trout, with and without additives. All lots were acceptable after one year at -18°C.

Fish mince (minced fish). Often, fish processors use machinery to separate fish meat from skin and bone, thereby increasing the yield of edible fish. Whittle et al (1983) compared the storage life at -15°C of fillet and mince blocks from different species. The quality of fish fingers prepared from the different raw materials was evaluated by means of a rather large panel, i.e. 42 staff members. The shelf life limit was the point at which 75% of the panel members still liked the product. Fillet blocks had PSL above 6 months at -15°C, while most mince blocks had no PSL, as the initial quality was not sufficiently high to meet the above criteria. The authors conclude that storage at -30°C is essential.

Moral et al (1983b) added protein protectors and antioxidants to sardine mince; also, one lot was washed with ice water. For vacuumpacked sardine mince PSL at -20°C was 4-8 months. Moral et al (1983a) stored vacuumpacked trout mince, with the same additives as above, at -12°C, -18°C, and -24°C. All lots maintained a good quality for one year, except the lot at -12°C without additives.

4. DISCUSSION

Product

COST-91 has included several experiments on the nature and quality of the raw material. For vegetables, the variety and stage of maturity have been studied, especially regarding methods for objectively determining the quality of the raw materials.

For fish, the seasonal variations have been found to influence the keepability very importantly.

A holding period prior to freezing has been known to influence PSL, often negatively, as the enzymatic activity created during this period (ageing) could reduce PSL. However, for chicken parts, a certain ageing was found to increase PSL.

The influence of the microbiological quality on PSL, still is an area where limited knowledge is available.

The influence of feeding on PSL of meat is another area where further studies would be welcomed.

Processing

Blanching of vegetables has been one of the main aspects studied by subgroup 3, and undoubtedly will be the subject of much research during the next years.

Mechanicalrecovery of meat and fish flesh is a very promising process, and it is important to study how a reasonably PSL can be achieved. This area is comparatively new, and further research is necessary. One of the interesting aspects is the storage life of frozen products with frozen mince as raw material.

Product formulation, e.g. the influence (positive or negative) of different additives on storage life has only be studied in a few projects, but is a very important area.

Packaging

In several studies, vacuumpackaging was used, and in some studies the influence of different packaging methods was studied. For fish, vacuum-packaging increased PSL less than expected in one project, while it about doubled PSL for lightly cured fish in another project.

Generally, a good packaging, i.e. a tight fitting packaging with low permeability and sufficient mechanical strength will increase storage life, expecially at rather warm temperatures, and especially for fatty foods.

TTT-PPP experiments

Our knowledge about PSL of different frozen foods have been greatly increased through the COST-91 projects.

For manufactured products such as fish fingers, pizza, and prepared meals, it is regrettable that so little information is available. The problem is complicated, as different manufacturers use different PPP-factors (different raw materials, different processes and recipes and different packagings), thus necessitating a number of studies for determining the influence of each factor and of different combinations of the PPP-factors.

Other frozen foods have been marketed for years, e.g. fruits and vegetables, and still the information about PSL is mainly based on the experience of industrialists. This is by no means a bad thing, but complicates matte concerning international considerations of the storage life (PSL) of differe foods.

Storage temperatures

In order to save energy, the possibilities of using warmer storage temperatures have drawn considerable interest in recent years.

Several studies have included this aspect and there is no doubt that for several frozen foods, PSL is rather independent of temperature. This means that it is not sufficient to determine PSL at for instance -18°C; knowledge of the TTT-characteristics is necessary, too. It might be possible to divide frozen foods in two or three categories (high - medium - low stability) and use different storage temperatures for each group, e.g. -12°C for products with high stability and -28°C for products with low stability. However, before being able to recommend this approach, more information is needed about TTT and PPP.

The TTT-curves in the Albany experiments were nearly always straight lines (in a semilogarithmic diagram), see fig. 2. The reason could be that fruits and vegetables normally behave like that, but another explanation could be the method used, see 2.1. Anyway, some COST-91 projects indicate that not all fruits and vegetables exhibit linear TTT-curves. TTT-studies on meat and poultry very often result in non-linear TTT-curves.

Future TTT-PPP studies should include the influence of the temperature conditions found in the real freezer chain, e.g. by placing samples in display cabinets for some time. The temperature conditions to be applied can be found in the time-temperature surveys which have been a substantial part of the projects in COST-91.

REFERENCES

Adams, J.B. & Robertson, A. (1983) Correlation of sensory and instrumental methods in the determination of colour in frozen green beans. COST-91, Athens.

Baardseth, P. (1978). Food Chemistry, 3, p. 271-282.

Bech-Jacobsen, K. & Bøgh-Sørensen, L. (1983) TTT-studies of different retail packed pork products. COST-91, Athens.

Bertelsen, G. & Bøgh-Sørensen, L. (1983) Colour stability of frozen ground beef. COST-91, Athens.

Bøgh-Sørensen, L. & Højmark Jensen, J. (1981) International Journal of Refrigeration, 4, No. 3, 139-142.

Bøgh-Sørensen, L., Højmark Jensen, J. & Jul, M. (1978) Konserveringsteknik (Food Preservation Technology) vol 1. DSR Forlag. Copenhagen.

Christensen, A.T. & Jessen, B. (1983) TTT-studies of cured fish products. COST-91, Athens.

Crivelli, G., Bertolo, G., Maestrelli, A. & Senesi, E. (1983) Suitability of some maturity index to determine optimum harvest time of green beans for freezing. COST-91, Athens.

Dalhoff, E. & Jul, M. (1963) Factors affecting the keeping quality of frozen foods. Proceedings of 11th International Congress of Refrigeration. Munich.

Dietrich, W.C., Lindquist, F.E., Miers, J.C., Bohart, G.S., Neumann, H.J. & Talburt, W. F. (1957) Food Technology, February 1957, p. 109-113.

Espinosa, J., Prestamo, G., Fuster, C. & Canet, W. (1983) TTT of frozen vegetables. Carrots. XVI intern. Congress of Refrigeration. Paris. C2, p. 611-616.

Fuster, C., Prestamo, G. & Espinosa, J. (1983a) Influence of treatments prior to freezing on the quality and stability of fruits and vegetables during frozen storage. COST-91, Athens.

Fuster, C., Prestamo, G., Canet, W. & Espinosa, J. (1983b) Detriment in cauliflower quality at different temperatures during frozen storage. XVI intern. Congress of Refrigeration. Paris. C2, p. 53-58.

Gormley, T.R. (1983) Acceptability of frozen white sliced bread. COST-91, Athens.

Guadagni, D.G. & Kelly, S.H. (1958) Food Technology, December 1958, p. 645-647.

Howgate, P. (1983) Measuring the storage lives of chilled and frozen fish products. Proceedings of 16th International Congress of Refrigeration. Paris. C2, p.434-442.

Houwing, H. (1983) Energy Saving in the frozen fish industry. COST-91, Athens.

Jimenez-Colmenero, F., Garcia-Matamoros, E. & Pelaez, M.C. (1983) Pretreatment of mechanically recovered meat for freezing and stability of mechanically recovered meat during frozen storage. COST-91, Athens.

Jul, M. (1983) The Quality of Frozen Foods. Academic Press (in press).

Katsaboxakis, K.Z. & Papanicolaou, D.N. (1983) The consequences of varying degrees of blanching on the quality of frozen green beans. COST-91, Athens.

Lindeløv, F. (1978) TTT-examination of frozen cured meat products. Ph.D. thesis (in Danish). Food Technology Laboratory. Technical University of Denmark.

Linko, P. & Karhunen, A. (1983) Quality considerations in freezing of dough and baked products. COST-91, Athens.

Ludovico-Pelayo, L., Hume, A. & Love, R.M. (1983) Seasonal variations in flavour change of cold-stored rainbow trout. COST-91, Athens.

Ludwig, J.W. (1983) The effect of the storage temperature at prolonged storage on the quality of deepfrozen French fries. COST-91, Athens.

Mikkelsen, K., Rasmussen, E.L. & Zinck, O. (1983). Retention of vitamin B_1, B_2, and B_6 in frozen meats. COST-91, Athens.

Moral, A., Tejada, M. & Borderias, A.J. (1983a) Stability of frozen trout. I. XVI Intern. Congress of Refrigeration. Paris. C2, p. 395-401.

Moral, A., Tejada, M. & Borderias, A.J. (1983b) Conservation de hachis de sardine en entreposage à l'état congelé. I. Hachis obtenus de sardine entière, et de sardine étêté et eviscéré. COST-91, Athens.

Olsson, P. (1981) In: Minutes of COST-91, subgroup 3, meeting in Karlsruhe.

Philippon, J. (1983) General aspects of fruit and vegetable blanching. COST-91, Athens.

Philippon, J., Rouet-Mayer, M-A. & Abbas, J. (1983a): Haricots Mange-Tout congelés sans blanchiment prealable. Relation entre la qualité et les facteurs temperature et durée d'entreposage. COST-91, Athens.

Philippon, J., Rouet-Mayer, M-A., Duminil, J-M. & Fontenay, P. (1983b) Influence du blanchiment, de l'enriettement des feuilles et de l'emballage sur la stabilité de l'arôme du persil congelé (in press).

Ristic, M. (1983) Shelf life of poultry parts in dependence of time of preparation. COST-91, Athens.

Rouet-Mayer, M-A., Philippon, J., Fontenay, P. & Duminil, J-M. (1983) Alteration de l'arôme et des chlorophylles du persil congelé non blanchi en fonction des facteurs temperature et durée de l'entreposage (in press).

Spiess, W.E.L. (1981) In: Minutes of COST-91, Sub-group 3, meeting in Karlsruhe.

Steinbuch, E., Spiess, W.E.L. & Grünewald, Th. (1977). Bull. Inst. Int. Refrig. (Annex 1977-1), p. 239-246.

Tejada, M., Borderias, A.J. & Moral, A. (1983) Stability of frozen trout. II. XVI Intern. Congress of Refrigeration. Paris. C2, p. 389-394.

Whittle, K.J., Young, K.W., Howgate, P., Craig, A. & Smith, G.L. (1983) Storage life of fish minces. COST-91, Athens.

INFLUENCE OF DIFFERENT FREEZING AND STORAGE CONDITIONS ON QUALITY AND CHARACTERISTICS OF BEEF CUTS

R. VERBEKE*, J. VAN HOOF** and H. LAUWERS***

Laboratory of Chemical Analysis of Food from Animal Origin*, Laboratory of Hygiene and Technology** and Laboratory of Veterinary Anatomy***, Faculty of Veterinary Medicine, University of Ghent, Belgium

Summary

Freezing, storage temperature and time conditions used in domestic freezing were shown to affect some quality aspects of polyethylene packed beef cuts. Some parameters such as WHC, MFI, extractable myofibrillar proteins, myoglobin and lipid oxidation were significantly altered in frozen-thawed cuts as compared with non-frozen cuts. However, freezing and thawing rates as used in the present study had no demonstrable effect on sensory quality characteristics (drip loss, colour, tenderness). This indicates that domestic freezing and thawing procedures were of minor importance for the eating quality of the meat. Time of storage has been proven to be the main factor enhancing chemical and physical deterioration of the muscle tissue. Histological and microbiological changes due to freezing and/or thawing were also observed.

1. INTRODUCTION

Reliable data concerning the effect of freezing, frozen storage and thawing as practiced in household circumstances on eating quality of retail meat cuts are scarce. Analogous to current industrial freezing and storage conditions high freezing rates combined with low storage temperatures (<-18°C) are generally assumed to be less detrimental for the meat quality as compared with slow freezing rates and storage temperatures above -18°C. To elucidate the effect of these parameters a comprehensive study including physical, chemical, biochemical, morphological and microbiological changes occurring during domestic freezing operations has been performed.

2. EXPERIMENTAL

Meat samples originated from a 33 months old steer (White-Red of East Flanders) with a slaughtered weight of 434 kg. Both carcass halves were chilled within 48 h after slaughter to a core temperature of +4°C. At that time, back ribs from the 6th thoracic through the 6th lumbar vertebra were severed from the carcass halves and further stored for 7 days at +2 to +3°C Nine days after slaughter both joints were deboned and stored under commercial conditions for a further 2 days at +2°C. After a total aging time of 12 days the joints were cut into 46 steaks (2 x 23) of approximately 3 cm thickness each with an average weight of 528 ± 92 g. Slicing of the meat occurred under laboratory conditions on a clean surface using non-sterile utensils.

Immediately after cutting and weighing of the slices, steaks were aerobically packed into 0.025 mm thick polyethylene bags and, after removal of the enclosed air by hand pressure, tied up. This type of packing was deliberately chosen in order to apply a procedure as currently used for household purposes.

Six reference steaks remained unpacked and were examined without freezing. The remaining 40 steaks were divided in 4 different groups at random and frozen at -12°C, -18°C, -24°C and -34°C respectively. After reaching the final temperature, the steaks were subdivided in 6 different groups (6 steaks each) according to the following freezing and storage combinations :

group	number of steaks	freezing temperature (°C)	effective freezing time (h)	storage temperature (°C)
1	6	-12	48	-12
2	6	-18	24	-18
3	6	-24	15	-18
4	6	-24	15	-24
5	6	-34	21	-18
6	6	-34	21	-24

Freezing was carried out in upright domestic freezing cabinets. After 2, 12 and 24 weeks of storage respectively, 2 samples from each subgroup were thawed-in-the-bag, either overnight (11-12 h), at +3°C or at room temperature (+20°C) for approximately 1 1/2 h. During freezing, frozen storage and thawing, core temperatures of steaks from each group were continuously measured by an inserted thermistor probe connected to an electronic YSI-telethermometer 42SC and recorded.

Physical, chemical, microbiological and histological examinations were performed on m. longissimus dorsi samples. pH-Values were measured by direct insertion of a combined electrode. The water-binding capacity (WHC) of the meat was evaluated using different methods : a) by weighing the amount of drip fluid collected after thawing of the slices, b) by determining the drip loss after centrifugation (15), c) by measuring the amount of water bound after equilibration of myofibril suspensions at pH 7, d) by weighing the cooking loss (75°C, 1 h) from the samples used for texture evaluation.

Surface appearance of the thawed samples was directly measured by the Göfo-value and the Y-value (EEL-photometer). The haempigment was extracted and measured as described by Hornsey (5). Percentage oxymyoglobin was obtained from the absorption spectrum of a water extract of the meat between 400 and 700 nm (18).

Meat tenderness was evaluated with an Instrom Tenderometer 1140, mounted with a Warner-Bratzler shearing device according to the EEC adopted standard procedure for sampling and measuring of meat texture (5).

The quality index was determined after homogenisation of minced tissue in a metaphosphoric/NaCl solution as described by Khan and Lentz (10). The filtrate was assayed for sulfhydryl groups using Ellman's reagent (16) and phenol-reagent positive materials by Folin-Lowry. The ratio of the two tests (µg glutathione-SH per g muscle/µg tyrosine-N per g muscle) has been termed the quality index (10).

Myofibrils were isolated from LD muscles and the myofibrillar index (MFI) determined as described by Olson et al (12). Myofibrillar proteins were extracted from the myofibrils during 10 min. with a) Guba-Straub solution (G&S extract, pH 6,5), b) Hasselbach-Schneider buffer (H&S extract) and centrifuged (17). The extraction yield is expressed as % of the total protein in myofibrils. Protein concentrations were measured using a microbiuret method standardized against bovine serum albumin.

SDS-polyacrylamide gel electrophoresis (7,5 % gel) of the extracts from myofibrils with the buffers were used according to the method of Weber and Osborn. The gels were stained with Coomassie Brilliant blue (0,2 %). After destaining, densitometry of the gels was performed with a Zeiss KM3 equipped with a scanner ZK3 at 610 nm.

The non-protein nitrogen of drip was obtained after precipitation with an equal volume of 30 % trichloroacetic acid. Fat was extracted from tissue by the method of Bligh and Dyer and the thiobarbituric acid number was determined as mg malonaldehyde/kg. The fatty acids, the peroxides and fatty acid composition of the fat tissue was determined by standard methods.

Microbiological examinations were carried out by the enumeration of total aerobic mesophilic (PCA, 72 h at 30°C, with overlayer) and psychrotrophic organisms (PCA, 10 days at +5°C), enterobacteriaceae (VRBG, 24 h at 30°C, with overlayer) as well as yeast and moulds (Glucose-yeast-OTC-gentamycin agar) according to Mossel (11).

For the histological examination frozen samples were collected from groups 1, 2, 4 and 6 both immediately after freezing, and after storage for 2, 12 and 24 weeks. Without thawing sections were obtained by the cryostate technique. These were immediately fixed to a glass slide at room temperature and stained with haematoxylin-eosin. Also, unfrozen samples were prelevated just before freezing and after the thawing procedure. These were quickly frozen at -120°C and further processed as the frozen samples.

3. RESULTS AND DISCUSSION

Average final pH values for the thoracic and lumbar parts of the LD muscle were pH 5.5 and 5.7 respectively.

Ranking of test samples according to storage temperature and time shows a significant increase in the amount of drip from thawed muscle from 2 weeks up to 24 weeks frozen storage. The amount of drip was inversily related to the WHC of the muscle and was statistically significant at the 1 % level (Table I). Similar observations were made by Khan and Lentz (10), using cuts of approx. 200 g. WHC, composition and amount of drip were not consistently affected by freezing rate and storage temperature nor by the thawing procedure. In contrast to the results reported by Awad et al. (1) we did not observe an effect of freezing temperatures on the amount and composition of drip. The protein content of drip (mg protein/g drip) did not change appreciably over the whole storage period. However, total sulfhydryl groups in the drip decreased significantly with increasing frozen storage time. In the non-protein fraction the phenol reagent positive materials (as tyrosin) increased from 2 weeks to 12 weeks of storage. The lack of a distinct variation in WHC in relation to freezing rates in the present experiment indicates that the used modes of freezing are slow to moderate (7), and suggests only negligible morphological variation between the frozen muscle tissue samples.

Histological examination of the frozen beef cuts, however revealed a conspicious alteration of the morphological structure due to the freezing process. Before freezing the muscle fibers have a regular shape and are surrounded by a small intercellular space, whereas frozen meat has a spongy appearance. Frozen fibers are smaller, shrunken and irregular, and the diameter of the intercellular spaces exceeds largely that of the fibers. These empty spaces represent the location of the ice cristals. No intracellular ice cristal formation is observed. The slow freezing procedure enables the migration of intracellular water to extracellular crystallization centres.

Especially because the sarcolemma remains apparently intact, the reverse phenomenon occurs during slow thawing, and the prefrozen morphology is restored. Neither the various freezing temperatures nor the different duration and temperature of storage change the described morphology. Furthermore, the microscopical study of the reference samples reveals no conspicious difference between unfrozen matured meat as compared with thawed beef after various freezing, storage and thawing conditions.

Additional study of samples, frozen at -120°C and at -45°C in isopentane, reveals intra- and extracellular ice crystal formation with higher risk of sarcolemma damage. Consequently, quick freezing does not lead necessarily to a better preservation of the structure. Although the present study differs from others in one or more experimental conditions, our results are largely analogous with the reported data (2, 3, 6, 13). However, alteration of morphology (6) during frozen storage caused by ice recrystallization was not observed.

In agreement with earlier observations (8) visual appreciation of the colour of the frozen steaks differed markedly between samples frozen at -12°C and those frozen at lower temperatures, the latter being characterized by a more bright appearance. However, differences in visual appearance of thawed steaks in relation to freezing temperature remained insignificant. Göfo and Y-values of the thawed steaks indicated a decline of lightness with increasing storage time. Freezer burn never occurred during the experimental period.

The combined effect of freezing, frozen storage and thawing results in a very significant ($p<0,001$) decrease of the percentage OxyMb in the beef cuts. In the frozen state the oxidation progresses rapidly up to a storage time of 12 weeks. The decrease in percentage OxyMb was similar at the different freezing and storage temperatures.

Table I : Effect of storage time on some chemical and physicochemical parameters of frozen bovine l. dorsi muscle.

Parameter	Storage time (weeks)			
	0	2	12	24
Drip (g/100g)	-	2.77(0.31)a	3.89(0.29)b	4.65(0.23)b
Spun drip	14.0(2.2)a	19.3(1.4)a	24.9(0.7)b	22.8(1.2)a
WHC (g/100g)	75.6(2.0)b	51.3(2.0)a	36.3(1.4)c	28.7(1.9)c
MFI	126 (4)b	98 (2)a	145 (3)b	189 (5)b
WB-shear	13.3(0.7)a	13.6(0.6)a	16.2(1.0)b	13.3(0.8)a
Quality index	1.90(0.12)b	1.60(0.05)a	1.75(0.08)a	1.48(0.04)c
Haem (mg/kg)	241 (4)b	203 (12)a	192 (3)a	199 (5)a
OxyMb (%)	95.9(1.6)b	86.4(1.1)a	77.3(1.0)c	76.0(0.7)c
Göfo	88.7(0.5)b	91.4(0.4)a	93.5(0.4)c	95.3(0.4)c
Myofibrils				
- G&S extract	7.5 (0.5)b	5.9 (0.1)a	6.1 (0.2)a	6.5 (0.4)a
- H&S extract	36.5(0.9)b	33.5(0.7)a	34.4(1.0)a	39.0(1.0)b

Samples were frozen at 12 days post-mortem and stored at different freezing temperatures (see experimental). Data show average values for twelve samples. Standard error of the mean is shown in parenthesis. Different index letter signifies that values are significantly different from 2 weeks frozen storage for at least $p<0,05$ (t-test analysis).

Freezing and frozen storage of bovine muscle during 2 weeks caused a considerable decline in different parameters as compared to unfrozen samples. MFI, extraction of the proteins from the myofibrils, quality index and the extracted haempigment decreased significantly. This parallel decline points to an appreciable denaturation of the myofibrillar proteins. Since SDS-PAGE electrophoresis of the extracted proteins did not show any significant difference in the relative composition of the proteinbands as compared to the non-frozen samples, it is suggested that all myofibrillar proteins are similarly affected. Awad et al (1) reported that total extractable actomyosin from muscle frozen at -4°C was not altered significantly by freezing up to 18 h. However, frozen storage over 2 weeks caused a considerable decline (-32 %) in the soluble actomyosin (1). This suggests that the first weeks of frozen storage is more detrimental to protein denaturation than freezing and thawing.

Further frozen storage of the muscle up to 24 weeks resulted in a steady and significant increase in the MFI. Parallel with the increase of MFI a slight rise was noted in the extractability of the protein from the myofibrils. The extractable sulfhydrylgroups did not change on further storage; the fall in quality index noted after 24 weeks of storage is caused by a significant increase in the soluble nitrogen. Since MFI is related to the degradation of myofibrillar protein during storage, an increase in protein breakdown products and in meat tenderness is expected (12). However, meat tenderness as measured with the WB-shearing device was not appreciably affected by storage time, nor by the freezing or thawing procedure. These findings agree with observations made by other investigators (4, 9). After 6 months of frozen storage average shear values were comparable to those of the non-frozen cuts (Table I).

The study of lipid alterations during frozen storage revealed a high variability in the results amoung the different beef cuts. The TBA-numbers increased with prolonged storage, the increase being 2 times higher with the cuts stored at -12°C as compared to -18 or -24°C. The peroxide value for lipid increased significantly from 2 weeks up to 24 weeks of storage.

Average total log counts for unfrozen steaks amounted up to 2.37 cfu/g for mesophilic and up to 1.37 cfu/g for psychrotrophic organisms. Enterobacteriaceae, yeasts and moulds remained below the detection level (<1 log/g). Mesophilic and psychrotrophic counts were significantly higher in frozen-thawed cuts (increase with 0.5-1.0 log units) as compared to the unfrozen cuts. Consistently higher counts in the samples thawed overnight at +3°C as compared to the samples thawed at room temperature for 1 1/2 h, suggest that proliferation occurred during the longer thawing process at 4°C. Freezing and storage temperature and conservation time did not significantly affect the microbiological status of the steaks. Enterobacteriaceae remained undetectable whereas yeasts and moulds slightly increased up to 1 to 2 log units/g.

REFERENCES

1. AWAD, A., POWRIE, W.D. and FENNEMA, O. (1968). Chemical deteriorations of frozen bovine muscle at -4°C. J. Fd. Sci. 33, 227-335.
2. BEVILACQUA, A.E. and ZARITZKY, N.E. (1982). Ice recrystallization in frozen beef. J. Fd. Sci. 47, 1410-1414.
3. BREHMER, H. (1972). Der Einfluss der Gefrierbehandlung auf die Gewebstruktur von Hähnchenmuskulatur. Arch. Lebensmittelhyg. 7, 150-156.
4. DRANSFIELD, E. (1974). Influence of freezing on the eating quality of meat. MRI symposium nr 3, 9.1.
5. EEC (1977). Recommended procedures for use in the measurements of meat qualities - Propositions for E.E.C. standard methods.
6. GOMA, M. and BIRO, G. (1970) Strukturelle Veranderungen gefrorenen Fleisches während der Gefrierlagerung. Fleischwirtsch. 50, 1075-1078.

7. HERRMANN, K. (1970). Tiefgefrorene Lebensmittel. P. Parey, Berlin, p. 38.
8. JAKOBSSON, B. and BENGTSSON, N. (1969). The influence of high freezing rates on the quality of frozen ground beef and small cuts of beef. Proceed. 15th European Meeting of Meat Res. Workers, p. 482.
9. JAKOBSSON, B. and BENGTSSON, N. (1973). Freezing of raw beef : influence of aging, freezing rate and cooking method on quality and yield. J. Fd. Sci. 38, 560-565.
10. KHAN, A.W. and LENTZ, C.P. (1977). Effects of freezing, thawing and storage on some quality factors for portion-size beef cuts. Meat Sci. 1, 263-270.
11. MOSSEL, D., BIJKER, P. and VAN DEN BROEK, M. (1977). Recommended routine monitoring procedures for the microbiological examination of foods of animal origin and drinking water. Vaco, Directorate for International Aid, Min. Foreign Affairs, The Netherlands.
12. OLSON, D.G., PARRISH, F.C. and STRACHS, M.H. (1976). Myofibrillar fragmentation and shear resistance of three bovine muscles during postmortem storage. J. Fd. Sci. 41, 1036-1041.
13. PARTMANN, W. (1973). Histologische Veränderungen in Rind- und Schweinefleisch sowie Schweineleber unter definierten Gefrier- und Auftaubedingungen. Fleischwirtsch. 53, 65-70.
14. PENNY, I.F. (1969). Protein denaturation and water-holding capacity in pork muscle. J. Fd. Technol. 4, 269-273.
15. PENNY, I.F. (1975). Use of a centrifuging method to measure the drip of pork longissimus dorsi slices before and after freezing and thawing. J. Sci. Fd. Agric. 26, 1593-1602.
16. SEDLAK, J. and LINDSAY, R.H. (1968). Estimation of total protein-bound and non-protein sulfhydrylgroups with Ellman's reagent. Anal. Biochem., 25, 192-205.
17. SUNG, S.K., ITO, T. and FUKAZAWA, T. (1976). Relationship between contractibility and some biochemical properties of myofibrils prepared from normal and PSE porcine muscle. J. Fd. Sci. 41, 102-107.
18. VAN DEN OORD, A.H.A. and WESTDORP, J.J. (1971). Analysis of pigments in intact beef samples. J. Fd. Technol. 6, 1-13.

INFLUENCE OF PACKAGING ON THE SHELF LIFE OF FROZEN CARROT, FISH AND ICE CREAM

P. AHVENAINEN and Y. MÄLKKI
Technical Research Centre of Finland,
Food Research Laboratory, Biologinkuja 1
SF-02150 Espoo 15, Finland

Summary

Shelf life of frozen carrot cubes, fillets of baltic herring, and of ice cream was studied when stored at -12, -15 and -18°C in vertical freezer cabinets, and in open display cabinets at -15°C, packed in 23 different types of plastic, cardboard and laminate packages. The packaging had the greatest effect on the shelf life of products stored in the top layer of an open display cabinet, where aluminium foil laminated and metallized packages gave the best results. The shelf life in these packages was 2...3 times that in packages without Al-layer. The packaging had significant effect also on preserving the quality of frozen foods stored in vertical cabinets, particularly of fatty foods, like fish and ice cream. The longest shelf life could be obtained in tight packagings of low oxygen and water vapour permeability. The shelf life of Baltic herring fillets in vacuum packagings was two times that in cardboard packages, for example. Nitrogen did not improve the preservation of the quality of carrot cubes and herring fillets compared with normal packagings or vacuum packagings, respectively. According to the study it is possible to increase frozen storage temperatures in retail stores if frozen foods are protected by aluminium foil laminated or metallized packagings.

1. INTRODUCTION

The significance of packagings on the shelf life of different frozen foods has not been studied very widely. Primarily the dependence of the quality of frozen fish and meat products on packaging materials and methods has been investigated to some extent (2,4,6,7). These investigations indicate that particularly fat fish, like trout, and also peeled shrimps, individually quick-frozen, and many meat products retain their quality in vacuum packagings of low oxygen and water vapour permeability up to two times longer than in a PE-bag or cardboard packages. Vacuum packaging has also a favourable effect on quality retention of some unblanched vegetables, resulting in a longer shelf life (8,9). According to Broman (3) ice cream preserves its quality better in aluminium foil laminated than in PE-coated cardboard packages. In aluminium foil laminated packagings the temperature of frozen products stored in the top layer of an open display cabinet has been shown to be 4...6 °C lower than in packagings without Al-layer (3,5).

In recent years attention has been also paid to possibilities of reducing the energy consumption of frozen storages. The simplest way would be to raise storage temperatures. The objective of this study was to investigate the significance of different packagings in improving the preservation of the quality of frozen foods and possibilities to increase storage temperatures in retail stores by improved packaging.

2. MATERIALS AND METHODS

The frozen foods tested were blanched carrot cubes, frozen individually in fluidized bed, fish fillets of Baltic herring, frozen in blocks in plate freezer, and ice cream. Both the freezing and packaging of products were made in industrial scale to get representative samples.

The significance of packaging for shelf life of frozen foods was studied in vertical freezer cabinets kept at three temperatures, -12 °C, -15 °C and -18 °C, on the average. The retailing conditions were simulated by storing in the top layer of open display cabinets with nominal temperatures of -15 °C under illumination. Reference samples were stored at -25 °C to -27 °C in an industrial storage.

The packagings tested included 23 different packagings or combinations of plastic, laminate, cardboard, aluminium foil laminated and metallized packagings. In a part of the packagings the air was replaced by nitrogen. In an open display cabinet the storage life of samples was tested in two different packagings; the packagings of carrot cubes were a white pigmented LDPE-(95 μm)-bag and a metallized PETP/PE (12/50)-bag, the packagings of Baltic herring fillets a vacuum deep drawn package (PA/PE-multilayer film, 175 μm) covered by an aluglass-coated cardboard box or without covering, the packagings of ice cream were a PE-coated and an aluminium foil laminated cardboard box. The packaging of reference samples was a LDPE (95 μm)-bag for carrot cubes, a vacuum deep drawn package covered by a PE-coated cardboard box for herring fillets, and a PE-coated cardboard box for ice cream.

The quality changes of the frozen foods during storage were followed principally by sensory evaluation, which was made for carrot cubes and herring fillets at intervals of 1 - 2 months and for ice cream at intervals of 1 - 6 weeks. The losses of weight were also determined.

The details concerning the materials and methods of this study are described elsewhere (1).

3. RESULTS AND DISCUSSION

Packaging had a distinct influence on the quality of carrot cubes stored in vertical cabinets only at temperatures above -15 °C (Table I). At -12 °C, polyamide/polyethylene-, polyester/polyethylene-and metallized polyester/polyethylene laminate bags retained the quality of carrot cubes fairly significantly or significantly better than polyethylene bags or some cardboard packages. Nitrogen did not lengthen the shelf life of carrot cubes compared with packagings without nitrogen (Table I). The packaging had the most significant influence on the storage life of carrot cubes stored in the top layer of an open display cabinet (Table I and Fig. 1). In the pigmented LD-polyethylene bag the quality deteriorated significantly faster than in the metallized package, the acceptability time in the latter being 2.5 times longer than in the former. This was evidently a result of very low oxygen permeability of metallized laminate at -15 °C, whereas polyethylene (95 μm) permeates oxygen about 200 ml/m^2 24 h atm still at -15 °C (1). Thus oxidation reactions occur catalyzed by UV-light although UV-transparency of pigmented polyethylene is low. Metallized polyester/polyethylene protected also better against the effect of light to increase the temperature. The temperature in the metallized bag was 1...2 °C lower than in the polyethylene bag.

The package and temperature had markedly greater influence on the shelf life of Baltic herring fillets stored in vertical cabinets than that of carrot cubes. In all packages the storage life was at -18 °C

Table I. The effect of packaging and storage temperature on the shelf life of carrot cubes stored in vertical cabinets and in an open display cabinet.

Package	Shelf life (months) until not saleable			
	Vertical cabinets			open display cabinet
	-12 °C	-15 °C	-18 °C	-15 °C
Cardboard packages	8.5-12	10.5-12.5	11.5-13.5	
LDPE-bag	8.5-10	>12.5	>13.5	3-5
Other PE-bags	7 - 8.5	>12.5	>13.5	
Met. PETP/PE-bag	>13.5	>12.5	>13.5	>11
PA/PE- and PETP/PE-bag	11 -13.5	>12.5	>13.5	
" " " with N_2	7 -13.5	>12.5	>13.5	
Reference	>13.5			

1.5...2 times longer than at -12 °C (Table II). The reference sample of fillets kept its quality very significantly better than fillets stored at -18 °C, except that compared with fillets packed in vacuum deep drawn package (bottom PA/PE-multilayer, 175 µm) the reference sample was only fairly significantly better.

Herring fillets stored in vertical cabinets at -12 °C to -18 °C, preserved their quality best just in the vacuum deep drawn package. This package was significantly better than the best cardboard package tested which was made of polyethylene-coated cardboard, but not heat sealed. In packagings with nitrogen, quality changes occurred slightly faster than in vacuum packagings, the difference being non-significant (Table II). Herring fillets stored in an open display cabinet retained their quality in the vacuum deep drawn package, covered with an aluglass-coated cardboard box, very significantly better than in the vacuum package without covering. The aluglass-box (metallized cardboard) protected very well from the effect of light. The temperature in that package was about 4 °C lower than in the package without covering.

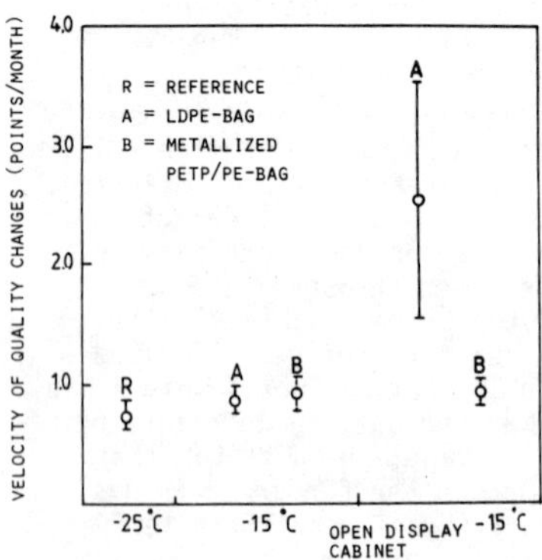

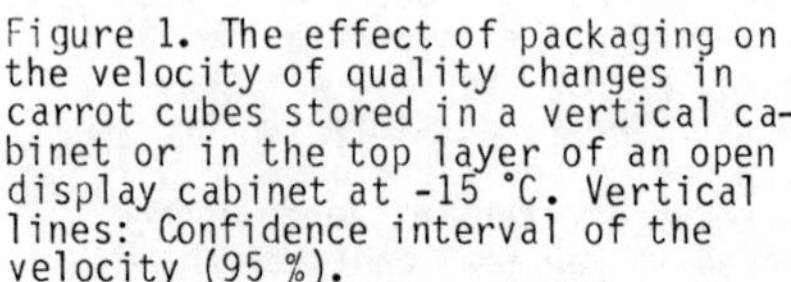
Figure 1. The effect of packaging on the velocity of quality changes in carrot cubes stored in a vertical cabinet or in the top layer of an open display cabinet at -15 °C. Vertical lines: Confidence interval of the velocity (95 %).

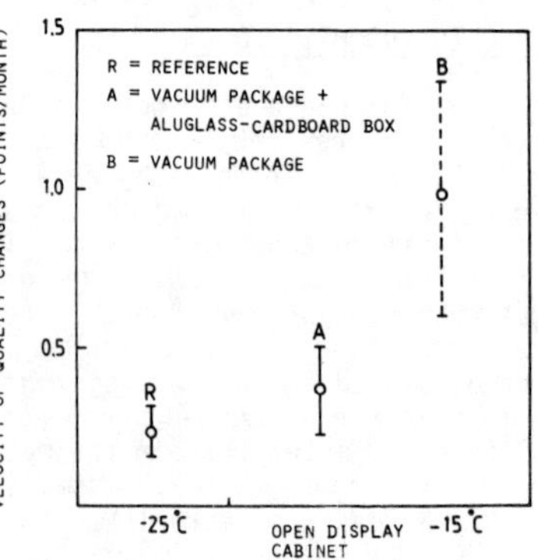

Figure 2. The effect of packaging on the velocity of quality changes in Baltic herring fillets stored in the top layer of the open display cabinet. Vertical lines as in Fig. 1.

Table II. The effect of packaging and storage temperature on the shelf life of Baltic herring fillets stored in vertical cabinets and in the top layer of an open display cabinet.

Package	Shelf life (months) until not saleable			
	Vertical cabinets			Open display
	-12 °C	-15 °C	-18 °C	cabinet -15 °C
Cardboard packages	4 - 5	4 - 5.5	6	
Vacuum deep drawn packages	6.5 - 8	8 - 11	10 - 13	6
Deep drawn packages with N_2	5 - 8	5.5 - 8	8 - 10	
Vacuum deep drawn package covered by metallized cardboard box				>11
Reference (-25 °C)	>13			

Herring fillets and carrot cubes had losses of weight principally only in cardboard packages. After storage of 6 months the losses of fillets were 1.4...5.3 % depending on packaging material and temperature.

The packaging was of importance also on the shelf life of ice cream stored in vertical cabinets. The quality of reference sample was quite good after a storage of 30 weeks (hedonic points 77 in total, max 100). At -12 °C and -15 °C the quality downgraded from first quality in less than one week and at -18 °C in 4-5 weeks independently of packagings. In the second quality grade ice cream kept not more than four weeks at -12 °C and -15 °C, differences influenced by packagings were insignificant. At -18 °C, the quality retained in the second grade about 15 weeks. In addition there was a significant difference between the packages. Among cardboard packages the hot-melt-coated cardboard box was best for quality, plastic packages were comparable to the hot-melt cardboard box.

With regard to the shelf life of ice cream in the third quality grade, still saleable, a significant difference in shelf lifes in different packages even at -12 °C (Table III) was observed. The best package for the quality of ice cream under these circumstances was the polystyrene box, and it was fairly significantly better than cardboard packages and HDPE box. The hot-melt cardboard box proved to be the best of the cardboard packages tested. It was better (but not significantly) than the PE-coated cardboard box used nowadays most of all as the packages of ice cream.

Table III The effect of packaging and storage temperature on the shelf life of ice cream stored in vertical cabinets and in an open display cabinet.

Package	Shelf life (weeks) until not saleable			
	Vertical cabinet			Open display
	-12 °C	-15 °C	-18 °C	cabinet -15 °C
Cardboard packages:				
Hot-melt-coated	8 - 10	11 - 16	22	
PE-coated	6 - 8	11 - 16	13 - 17	10
Latex-coated	6 - 8	7 - 10	22	
WR-coated	3 - 4	4 - 5	8 - 10	
Al-foil laminated				16 - 18
Plastic packages:				
PS-box	16	22 - 26	>30	
HDPE-box	16	19	22 - 26	

In open display cabinet at -15 °C the storage time could be extended by using aluminium foil laminated cardboard package by ca. two months, as compared to the conventional PE coated cardboard package. (Table III). The effect of the improved package remained smaller as compared to that for carrot cubes and fish fillets, evidently due to partial melting of the ice cream at this temperature.

4. CONCLUSIONS

As in the previous studies, the influence of packaging was of greatest importance to the shelf life of fatty foods such as fish. The best for these products are packages of low oxygen and water vapour permeability. The packaging must also cover the products tightly. This allows a shelf life in retailing temperatures up to two times that in permeable packages. Also the shelf life of vegetables depends significantly on packaging, particularly if the storage temperature is high, or the frozen foods are subjected to light and fluctuating temperatures.

The packaging has the greatest significance on the quality of frozen foods stored in an open display cabinet. Oxygen permeability, water vapour transmission and transparency of packaging material intended for frozen foods stored in retailing conditions must be very low. Metallization seems to protect as well as aluminium foil from the effects of light and fluctuating temperatures.

It is possible to increase the storage temperatures of frozen foods other than ice cream in retail stores if they are protected by aluminium foil laminated or metallized packages. In these packages, carrot cubes and herring fillets stored in an open display cabinet at temperature -15 °C kept their quality nearly as the well as references of these products stored in an industrial storage at -25 °C.

REFERENCES

1. AHVENAINEN, R., MÄLKKI, Y. and SALLINEN, P., Pakkauksen vaikutus pakasteiden säilyvyyteen (Influence of packaging on the shelf life of frozen foods), Research Reports 199, Technical Research Centre of Finland, Espoo 1983. 87+12 pp.
2. BRAMSNAES, F. & SÖRENSEN, H.C., Vacuum-packed frozen fatty fish. Bull. Intl. Inst. Refrig. Annex 1960-3, p. 281.
3. BROMAN, U-I., Pakkauksen ja säilyttimen vaikutus jäätelön säilyvyyteen vähittäiskaupan myymäläolosuhteissa. EKT-sarja 488, University of Helsinki, Department of Food Chemistry and Technology, Helsinki 1979. 90 pp..
4. BØGH-SÖRENSEN, L. & JENSEN, J.H., Factors affecting the storage life of frozen meat products, Int. J. Refrig. 4(1981) No 3, 139 - 142.
5. HAWKINS, A.E., PEARSON, C.A., RAYNOR, D., Advantages of low emissivity materials to products in commercial refrigerated open display cabinets. Austr. Refrig., Air Cond. Heat. 29(1975) No 2, 15 -25.
6. LINDSAY, R.C., The effect of film packaging on oxidative quality of fish during long-term frozen storage. Food Prod. Dev. (1977) No 10, 93 -96.
7. LÖNDAHL, G., DANIELSON, C. E., Time-temperature-tolerances for some fish and meat products, Bull. Intl. Inst. Refrig. Annex 1972-2, p.295.
8. STEINBUCH, E., Technical note: Quality retention of unblanched frozen vegetables by vacuum packing. I Mushrooms. J. Food Technol. 14(1979), 321 - 323.
9. STEINBUCH, E., Technical note: Quality retention of unblanched frozen vegetables by vacuum packing. II Asparagus, parsley and celery. J. Food Technol 15(1980), 351 - 352.

INFLUENCE DES TRAITEMENTS PREALABLES A LA CONGELATION SUR LA QUALITE DES FRUITS ET LEGUMES

J. ESPINOSA
Instituto del Frío. Ciudad Universitaria. Madrid -3 (Espagne)

Effects of the treatments prior to freezing on the quality and stability of fruits and vegetables

SUMMARY

The present survey sets forth the chemical products most commonly used in the operations involved in the processing of frozen vegetables as well as in the pre-freezing treatments with a view to improving the organoleptic properties and securing a higher stability of these vegetables.

1. CONSIDERATIONS PREALABLES

Parmi les procédés de conservation des denrées d'origine végétale c'est la congélation qui maintient en général les caractéristiques organoleptiques et nutritionnelles les plus proches de celles du produit frais au moment de la consommation. Au cours de l'élaboration des produits végétaux congelés on applique fondamentalement des traitements thermiques (blanchiment et congélation) avec à peine l'emploi d'additifs.

Le Manuel de Procédé du Programme Conjoint FAO/OMS concernant les Normes Alimentaires (1) dit comme suit: "L'emploi d'additifs alimentaires n'est justifié que quand ils accomplissent l'un ou plus d'un des objectifs signalés dans les points (a) à (d) et quand ces buts ne peuvent être atteints par d'autres moyens économiquement et technologiquement applicables et ne présentent pas de risques pour la santé du consommateur;

a) Conserver la qualité nutritionnelle de la denrée alimentaire; une diminution intentionnelle de la qualité nutritionnelle d'un aliment serait justifiée dans les circonstances indiquées sous le point (b), ainsi que dans d'autres circonstances où la denrée ne constitue pas un composant important d'un régime alimentaire normal;

b) fournir les ingrédients ou constituants nécesaires pour les denrées fabriquées à destination de groupes de consommateurs qui ont des besoins diéthétiques spéciaux;

c) élever l'aptitude à la conservation ou la stabilité d'une denrée ou en améliorer les propriétés organoleptiques à condition que cette dose n'altère pas la nature, la substance ou la qualité de l'aliment au point de tromper le consommateur;

d) fournir un appui dans la fabrication, élaboration, préparation, traitement, empaquetage, transport ou entreposage de la denrée, à condition que l'additif ne soit pas utilisé pour déguiser les effets de l'emploi de matières prèmieres défectueuses ou de pratiques (y compris celles non hygiéniques) ou de techniques indésirables au cours de n'importe quelle de ces opérations".

La plupart des produits chimiques employés dans l'élaboration de denrées végétales congelés répondent directement au but indiqué dans le point (c) e indirectement a celui indiqué sous (d).

Quant au mode d'application de ces produits chimiques, ceux-ci sont dissous dans l'eau de blanchiment, ce qui est fondamentalement le cas des légumes, ou incorporés à un sirop, comme c'est, en général, le cas des fruits.

La plupart des efforts, qui d'ailleurs s'accordent logiquement avec les données rencontrées dans la bibliographie ont été dirigés vers l'application des produits chimiques aux espèces végétales qui, pour des raisons diverses, ne présentent pas une bonne aptitude à la congélation.

Les traitements à base de produits chimiques sont spécifiques pour chaque produit quant aux paramètres qui définissent ces traitements: dose, durée, pH du milieu; il n'existe pas de recettes universelles.

2. PRODUITS FRAIS

Il n'y a que très peu de renseignements concernants les traitements du produit frais avant le comencement du processus d'élaboration. Un traitement des hricots verts avec atmosphères enrichies de dioxyde de carbone à la température ambiante, exerce une action profitable sur la qualité du produit congelé. On sait que l'exposition des produits d'origine végétale au CO_2 comporte une élévation du pH et un décroissement de l'acidité qui agit indirectement sur la couleur et la texture des tissus, surtout pendant les opérations d'échauffement /2/.

3. OPERATIONS PRELIMINAIRES

En vue d'obtenir des produits d'une utilisation facile et susceptibles d'être soumis, dans les meilleurs conditions, aux traitements thermiques de blanchiment et congélation, les fruits et légumes subissent un nombre d'opérations, telles que: nettoyage, lavege, égrenage, triage, calibrage, épointage, découpage, pelage.

On emploie des produits chimiques pour le lavage, en additionant l'eau, généralement, de chlore sous forme gazeuse ou d'hypochlorite de sodium. Une teneur de l'eau en chlore libre de l'ordre de 5 à 10 ppm n'influe pas sur la saveur du produit ni exerce une action corrosive sur les matériaux des chaînes d'élaboration /3/.

Pour le pelage par voie chimique des produits qui le réclament on emploie de la soude caustique.

Le temps de séjour dans le bain, la température de celui-ci et la concentration de soude sont les paramètres qui définissent l'opération et qui sont également en fonction du type de produit.

4. MODIFICATION DU pH DE L'EAU DE BLANCHIMENT

On sait que les enzymes et les micro-organismes sont moins résistants à la chaleur dans un milieu acide.

Conséquemment il existe la possibilité de réduire le temps et/ou la température de blanchiment en abaissant le pH de l'eau. Ainsi, par exemple, on a employé, avec des résultats positifs, l'acide citrique dans le cas de la pomme de terre et de l'artichaut.

L'action de l'acidification du milieu peutêtre renforcée si l'on ajoute un agent séquestrateur tel que sel sodique de l'acide éthylen-diamino-tétracétique (EDTA), bien que ce procédé ne soit pas faisable dans les pays où l'emploi de ce produit est interdit.

Néanmoins, l'acidification ne peu pas être appliqué dans tous les cas parce qu'elle favorise la transformation des chlorophylles en phéophytines, ce qui exerce une influence négative sur la couleur des légumes verts.

Par ailleurs, l'addition de solution de CO_3Na_2 à 5% ou de NaCl à 10% à l'eau de blanchiment, provoque le ralentissement de l'inactivation thermique des enzymes dans le cas des haricots verts; en voie de conséquence, elle prolonge la durée du traitement/4/. En général, l'utilisation d'additifs dans l'eau de blanchiment en but d'améliorer la retention de substances solubles, la texture ou la capacité de conservation des végétaux congelés, devra faire l'objet d'une analyse de l'influence de ces additifs sur le temp nécessaire pour arriver à l'inactivation des enzymes.

5. CHANGEMENTS DE COULEUR

Les fruits (tels que pêches, abricots, prunes, poires, pommes, etc.) et légumes (tels que champignons, pommes de terre, aubergines, etc.) dont la composition comprend des composés diphénoliques, (acide chlorogénique, catequines, etc.) incolores, en présence de l'enzyme polyphénoloxydase réagissent avec l'oxygène gazeux donant lieu, d'abord, à la formation de quinones, également incolores, puis à la formation de polymères bruns, d'accord avec la réaction suivante, très schématisée:

OH OH (Incolore) —O-diphénoloxydase, $+ \frac{1}{2} O_2$→ O O (Incolore) —Polymérisation→ Composés Colorés

Pour empêcher le brunissement enzymatique on peut employer des agents réducteurs tels que l'acide ascorbique qui agit comme un antioxydant réduisant les quinones aux composés phénoliques originaux; toutefois, ce procédé s'avère inefficace dès que la polymérisation aura eu lieu.

Un autre additif amplement employé c'est l'anhydride sulfureux, générallement sous forme de métabisulfite de sodium ou de potassium. L'anhydride sulfureux entrave de façon irréversible la réaction avec les quinones pour former des produits d'addition non colorés et empêche la condensation ultérieure des quinones. Cependant, l'emploi de SO_2 présente les inconvénients suivants: il peut donner lieu à des altérations de la saveur, il détruit la vitamine B1, l'emploi de ce produit n'est pas permis dans certains pays et il est dangereux pour la santé s'il dépasse les limites normales de tolérance.

Le blanchiment classique est le procédé généralement employé pour inactiver les enzymes. Toutefois, l'application de cette méthode est accompagnée d'un nombre d'inconvénients, c'est pourquoi certains auteurs considèrent ce traitement thermique comme un mal nécessaire. Pour la presque totalité des fruits et pour certains légumes (champignons, aubergines, et tomates, par exemple), les altérations de la saveur et de la texture et les pertes de substances solubles ont imposé la nécessité de rechercher des méthodes alternatives pour bloquer l'activité enzymatique.

Ainsi, dans les fruits, on a employé le saccharose incorporé directement au fruit ou sous forme de sirop comme agent protecteur en raison de son action antioxydante indirecte du fait qu'il expulse l'air des tissus cellulaires, en même temps qu'il renforce la saveur et l'arôme des fruits.

On a la possibilité de remplacer le saccharose et le sirop de glucose (formes déconseillées aux personnes souffrant de diabète) par le fructose, la xylite et la D-sorbite /5/, pour répondre au but mentionné dans le point (b) concernant l'emploi d'additifs alimentaires.

La congélation de mélanges de fruits avec sirops est employée de moins en moins en raison des inconvénients qu'elle présente.

Actuellement on préfère le traitement par immersion pendant un certain temps dans des sirops à diverses concentrations, contenant composés antioxydants ou réducteurs pour renforcer l'action du sucre.

Les traitements préalables à l'élaboration de tranches de fruits réfrigérées (le cas de la pomme) doivent être différents de ceux qu'on applique à un produit destiné à la congélation, du fait que celle-ci détruit l'organisation cellulaire, ce qui permet les enzymes et le substratum d'entrer en contact.

Dans le premier cas, on n'aura besoin que de traitements superficiels; par contre, dans le cas de la congélation, des traitements de pénétration seront nécessaires.

En employant des proportions convenables de chlorure de sodium et d'anhydride sulfureux dans une solution d'une valeur pH déterminée, la texture, la saveur et la couleur des tranches de pomme réfrigérées peuvent être équilibrées, pour obtenir un produit de bonne qualité après l'entreposage /6/. En employant à la fois l'acide ascorbique, le chlorure calcique et l'anhydride sulfureux, on n'obtient pas de meilleurs résultats qu'avec les combinations binaires de chlorure calcique, soit avec l'anhydride sulfureux soit avec l'acide ascorbique /7/.

Pour obtenir, au moment de la décongélation, des tranches de pomme avec d'une texture, saveur et couleur les plus semblable à celles du produit frais, un traitement consistant à immerger le produit pendant 17 heures, soit dans un sirop contenant 30% de sucre, 0,2% d'acide ascorbique et 0,3% de chlorure calcique, soit dans un sirop de la même concentration avec 0,02% d'anhydride sulfureux et 0,5% d'acide malique (pH = 2,4) donne de bons résultats et, bien que le procédé soit lent et ennuyeux, l'amélioration de la qualité du produit peut en justifier l'emploi /8/.

Lorsque la concentration ionique s'accroît, le brunissement est inhibé progressivement. L'emploi de chlorures de sodium et de calcium (s'est avéré efficace pour empêcher le brunissement enzymatique, bien que leur mode d'action n'a pas encore été élucidé. Ainsi, dans le cas de tranches de pomme réfrigérées, l'immersion préalable pendant 30 minutes dans un sirop à 30º Brix, additionné de 0,5% de chlorure sodique et 0,28% de chlorure calcique donnait de bons résultats permettant d'entreposer les tranche pendant 4 semaines à 0ºC. C'est l'ion chlore, et non pas le cati calcium, qui joue le rôle principal comme inhibiteur du brunissement /9/.

Les traitements préalables par $NaHSO_3$ (60 à 80 ppm de S acide citrique (0,5 à 1%), $SnCl_2$ (100 à 200 ppm) ont donné d bons résultats en ce qui concerne la stabilisation de la couleur des fraises, récoltées mécaniquement, congelées et destinée à subir une seconde transformation. Le SO_2 et le $SnCl_2$ fournisse un milieau réducteur qui empêche la décomposition des anthocyanes et de l'acide ascorbique. Le traitement par acide citrique agit en abaissant le pH, inhibant ainsi la dégradation enzymatique /10/.

D'autres auteurs estiment que, dans le cas des fraises congelées pour la consommation directe, un traitement par immersion pendant 5 minutes dans un bain contenant 0,2% de $K_2S_2O_5$ ne présentait pas d'avantages sur le témoin quant à la texture, la saveur et l'aspect /11/.

Pour beaucoup de légumes l'emploi de sucre comme antioxydant n'est pas possible, et le procédé employé est l'immersion dans des solutions qui contiennent les produits actifs.

On présente ci-dessous quelques exemples.

Le traitement préalable par immersion pendant cinq minutes dans une solution de $K_2S_2O_5$ à 0,35% et d'acide citrique à 0,2% remplace avantageusement le blanchiment de tranches d'aubergine pour éviter le phénomènes de brunissement de la pulpe /12/.

La congélation préalable à une opération classique de séchage améliore considérablement la stabilité des caroténoides et des lipides dans le cas des carottes déshydratées. L'immersion des carottes dans une solution contenant 3% de ClNa et 0,2% métabisulfite sodique pendant 20 minutes avant la congélation et le séchage ultérieur, a réduit significativement la formation de composés bruns, aussi bien durant le processus de séchage lui-même, qu'au cours de l'entreposage ultérieur, et la couleur, la saveur, la texture et les caractéristiques de reconstitution ont été meilleurs que dans les témoins non traités /13/.

L'anhydride sulfureux agit comme un antioxydant, reduisant ainsi les pertes d'acide ascorbique au cours de l'entreposage de piments congelés, tant préalablement blanchis que non blanchis.

De même, la rétention de pigments, y compris la chlorophylle, est plus important dans les échantillons traités par anhydride sulfureux /14/.

L'un des traitements alternatifs du blanchiment des champignons c'est l'emploi de l'anhydride sulfureux.

Dans le cas d'un produit congelé on obtient les résultats acceptables par immersion pendant 5 minutes dans un bain contenant 0,2% de $K_2S_2O_5$ /15/. Pour champignons destinés à être lyophilisés, l'immersion dans une solution de $Na_2S_2O_5$ contenant 200 ppm de SO_2, pendant 10 minutes, influe favorablement sur la qualité finale du produit /16/.

On peut également intervenir par voie génétique pour pallier les inconvénients que présente le brunissement enzymatique, par l'obtention de variétés ayant une faible teneur en substratums phénoliques.

6. ALTERATIONS DE LA TEXTURE

Les traitements thermiques qui constituent le processus d'élaboration des produits végétaux congelés exercent une très forte influence sur la structure du produit, causant d'importantes dommages à sa texture, qui est un attribut de la qualité très important.

Pendant l'opération de blanchiment, se produit une solubilisation des composés pectiques intercellulaires qui occasionnent une perte d'adhésion et de cohésion entre les cellules. La présence de cations divalents (Mg^{2+} et Ca^{2+}) provoque la formation de pectates insolubles, et conséquemment, corrige les effets négatifs du blanchiment sur la texture, lesquels s'aggravent ultérieurement au cours de la congélation.

Les traitements par blanchiment échelonné ou "stepwise blanching", composé de deux phases dont la première comprend la combinaison basse température-temps prolongé (BT-TP) suivie de refroidissement, et la seconde la combinaison température élevée-court temps (TE-CT), améliorent le comportement rhéologique de la texture et la structure du produit /17/. Dans le traitement préalable l'enzyme pectin-méthylestérase est activée laissant en liberté des groupes acido-carboxyliques qui forment des sels avec les cations divalents. La formation de ces sels unissent des molécules pectiques adjacentes, donnant pour résultat une plus grande fermeté de la structure du produit /18/.

De tout ce qui vient d'être exposé, on déduit que l'activation de la pectin-méthylestérase et l'addition de sels de calcium, sont deux procédés possibles pour résoudre les problèmes de texture qui se présentent dans les produits végétaux congelés.

D'ailleurs on a constaté que le chlorure calcique produit un effet synergique avec l'acide ascorbique et l'anhydride sulfureux pour empêcher le brunissement.

Dans le cas des tranches de mangue congelées, l'immersion préalable à la pasteurisation dans un sirop à 20º Brix contenant 20% de $CaCl_2$, avait une action positive et marquée sur la texture, laquelle, d'autre part, n'était pas influencée par les vitesses de congélation essayées /19/.

En vue d'améliorer la texture de certains fruits qui la perdent en grande partie après décongélation comme c'est le cas de la fraise et le melon par example, on a appliqué des traitements chimiques préalables à la congélation, à base de collöides.

Les pectines ayant une faible teneur en groupes méthoxyle, alginates et extraits d'agar-agar appliqués aux proportions de 0,1% a 0,4% du fruit, dispersant les collöides dans le sirop, ont un effet positif du fait qu'ils augmentent considérablement le poids égoutté, ainsi que l'aspect de fraises, après décongélation /20/.

L'immersion de boules de melon dans un sirop d'une concentration de sucre égale à la concentration totale de solides solubles du produit frais, et qui contenant 0,125% de pectines modifiées par poids de fruit, améliorait la texture du produit au moment de la décongélation /21/.

7. DECONGELATION

Dans les fruits congelés sans traitement antioxydant préalable on peut éviter le brunissement, au moment de la décongélation, en les immergeant dans une solution (eau ou sirop de sucre) contenant un réducteur, par exemple acide ascorbique.

Il existe des procédés tels que par exemple, un traitement thermique comme s'est le cas des purées de pêche et d'abricot obtenues à partir de fruits congelés entiers qui sont pasteurisés au moment de la réduction /22/.

8. REFERENCES

1. ANONIMO (1975): Comisión del Codex Alimentarius. Manual de Procedimiento. Programa Conjunto FAO/OMS sobre Normas Alimentarias.

2. BUESCHER, R.W. and BROWN, H. (1979). Regulation of frozen snap bean quality by postharvest holding in carbon dioxide enriched atmospheres. Journal Food Science, 44, 1494-1497.
3. PHILIPPON, J. (1975). La surgélation des denrées d'origine végétale. Bulletin Technique d'Information du Ministère d'Agriculture, nº 296, 81-85.
4. THOMOPOULOS, C. (1975). Influence de certains sels de sodium sur la durée du blanchiment des haricots verts. Ind. Aliment. Agric., 92, 531-534.
5. STOLL, K. (1979). Essais de l'évacuation de l'air et de l'usage de différents formes de sucre sur les qualités organoleptiques des produits surgelés. Comptes rendus XVe Congrès Int. Froid, Vol. III, 789-892.
6. PONTING, J.D., JACKSON, R. and WATTERS, G. (1971). Refrigerated apple slices: Effects of pH, sulfites and calcium on texture. Journal Food Science, 36, 349-350.
7. PONTING, J.D., JACKSON, R. and WATTERS, G. (1972). Refrigerated apple slices: preservative effects of ascorbic acid, calcium and sulfites. Journal Food Science, 37, 434-436.
8. PONTING, J.D. and JACKSON, R. (1972). Pre-freezing processing of Golden Delicious apple slices. Journal Food Science, 37, 812-814.
9. PHILIPPON, J., ROUET-MAYER, M.A. et BROCHIER, J.J. (1978). Conservation par réfrigération de tranches de pommes destinés à l'industrie. Problèmes sanitaires et prévention des brunissements enzymatiques. Bull. Inst. Int. Froid. Annex 1978-2, 29-39.
10. SISTRUNK, W.A., MORRIS, J.R. and KOZUP, J. (1982). The effect of chemical treatments and heat on color stability of frozen machine-harvested strawberries for jam. J. Am. Soc. Hortic. Sci., 107 (4), 693-697.
11. FUSTER, C., PRESTAMO, G. and CANET, W. (1982). Effects of pretreatments and freezing on different strawberry varieties. Bull. Inst. Int. Froid, Annex 1982-4, 350-355.
12. CRIVELLI, G., MAESTRELLI, A. et BERTOLO, G. (1980). Comportement des légumes traités par congélation rapide. Aubergines. Rev. Gén. Froid, 70 (12), 625-628.
13. ARYA, S.S., NATESAN, V. and PREMAVALLI, K.S. (1982). Effect of pre-freezing on the stability of carotenoides in unblanched air-dried carrots. Journal Food Technology, 17, 109-113.
14. RAHMAN, F.M.M. and BUCKLE, K.A. (1981). Effects of blanching and sulphur dioxide on ascorbic acid and pigments of frozen capsicum. Journal Food Technology, 16, 671-682.
15. PRESTAMO, G. and FUSTER, C. (1982). Influence of various treatments and blanching on the quality of frozen mushroom. Bull. Inst. Int. Froid, Annex 1982-4, 290-295.
16. FANG, T.T., FOOTRAKUL, P. and LUH, B.S. (1971). Effects of blanching, chemical treatments and freezing methods on quality of freeze-dried mushrooms. Journal Food Science, 36, 1044-1048.
17. STEINBUCH, E. (1976). Technical Note: Improvement of texture of frozen vegetables by stepwise blanching treatments. Journal Food Technology, 11, 313-316.
18. BARTOLOME, L.G. and HOFF, J.E. (1972). Firming of potatoes: brochemical effects of pretreating. Journal Agric. Food Chem., 20, 266-270.

19. COOK, R.D. et al. (1976). Studies of mango processing. II. Deep freezing of mango slices. Journal Food Technol, 11, 475-484.
20. WEGENER, J.B., BAER, B.H. and RODGERS, F.D. (1951). Improving quality of frozen strawberries with added colloids. Food Technology, 2, 76-78.
21. RIO, M.A. and MILLER, M.W. (1979). Effect of pretreatment on the quality of frozen melon balls. Proced. XVth Int. Congr. Refrig., Vol. III, 923-926.
22. PHILIPPON, J. et ROUET-MAYER, M.A. (1973). Décongélation des fruits à noyau. Prévention des brunissements enzymatiques. Rev. Gén. Froid, 64 (5), 487.493.

STORAGE LIFE AND EATING-RELATED QUALITY OF NEW ZEALAND FROZEN LAMB: A COMPENDIUM OF IRREPRESSIBLE LONGEVITY

R.J. WINGER

Meat Industry Research Institute of New Zealand (Inc.),
P.O. Box 617, Hamilton, New Zealand.

SUMMARY

The eating quality of roasted New Zealand export lamb -- treated with the new electrical stimulation procedure, processed to strict international hygiene requirements and wrapped in plastic -- is essentially unchanged for at least 2 years' storage at -10°C. But the storage life of lamb is determined by many factors apart from frozen storage temperature. One day of chilled storage prior to freezing reduces subsequent frozen storage life by about 25%. Animal-to-animal variability can result in differences as great as 50%. Packaging also plays an important role.

Of the microbiological changes in frozen meat, yeasts which can grow at -5°C cause no organoleptic change in the meat. Some moulds are also capable of slow growth at this temperature. There is, however, absolutely no microbiological growth on meat at temperatures below -8°C. Therefore the limit to the storage life of properly frozen-stored meat is not related to microbiological growth.

Appearance, colour and cooking aromas are quality attributes limited to the surface layers of the meat. These attributes have no reliable or consistent relationships to the eating quality of lamb which, in fact, is influenced predominantly by the bulk muscle tissue and sub-surface fat.

1. INTRODUCTION

This overview focuses on the freezing preservation of lamb and the factors that influence its eating quality. It does not review all meats because, in my opinion, results from studies on other mammalian meat (beef, pork) do not relate well to lamb.

Meat "quality" is a nebulous term which relates to a consumer's subjective response to meat -- a response involving at least visual, tactile, olfactory and gustatory senses. Meat quality includes such factors as appearance, colour, aroma, texture, flavour and juiciness. Every person responds differently to any one food product, although consumers with similar backgrounds tend to have similar responses to certain foods. To complicate this picture, we have found that consumers cannot be relied on to describe the factors they consider relate to meat quality. Subjective responses can also be open to unwitting misinterpretation by individuals, who, while evaluating quality, tend to incorporate their personal beliefs and biases. Thus, the real description of the term "quality" for any group of people can be investigated only in an impartial scientific manner, such as by the use of consumer taste panels to evaluate eating quality.

1.1 Commercial processing practices in New Zealand

Processing practices from animal slaughter to the entry into freezer stores have a significant influence on the eating quality of meat. The most important quality attribute influenced during this time is that of tenderness. The emphasis given to hygiene and the widely accepted belief that meat should be cooled as quickly as possible after animal death has led to the improved efficiency and high capacity of modern refrigerating plants. This has a disastrous effect on meat tenderness. The temperature history of meat in the prerigor period far outweighs all other factors in determining its tenderness and rapid cooling is the worst thing that can be done. The details related to processing practices and the eating quality of lamb have been extensively reviewed (2, 12) and will not be elaborated further in this paper.

1.1.1 Conditioning

This process involves holding newly killed and processed animals at a sufficiently high chill temperature to ensure the muscles are sufficiently close to *rigor mortis* for shortening (and as a result toughening) to be negligible during subsequent refrigeration. A typical conditioning process would involve chilling at 13°C for 24 hr, with an air movement between 0.25 to 0.75 m/sec and a relative humidity controlled at 80 to 85% (12). Modern practices usually involve a colder chill temperature (e.g., 10°C).

It is worth mentioning that the average New Zealand meat factory processes about 10,000 lambs per day (one every 2 sec) and some as many as 20,000. Conditioning of the total production in these factories would require very large controlled chilling rooms, which is a very expensive consideration.

1.1.2 Accelerated conditioning (AC)

In the early 1970s researchers at MIRINZ developed the commercial application of electrical stimulation: a process designed to speed up the muscles' progress into *rigor mortis* (3). Even after electrical stimulation, however, muscles are still capable of cold-shortening and thaw-shortening and hence they can become unacceptably tough to eat. Thus electrical stimulation is followed by a period of chilling (minimum of 2 hr *post mortem* at 7°C) and a controlled freezing rate (12 to 14 hr for deep leg temperature to reach -4°C). All carcasses must be frozen below -12°C before transfer to freezer storage. This entire process involving electrical stimulation, chilling and controlled freezing is called "accelerated conditioning" --or AC. One important benefit of AC is that processing times from animal slaughter to entry into freezer stores are reduced from the 2 days in the conditioning process to a 24 hr cycle in the AC process.

1.1.3 Packaging practices

During the season to 30 September 1982 a total of 31 million lambs were slaughtered for export. Of these about 82% were exported as intact carcasses and the remainder were cut into primal cuts and/or boned and exported in that form. Almost all (99.7%) of this lamb was exported frozen.

Most lamb carcasses are wrapped solely in "stockinet" -- a cotton material which, for practical meat quality purposes, is no protection at all. A small proportion of carcasses were wrapped with loose-fitting plastic bags over which the stockinet was placed. The shrinkwrapping of entire lamb carcasses in specially designed moisture-impermeable plastic bags is also commercial feasible now. All lamb cuts are individually wrapped usually in oxygen-permeable, moisture-impermeable plastic film. Vacuum-packaging or heat-shrinking help ensure an intimate contact between film and meat surfaces.

All carcasses are usually frozen naked or wrapped in stockinet bags. Thus there is usually a considerable weight loss in carcasses from slaughter to their entry into freezer stores (13, 17). For conditioned carcasses, weight losses average 3.5% to freezer store entry. Average weight loss for AC carcasses is about 2.3%. Weight losses in chilling depend upon air humidity and air velocity, whereas loss during freezing is determined primarily by carcass weight and fat cover.

Weight loss of stockinet-packaged carcasses during freezer storage averages 0.4% per month at -15°C for carcasses with a moderate fat cover and 0.6% per month at -15°C for light weight carcasses with a thin fat cover. Carcasses or cuts packaged in tight fitting plastic film have essentially no weight loss during frozen storage. Loose fitting bags, however, do allow some desiccation to occur from the meat with the resulting formation of ice within the bag.

1.2 The freezing preservation of meat. What changes occur during frozen storage?

Not all eating quality attributes change during frozen storage. For example, if lamb is properly processed before freezing, both the scientific literature and experiences at MIRINZ indicate that meat tenderness will not change significantly during frozen storage. Cooking aromas and juiciness of lamb have received essentially no research interest to date and hence their changes during frozen storage cannot be documented. This paper will be confined to the eating quality attributes of lamb and not appearance or colour. It is worth noting, however, that raw meat appearance and colour, crucial as they may be toward the consumers' short-term buying habits, bear no reliable or consistent relationships to the eating quality of the product. Therefore this paper considers only flavour changes in lamb during frozen storage. In this regard it is commonly believed that the limit to the frozen storage life of properly packaged lamb is the development of "rancid flavour" caused by oxidative deterioration of the fat.

1.3 Microbial considerations for frozen storage

A discussion of the quality of meat is incomplete without due consideration being given to microbiological changes that may occur. The limit to the storage life of **chilled** meat is directly related to microbiological growth. One cannot, however, extrapolate these findings to **frozen** meat. Thus, the importance of microbiological growth which may occur on meat below its freezing point is of great interest.

1.3.1 Physical properties of meat which affect microbiological growth

Typically meat has a freezing point of about -1°C. Of direct significance to microbiological growth are the considerations related to the water activity (a_w) of the meat surface. In chilled meat the a_w is determined by the osmotic properties of the surface layers of the meat. A normal a_w of unfrozen (chilled) meat is about 0.99. In frozen meat the a_w is solely determined by the temperature of the meat: osmotic constituents have no influence at all. a_w of frozen meat is identical to the a_w of pure ice at the same temperature (Table 1).

Table 1. Physical properties of frozen meat.

Temperature °C	a_w	% water as ice*
-5	0.95	74
-10	0.90	83
-20	0.82	88

* eutectic temperature -40°C with 89% water as ice (20).

1.3.2 Bacterial growth

Spoilage of red meats at chill temperatures is usually caused by bacteria, as their rapid growth rates preclude development of the slower growing yeasts and moulds. Psychrotrophic species of Pseudomonas characteristically dominate the spoilage floras along with strains of Moraxella, Acinetobacter, Lactobacillus and certain genera of the family Enterobacteriaceae (8). These bacteria, however, assume little significance in the flora of frozen red meats. Although reports of bacterial growth at -1.5°C are common, there are few reliable reports detailing significant bacterial growth below -3°C. This lack of growth probably occurs because of the reduction in water activity that accompanies freezing, as the common meat spoilage bacteria characteristically exhibit marked sensitivity to reduced water activity (11) and this sensitivity can be expected to be enhanced as their minimum growth temperatures are approached.

1.3.3 Mould growth

At least four distinct forms of mould growth have been recognized on frozen meat: "black spot", "white spot", "whiskers" and "blue-green mould".

Although minimum growth temperatures for moulds between -10 and -12°C have been reported, much of the data can be questioned on the grounds that they were obtained under conditions where accurate temperature control for long periods could not be assured.

Recent research conducted at MIRINZ has provided accurate data on the limiting growth conditions of the moulds capable of growth on frozen meat. Historically the most important mould growth on meat is "black spot", a condition characterized by the penetration of the black hyphae into the underlying tissues. "Black spot" is now known to be caused by at least four species of fungi, Cladosporium cladosporioides being the most common (6). On a nutrient agar medium, containing just sufficient glycerol to prevent freezing, the "black spot" moulds exhibit growth temperature minima of about -5°C (7). These organisms are characteristically moderately xerotolerant. The major causative species of "white spot", Chrysosporium pannorum, also shows similar limits of xerotolerance and minimum growth temperature (14). Growth rates of these moulds at -5°C are exceedingly slow and colonies are barely visible after 6-8 months storage. Growth rates of these moulds increase with increasing storage temperatures, but at temperatures above about -2°C bacterial growth rates are normally such that these moulds cannot compete and the meat spoils before significant mould growth is observed. However moulds, because of their greater xerotolerance, have a distinct growth advantage under conditions where moderate surface drying of the meat has occurred. As fungal growth at -5°C is extremely slow, it would appear that "black spot" and "white spot" mould growths normally develop on meat when surface temperatures have been in the range -2 to about +5°C, where some amount of surface desiccation controls bacterial growth. Similar levels of desiccation are also needed to permit development of a Penicillium species responsible for "blue-green" mould forms, although this species does not develop below -2°C (14).

The fourth mould group, "whiskers", caused by the moulds Thamnidium elegans and Mucor racemosus does not fit this pattern. Both species show a relatively low level of xerotolerance which is only slightly greater than bacteria and they do not grow readily on frozen substrates. However at temperatures near freezing or where fluctuating temperatures limit bacterial proliferation these moulds exhibit markedly faster growth rates than the other mould species and in fact would appear to compete successfully with the bacteria (14).

1.3.4 Yeast growth

Recent storage trials of frozen lamb at MIRINZ have revealed that substantial yeast growth can occur at -5°C. Although mould colonies do not appear until at least 35 weeks' storage at this temperature, yeasts begin to appear in the flora after 10 weeks' storage. One species, Cryptococcus laurentii consistently becomes the dominant species making up more than 90% of the flora. This species, initially comprising only 0.1% of the total micro flora, grows comparatively rapidly reaching maximum numbers of $10^6/cm^2$ with 20 weeks' storage at -5°C (15). The dominating growth of yeast at -5°C has previously been observed in frozen chickens (21). Work in progress would indicate that this species can develop at temperatures as low as -6.5°C but to date no growth has been observed at -8°C.

1.3.5 Overall microbiology of frozen lamb

The extensive research carried out by MIRINZ in the field of microbial growth at sub-freezing temperatures clearly indicates that meat or meat products stored at product temperatures below -8°C will not support any microbial growth.

It can, therefore, be stated that provided meat is safe and wholesome when frozen it will remain so indefinitely in proper frozen storage at temperatures below -8°C -- barring any accident. The limits to the storage of meat that is properly frozen and stored are physical, chemical or biochemical changes which are unrelated to microbiological proliferation. Therefore, frozen storage life is limited by changes in appearance or taste unrelated to microbiological activity. There is no possibility of any health hazard developing with prolonged storage of frozen meat.

2. FROZEN STORAGE LIFE OF LAMB - A REVIEW OF THE LITERATURE

The earliest report on the palatable storage life of frozen lamb was published in 1931 (10). In this study, lamb was frozen 15 min after dressing. At -10°C the meat and fat remained of an "excellent" flavour after 18 months of storage. A later study (4) evaluated the storage life of lamb chops wrapped in "Riegel's Moisture-proof Vegetable Parchment". The lamb was aged at 0°C for 7 days prior to chops being cut and frozen. Storage life of these samples was 4 months at -9°C, 5 months at -12°C and more than 14 months at -18°C. In 1947 the maximum storage life of lamb chops wrapped in cellophane was estimated to be less than 17 weeks at -10°C and 35 weeks at -18°C (19). It is not clear from the paper whether this limit to storage life was loss of colour, appearance or palatability. In a paper presented to the American Institute of Food Technologists, the storage life for unnamed samples of lamb was given as about 14 weeks at -12°C and 28 weeks at -18°C (23). Vacuum-packaged lamb loins which had been aged for 8 or 13 days at 1°C prior to freezing remained acceptable for 48 weeks at either -8 or -10°C. Unpackaged lamb was considered unpalatable within 12 weeks at -8 to -10°C or within 30 weeks at -18°C (9). In 1974 it was reported that there were only minor changes in the eating quality of New Zealand lamb stored in stockinet at -12°C under commercial conditions for up to 24 months (16). Lamb with increased levels of polyunsaturated fat could readily be stored in oxygen-permeable film at -10 to -13°C for up to 22 weeks without a significant change in flavour acceptability (18). "Normal" Australian lamb packaged in oxygen-permeable film had a high quality life (HQL) of 29 weeks at -10°C, and of more than 12 months at -20°C (1). Finally, steaks cut from excised muscles from one-year-old Merino rams and stored in polyethylene bags at -20°C were found to have a storage life of less than 3 months (5).

Clearly these studies give an enormous range of values for lamb. For example, considering -10 to -12°C only, storage life ranges from 12 weeks (9) to nearly 2 years (16). There may be many possible reasons for these different values. Unfortunately, these research reports have not included enough experimental information to adequately assess the work and therefore one cannot reach useful or justifiable conclusions about the storage life of lamb.

The published reviews on frozen storage of meat, which will not be referenced here, have used one or more of these published papers as their data base for lamb.

3. MIRINZ RESEARCH ON FLAVOUR CHANGES DURING FROZEN STORAGE

3.1 Sensory techniques

Three different sensory techniques have been used at MIRINZ to study changes in the eating quality of sheepmeat during frozen storage.

3.1.1 Highly trained, analytical taste panels. There are two separate panels: one trained to detect the unique "rancid" flavour in sheepmeat, the second to detect subtle flavour changes that occur before the rancid flavour becomes apparent. The details of these two panels have been described (24, 28).

3.1.2 "In-house" consumer panels. These tests involve feeding roasted meat to a 40-60 member panel consisting of MIRINZ personnel. Although these tasters are not selected or trained for any particular ability, they have tasted a lot of meat over the last 3-4 years and have become a very sensitive panel. This panel scores acceptability of aroma, texture, flavour, juiciness and their overall impression.

3.1.3 General-public consumer test. This involves asking 80-120 members of the general public for their attitude toward two samples of meat: they score overall acceptability only, as this is the only reliable information they can supply.

3.2. Results on frozen storage of New Zealand sheepmeat

3.2.1 Flavour changes during frozen storage

The mincing of lamb samples, incorporating 25% by weight of fatty tissue and then cooking using a "boiling-frying" procedure (28) has turned out to be a very reliable and sensitive method for analytical panel work in following frozen-storage flavour changes. As an example of the development of flavours in frozen-stored sheepmeat, data are presented from a storage trial on minced, then frozen sheepmeat stored at -5°C (Fig.1). The true rancid flavour was not evident until 13 weeks' storage in this trial, but several other significant flavour changes had occured before that time.

FIGURE 1.

Development of flavours during frozen storage of minced lamb at -5°C.

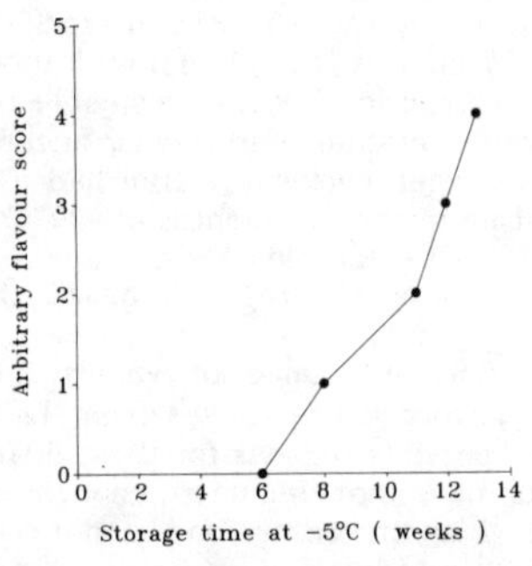

One unit on the ordinate represents a significant difference between samples ($p<0.01$) using triangle tests with 24 panelists. (After 24.)

The tasters on the rancid flavour panel, in the main, are not very sensitive to these pre-rancid flavours. Tasters on the other trained flavour panel have little difficulty in detecting these subtle frozen storage flavours. They can also correctly define the relative intensity of these flavours and can thus give a good and reliable estimate of the "length of frozen storage" a sample has been subjected to.

3.2.2 Accelerated frozen storage trials

It has been well recognised that many factors, apart from the temperature of storage, may have a significant influence on the frozen storage life of foods. These factors have been collectively termed "product, processing and packaging" -- or PPP.

In order to evaluate these factors, we have used an accelerated storage treatment. Samples are stored at -5 ± 1°C. The first discernable rancid flavour develops within a reasonable time period -- generally after 20 to 40 weeks of storage.

One of our concerns was that of microbial growth on the meat surfaces and how this may influence meat flavour. The specific nature of microbiological growth on frozen meat is discussed earlier in this paper. We have found no significant organoleptic effect caused by the growth of yeasts at -5°C (25, 26). We have just finished evaluating any flavour changes that may be caused by common moulds found on frozen meat, recognizing it usually takes longer than 40 weeks at -5°C to produce significant mould growth on lamb. Even after 45 weeks of storage at -5°C we have found no flavour changes in the meat because of mould growth. We expect to publish this work in 1984. Under the experimental conditions used in our accelerated storage trials it appears safe to say that microbiological growth on frozen meat stored at -5°C for up to about 45 weeks of storage does not cause any significant flavour change in the meat samples.

We have not made any in-depth studies related to the effects of packaging on storage life of lamb. In our early work we compared anaerobic and aerobic packaging (27). The anaerobic system resulted in an extension of about 20% in storage life. In a chilled meat oxygen-impermeable vacuum package, residual oxygen is scavenged from the air by the meat and microbiological flora, essentially creating a total anaerobic environment. This is not the case for frozen meat and the slightest trace of oxygen in the package negates the effect of good anaerobic packaging. As it is not commercially feasible to routinely produce complete anaerobic conditions in any plastic pack of frozen meat, the practical use of anaerobic packaging for frozen meat storage is questionable. We have not conducted any specific tests on the comparison between plastic-wrapped and stockinet-wrapped lambs. However, our experience indicates that desiccated lamb meat (e.g., from stockinet-wrapped lamb) has an objectionable cooking odour and the desiccated layers of meat have an unpleasant taste. If these desiccated layers are trimmed away before cooking the objectionable cooking odours and unpleasant tastes largely disappear. Thus, a moisture-impermeable plastic wrap over frozen meat results in less weight loss due to desiccation, better maintenance of good colour and appearance and apparently less objectionable cooking aromas and external tissue flavours of stored meat.

The effects of some processing treatments have also been evaluated. The most significant effect studied to date is that of the period of chilling given a carcass prior to freezing. A 24 hr chill at 0°C reduces the frozen storage of lamb by about 25% compared to the AC treatment, which involves a 2 hr post-mortem chill. It takes no imagination to recognize that the longer the pre-freeze chill period, the shorter the subsequent frozen storage life of lamb. This may explain much of the very short frozen storage life data recorded in the early literature, because many of these experiments involved chilling meat for as long as 7 to 13 days prior to freezing.

One problem always apparent in experiments involving meat animals is that of animal variability. For example, at -5°C one experiment indicated a storage life of about 20 weeks, with all samples becoming rancid over a range of about 5 weeks. A repeat experiment involving identical processing, packaging and frozen storage conditions found the storage life to be more than 40 weeks, once again with all samples becoming rancid over about a 5-week period. The only difference between experiments was the animals used. The reasons for the extreme differences in storage life are not known. Many factors may have an influence here -- e.g., season, animal age, sex, plane of nutrition, breed. It should be noted that lambs (ovine species) have specific enzymes in their intestines which destroy carotenoids (22). Thus variations in storage life for sheepmeats may not be caused by differences in the common naturally-occuring antioxidants (carotenoids).

Excessive antemortem animal stress results in the meat from those animals being of high ultimate pH. Contrary to the case for beef, we have found absolutely no significant influence of the ultimate pH of lamb muscle on either basic meat flavour acceptability or the pattern, rate and extent of flavour changes that occur during frozen storage at -5°C (25).

It must be emphasized here that there is a very poor relationship between raw meat colour/appearance and any aspect of organoleptic quality of cooked meat. I have seen meat looking almost green which, when cooked, panelists have described as some of the best meat they have eaten. Conversely some meat of excellent raw appearance has been inedible. In short, there is no reliable comparison between raw meat appearance and cooked meat quality. Methods used to improve raw meat colour (eg. anaerobic packaging) have not resulted in significant alterations in the normal pattern of flavours that develop during frozen storage. The control of meat appearance and colour, therefore, involves control of reactions at the surface of the meat. These controls do not necessarily affect the bulk of the meat.

3.2.3 Time-temperature tolerence (TTT)

A major storage trial was begun in April 1981 involving 132 lambs processed by 2 different methods (conditioning vs. AC) and stored at three different temperatures (-10, -15, -20°C). In this experiment all samples were frozen as intact carcasses, then cut. The loins only were used and these were wrapped in 'Cryovac D-film' (an oxygen-permeable plastic film). The original concept was to wait until the samples became "rancid" and record that unique specific point as the endpoint for this experiment. To date -- some 120 weeks (2 years 4 months) into this trial, we do not have any signs of rancid flavour at either -15 or -20°C and only indications that samples stored at -10°C may be near their endpoint.

To obtain some useful comparative data I have evaluated half of these samples at each storage temperature using a trained flavour panel to describe the "frozen storage life" of these samples. The data are presented in Figure 2. It can be stated that samples stored at -20°C are still considered almost fresh. Samples at -15°C are slightly altered and those at -10°C are close to becoming "rancid".

The effect of two different processing treatments can also be seen in the -10°C data of this figure. The equivalent average storage life value for these two treatments is 60 weeks at -10°C for conditioned meat, and about 80 weeks for the AC meat. In short, the average storage life for conditioned meat appears to be approximately 75% that of the AC product. Useful TTT data are not yet available because samples have not yet been stored long enough at either -15 or -20°C.

FIGURE 2.

Frozen storage life scores of plastic-wrapped frozen lamb loins.

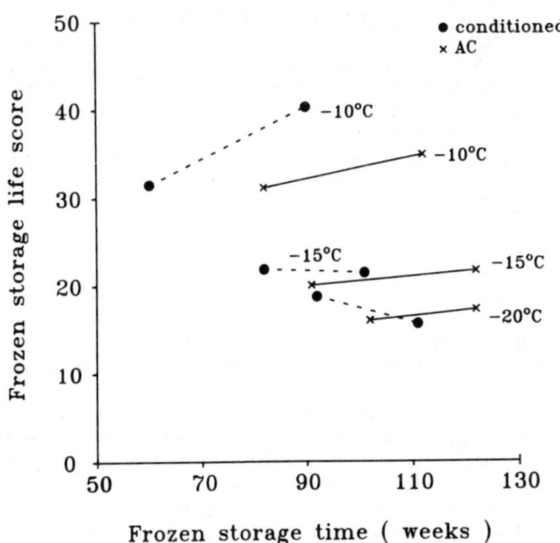

Frozen storage life was measured on a 15-cm unstructured line scale (0 = "fresh"; 100 = "rancid"). Each point represents the mean of 10 panelists tasting 10 (AC) or 11 (conditioned) samples, each from a different animal.

We have conducted several tentative experiments involving consumer testing of meat stored at -5°C for various periods. Using in-house (MIRINZ) consumer taste panels we have found that lamb which the analytical panel considers "rancid" following storage at -5°C is, when roasted, judged to be indistinguishable from paired samples stored at -35°C for the same period of time (25). This relationship would appear to be consistent over the experiments conducted to date. We have extended this to a mass consumer panel in two separate experiments. One experiment involved a comparison between legs of lamb stored at either -5 or -35°C for 25 weeks. The rancid flavour panel considered most of the legs stored at -5°C to be rancid and those at -35°C to be fresh but the consumer panel could not distinguish between the two samples.

The second consumer panel involved comparison between freshly frozen lamb and randomly selected carcasses of lamb stored for 12, 18 and 30 months in commercial freezer stores around New Zealand. Only a few carcasses stored for such a long term were found, but the data indicate that New Zealand consumers could not significantly distinguish freshly frozen meat from long-term frozen stored meat. This supports the analytical panel data in that we have found a storage life (to rancid flavour development) of more than 2 years at -10°C or colder. It should be noted that in this latter experiment, the lamb carcasses lamb that had been stored for 30 months were wrapped in stockinet. Cooking aromas were most unpleasant and the outer, desiccated layers had an unacceptable flavour.

The exterior layers were removed after cooking and it was the interior meat that was acceptable to the consumers.

3.3. Random selection and tasting

From time to time, cartons of primal lamb cuts or carasses of lamb stored for a prolonged period in commercial cold storage (exact product temperature history unknown) have been tasted. We have encountered some samples which, when roasted, have had flavours that were unacceptable to the MIRINZ in-house consumer panel. The flavours of these samples are not typical frozen storage flavours, yet we do not appear to encounter them in fresh lamb. This indicates a complex flavour interaction in some samples of lamb during frozen storage. To date we have had no indication as to the cause of these flavours, but this does add a very important, albeit confusing picture to the study of the frozen storage of lamb.

4. CONCLUSIONS

Storage of lamb at product temperatures below -8°C will ensure that there is absolutely no microbiological growth of any kind. Provided that the food is safe and wholesome when frozen, the frozen storage life of meat -- barring any accident -- is limited by changes in appearance and taste and not by spoilage or hazard to health.

Appearance and colour changes are surface-related phenomena which occur on prolonged frozen storage of meat under aerobic and/or drying conditions. Packaging of the lamb product has a critical influence on the appearance, colour and degree of desiccation which occurs. Appearance and colour are important for the short-term immediate consumer purchase of the product. Desiccated portions of meat are responsible for many unpleasant cooking odours and these cooked, desiccated portions have an unacceptable flavour. There is, however, no reliable, consistent, or useful relationship among raw meat appearance or colour, cooking aromas and/or meat flavours or any other organoleptic property of cooked lamb.

Frozen AC lamb, cut into primal cuts and stored under moisture-impermeable, oxygen-permeable plastic film will not develop a "discernible rancid flavour" for about 9 months at -5°C and at least two years of storage at -10°C. With consumer testing, the "first discernible rancid flavour" does not detract from the flavour acceptance of roasted meat. Hence to the average New Zealand consumer the practical storage life of this lamb at -10°C is in excess of two years.

Animal variability can cause enormous differences in results from one experiment to the next. Fortunately, animal variability is small within any one group of animals from one farm at any one time of the year. Processing procedures such as prolonged chilling prior to freezing have a decisive and dramatic influence on subsequent frozen storage life: a 24 hr post-mortem chill period (0°C) prior to freezing can reduce the frozen storage life by about 25% compared to AC meat.

From a scientific viewpoint I question the appropriateness of "rancidity" or rancid flavour as the limit to the frozen storage life of sheepmeat. Clearly, if rancidity is an endpoint, our packaged sheepmeat has a storage life of many years at temperatures below -10°C. However, other factors such as cooking odours (especially with sheepmeats) may be a more important criterion for consumer acceptance of this product. The real limit to the storage of any frozen food is purely a subjective consumer response: it is vital to ensure the appropriate response is studied.

There is a disturbing trend toward regulations concerned with stipulating maximum times that meat can remain in frozen storage before sale to customers ("expiry dates"). Codex Alimentarius defines the need for regulations as the protection of the health of consumers and the ensuring of fair trade practices in the food industry. Regulations are necessary. If carefully devised they should achieve their aim of satisfactory consumer protection. Over the last century of New Zealand's trade in frozen meat we have seen the promulgation of many regulations, the majority of which have been soundly based and aimed at ensuring a wholesome, hygienic and hazard-free product. They encompass regulations on animal health, ante- and post-mortem inspection requirements, carcass treatments and processing, freezing, packaging and storage conditions. However with expiry dates the assumption is that storage **time alone** is a suitable criterion for assuring the quality of a frozen food product. That is blatantly incorrect! **Storage time** is absolutely no guide to any aspect of hygiene, safety, appearance or eating quality of frozen meat. Many regulations suitable and necessary for chilled and perishable food products have little, if any, application for assuring the quality of frozen foods. Frozen storage "expiry dates", although useful as an indication of quality, should never be regulated for. This is especially true if we recognize there are no standard methods of analysis and scientists and technologists often cannot reach a suitable consensus as to an appropriate figure for storage life. In a commercial situation where storage temperatures from manufacture to retail can range from near 0°C to -20°C and lower, what does a regulation involving an expiry date which has been found on storage at -10°C mean? What protection does it give? We must not allow such deception to continue or evolve further.

Finally it is clear that any attempt to compile TTT data on any food, must involve a critical selection of data. In the meat industry since 1970 there have been very significant changes in animal breeds and genetic engineering as well as packaging technology. Processing procedures of the 1970's are very different from earlier years, and the practices of the 1980's are revolutionary. I strongly suggest, therefore, that all research into the storage life of lamb that has been conducted prior to 1970 -- irrespective of the quality of the work performed -- should be completely ignored in compiling TTT data. This current paper has clearly shown that many factors play a crucial role in determining the storage life of lamb. Thus a sound indication of storage life requires a complete and detailed description of the PPP factors used and their control.

REFERENCES

1. Bremner, H.A., Ford, A.L., Macfarlane, J.J., Ratcliff, D. and Russell, N.T. (1976). Meat with high linoleic acid content: Oxidative changes during frozen storage. J. Food Sci. 41: 757-761.
2. Chrystall, B.B. (1984). Procedures for slaughtering sheep in New Zealand. In: Milk and Meat. (H.R. Cross, ed). World Animal Science Series, Vol. 23.
3. Chrystall, B.B. and Devine, C.E. (1982). Electrical stimulation in New Zealand. In: Meat Science and Technology: International symposium proceedings. National Livestock and Meat Board, Illinois. Pp 115-136.
4. Du Bois, C.W., Tressler, D.K. and Fenton, F. (1940). Influence of rate of freezing and temperature of storage on quality of frozen meat. Proc. 1st Food Conference, Institute of Food Technologists 1: 167-179.
5. El-Wakeil, F.A., El-Banna, H.M., Abdallah, N. and El-Magoli, S.B. (1982). Effect of freezing, frozen storage, and cooking on the chemical changes and quality characteristics of lamb meat. I. Amino acids and quality characteristics. 28th European Meeting of Meat Research Workers, Madrid. Paper 3.07. Page 138-141.
6. Gill, C.O., Lowry, P.D. and Di Menna, M.E. (1981). A note on the identities of organisms causing black spot spoilage of meat. J. Appl. Bact. 51: 183-187.
7. Gill, C.O. and Lowry P.D. (1982). Growth at sub-zero temperatures of black spot fungi from meat. J. Appl. Bact. 52: 245-250.
8. Gill, C.O. and Newton, K.G. (1978). The ecology of bacterial spoilage of fresh meat at chill temperatures. Meat Sci. 2: 207-217.

9. Hiner, R.L., Gaddis, A.M. and Hankins, O.G. (1951). Effect of methods of protection on palatability of freezer-stored meat. Food Technol. 5: 223-229.
10. Lea, C.H. (1931). A note on the changes in the fat of frozen mutton. J. Soc. Chem. Ind. 50: 409T-410T.
11. Leistner, L., Rodel, W. and Krispien, K. (1981). Microbiology of meat and meat products in high- and intermediate-moisture ranges. In: Water Activity: Influences on Food Quality. Ed. L.B. Rockland and G.F. Stewart. Academic Press, New York. Pages 855-916.
12. Locker, R.H., Davey, C.L., Nottingham, P.M., Haughey, D.P. and Law, N.H. (1975). New concepts in meat processing. Advan. Food Res. 21: 157-222.
13. Longdill, G.R. and Pham, Q.T. (1982). Weight losses of New Zealand lamb carcasses from slaughter to market. Proc. Int. Inst. Refrig. Conf., New Zealand. I.I.R. France. Pages 125-131.
14. Lowry, P.D. and Gill, C.O. (1983). Temperature and water activity minima for growth of spoilage moulds from meat. J. Appl. Bact. (in press).
15. Lowry, P.D. and Gill, C.O. (1984). The development of a yeast microflora on frozen lamb stored at -5°C. J. Fd Prot. (in press).
16. Mawson, R.F., Collinson, B.R., Carse, W.A. and McLeod, K. (1974). Changes in frozen lamb stored at -12°C in stockinet wraps. Synopsis. Meat Ind. Res. Inst. N.Z. Publication number 419.
17. Pham, Q.T., Durbin, J.R. and Willix, J. (1982). Survey of weight loss from lambs in cold storage. Proc. Int. Inst. Refrig. Conf., New Zealand. I.I.R., France. Pages 117-123.
18. Purchas, R.W. and Barton, R.A. (1975). The effect of length of frozen storage on the palatability of lamb with elevated levels of polyunsaturated fat. Food Tech. in N.Z. 10(5): 7, 9.
19. Ramsbottom, J.M. (1947). Freezer storage effect on fresh meat quality. Refrig. Eng. 53: 19-23.
20. Riedel, L. (1957). Kalorimetrische Untersuchungen uber das Gefrieren von Fleisch. Kaltetechnik 9(2): 38-40.
21. Schmidt-Lorenz, W. and Gutschmidt, J. (1969). Mikrobielle und sensoriche Veranderungen getrorener Brathahnchan und Poularden bei Lagerung im Temperaturbereich von -2.5 bis -10°C. Fleischwirtschaft 49: 1033-1041.
22. Singh, H. and Cama, H.R. (1975). Metabolism of carotenoids. J. Scientific and Industrial Res. 34: 219-230.
23. Wiesman, C.K. (1947). Factors influencing quality of frozen meats. Ice and Refrig. 112(4): 21-24.
24. Winger, R.J. (1983). Selection of judges and assessment of storage-related flavours in frozen sheepmeats. J. Food Technol. (in press).
25. Winger, R.J. and Duganzich, D.M. (1984). Eating quality of frozen-stored sheepmeat with different ultimate pH values (in preparation).
26. Winger, R.J. and Lowry, P.D. (1983). Sensory evaluation of lamb after growth of yeasts at -5°C. J. Food Sci. (in press).
27. Winger, R.J. and Pope, C.G. (1981a). Processing-induced changes in frozen storage life of lamb. Int. J. Refrig. 4: 335-339.
28. Winger, R.J. and Pope, C.G. (1981b). Selection and training of panelists for sensory evaulation of meat flavours. J. Food Technol. 16: 661-669.

HEAT SHOCK TREATMENT FOR VEGETABLES TO BE FROZEN AS AN ALTERNATIVE FOR BLANCHING

E. STEINBUCH
Sprenger Institute, Wageningen
The Netherlands

Summary
The application of heat shock treatments, in stead of normal blanching procedures, might be feasible to leguminous vegetables, such as beans and peas, in order to retain flavour and colour during frozen storage. Exposure to boiling water or condensing steam during 5-15 seconds gave satisfactory results, notwithstanding the residual activity of peroxydase and even catalase and lipoxygenase. Beans, treated by this method indicate a natural firm texture which differs highly from the softness of conventionally blanched green and wax beans.

1. INTRODUCTION

Considering the history of the production of frozen foods a remarkable fact can be observed regarding the difference between vegetables and the other food products. While meat and fish appeared to be very suitable for freezing, vegetables were originally not considered as an excellent raw material to be frozen, on account of serious quality deterioration. During storage at even -20°C discolourations and off-flavours cause dramatic reduction of appearance, attractiveness and edibility, leading to total unfitness for human consumption.

However, this inconvenience of freezing vegetables is solved by the invention of the blanching process. Blanching, defined as a heat treatment at 90-98°C, during a relatively short time, depending on the nature and the size of the concerned vegetable, completes the inactivation of those enzymes being harmful to quality retention.

Actually, the disadvantages of conventional blanching procedures, in regard to energy consumption, waste water production and undesired texture changes in the frozen vegetables require a renewed approach of this heat treatment. In particular, the unfavourable texture changes of some vegetables, into softness, toughness and rubberness stimulate the search for alternative treatments.

Possible alternative treatments, which might be considered, are based on the removal of oxygen from the tissue, in connection with the suggestion of Aylward and Haisman (1969), that the quality decrease of unblanched frozen vegetables may be mainly oxidative (2). The application of vacuum packing resulted in the indication that the enzym systems in mushrooms (Steinbuch, 1978) are much more oxygen-depending than those in frozen spinach (Birnbaum et al., 1979) (6, 3). Vacuum treatments do not contribute to stabilization of chlorophyll in unblanched frozen green vegetables. Therefore, heat shock treatments are applied in order to inactivate the chlorophyll converting enzymes in the surface of the product such as green beans. The results of those experiments were rather surprising: both colour maintenance and tenability increased without texture damage (Steinbuch 1980) (8).

2. MATERIAL AND METHODS

For these experiments a variety of vegetables, intended for normal consumption, whether fresh or frozen, are selected, as green and wax beans, garden peas, leek and sprouts.

After proper cleaning and washing, preparation, as cutting and size-grading, the vegetables were exposed to various heat treatments. The heat treatments were carried out by soaking the product in an abundant amount of boiling water during a certain time. Occasionally the products were exposed to condensing steam, when limitation of leaching of soluble substances was required.

After being rapidly cooled with tap water, the vegetables, whether packed or unpacked, were frozen in a blast freezer at -40°C and stored at -18 to -20°C.

After being stored during various times (1 week - 2 weeks - 1 month - 2 months - 6 months - 12 months) samples are sensorically evaluated for appearance, colour, flavour and texture. The sensorical analysis has been executed by a panel of 3-5 experts.

Methods used for the analysis of both the enzyme activities in the raw material and the residual enzyme activities in the processed product are described earlier (Steinbuch et al., 1979) (7).

3. RESULTS AND DISCUSSION

The application of heat shocks, whether in hot water or with condensing steam, during varying times has been deliberately executed frequently in some years, in order to investigate the interrelation between the exposure time to heat, the residual activity of relevant enzymes and finally the product quality retention, determined by the presence or absence of discolourations and off-flavours. Table I shows the effect of the duration of heating on the residual activity of enzymes in green beans.

Table I: The relation between heating time and residual enzyme activity in green beans

heat treatment (± 98°C)	lipoxygenase		catalase		peroxydase	
	1980	1981	1980	1981	1980	1981
no	100%	100%	100%	100%	100%	100%
2.5 s (w)	95		68		81	
5 s (w)	77	47	44	82	69	74
10 s (w)	20	15	21	41	61	72
15 s (w)		1.1		34		52
20 s (w)		0.5		21		43
20 s (st)		0.2		-		2.4
3 min. (w)	0.2	0.13	-	-	1	0.8

(w) = water (st) = steam

The thermal enzyme inactivation values in green beans in both 1980 and 1981 are rather comparable between certain limits. Differences, in regard to variety, maturity and climatic conditions might be the reason of slight variations in the heat-resistance of the investigated enzymes. Heating with steam results in a higher enzym inactivation in comparison with boiling water. The relatively high heat transfer of condensing steam might be the cause, in spite of gases, released from the tissue during heating.

Results of the heat shock treatment on the quality of the green beans as shown in table II, indicate a required heating time of 10 sec. for the

Table II: The relation between the heat shock treatment and the quality of frozen green beans

heat treatments in water (98°C)	after 3 months			after 7 months			after 12 months		
	colour	flavour	texture	colour	flavour	texture	colour	flavour	texture
no	discoloured	strong off-flavour	good	discoloured	strong off-flavour	good	discoloured	strong off-flavour	good
2.5 s	good	slight off-flavour	good	good	off-flavour	good	discoloured	off-flavour	good
5 s	good	good	good	good	good	good	discoloured	off-flavour	good
10 s	good	good	good	good	good	good	good	good	good
3 min.	good	good	good	good	good	soft	good	good	soft

retention of flavour and colour as well as texture. If the beans were exposed to heat during 2,5-5 s a gradual decrease of colour and flavour can be observed within one year.

The application of heat shock treatments with wax beans (table III) results in residual enzym activities, comparable with those of green beans. Although the lipoxygenase-content of fresh wax beans was only 20% of that of fresh green beans.

Table III: The relation between the heat treatment, residual enzym activity and quality retention of wax beans

heat treatment in water (98°C)	residual enzyme activity			quality evaluation after one year		
	lipoxygenase	catalase	peroxydase	colour	flavour	texture
no	100%	100%	100%	discoloured	strong off-flavour	very firm
5 s	20	68	75	yellow	off-flavour	tough
10 s	17	64	72	yellow	off-flavour	firm
15 s	13	49	23	yellow	good	firm
20 s	7	5	4	yellow	good	less firm
30 s	2	3	4	yellow	good	soft
2 min.	1	-	1	bright yellow	good	very soft

As observed from the evaluation of three quality aspects, a heat shock time of 15 seconds appears to be required. Shorter heating times (5-10 seconds) result in off-flavour and undesired textures during a one year storage, while only the unblanched sample show faded pale discolourations.

Loss of the original texture, as defined by increasing softness, is strongly related with the application of prolonged heating times ($\geqslant$ 20 seconds).

Both colour and flavour of frozen peas can be retained by the application of very short heat shock times as shown in table IV. After a heat treatment of 10 s both flavour and colour are well retained. However, the various pea samples show only slight texture differences. Heating with condensing steam results in an enzyme inactivation to a higher degree than it was mentioned earlier in regard to green beans, when compared with soaking in boiling water.

Besides, the degree of inactivation of lipoxygenase, derived from table IV, indicates a less heat resistance of this enzyme, in comparison to thermal inactivation data of Svensson et al. (1972) and Ohlsson et al. (1974) (9, 5).

Table IV: The relation between the heating time, residual enzyme activity and quality retention of frozen peas

heat treatment in water (98°C)	residual enzyme activity			quality evaluation after one year storage	
	lipoxygenase	catalase	peroxydase	colour	flavour
no	100%	100%	100%	dis-coloured	strong off-flacour
2.5 s	80%	36%	65%	dis-coloured	off-flavour
5 s	62%	28%	52%	dis-coloured	good
10 s	6%	2%	34%	good	good
15 s	1%	0.3%	23%	good	good
3 min.	-	-	0.3%	good	good

The investigation about the application possibilities of heat shocks, in stead of blanching vegetables for freezing are moreover executed with leek and Brussels sprouts. Observations on the quality retention of these frozen vegetables indicate considerable differences in the behaviour of flavour properties. White, non-chlorophyll containing leek could be heat-shocked in order to obtain a satisfactory storage life. On the other hand both green coloured leek and Brussels sprouts require a thorough blanching treatment, in order to avoid any undesired quality alterations during frozen storage of at least one year. Complete peroxydase inactivation appears to offer a guarantee for flavour retention of frozen sprouts, illustrating the inapplicability of heat shocks for these kind of vegetables.

Overlooking the effects of heat shock treatments on various vegetables for freezing, comprehensive differences in "enzyme inactivation - quality retention" relations can be ascertained (Table V).

Table V: The relation between the maximal residual enzyme activity and the minimal high quality of frozen vegetables

product	residual enzyme activity		
	lipoxygenase	catalase	peroxydase
peas	10%	10%	40%
beans	20%	20%	50%
sprouts	0	0	slight

Figures in table V indicate the level of activity of various enzyme systems, determinative for the keeping quality of frozen vegetables. The lipoxygenase content of fresh sprouts is considered to be neglectable. On the other hand, the occurrence of this enzyme in leguminous vegetables relates highly with the presence of the green colour. Lipoxygenase is presumed to be involved in the formation of rancidlike off-flavours in unblanched frozen vegetables (Arens et al., 1973) (1). However, a residual activity of 10-20% might not be detrimental for flavour retention.

Therefore, it can be questioned, if there is any unknown enzyme system, which determines actually quality retention of frozen vegetables. On the other hand, the favourable effect of heat shock treatments, during fixed times, depending on kind and size grade of the product might be explained by assuming plural thermal consequences for vegetable tissues:

1. Increasing enzyme inactivation. Moreover, it is unknown, which relation exists between the residual activity, determined in an artificial substrate and the actual ability of the enzyme to function in the product.
2. Partial protein coagulation.
3. Reducing the air (oxygen) content, which might limit the activity of oxygen-consuming enzymes.
4. Due to the limited heating of the inner part of the product to ca. 50-60°C, texture is highly maintained and not softened by hydration of cell wall components (Van Buren, 1979) (10).

4. ACKNOWLEDGEMENT

The author is indebted to Prof. W. Pilnik for supporting discussions and Mr. W. Klop and Mr. J.E. Robbers for skilful enzyme analysis.

5. LITERATURE

1. Arens, D., Laskawij, G. and Grosch, W. (1973). Lipoxygenase aus Erbsen. Lebensm. Unters.-Forsch., 151, 162.
2. Aylward, F. and D.R. Haisman (1969). Oxidation systems in fruits and vegetables: their relation to the quality of preserved products. Adv. Food Res., 17, 1-76.
3. Birnbaum, N.R., J.R. Hicks, M.H. Tabacchi and P.E. Brecht (1979). Evaluation of evacuated packages as an alternative to blanching for frozen spinach. J. Food Sci., 44, 404-406.
4. Klop, W. (1974-1975). Enzyme inactivation and reactivation during the processing of horticultural produce. Annual Report, Sprenger Instituut, 15.
5. Ohlsson T. and S. Svensson (1974). Calculation of heat transfer and enzyme inactivation in blanching of peas. Proc. IV Int. Congress Food Sci. and Technol., Vol. IV, p. 380-389.
6. Steinbuch, E. (1978). Factors affecting quality and shrinkage losses of processed mushrooms. Mushr. Sci., 10. Part II, p. 759-766.
7. Steinbuch, E., R.A. Hilhorst, W. Klop, J.E. Robbers, W. Rol and R.G. van der Vuurst de Vries (1979). Quality changes in frozen Brussels sprouts during storage. I. Sensory characteristics and residual enzyme activities. J. Food. Technol., 14, 289-299.
8. Steinbuch, E. (1980). The effect of heat shocks on quality retention of green beans during frozen storage. J. Food Technol., 15, 353-355.
9. Svensson, S.G. and C.E. Eriksson (1972). Thermal inactivation of lipoxygenase from peas. I. Time-Temperature relationships and pH-dependence. Lebensmitt.-Wiss. u. Technol., 5. p. 118-123.
10. Van Buren, J.P. (1979). The chemistry of texture in fruits and vegetables J. Text. Stud. 10, 1-23.

THE INFLUENCE OF THE DEGREE OF BLANCHING ON THE QUALITY OF FROZEN VEGETABLES

K.Z. Katsaboxakis
Institute of Technology of Agricultural Products
Athens Greece

Summary

Blanching conditions and mainly the degree of thermal treatment given, influences the quality of frozen vegetables during frozen storage. Texture degradation increases with the degree of blanching in the majority of vegetables and their quality could be improved considerably if this operation was reduced to a minimum possible and so retain to the maximum extent their natural characteristics. The degree of blanch treatment also influences the rate of chlorophyll deterioration during processing as well as during frozen storage. Unblanched green vegetables show a constant rate of total chlorophyll loss during frozen storage. Overblanching results in increased total chlorophyll loss after a certain lag storage period. Peroxidase inactivation by blanching is not necessary in all vegetables and their varieties. For many vegetables the best sensory quality is achieved prior to complete inactivation of peroxidase. This suggests that further research is needed on identifying the enzyme systems mainly responsible for off flavour development and not allowing them to remain active in the frozen product.

1. INTRODUCTION

Freezing improves the storage life of vegetables by retarding enzyme activity, chemical changes and microbial growth. However since 1930's it has been recognized that enzyme systems remain active even at sub-zero temperatures (4) and that freezing alone does not completely prevent off flavour development, colour and texture deterioration in frozen vegetables. It was also established many years ago that production of off-flavours in frozen vegetables could be prevented by blanching, a brief heat treatment of the raw material in steam or hot water prior to freezing and frozen storage. Although the most important effect of blanching is that of inactivation of the naturally occurring enzymes in the vegetable tissue, it also reduces the level of infection by microorganisms and improves the visual green colour of vegetables as a result of the removal of gases from the vegetable surface and from the intercellular spaces. Blanching finally may improve flavour by expelling gases and other volatile degradation products, formed during the post harvest interval.

The adverse effects of blanching are mainly the permanent modification of the cellular stucture in the vegetable tissue,the solubilization and/or destruction of some nutrients and vitamins in the blanching media and the conversion of green chlorophylls to yellow green pheophytins (10,15). Alternatives replacing conventional blanching have been studied (6,17). Vacuum packaging for example was found not effective as blanching for long term frozen storage periods. It was also observed (8,18) that some vegetables can tolerate an amount of residual peroxidase activity after blanching without significant loss of quality during frozen storage. Conclusively blanching is necessary as a part of the preparation for freezing preservation of vegetables and its optimization should be studied in each case.

2. EFFECT OF BLANCHING ON THE CELLULAR STRUCTURE AND THE PHYSICAL CHARACTERISTICS OF THE VEGETABLE TISSUES.

Vegetable tissues are living materials and their properties that reflect the freshness and turgidity depend largely upon the structural arrangement and chemical composition of the cell wall and of the intercellular spaces where pectic substances are the primary constituents. Heat given by blanching kills the cells breaks down pectic substances and causes irreversible changes in the cell structure and the physical characteristics of the vegetables tissues. Figure 1 shows diagrammatically the main effects of blanching on a typical plant cell.

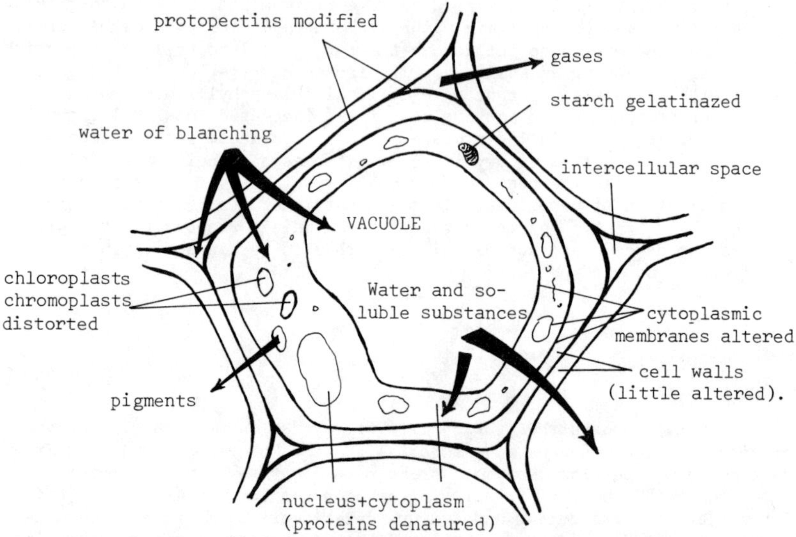

Fig.1. Diagram showing the main effects of blanching on a generalised plant cell.

The disruption of the cytoplasmic membranes increases their permeability water enters the cells and the intercellular spaces expelling gases and other volatile products. Proteins become denatured and a loss of soluble nutrients (vitamins, sugars and minerals in perticular) occurrs at the same time. Chloroplasts and chromoplasts become swollen during blanching and distorted and carotens or chlorophylls become more or less diffused throughout the cell and in the blanching media.

An appreciable amount of research has been carried out on studying the effect of blanching and freezing on the texture and other quality characteristics of frozen vegetables.In the case of green beans it was found (8) that while in unblanched samples cell wall damage was randon in blanched and frozen,damage had a radial pattern. It was also found (9) that blanching conditions affect sloughing of beans, that is a tendency for the skins to become loose and detach away from the main part of the pod. Water blanching at 97°C of carrot root tissue and pea seeds indicated (16) that turgor loss during this operation results in contraction of the cells (depending on elasticity of the cell walls and perhaps in peas on the degree of starch gelatinization) with expression of some cell contents, followed by diffusion of solutes from the cells. In the case of mushrooms conventional blanching in water,sufficient to inactivate a highly active endogenous enzyme system cause an appreciable amount of shrinkage and results in a 28-30% loss by weight (2). The texture of blanched and normally frozen mushrooms is also unacceptably tough. A modified blanching process was developed at Pensylvania state university USA that involves a soak of mushrooms in water (20 min) followed by chill storage at 2-4°C (for 24,48 or 72 hr) followed by a further soak (2hr). The mushrooms are then blanched and frozen or canned conventionally. It was shown (11) that the above process in combination with other treatments is very effective in reducing shrinkage in canned or frozen mushrooms.

It is quite evident from the above findings that texture degradation increases with the degree of blanching in the majority of vegetables and their quality could be improved considerably if this operation was avoided, replaced by another or limited to an optimum level. Work at various institutes is being carried out at that directions.Finally it should be pointed out that there are a few vegetables that suffer severe reduction in texture and overall quality if blanched prior to freezing and which are generally acknowledged to maintain their quality during frozen storage (at -18°C) without blanching. It has been found (13) that the quality of unblanched frozen tomatoes, red and green peppers, cabbage, diced celery cucumbers, leeks, onion, diced carrots, parsley and some oil containing herbs was better than that of blanched counterparts after 12 months frozen storage.

3. EFFECT OF BLANCHING ON CHLOROPHYLLS AND SOME OTHER NUTRIENTS

Chlorophylls.Colour is the primary quality attribute by which the consumer assesses natural and processed food. The colour of a processed food is often expected to be as similar as possible to the natural ptoduct. Chlorophylls are mainly responsible for colour in green vegetables (beans, peas, spinach, okras etc). Time-temperature conditions during blanching of green vegetables were found to influence the rate of chlorophyll deterioration during processing as well as during frozen storage (12,19,20).
As shown in figure 2 an amount of total chlorophyll is lost during blanching converted to pheophytin,the conversion being proportional to the degree of heat treatment applied during blanching. At the initial steps of blanching chlor.a appears to be more sensitive than chlor.b but after the 4th minute of blanching increased destruction rate was observed and for chlor.b.Chlor.a/b ratio decreases contineously as blanch time increases from 0 to 4 min and then increases again due to the change in the destruction rate of chlor.b. Similar results were also obtained by other workers (10,19, 20). It has been also established that changes in chlorophylls continue during frozen storage and that these changes are depended largely upon the amount of residual enzyme activity and/or the degree of heat treatment given during blanching.As it can be seen in figure 3 in unblanched samples a constant rate of total chlorophyll loss occurrs during the entire storage period. The loss of total chlorophyll diminished in samples blanched from 30 to 60 secs. Samples blanched from 2 to 3 min showed no significant loss in total chlorophyll for the first 6 months of frozen storage, compared to samples with lower heat treatment. However after that storage period a

progressive increase was observed in the rate of chlorophyll loss. This loss rate was increased in samples of beans received more prolonged thermal treatment. Obviously thermal treatment higher than that required for peroxidase inactivation results in increased loss rate in chlorophyll after a cartain lag storage period due to oxidation and related to the process of fat peraxidation as proved by other workers (19,20).

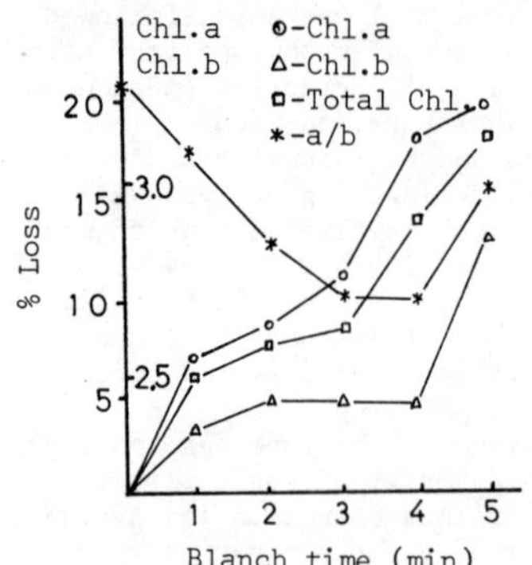

Fig.2. Effect of blanch time (98°C) on % loss of chlor.a, chlor.b, total chlor. and chl.a/b ratio (Katsaboxakis, 1983)

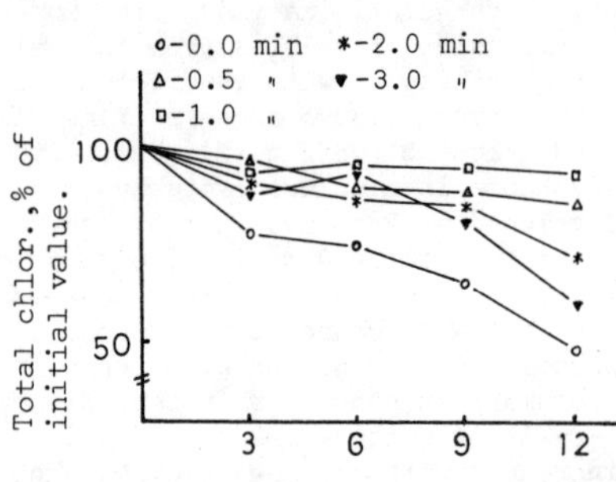

Fig.3. Total Chlor.,% of initial value of green beans blanched for 0, 0.5, 1, 2 and 3 min, during frozen storage at -18°C. (Katsaboxakis).

Changes in chlorophylls influence significantly the visual green colour of vegetables as demonstrated in table I where the mean sensory scores of colour enaluation of frozen green beans, blanched from 0 to 3 min and stored for 0,3,6 and 12 months at -18°C are presented. A statistically significant difference was found for mean colour scores between unblanched and blanched samples. A significant also difference can be observed between samples blanched from 30 to 60 secs and samples subjected to a prolonged heat-treatment.

Table I Mean sensory scores for colour evaluation* of frozen green beans blanched from 0 to 3 min, stored for 0,3,6 and 12 months at -18°C.

Storage	Blanch time (sec)							F	LSD
-18°C	0	30	60	90	120	150	180	test	P=0.05
0	3.22	6.25	4.77	5.45	4.48	4.28	3.65	**	0.671
3	3.63	6.13	4.94	4.91	4.66	3.50	3.30	**	0.676
6	3.56	5.76	5.13	5.20	4.93	3.50	2.93	**	0.788
12	2.86	4.63	4.76	4.50	3.83	3.73	2.90	**	0.783

* Green colour intensity and uniformity was scored in a 7 point scale. (Katsaboxakis, 1983).

Ascorbic acid and other nutrients. The effect of the duration, temperature and type of blanching on the nutrient and mineral composition of vegetables has been studied by many workers (1,14,16). Steam blanching has been generally recognized to have less effect in nutrient and minerals than water blanching. Water-blanched brocolli was lower in solids, as h,P,K and

Ascorbic acid than steam-blanched samples (14). The loss in total ash is primarily due to losses in potassium which is more soluble and pressent in comparatively large quantities.

Figure 4 shows the effect of blanch time at 98°C (±2°C) on L-ascorbic acid retention of green beans during frozen storage.

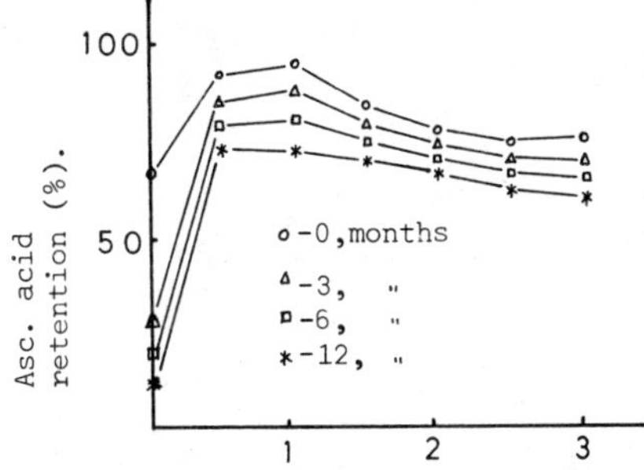

Fig.4. Effect of Blanch time at 98°C (±2°C) on Ascorbic acid retention of green beans stored for 0.3.6.12 months at -18°C. (Katsaboxakis,1983).

It can be seen from this figure that blanching has a definite effect on ascorbic acid retention. Unblanched samples showed a 75% loss of vitamin content during the first 3 months of frozen storage and almost complete loss 6 months later. L-ascorbic acid levels were higher in samples which had received 30 to 60 secs blanch treatment during the entire frozen storage period. Obviously a minimum thermal treatment given by blanching is required for vitamin retention during frozen storage and this depends mainly on the particular vegetable and its variety. This minimum level of the thermal process does not nenessarily results in complete peroxidase inactivation. Finally it should be pointed out that in the case of frozen vegetables ascorbic acid content is a valuable and sensitive index for the evaluation of the thermal process and of the frozen storage conditions.

4. ENZYMES AND SENSORY QUALITY.

Steam or hot water blanching of vegetables is a prefreezing process of primary importance because heat inactivates the naturally occurring enzymes which are responsible for quality losses during frozen storage. Peroxidase is the most heat-stable enzyme present in the majority of vegetables and its inactivation is very commonly used for determining the adequacy of the blanching process. The activity of peroxidase can be easily determined in the Laboratory and factory, by rapid tests developed for this purpose. Although good correlation of peroxidase inactivation with quality stability in frozen vegetables has been found, more recent studies have demonstrated that a significant proportion of enzyme activity may be left after blanching without detectable loss in quality during frozen storage (7,18).

As shown in figure 5,30 secs blanch treatment at 98°C (±2°C) prevents effectively off flavour development in green beans, for 12 months frozen storage at -18°C. This treatment leaves approximately 7% of the level of peroxidase activity in the unblanched samples and indicates that in green beans the best sensory quality is achieved prior to complete inactivation of peroxidase.

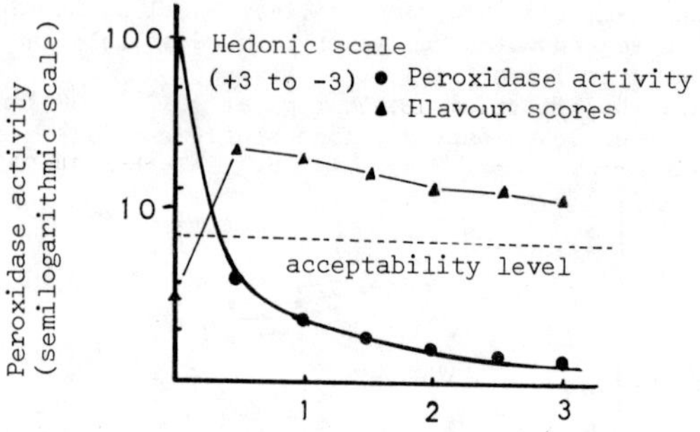

Fig.5. Effect of blanch time at 98°C (±2°C) on Peroxidase activity and flavour score of green beans stored for 12 months at -18°C (Katsaboxakis,1983).

Conclusively attempts to minimize peroxidase activity by blanching does not seem necessary in all vegetables and the test for determining peroxidase activity may not be appropriate in every case. This suggests that further research is needed on identifying the enzymes mainly responsible for off flavour development and not allowing them to remain active in the frozen product. Among other lipolytic enzymes lipoxygenase is generally considered to be more closely related to the loss of flavour acceptability throught lipid oxidation, than peroxidase. Off odour in frozen raw or insufficiently blanched vegetables have generally been attributed to the formation of volatile carbonyl compounds (3,5). The reactions involved are principally enzymic and may result in the formation of ethanol by anaerobic fermentation of pyruvate or of a variety oxidation products of polyunsaturated fatty acids of the plant tissues. It has been reported (5) that hexanal formation, which is the major product of linoleic acid oxidation in blanched (100°C for 2 min) peas and stored at -5°C, occurrs simulteneously with sensory detection of off-flavours. Hexanal is formed either enzymatically (lipoxygenase) or through autoxidation in processed samples, even during frozen storage. Ethanol is also especially interesting and although both are formed during post-harvest storage, hexanal was found alone, during frozen storage. It is obvious that the relative proportion of these compounds in a deteriorated frozen vegetable might be an indication whether deterioration occurred before or after processing. Lipoxygenase is much more easily inactivated than peroxidase (3) and its inactivation could be taken as determining of the adequacy of the blanching process.

REFERENCES

1. ABRAMS, C.I. (1975). The ascorbic acid content of quick frozen sprouts. J.Fd. Technol. 10:203
2. Mc ARDLE, F.J. and CURWEN, D.(1962). Some factors influencing shrinkage of mushrooms during processing. Mush. Sci., 5:547
3. AYLWARD, F. and HAISMAN, D.R. (1969). Oxidation systems in fruits and vegetables. Adv.Food Res. 17:1.
4. BALLS, A.K. and LINEWEAVER, H. (1938). Action of enzymes at low temperatures. Fd Res. 3:57.
5. BENGTSSON, B.L. et al. (1967). Hexanal and Ethanol formation in peas in relation to off-flavor development. Fd Technol. 21:458.
6. BIRNBAUM, N.R. et al (1979).Evaluation of evacuated packages as an alternative to blanching for frozen spinach J.Fd Sci. 44:404.

7. BÖTTCHER, H. (1975). Enzyme activity and quality of frozen vegetables. I. Residual activity of peroxydase. Nahrung, DE., 19:173
8. BROWN, M.S. (1967). Texture of frozen vegetables:Effect of freezing rate on green beans.J.Sci. Fd Agric. 18:77.
9. BUREN, J.P. et al.(1960). Influence of blanching conditions on sloughing, splitting and firmness of canned snap beans. Fd Techn. 14:233.
10. DIETRICH, W.C. et al. (1959b). Time-Temperature-Tolerance of frozen foods. XVIII. Effect of blanching conditions on colour stabitlity of frozen beans. Fd Technol. 13:258.
11. GORMLEY, T.R. and WALSHE,P.E. (1982). Studies on methods reducing shrinkage in canned and frozen mushrooms.COST 91, Sub-group 3, Freezing and Thawing, Athens final seminar Nov. 14-18.
12. KATSABOXAKIS, K.Z. and PAPANICOLAOU, D.N. (1983). The consequences of varying degrees of blanching on the quality of frozen green beans.COST 91, Sub-group 3, Freezing and Thawing, Athens final seminar Nov.14-18.
13. KOZLOWSKI, A.V. (1979). Is it necessary to blanch all vegetables before freezing? Quick frozen fds international 20:83.
14. ODLAND, D. and EHEART, M.S. (1975). Ascorbic acid, mineral and quality retention in frozen broccoli blanched in water, steam and ammonia-steam. J.Fd Sci. 40:1004.
15. PHILIPPON. J. (1969). La congélation des haricots vert. Effect combinés du temps et de la température pendant la préparation et la congélation sur la qualité finale du produit. Rev.Gén. Froid No 1:101.
16. SELMAN, J.D.and ROLFE E.J. (1979). Effects of water blanching on pea seeds. I. Fresh weght changes and solute loss. J.Fd Technol. 14:493.
17. STEINBUCH, E. (1979). Quality retention of unblanched frozen vegetables by vacuum packing. I.Mushrooms J.Fd Technol. 14:321.
18. STEINBUCH, E. (1980b). The effect of heat shocks on quality retention of green beans during frozen storage. J.Fd Technol. 15:353.
19. WALKER, G.C. (1964a). Color deterioration in frozen green beans (Phaseolus vulgaris). J.Fd Sci. 29:383.
20. WALKER, G.C. (1964b). Color deterioration in frozen green beans.2. The effect of blanching. J.Fd Sci. 29:389.

THE THEORY AND PRACTICE OF FOOD THAWING

S.J. James & C. Bailey
Agricultural and Food Research Council, Meat Research Institute,
Langford, Bristol, UK, BS17 8DY

1. INTRODUCTION

With the exception of a few types of frozen desserts, frozen foods must be thawed before they are eaten. In many cases thawing is a simple process and presents little difficulty to the consumer. Frozen vegetables can be heated directly from a frozen state in boiling water; beefburgers, individually frozen sausages and fish fingers can be fried or grilled without preliminary thawing. More and more commonly, pre-cooked pies and individual meals are being prepared so that they can be directly warmed in microwave ovens. The only occasion that domestic consumers meet a thawing problem is when they are presented with large frozen joints of red meat or whole poultry. A typical 6 kg joint can take a day to thaw at normal ambient temperatures (20°C) and 3 to 4 days in a refrigerator at a nominal +1°C. Numerous food poisoning cases have been traced to the inadequate cooking of incompletely thawed turkeys or chickens.

In commercial operations, the thawing requirement is scaled up and it is product size that produces the majority, if not all of the problems encountered. A growing proportion of the meat consumed in the United Kingdom is frozen in carcass or boned-out form, either within the country or before importation. The majority of fish caught in distant waters or imported has been plate-frozen at sea. This frozen meat or fish has to be thawed before it can be processed for either refreezing in consumer packs or for direct sale in its thawed state. Medium to large processing companies will need to thaw between 10 and 100 tonnes of frozen product per day which presents problems in handling and in the provision of physical space for the operation.

If the individual items to be thawed were small, then the problem would be minimal since the short thawing times required could be achieved in small continuous thawing plants. However, fish is commonly frozen in 50 x 100 x 10 cm thick blocks weighing about 40 kg, beef in quarters up to 30 cm thick and 40 to 80 kg in weight, lamb and pigs as whole carcasses weighing from 15 to 25 kg and 50 to 100 kg respectively, and boned-out meat in 25 kg slabs nominally 15 cm thick. In the dairy industry, butter from Australia and New Zealand is imported in frozen slabs 30 x 60 x 30 cm thick. For reasons that will be explained later the thawing cycle for most if not all of these products must be measured in days rather than hours. The logistical problems are therefore intensified by the duration of the thawing operation.

Thawing is often considered as simply the reversal of the freezing process, and in many ways it is. However, inherent in thawing is a major problem that does not occur in the freezing operation. The majority of the bacteria that cause spoilage or food poisoning are found on the surfaces of meat and fish. During the freezing operation surface temperatures are reduced rapidly and bacterial multiplication is severely limited, with bacteria becoming completely dormant at below -10°C. In the thawing operation these same surface areas are the first to rise in temperature and bacterial multiplication can recommence. On large objects subjected to long uncontrolled thawing cycles, surface spoilage can occur before the centre regions have fully thawed.

For these and other reasons, thawing is not a simple operation. In this paper the theoretical considerations that govern thawing are discussed, followed by descriptions of the more important methods for thawing foodstuffs together with experimental data on thawing times for a range of products. The paper concludes with a brief discussion of the factors that need to be considered when choosing a thawing method for a particular product. Since most commercial thawing is carried out on either fish or meat only these two foodstuffs have been specifically considered. However the majority of the points discussed would apply equally to any high water content product.

2. THEORETICAL CONSIDERATIONS

Thawing methods can be divided into two groups depending upon whether heat is supplied to the frozen foodstuff by conduction through the surface or by electro-magnetic radiation. Since the substantial majority of commercial systems operate using the conductive method, its theory and problems will be examined in detail and particular aspects of electromagnetic thawing will be considered in the section on methods.

Before the factors that determine thawing time can be considered, the thawing process itself needs to be clearly defined. Freezing of a high water content foodstuff such as meat or fish takes place over a range of temperatures rather than at an exact point, because as freezing proceeds the concentration of solutes in the fluid portion steadily increases and progressively lowers the freezing temperature. Thawing simply reverses this process and the temperature at the centre of a meat cut would follow the path shown in Figure 1. The relationship between the percentage of water that is frozen, the enthalpy of lean meat, and temperature is also given. For the thawing process to be complete no ice should remain and the minimum temperature has to be above -1°C (Figure 1). To thaw 1 kg of meat from a starting temperature of -40°C would require the addition of 300 kJ of energy if the meat was very lean, falling to about 180 kJ if very fat. Similar figures for fish range from 290 kJ/kg for white fish to 230 kJ/kg for oily fish which have a high fat content.

Frozen meat that requires boning or fish filleting have to be completely thawed. However an increasing proportion of meat is boned before freezing and if it is subsequently used in products such as pies, sausages, etc., it can be cut by machine in a semi-frozen (tempered) state. Similarly, blocks of whole fish can be separated in a tempered state, and the individual fish thawed with comparative ease before further processing. The range of temperatures used for tempering is not strictly defined and varies between products. For meat and white fish -2° to -7°C is the usually accepted range. In bacon and other cured foodstuffs, the temperatures depend upon the salt content and can be as low as -9° to -13°C. To temper meat from -40°C to an average temperature of -4°C requires a heat input of approximately 100 kJ/kg, only one third of that required for complete thawing. Tempering is therefore a much less energy-intensive operation than thawing and is simpler and quicker; but should not be confused with thawing.

The temperature dependence of thermal conductivity in high water content foodstuffs also aids the tempering process. Because conductivity of both frozen white fish and lean meat is 3 times that of the thawed material, the rate of heat flow to the centre is high; however when thawing commences the surface rises above the initial freezing point, an increasing thickness of poorly conducting material extends from the surface into the foodstuff reducing the rate of heat flow into the centre of the material, and substantially increasing the time required for thawing.

The duration of a thawing process depends upon a number of factors related to the product, the thawing method and the environmental conditions used. The dimensions and shape of the product, particularly the maximum thickness and its thermal properties, are of principal concern. The total amount of energy that must be introduced into the product is equal to the enthalpy change (ΔH) between the initial temperature and the average temperature required within the material after thawing. The important physical property is thermal conductivity (k) the effect of which has already been considered, whilst the main environmental factors are the temperature of the thawing medium (Ta) and the surface heat transfer coefficient (h) which is a function of the shape and surface condition of the product, the thawing medium used and its velocity. Except for very simple configurations, h cannot be derived theoretically and must be measured experimentally. Few such measurements have been made for the thawing of foodstuffs (1,2), but typical ranges of h for the main thawing systems are:-

Air - free convection	5 to 15 $W/^2{}^\circ C$
Air - forced convection	10 to 70 "
Water	100 to 400 "
Vacuum steam heat	150 to 1000 "
Plate	100 to 300 "

In air thawing, h is not constant and is a function of relative humidity (3). In the initial stages, water vapour condenses onto the frozen surface, immediately changing to ice. This is followed by a stage where vapour condenses in the form of water until the surface temperature is above the dew point of the air and all condensation ceases. The varying rate of condensation produces substantial changes in the value of h during the thawing process.

Analysis of the pattern of heat flow during thawing is difficult due to the temperature dependence of the thermal properties of high water content foodstuffs. Consequently, most formulae designed to calculate thawing times only cover simple geometric shapes such as infinite and finite slabs and cylinders, and spheres. Thawing times of more complicated irregular shapes are handled by relating these shapes to a simple geometric analogy.

Plank (4) provided one of the earliest analyses of the problem of heat conduction involving phase change, and produced an equation which has been the basis for most freezing- and thawing-time calculations over the past 50 years, although using the equation involves numerous initial assumptions rarely met in practice. He considered the freezing of an infinite slab and assumed that:-

a. The material is initially unfrozen but uniformly at its freezing point.
b. The material has a definite freezing point.
c. At any time steady state conditions exist.
d. The thermal conductivity will be constant in the frozen phase.

Under these conditions the freezing time t was given by:-

$$t = \frac{\rho L}{\Delta\theta}\left[\frac{X}{2h} + \frac{X^2}{8k}\right] \qquad (1)$$

where ρ = density of the block, L = latent heat, X = thickness and $\Delta\theta$ = temperature difference between the ambient and the freezing point of the

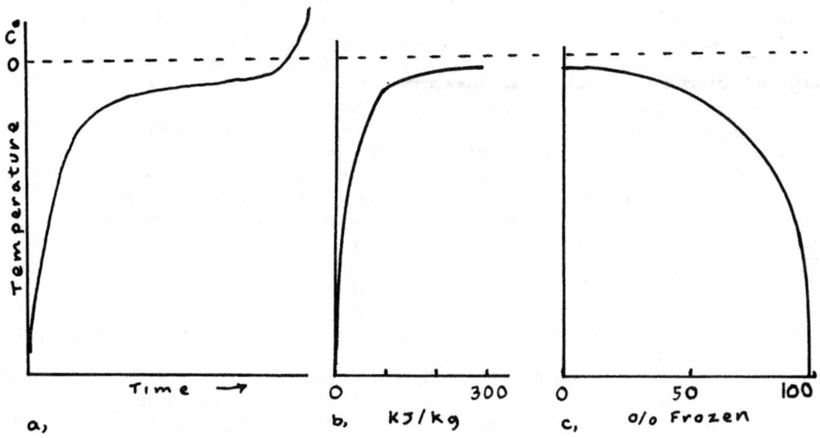

Fig. 1. (a) Typical thawing curve.
(b) Enthalpy/temperature, lean meat.
(c) Percent water frozen, lean meat.

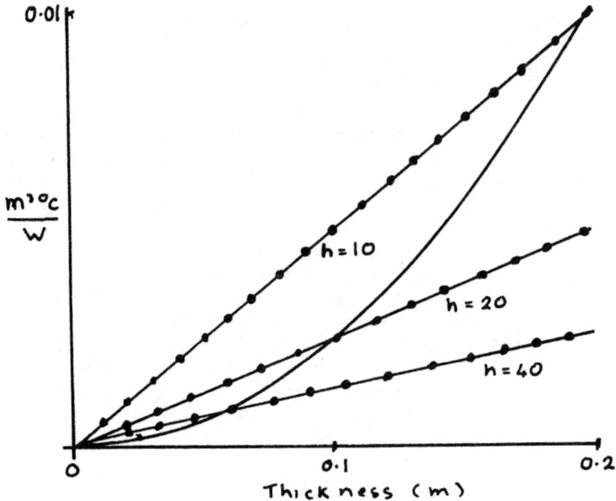

Fig. 2. (●—●) X/2h and (——) $X^2/8k$ against thickness (X).

material.

This basic form of the equation can equally be used for the determination of thawing time. There have been numerous modifications to the basic equation both to extend it to other shapes and to overcome the two initial assumptions (a and b above) (5,6). However, the simple form of equation 1 given above allows clear, if imprecise, estimates of the relative importance of the factors.

The temperature of the thawing medium is of primary importance in all systems since thawing time is universely proportional to the temperature difference between the final thawing point (nominally -1°C) and the temperature of the medium. A 5°C increase in thawing temperature from 4° to 9° would halve the thawing time. The relative effect at higher temperatures, is much smaller, eg increasing from 24° to 29°C produces a 20 per cent reduction in time.

Thawing time is directly related to the latent heat (change in enthalpy) requirement. Since water content is the main factor determining the change in enthalpy, high water content foodstuffs, which also have high densities, tend to have long thawing times.

By plotting the last two terms $X/2h$ and $X^2/8k$, against slab thickness (Figure II), assuming $K = 0.5$ W/m°C, which is approximately correct for lean beef and white fish, their relative importance becomes clear. Increasing the value of h has a substantial effect on the thawing time of thin materials (<10 cm) and at low initial h values (h <20 W/m²°C). The effect of increasing h is far less important as thickness increases because conductivity then becomes the factor controlling thawing rate. In situations where the product is packaged, solid and corrugated fibreboard, and the air gap often produced between the packaging material and the frozen foodstuff, act as an insulating barrier to heat flow and considerably reduce the effective heat transfer coefficient at the surface of the product. In such cases, increasing the surface heat transfer coefficient has only limited effect on thawing time.

3. QUALITY AND MICROBIOLOGICAL CONSIDERATIONS

There are few published data relating thawing processes to the palatability of meat and with meat, eating quality is generally independent of thawing method (7). However, two reports indicated that cooking directly from the frozen state produced less juicy lamb rib-loins (8), and less tender beef rolled rib joints (9) when compared with meat that had been thawed before cooking.

The main detrimental effects of freezing and thawing meat is the large increase in the amount of proteinaceous fluid (drip) released on final cutting, yet the influence of thawing rate on drip production is not clear. There was no significant effect of thawing rate on the volume of drip in beef (10,11) or pork (11). Several authors (12,13,14) concluded that fast thawing rates would produce increased drip, while others showed (15,16,17) the opposite. Thawing times from -7 to 0°C of less than 1 minute or greater than 2000 minutes led to increased drip loss (18). The results are therefore conflicting and provide no useful design data for optimising a thawing system.

The principle criteria governing quality of thawed foods are therefore normally the appearance and bacteriological condition. These are major factors if the product is to be sold thawed but are less important if the food is destined for processing and heat treatment.

Bacteriological problems can arise during thawing of food in bulk since while the centre is rising slowly to 0°C, the exterior surface may be held at 10-15°C for many hours, or even days, during which time extensive growth of spoilage organisms can occur. The time required to reach the level of bacterial concentration regarded as 'spoiled' is dependent, among other factors, upon the number of bacteria initially present. Since freezing and frozen storage have little effect on the number of viable bacteria present, material of poor bacteriological quality before freezing is likely to spoil more quickly during thawing (19). The use of high thawing (>10°C) temperatures for carcass meats tends to lead to large increases in bacterial numbers (20,21).

It is often asserted that thawed food is more perishable than fresh or chilled produce, but experiments (22,23) failed to demonstrate any difference of practical significance between the growth of meat spoilage organisms on fresh or thawed slices of meat.

4. THAWING SYSTEMS

4.1 Conduction

4.1.1. Air Thawing Air thawing systems transfer heat to the frozen material by conduction through the static air boundary layer at the product surface and the rate of heat transfer is a function of the difference in temperature between the product and the air and the air velocity.

4.1.1.1. Still air Single frozen fish, blocks of fish, or thin blocks (<10 cms) of meat can be thawed overnight at room temperature and provided the surface of the product does not become too dry, the thawed product can be perfectly acceptable. Air temperatures should not be greater than 15°C.

For thicker materials still air thawing is not recommended, since thawing times extend to days, rather than hours, and the surface layers may become warm and spoil long before the centre is thawed. Still air thawing is practicable only on a small scale, because considerable space is required, the process is uncontrolled and the time taken is often too long to fit in with processing cycles. The sole advantage is that little or no equipment is required.

In still air blocks (10 cm thick) of whole cod require 20 hours to thaw at 15°C, while individual cod (also 10 cm thick) require 8 to 10 hours (24).

4.1.1.2. Moving Air The majority of commercial thawing systems use moving air as the thawing medium. Not only does the increased h value produced by moving air result in faster thawing but it also produces much better control than using still air. Control of relative humidity is important with unwrapped products to reduce surface desiccation and increase the rate of heat transfer to the foodstuff, 85 to 95% RH being recommended for meat (21) and near saturation for fish (24).

With 250 g slabs of meat (25) weight loss was a function of temperature, velocity and RH. In all cases, increasing the air temperature, or decreasing the air velocity, produced a decrease in percentage weight loss at 85-88% RH or an increase in weight gain at 95 to 98% RH. Changes ranged from a 2.5% per cent loss at 5°C, 5 m/s, 85-88% RH to a 0.5% weight gain at 25°C, 1 m/s, 95-98% RH.

In general for fish it recommended that air velocities near 6 m/s with a maximum temperature of 20°C are used (24). Since frozen meat is usually in thicker pieces, and thawing times are consequently longer, lower air velocities (0.5 to 1.0 m/s) and temperatures (8° to 15°C) are advised to minimise increases in bacterial numbers (7).

4.1.1.3. Two stage Air Two stage air thawing has often been proposed as a means of shortening the thawing process. In the first stage a high air temperature is maintained until the surface reaches a predetermined set temperature, thus ensuring a rapid initial input of energy. The air temperature is then reduced rapidly and maintained below 10°C until the end of the thawing process. Heat flows from the hotter surface regions to the centre of the frozen foodstuff lowering the surface temperature to that of the ambient air. Since this temperature is below 10°C, and the overall thawing time is short, total bacteria growth is small. A patent (26) has been taken out on a two stage thawing system using almost saturated air between 35° and 60°C, followed by air between 5° and 10°C after the surface temperature of the product has reached 30° to 35°C. The first stage normally takes 1 to 1.5 hours, the second 15 to 20 hours and it is claimed that weight loss is low and drip loss minimal.

4.1.2. Immersion/Plate-contact In this method (27) blocks of the frozen material are placed in a vertical position between thin heat-exchange plates with internal channels through which heated water at a constant temperature is circulated. The whole system is immersed in still water to transfer heat from the plates to the blocks. This water is changed regularly to prevent accumulation of bacteria. This removes the requirement found in most plate contact systems for a method of applying pressure to the plates to achieve close contact needed to promote good heat transfer. A particularly good application of this method is for blocks made up of small individual items with large air spaces which normally lead to poor heat conduction. 8 cm thick blocks of whole shrimp with approximately 50% of free air can be thawed within an hour if a circulated water temperature of 25°C is maintained.

4.1.3. Water Thawing The mechanism of heat transfer in water is similar to that in air, but because the heat transfer coefficients obtained are considerably larger, the thawing times of thinner cuts are effectively reduced. However, there are practical problems that limit the use of water thawing systems. Boxed or packaged goods (unless shrink-wrapped or vacuum-packed) must be removed from their containers before they can be water thawed. Blocks made up of small pieces break up readily in water. This may be an advantage with individual items such as shrimps or small fish, but meat blocks are commonly comprised of a mixture of large and small pieces and it is undesirable for these small pieces to become dispersed.

Since heat transfer coefficients are high in water thawing (typically 100 $W/m^2 °C$) the effect on thawing time of practically feasible changes in the speed of water movement is small. Increasing the water speed from 0.006 m/s to 0.23 m/s at 10°C produced a reduction from 20.1 to 20 hours with 7 kg legs of pork (21). When thawing fish blocks (28) in large tanks problems were encountered with temperature stratification at 0.46 m/min while at 1.2 m/min other authors encountered problems with frothing (29).

4.1.4. Vacuum-steam thawing A vacuum-steam thawing system operates by transferring the heat of condensing steam under vacuum to the frozen product. Theoretically a condensing vapour in the presence of a minimum amount of a non-condensible gas can achieve a surface film heat transfer coefficient far higher than that achieved in water thawing. The principle of operation is that when steam is generated under vacuum, the vapour temperature will correspond to its equivalent vapour pressure. For example, if the vapour pressure is maintained at 1706 N/m^2, steam will be generated at 15°C and will condense onto any cooler surface such as a frozen product. The benefits of latent heat transfer can be obtained without the problems of cooking which would occur at atmospheric pressure.

With thin materials, thawing cycles are very rapid enabling high daily throughputs to be achieved. The advantage of a high h value becomes less marked as material thickness increases and beef quarters or 25 kg meat blocks require thawing times permitting no more than one cycle per day. Under these conditions the economics of the system and the largest capacity unit available (10-12 tonnes) severely restrict its application.

4.2 Electrical methods In all of the methods described above the rate of thawing is a function of the transfer of heat from the thawing medium to the surface of the meat, and the conduction of this heat into the centre of the carcass or cut. In theory, electrical systems should overcome these problems because heat is generated within the material and the limitations of thermal conductivity are circumvented. In such systems the kinetic energy imparted to molecules by the action of an oscillating electromagnetic field, is dissipated by inelastic collisions with surrounding molecules and this energy appears as heat. Thus electromagnetic radiation may be used to heat foodstuffs.

Three regions of the electromagnetic spectrum have been used for such heating:-

Resistive	50 Hz
Dielectric	0.01 to 0.1 GHz
Microwave	0.915 and 2.45 GHz

Since thawing using the dielectric region has proved more practical than either resistive or microwave it is considered first and in more detail.

4.2.1. Dielectric During dielectric thawing, heat is produced in the frozen foodstuff as a result of dielectric losses when a product is subjected to an alternating electric field. In an idealized case of dielectric heating the foodstuff, consisting of a regular slab of homogeneous material at a uniform temperature, is placed between parallel electrodes and no heat is exchanged with its surroundings. When an alternating e.m.f. is applied through the electrodes the resulting field in the slab is uniform, so the energy and the resultant temperature rise is identical in all parts of the food (30).

In practice this situation rarely applies. Foodstuffs are not generally in the shape of perfect parallelepipeds: most foods consist of at least two components ie fat and lean, crust and filling, etc: during loading frozen foods pick up heat from the surroundings, the surface temperature rises and the dielectric system is not presented with the uniform temperature distribution required for even heating.

The dielectric properties of a material are specified by the loss angle δ, the real and imaginary permittivities ε' and ε'' and the conductivity. When a sinusoidal e.m.f. of angular frequency w is applied, dielectric theory shows (31) that the power per unit volume 'P' absorbed by the dielectric material is given by:

$$P = \frac{J^2 \sin 2\delta}{0.55 \times 10^{10} \times 2 f \varepsilon_\phi} \qquad (2)$$

where J is the r.m.s. current density in the material and $f = w/2\pi$ and ε_ϕ is the relative permittivity ($\varepsilon_\phi = \varepsilon'/\varepsilon_0$, $\varepsilon_0 = 8.85 \times 10^{-12}$F/m).

Values of $\sin 2\delta/\varepsilon_\phi$ for beef at 35 MHz are (32)

	Frozen	Thawed
Lean	0.09	0.01
Fat	0.13	0.14

and using these values in equation 2 indicates that any thawed areas will preferentially absorb energy leading to runaway heating. Similar problems are found with other foodstuffs.

By using a conveyorised system to keep the product moving past the electrodes and/or surrounding the material by water, commercial systems have been produced for blocks of oily fish (33) and white fish (34). Successful thawing of 13 cm thick meat, and 14 cm thick offal, blocks have also been reported (30), but the temperature range at the end of thawing (44 mins) was stated to be -2° to 19°C and -2° to 4°C respectively, and the product may not have been fully thawed.

To overcome runaway heating with slabs of frozen pork bellies workers (35) have tried coating the electrodes with lard, placing the bellies in oil, water and saline baths and wrapping the meat in a cheesecloth soaked in saline solution. Only the last treatment was successful but even that was not deemed practical. Further experiments used specially shaped copper coil electrodes in which bellies were thawed from -20°C to a maximum temperature of 5°C in 40 seconds. The main practical problem was considered to be the limited capacity of any commercial plant since oscillators were only 50% efficient and 1 kW would only thaw 22 kg per hour.

Dielectric systems have been used successfully to thaw 14 kg cartons of both frozen eggs and fruit in approximately 15 minutes using a power of 2 kW (36). Another interesting application of high frequency thawing is that of Swiss veterinary inspectors (37) who wished to thaw a number of 25 kg meat blocks from each lorry load of imported frozen meat to check its quality. Normal convection systems required 2 to 3 days, produced high drip losses and resulted in possible contamination. A 25 kW generator is now used and is reported to thaw 20 blocks from -15°C to 5°C in 1.5 hrs. Problems with runaway heating are not discussed and may not be particularly important in this limited application. However the energy requirements of a large commercial thawing system would present considerable practical and economic problems.

4.1.3. <u>Resistive thawing</u> A frozen foodstuff can be heated by placing it between two electrodes and applying a low voltage at normal mains frequency. As the electric current flows through the material it becomes warm. Frozen meat and fish at a low temperature do not readily conduct electricity, but as they become warmer their electrical resistance falls, a larger current can flow and more heat is generated within the product. In practice the system is only suitable for thin (5 cm) homogeneous blocks such as catering blocks of fish fillets, liver, crab meat etc. since current flow is very small through thick blocks and inhomogeneities lead to runaway heating problems.

A hybrid resistive thawing system has been produced for fish in which the block is partially thawed in water or between hot plates before the current is switched on (27).

4.1.4. <u>Microwave thawing</u> Despite a widespread belief to the contrary, there is no published information on a commercially successful microwave thawing system. Whilst the heating of frozen meat by microwave energy is potentially a very fast method of thawing, its application is constrained by thermal instability. At its worst, parts of the food may be cooked whilst the rest is substantially frozen. This arises because the absorption by frozen food of electro-magnetic radiation in this frequency range, increases as the temperature rises, this dependence being especially large at about -5°C, increasing as the initial freezing point is approached. If for any reason during irradiation a region of the material is slightly hotter than its surroundings, proportionately more energy will be absorbed within that region and the original difference in enthalpy will be increased. As the enthalpy increases so the absorption increases and the unevenness of heating worsens at an ever increasing rate. Below the initial freezing point the temperature increase is held in check by thermal inertia since for a given energy input the temperature rise is inversely proportional to the thermal capacity. If irradiation is continued after the hot spot has reached its initial freezing point, the temperature rises at a catastrophic rate.

This runaway heating can be reduced by lowering the power density (and thereby increasing the processing time) to allow thermal conduction to even out the enthalpy distribution through the material. Order of magnitude calculations have been made (38) of the times involved for thermal stability and showed that whereas below -5°C and above 0°C relatively uniform rapid heating is possible, at temperatures between -5°C and -1°C, where the major enthalpy change on thawing occurs, conditions are unstable unless prohibitively long processing times are employed.

Since the main instability tends to occur at the surface, attempts have been made to cool the surface during thawing using air and liquid nitrogen (39). Although experimentally successful the systems are not economically viable. A hybrid microwave/vacuum system, in which boiling of surface water at low temperature is used to cool the surface, thawed 15 cm thick cartoned meat in 1 to 2 hours without runaway heating, but problems of control and cost would appear to limit the commercial use (40).

5. EXAMPLES OF THAWING TIMES

Table 1 shows typical thawing times for a range of products in different thawing systems.

6. CHOICE OF THAWING METHOD

There is no simple guide to the choice of an optimum thawing system for a company, or even for a particular foodstuff. A thawing system should be considered as one operation in the production chain. It receives frozen material, hopefully, within a known temperature range and of specified microbiological condition. It is expected to deliver that same material in a given time in a totally thawed state. The weight loss and increase in bacterial numbers during thawing should be within acceptable limits, which will vary from process to process. In some circumstances, eg direct sale to the consumer, the appearance of the thawed product is crucial, in others it may be irrelevant. Apart from these factors the economics and overall practicality of the thawing operation, including the capital and running costs of the plant, the labour requirements, ease of cleaning and the flexibility of the plant to handle different products, must be considered.

The Torry Research Station has produced general guidelines for systems suitable for fish (24). For continuous thawing of large blocks of whole fish the choice lies between air blast thawing and the dielectric systems. Smaller quantities of similar blocks can be thawed in a simple batch air thawer or overnight in slowly moving air. Water thawing is practicable for small quantities of whole fish and electrical resistance is best for catering blocks of uniform consistency.

The choice of a practical thawing system for meat tends to be more restrictive. Standard 25 kg, 15 cm thick cartoned blocks can only be thawed in air blast systems. Large quantities of beef quarters or lamb

Table I: Thawing times from -30° to 0° for a range of products using different thawing methods.

Thawing system/ product	Size	Conditions/Thawing time (Typical range)				Ref.
AIR						
Beef Hind-	70 kg	5°C, .25 m/s	115 h	30°C, 5 m/s	28 h	20
quarters	50 kg	"	90 h	"	22 h	20
Beef block						
Unwrapped	15 cm	5°C, .25 m/s	75 h	30°C, 5 m/s	17 h	41
Corr. carton	15 cm	"	95 h	"	24 h	41
Pork legs	7 kg	10°C, .25 m/s	28 h	30°C, 5 m/s	12 h	21
	3 kg	"	15 h	"	5 h	21
Chicken	8 week	3-5°C, 0 m/s	46 h	26-28°C, 0 m/s	7 h	17
Fish blocks						
Cod	11.4 cm	16°C, 7.6 m/s	5.9 h	24°C, 7.6 m/s	4.4 h	42
Sardine	10 cm	4°C, 0 m/s	72 h	20°C, 2 m/s	8 h	43
WATER						
Pork legs	7 kg	10°C, 006 m/s	20 h	30°C, .023 m/s	10 h	21
	3 kg	"	9 h	"	4 h	21
Turkeys	3.5 kg	21-27°C 0 m/s	4 h	44-46°C, 0 m/s	2.1 h	44
Fish blocks						
Cod	11.4 cm	16°C, .02 m/s	5.3 h	27°C, .05 m/s	3.4 h	42
Sardine	7.5 cm	5°C, 1 m/s	1.2 h	25°C, 1 m/s	0.3 h	46
PLATE						
Fish blocks						
Whitefish	10 cm	5.8°C	6.7 h	20.4°C	4.5 h	47
VACUUM						
Beef Hind	70 kg	5°C	87 h	30°C	20 h	20
-quarters	50 kg	"	68 h	"	20 h	20
Beef block						
Unwrapped	15 cm	5°C	40 h	30°C	12 h	41
Pork legs	7 kg	10°C	19 h	30°C	9 h	21
	3 kg	"	9 h	"	5 h	21
Chickens	1.3 kg	27°C	2 h			47
Fish blocks						
Whitefish	10 cm	15°C	6.6 h	30°C	4.5 h	46
Tuna	10 cm	10°C	3.8 h	25°C	2.2 h	48
DIELECTRIC						
Beef block	16 cm	4-7 KW, 13 MHz	2 h			49
Fish blocks						
Hake	5.8 cm	0.75 KW, 40 MHz	50 min			50
Herring	8.3 cm	1.5 KW, 40 MHz	13 min			51
MICROWAVE						
Cod/Herring	7.5 cm	20 KW, 245 MHz	10-15 min			39

or pig carcasses, are again restricted to air thawing systems. If the microbiological quality and appearance are important then air conditions of 10°C, 0.25 m/s and 85% RH are near optimal. However, thawing times are long, e.g. 2 to 3 days for beef quarters. Smaller quantities can be handled in vacuum-thawing plants. Water thawing is possible for vacuum packaged primal meat cuts, and pork joints that are to be subsequently cured. To our knowledge no electrical thawing methods are being used commercially on meat or meat products.

REFERENCES

1. ARCE, J. and SWEAT, V.E. ASHRAE Trans. (1980). 86 (2) p235.
2. VANICHSENI, S. (1971). Meat Research Institute of New Zealand Report No. 233.
3. JAMES, S.J. and BAILEY, C. (1982). Proc. 28th Eur. Meeting Meat Res. Wkrs., Madrid. 1 (3.16), p160.
4. PLANK, R. (1913). Z. ges. Kalteindustr. 20 (5) p109.
5. CLELAND, A.C. and EARLE, R.L. (1976). B.I.I.F., Annex 1, Melbourne.
6. HAYAKAWA,K.I. (1977). Adv. Fd. Res. 23 p.74.
7. BAILEY, C. (1975). Proc. 6th Eur. Symp. 'Food - Engineering and food quality', Cambridge, U.K. Pub. Soc. Chem. Ind., London, p.175.
8. WOODHAMS, P.R. and SMITH, R.A. (1965). Meat Research Institute of New Zealand Report No. 99.
9. JAMES, S.J. and RHODES, D.N. (1978). J. Sci. Fd. Agric., 29, p.187.
10. EMPEY, W.A. (1933). J. Soc. Chem. Ind., 52, p235.
11. CIOBANU, A. (1972). Lucrari Cercetare R., 9, p63.
12. CUTTING, C.L. (1974). Refrig. Air Con., 77, p45.
13. LOVE, R.M. (1966). In Cryobiology, Academic Press, London.
14. RICHARDSON, A.S. and SCHRUBEL, R. (1908). Cited in Cryobiology.
15. DEATHERAGE, F.E. and HAMM, R. (1960). Fd. Res. 25, p623.
16. FINN, D.B. (1932). Proc. Royal Soc., 111, p396.
17. SINGH, S.P. and ESSARY, E.O. (1971). Poultry Sci., 50, p364.
18. JAMES, S.J., NAIR, C. and BAILEY, C. (1983). COST 91 Terminal Session.
19. ROBERTS, T.A. (1974). Meat Freezing: Why and How? Meat Research Institute Symp. No. 3 p8.1.
20. JAMES, S.J. and CREED, P.G. (1980). Int. J. Refrig. 3 (4) p237.
21. BAILEY, C., JAMES, S.J., KITCHELL, A.G. and HUDSON, W.R. (1974). J. Sci. Food Agric. 25, p81.
22. KITCHELL, A.G. and INGRAM, M. (1956). Annls. Inst. Pasteur., Lille, 8, p121.
23. KITCHELL, A.G. and INGRAM, M. (1959). B.I.I.R. 30. p866.
24. Anon. Torry Research Station Advisory Note No. 25.
25. ZAGRADZKI, S., NIEDZIELSKI, Z. and KULOGAWSKA, A.. Acta Alimentaria Polonica (1977) 3. (27) 1 p11.
26. British Patent No. 1355164 (1974).
27. Danish Ministry of Fish Research Laboratory Annual Report 1965. p.42.
28. HEWITT, M.R. (1969). Fishing News 4, p201.
29. MacCALLUM, W.A. and ELLIS, D.G. J. Fish Res. Bd. Canada 1964 21 (1), p115.
30. Sanders, H.R. (1961). J. Fd. Technol., 1, p183.
31. von HIPPEL, A.R. (1954). Dielectric Materials and Applications, p3. The Technology Press of M.I.T.; Wiley, New York.
32. BENGTSSON, N.E., MELIN, R., REMI, K. and SODERLIND, S., (1963). J. Sci. Fd. Agric. 14, p592.
33. JASON, A.C. and SANDERS, H.R., (1962). Fd. Technol. p101.

34. JASON, A.C. and SANDERS, H.R., (1962). Fd. Technol. p107.
35. SATCHELL, F.E. and DOTY, D.M., (1951). American Meat Institute Foundation Bulletin No. 12, p95.
36. CATHCART, W.H. and PARKER, J.J., (1946). Food Research 11 p341.
37. von HEEREN, H., (1964). Kalte Rundschau 2.1 p7.
38. JASON, A.C. (1974). Meat Freezing; Why and How? Meat Research Institute Symp. No. 3. p43.31.
39. Les Micro-ondes Industrielles (undated). L.M.I. Gigatrons in the fish industry. L.M.I., Epone, France.
40. JAMES, S.J. (1983). COST 91 Terminal Session. Athens.
41. CREED, P.G. & JAMES, S.J. (1981). Int. J. Refrig. 4 p355.
42. MERRITT, J.H. & BANKS, A. (1964). Bull. Int. Inst. Refrig., Annexe 1964-1, p65.
43. CREPEY, J.R. & HAN-CHING, L. (1979). Proc. 15th Int. Cong. Refrig., Venice, Comm. C2, Paper 60.
44. KORSLUND, M.K. & ESSARY, E.O. (1971). Poult. Sci. 50 p1790.
45. OHMORI, H., NAKAMURA, K., HORI, T. and YAMAMURA, M. (1981). Int. J. Refrig. 4, p27.
46. LORENTZEN, G. (1969). Kjole. Tek. og Frys. (Norway) 21 (1) p3.
47. CHRISTIE, R.H. and JASON, A.C. (1975). Proc. 6th Eur. Symp. 'Food Engineering and food quality'. Cambridge, UK Pub. Soc. Chem. Ind., London.
48. MIKI, H. and NISHIMOTO, J-I. (1975). Mem. Fac. Fish., Kagoshima Univ. 24 p161.
49. SCHLUSSELBURG, G. (1974). Fleischwirtschaft 54 p672.
50. LANG, O. (1961). Kalte 41 (10) p615.
51. JASON, A.C. and SANDERS, H.R. (1959). Proc. 10th Int. Cong. Refrig., Copenhagen, 3 p19.

LOW POWER MICROWAVE THAWING OF ANIMAL FOODS

Thomas Ohlsson
SIK-The Swedish Food Institute
Box 5401
S-402 29 Göteborg
Sweden

SUMMARY

The thawing of animal foods requires much energy to be added to the frozen product, at the same time as the surface temperatures have to be kept low. The low heat conductivity of thawed animal foods makes thawing times long when using conventional techniques. This can be overcome by using microwaves that have an ability directly to heat the interior of foods. There are however problems with uneven temperature distribution during rapid microwave thawing of large blocks of meat and fish. The effect on the thawing of different processing parameters was studied by means of a computer calculation program for microwave thawing. The calculations show that processing in less than one hour is only possible for thicknesses below 50 mm. In such a short time thicker blocks can only be partially thawed or tempered which also is the industrial practice. With longer thawing times (5-10 hours) also thicker blocks (150-250 mm) can be successfully thawed at 915 MHz.

INTRODUCTION

With conventional methods, such as air thawing, it often takes days to thaw frozen blocks of animal foods. To limit bacteriological growth the temperatures at the surfaces must be kept low. This limits the temperature gradient needed for the heat transfer from the surface to the interior to 10-20°C. Together with the low thermal conductivity of thawed animal foods this results in the long thawing times.

Microwave heating offers a method to overcome these limitations. In microwave heating the energy is transferred directly to the interior of the foods. The heating mechanism is based on interaction between the rapidly oscillating electrical field and the remaining not frozen water molecules and therein dissolved salts. However, the penetration ability of microwaves is limited, as shown in figure 1, (1, 2). It illustrates the dramatic shift in the penetration depth of the microwaves during the thawing process. Low penetration depth in already thawed food will lead to absorption of the microwaves in the already thawed part resulting in rapid temperature increases on the surfaces. This so called run-away microwave heating effect is illustrated in figure 2. Consequently industrial microwave thawing of frozen blocks of meat is limited to tempering to about -2°C, where approximately 50 % of the water remains frozen. In the industry microwave thawing is done using 915 MHz microwave heating units of 30 kW. (3). However, there is an industrial need for well controlled complete thawing of frozen animal foods, e.g. for meat carcasses that are to be cut down into retail portions.

METHODS

In the calculation program the heat transfer was calculated using the Fourier heat transfer equation for an infinite slab. For temperatures below the freezing point, the changes in enthalphy with time were calculated. A subroutine was used to convert from enthalphy to temperature and vice versa. Another subroutine gave the temperature dependent factor λ/ρ as a function of the enthalphy. For the numerical solution a finite difference approximation of the Fourier equation was used.

The microwave heat transfer was calculated using the method presented by Ohlsson & Bengtsson (4). This involves calculating the microwave power density for each time and space increment, using an exponential equation. The absorption factor in the equation for the microwave heating had to be determined for each time and space increment using a subroutine due to the high temperature dependence of the dielectric properties, taken from (1 and 2). Calculations were performed both for 915 and 2450 MHz. The surrounding air was -20 or +4°C, with a surface heat transfer coefficient of 5 W/m^{2},K.

The calculation program has previously been used in simulations of thawing in institutional and domestic microwave ovens (5). These calculations showed that for good accuracy ($\pm$5% of change in enthalphy) between calculated and experimental temperature profiles, the number of space increments should be large. (5).

RESULTS AND DISCUSSION

In the evaluation of the results it is important to remember that the temperature profiles are calculated and not measured. Also, the temperatures represent the center line temperature of a large slab of food. Corners and edges, that often show the highest temperatures (6), are not taken into account using this one-dimensional calculation method.

For slow thawing ($>$ 5-10 h) the maximum surface temperatures are limited by the risks for bacterial growth to +8 - 10°C. For fast thawing ($<$ 2-3 h) the maximum surface temperatures are more limited by losses in functional properties of the animal foods, such as water holding capacity. For meat, temperatures above approximately +20°C should be avoided. In this evaluation, surface temperatures below +5 at slow and +10°C at fast thawing have been considered to be accepted, taking the previously discussion overheating of corners and edges into consideration.

The difficulties of complete thawing of thick blocks of frozen foods are illustrated in figure 2. It is impossible to thaw 100 mm blocks at 2450 MHz successfully within 1 hour's thawing time. The calculations showed that even 50 mm blocks cannot be thawed acceptably within 1 hour. However, tempering to about 55% remaining ice content of 100 mm blocks seems to be feasible. Figure 3 shows the temperature profiles for thawing to 70% (14 minutes) and 40% (28 minutes) remaining ice content. Using a surrounding air temperature of -30°C instead of +5°C only meant a small improvement on the center line temperature profile. The effect on corner and edge overheating tendencies can be expeced to be larger.

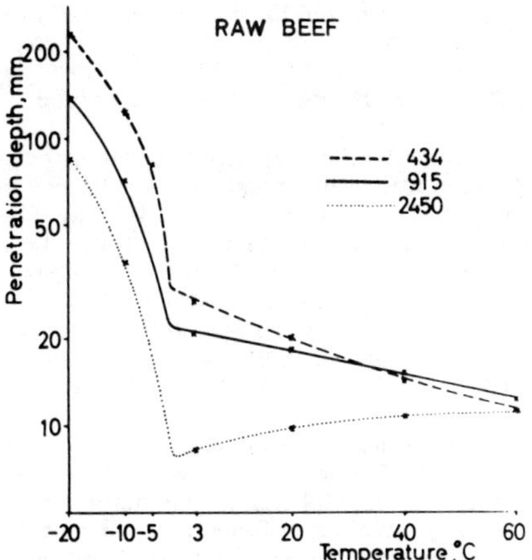

Figure 1 Penetration depth in raw beef as a function of temperature and frequency

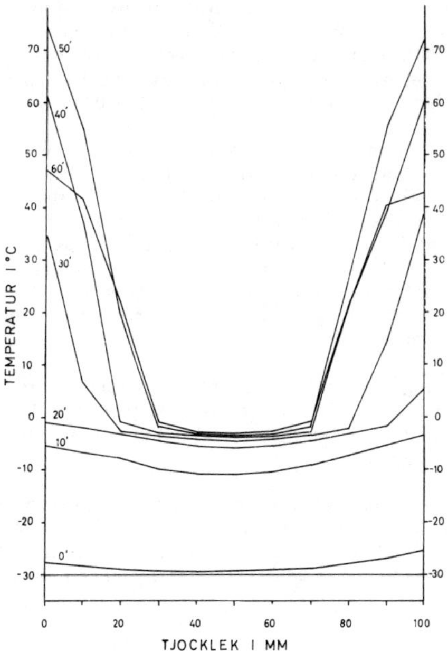

Figure 2 Calculated temperature profiles in a 100 mm block of fish after 20, 40 and 60 minutes of microwave thawing at 2450 MHz.

The better penetration depth of 915 MHz means that 150 mm blocks can be successfully microwave tempered to 40% remaining ice content in 20 minutes (corresponding to a power density of 1.0 W/cm^2) with a maximal surface temperature of 10^oC. This is also the dominating method used in the meat industry.

Even at 915 MHz complete thawing within 1 hour is not possible for 100 or 150 mm blocks. For 50 mm blocks, however, the calculations indicate a maximum surface temperature of $+13^oC$, which may be conditionally acceptable. A detailed inspection of the temperature profile development during the microwave thawing process shows that there are rather distinct limits in remaining ice contents, where the surface temperatures start to "run-away". At 0.5 W/cm^2 the run-away starts at about 45% remaining ice content. This corresponds to approximately 35 minutes. At 1.0 W/cm^2 the "run-away" heating starts at approximately 35% remaining ice content, corresponding to about 25 minutes microwave heating time.

To reach good thawing results, reduced power levels are used in domestic and institutional microwave ovens. Therefore, we decided to run a second series of calculations to investigate the possibilities of complete thawing at lower microwave power level longer time. We chose 8 hours as our calculation standard time. The prolonged time will allow the thermal conductivity to equilibrate the unevenness created by the microwaves.

The calculations have demonstrated that complete thawing of frozen blocks of 100-250 mm is possible if thawing times of up to 8 hours are permitted. Figure 4 shows the temperature profiles for low power microwave thawing of a 150 mm block at 2450 MHz and in an air temperature of $+4^oC$. The surface temperatures are slightly too high but with somewhat shorter exposure to lower power density or -20^oC air temperature, the surface temperatures were acceptable. Also, a two-hour equilibration period after the microwave heating period meant a rapid decrease of the surface temperatures, as shown in figure 4.

Again -20^oC instead of $+4^oC$ air temperature means a rather limited improvement for the centerline temperatures, but a much larger effect is expected for corner and edge temperatures. Our calculations also show that at 915 MHz it is possible completely to thaw 200 mm thick blocks in 8 hours (at 0.07 W/cm^2), with surface temperatures below $+8^oC$ using -20^oC air temperature. The temperature profile after 8 hours of thawing at 915 MHz is very similar to the 10 h profile in figure 4. For 150 mm blocks surface temperatures below $+5^oC$ are reached even at $+4^oC$ air temperature. For larger thicknesses, insulation of the surfaces markedly lowers the maximum surface temperature.

Although of limited extention, these calculation has demonstrated the advantages of low power microwave thawing of blocks of frozen animal foods of thicknesses up to at least 200mm in maintaining surface temperatures low while allowing complete thawing.

An interesting aspect of the low power microwave thawing is that the installed microwave power can be kept low, e.g. for 15 ton units only 10kW of microwave power needs to be installed for an 8-hour thawing cycle. This also enables better utilization of the units, as the thawing can be done overnight.

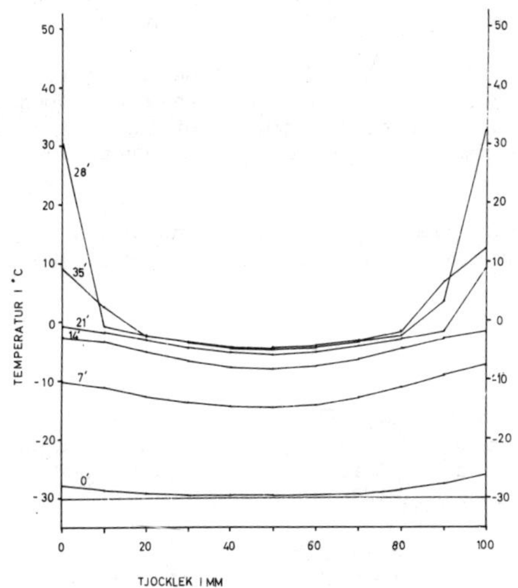

Figure 3 Calculated temperature profiles in a 100 mm block of fish after 14 and 28 minutes of microwave tempering at 2450 MHz.

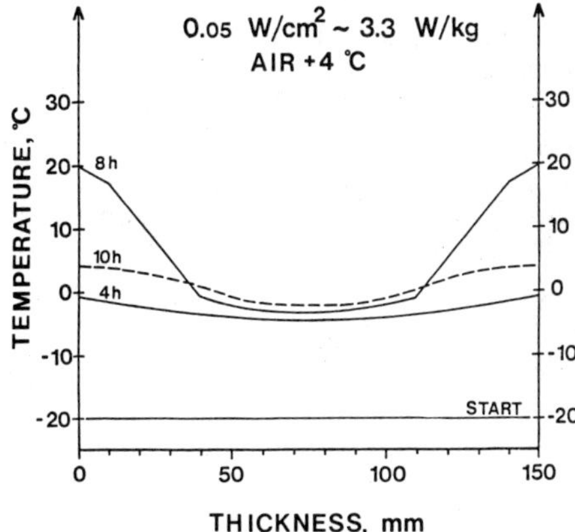

Figure 4 Calculated temperature profiles in a 150 mm block of fish after 4 and 8 hours of low power microwave thawing at 2450 MHz. The broken line represents the temperature profile after 2 hours of temperature equilibration after a 8 hour thawing. It also represents the profile after 8 hours of thawing using 915 MHz.

This method for complete thawing can be used both for meat and fish for further industrial processing and for distribution as whole cuts.

Although slower than the present short time microwave tempering, the low power microwave thawing is many times faster than conventional methods. It also offers possibilities to minimize the effects of too long holding periods at temperatures, where risks are high for bacteriological growth or losses in functional properties.

Some limited experiments with low power microwave thawing have verified that low surface temperatures are maintained throughout the complete thawing operation.

REFERENCES

1 Bengtsson, N.E. and Risman, P.O., 1971
Dielectric properties of food at 3 GHz as determined by a cavity perturbation technique II Measurements on food materials.
J. Microwave Power 6:2 p.107

2. Ohlsson, T. and Bengtsson, N.E., 1974
The frequency and temperature dependence of dielectric food data as determined by a cavity perturbation technique.
J. Microwave Power, 9 p. 129

3. Bezanson, A., 1975
Thawing and tempering frozen meat.
Proceedings of the Meat Industry Research Conf., March 20-21, 1975

4. Ohlsson, Th. and Bengtsson, N.E., 1971
Microwave heating profiles in foods.
Microwave Energy Applications Newletter 4:6p.3

5. Ohlsson, Th., 1975
Temperature profiles in microwave heating.
Computer simulations and comparisons to experiments.
SIK Service Series No 508 (In Swedish)

6. Ohlsson, Th., 1982
Fundamentals of Microwave Cooking
Microwave World

TIME-TEMPERATURE SURVEYS IN THE FROZEN FOOD CHAIN

W.E.L. Spieß, D. Folkers,
Federal Research Centre for Nutrition, D-75oo Karlsruhe

1. Introduction

The conditions under which deep frozen food passes the frozen food chain have been subject to research in many countries all over the world. The studies demonstrate that there are differences, but also many similarities in the various countries; existing problems in particular are more or less the same /1, 2, 8, 9, 11, 12, 14, 15/.

Some of these problems are illustrated in this paper which summarizes several studies carried out at the Federal Research Centre for Nutrition, Karlsruhe. The paper focusses on the structure of the frozen food chain, on temperatures and residence times observed for the entire chain as well as for specific links. Details and consequences will be discussed in a later communication.

2. Structure of the frozen food chain

The frozen food chain in the Federal Republic of Germany may be broken down to four levels connected by transport processes:

- Level 1: processing of raw material and limited storage periods in plant coldstores
- Level 2: storage in central coldstores
- Level 3: storage in distribution and local coldstores, including facilities at wholesale organizations
- Level 4: storage at retail outlets in freezer units.

There are of course exceptions from this model: products may be stored exclusively in a plant coldstore or be transported directly from plant to distribution coldstores(see Fig. 1).

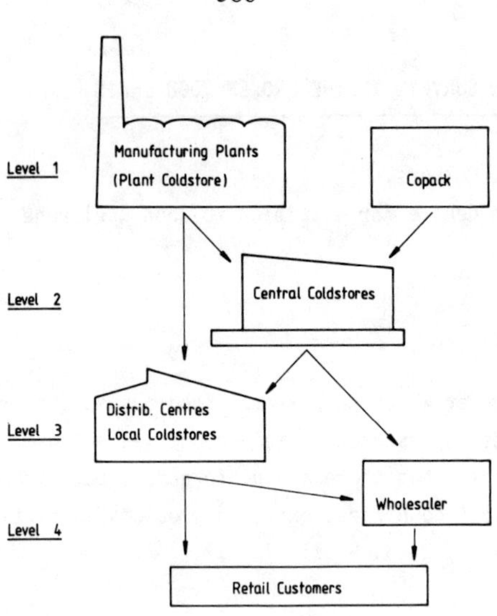

Fig. 1 The frozen food chain
Distribution flow chart for deep frozen food

3. Temperatures in the frozen food chain

3.1 Storage

Level 1: freezing and plant coldstores

Products intended for deep freezing are usually frozen directly after production. Freezing is carried out in such a way that the product temperature, after an equilibration time, settles below -18°C. Equilibration takes place in freezers in case of small product particles, and in frozen storage facilities for larger particles. Air temperature in plant coldstores is usually at -24°C. The products mostly remain in plant coldstores until they are cooled down to this temperature.

Level 2: central coldstores

Temperatures were observed in 2 coldstores selected to represent a universe of about 8o larger coldstores.

All temperatures observed in products were below -24°C, the lowest being -28°C. The relevant trade association reports this to hold for 87 % of commercial coldstores. Central coldstores with temperatures above -24°C are intended for products other than deep frozen, e.g. fruit juice concentrates etc (see Table 1).

Table 1: Total capacity of central coldstores in the Federal Republic of Germany by temperature range

Temperature range °C	Capacity cbm	%
at least 0	1o5,4o9	3.2
at least -12	9,464	0.3
at least -17	42,674	1.3
at least -24	271,1o8	8.4
below -24	2,812,549	86.8
Total	3,241,204	100.0

Source: Fachverband der Kühlhäuser und Eisfabriken e.V., Bonn (Date: February 15, 1983)

Temperature controls in the coldstores are strict since specified temperatures are contracted between store operators and product owners.

Level 3: distribution and local coldstore

Distribution and local coldstores are generally operated at the same temperatures as central coldstores. The average temperature may be slightly higher than in the latter, however, due to more frequent deliveries to and from the store. Fluctuations up to -18°C may occur near doors and at places where newly delivered products which warmed up during transport are kept.

Level 4: Retail freezer units

At the final point of sale frozen foods must be well displayed and easily accessible for consumers. Maintaining low product temperatures in retail freezer units during business hours therefore is most difficult, no matter whether open freezer chests or closed freezer cabinets are used.

In a representative study of 1976/77, product temperatures were measured in the top layer of retail freezer chests in more than 3oo retail shops all over the Federal Republic of Germany. The temperatures varied between -1°C and -32°C. Only 46 % of the products showed temperatures up to -15°C as required by official regulations, i.e. more than half of the samples were above the tolerable limit. The distribution of product temperatures is given in Table 2 /6/.

Table 2: Product centre temperature of deep frozen food in open retail freezer cabinets

Temperature range °C	Samples (n=877) %
at least -12	30
below -12 to -15	24
below -15 to -18	19
below -18 to -35	27
Total	100

The centre temperatures of important products are about normally distributed with a mean temperature of -15°C and a standard deviation of ±2,5°C (see Fig. 2).

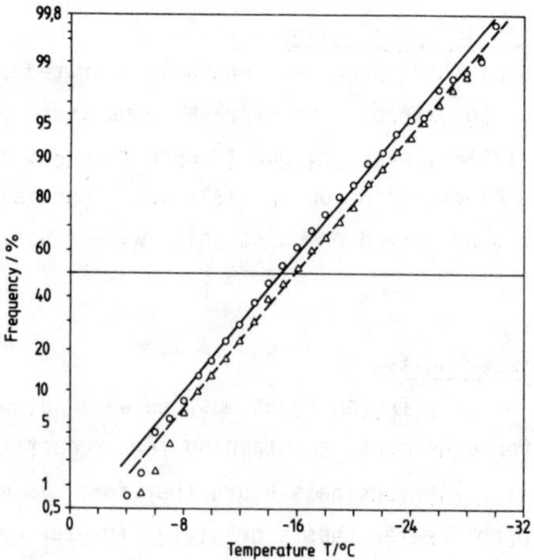

Fig. 2 Cumulative frequency distribution of product temperatures of fish
○ centre temperature; △ surface temperature

Surface temperatures were often below centre temperatures, because most measurements were taken during mornings when product temperatures were again decreasing after a temporary rise through defrosting.

Continuing observations of the frozen food chain suggest that the 1977/78 results are still valid in 1983. The reasons for the relatively high temperatures in retail outlets are manifold and have been illuminated by several studies. According to these a wide distribution of temperatures must be expected even under laboratory conditions in a completely filled open freezer chest (Fig. 3). Measurements according to DIN 8954 show a temperature range for the warmest pack (curve A) of -15°C to -12°C. The average temperature of top layer packs (curve B) varies from -17°C to -14°C, that of bottom layer samples (curve D) from -25°C to -2o°C. Defrosting usually causes a pronounced rise in product temperatures which is compensated again only after several operating hours at normal schedule. If the cabinet is only partially loaded, the average temperature is generally lower, and temperatures are less stable /4/.

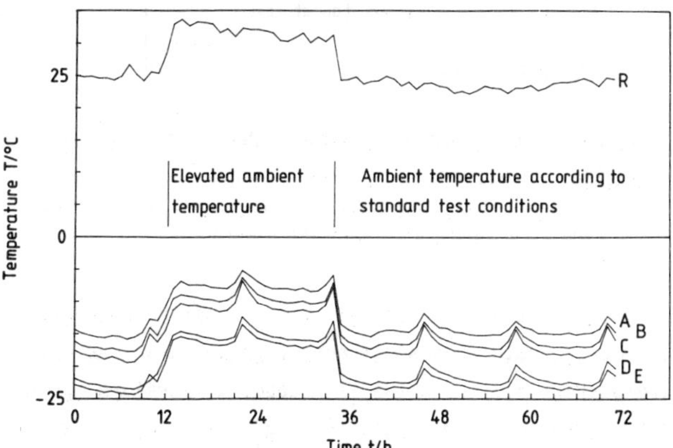

Fig. 3 Product temperature in a retail freezer cabinet operated under standardized conditions (DIN 8954).

A temperature of the warmest pack in the top layer
B mean temperature of the top layer
C temperature of the coldest pack in the top layer
D mean temperature of bottom layer
E temperature of the coldest pack in the bottom layer
R room temperature

Product temperatures may also be raised by sales activities and by above-average temperatures around the retail unit. Environmental temperature increase will immediately raise product temperatures unless they are compensated by the refrigeration equipment. Any purchasing activity

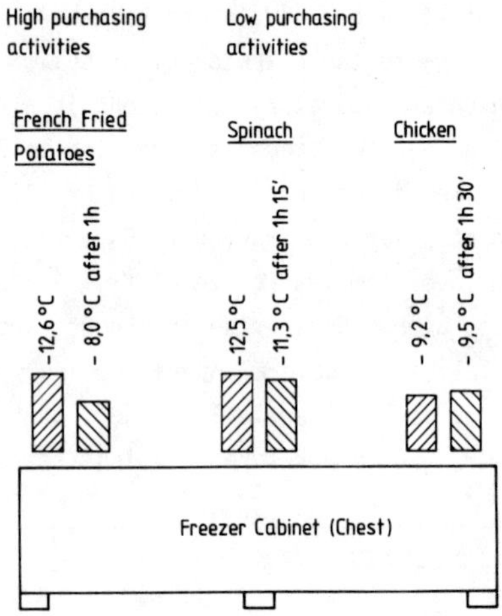

Fig. 4 Product temperatures in a retail freezer cabinet operated under practical conditions in a retail shop

Increase in the mean product temperature (centre) in the top layer during business hours (example)

reduces the effect of the layer of cold air protecting products against warmer environments, both in case of open freezer chests as well as of closed cabinets (Fig. 4). Increases in product temperature can be reduced by using night covers on open chests (Fig. 5). Other measures of keeping product temperatures at a low level are installing the unit at a place with little air draught and without direct heat radiation into its opening. Shopkeepers should also be concerned about proper maintenance of aggregates, observe maximum loading marks and take care that the cold air curtain functions properly. Designers of freezer units may help to avoid excessive increases in product temperatures by reducing the dimensions of openings to remove the merchandise from the chest and by providing adequate refrigeration capacity and insulation. In defrosting

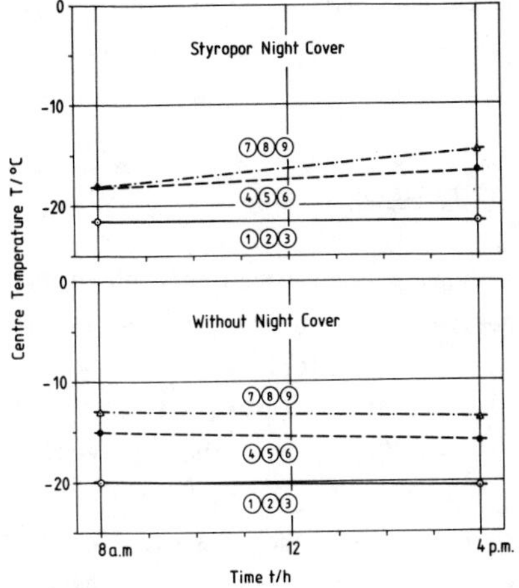

Fig. 5 Product temperatures in a retail freezer cabinet operated under practical conditions in a retail shop - influence of night cover

processes heat should exclusively be directed to sites where defrosting is needed and not to the products stored /1/.

Table 3: Quality of deep-frozen foods

Quality	Total (n=803) %	Chicken (n=270) %	Fish (n=273) %	Vegetables (n=26o) %
good	81	87	71	85
medium	16	12	23	13
poor	3	1	6	2
Total	100	100	100	100

Studies on the quality of products sampled for temperature measurements yield no direct relation between product temperature observed at storage in the retail unit and product quality. In the above-mentioned nationwide survey yielding 54 % of samples warmer than -15°C only 3 % were of insufficient quality(see Table 3).

3.2 Transport and distribution

Deep frozen food is usually transported between coldstores in trucks sufficiently equipped for maintaining temperatures below -18°C. A rise in temperature above the tolerable limit has to be expected only in cases of technical failure of the refrigeration equipment or extremely high ambient temperatures. For maintaining the required temperatures the load should be properly arranged in the vehicle so that the cold air can pass through the stacks. Adequate air flow is guaranteed best if the packages are placed on palettes. The use of palettes also simplifies loading and unloading of transport vehicles so that the danger of product temperature increases at high ambient temperatures is minimized /3/.

The ramps over which the products are transported between coldstore and vehicle are a problem area in most coldstores. If the deep-frozen food palettes are placed on open ramps for longer periods, temperatures at product surfaces may rise considerably, particularly in summer. Therefore dock shelters in loading areas are advisable, especially when sensitive products are transported /1o/.

Similarly,distribution of deep-frozen food from coldstores to retail shops involves the danger that the products are exposed to higher temperatures, mainly during unloading. Measurements of the product temperature at the beginning and the end of distribution tours with modern distribution vehicles (equipped with eutectic plates charged overnight at the local cold store) showed that even under favourable conditions a pronounced increase of product temperatures has to be encountered during distribution.
Mean product temperatures obtained during a series of measurements in summer 1977 were -25.4^{o}C at the beginning and -17,5^{o}C at the end of the tour. Fig 6 shows the frequency distributions of product temperatures (68 door openings) at the beginning and at the end of tours /3/.

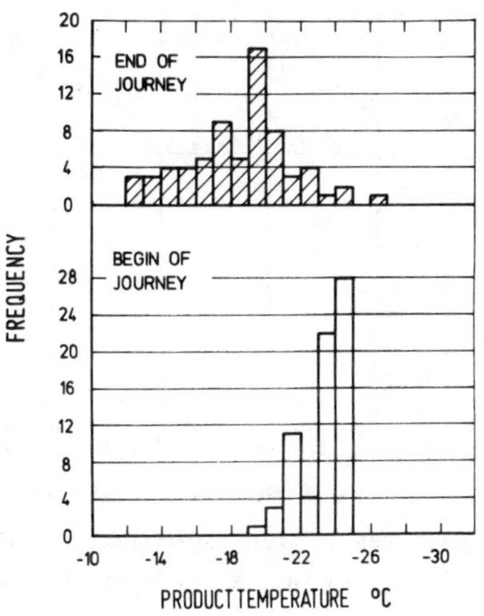

Fig. 6 Frequency of product temperatures at the beginning and at the end of distribution tours.

Mean temperatures:

Ambient air temp.: t_a = +24.2°C

Air Temp. in vehicle begin: t_b = -21.7°C end: t_e = -11.5°C

Prod. Temp.: begin: t_b = -25.4°C end: t_e = -17.5°C

In 1982 and 1983 temperature during distribution was controlled with a newly developed measuring device. This confirmed the results obtained in 1975 and specified that temperatures increase immediately after starting the distribution process. If doors are opened for shorter periods, for instance, by presorting the products to be distributed, temperature increases are less pronounced.

Similarly, extended distribution processes involving reloading from large long-distance trucks into smaller vehicles for local distribution were monitored. Measurements made in the top layers of palettes showed that the product temperature adjusts very fast to the environment; therefore products should be kept well away from high temperature environments (Fig. 7).

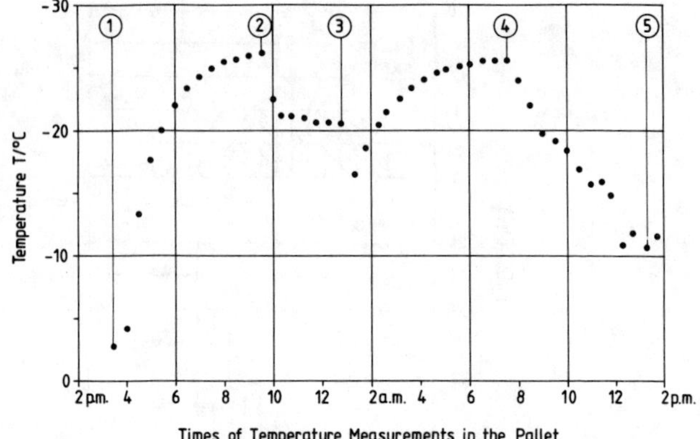

Fig. 7 Product temperatures in transit/distribution vehicle during distribution: measured in the top layer of a palette (spinach)

1 insertion of measuring device into palette at distribution coldstore
2 loading of transit vehicle at distribution coldstore
3 reloading into distribution vehicle at local coldstore
4 start of delivery tour
5 unpacking of measured palette

Apart from the requirement that trucks transporting deep-frozen food be adequately equipped and insulated, the following points should be observed:

- the temperature of the load at the beginning of the tour should be about -24°; much lower temperatures are not economical under the conditions described
- to admit as little heat as possible, the frequency and length of door openings should be reduced to an absolute minimum; this can be arranged by prepacking goods in the depot and loading according to the delivery route.

4. Residence time of frozen food in the cold chain

The date of packing is coded on most deep frozen foods. This date often coincides with the date of production. Therefore, the interval between packing and collection of samples at a given level of the frozen food chain can be used to determine the residence time of samples in the chain so far.

Residence times in specific links of the cold chain, however, can only be controlled by marking product samples before they enter the link under observation.

At the Federal Research Centre for Nutrition, both methods to determine residence times are employed. Total residence time up to level 4 of the frozen food chain was determined at retail shops. In separate studies, residence times in central coldstores and in retail cabinets were evaluated by date-marking product samples.

Products containing meat or fish have been subject to open date-marking regulations which are currently being extended to the whole product range. Most manufacturers meet this by printing a sell-by date on the pack, i.e. the date by which the product should be sold. The interval between collection and sell-by date - the sell-by reserve - yields another indicator of the time which products spend in the frozen food chain. Results on total residence time and on sell-by reserves are therefore compared in a survey conducted by the Federal Research Centre at the retail stage.

Total residence times between levels 1 and 4

Following extended pilot studies in the community of Karlsruhe, a representative nationwide survey was conducted in the Federal Republic of

Table 4: Sampling frame

Stage	Sampling unit	Sampling method	Sample size
1	electoral district	stratified random sample	700
2	food retail outlet	random route	1036
3	deep frozen product	random selection	4753

Germany in 1981 to determine total residence times for 1o selected deep frozen foods. A multistage sample was designed for this purpose /5/(Tab. 4).

Due to stratification, stage 2 sample of shops corresponds with the regional distribution of both households and total food retail outlets (Tab. 5).

Table 5: Regional distribution of households and food retail shops in the Federal Republic of Germany

State	Households (Basis: 24.2 mill.) %	Shops (Basis: 95.000) %	Shop sample (n = 1.036) %
Schleswig-Holstein, Hamburg, Lower Saxony, Bremen	20	19	18
North Rhine-Westphalia	28	26	27
Hesse, Rhineland-Palatinate, Saar	16	18	17
Baden-Württemberg	15	16	15
Bavaria	17	20	21
Berlin	1	1	2
Total	100	100	100

Source: Nielsen-Lebensmitteleinzelhandelsindex (Date: 1.1.198o) and own survey (Sample Institut, Köln, 1981)

In each shop sampled at stage 2 one item of each product on the list was selected if the product was available. The product list comprised 4

Table 6: List of products

Product	Pack sizes/ Types	Product variants
Meat loaves	all	incl. Farmers' Minced Meat steaks, excl. steaklets
Pizza	all	with meat or sausage, e.g. "Napoli"
Fish sticks	10 pieces	all
Cod fillet	all	all
Peas	300 g	including young peas
Green beans	300 g	all
Spinach with cream	45o g	all
Strawberries	about 300 g	all
French fried potatoes	cardboard boxes	all
Puff Pastry	all	all

meat products (ready-to-heat), 2 fish products, 3 vegetable products, 1 fruit product, 1 potato and 1 pastry product (see Table 6).

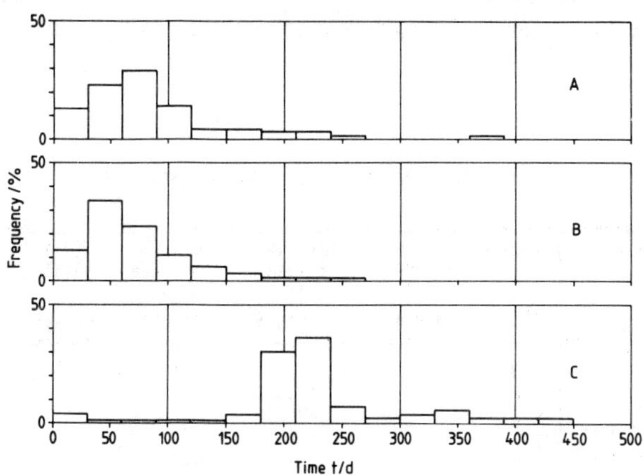

Fig. 8 Frequency of residence times in the cold chain

A cod fillet
B fish sticks
C spinach with cream

Empirical residence times scatter considerably. Their distribution is usually skewed to the left, i.e. the majority of product samples passes the chain fairly fast whereas a small proportion stays relatively long in the chain. The pattern varies among products due to type and season of harvest, of catch and packing, with spinach appearing as an exception to the rule (Fig. 8).

The following table summarizes the major aspects of residence time distributions for all lo products. For the average product, 5o % of its samples pass the frozen food chain in less than 4 months after packing. The period required to sell 9o % varies from 5 months for fish sticks to 14 months in the case of strawberries (see Table 7).

Table 7: Parameters of empirical residence time distributions

Product	Samples (abs.)	Mean residence time (months)	Minimum residence time (months)	Maximum residence time (months)	50 % point (months)	90 % point (months)
Meat loaves	372	4	1	26	3	8
Pizza	550	3	1	16	3	6
Fish sticks	661	3	1	16	2	5
Cod fillet	328	3	1	≥ 42	3	7
Peas	472	4	1	17	3	8
Green beans	355	4	1	29	3	7
Spinach with cream	658	7	1	21	8	11
Strawberries	293	5	1	22	4	14
French fried potatoes	63	4	1	14	3	7
Puff pastry	360	5	1	18	4	9
Total	4.112	4	1	≥ 42	4	8

Samples were not only checked for their dates but also for manufacturer, trade mark or brand to account for supply-based differences in residence times. Moreover, shop-related characteristics were recorded to establish factors potentially influencing the market for deep frozen foods (see Table 8).

Table 8: Influencing factors

Factor category	Factor group	Factor
Product-related	Manufacturer	Manufacturer/Trade mark
Shop-related	Location	State Community size Location in community
	Type of shop	Kind of service/No. of tills Assortment Social group/Area
	Frozen food storage facilities	Existence Capacity (equipment length) Icecream storage facilities

The influence of these factors on residence time and sell-by reserve was statistically tested (Chi^2 test, α = o,o1). Accordingly shop location is of limited influence on availability level and residence time of deep frozen foods.

Table 9: Residence times of fish sticks by shop size

Residence times (months)	Total (n=661) %	Shops with 1 till (n=314) %	Shops with 2 and more tills (n=347) %
1	15	11	18
2	37	31	42
3	24	28	20
4	12	15	8
5	6	6	5
6	3	3	3
at least 7	4	7	2
Total	100	100	100

The type of shop was found to be a major determinant. Deep frozen foods were significantly more frequently found in larger than in smaller shops, the size of shops being evaluated according to the number of tills. Empirical residence times are shorter for deep frozen foods sampled in larger shops,to some extent also for samples from shops offering a wide assortment of food products. Table 9 illustrates such differences in residence times for fish sticks, a product which is at stock in only 51 % of shops without self-service, but in 97 % of larger self-service outlets with at least five tills.

Similarly, shop storage capacity for frozen foods influences both the frequency of such foods being at stock and the speed at which they pass through the frozen food chain. Puff pastry, for example, was found in just 15 % of storage units below 2 m in length but in 76 % of units showing a length of 1o m and more. Its residence time pattern is equally affected by storage capacity(see Table 1o).

Table 10: Residence times of puff pastry by storage capacity for deep frozen foods

Residence times (months)	Total (n=360) %	Units of up to 3 meters length (n=87) %	Units of 3 meters length and more (n=272) %
up to 2	12	10	13
3	21	9	25
4	29	29	30
5	6	3	6
6	10	11	9
at least 7	21	37	17
Total	100	100	100

Although the frequency of such goods being at stock and residence time patterns vary significantly among products, it holds for all products that the influence of shop-related characteristics on both is similar in its tendency while different in intensity.

Residence times are not the only indicator of time patterns in the frozen food chain. For products containing meat or fish sell-by reserves were established for comparison. Positive values of sell-by reserve data show that samples were collected before their sell-by date whereas negative values indicate that the sell-by date was exceeded.

Such data are of particular interest because they confirm that only a very small proportion - 2 % on the average - of deep frozen food sampled at the retail stage is beyond its sell-by date, i.e. eventually below the standard defined as "good commercial quality". More detailed analyses would show that 94 % of all samples have more than 2 months until their sell-by date is reached.

Table 11: Parameters of empirical sell-by reserve distributions

Product	Samples (abs.)	Mean sell-by reserve (months)	Minimum sell-by reserve (months)	Maximum sell-by reserve (months)	Samples beyond sell-by date %
Meat loaves	378	+ 7	- 16	+ 13	5
Pizza	562	+ 6	- 5	+ 15	1
Fish sticks	731	+ 8	- 7	+ 17	1
Cod fillet	429	+ 8	- 34	+ 15	3
Total	2.100	+ 7	- 34	+ 17	2

If sell-by dates were set at a fixed distance from the packing date, empirical sell-by reserves would immediately follow from the observations on residence times. However, the empirical patterns of these two time indicators do not correspond perfectly (see Table 11). Empirical sell-by dates range from 4 to 18 month after packing, depending on the type of product, product quality, and supplier/brand. Sell-by dates are also influenced by shop-related characteristics comparable to residence times. It is therefore advisable to use both indicators in their respective contexts.

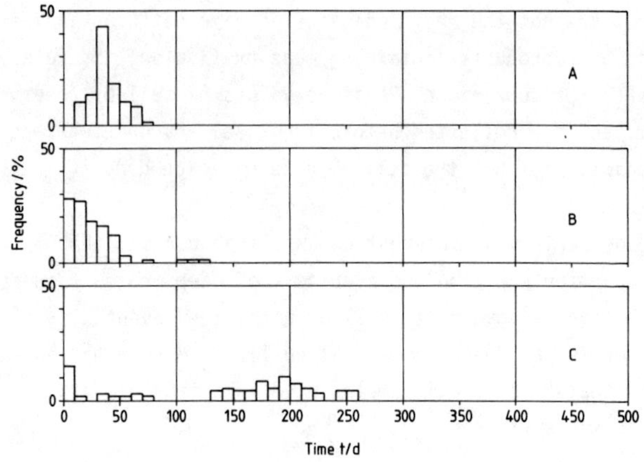

Fig. 9 Frequency of residence times in central coldstores

A cod fillet
B fish sticks
C spinach with cream

Level 2:

To obtain information on the residence time of deep frozen food in central coldstores, 2 of these were selected with the help of the German deep freezing industry. They may be regarded as typical for central coldstores in the Federal Republic of Germany. One was situated in a region of industrial concentration in North Rhine-Westphalia, the second in Baden-Württemberg. The observation period was 9 month (12.8o - 8.81). Residence times of 1o selected products were recorded in such a way that the date of entry into the store and the date of leaving the store were recorded on the palette.

The results for all products investigated showed residence time patterns similar to those in the entire chain; however, the residence times were much shorter, and more narrowly distributed(see Fig. 9).

Level 4:

Residence times in retail freezer chests could, for technical reasons, be determined in only 2 freezer units (cabinet and chest) of one retail shop in Mannheim. The sample is of course selective and far from being representative. Its function is limited to a control and confirmation of results obtained from comparisons of total residence times with those in the first two levels of the chain.

5. Summary and conclusions

The studies carried out by the Federal Research Centre for Nutrition, Karlsruhe, on temperatures and residence times of deep frozen food in the frozen food chain show that such food is stored at temperatures below -24^{0}C for a large proportion of the time it spends in the chain. This temperature limit is continuously observed in central coldstores as well as in distribution and local coldstores. Problems arise primarily during distribution and sale. A temporary increase during distribution is of little consequence if the product is cooled down again immediately after delivery. Serious quality losses may occur if products are exposed to higher temperatures in retail freezer units over longer periods. Practical experiences in the retail trade have shown, however, that products with long residence times frequently rest in the bottom layers of chests where temperatures are below average.

Representative time studies confirm that the bulk of deep frozen food - particularly of sensitive products in large, well equipped outlets - passes the frozen food chain within a few months. Sell-by dates prescribed by producers are rarely exceeded. For these reasons it seems realistic to assume good product quality even in cases where products are exposed to somewhat higher temperatures in parts of the frozen food chain. To ensure good quality of all products, however, it seems advisable to calculate acceptable storage lives of deep frozen food under the assumption of relatively high product temperatures during residence at the retail stage.

REFERENCES

1. BØGH-SØRENSEN, L.; BRAMSNAES, F. - The effect of storage in retail cabinets on frozen food. Freezing, frozen storage and freeze-drying; Commissions C1-C2, Karlsruhe, 1977. International Institute of Refrigeration, Paris, 1977-1.
2. BØGH-SØRENSEN, L.; BECH-JACOBSEN, K. - Time-temperature in the freezer chain in Denmark. Proceedings of the 16th International Congress of Refrigeration, Paris 1983.
3. BORNSCHLEGEL, A. - Temperature conditions during long-distance distribution of perishable food products. Freezing, Frozen storage and freeze-drying; Commissions C1-C2, Karlsruhe, 1977. International Institute of Refrigeration, Paris, 1977-1.
4. DEUTSCHES INSTITUT FÜR NORMUNG E.V. DIN 8954 - Offene Verkaufskühlmöbel; Blatt 4, Temperaturprüfung. Beuth-Vertrieb; Berlin, Köln.
5. FOLKERS, D.; SPIESS, W.E.L. - Zum Verweilzeitverhalten tiefgefrorener Lebensmittel in der Tiefkühlkette. Temperatur Technik 20 (2) 8 - 11; 1982.
6. FRANCK, R.; NÜSE, K.-H. - Deutsches Lebensmittelbuch. Carl Heymanns; Köln, Berlin, Bonn, München 1982.
7. GAC, A.; GAUTHERIN, W. - Amélioration du fonctionnement des meubles ouverts de vente au détail des produits surgelés. Cooling, freezing, storage and transport: biological and technical aspects; Commissions C2-D1-D2, Budapest, 1978. International Institute of Refrigeration, Paris, 1978-2.
8. Joint ECE/CODEX Alimentarius Group of Experts on Standardisation of Quick Frozen Foods. Report of the working Group on Temperature and Quality of Quick Frozen Foods. Joint FAO/WHO Food Standardsprogramme CX/OFF 78/3; Rome 5. Okt. 1978.
9. JUL, M. - The intricacies of the freezer chain. Refrigation of perishable products for distant markets; Commissions C2-D1-D2-D3, Hamilton NZ, 1982. International Institute of Refrigeration, Paris, 1982-1.
10. LÖNDAHL, G. - How to maintain a sufficiently low temperature in frozen food distribution. Freezing, frozen storage and freeze-drying; Commissions C1-C2, Karlsruhe, 1977. International Institute of Refrigeration, Paris, 1977-1.
11. MEFFERT, H.F.TH. - Maintaining quality of perishable produce in cold chain by proper time-temperature conditions. Recent developments and trends in refrigerated transport; Commission D2, Vienna, Austria, 1978. International Institute of Refrigeration, Paris, 1978-4.
12. MIDDLEHURST, J.; RICHARDSON, K.C.; EDWARDS, R.A. - Handling distribution and retailing of frozen foods. Food Technology in Australia; 24(11) 560, 561, 563-564, 567, 569-571; 1972.
13. SPIESS, W.E.L.; WOLF, W.; WIEN, K.J.; JUNG, G. - Temperature Maintenance during the local distribution of deep frozen food products. Freezing, frozen storage and freeze-drying; Commissions C1-C2, Karlsruhe, 1977. International Institute of Refrigeration, Paris, 1977-1.
14. STOREY, R.M. - Some observations on temperature and quality in the UK frozen food chain. COST 91, Concluding - Seminar; Athens, 1982.
15. WINGER, R. - The frozen food chain in New Zealand and shipment of frozen foods from New Zealand to Great Britain. Personal Communication 1983.

SOME OBSERVATIONS ON TEMPERATURE AND QUALITY IN THE UK FROZEN FOOD CHAIN

J. WIGNALL , D. POTTER and R. M. STOREY
Humber Laboratory, Ministry of Agriculture, Fisheries and Food

Summary

The paper outlines the scope of a nationwide survey of the temperature and quality of selected frozen foods at successive stages in the distribution chain from production to retail display. The mean temperature of samples on retail display was found to be of the order of -15°C, to be -19°C in secondary cold stores and -23°C in primary cold stores. The mean time in distribution was of the order of 96 days. Details of quality changes are also given; for example 97% of the fish samples had either none or only slight deterioration due to cold storage. Much analysis of the data from over 8000 individual observations remains to be completed and will be published in due course.

1. INTRODUCTION

Technological research cannot be divorced from the industry which it serves and whereas many problems are self evident, there can be others which only become apparent when the technologist works closely with industry. With this in mind the Humber Laboratory, an outstation of the UK Ministry of Agriculture, Fisheries and Food's Torry Research Station, in conjunction with the Ministry's Food Science Division and The Campden Food Preservation Research Association planned a comprehensive survey of temperature and quality profiles of frozen food in the distribution chain.

It is hoped that the survey would achieve a number of objectives such as the collection of factual data on time temperature histories currently obtaining in the frozen food chain, their effects on any cold storage deterioration, and the identification of any problem areas.

In order to undertake a survey of such magnitude it was necessary to obtain the full co-operation of all sections of industry, via their respective Trade Organisations. The success of the survey is in no small measure due to the excellent co-operation of the UK industry and to whom must be given our warmest thanks.

A practical and hopefully realistic approach was made in the planning of the survey. The frozen food products chosen are those most commonly consumed in the UK. Meat products with the exception of poultry had to be excluded simply because facilities for the determination of quality were not available to us. The products chosen were:-

Fish
Brussels sprouts
Peas
Chips
Chickens

which represent a large proportion of the more important products offered for sale in the UK. The fish was sub-divided into uncoated products, coated products and fish in sauce. Preference was given to cod products prepared where possible from UK caught fish, haddock being substituted when necessary. Only UK produced vegetables and chickens, even more specifically water chilled, non polyphosphated ones in the popular range 1.4 to 1.8 kg (3 to 4 lb) were sampled.

The quality of the products was measured by trained taste panels, the sensory techniques used being designed to be as objective as possible. The main aim was the determination of the degree of change due to cold storage, which can be regarded as the product's own integral of it's previous time temperature history.

2. OBSERVATIONS

It is desirable to make as many observations as possible for subsequent correlation. For example manufacturers' date codes were used, with their co-operation, during both the retail and wholesale sampling to determine residence times.

Temperatures were measured to $\pm$ 0.1°C, the reading rounded to the nearest degree, with accurate electronic thermometers in accordance with Codex Alimentarius recommendations, usually by drilling and inserting a pre-cooled probe into selected products whilst still in their original positions wherever possible. These sampling positions were those recommended by Codex Alimentarius; additional samples were taken to supplement these data.

All positions were carefully coded and recorded. Cabinet air temperature were measured and, in addition, numerous non-destructive product temperatures were measured by placing the thermometer probe between retail packs. All the samples into which thermometers had been inserted were placed into cooled insulated containers and purchased. In order to reduce to an insignificant minimum any changes in quality post sampling, samples were transferred as rapidly as possible to a liquid nitrogen cooled cabinet mounted on a lorry. The samples were held at -25°C for upto 4 days and then transported to the laboratory and stored at -30°C. Tasting was usually completed within 2 weeks of sampling.

Whilst an appraisal of some of the quality, temperature and residence time data are given later, much more information has been collected which has yet to be analysed. For example, type of freezer, volume of cabinet occupied, use of night covers, frequency and time of delivery of goods, etc. were all noted for future study.

The selection of retail sampling points and the frequency of sampling at each point was based on the throughput of each of 4 types of retail outlet; hyper/supermarkets, national chains of food retailers, frozen food centres and small individual shops, in order to obtain as representative a picture as possible (1). Further, the selection of the region to be studied was made to take into account possible differing consumption patterns related to type of area eg. rural, towns and inner cities. The survey has now been completed in the 5 regions, Kingston-upon-Thames, Bath, Birmingham, Kilmarnock and Newcastle, selected to represent different types of communities (2). Four regions were revisited during opposite seasons and one during the same season, 1 year later. Data from the 10 individual retail surveys are currently being analysed. The data given later are a summary of seven of these surveys, in which some 1800 sample temperatures and 2400 non destructive temperature measurements were made and some 900 date codes identified.

Surveys of temperature, time in distribution, quality of the product and details of some of the cold stores were carried out in "feeder" or secondary cold stores supplying each region visited. About 1100 samples were taken from 66 secondary cold store visits in this part of the chain. Further back in the chain, visits were made to 24 primary cold stores (ie. producer's cold stores, or cold stores in which products were stored after packing, but before further distribution) from which about 400 samples were taken.

Thus data have been obtained from the whole of the distribution chain from packing to retail sale and, where possible, distinct chains of distribution were followed.

3. RESULTS

3.1 Retail

3.1.1 Temperature

It must be stressed that the information given below is of a preliminary nature and is presented as an indication, only, of the type of data provided by the survey. Until full statistical analysis of all the data collected is completed, it would be unwise to draw any firm conclusions. Many data have been considered together to provide an overview and to simplify their presentation in this paper; due to rounding the sum of percentage distributions is not necessarily 100%.

The results given for the retail part of the survey are based on about 70% of the data collected. Table I is a summary of product and air temperatures and non-destructive product temperature measurement in retail display cabinets.

TABLE I Temperature distribution in retail cabinets

Temperature range °C	Percentage of sample falling within range		
	Products *(1826)	Non destructive *(1220)	Cabinet air *(1168)
-24 and colder	6	9	12
-23 to -18	28	35	34
-17 to -12	43	41	37
-11 to -6	20	14	15
-5 and warmer	2	1	2
Mean	-15.4	-16.4	-16.9

(*Figures in brackets are numbers of observations).

During the survey it was found that the vast majority of cabinets was of the chest type and whilst a small proportion was covered, the results presented represent the mean for all observations in the chest cabinets surveyed. Very few vertical cabinets, either open or closed, were found during this survey. Detailed analysis to study the effects of covers and the differences between types of cabinets is in progress.

As is known there is generally a temperature gradient within cabinets. The average temperature distribution found is shown in Table II; measurements were not made above the load line. Further, as might be expected, the range of the mean temperatures found in the upper layers for each survey was greatest, and was least in the bottom layers, the highest temperatures being found in the upper layers.

3.1.2 Taste panel evaluation

As mentioned in the introduction the major reason for tasting the products surveyed was to determine the degree of any changes due to cold storage. These are more easily identified in fish than in some other products, for example certain vegetables. Cold storage changes subsequent to freezing can be, to some extent, dependent on the pre-freezing history of the product. It is very difficult for example to separate the effects of cold storage on colour, flavour and texture of vegetables without prior knowledge of the values of the measured parameters before storage.

TABLE II Temperature distribution within cabinet

Temperature range	Percentage of samples falling within range		
	Top **(547)	Middle **(644)	Bottom **(632)
-24 and colder	1	6	11
-23 to -18	14	32	37
-17 to -12	47	46	36
-11 to 06	33	15	15
-5 and warmer	5	1	1
Mean	-12.8	-16.1	-17.0
*Range of means	-10.2 to -14.0	-13.0 to -17.3	-16 to -18.2

*Range of mean values found in each area visited
**Numbers in brackets are numbers of observations

The effects of 'PPP' factors such as processing (eg. blanching of vegetables) and packaging are clearly important when considering rates of change during cold storage. For this reason the fish products were divided into the three groups to which reference has already been made. Cold storage changes in white fleshed species such as cod and haddock are manifested by the development of characteristic cold storage flavour and of some toughening.

Table III is a summary of the preliminary analysis of the taste panel results to date. The majority of the products were found to have only little or moderate development of cold store flavour and toughening. The significance of the differences found between products is not yet clear, but is most probably related to preparation and packaging. Potential masking effects on taste panel evaluation due to the various coatings and the addition of sauces, which were removed prior to the products being cooked in a steam oven, has not yet been separated from any inherent differences between products. Thus it would be irresponsible to draw

TABLE III Cold storage flavour and texture

Parameter	Percentage of samples falling within grades		
	Uncoated *(273)	Battered and breaded *(312)	In sauce *(201)
Cold storage flavour			
1. None to just detectable	57	51	77
2. Slight to moderate	37	45	22
3. Moderate to strong	6	4	1
Texture			
1. Normal to slightly chewy	74	84	67
2. Chewy to tough	25	16	33
3. Very tough	1	0	0

*Numbers of samples in brackets

too many conclusions from the tasting data at this stage, which should be regarded only as a further indicator of the scope of the survey.

The flavour and texture of both red (thigh) and white (breast) poultry meat were examined by a taste panel. Changes in flavour attributable to cold storage were found to be negligible; 93% of the samples (148) of white meat and 97% of the samples of red meat had no or only just detectable changes in flavour attributable to cold storage. There appear to be slightly greater changes in texture, but in no case were these marked and hence readily detectable. In general, the effects of cold storage were found to be small.

3.1.3 Time in distribution

The time in distribution from packing to sampling was determined from the manufacturers' date codes, with their co-operation. This was determined for individual products except chickens which are not date coded and is summarised in Table IV.

Thus, based on the computed average residence time, 61% of all products were 90 days or less in distribution, 77% less than 120 days, 86% less than 150 days, with an overall average of 96 days elapsing between packing and sampling.

A further analysis of the data for all products is given in Table V in which time from packing is related to position in cabinet at the time of sampling.

The data in Table V suggest that the time from packing to sampling of products from the middle layer is shortest and is longest for the bottom samples; this observation will be referred to later.

TABLE IV Time in distribution

Time from packing (days)	Percentage of samples falling within each time range		
	Fish *(472)	Vegetables *(420)	All products *(892)
1 - 30	11	8	10
31 - 60	29	26	28
61 - 90	22	24	23
91 - 120	17	14	16
121 - 150	8	11	9
151 - 180	4	4	4
Over 180	10	11	11
Mean time (days)	94	100	96

*Figures in brackets are numbers of samples

TABLE V Effect of position of sample on time in distribution

Time from packing (days)	Percentage of samples in each time range		
	Top *(263)	Middle *(316)	Bottom *(312)
1 to 30	11	11	8
31 to 60	29	27	27
61 to 90	23	25	21
91 to 120	12	18	17
121 to 150	9	10	8
151 to 180	5	2	5
Over 180	11	6	16
Max. average time (days)	136	107	146
Min. average time (days)	67	67	75
Average mean time (days)	95	88	107

*Figures in bracket are numbers of samples.

3.2 Primary and secondary cold stores

3.2.1 Temperature and residence time

The data given in this section are based on observations made during visits to 14 primary and 30 secondary cold stores, and represent less than 50% of the data collected in the whole survey. Table VI is a summary of the data processed to date. In total some 90 cold store visits have been made, 24 to primary and 66 to secondary stores. The numbers of individual observations were proportionately fewer than those obtained in the retail part of the survey.

TABLE VI Summary of cold store data

	Store temperature (°C)		Product temperature (°C)		Time from packing (days)	
	Average	Range[(a)]	Average	Range[(b)]	Average	Range[(c)]
Primary (14)	-24.3	-17 to -29	-22.7 (137)	-13 to -29	50 (109)	4 to 437
Secondary (30)	-21.7	-13 to -25	-19.3 (354)	-12 to -28	69 (223)	6 to 357

(a) Range of average temperature in each cold store
(b) Range of individual product temperatures
(c) Range of time in distribution of individual samples

Figures in brackets are numbers of samples upon which data are based

All cold stores which supply directly to retail outlets have been included under the heading of secondary cold stores. Some distribution chains are complex and warrant individual analysis; Table VI is presented to indicate the average conditions found. It can be seen that the average temperature of all products in the primary cold stores was -22.7°C and ranged from -13 to -29°C. Care must again be exercised in the interpretation of the data which include that from cold stores used to store only a single product. The averages are not weighted with respect to volume. Care was taken however not to include samples which had just been packed and which might produce an unfair bias in the results.

The average temperature of products in secondary cold stores was slightly warmer at -19.3°C, but again this figure is not a weighted mean, each sample being given an equal weighting. As for data previously discussed, statistical analysis is in progress, the information given cannot be considered as a precise indication of the current situation.

The average time from packing to sampling was 50 days in primary cold stores and 69 days in secondary stores, which suggests, again on average, that the residence time in secondary cold stores was of the order of 19 days. If the average time in distribution on sampling at the secondary stores is subtracted from the times found in retail cabinets (Table V), the differences in residence time found in the top, middle and bottom layers of product is more clearly differentiated. These become 26, 19 and 38 days in the top, middle and bottom layers respectively. Neither the reasons for these apparent differences, which probably include variations in stock rotation and the possibility of customers avoiding the top layer, nor their importance, have yet been determined.

There are, as expected in any survey of a real situation, outliers and some indication of these can be seen in the ranges of the data given in Table VI.

3.2.2 Taste panel evaluation

As for the retail part of the survey the data on fish have been selected from that obtained in total to indicate the scope of the changes. Since on average the cold storage changes are small it is necessary to discuss the taste panel results in slightly greater detail to indicate the order of the changes occurring during distribution.

Tasting was carried out by a trained panel using scoring systems similar to those given below. Other parameters were determined and only two factors are considered here for illustration.

Cold storage flavour

Flavour intensity	Score
None	0
Just detectable	1
Slight	2
Moderate	3
Strong	4
Very strong	5

Texture
(Toughness on chewing)

Toughness	Score
Sloppy	-1
Fresh fish	0 1
Chewy	2
Tough	3
Very tough	4 5

The mean taste panel scores for samples from the three stages in the distribution chain are given in Table VII.

TABLE VII Summary of taste panel observations on fish samples

Stage	Cold store flavour		Texture	
	Mean	Range	Mean	Range
Primary (93)*	1.1	0 to 2.7	1.4	0 to 3.3
Secondary (190)	1.1	o to 3.0	1.3	0 to 3.4
Retail (788)	1.5	0 to 4.5	1.7	0 to 4.5

*Figures in brackets are numbers of samples included in this summary.

As can be seen in Table VII on average there is very little differenc between samples from the primary and secondary cold stores. There are more readily detectable differences in the samples from the retail survey however, as might be expected.

4. CONCLUSIONS

Although time consuming, the survey has proved to be a valuable exercise. Much has been learnt about the current situation in the UK industry, from whom excellent co-operation was received. It is hoped that the final analysis will demonstrate any problem areas requiring further attention. Much detail has had, of necessity, to be omitted from this account, and firm conclusions cannot be drawn until all the analyses and inter-relations have been determined.

REFERENCES

1. Birds Eye Annual Review (1980), Birds Eye Walls Ltd., Walton on Thames, Surrey.

2. Ministry of Agriculture, Fisheries and Food. Household Food Consumption and Expenditure. Annual Reports, HMSO, London.

TIME-TEMPERATURE SURVEY OF THE FROZEN SHEEPMEAT TRADE BETWEEN NEW ZEALAND AND THE UNITED KINGDOM

R.J. WINGER

Meat Industry Research Institute of New Zealand,
P.O. Box 617, Hamilton, New Zealand.

Summary

The paper summarizes the time-temperature history of frozen lamb exported from New Zealand to the United Kingdom in 1979. Lamb slaughtered and frozen at licensed export factories in New Zealand was stored for an average of about 6 weeks at -14.5°C before being transferred to containers for shipping. These containers were maintained at about -17°C on board ship for 4 to 5 weeks on average. There are three major distribution cold chains for this meat in the U.K., and the conditions in each chain are outlined briefly in this report.

1. INTRODUCTION

This paper summarizes the results of a survey conducted in 1979 by the Meat Industry Research Institute of New Zealand on the cold distribution chain for frozen lamb between the factory in New Zealand (N.Z.) and retail in the United Kingdom (U.K.) (1).

2. OVERVIEW OF THE DISTRIBUTION SYSTEM

To study the time-temperature history of lamb in each part of the cold distribution chain, the relative product flows within the chain must be known. Some of the data presented are only approximate, and all apply to the calendar year 1979.

N.Z. lamb for the U.K. market is frozen predominantly in carcass form within 1 to 3 days of slaughter. Once frozen, the carcasses are usually stored for several weeks before being transferred to containers and transported to the nearest port. At the wharf the containers are mechanically refrigerated from arrival until loading onto the ship. They are refrigerated continuously during their journey to the U.K.

In 1979 the U.K. imported 15.4 million whole carcasses of N.Z. lamb and the equivalent of 1 million carcasses in cut form. Of that total, 86% was shipped in containers, the remainder in conventional vessels. About 21 000 containers would have been used.

In the U.K., 95% of the containers were discharged through Tilbury, and the remaining 5% through Liverpool. All conventionally shipped lamb was discharged through Avonmouth.

The lamb was imported into the U.K. by 19 companies (also called primary wholesalers), who sold the lamb in one of three ways:

(i) Direct ex-ship (or ex-quay). The lamb was sold to 'multiples' (butcher and supermarket chains with many retail outlets) and also to other wholesale companies. The importer reduced his handling and inventory costs by transporting the container from the quay to the buyer.

(ii) Through the importers' own wholesale depots. The primary wholesalers owned some 250 depots and central meat market (CMM) stalls (such as in Smithfield) both of which sold a full range of fresh and frozen meat, mainly to independent butchers. The depots and stalls received lamb either in containers direct from the ship or from public cold stores.

(iii) Delivered ex-store. When an importer was unable to sell newly arrived lamb at a reasonable price, he was forced to store it until the supply and demand balance was restored and the wholesale price rose. Importers usually tried to sell the lamb on arrival rather than incur the extra costs of storage.

A survey of importers handling 63% of the N.Z. lamb imported in 1979 indicated that 34% of this lamb was stored in public cold stores for some period. A survey of a larger sample of importers indicated that about 49% of imports was sold direct ex-ship. The remaining 17% would have gone to the importers' own wholesale depots (Fig. 1).

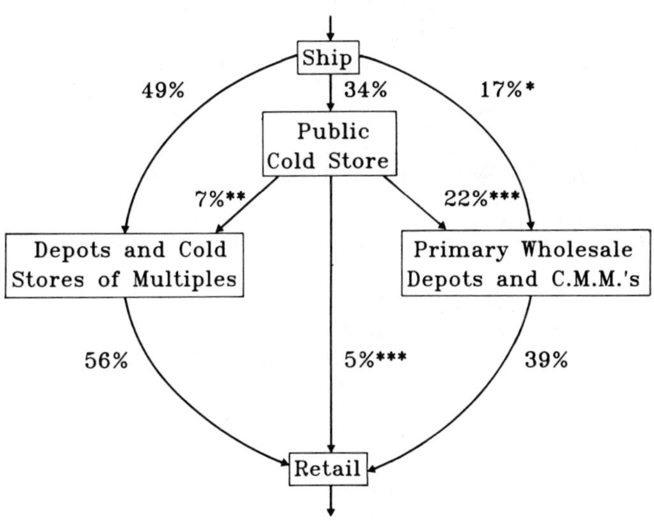

* calculated by difference
** estimate based on a limited section of the trade
*** educated guess

Figure 1. The movement of New Zealand lamb from ship to retail in the United Kingdom.

The sale of N.Z. lamb through different types of retail outlets in 1979 was as follows:

Multiples/grocers (i.e., supermarkets)	36.5%
Freezer centres	9.2%
Butchers	49.0%
All other outlets	5.3%

Almost all N.Z. lamb sold by supermarkets and freezer centres would be frozen. A significant portion of butchers' sales would be frozen freezer packs, and probably half the lamb sold through other outlets would be frozen. The rest would be thawed prior to sale.

3. RESULTS AND DISCUSSION

3.1 Time-temperature history of the distribution chain

3.1.1 Limits of using the time-temperature history

The condition of stockinet-wrapped carcasses is affected not only by storage temperature but also by the relative humidity, air velocity and temperature fluctuations in the store. Although these other variables are not included in this analysis, they are important and at times their effects can far outweigh the effect of the mean storage temperature.

3.1.2 Experimental outline

The storage data were collected using the following sampling scheme:

3.1.2.1 Cold stores in N.Z.

Eight factories, processing 20% of the total N.Z. lamb production, were sampled.

3.1.2.2 Containers/shipping

A sample of 117 containers was identified and the time-temperature histories of these containers were followed from loading at the factory in N.Z. to unloading at Tilbury container terminal in the U.K.

3.1.2.3 Storage in the U.K.

As much information as possible was obtained from records of companies importing 63% of N.Z. lamb, from spot checks of a sample of stores and from verbal discussions with store managers. These latter data were of a cursory nature.

3.1.3 Storage time results

3.1.3.1 Time spent in N.Z. factory cold stores

Storage time data for the 1978/79 season are presented in Figure 2.

The average storage time was about 6 weeks, and 90% of all product was moved within 22 weeks.

3.1.3.2 Time spent in transit from N.Z. factory to ship

Transit times from works to ship are shown in Figure 3. The average transit time was 9.8 days. This comprised 1.5 days between loading at the works and connection to the terminal refrigeration, 1.6 days on mechanical refrigeration to lower the container temperature toward -20°C, and 6.7 days controlled refrigeration at -20°C before disconnection and loading on board ship.

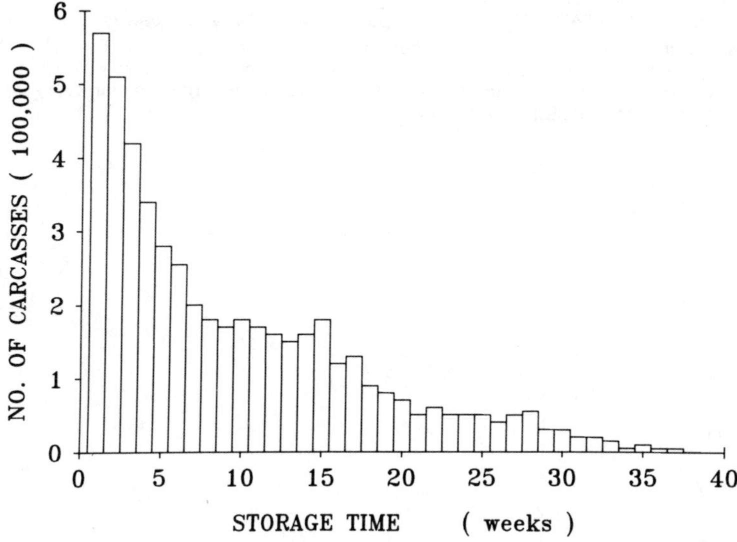

Figure 2. Length of time lamb carcasses spent in N.Z. cold stores.

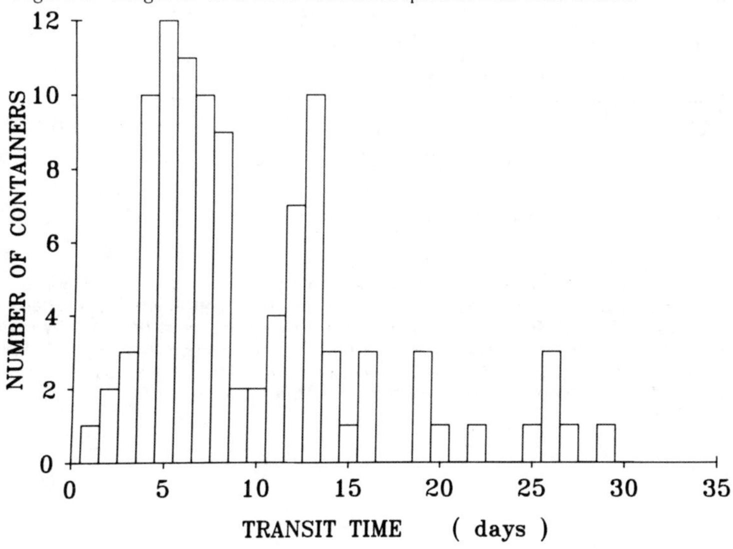

Figure 3. Transit times for containers from works to ship.

3.1.3.3 Time spent on board ship

The length of time the container was on board ship was taken from the ships' logs, which recorded the time the container was on refrigeration (Fig. 4). The average voyage time was 4.4 weeks, and 90% of all containers arrived within 6.5 weeks. The long voyage times were due to the vessel either being delayed or having many intermediate ports of call.

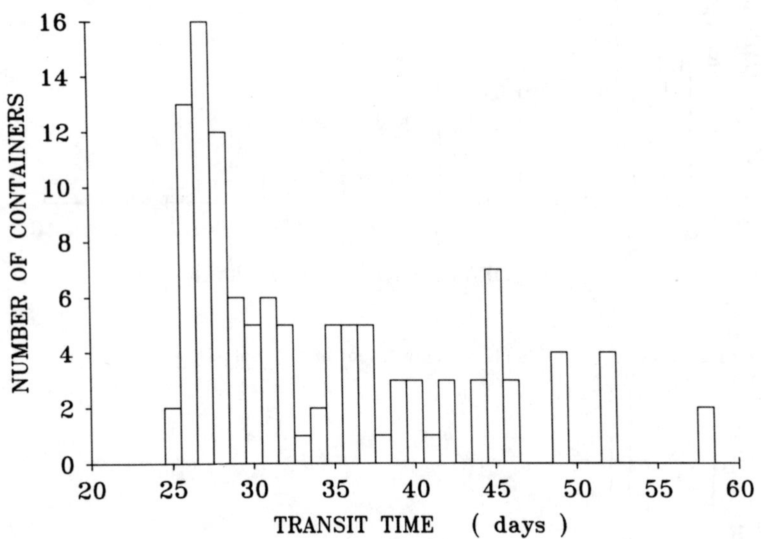

Figure 4. Distribution of transit times on ship.

3.1.3.4 Tilbury container terminal

In 1979, 81% of all N.Z. lamb shipped to the U.K. passed through Tilbury container terminal. The terminal had capacity for 1456 containers on mechanical refrigeration units. Refrigeration for additional containers was provided by dump charging of liquid nitrogen. The time distribution of containers holding N.Z. lamb at this terminal is shown in Figure 5. The average holding time was less than 1 week, with 90% of all containers being moved within 1.5 weeks.

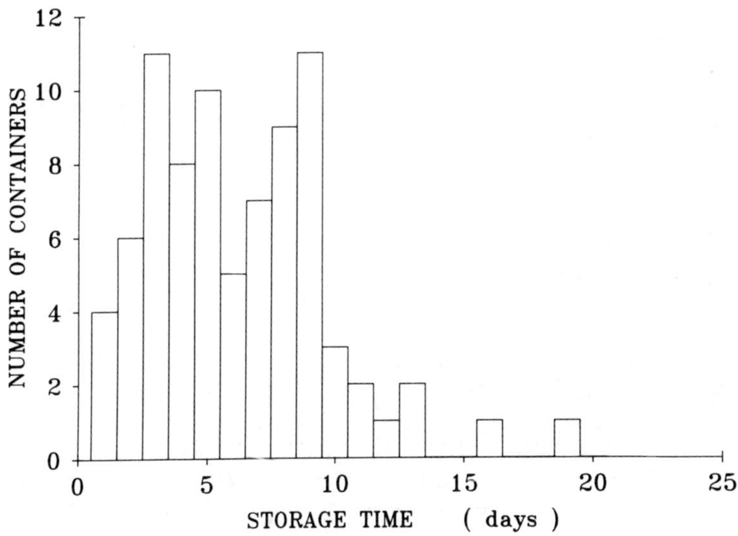

Figure 5. Distribution of holding times at Tilbury container terminal.

3.1.3.5 Public cold storage in the U.K.

Lamb was stored in public cold stores mainly by the importers. Because of the unusual market situation for lamb in 1979, an estimated 5% more lamb than normal was stored. Thus in a 'normal' year about 29% of N.Z. lamb imports might pass through public cold stores. Storage times for N.Z. lamb in public cold stores are given in Figure 6. Storage-time data for 2.74 million lamb carcasses were obtained. The average storage time was about 11 weeks, with 90% of the product being moved within 34 weeks.

3.1.3.6 Secondary storage

This part of the distribution chain includes cold stores associated with wholesale depots, central meat markets, distribution centres of multiples, and cutting plants. The survey assumed that all N.Z. lamb passed through these stores, although in Figure 1 five percent of lamb imports is shown by-passing this secondary storage.

Data were obtained from 16 representative secondary cold stores (Fig. 7). This information was obtained verbally from managers and foremen, and thus may not be as reliable as data obtained by personal measurement. The average storage time was about 4 weeks, with 90% of the product being moved within 10 weeks.

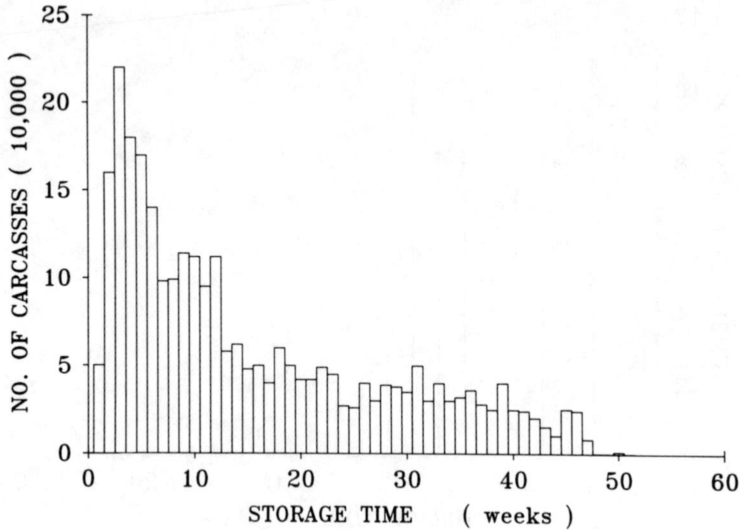

Figure 6. Length of time lamb carcasses were held in U.K. public cold stores.

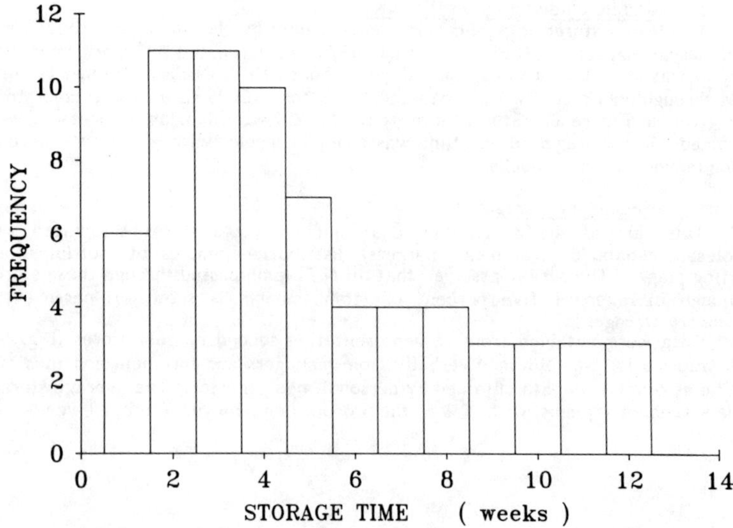

Figure 7. Holding times of carcasses in secondary cold stores.

3.1.4 Storage temperature distribution

The storage temperatures of the cold stores in N.Z. and the U.K. are presented in Figure 8.

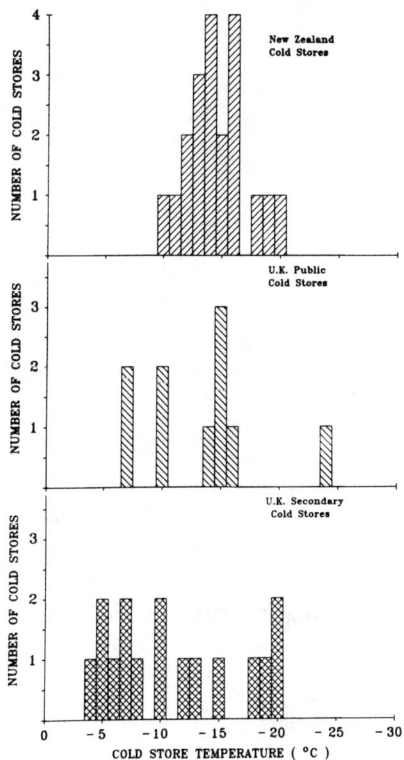

Figure 8. Frequency distribution of storage temperature for N.Z. lamb.

The average temperature in N.Z. cold stores was -14.5°C; the average in U.K. public cold stores was -14°C; and the average in U.K. secondary stores was -10°C. Cold store temperatures became more diverse as the product progressed along the cold chain (Fig. 8).

Temperatures in containers from factory loadout to the N.Z. container terminal were extremely varied. At these terminals the containers were mechanically refrigerated for 1 to 2 days to -20°C. Return air temperatures measured at the onset of this refrigeration were sometimes as high as -6°C. Some of these high temperatures were associated with lengthy transit times from factory to terminal or with long loading times under adverse conditions. A product temperature warmer than -4°C could result in deformation of the carcasses (2).

On the ship, storage temperatures ranged from -15 to -20°C within about 4 days of loading and remained constant (±2°C) during the voyage. Average storage temperature on the ship was -17°C. During unloading at Tilbury, 90% of the product was kept colder than -13°C. The average product temperature at unloading was -16°C.

3.1.5 Average time-temperature conditions

The average time-temperature conditions for the frozen lamb trade in 1979 from N.Z. to the U.K. are given in Figure 9.

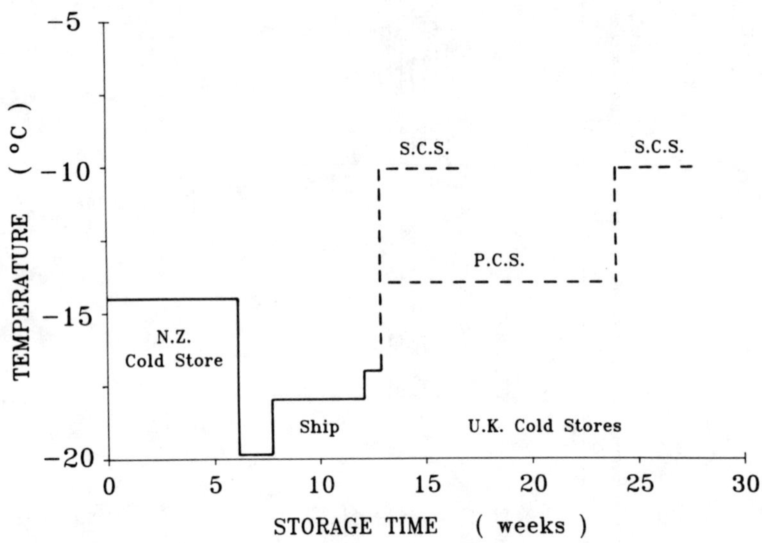

S.C.S. - secondary cold stores
P.C.S. - public cold stores

Figure 9. Average time-temperature history for N.Z. lamb from works to retail.

REFERENCES

1. DURBIN, J.R. (1980). The cold distribution chain and lamb quality in the U.K. Meat Ind. Res. Inst. N.Z. Publ. No. RM. 124.
2. HAUGHEY, D.P. (1970). Transportation of frozen meat in insulated containers within New Zealand. Meat Ind. Res. Inst. N.Z. Publ. No. 180.

NUTRITIONAL CONSEQUENCES OF FOOD FREEZING

Orla Zinck
National Food Institute
Copenhagen

SUMMARY

In a classical sense the nutrition situation in Europe is good. It is even difficult to identify cases of the classical nutrient deficiency diseases among people in the industrialized nations today. The major nutritional problems are related to affluence and affecting very large population segments indeed. The change in nutritional value during freezing is of minor importance compared to changes in eating habits. Generally the changes in consumption patterns may be modest from one year to the next, but over a longer period they become drastic.

The present consumption of frozen foods is generally so low that it is of no nutritional importance for the public in general - even if all the nutrients in the frozen products had been lost. But generally frozen products retain more nutrients than for instance canned products.

For special groups where frozen foods supply a large proportion of the total food intake losses in nutrients are, however, very important. Homeliving elderly participating in "meals-on-wheels" based on frozen foods is such a group. For this group it is of importance that the products have a very high acceptability and palatability so they do not reject it and by doing this deprive themselves of important nutrients.

1. INTRODUCTION

When evaluating the nutritional consequences of food freezing one has to relate the changes in nutritional value during freezing of foods to the nutritional problems of those who eat the foods and also to the importance of frozen foods in the diet. The use of frozen food is only of importans in the industrialized countries. It is the nutritional problems of these countries we have to consider. Also one must have in mind the alternatives to freezing and compare with the changes in nutritional value during other forms of food preservation. It is neither practical nor realistic to look at one process without some kind of reference.

The nutritional losses during the various preservation processes have to be seen in relation to the changes in nutrient intakes due to changes in consumption patterns.

2. THE NUTRITIONAL PROBLEMS TODAY

2.1 Overconsumption and lack of nutritients. In a classical sense the nutrition situation in Europe is good. It is even difficult to identify cases of the classical nutrient deficiency diseases among people in the industrialized nations today. The few cases found are either associated with clinical diseases, alcoholism or with the special conditions of the migrant workers. There are, however, "modern" nutrition problems affecting large segments of the populations in the rich countries, namely problems associated with overconsumption of food and alcohol (and tobacco).

The only well proved nutrient deficiency in Europe is goiter, which is endemic in certain areas due to lack of the mineral iodine in the soil.

The major nutritional problems are related to affluence and affecting very large population segments indeed. Thus, there is no reason for immediate concern about lack of nutrients in the European diets. Fiber is, however, lacking in many western diets because of a relatively high intake of refined foods. Fiber is of great nutritional importance, but not classified as a nutrient in the classical sense.

Based on data from food balance sheets there is no lack of nutrients in the European diets. But there are some groups at risk in certain segments of the population. The vitamins and minerals that these groups might be lacking are: vitamin A, vitamin B_1, B_6, folate,vitamin C, iron and calcium (15). These are the nutrients one has to study during food freezing. At the same time one has to know how much frozen food these groups consume.

The consumption figures are very difficult to get. For instanse in Denmark we do not know the food intake of the varies populationgroups. From a pragmatic point of wiew it is more feaseble to study what happens to the nutrients during freezing. In the modern societies, however, the eating patterns are changing very rapidly and the nutritionists will have to constantly follow the changing nutrient intake of the various population groups.

2.2 Changes in nutritient intakes. These changes may be caused by different developments. Usually nutritionists are most concerned about the losses in nutritive value during processing of foods. These losses may be considerable and of great nutritional importance since our food supply is processed to an increasingly larger extent. Very often it is overlooked that the consumption patterns are changing drastically at the same time with an effect on the nutrient intake of the same order of magnitude.

2.2.1 Changes in nutrient content of foods. The modern food supply system is caracterized by large-scale, centralized production. This tendency is not limited to the processing sector. It is also true for agriculture, where we for instance become more and more dependent on a very small number of high-yielding plant varieties. The increasingly uniform way of production, processing and distribution makes the situation vulnerable from a nutritional point of view. If a drastic change occurs in the nutrient content in a food which is an important supplier of a nutrient it might affect the nutrient intake of a whole population. In many European countries potatoes for instance supply 20-40% of the vitamin C in the diets. If a new potato variety with a low vitamin C content were introduced it might seriously affect the vitamin C intake. Therefore, the developments in the whole food production system have to be followed.

Due to this the "COST 91 - sub-group 4 - nutrition" recommends that concerning critical nutrients one should take into accounts (16):

- whether or not food items contribute substantially to nutrient supply of the general population or of a population group at risk,
- whether or not methods exist for determination of the nutrients under consideration,
- whether or not diminished intake of the nutrients in question may create serious problems in a given population group.

Main emphases should be given to vitamins, minerals and dietary fibre, according to the following list:

- Vitamins: (thiamin, riboflavin, pyridoxine, folate, ascorbic acid, vitamin D, and vitamin A including -carotene).
- minerals and trace elements: (iron including iron/ascorbic acid relationship, iodine and calcium).
- Dietary fibre

2.2.2 Changes in consumption patterns. The developments in the modern food supply system cause changes not only in the nutrient content of foods but also in the amounts consumed. Products that were previously seasonal may now be obtained all year and new products appear on the market. Our food consumption changes with our changing life-styles. For example the increasing number of women who are working outside the home. The need for convenience foods. Also changes in technology i.e. in the home. In Denmark we have had an increase in home freezer from 100.000 to 1.500.000 from 1960 to 1980. In the same period we also had a 10 times increase in consumption of frozen food.

Generally the changes in consumption patterns may be modest from one year to the next, but over a longer period they become drastic. Figure 1 shows an example of the changes in consumption over three decades in Denmark from 1950/52 to 1980/82. Similar changes have taken place in other European countries.

Figure 1. Change in food consumption in Denmark (Food balance sheets) (8).

	Average food consumption (kg/year)		Change in % 50/52–80/82
	1950/52	1980/82	
Cereals	103	70	÷ 32
Sugar	37	42	+ 16
Potatoes	123	68	÷ 45
Cabbage	20	11	÷ 45
Other vegetables	49	45	÷ 8
Fruits and berries	57	47	÷ 17
Meat, fish, eggs and chese	86	106	+ 23
Milk products	173	166	÷ 4
Butter, marg., fat	25	30	+ 21
Alcohol	3	10	+188

These changes in consumption patterns have drasticly changed the energy-distribution in the diet. The percentage of energy from fat has increased from 33 to 40 and the percentage of energy from carbohydrate has decreased from 53 to 43, while protein has been constant at 12%.

In spite of the drastic changes the vitamin and mineral intakes have generally increased or stayed constant. The only exceptions are vitamin C, calcium and iron where there have been a minor decrease (see fig. 2).

Figure 2. Changes in vitamin and mineral intake in Denmark (Food Balance Sheets) (8).

	1955	1981	Change in %
Retinol mg	1,0	1,4	+40
Ascorbic acid mg	118	111	÷ 6
Niacin equivalents mg	34,6	40,5	+17
Thiamin mg	2,4	2,5	+ 4
Riboflavin mg	2,7	2,9	+ 7
Pyridoxin mg	2,4	2,7	+13
Cobolamin µg	8,5	11,4	+34
Folate mg	99	99	0
Calcium g	1,6	1,5	÷ 6
Iron mg	20	19	÷ 5

The decrease in vitamin C intake is due to the decrease in potato consumption.

Generally the vitamin and mineral content of the Danish diet has not decreased. There is a special problem concerning folate. We do not know the real intake of folate. There are not good methods for determination of folate in foods. Similar problems exist with pyridoxin but not to same extend.

3. CHANGES IN NUTRITIONAL VALUE DURING FREEZING

Freezing is usually considered one of the most favourable preservation methods from a nutritional point of view. There is, however, no guaranty that the end result will be acceptable. For the consumer it is the intake that counts. Therefore one has to consider the entire process, and not only the freezing process itself, for instance: blanching, freezing, storage, thawing and preparation.

Figures 3 and 4 illustrate typical losses of vitamin C in vegetables during the freezing process. The importance of the storage temperature is also shown. Two months of storage at -6°C will destroy all vitamin C, where as pratically all of the vitamin C is retained after one year if the storage temperature is kept at -18C. Vitamin C is the most susceptible vitamin and can be used as an indicator of the influence of the treatment.

Figure 3. Loss of vitamin C from various vegetables during the entire freezing process. (5).

Product	Typical Amount of Vitamin C in Fresh Product (Mg/100Gm)	Loss of Vitamin C During 6-12 Months at -18°C (Mean % and Range)
Asparagus	**33**	**10 (12-13)**
Beans, green	**19**	**45 (30-68)**
Beans, lima	**29**	**51 (39-64)**
Broccoli	**113**	**49 (35-68)**
Cauliflower	**78**	**50 (40-60)**
Peas, green	**27**	**43 (32-67)**
Spinach	**51**	**65 (54-80)**

Figure 4. Loss of ascorbic acid in frozen peas. (13)

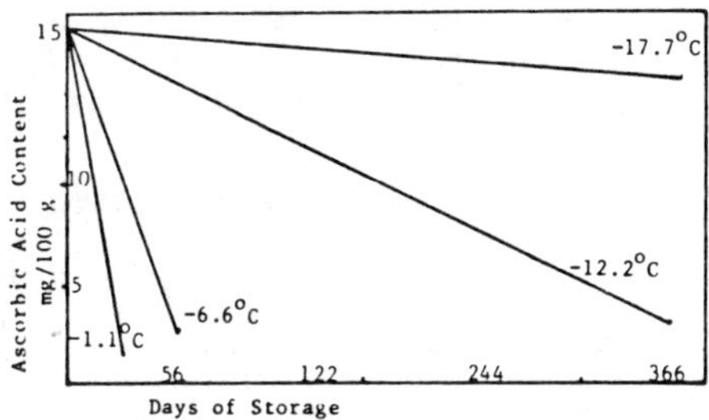

In figure 5 is shown the losses of several vitamins in fruits and vegetables during the freezing process compared to canning. The losses during freezing are usually smaller than those during canning, but it has to be noted that they may be considerable - up to 3/4 of vitamin C and 2/3 of B_1 and B_2 during the freezing process.

Figure 5. Comparative losses of vitamins from fruits during various methods of preservation. (5)

Method of Preservation	Number of Fruits Examined		Loss of Vitamins as Compared to Values of Fresh Products A (%)	B-1 (%)	B-2 (%)	Niacin (%)	C (%)
Frozen, not thawed	8	Mean	37	29	17	16	18
		Range	0-78	0-66	0-67	0-33	0-50
Canned, solids and liquid	8	Mean	39	47	57	42	56
		Range	0-68	22-67	33-83	25-60	11-86
Dried, uncooked	4	Mean	6	55	0	0	39
		Range	0-18	11-90	—	—	0-65

The results for retention of nutrients during freezing of animal tissue are confusing especially regarding the loss of B-vitamins. Sometimes there appears to be considerable <u>increases</u> in the vitamin content which are difficult to explain. The bulk of work concerning retention of B-vitamins in frozen meats was done in the fifties. Different studies of the retention of B-vitamins in the same muscle stored at the same temperature in the same period have produced different results. Lehrer (7), Lee (6) and Westerman (17) found a retention of 85-112% for thiamine, 56-106% for riboflavine in pig longissimus dorsi stored at -18C for 180 days.

Mikkelsen et al (9) found that retentions (%) after preparation and one year storage at -12C or -24C were within the following range: For pork loin B_1 50-85, B_2 90-112 and B_6 48-88. For ground meat B_1 77-78, B_2 84-85 and B_6 107-122. For "medister" sausage B_1 83-87, B_2 90-93 and B_6 98-115. For vienna sausage B_1 79-82, B_2 66-67 and B_6 115-133.

It is difficult to explain the increases in vitamin content during frozen storage. Possibly the changes during frozen storage make the vitamins easier to analyse using the present methods.

4. NUTRITIONAL CONSEQUENCES OF FOOD FREEZING

The consumption of frozen foods have increased 10 times from 1960 to 1980 in several European countries (18). There are big variations among the countries with the highest consumption in Sweden (25 kg/year) and one of the lowest in Italy (3,7 kg/year). Denmark is close to Sweden with a consumption of 23 kg/year (2).

This increase in consumption has of course not been possible without the use of home freezers. In the same period the number of home freezers in Denmark has increased from 100.000 in 1960 to 1.500.000 in 1980.

The frozen products that are most commonly sold are vegetables, meats (especially chicken) and fish. There are considerable variations from country to country, but normally vegetables are either first or second.

The various nutrient that might be important for some vulnerable groups are:

Protein: In general the diets in Europe are rich in protein. The proportion of animal protein is significant. Freezing can result in some moderate denaturation and affect protein availability, but the changes in proteins due to freezing are not of nutritional concern. As shown by Moreiras-Varela et al (11) cooking, freezing and reheating has practically no effect on the nutritive value of food protein. They base their evaluation on Diet Growth Index (DGI), Net Protein Utilization (NPU), True Digestibility (TD), and Biological Value (BV).

Retinol. The Danish population seems to be well supplied with this vitamin. Even considerable damage, which might be caused by freezing, appears not be of nutritional significance.

Thiamin. Meat is on of the most important sources of thiamin. The average diet contains an ample amount of thiamin. Even a considerable loss of that part of the thiamin which is derived from meat would hardly have any nutritional significance.

Riboflavin. The situation for this vitamin is much the same as for thiamin.

Pyridoxin. The Danish diet seems modestly supplied with this vitamin. Especially among the elderly (3), there may be symptoms of pyridoxin deficiency. Also, the use of oral contraceptives may result in an increased requirement. About 40% of the pyridoxin intake is derived from meat. Thus, any loss caused by processing could have substantial, adverse nutritional significances. It is, however, worth stressing that the analysis of pyridoxin causes difficulties. Thus, the intake is not known with any high degree of accuracy. Also, doubts exist with regard to requirements. Nevertheless, the effect of freezing of meat on this vitamin is worth close scrutiny.

But the decrease in pyridoxin content in meat during freezing seem to be too small to be of importance.

Niacin. The diet in Denmark seems amply supplied with niacin. A significant amount is derived from meat, but even a substantial loss of this amount during freezing will not be likely to lead to any deficiency situation.

Cobolamin. The average diet seems very well supplied with this vitamin. Even though a major part, app. 65% of RDA, is supplied by meat, a substantial loss would not render the diet deficient with regard to this vitamin. A concern is often expressed that a diet without animal products may easily be deficient in vitamin B_{12}. However, even for vegans this is a rare occurrence.

Calcium. The diet seems generally well supplied with calcium and the main bulk of the calcium comes from milk and bread. Loss of calcium during thawing would not be of nutritional significance.

Iron. It is known that anaemi is quite frequent in Western societies. About half of the iron intake is derived from meat. Therefore, any loss of this mineral due to freezing and thawing would have serious consequences. In addition, the iron in meat is known to be biologically easily available, a factor which further stresses the importance of avoiding any losses of this mineral during freezing and thawing of meat. However, freezing does not affect iron availability and drip from meat is normally limited to a few per cent. Freezing of meat is unlikely to have any significant consequences as far as iron is concerned.

Ascorbic acid. Vegetables and fruits are the main suppliers of ascorbic acid. The loss of ascorbic acid due to freezing seems to be determined by the storage temperature. The losses might be of nutritional importance. But changing patterns of consumptions seems to be a more important factor in determing the intake.

Generally the importance of frozen foods in the diet is limited. One may exclude all frozen foods from the average diet and still have a sufficient intake of the various vitamins and minerals.

5. PALATABILITY AND ACCEPTABILITY OF FROZEN FOOD

From a nutritional point of view food is only of importance if it is eaten. Therefore the palatability and the acceptability of the various products are essential.

Generally frozen foods have both a high palatability and acceptability. Sanderson-Walther (14) mentioned that a survey of consumers attitudes towards frozen foods showed that taste and apperance seems to be more important for frozen foods than for both canned and dried.

Millross et al (10) showed that schoolchildren preferred precooked frozen food reheated at the school to food transported hot from a central kitchen located some distance away due to a better taste. Also frozen reheated food is normally of better nutritionally quality.

Not all frozen products in a meal may be reheated with an acceptable result (1). Skjöldebrand et al (16) recommend that some vegetables and especially potatoes have to be cooked from the raw state. This is in accordance with Mikkelsen (8) who conclude that the eating quality of reheated frozen potatoes is lower than both chilled and reheated and potatoes that have been kept heated for 2 hours. The problem with frozen potatoes is that the center gets soft after reheating due to retogradation, changes in texture and structure. The eating quality of the end result is poor.

The use of frozen potatoes in the "meals-on-wheels" system might be of nutritional importance. The people normally recieving meals-on-wheels are elderly. It is a group who is at risk concerning vitamin C. Potatoes is the most important source of vitamin C - supplying 20-40%. If people stop eating them due to poor quality this would seriously affect the intake of vitamin C of this vulnerable group. From a nutritional point of view it is therefore important to develop better processing methods for frozen potatoes. And if this is not possible then to find alternative solution in the cook-freeze system. The people that take part in "meals-on-wheels" do not normally have other choises than to eat the food offered or reject it.

6. CONCLUSION

The change in nutritional value during freezing is of minor importance compared to changes in eating habits. The present consumption of frozen foods is generally so low that it is of no nutritional importance for the public in general - even if all the nutrients in the frozen products had been lost. But generally frozen products retain more nutrients than for instance canned products.

For special groups where frozen foods supply a large proportion of the total food intake losses in nutrients are, however, very important. Homeliving elderly participating in "meals-on-wheels" based on frozen foods is such a group.

For this group it is of importance that the products have a very high acceptability and palatability so they do not reject it and by doing this deprive themselves of important nutrients.

REFERENCES

1. Anon. (1982). Schulverpflegung, mit speisen aus eigener zubereitung und industrieller herstellung. Mischküche - stufe 1. Bundesministerium für Ernährung, Landswirtschaft und Forsten und der Bundesforschunganstallt für Ernährung.

2. Dybfrostinstituttet. (1983). Copenhagen. Personel communication.

3. Elsborg, L. (1982). Vitamin og mineralindtagelse. De ældres kost. Husmandsforeningernes Husholdningsudvalg, Aarhus og Foreningen af Jyske Landboforeninger, Skanderborg. Denmark.

4. Emborg, S. (1983). Levnedsmiddelstatistik og fødevareforbrug. National Food Institute. Copenhagen.

5. Harris, R.S. and Karmas E. (1978). Nutritional evaluation of food processing. Second edition. AVI Publishing Co. Westport.

6. Lee, F.A., Brooks, R.F., Pearson, A.M., Miller, J.I. and Wanderstock, J.J. (1954). Effect of rate of freezing on pork quality. J. Am. Dietet. Assoc. 30, 351-354.

7. Lehrer, W.P., Jr., Wiese, A.C., Harvey, W.R. and Moore, P.R. (1951). Effect of frozen Storage and subsequent cooking on the thiamin, riboflavin, and nicotinic acid content of pork chops. Food Res. 16, 485-491.

8. Mikkelsen, B. (1983). Quality of potatoes in meals-on-wheels. M.Sc. Thesis, Department of food Preservation, Royal Veterinary and Agricultural University, Copenhagen.

9. Mikkelsen, K., Rasmussen, E.L. and Zinck, O. (1983). Retention of vitamin B_1, B_2 and B_6 in frozen meat. Concluding seminar on Thermal Processing and Quality of Foods. Cost-91, Athens.

10. Millross, J., Spekt, A., Holdsworth, M.M. and Glew, G. (1973). The utilisation of the cook-freeze system for school meals. University of Leeds.

11. Moreiras-Varela, O., Ruiz-Roso, B. and Varela, G. (1983). Influence of the CFR (cooking-freezing-reheating) system on the nutritive quality of food protein. Concluding seminar on Thermal Processing and Quality of Foods. Cost-91, Athens.

12. Nair, B.M. (1983). The effect of freezing and canning on the vitamin content of green peas. Food and Nutrition. vol. 9 - no. 2. FAO.

13. Pietrzik, K. (1983). Nutrients considered to be worthy of examinations in processed food. Concluding seminar on Thermal Processing and Quality of Foods. Cost-91, Athens.

14. Sanderson-Walker, M. (1983). The application of research data in the quick frozen food industry. Concluding seminar on Thermal Processing and Quality of Foods. Cost-91, Athens.

15. Schierf, G., Wolgram, G. (1982). Mangelernährung in Mitteleuropa. Wiss. Verlagsgesellschaft, Stuttgart.

16. Skjöldebrand, C., Ohlsson, Th., O'Sullivan, K. and Turner, M. (1983) Reheating of food in catering. Concluding seminar on Thermal Processing and Quality of Foods. Cost-91, Athens.

17. Westerman, B.D., Oliver, B. and Mackintosh, D.L. (1955). Influence of chilling rate and frozen storage on B-complex vitamin content of pork. Agricultural and Food Chemistry 3(7), 603-605.

18. Åström, S. og Bäch, L. (1979). Den internationelle djupfrystmarknaden. Livsmedelsteknik nr. 3.

THE APPLICATION OF RESEARCH DATA IN THE QUICK FROZEN FOOD INDUSTRY

M. SANDERSON-WALKER
Field Quality Audit Manager
Birds Eye Wall's Ltd. (UK).

Summary

The consumer is a tough customer who demands value for money, and it is necessary to defend quality standards. Market Research has shown that taste and nutritional value are rated very highly. A prudent manufacturer has regard to consumer needs and ensures that packets are artworked so that an informed choice can be made. Products must also be designed within the Time Temperature Tolerance factors relevant to the distribution chain, whilst also taking account of constraints/opportunities presented by new sources of supply, packaging innovation, and novel processes. Some changes in distribution handling are considered as well as the latest trends in product development. It is concluded that major advances in the future will come from a mixture of scientific disciplines working together. There is also a need for discussions with retailers on the importance of good handling and temperature control. Manufacturers need to be involved in future debates upon nutrition, and the storage lives of products.

1. INTRODUCTION

"During 1982 the UK retail quick frozen food market reached £1182M, a growth of 15%. The growth owed little to price increases, and such rises in quick frozen foods were less than in other foods. It has been a remarkable feat of the industry to have expanded right through the present recession. (Sales to the catering industry were still affected by the recession and fell by nearly 4% to £228M).

The housewife with her shopping basket is a tough customer who demands and receives exceptional value for money. Whatever happened to the view that she was a poor defenceless creature in need of protection from price controls?

It is no secret that the intense competition in frozen foods, in the UK, is putting pressure on the margins of both manufacturers and retailers. In our quest to cut costs the industry has to be careful not to jeopardise product quality.

At Birds Eye we have been vigilant in the defence of our own high quality standards, for this is the rock on which our branded business is founded. We sometimes worry that not everyone in the industry is as concerned about quality as we are. It will be a pity if the high reputation of the industry were to be jeopardised by a willingness to cut corners on quality so as to get down to a price". - D. Angel (1).

2. CONSUMERS ATTITUDE TO FOOD ESPECIALLY QUICK FROZEN FOODS

A survey in February 1983 by Gallup on behalf of Birds Eye studied the consumer's attitude to food values (2). These showed overwhelmingly that the majority of people (over 80%) rate "taste" as very important when selecting food and, over 60% of them give "nutritional value" a similar importance. At the same time, "value for money" gets stronger ratings than "price" alone. - Table 1.

Table 1. Reasons for Selecting The Foods We Eat

	Very Important	Quite Important	Not very Important	Not at all Important	Don't Know
Taste	83%	14%	1%	0%	0%
Value for Money	67%	26%	5%	1%	1%
Nutritional Value	62%	27%	9%	2%	1%
Price	55%	33%	10%	2%	1%
Family influence/ preference.	52%	30%	10%	5%	3%
Convenience	24%	41%	29%	5%	0%

The same survey shows that frozen foods are given the best ratings for "taste", "nutritional value", and "value for money", when compared with the other main types of purchased foods. They are overwhelmingly preferred for their "freshness" and "appearance". - Table 2.

Table 2. Relative Merits of Preserved Foods

	Frozen	Canned	Dried	Don't Know
Taste	56%	30%	7%	7%
Appearance	62%	28%	3%	7%
Ease of Preparation	40%	52%	3%	5%
Nutritional Value	51%	20%	11%	18%
Value for Money	47%	29%	10%	13%
Freshness	71%	14%	5%	10%

There seems no reason to doubt, therefore, that it is genuine consumer choice which has kept up the growth rate of the quick frozen food industry.

3. THE COLD CHAIN

Having studied the needs of the consumer it is necessary to define the chain, producing and distributing products to the consumer. The distribution aspects have generally been called the "cold chain" but in recent years policy changes within the EEC have led to changes in the supply position for some raw materials in the UK. The most profound is that much cod is now obtained as a pre-formed frozen filleted block from other Member States (and Norway) whereas it used to be landed as "fish on ice" from trawlers. These blocks are then converted into products. There have been changes in the supply position of other materials needed for products. Their procurement is a specialist task, which requires the support of the food technologist.

Any consideration of the cold chain has also to consider the supply chain to the factory to ensure that the "time temperature tolerance" factors and the "product processing packaging" factors (TTT and PPP) have been taken into account at all stages.

This extended cold chain can be described as shown in Table 3.

Table 3. The Extended Cold Chain

STAGE	NEEDS
Frozen/non-frozen materials plus packaging.	Availability at the right time/ quality/cost.
Factory	Production at the right time/ quality/cost/quantity using correct processes.
Factory Cold Stores	To handle frozen bulk supplies and stock awaiting distribution.
Distribution	Easily identified and handled packs on pallets/in cases/units.
Retailers	Good products/right costs/ availability/good merchandising.
Consumers	Quality/value for money/convenience/ information.

The main factors which most affect quality are:-

Initial quality of raw materials/ingredients.
Processing conditions.
Packaging integrity.
Stock rotation.
Temperature control.

There must, therefore, be an interaction of different disciplines, including research, in order to develop a product and distribute it successfully to the consumer.

4. CURRENT TRENDS IN PRODUCT INNOVATION

New food ideas do not come easily. Often they are expensive; sometimes a stroke of creativity gives a break through. Some examples of current trends are shown in Table 4.

Table 4. Current New Product Trends.

PRODUCT	COMMENT
Ovenable Batters.	For coating fish steaks and fingers.
Crispy Crunch Crumb.	For fish steaks and fingers.
Steakhouse Grills.	A beef product designed to fill the gap between the burger and the butcher's steak.
Pastry sheets.	A simple concept that required new equipment to be developed.
Supermousse.	New Flavours and Toppings.
Superwhip.	A whipped creamy topping.
Arctic Gateaux.	A combination of ice cream/sauce/sponge.
Potato Waffles.	A waffle made from moulded potato and pre-fried.
Mini Waffles.	A development of the Waffle.

These represent a mixture of added value improvements and innovations. It would be understandable to believe that there is little more room for development but this is far from true. There is still room for more innovation.

Currently, there is a continuing interest in products which can be cooked in a microwave oven, and an increasing interest in nutritional values.

Each of these changes has and does represent a challenge in itself as research had developed new materials and processes which allowed products to be developed and marketed. At the same time, research workers were investigating the stability of frozen products when stored and defining such terms as Time Temperature Tolerance and Q10 which are relatively well understood today (3). However, because of the effects of the PPP factors it is difficult to define storage lives as can be seen by considerably different recommendations in the literature.

5. APPLICATION OF RESEARCH FINDINGS IN THE AREA OF STORAGE AND DISTRIBUTION

A prudent manufacturer has regard to the TTT factors and ensures that adequate storage tests at different temperatures are conducted on products from initial production (4,5).

Most storage tests are carried out at constant temperatures but in

1973 the effects of fluctuating temperatures on quick frozen food were published in a research paper by Hawkins, Pearson and Raynor (6). Again a prudent manufacturer will store products under fluctuating temperature conditions such as those found in a retail display cabinet, to ensure adequate stability and practical storage life. In addition, abuse tests will be carried out to assess the product's ability to withstand abnormal treatment.

Manufacturers need a computerised model of the distribution chain based upon known parameters, so that product design can be optimised. Software is available in some Companies which allows packaging designs to be optimised. Similarly a system is needed which takes the TTT factors and if possible the PPP factors into account. Such a system might be statistically based so that degrees of confidence or utility can be defined.

6. EXAMPLE OF A COLD CHAIN

The UK Association of Frozen Food Producers in its Code of Practice (7) defines a cold chain for the United Kingdom. This can be represented in simple form by Figure 1.

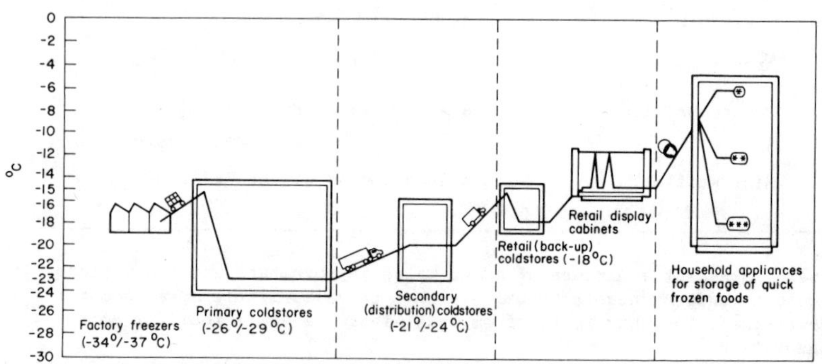

Fig.1 - Temperature profile of the cold chain

This embodies the principle of a more or less controlled rise in temperature from manufacture to consumer (4). Some small heat gain is inevitable when product transfers occur, for example between cold stores/ vehicles and vehicles/shops.

My Company audits temperatures of products through those parts of the cold chain under its control. The increasing complexity of UK distribution is leading to the development of Regional Distribution Centres rather than small local depots. These are designed to operate with air temperatures of -26°C rather than -21/-24°C for the depots shown in Figure 1.

7. CHANGES IN DISTRIBUTION HANDLING

The new cold stores contain narrow aisle racking for storing pallet loads. Selection is by use of turret trucks which are guided down the aisles. In addition, there are order packing/assembly areas, covered loading banks and port doors.

New types of trunker vehicles for bulk deliveries are being developed with such innovations as:-

Eutectic tubes beneath the roof.
Hydraulic drives to the refrigeration plant from the back axle.
Standby diesel generators beneath the body.
Electric plug-in capacity when stationary.

These items help to overcome the "noise" factor present in some vehicle refrigeration systems which can cause complaints on environmental grounds.

Radial distribution vehicles have also been developed which can carry pallets/trolleys/bulk packed orders to the stockists. Again, they have improved temperature holding capacity with the eutectic tubes beneath the roof being recharged when the vehicle engine is running.

Many wholesalers, and retailers who operate their own general traffic system, use insulated containers such as Thermotainers for the movement of stock from their cold stores to their stockists. Pre-cooling of the containers prior to filling is very important and temperatures can be maintained by placing a cooled eutectic plate or drikold above the packs.

8. RETAIL STORAGE AND DISPLAY

The International Standard for Retail Display Cabinets has been converted into a revised version of British Standard 3053 (1983) for UK use. The temperature requirements are very similar to the previous version of British Standard 3053 (1971).

The problem of the temperature performance of open top display cabinets is well known. A Random survey of top layer product temperatures in about 400 cabinets, some years ago, gave a slightly skewed normal distribution with a mean of about -15°C and a standard deviation of about 3.3. This was significantly warmer than the ISO/BSI test standards, and it is believed that the position is similar today.

Some ten years ago Hawkins, Pearson and Raynor (6) investigated the problems of infra-red radiation in shops and the effects at the load limit line of refrigerated display cabinets. They showed that reflective night blinds and specially designed reflective ceilings would result in colder product temperatures, together with better quality maintenance. There seems to have been only limited exploitation of their findings and only a few reflective night blinds are found in shops.

Cabinets with lids or doors can benefit from the use of Donnelly glass (8). This is a patented process which provides glass with an infra-red reflecting film laminated to it on the underside. Visible frequencies are transmitted but it allows energy to be saved if product temperatures are kept at the same level. Colder temperatures can be achieved for the same power settings. In addition, condensation is reduced and a year's experience has been gained so far with the material having good durability.

The writer agrees with the views expressed by Jul (9) "that the greatest risk of quality deterioration is in the retail cabinet. There appears to be little chance of altering the situation" - without a programme of information and advice which has a positive result. This is a major task! Therefore it becomes a matter of serious concern to make certain that the turnover in the retail stores is as rapid as possible.

9. QUALITY LOSS/AGE OF STOCK

It must be stressed that the quality of quick frozen foods is generally acceptable to the consumer. If a defective item occurs, it will be discovered eventually by the consumer; as this is the 100% sampling scheme which rarely misses and which is the ultimate measure of the quality of the product. Brand loyalty is strong and the consumer assumes that the manufacturer is likely to react to complaints. A sympathetic, positive and rapid response to a consumer's complaint often converts an antagonist into a protagonist for a Company and its products. An analysis of the complaints known to my Company which are caused by a failure of the control of temperature (or time) at any link in the cold chain have shown a dramatic fall over the last fifteen years (by about 70%).

It is found that stock leaving depots has a log mean age of 63 days (Sd (log) 0.23) and the age of packs when eaten has been estimated to have a log mean of 89 days (Sd (log) 0.22). It should be noted that the age at the time of eating includes storage at the retailer and at home. (Some caution has to be expressed as the figures include items packed only once a year into cartons, e.g. broccoli. In addition, many packs are multi-portion and are only partially consumed on any one date).

Thus about 50% of packs are less than two months old leaving depots, and almost 95% are less than five months old. Surprisingly, 50% are consumed within three months of packing and 95% within eight months.

A small sample of temperature/time abused stock has been examined and it had a log mean of 145 days (Sd (log) 0.24). Surprisingly, 91% of the samples were less than 1 year old with much of the older items being Vegetables rather than Fish/Meat/Cakes & Desserts.

Thus lower quality items which had apparently suffered serious quality loss characteristic of excessive time-temperature exposure, had time scales of the same order of magnitude as of normal items, even though the log mean is somewhat greater. The quality of the products was known to be satisfactory at the time of production, and failures occur as random events. Intrinsic quality and packaging is similar for all samples so a factor which is not common is needed. This is most likely to be the retail display cabinet/retail storage. The variability of storage temperatures at this stage has already been noted.

The results of the survey work being carried out by Storey and Spiess (10,11) will be of great interest in the evaluation of the age profile of stock in the retail cabinets.

It can be argued that the consumer may store products incorrectly, and this may be true in some instances. The vast majority of consumers handle our branded products carefully and our packs carry storage/handling instructions on them including "star marking". Freezers and frozen food storage compartments of household refrigerators are also marked in this way. The producer defines the time of recommended storage for the relevant "star marking".

It is of considerable concern to learn that automatic defrosting household freezers and fridge/freezers are being designed which may have an effective integrated average storage temperature warmer than currently defined by star marking.- See Figure 2. Although the simple arithmetic mean temperature may be adequate,the effect of the Q10 has to be considered as well as the peak temperature fluctuation. In such circumstances a prudent manufacturer may have no option but to reduce the recommended storage time in the star marking information panel on the pack.

10. CONSUMERS AND MANUFACTURERS NEW NEEDS FOR INFORMATION

The Gallup Poll results (2) indicated the consumers interest in nutrition. In 1979 we introduced nutritional labelling. We tried to make the information on packs as easy as possible to understand to help the consumers with their choice. We give nutrients in easily identifiable amounts of food such as 1 fish finger or 1 packet of fish in sauce.

We have not in general gone for nutrients per quantity by weight or portion size. My sixteen year old son and my eighty one year old mother seem to disagree about portion size.

FOOD FACTS

One cod steak provides at least one fifth of the protein recommended daily for most people.
To keep the total fat and calorie content down – simply grill.

Each cooked cod steak contains

	Grilled	Shallow fat fried	Deep fat fried
Protein	14 g	14 g	14 g
Fat	9 g	12 g	14 g
Carbohydrate	13 g	13 g	13 g
Calories	185	210	230
Calories per oz	51	64	84

TO STORE

'Star Marked Frozen Food Compartments		Food Freezer (****)	3 months
(***)	3 months	Ice Making Compartment	3 days
(**)	1 month	Cool Place	12 hours
(*)	1 week		

In case of dissatisfaction please return quality code panel to address below.

Quality Code
RS

Birds Eye Wall's Ltd., Walton-on-Thames, Surrey, England.
Produce of the United Kingdom.

Fig.2 - Example of information on packs

The important feature is that the information provided refers to the product after it has been cooked in the home. - See Figure 2. Frankly we do not think consumers will all beat a path to our door because of what we have done. We do seek to provide the means for people to choose better for themselves.

As manufacturers we need feed-back about our products, and we have progressively introduced a quality code panel on the packs should the consumer decide to express comments. - See Figure 2. The panel contains the factory production code impressed or printed during pack sealing. We are now including two alpha numeric symbols printed in, or near, the panel which are identical to our computer code identification of the product. Thus we have a simple way of identifying the product/pack size and production location/ date. This is already starting to provide feed-back information to management.

11. THE FOOD CONTROL SYSTEM

The early products which were developed and which were successful were largely "commodity" items. The processes were labour intensive and as stated earlier the packaging was generally of the carton type. Today most processes are much more mechanised and the whole system interlocks with each step in the process being dependent upon its predecessor. Thus, raw material handling/processing/dispensing/packing/freezing/palletising, can be in one continuous production line. A breakdown or failure in any step of the operation can cause the whole operation to stop unless adequate "buffering" is possible.

New products have, when fully developed, to fit into this mechanised environment so that there are many disciplines needed to achieve an acceptable operation. The "food control" system has to be designed to prevent or highlight changes which can occur over a long period. Each small increment or change may be insignificant in itself but overall the process, and product, can change dramatically.

Even though a process has been well designed there is often a need for applied research effort to optimise it or create systems which can keep it in control. Again, changes in the raw material supply position can cause research effort to be applied to give techniques of measurement, and handling, to allow such items to be used.

12. CONCLUSIONS

The consumer is a tough customer who demands and receives value for money and it is necessary to defend quality standards. Market research has shown that customers rate taste and nutritional value very highly and that value for money gets stronger ratings than price alone. A prudent manufacturer will ensure that he has regard to the PPP and TTT factors, to ensure that the products reach the consumer at the right quality.

In a recently published Paper, Strachan (12) when discussing the optimised quality in frozen foods, stated -:

"An anonymous cynic (or was he a sage?) has described development as the work you know you want and research is that which you don't yet realise you want. It was his personal feeling that significant advances will not come from traditional food technology. That has an important part to play in implementing current practice and describing the symptons of new problems. He thought that we should be looking to less obvious disciplines to guide us through the step of new insight and understanding which can set our applied efforts on a fresh road.

The example quoted of radiant heat absorption amplifies the insight provided by a team of basic physicists who chose to turn their attention to the foods area. This was not meant to discourage those of us already associated with food technology. Rather, it should challenge all of us in universities, institutes and industrial research to widen our interpretation of the boundaries of our traditional disciplines. We need the contribution of basic physicists, chemists, rheologists and physiologists to quantify our materials and their response to extreme physical conditions. He hoped that they will recognise that the food industry is a fertile and wide open field in which to seek their new frontiers".

The writer of this Paper endorses the above conclusions and in addition recommends that :-

1. Manufacturers must ensure that they are able to engage with confidence in the debate on nutrition which is likely to develop increasingly in the next few years, and we believe we are well placed to do so.

2. Any redefinition of the optimum time and temperature profile of products through the distribution chain, must have regard to the technical, commercial and practical constraints which are involved.

3. Should more stringent regulations be considered for frozen foods they should not be applied without establishing that the very high additional capital and energy costs are justified (9).

Much of the discussion has mentioned the need for the active support of retailers/stockists in maintaining product quality. There is, therefore, a need for discussions with retailers to improve handling and temperature control. The efforts of Research, Development, Manufacturing, and Distribution can be nullified by inadequate retail storage.

As a manufacturer we intend to play our part by providing information and advice to retailers and enforcement officers.

REFERENCES

1. D. ANGEL. The Quest for Values. Birds Eye Wall's Ltd. (1983).
2. SOCIAL SURVEYS (GALLUP POLL). February (1983) on behalf of Birds Eye Wall's Ltd.
3. Conference on Frozen Food Quality, Albany California. November (1960).
4. M. SANDERSON-WALKER. Time-temperature monitoring and quality inspection for Quick Frozen Foods - Design factors, IIR Budapest (1978) and International Journal of Refrigeration March (1979).
5. M. SANDERSON-WALKER. Time-temperature monitoring and quality inspection for Quick Frozen Foods - Practical considerations IIR Budapest (1978) and International Journal of Refrigeration March (1979).
6. HAWKINS PEARSON and RAYNOR. The Advantages of Low Excessivity Materials to Products in Commercial Refrigerated Open Display Cabinets - Institute of Refrigeration - (UK), March (1973).
7. UK ASSOCIATION OF FROZEN FOOD PRODUCERS. Code of Recommended Practice for the Handling of Quick Frozen Foods - (1978).
8. Donnelly Glass - Information from MR.M.S. WARE, 35 Birchwood Grove, Hampton, Middlesex, TW12 3DU.
9. These phrases appeared in a paper by M. JUL entitled The Intricacies of the Freezer Chain. IIR Hamilton (1982) which also appeared in the International Journal of Refrigeration vol.5 No.4 July 1982 pp 226-230, published by Butterworth Scientific Ltd., Guildford, Surrey, UK.
10. R.M. STOREY. Some observations on temperature and quality in the UK QFF chain - COST 91 Athens (1983).
11. WEL SPIESS. Residence time behaviour of ten selected products in the Federal Republic of Germany - COST 91 Athens (1983).
12. P.W. STRACHAN. Optimised Quality in Frozen Foods - Institute of Food Science & Technology (UK) proceedings June (1983).

THE CONSUMERS' VIEW OF FROZEN FOODS

J A SALMON
The KMS Partnership Ltd.

Summary

There is often no relationship between what people think about food and what they eat. But attitudes probably are important and the food producer should be aware of them. In general, all preserved foods, including frozen, are a long way behind 'fresh' in the consumer's mind when she thinks about nutritional value, taste, appearance and value for money. But frozen foods do better than canned or dried. Already chilled foods are becoming more popular in many European countries and are seen as almost identical to 'fresh'. If frozen food producers believe that their products are 'misunderstood' and that this is harming sales, new methods of conveying the truth will need to be found.

1. WHO KNOWS WHAT PEOPLE THINK?

Much is heard from Consumer Groups in Europe about the consumer's need, or want, for more information about the food she buys. These Groups do their best to represent consumers, but they do not always have access to up-to-date research studies which reflect the spread of opinion. This paper is based on consumer research carried out, mainly in UK, between 1978 and 1983. Therefore, it records what people think about certain issues, what they believe. That is not necessarily any indication of what they buy or do. Some food producers take the view that it is only consumers' actions that matter, what they think is irrelevant. The debate about the relationship between attitudes and behaviour is an endless one, but it is difficult to understand how any food producer could be happy with a situation in which many of his potential, or actual, customers have negative views about his goods, especially when those views are not supported by the facts.

2. FROZEN FOOD OR JUST FOOD?

Consumer attitudes to frozen foods need to be approached from two points of view. Firstly, what do people think of the actual process of freezing. Secondly, what do they think about food in general and how well do they think frozen foods now available measure up to their expectations?

3. HOW MUCH INFORMATION?

There is little doubt that more people, women in particular, are becoming more aware of what they eat. They are asking magazines and food producers for nutrition and ingredient information. Judging by the demands of some Consumer Groups, it seems they believe that the more information the consumer is given, the happier she will be. That simply is not true. When various methods of presenting nutrition information were tested in 1978 (1) the very comprehensive American style label received a definite 'thumbs down'. An overly simplified pictorial representation was not liked either. The majority of consumers who wanted nutrition information liked and understood a small amount of numerical nutrition data. Clearly a great deal of thought needs to be given to the amount of information that is given and the way it is presented.

It may be true that consumers should have some information they don't ask for, but if a manufacturer does give it, he will need to spend a lot of money, time and effort explaining why the information is necessary, how it will benefit the user, what it means and how it should be used.

It is important to realise that consumers don't always make systematic comparisons between frozen, canned and dried foods. As more sophisticated frozen foods are developed, each with many ingredients, there are no obvious comparisons with canned or fresh to make. People tend to make value judgements by assessing how well a food fits a whole series of parameters including taste, appearance, portion size, value for money, suitability for their life style and meal occasion, and sometimes nutritional value too.

What do people think about frozen foods? in relation to their freshness, taste and value for money? Mike Sanderson Walker's paper showed that, by these parameters, frozen foods outstrip canned and dried. That seems to say that people have a positive view about frozen foods, and sales in many countries are healthy enough to support this. But what about that rather important group of foods known as 'fresh' or 'market bought'? Many market researchers say if you do include them in questions of this kind, they so overshadow frozen, canned and dried that you can't get any meaningful views about preserved foods. The fact is that when you do ask how people rate fresh, frozen, canned and dried foods for taste, appearance etc. (2) 95% say fresh is best for taste, 92% say fresh is best for nutritional value, 84% say its best for appearance and almost three quarters think fresh foods give the best value for money.

Even on the matter of ease of preparation, only about half the women interviewed thought frozen foods the easiest to prepare. One in three said fresh foods were most convenient and fewer thought canned foods most convenient.

The message here is clear. Frozen food producers shouldn't rely on frozen foods being the obvious choice because people want convenience as well as good quality. Already fresh foods rate well on convenience and it only needs a few well orchestrated compaigns to hammer the point home for preserved foods to feel the effect.

NUTRITION. Practically every nutritionist and food scientist knows that freezing itself has no effect on any nutrient. And as long as pre - and post - freezing storage and heating are carried out according to good practice, by the time foods are ready to be eaten, frozen food is, by and large, as good as the so-called fresh equivalent.

You know. I know. But what about all the other people who buy food? Many are not convinced, consumers don't put frozen foods anywhere near fresh. They are adament that a food which has been 'processed' can't be as good for you as untouched fresh food. (3) That view is endorsed by much that is written in the popular press. Ask people to explain what is different and they can't be explicit. They offer a statement such as "I don't know what they (the manufacturers) do, but something must happen to nutrients when they process the food". (3) A particularly vivid illustration of this attitude came in some research carried out in 1978 (4) 97% of housewives thought fish fillet from the fishmonger was very good for them. Only 37% thought the same about frozen fish fillets. In fact the difference in nutritional value is probably not measureable.

Ever since commercial food freezing began, those with a vested interest have been trying to persuade people that freezing doesn't have any significant effect on nutrients. One is forced to say that those efforts have not been very successful. The question is "Does it matter?" If you believe it does, then clearly new ways to convey the truth must be found.

Now that many of the frozen foods in the supermarkets are complex recipe products, people are beginning to ask 'How much meat is there in a beefburger?' 'How much fish in the sole provencal?' We are getting to the stage where the overall nutritional value of many of the frozen foods in the cabinet is much more dependent on the proportions of various ingredients than on the effects of blanching, freezing, distribution and storage. It is however, that end-cooking in the home, that can have a profound effect on some vitamins!

Food quality/freshness There's little doubt that, in recent years, the range of food offered to customers has improved, and many in the food industry would say its quality is better too. In judging food quality many customers rate food freshness as the most important component. They put it above price and acknowledge ingredients. Shoppers like to be able to assess the quality of foods before they buy. When they're shopping for fresh carcase meat, fruits, vegetables and cheeses they can see the product, very often feel it and sometimes smell it. These may not be very accurate measures of eating quality, but they are certainly better than nothing. But even these are not available to people buying frozen foods. They can't feel or smell the product and even if they can see it they are no wiser. Buying frozen foods has been described as "a bit of a lottery", people are dependent on the manufacturer, his reputation and the information he supplies.

It's not surprising then that, for many people, the best means of judging the feshness of frozen, canned or dried food is to know the age of the product. In pre-freezing days age was a rough indication of freshness. For frozen foods, age alone is not. But the message has not been understood by many people. They simply can't understand why temperature is so important. Even less can they understand that, if it is important, why it can't be controlled. Then, surely, they can have some sort of date on the packet.

When asked about packaged food in general, not just frozen, 42% of people said they wanted some indication of the age of the product (5). They called it 'eat-by date', 'sell-by' or 'best before date'. To put this into context a quarter said ingredient listing was essential to them and a mere 4% mentioned spontaneously a need for nutrition information.

Star Marking In an attempt to integrate time and temperature at the household end of the distribution chain, 20 years ago the frozen food industry and refrigerator manufacturers developed star marking. A group of home economics teachers who, of all people, ought to know the generally recommended storage times at different temperatures, were asked questions to test their knowledge (6) 66% did not get it right. Perhaps less surprising was that the same number didn't know acceptable product temperatures in a retail display cabinet.

The frozen food industry in many countries has run educational campaigns to tell people about the production, storage and handling of frozen foods. In some countries these efforts have now virtually stopped, partly as an economy measure, partly because companies take the view "we've done the education bit". They forget that new customers come along every day. We have a new generation who haven't received the kind of education the frozen food industry lavished on their parents.

Perhaps it doesn't matter to the industry that freezing technology isn't understood too well, that people have the wrong ideas about the nutritional quality of the food. If it does, some thought must be given to action programmes to put matters right.

There are lobbies in different countries for nutrition information on packs, for open date marking and more comprehensive ingredients declarations. Frozen food producers may not like them. Sometimes there are good reasons why requests are just not practicable. But there are only two options open to the industry. Either it bows to the pressure or it convinces the lobbyists and consumers that their demands are not workable. The one thing the industry cannot do is ignore these people. They won't go away!

Food in general So much for frozen foods as such. Frozen food manufacturers need to be aware of consumer attitudes to food in general. After all, freezing is really only a convenient way of distributing food in good condition. There aren't too many products that are intended to be eaten frozen.

Food Additives Information in the press on radio and television in almost every western country would leave many people believing that all emulsifiers, flavourings, colourings and other additives are a sure route to death, or at least serious ill health. Since most of the information people have about food comes from the mass media, one might expect these views to be played back through market research, with a high proportion of people expressing concern about additives. In 1982 people were given a long list of food ingredients and asked whether they were concerned about any of them. (7) Half said they didn't have a view one way or the other. A quarter expressed concern about additives in general. Within that group preservatives and colouring caused most anxiety. Sugar levels in food bothered 10% and salt only 30%

This year we asked a different question. Would people be prepared to spend a few pennies more to get food without additives. Half said they would, a quarter would not. And almost half said they would be prepared to shop a bit more often if they had to, to get food without additives. (8)

Now that more medical research is being carried out into food additives and allergies, it is likely that more will be written about them in the popular as well as medical press. For years the food industry has been trying to convince customers that food additives are thoroughly tested, and are safer than some natural foods. People are still not persuaded.

The truth is one thing. Quite another, and more likely to affect sales, is peoples perception of what additives will do. Frozen food producers are at least in a better position than others to reduce their usage of additives if they think it would be a good move. But it won't do the company any good at all unless people are made aware of the action.

There is little hard data yet on the in-roads chilled foods are making, but my own company is currently conducting perception analysis of chilled and fresh food compared with frozen, canned and dried. The current view is that the methods of extending the shelf life of chilled foods means they can be kept at home for at least one week. And this is all most people want. If these foods are indeed seen as being fresher, more natural than frozen, it is likely that their sales will grow dramatically over the next few years. Many retailers agree. They are increasing their chill capacity. The old argument that frozen foods can be kept for 3 months is not as appealing as it used to be. People are changing their huge freezers for smaller capacity fridge/freezers (9). They are buying medium (2 lb) packs of food not large 5 & 10 lb sizes. They actually like shopping once a week.

Health in general, Slimming in particular? The USA has seen a massive expansion in the market for frozen low calorie meals during the past 2 years. They have been successful for two reasons. People want to be slim but they also want to eat good food while they get thin. Up until the Stouffers Lean Cuisine range, and similar products from other manufacturers, the foods designed for slimmers were pretty dreadful to taste.

Next year the UK market will probably see three or four companies launching frozen meals for slimmers. As long as the manufacturers realise that the UK consumer is not the same as her American cousin, they may be successful. If they are not it will be because the name of the product, or its formulation or market positioning is wrong, not because there is no demand. Mind you the UK consumer is cynical about 'food for slimmers'. It is not unusual to hear them described as"a rip-off". Foods which taste very good and look like meat/fish/vegetables are less likely to receive this kind of reaction than biscuits etc. But even now we are beginning to see in UK a move away from slimming as the only goal, towards a broader appreciation of health. Slimming is beginning to look like a negative approach involving self denial. The magazines (we have 5 major magazines devoted to slimming) and consumers are moving towards a positive approach to better health, which includes a better total diet, more exercise and less stress. This year we asked people whether they cared about the kind of food they ate (10) 42% admitted they ate what they liked, but 28% said they were concerned about their food because of their health. Only 4% said they were concerned because of their appearance. When it comes to consumers implementing dietary change - eating more filling, less fatty food, more whole-meal cereals, a majority of people interviewed last year said they were trying to change. So far we have seen only small changes in the amounts of these foods actually eaten. All the figures used here are based on what people said about their beliefs/attitudes. They are a guide to what people might want and might buy. But they are by no means infallible. The one thing about people's buying and eating habits that doesn't change is their unpredictability. At the end of the day, a manufacturer has no choice but to market a food and see what happens. He has to remember that the people who buy his foods are not highly trained taste pannelists, and they may not always agree with him about what good quality is. In selling anything, the potential buyer is always right. If a manufacturer doesn't accept that he'll go out of business - albeit fighting, as he goes down.

But the frozen food industry is better placed than most to take advantage of the new markets created by the consumer's growing awareness of and interest in her food.

REFERENCES

1. Nutrition Information on Packs 1974. Food & Drink ResearchLtd.
2. Food Values Study 1982. Gallup.
3. A Survey of Attitudes to Foods ' Health part 1 1978. Kraft Foods.
4. Food & Cooking Opinions Survey 1978. RBL.
5. Food & Cooking Opinions Survey 1978. RBL.
6. Jenny Salmon - Unpublished data.
7. Food Values Study 1982. Gallup.
8. Food Survey 1983. MAS Survey Research Ltd.
9. Birds Eye Wall's Annual Review.
10. Food Survey 1983. MAS Survey Research Ltd.

SHELF LIFE OF POULTRY PARTS IN DEPENDENCE OF TIME OF PREPARATION

M. RISTIĆ
Federal Meat Research Centre, Institute for Meat Production and Marketing, Kulmbach, Fed. Rep. of Germany

Summary

1600 Broilers of Lohmann breed were slaughtered at the age of six weeks. One half of the carcasses was cooled in water, the other one in air. After 1 - 8 days of storage the carcasses were cut and frozen at -15°C and -21°C up to 20 months. Every 2 months the breast and leg muscles were studied with respect to: Freezer burn, drip loss in percent, peroxide number, total microbial count, taste, total impression, grill loss in percent and objective tenderness. The chilling with water resulted in more favourable values in the sensoric evaluation in the leg muscles. Leg muscles could be stored for 12 months. Water chilling and freezing at -15°C exhibited a better taste in the breast muscles. Cold storage at 3 and 6 days showed the best taste in the unfrozen material. Breast muscles could be stored for 16 months.

1. INTRODUCTION

In dependency of the condition at offer the freshly slaughtered broiler will be either water or air chilled. A center temperature between -1°C to +4°C will be reached by this chilling. The best storage duration of the fresh broiler carcasses is one week (RISTIĆ and HECHELMANN, 1975; RISTIĆ, 1976). A maturity takes also place with poultry meat, however it goes faster than at other species and it takes 2 to 3 hours (LEE BANG et al., 1978; HEINZ, 1976; RISTIĆ et al., 1980; WHITING et al., 1975). Aim of this investigation was to find out the storage capability of produced poultry parts from freshly stored broiler after finished storage period of 1 to 8 days at various chilling; storage duration was 20 months at the most.

2. MATERIAL, TREATMENT, METHOD AND CHARACTERISTICS

The material consisted of 1600 broiler (= 3200 parts) of the Lohmann origin, slaughtered after 6 weeks of fattening at floor-keeping. Carcasses had been dressed ready-to-grill and belonged to grading class A with a weight of 1000 grams. After finished water chilling (30 min at +10°C and +5°C resp.) and air chilling (3 hours at -30°C) as well as additional storage of the fresh broiler (1 - 8 days) the dissection was carried out. After freezing of the primal cuts - breast and leg - they were stored at -15°C and -21°C resp. up to 20 months for further investigations. The material was taken out of storage at intervals of 2 months. Recorded were the following characteristics to the different periods of storage.

The test was planned four-factorial-orthogonal. First the data were calculated seperately with a three-factorial analysis of variance for each chilling procedure, afterwards for the total material with the Least-Square-Analysis according R. HARVEY as four-factorial, whereby the significance was checked by means of the F and t-test.

characteristics	time of measurement months	repeating of each sample	procedure resp.appartus
freezer burn	every 2 months	each	visual
drip in %	"	"	difference of weight
peroxide number	"	2 times	DFG-method according D.H.WHEELER
total count	"	"	surface inoculation procedure
sensory	"	6 times	sematic-numeric scale from 6 to 1
grill loss in %	"	1 time	difference of weight
objective tenderness	"	2 times	Instron (model 1140)

3. RESULTS

3.1 Leg muscles

Generally the leg muscles showed a shorter storage capability than the breast muscles. This was caused by faster drying-out, loss of flavor and rancidity of the leg muscles. During the course of the storage period the scores for flavor and overall impression of the leg muscles declined for the water chilled as well as for the air chilled leg muscles (table 1). The storage temperature did not influence at all. Air chilling led to lower scores for flavor. The 6-days storage period of the fresh broilers revealed the best evaluation of the leg muscles. Similar changes occured while scoring overall impression. Smaller loss of grilling was found at storage temperature of -15°C, shorter storage period of the fresh broiler as well as air chilling of the leg muscles. The energy, as a measure for the objective tenderness, was influenced by all four factors of influence. A better tenderness of leg muscles as well as of parts was reached with increasing time of storage. The lower temperature of storage at -21°C led also to a better tenderness.

3.2 Breast muscles

Breast muscles showed a significantly higher freezer burn at higher storage temperature of -15°C and air chilling (table 2). The freezer burn decreased for one score after 16 months storage time at air chilling and after 18 months storage time of water chilling. The weight loss after thawing (drip in %) recorded for each pack (2 breasts and 2 legs) was in average lower at air chilling (1,07 : 1,48). The storage temperature of -15°C showed better drip values. With increasing period of storage of the fresh broiler as well as of the parts an increase of the peroxide number was found, determined in the abdominal fat. The peroxide numbers increased considerably after 10 months of storage. Deeper storage temperatures of -21°C reached an approximately 100 % lower peroxide number (F = 660,62***). Air chilled broiler showed slightly better peroxide numbers. The total count was strongly influenced by the effects of chilling, and period of storage of the fresh broiler. The one day storage as well as the air chilling led to lower results. The total count decreased during the course of storage.

The changes of the flavor scores were dependent on all four causes of variance. Water chilling as well as higher storage temperature revealed better evaluation scores. Evaluation scores decreased with increasing storage periode. Fresh material of 3 days, resp. 6 days of storage was scored the best for flavor. Overall impression was scored one score lower after 20 months at water chilling, and after 14 months at air chilling. Other pattern of influence was identical in its changes to those as with the flavor. The grill loss of the breast muscles decreased during storage, they were better in the area of temperature of -15°C. Air chilling led to lower grill loss. The air chilled samples of the breast showed lower energy values, i. e. better tenderness. Up to 10 months of storage the energy values were higher than after 10 months, and the tenderness was positively influenced up from 12 months onwards.

REFERENCES

HEINZ, G., 1976: Warmzerlegung von Fleisch. Die Fleischwirtschaft 56, 1713. - RISTIĆ, M. und H. HECHELMANN, 1975: Einflüsse der Brühtemperatur und des Kühlverfahrens auf den Keimgehalt frischer Masthähnchen. Jahresbericht der Bundesanstalt für Fleischforschung, C 15. - RISTIĆ, M., 1976: Einfluß des Schlachtvorganges auf die Lagerfähigkeit frischer Masthähnchen. Archiv f.Geflügelk.40, 52-55. - WHITING, R.C. und J.F. RICHARD, 1975: Thaw rigor induced isometric thension and shortening in broiler-type chicken muscles. J.of Food Sci.40, 960. - YU BANG LEE and D.A.RICKANSRUD,1978: Effect of temperature on shortening in chicken muscle. J.of Food Sci.43, 1613.

TABLE 1: Mean values of sensory scores, grill loss as well as objective tenderness - leg muscles -

Cause of variance		n	flavor wc	flavor ac*)	overall impression wc	overall impression ac	grill loss % wc	grill loss % ac	energy wc	energy ac
Storage duration (A)	1	160	4,4	3,9	4,6	4,3	26,1	24,3	91,8	62,4
	2	160	4,0	3,9	4,4	4,2	25,6	25,3	55,1	53,3
	3	160	4,0	3,4	4,2	3,9	26,7	24,9	61,7	51,0
	4	160	3,3	3,6	3,7	3,8	26,9	24,7	45,7	49,7
	5	160	3,2	3,0	3,6	3,3	28,4	24,7	46,4	46,6
	6	160	3,4	2,8	3,8	3,2	26,8	25,8	46,1	47,9
Storage temperature (B)	1	480	3,7	3,4	4,0	3,8	26,3	24,2	59,5	54,1
	2	480	3,7	3,4	4,0	3,8	27,2	25,6	56,1	49,5
St.duration of fresh broiler (C)	1	240	3,7	3,3	4,0	3,7	26,2	24,0	64,3	55,9
	2	240	3,7	3,4	4,0	3,8	27,4	23,9	65,9	53,4
	3	240	3,9	3,6	4,2	3,9	26,9	25,3	51,3	48,3
	4	240	3,6	3,4	3,9	3,8	26,6	26,6	49,6	49,6
Total		960	3,7	3,4	4,0	3,8	26,8	25,0	57,8	51,8

Cause of variance	DF	F - value flavor	F - value overall impression	F - value grill loss %	F - value energy
Chilling media	1	134,91***	120,05***	68,73***	37,25***
St. of fresh broiler	3	18,34***	18,71***	8,10***	35,72***
St. temperature	1	-	2,84	25,23***	16,70***
St. duration of parts	5	225,54***	237,78***	3,66**	93,53***
Rest	949				

*) Chilling media: wc = water chilling, ac = air chilling

(A) 1 = 2 months, 2 = 4 mo., 3 = 6 mo., 4 = 8 mo., 5 = 10 mo., 6 = 12 mo., 7 = 14 mo., 8 = 16 mo. 9 = 18 mo., 10 = 20 mo.

(B) 1 = -15°C, 2 = -21°C

(C) 1 = 1 day, 2 = 3 days, 3 = 6 days, 4 = 8 days

TABLE 2: Mean values of the freezer burn, drip, peroxide number as well as total count

Cause of variance		n	freezer burn		drip in %		peroxide number		total count	
			wc	ac	wc	ac	wc	ac	wc	ac
Storage duration	1	160	3,98	3,96	1,98	1,05	3,12	2,99	6,02	5,81
	2	160	4,00	3,96	1,22	0,86	3,57	3,13	6,06	5,48
	3	160	3,99	3,95	1,13	0,70	6,34	4,57	6,19	5,52
	4	160	3,96	3,99	0,94	0,57	5,48	5,34	5,73	5,30
	5	160	3,90	3,91	0,90	1,30	13,02	9,95	5,73	5,23
	6	160	3,80	3,53	1,53	1,44	10,12	11,73	5,85	4,86
	7	160	3,74	3,41	1,64	1,56	24,28	17,89	5,18	5,02
	8	160	3,30	2,55	2,13	1,24	16,83	15,72	5,24	5,04
	9	160	2,46	2,26	1,47	1,06	15,38	15,71	4,60	4,96
	10	160	2,70	2,06	1,84	1,21	15,35	12,94	5,33	4,79
Storage temperature	1	800	3,34	3,00	1,37	1,01	14,93	14,00	5,42	5,27
	2	800	3,83	3,72	1,59	1,13	7,77	5,99	5,77	5,13
St.duration of fresh broiler	1	400	3,75	3,52	1,52	1,02	7,65	7,78	5,15	4,76
	2	400	3,46	3,41	1,57	1,11	11,33	8,43		
	3	400	3,57	3,37	1,50	1,11	13,21	11,49		
	4	400	3,56	3,14	1,32	1,05	13,20	12,29	6,04	5,64
Total		1600	3,58	3,36	1,48	1,07	11,35	10,00	5,59	5,20

Cause of variance	DF	F - value			
Chilling media	1	125,03***	65,94***	20,93***	90,32***
St. of fresh broiler	3	35,01***	1,78	62,87***	459,11***
St. temperature	1	933,12***	11,43***	660,62***	5,84*
St. duration of parts	9	439,70***	15,59***	174,83***	33,62***
Rest	1585				

TABLE 3: Mean values of sensory scores of the grill loss as well as tenderness - breast muscles -

Cause of variance		n	flavor		overall impression		grill loss %		energy	
			wc	ac	wc	ac	wc	ac	wc	ac
Storage duration	1	160	4,2	3,9	4,2	3,9	25,2	24,4	117,5	105,0
	2	160	3,6	3,7	3,7	3,8	27,1	24,5	84,4	80,7
	3	160	4,0	3,9	3,9	4,0	25,0	23,5	100,5	68,2
	4	160	3,8	3,9	3,8	3,9	25,8	23,2	101,4	71,5
	5	160	3,7	3,6	3,8	3,7	25,3	22,6	90,4	73,4
	6	160	3,7	3,6	3,8	3,7	24,8	22,1	78,1	65,6
	7	160	3,0	3,7	3,2	2,8	25,0	23,5	70,3	67,7
	8	160	3,7	3,6	3,7	3,8	23,8	20,6	81,5	70,7
	9	160	3,8	3,7	3,9	3,8	22,7	20,0	74,1	72,5
	10	160	3,0	2,7	3,2	2,9	24,4	22,7	89,5	74,6
Storage temperature	1	800	3,7	3,6	3,8	3,7	24,8	22,4	86,5	69,2
	2	800	3,6	3,5	3,7	3,6	25,1	23,1	91,1	80,8
St.duration of fresh broiler	1	400	3,6	3,5	3,7	3,6	25,0	22,4	91,0	74,5
	2	400	3,6	3,7	3,6	3,7	25,7	22,9	90,7	76,9
	3	400	3,8	3,6	3,9	3,7	24,6	22,8	85,7	71,8
	4	400	3,6	3,4	3,7	3,4	24,4	22,8	87,8	76,8
Total		1600	3,7	3,5	3,7	3,6	24,9	22,7	88,8	75,0

Cause of variance	DF	F - value			
Chilling media	1	47,18***	29,43***	210,17***	201,33***
St.of fresh broiler	3	40,92***	37,91***	5,02**	5,10**
St. temperature	1	31,29***	47,08***	10,80**	69,95***
St. duration of parts	9	225,85***	182,67***	27,84***	59,71***
Rest	1585				

PRE-TREATMENT OF MECHANICALLY RECOVERED MEAT FOR FREEZING AND STABILITY OF MECHANICALLY RECOVERED MEAT DURING FROZEN STORAGE

F. JIMENEZ-COLMENERO, E. GARCIA-MATAMOROS, and M.C. PELAEZ
Instituto del Frío, Ciudad Universitaria, Madrid-3, Spain

Summary

The preservation of mechanically recovered meat (MRM) in cold storage is basically limited by microbial development and the oxidation of lipids. Various methods have been tried to reduce these two processes. A vanguard technological method that can help solve these two problems consists of washing the MRM. The characteristics of the product after washing (composition, microbiological quality, and physico-chemical properties) and the effect of freezing and frozen storage on the microbiological aspect and on the stability of washed MRM have been determined with a view to evaluating the possible advantages of this method. Washing affords advantages, such as enhanced microbiological quality, deflavouring, bleaching, and the elimination of fats and other causes of the alteration of fats, inasmuch as losses in the composition and in the physical and chemical properties of the product can be compensated for. During cold storage, the action of washing generally causes a more marked decrease in the number of microorganisms, and this effect is stronger in the case of sulphite-reducing anaerobes and pathogenic staphylococci. The MRM exhibits a lower emulsifying capacity than the control sample, a condition maintained throughout storage. Oxidative rancidity shows similar results.

1. INTRODUCTION

Mechanically recovered meat (MRM) offers advantages both from an economic standpoint (production of MRM from cattle, sheep, pigs, and poultry in 1979 amounted to more than 63 000 t, with a value approaching 1 500 million pesetas) and the nutritional standpoint. However, due to its composition, physical structure (high surface/volume ratio) and high pH value, MRM is an ideal medium for the proliferation of microorganisms. This considerably restricts its frozen storage life and can create problems in the use of this meat for human consumption.

Due to its low stability, MRM must either be rapidly chilled (less than 4 ºC) and used within 48 hours after production or frozen if longer storage periods are required. In an attempt to improve the microbiological quality of MRM, a number of methods have been tried, among which are thermal treatment of the raw material, the washing of poultry carcasses in antiseptic solutions (1), holding MRM at high temperatures for short periods (pasteurization) (2), and the addition of lactic-acid bacteria (3), etc.

The washing of MRM is a technical process which addresses these problems. It acts to enhance the stability of the product by bringing about a reduction in the initial level of microorganisms present in the meat (still further reduced if antiseptics are added to the wash water), in the amount of enzymes (both intrinsic and bacterial in origin) present, as well as in the concentration of hemoproteins. Furthermore, washing also results in the bleaching and deflavouring of MRM, which means that it can incorporated more readily and in greater quantities into those products into which it is finally processed, and/or that it can be employed in other products in which the sensory properties of MRM would otherwise restrict its use.

The objective of the present experiment was to assess the effect of washing on MRM by analyzing the properties of the final product and relating them to the processing conditions it had undergone, as well as to investigate the effect of freezing and frozen storage on the microbiological characteristics and on the stability of washed MRM.

2. MATERIALS AND METHODS

Following conventional manual deboning, mechanical deboning took place using the backbones of pigs kept for 72 hours at 0 ºC. The MRM was prepared in a Protecon MRS-40P deboner at a pressure of 100 atm. The increase in the temperature of the meat during mechanical separation was 7 ºC. The MRM was then kept at 0 ºC until use.

In order to investigate the effect of washing conditions on the characteristics of the MRM, the meat was washed by mixing it with ice water, with occasional stirring. The MRM/water ratios tested were 1/2 and 1/5, with washing times of 0 and 15 minutes. Dewatering was effected by placing the aqueous MRM suspension in a cloth bag made of a coarse muslin-like fabric and centrifuging in a centrifugal dewaterer until the continuous flow of liquid draining off stopped. The characteristics of the meat after washing (composition, microbiological population, and physicochemical properties) were studied as described by Jiménez-Colmenero and García-Matamoros (4).

The following experimental procedure was applied in order to determine the effects of freezing and frozen storage on the stability and microbiological characteristics of MRM subjected to a washing process. The MRM was divided into two lots, a control lot and another which underwent washing. The washing process was effected by mixing 4 kg of MRM with 15 l of ice water for 20 min with occasional stirring. The wash water was removed by placing the aqueous suspension in a cloth bag and pressing the contents until the liquid draining off stopped flowing easily. The MRM yield under the experimental conditions came to 88 %. Both the washed MRM and the control lot were placed on trays holding about 500 g each and then frozen in a blast freezer (air flow: 5 m/s) at -20 ºC. The frozen MRM was vacuum-packed and stored at -12 ºC until the end of the experiment, which lasted three months. The composition, bacteriological analysis, and oxidative rancidity were as described by Peláez et al. (5).

Except in the case of chemical composition, the analyses described were performed on fresh MRM and on frozen MRM immediately following freezing and after three months in frozen storage at -12 ºC. The degree of significance of the results was determined by applying analysis of variance using an F test.

3. RESULTS AND DISCUSSION

3.1 Effects of washing

The MRM yield after washing together with the results of the analyses of the moisture, total nitrogen, fat, and ash contents, as well as the loss of dry matter by substance, are presented in Table I. MRM loss during washing increases with the MRM/water ratio and decreases with washing time. With regard to the MRM/water ratio, these results may be due to the higher solubilization of certain substances (ash and total nitrogen) and to fat loss; the results in relation to the washing time may be due to protein insolubilization during the washing process, as shown in Table II. Washing removes a large part of the protein fraction, which is replaced by water, giving rise to a higher moisture content in washed MRM than in its unwashed counterpart. Washing of MRM results in losses of some 50 % in total nitrogen x 6.25 and of from 44 to 72 % in the ash content. Fat loss is much lower.

TABLE I. Proximate analysis of MRM

Analysis	Lot			
	Unwashed	1/2 (0')	1/5 (0')	1/5 (15')
Moisture content (%)	56.63	62.82	57.97	59.44
MRM loss (%)	-	17.34	24.25	19.82
N x 6.25 (%)	13.36	7.04	8.37	8.17
N x 6.25 loss (%)*	-	49.77	52.07	46.25
Fat (%)	27.20	28.51	33.30	29.33
Fat loss (%)*	-	0	6.34	5.11
Ashes (%)	0.95	0.40	0.36	0.33
Ash loss (%)*	-	44.21	71.57	69.47

* The percent loss was calculated on the basis of the post-wash yield and final composition of the MRM

Washing results in a substantial loss of minerals (4), mainly dependent on the MRM/water ratio.

TABLE II. Properties of MRM

	Lot			
	Unwashed	1/2 (0')	1/5 (0')	1/5 (15')
pH	6.58	6.61	6.87	6.78
Protein solubility (%)	67.08	62.16	62.83	54.88
WHC	2.45	18.5	10.1	13.42
Cooking drip loss (%)	19.69	37.46	30.73	31.68
EC	44.15	40.15	41.49	38.85
VGC	$2.38.10^7$	$1.09.10^7$	$8.42.10^6$	$8.99.10^4$

WHC - Water holding capacity
EC - Emulsifying capacity (ml of oil/g of MRM)
VGC - Viable germ count

Washing the MRM alters the physico-chemical properties of the meat, so that the pH rises, decreasing the soluble protein, which, together with other factors, brings about a reduction in the water holding capacity and in the emulsifying capacity (Table II). These alterations are dependent on the washing time and/or the MRM/water ratio (4). The washing enhances the microbiological quality, which improves in direct proportion to the amount of water used, becoming even more pronounced as washing time is increased (Table IV), since microorganisms are washed away.

From the results of sensory analysis by the panel of tasters (4), it can be seen that washing considerably affects the sensory properties of the

MRM. With regard to taste, all the lots were rated average to good; washing lowers the MRM rating for this property, basically as a result of the loss of soluble substances. As for the remaining two properties tested, washing does not affect the perception of texture, but it does affect the overall acceptability, which is lowered by washing.

On the basis of the results obtained, washing leads to a reduction in the sensory properties of the product, so that the taste rating, for example, is adversely affected because of deflavouring; on the other hand, this does afford certain technical advantages.

3.2 Freezing and frozen storage of MRM

It has already been seen that, as a result of the washing process, a new MRM with a different composition is obtained; this MRM has a higher water content and smaller percentages of protein, fat, and ash (5). The removal of soluble nitrogenous compounds in the wash water causes losses of 42 % of the protein and 53 % of the ash, fractions which are replaced in the washed MRM with water (a gain of 11.9 % compared to the control lot). The fat loss of around 7 % may in part be the result of elimination with the wash water and in part by adherence to the walls of the draining device used when removing the water.

TABLE III. Stability of MRM during treatment*

	MRM	Initial	Frozen	Stored
EC g oil/g MRM	Control	45.0^{a}	45.8^{a}	39.3^{b}
	Washed	34.6^{a}	32.2^{a}	30.8^{b}
TBA malonald./100 g MRM	Control	0.85^{a}	1.06^{b}	3.56^{c}
	Washed	0.48^{a}	0.74^{b}	2.45^{c}

* Different letters on the same line indicate significant differences (P less than 0.05)

Washing, as already stated, brings about a decrease in the emulsifying capacity of the MRM (Table III) due to a conjunction of several factors, but basically to the reduction in the protein fraction making up the continuous phase of the emulsion. The freezing process has no influence on the EC, since there are no significant differences for MRM before and after freezing, in either the control lot or the washed samples. As can be seen in Table III, frozen storage for three months reduces the MRM's EC, probably due to the alterations in the characteristics of the proteins taking place during the storage period.

Washing eliminates part of the malonic aldehyde originating from the auto-oxidation of lipids, since the initial TBA levels are reduced by 56 % (Table III). Some oxidation of lipids occurs during freezing, though the control MRM has the highest rancidity index throughout the period studied (freezing and frozen storage). Despite the fact that washing also lowers the amount of lipids present and the amount of catalysts that tend to bring about oxidation of the lipids (Table II), the rancidification rate during freezing is higher in the washed MRM. As shown in Table III, this rate is equal in both lots during the frozen storage time tested.

The results of the bacteriological analyses carried out are presented in Table IV.

TABLE IV. Changes in the microbial population during treatment*

	MRM	Initial	Frozen	Stored
Revivificable aerobes	Control	7.80 a,1	7.79 a	7.30 b
	Washed	7.40 a,2	7.28 b	6.92 c
Pseudomonas	Control	7.72 a,1	7.61 a	7.38 b
	Washed	7.39 a,2	7.19 b	6.87 c
Staphylococcus Coag.+	Control	2.97 a,1	2.40 b	1.00 c
	Washed	2.78 a,1	2.00 b	0.01 c
Clostridium	Control	2.95 a,1	3.03 b	2.06 b
	Washed	2.72 a,1	2.97 a	1,00 b
Salmonella	Control	+	+	+
	Washed	+	+	+
Escherichia coli	Control	4.04	3.36	2.17
	Washed	4.04	4.66	2.17

* Different letters on the same line indicate significant differences (P less than 0.1)
Different numbers in the same column indicate significant differences (P less than 0.1)
Microbe values expressed as the log no. of germs/g of MRM
The presence of Salmonella was determined in 25 g of sample

The initial values found for the total number of viable germs are very high compared to those found in mechanically recovered meat by other authors (6). The genus Pseudomonas, which makes up most of the deteriorating psychrotrophic flora in the MRM, is found in very large amounts, which can be explained by the high frequency of these microbes in the environment and because it is an aerobic germ which grows readily in an alkaline medium such as MRM. In addition to the high pH value of the MRM (pH 6.6), this type of meat is a suitable medium for the multiplication of bacteria due to its large surface area. The values found are in agreement with those reported by Gardner (6).

Washing significantly reduces all the microbial groups investigated, except for the sulphite-reducing group, the coagulase-positive staphylococci, and E. coli.

Except for the coagulase-positive staphylococci, the freezing process had no significant effect on any of the microbial groups investigated in the case of the unwashed MRM. This effect on staphylococci is an explainable phenomenon in view of their well known sensitivity to low temperatures. Buckley and Kearney (7) also did not find any significant differences in total germs immediately after freezing, which is predictable when proper freezing procedures are applied. However, this phenomenon does not occur in the washed MRM, as significant differences in total germs, Pseudomonas, and staphylococci were found. It would seem that, when washing treatments are carried out, the new conditions created contribute to heavier destruction of the total flora due to thermal shock.

During the frozen storage period there was a significant reduction in all the microbial groups investigated. This agrees with the results obtained by other authors (6). The reduction was proportionally higher in the washed MRM for most of the microbes considered. The bacterial decrease is much more pronounced for the sulphite-reducing anaerobes and the coagulase-positive staphylococci; this latter group may even disappear in the washed MRM. In the case of chicken MRM stored at -23 ºC for 46 weeks, Lilliard (8) reported significant reductions in Clostridium cells and spores. In the present experiment this reduction is also much more apparent in the case of washed MRM.

Significant reductions in the number of Escherichia coli per gram of sample were found during the frozen storage period, but no significant differences were found on freezing.

Salmonella was found to be present in all the lots in 25 g of sample, and neither washing nor freezing seemed to have any effect on this bacterium. There is, however, some controversy on this point, as other authors have reported this microorganism's sensitivity to low temperatures.

REFERENCES

1. DESCHAMPS, B.F.M. La séparation mécanique des viandes de volailles. Thesis. Tolouse Univ. (19788).
2. MAST, M.G. and J.H. MACNEIL. J. Food Sci. 45, 641. (1980).
3. RACCACH, M. and R.C. BAKER. Poultry Sci. 58, 144. (1979).
4. JIMENEZ-COLMENERO, F. and E. GARCIA-MATAMOROS. Proc. 27th European Meeting Meat Research Workers (1981) 351.
5. PELAEZ, M.C., E. ARROYO, and F. JIMENEZ-COLMENERO. Preprints 16th International Congress of Refrigeration C-2. Paris 1983.
6. GARDNER, C.A. Proc. 27th European Meeting Meat Research Workers. 629. (1981).
7. BUCKLEY, J. and L. KEARNEY. Proc. 21st European Meeting Meat Research Workers. 231. (1975).
8. LILLARD, H.S. Poultry Sci. 56, 2052. (1977).

SEASONAL VARIATIONS IN FLAVOUR CHANGE OF COLD-STORED RAINBOW TROUT

Lilia Ludovico-Pelayo, Alan Hume and R. Malcolm Love
Torry Research Station,
Aberdeen, Scotland

Summary

The flesh of fish gradually acquires an undesirable flavour and odour during frozen storage, because of the action by atmospheric oxygen on unsaturated fatty acids. Since the proportion of these varies seasonally, there is scope for seasonal variation in the development of off-flavour. In the present work, the lipid composition of rainbow trout (*Salmo gairdneri* Richardson) was studied monthly from January to July. The overall degree of unsaturation of the fatty acids decreased as warm-weather feeding occurred, and less cold-store flavour developed. Artificial starvation of the fish resulted in an increased proportion of polyunsaturated fatty acids in the body lipids, and refeeding caused a decrease. However, starving fish containing most C22:6 did not develop detectably greater cold-store flavour than those containing less.
Refeeding the starving fish gave rise to a smaller proportion of C22:6 but detectably greater rancidity.
Frozen rainbow trout therefore behave differently from frozen cod.
Possible reasons are discussed.

INTRODUCTION

The quality of foodstuffs is sometimes affected more by seasonal variation than by variations in the manner of processing (1). Ross & Love (2) showed that the seasonal influence extended even to the quality of cold-stored cod (*Gadus morhua* L.), since the off-flavour which develops at low temperatures is the result of the atmospheric oxidation of polyunsaturated fatty acids (3) which themselves vary at different times of the year. Starvation reduces the proportion of polyunsaturated acids in cod (2), and it was found possible to reduce greatly the development of off-flavour by starving them before killing and freezing. The principal polyunsaturated fatty acid of cod flesh is C22:6ω3, which is more concentrated in the phospholipid fraction than in the triacylglycerols (4). Variations in C22:6 and in the proportions of phospholipids are therefore of particular interest in this work.

The recent upsurge in the cultivation of salmonids is leading to an increase in the practice of cold-storage to maintain a steady supply of fish to the market despite gaps between the periods of slaughtering.

The purpose of the present investigation was to study the reactions of rainbow trout (*Salmo gairdneri* Richardson) flesh to cold storage, and ascertain whether any seasons give rise to fish which is less suitable, or more suitable, for frozen storage.

1. EXPERIMENTAL

Rainbow trout approaching marketable size (220-250 g) were obtained on the last week of each month from a fish farm. They had been reared in earth ponds supplied with river water. Starvation lasted for 10 weeks at 8-12°C, and refeeding was done for 8 weeks using a trout grower pellet (no. 7, Fulmar, Marine Harvest Ltd, Edinburgh). Fillets were sealed in polyethylene bags and stored at -15° for 6 weeks, after which they were transferred to liquid nitrogen until sensory evaluation was done.

Water content was determined by drying at 100° for 5 days, and total lipids were extracted by the method of Hanson and Olley (5). Fatty acids were methylated by the method of Thomson (6), and determined by gas chromatography (Hewlett-Packard 5880 A-series). Lipid class separation was carried out by thin-layer chromatography on silica gel plates (Merck, 250 μ thickness). Spots were identified by comparison with pure standards run on the same plate, and scanned by a Vitatron Densitometer (TLD-100, Holland) connected to a Hewlett-Packard (HP 3380A) recorder-integrator. The method is a standard one, but the results should be read only in a comparative way, rather than strictly quantitatively (Table 1). Cold-store flavour of the steam-cooked muscle was assessed by a trained panel of 6 people, using a scale developed for evaluating frozen fish (7).

2. RESULTS

The work falls into two sections comprising, firstly, a monthly survey from January to July both of flesh lipid composition and off-flavour development after cold storage, and secondly, complete starvation for 10 weeks followed by refeeding for 8 weeks, the composition of the flesh and off-flavour again being monitored.

2.1 Seasonal Survey

Table 1 shows the composition of trout 'white' muscle (usually known as mosaic muscle in salmonids) in different months. The proportion of triacylglycerols increased steadily from April, while free fatty acids became drastically reduced. Phospholipids and cholesterol showed no obvious trends.

TABLE I
Seasonal Variation in the White Muscle Lipid Fractions of Rainbow Trout

		Percentage of total lipids				
Month	TL (%)	PL	FFA	pAG	CH	TG
Jan	2.1	33	12	4	12	39
Feb	1.7	33	17	2	11	37
Mar	1.7	43	14	0.7	11	31
Apr	1.7	37	7	1	12	43
May	1.6	37	3	0.3	14	46
Jun	1.4	32	1	0.5	19	48
Jul	1.7	35	1	1	13	50

TL: Total lipid   PL: Phospholipid   FFA: Free fatty acids
pAG: Partial acylglycerols   CH: Cholesterol   TG: Triacylglycerols

Fig. 1 shows that the proportion of polyunsaturated fatty acids in the total lipids falls appreciably in June, and more still in July, presumably in consequence of the build-up of triacylglycerols (poor in polyunsaturates) in the warmer months of higher dietary intake. Fig. 2 shows the corresponding taste panel results, a rise in the off-flavour from January to April, followed by a sharp fall from May onwards.

There is enough information here for us to conclude that, as in cod, there is a seasonal variation in the cold-store flavour of trout. There is, however, no straightforward correlation with polyunsaturation (which declines a month later than cold-store flavour) or with the proportion of phospholipids, which change little.

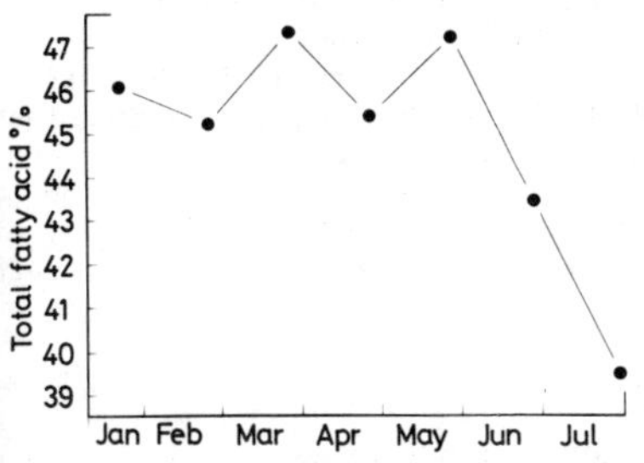

Fig. 1: Seasonal changes in the polyunsaturated fatty acids of rainbow trout muscle.

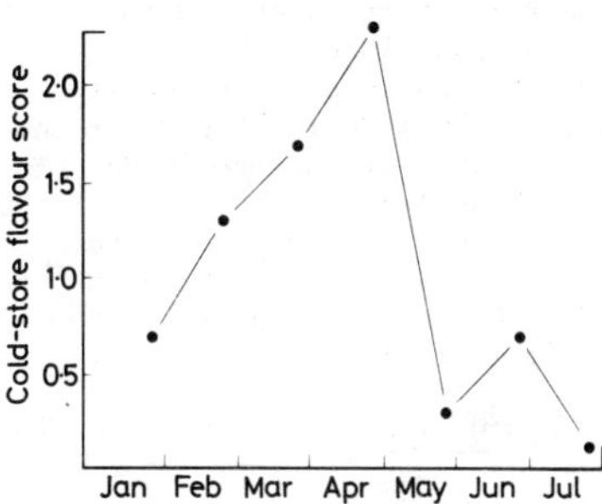

Fig. 2: Seasonal changes in the flavour developed by trout muscle after a fixed period of cold storage (flavour assessed after cooking by a panel of tasters)

2.2 Starvation

The muscle of starving cod shows a steadily increasing water content, since water substitutes, at least partially, for the structural protein which is being metabolised for energy purposes (8). Polyunsaturated fatty acids are reduced (2), resulting in less off-flavour being formed on cold storage. In trout (present work), the water content of the flesh also increases during starvation, but here the polyunsaturation increases (Fig. 3). Taste panel

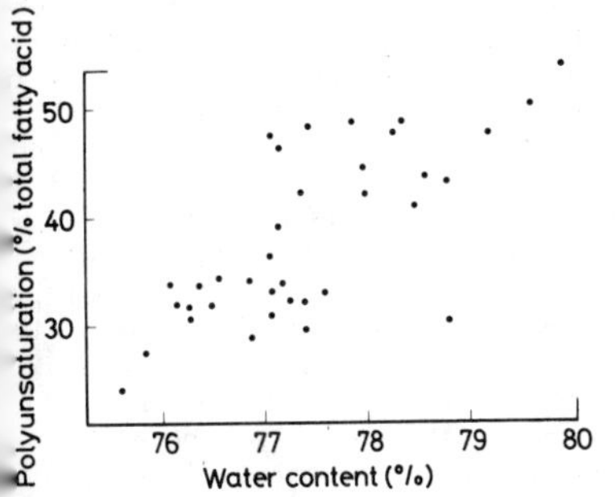

Fig.3: Polyunsaturated fatty acids in the lipid of trout white muscle after starving to different extents. Water content increases as starvation is prolonged.

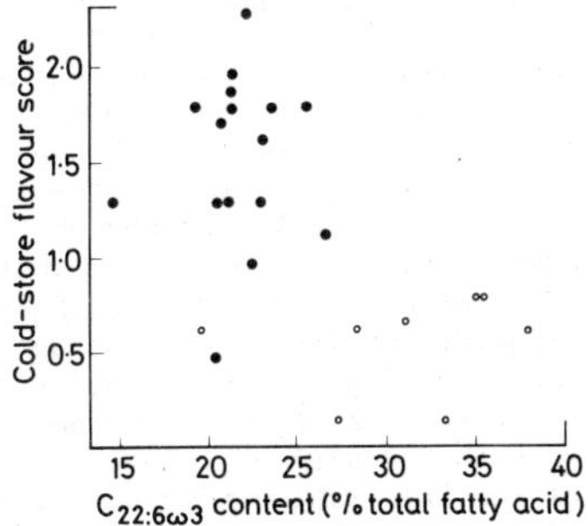

Fig.4: Cold-store flavour in trout muscle in relation to the proportion of C22:6ω3 fatty acid present.
O = Starving
● = Re-fed

scores of the cooked, cold-stored product (Fig. 4) showed no correlation with the quantity of C22:6 present, in contrast with the findings using cod. Refed fish, all with relatively low proportions of C22:6, gave rise to considerable cold-store flavour, but starved fish developed little of it despite a wide range of C22:6 in their muscle. The proportion of total polyunsaturated fatty acids rose during starvation and fell during refeeding.

It was found that 89.3% of the white muscle lipids utilised during 10 weeks of starvation were neutral lipids (mostly triacylglycerols), 9.4% were free fatty acids, while only 1.3% of the lipids removed by starvation were phospholipids. The changes in the lipid composition of trout flesh are therefore basically the result of influx or efflux of triacylglycerols. Table 1 shows that the proportion of triacylglycerols in trout flesh usually exceeds that of phospholipids. The phospholipids of cod muscle, on the other hand, have been shown (4) to lie between 84 and 91% of the total lipids, and to fall to about 76% in severe starvation, so this must be regarded as the 'dynamic' lipid fraction in that species. Triacylglycerols of cod muscle, at 0.8 to 2% of the total (4), are an almost negligible fraction, and are indeed little affected by starvation.

3. DISCUSSION

These results do not agree with the earlier findings on cod (4), because here a net increase in unsaturation of the fatty acids occurs during starvati while in cod there is a decrease. There is also a contradiction within the present results, since, while a seasonal decrease in unsaturation corresponds approximately with a decrease in cold-store flavour, the increase in unsaturation caused by artificial starvation does not engender a greater cold-store flavour, but less. Indeed, in such starving fish the degree of unsaturation obviously bears no relation to flavour (Fig. 4).

The compound actually detected by the tasters and identified as 'cold-store flavour' is cis-4-heptenal (3), which results from the oxidation of ω-3 polyunsaturated fatty acids, of which C22:6ω3 is the most important in these fish, so the decrease in the cold-store flavour with an increase in unsaturation is unexpected. Perhaps the oxidation of unsaturated fatty acids in the triacyl glycerols is more important in producing cold-store flavour than is the oxidation of those in phospholipids.

It should be borne in mind that there is no real increase in any component, but only in relative terms. All lipid fractions are consumed to some extent during starvation, but the phospholipids are consumed hardly at all in trout. They therefore become more important as the less-unsaturated triacyl glycerols are removed from the muscle tissue during starvation, and the overall picture is of a more unsaturated lipid remaining.

Triacylglycerols and phospholipids are themselves composed of several constituents. Ross (4) showed that phosphatidyl ethanolamine, one of the constituent phospholipids, contained much more C22:6 than did, for example, phosphatidyl choline. He also demonstrated considerable variations occurrin between individual phospholipid fractions as starvation progressed, though they varied in a haphazard manner and not in step with the starvation.

In the present work, well-fed fish caught in the summer months showed increased triacylglycerols in the muscle and reduced cold-store flavour (Figs 1 and 2) while fish refed after starvation show increased cold-store flavour. Other studies have shown that refeeding after complete starvation produces for a while some unusually high levels of various substances, e.g. carbohydrates (9) as a sort of over-compensation ('over-shoot'), and here it is feasible that the restored lipid is in some way different from that of fi feeding steadily.

Ross (4) showed by histological sections that the lipid stored for energy purposes in herring muscle (presumably mostly triacylglycerols) occurred as globules surrounding the individual muscle cells. It is possible that such lipids protect the phospholipids, which occur mostly within the cell and cell-wall, from atmospheric oxidation, by their physical distribution. This could explain why better-fed trout in June and July exhibit less cold-store flavour than those killed in the spring.

All these attempts at explanation are speculative. What we can say definitely is that the presence of relatively large quantities of triacylglycerols (which also do contain some polyunsaturated fatty acids) in trout flesh gives rise to behaviour in the cold-store which is quite different from that of cod, which contains phospholipid almost exclusively.

4. ACKNOWLEDGMENT

The work described in this paper formed part of the programme of the Ministry of Agriculture, Fisheries and Food.

5. REFERENCES

1. LOVE, R.M. (1980). The Chemical Biology of Fishes, Volume 2, 943 pp. Academic Press, London
2. ROSS, D.A. and LOVE, R.M. (1979). Decrease in the cold store flavour developed by frozen fillets of starved cod (*Gadus morhua* L.). J. Food Technol. 14, 115-122
3. HARDY, R., McGILL, A.S. and GUNSTONE, F.D. (1979). Lipid autoxidative changes in cold stored cod (*Gadus morhua*). J. Sci. Food Agric. 30, 999-1006
4. ROSS, D.A. (1977). Lipid metabolism of the cod, *Gadus morhua* L. Ph.D Thesis, University of Aberdeen, Scotland
5. HANSON, S.W. and OLLEY, J. (1963). Application of the Bligh and Dyer method of lipid extraction to tissue homogenates. Biochem. J. 89, 101P
6. THOMSON, A.B. (1980). A comparison of different methods for the preparation of fatty acid methyl esters from fish lipids. Torry Document 1457 (Torry Research Station, Aberdeen, Scotland)
7. BAINES, C.R., CONNELL, J.J., GIBSON, D.M., HOWGATE, P.F., LIVINGSTON, E.I. AND SHEWAN, J.M. (1969). A taste panel technique for evaluating the eating quality of frozen cod. In: Freezing and Irradiation of Fish. Ed. R. Kreuzer. Fishing News (Books), London, 361-366
8. LOVE, R.M. (1970). The Chemical Biology of Fishes, Volume 1, 547 pp. Academic Press, London
9. BLACK, D. (1983). The metabolic response to starvation and refeeding in fish. Ph.D. Thesis, University of Aberdeen, Scotland

CONSERVATION DE HACHIS DE SARDINE EN ENTREPOSAGE A L'ETAT CONGELE. I. HACHIS OBTENUS DE SARDINE ENTIERE, ET DE SARDINE ETETE ET EVISCERE

A. MORAL, M. TEJADA et A.J. BORDERIAS
Instituto del Frío (CSIC). Ciudad Universitaria. Madrid-3 (Espagne)

SUMMARY

The practical objective of this research work is to improve the technology intented for obtaining sardine minces and achieve a maximum practical storage life of this product so as to obtain a raw material suitable for facilitating the introduction into the market of a wide variety of food products based on sardine minces, thus finding an outlet for a considerable economic potential which is at present wasted on our coasts.

The mince was prepared from both whole sardines (Lots E) and gutted and headless sardines (Lots D). Protein protectors and antioxydants were added to the product. One lot (DL) was washed with ice water. The results obtained by means of objective texturometric and sensory tests show that these mince, if, vacuum-packed, can be usable till after 4-8 months of storage at -20ºC.

1. INTRODUCTION

On estime que, dans les prochaines années, l'Espagne devrait prêter attention à l'utilisation intégrale des débarquements de sardine et, en général, des petites espèces pélagiques.

La sardine représente 35,84% des debarquements de poissons du littoral et de haute mer, les clupéides 45,97% et les petits pélagiques (clupéidés, maquereau et carangiformes) 56,05%. Tout de même, la valeur en pesetas représentée par ces poissons est de 7,12%, 16,6% et 27,11% respectivement /1/. Ce fait porte a penser que les études dirigées à obtenir une utilisation plus diversifiée de ces types de poisson peuvent avoir une répercussion sur le développement technologique et socio-économique du secteur.

Le but pratique de ces travaux de recherche c'est d'améliorer la technologie destinée à obtenir hachis de sardine et atteindre une durée de conservation pratique maxima de ce produit en vue d'obtenir une matière première susceptible de faciliter la présentation au marché d'une grande diversité de denrées alimentaires à basse de hachis de sardine, assurant ainsi l'écoulement d'un grand potentiel économique gaspillé dans nos côtes.

2. MATHERIEL ET METHODES

Pour l'expérience décritee içi, on a employé des sardines (Sardina pilchardus, Walb.) du jour, provenant du marché de Castellón (Espagne), pêchées en mai. On a procédé à l'obtention de hachis à l'aide d'une machine d'extrusion Baader 694, munie d'un cylindre à orifices d'un diamètre de 3 mm. Une partie du hachis a été obtenue de sardine entière (lots E), et une autre partie a été extrait de sardine étêtée et éviscérée (lots D). Les sardines se trouvaient à l'état de *rigor mortis* au moment d'être soumises aux opérations de transformation.

Une portion du lot D a été lavée pendant 10 minutes à l'eau refroidie par glace au rapport hachis-eau de 1/5, en agitant de temps en temps /3/. Ensuite l'eau a été décantée. Puis le hachis a été mis dans une enveloppe en gaze et pressé avec la main pour extraction de l'eau restante.

A partir du hachis obtenu on a préparé les lots suivants:

		Protecteurs protéiques	Antioxydants	Lavé
	ESS			
E (hachis de sardine entière)	ESA	+	+	
	ESAO		+	
	ESP	+		
D (hachis de sardine étêté et éviscérée)	DS			
	DC	+	+	
	DL	+	+	+

Tous les lots, y compris les témoins, ont été homogénéisés dans une mélangeuse afin d'assurer une distribution uniforme des protecteurs protéiques, des liants d'eau et des antioxydants.

La formule adoptée pour les ingrédients (protecteurs protéiques et liants d'eau) contenait: glucose à 5%, glutamate à 1% et polyphosphate à 0,2%; celle employée pour les antioxydants était constituée par butyl-hydroxyaniso (BHA) à 0,01% et butyl-hydroxy-toluène (BHT) à 0,01% par rapport à la graisse du hachis. La pâte on a congelé dans une armoire à plaques horizontales jusqu'à ce que la température du centre thermique s'est abaissé à -20ºC. Les blocs de hachis congelé ont été emballés sous vide en sacs de CRYOVAC B13-ITS, puis entreposés en chambre froide à -20ºC pendant 12 mois.

On a déterminé les rendements des différents échantillons. Pour la réalisation de l'analyse sensorielle, on a utilisé des échelles à intervalles de cinq points qui permettaient de déterminer les paramètres de goût, rancidité et texture. Ainsi, pour le goût on a établi l'échelle suivante: 5, très bon; 4, bon; 3, passable; 2, limite; 1, mauvais. Pour la texture: 5, normale; 4, un peu molle, ou un peu dure; 3, molle ou dure; 2, très molle, ou très dure; 1, immangeable. Pour la rancidité: 5, nulle; 3, produit un peu rance; 1, produit très rance.

L'analyse sensorielle a été effectuée dans une salle de dégustation par un jury de 6 personnes moyennement entraînées, choisies parmi le personnel du laboratoire.

Par ailleurs, on a effectué un test d'acceptabilité avec intervention de 33 dégustateurs non entraînés choisis parmi le personnel de l'Instituto. Ils étaient priés de répondre à une échelle hédonique allant de 9 points (extrêmement désagréable) à 1 point (extrêmement agréable) et on leur demandait également si la consommation du hachis devenait plus (1), également (2) ou moins (3) ennuyante que dans le cas de la sardine entière.

L'analyse instrumentale de la texture a été effectuée au moyen d'un texturomètre Instron 1140 en mesurant la résistance au cisaillement à l'aide de la cellule de Kramer et adoptant la hauteur maxima du pic comme mesure de la dureté. On a utilisé également la compression pour déterminer la fermeté, l'élasticité et la cohésivité suivant, dans tous les cas, la méthodologie employée par Borderías et coll. /3/

Comme analyse statistique pour déterminer la signification des différences entre les divers traitements et dans chaque traitement au cours du temps, on a effectué l'épreuve de Tuckey /4/ après avoir constaté, lors de l'analyse de variance, qu'il existait une interaction aussi bien pour les traitements que pour les contrôles.

3. RESULTATS ET DISCUSSION

Les hachis ont été à partir de sardines d'une taille moyenne de 14,5 cm. Le rendement en hachis de la sardine entière a été de 87%, tandis que dans le cas des poissons étêtés et éviscérés il a été de 67,7%. Le hachis obtenu à partir de sardine étêtée et éviscérée soumise à lavage par eau de fusion de la glace a accusé une perte par lavage de 28%, ce qui à peu près d'accord avec les pertes par lavage d'autres espèces telles que le chinchard. /2/

Les valeurs obtenus de l'analyse instrumentale de la texture sont consignées dans le Tableau 1. La résistance au cisaillement mesurée par la cellule de Kramer (dureté) sur échantillons crus, indique que le lot DS s'avère, pendant tout le temps d'entreposage, plus dur que le lot ESS et accuse une tendance à accroisser la dureté à la fin de l'entreposage, ce qui correspond à une plus grande augmentation d'agrégation détectée au moyen de l'analyse de solubilité protéique. l'effet des protecteurs protéiques est plus évident dans les lots D que dans les lots E, et on constate, en outre, un amollissement significatif causé par la seule addition de ces additifs. Le lavage (DL) augment la dureté dès le début de l'entreposage, malgré la préalable incorporation de protecteurs protéiques au produit. Nous n'avons pas trouvé de corrélation entre ces résultats et la solubilité protéique. Après cuison, on constate une augmentation très évidente de la dureté dans les lots D, tandis que dans les lots E les valeurs de dureté restent invariables. Ce phénomène pourrait être attribué à la différente teneur en lipides des échantillons. La variation des valeurs obtenues des différents contrôles peut être attribuable à la difficulté que présente la standardisation des échantillons cuits.

On observe que, de même que dans le cas de la dureté, les échantillons D accusent une plus grande fermeté que les lots E. Etant donné qu'à l'aide de la cellule de Kramer on analyse principalement la dureté de la fibre musculaire, tandis que l'analyse de la fermeté donne une combinaison de dureté des particules et de cohésion entre elles, on peut affirmer, en général, que les lots D ont des particules plus dures et/ou plus cohésives. Avec cet indice également, on détecte dans les échantillons D l'effet ramollissant des protecteurs protéiques. De même que dans le cas de la dureté, le lot lavé accuse une plus grande fermeté, qui augmente nettement tout au long du temps d'entreposage.

De même que pour les autres paramètres de texture, une plus grande élasticité initiale des lots D est également évidente; toutefois cette augmentation au cours de la conservation est plus accusées dans les lots E. Les protecteurs protéiques stabilisent l'élasticité au cours du temps, et cet action est plus évidente dans les lots D. Le lavage permet d'obtenir un produit plus élastique dès le début, qui maintient des valeurs semblables tout au long de la période d'entreposage.

On constate que les valeurs de cohésivité augmentent légèrement au cours de l'entreposage; il en est de même pour les mesures de texture effectuées, mais les protecteurs incorporés ont un effet plus faible sur la stabilisation au cours de l'entreposage. On observe aussi que la cohésivité est plus forte dès le début dans les lots D, et qu'elle est plus accusée, dès le commencement, dans le lot lavé et ne subit pas de variations au cours du temps d'entreposage.

De l'analyse instrumentale de texture on peut conclure que dans les lots E la myofibrille (fibre musculaire) ne subit pas de grandes variations d'après les valeurs de résistance au cisaillement et de solubilité protéique. Toutefois, l'augmentation de la fermeté, l'élasticité et la cohésivité indiquent que d'autres composants protéiques du hachis se modifient au cours du temps de conservation. Il est évident que les protecteurs protéiques exercent une action non seulement sur la protéine myofibrillaire, mais aussi sur ces composants.

Tous les paramètres de texture étudiés sont initialement plus élevés dans les lots D que dans les lots E, bien que les variations au cours du temps soient moins manifestes. Ceci est attribuable à la différent quantité et qualité des lipides et protéines existant entre les deux lots, ce qui devient évident surtout après la cuison des échantillons. L'action des protecteurs protéiques sur la protéine myofibrillaire se manifeste nettement dans les valeurs de dureté, fermeté et élasticité obtenues, mais ils n'exercent aucune action sur les composants qui agissent comme forces de cohésion.

Le lavage entraîne la perte des composants du hachis qui se modifient au cours de l'entreposage, c'est pouquoi le lot lavé présente des valeurs initiales élevées, mais stables, pendant tout le temps de conservation.

L'observation du Tableau 2 permet de voir que les valeurs de dureté obtenues par analyse sensorielle concordent, en général, avec celles obtenues par analyse instrumentale sur le hachis cuit. L'échantillon lavé (DL) accuse, au départ, une augmentation de la dureté, mais celle-ci se stabilise au cours de l'entreposage. Dans les lots E on trouve une texture qui doit être considérée comme trop molle lorsque le hachis est consomé de la façon décrite, et ceci est plus évident dans les lots additionnés de protecteurs protéiques.

En ce qui concerne l'évaluation du goût (Tableau 2), on constate que le lot DC et surtout le lot DL, ont une qualification plus basse que le témoin (DS), à cause de la saveur douce apportée par le sucre incorporé. Dans le lot DL, on perçoit plus nettement le goût doux, du fait qu'il est lavé. De toute façon, ce goût n'est pas du tout désagréable. Dans les lots E, où le goût de sardine est plus fort, le goût doux n'est pas si évidente. D'autre part, pour le consommateur espagnol le goût fort de sardine, qui disparaît au lavage, influe défavorablement sur la qualité du produit, bien que ce soit précisément cette perte de goût qui le rend apte pour les consommateurs qui n'aiment pas le goût fort de poisson, ou approprié à être utilisé en d'autres produits, du fait que le lavage ne semble pas endommager les propriétés fonctionnelles de la protéine; au contraire, il les améliore même, dans certains cas.

Les goûts de rance apparaissent dans le 4ème mois d'entreposage dans quelques lots du groupe E et dans le groupe DS. Dans le lot ESS ce goût est perçu jusqu'à la fin de l'entreposage, et dans le rest des lots on ne le détecte plus, pratiquement, jusqu'au dernier contrôle. Skachkov et Yudine /5/ le constataient bien avant que nous. La raison de ce fait pourrait être l'emploi de l'emballage sous vide, méthode qui n'avait pas été appliquée dans les autres expériences et qui n'est pas coûteuse aujourd'hui. Hiremath /6/ avait déjà signalé cette possibilité en vue d'empêcher le rancissement de la sardine.

Le test d'acceptabilité effectué par 33 membres du personnel de l'Instituto del Frío sur l'échantillon ESS fraîchement préparé et divisé en portions a eu pour résultat 3,27, ce qui signifiait l'acceptation du produit et que la consommation réitérée de ce hachis ennuie un peu plus. Bien que l'on ne l'ait pas analysée, nombreux membres du jury ont fait observer que la couleur était, à leur avis, peu attrayante.

4. CONCLUSIONS

- La méthode d'extrudage de la sardine entière semble être très convenable du point de vue de l'économie de main-d'oeuvre et du rendement obtenu (87% contre 68% de l'extrudage de la sardine étêtée et éviscérée).
- On ne constate de rancidité dans aucun de échantillons jusqu'au 4ème mois d'entreposage, mais à partir de cette date elle se présente dans les lots E non additionnés d'antioxydants. Les lots D ne présentent pas de goût de rance jusqu'après huit mois d'entreposage.
- En ce qui concerne la texture, les échantillons D s'avèrent plus durs, fermes et élastiques que les échantillons E, probablement à cause d'une plus grande quantité de protéines myofibrillaires. Les protecteurs protéiques essayés stabilisent les paramètres de texture étudiés, à l'exception de la cohésivité, sur laquelle ils ne semblent avoir guère d'influence.
- Le lavage augmente la dureté, la fermeté, l'élasticité et la cohésivité dès le début de l'entreposage, puis les valeurs restent inaltérées pendant la conservation.
- La quantité de glucose ajoutée (5%) s'avère excessive à l'égard du goût parce que, surtout dans les lots D, la saveur douce qu'elle produit diminue l'acceptabilité.
- Le hachis extrait de sardines entières est un produit apprécié par le consommateur espagnol, même s'il est consommé tel quel en morceaux enrobés et frits.
- L'échantillon lavé perd dans une large mesure son goût caractéristique de sardine, et cela peut être intéressant pour la préparation de certains produits.

REFERENCES

1. ANUARIO DE PESCA MARITIMA. Dirección General de Pesca Marítima. Ministerio de Agricultura, Pesca y Alimentación. (1981)
2. M. TEJADA, A.J. BORDERIAS et A. MORAL. (1981). Effects of washing of horse mackerel (Trachurus trachurus L.) minces on the removal of substances detrimental to preservation in cold storage. Advances in the refrigerated treatment of fish. International Institute of Refrigeration. Boston, pp. 371-376.
3. A.J. BORDERIAS, M. LAMUA et M. TEJADA. (1983). Texture analysis of fish fillets and minced fish by both sensory and instrumental methods. J. Fd. Technol., 18, pp. 85-95.
4. J.W. TUCKEY (1961), cité en Statistical methods: Applied to experiment in Agriculture and biology. 5e ed. G.W. SNEDECOR. The Iowa State University Press. pp. 251.
5. V.P. SKACHKOV et O.P. YUDINA. Storage of comminuted meat from small fish . Rybnoe Khozyaistvo, nº 6, 54-55 (1975) (dans FSTA, R0213, 76-04.).
6. G.G. HIREMATH. (1973). Prevention of rancidity in frozen fatty fishes during cold storage. Univ. Agric. Sci. Fish. Coll., 27, pp. 20-24 (dans FSTA, R0625, 74-12).

TABLEAU I

Analyse instrumentale de la texture

Jours d'entre-posage	Echantillon	Dureté (Kg/g échantillon) Cru	Cuit	Force (g)	Elasticité (cm)	Cohésivité
	ESS	0,26a/x	0,31a/x	36	0,87a/x	0,19a/x
	ESA	0,30a/xy	0,20a/x	60,7a/x	0,63a/y	0,19a/x
	ESAO	0,27ab/x	0,27ab/x	60,7a/x	1,04a/x	0,20a/x
Après congélation	ESP	0,27a/x	0,17a/x	51,3ab/x	0,67a/y	0,18a/x
	DS	0,37a/y	1,25a/y	132,7a/y	1,82a	0,44a/y
	DC	0,27a/x	0,91a	80a/x	1,63ab	0,35a
	DL	0,42a/y	1,25a/y	143,3a/y	2,25a	0,48ab/y
	ESS	0,34a/x	0,25a/x	56,7a/x	0,83a/x	0,16a/x
	ESA	0,22a/y	0,20a/x	45a/x	0,70a/x	0,19a/x
	ESAO	0,30ab/xy	0,16a/x	55a/x	0,87a/x	0,21a/xy
30	ESP	0,23a/y	0,15a/x	60a/x	1,06b	0,24a/y
	DS	0,36a/x	1,35a/y	127,7a	1,77a	0,40a/z
	DC	0,26a/y	1,02ab/y	80a	1,27a	0,35a
	DL	0,45a	1,27a/y	196,7b	2,08a	0,41a/z
	ESS	0,33a/xy	0,18/x	58,3a/x	1,03a/x	0,20a/x
	ESA	0,21a/x	0,31b/x	50a/x	0,76a/x	0,19a/x
	ESAO	0,22a/x	0,32b/x	53,3a/x	0,90a/x	0,19a/x
60	ESP	0,20a/z	0,20a/x	55a/x	0,80a/x	0,19a/x
	DS	0,39a/yz	1,31a/y	138,3ab	1,85a/y	0,40a
	DC	0,30a/xy	1,34b/y	111,7	1,77b/y	0,32a
	DL	0,47a/yz	2,49b	176,7ab	2,27a	0,50ab
	ESS	0,32a/xy	0,28a/x	76,6b/x	1,20ab/x	0,49ab/xy
	ESA	0,28a/xy	0,19a/x	48,3a/x	0,76a/x	0,37b/x
	ESAO	0,30ab/xy	0,18a/x	47,3a/x	1,50b/x	0,36ab/x
90	ESP	0,24a/x	0,22a/x	55a/x	1,10b/x	0,87b
	DS	0,39a/y	1,63a/y	116,6a	2,20b/y	0,55ab/xy
	DC	0,19b/x	0,81a	67,3a/x	1,55ab/y	0,54b/xy
	DL	0,44a/y	1,40a/y	195b	1,95a/y	0,62b/y
	ESS	0,30a/x	0,33a/x	78b/x	1,50b/x	0,30ab/x
	ESA	0,24a/x	0,22a/x	85/x	1,21b/x	0,26a/x
	ESAO	0,35b/xz	0,34b/x	91,7b/x	1,37ab/x	0,30ab/x
120	ESP	0,22a/xy	0,19a/x	71,7/x	1,13b/x	0,24a/x
	DS	0,41ab/z	1,42a/y	178,3b	2,35b/y	0,44a/y
	DC	0,12b/y	1,26b/y	90a/x	1,30a/x	0,31a/x
	DL	0,44a/z	1,86ab	231,7b	2,27a/y	0,46a/y
	ESS	0,31a/x	0,33a/x	63,3a/x	1,33b/x	0,55b/x
	ESA	0,19a/y	0,37b/x	46,6a/x	0,80a/y	0,36b/x
	ESAO	0,30ab/x	0,68/xy	46,7a/x	1,53b/x	0,44b/x
180	ESP	0,27a/y	0,14a	41b/x	0,93b/y	0,61b/x
	DS	0,39a/xz	2,52	103,3a/y	2,09ab/z	0,55ab/x
	DC	0,25a/y	1,06ab/xy	76,7a/xy	1,40ab/x	0,53b/x
	DL	0,45a/xz	1,85ab	198,3b	2,20a/z	0,45a/x
	ESS	0,19/x	0,27a/x	81,7b/x	1,97/x	0,42ab/x
	ESA	0,26a/y	0,19a/x	56,6a/x	1,26b/y	0,39b/x
	ESAO	0,31ab/y	0,36b/x	85b/x	1,90b/x	0,47b/x
360	ESP	0,15/x	0,23a/x	53,3a/x	1,37/y	0,25a/x
	DS	0,46b	1,21a	126,7a	2,13b/x	0,64b/x
	DC	0,24a/y	0,75a	75a/x	1,77b/xy	0,51b/x
	DL	-	-	-	-	-

Les moyennes de même lot pendant la conservation suivies de la même lettre (a, b, c, etc.) indiquent absence de différences significatives ($p > 0,05$). Les moyennes, pendant la même période, des différents lots, suivies de la même lettre (x, y, z) indiquent absence de différences significatives.

TABLEAU II
Résultats de l'analyse sensorielle

Jours d'entreposage	Echantillon	Goût X̄	Goût CV	Rancidité X̄	Rancidité CV	Texture X̄	Texture CV
	ESS	3,9	4	5	0	4,4(B)	17
	ESA	3,5	14	5	0	3,8(B)	10
	ESAO	3,6	13	5	0	3,1(B)	34
30	ESP	3,6	14	5	0	3,6(B)	21
	DS	3,7	36	5	0	4,2(D)	23
	DC	3	33	5	0	5	0
	DL	2,2	34	5	0	4,6(D)	11
	ESS	3,7	13	5	0	4,1(B)	21
	ESA	3,3	13	5	0	4 (B)	17
	ESAO	3,8	19	5	0	4,4(B)	20
60	ESP	3,2	13	5	0	3,7(B)	25
	DS	3,7	29	5	0	4,3(D)	17
	DC	3,7	25	5	0	4,6(D)	17
	DL	3,4	15	5	0	4,3(D)	18
	ESS	3,2	36	5	0	4,2(B)	24
	ESA	3,3	15	5	0	4,3(B)	22
	ESAO	2,8	46	5	0	5	0
90	ESP	2,8	45	5	0	4 (B)	0
	DS	4	0	5	0	4,2(D)	26
	DC	2,7	51	5	0	4,3(D)	18
	DL	2,6	35	5	0	4,6(D)	19
	ESS	3,2	23	4,8	10	3,7(B)	25
	ESA	3,8	11	5	0	4,2(B)	10
	ESAO	2,8	29	4,8	10	4,5(B)	22
120	ESP	3,6	15	4,4	20	4,4(B)	20
	DS	3,7	13	4,1	20	3,9(D)	17
	DC	3,3	24	5	0	4,3(D)	18
	DL	2,7	18	5	0	2,7(D)	18
	ESS	3,4	25	4,5	10	4,5(B)	12
	ESA	4,4	15	5	0	4,3(B)	18
	ESAO	3,5	15	5	0	4 (B)	27
180	ESP	3,5	15	5	0	4,2(B)	19
	DS	3	38	5	0	4,3(D)	18
	DC	3	27	5	0	4,6(D)	11
	DL	3	25	5	0	4,1(D)	18
	ESS	3,5	15	4	20	4 (B)	24
	ESA	3,5	15	5	0	3 (B)	42
	ESAO	3,3	15	5	0	3,8(B)	10
240	ESP	3,6	15	4,8	10	4 (B)	16
	DS	3,9	19	5	0	4,1(D)	15
	DC	3,6	16	5	0	4,4(D)	12
	DL	3,4	16	5	0	4,4(D)	12
	ESS	3	23	3,8	28	4,6(B)	11
	ESA	2,8	15	4,7	10	3 (B)	33
	ESAO	3,6	15	4,4	20	5	0
360	ESP	3,8	22	4,2	22	4 (B)	34
	DS	4,2	10	4,8	10	5	0
	DC	3,4	26	4,4	20	4,8(D)	9
	DL	-	-	-	-	-	-

(B): Mou. (D); Dur. X̄:moyenne. CV: coefficient de variation.

INFLUENCE OF TREATMENTS PRIOR TO FREEZING ON THE QUALITY AND STABILITY OF FRUITS AND VEGETABLES DURING FROZEN STORAGE

C. FUSTER, G. PRESTAMO and J. ESPINOSA
Instituto del Frío. Ciudad Universitaria. Madrid-3 (Spain)

Summary

The products studied have been: mushroom, apple slices and strawberry. Influence of various treatments, blanching and packaging (atmospheric pressure and under vacuum) on the quality of frozen mushroom has been investigated. Blanching and different chemical treatments have been used to inhibite the o-diphenol-oxidase activity. The samples packed under vacuum show better appearance and colour. Blanched mushroom has a fibrous and spongy texture and the colour is too white. The best results were obtained when dipping in metabisulfite. - This treatment decreases strongly the o-diphenoloxidase activity and gives gives a product with the best texture and colour. Effects of - different treatments and vacuum on the texture and colour of frozen apple slices (Golden Delicious and Rennet varieties) have been studied. Previous assays were made to select the pre-freezing processing. Calcium and sulfite treatment has been found to be the best, it is - very effective in firming the apple slices and maintaining the colour. Effects of pretreatments and freezing on different Spanish strawberries varieties have been investigated. Sugar and salts treatments have been applied and they do not have great influence on the product texture. It is not necessary to apply any treatment prior to freezing. Among the studied varieties, Cambridge variety is considered the best for freezing.

1. INTRODUCTION

Influence of treatments prior to freezing on the quality and stability of fruits and vegetables during frozen storage has been studied in mushroom, strawberries and apple slices.

1.1. Mushrooms

After thawing frozen mushrooms undergo a strong browning and texture loss. Browning is caused by enzymes which catalyze the oxidation of phenolic substances. Phenoloxidase catalyzes the oxidation of phenolics to quinones, which then condense to form melanins. Sulfite interfers with this condensation by combining irreversibly with the quinones to form colorless addition products.

It also gradually diminishes the ability of the enzyme to oxidize o-diphenols.

Ascorbic acid acts as an antioxidant, reducing the quinones to original phenolics. Blanching is used to control the activity of mushrooms enzymes. However, under-or over-blanching can cause poor quality mushroom products.

The texture of blanched and normally frozen mushrooms indicates unfavourable thoughness. Experiments with unblanched mushrooms confirm enzyme-induced defects with regard to colour and flavour after storage at -20ºC. Therefore, the retention of texture of these unblanched frozen mushrooms has indicated the necessity of investigating alternative technological treatments, which might avoid the harmful action of enzymes in the unblanched frozen product. Since the oxygen dependence of most enzymes is generally known, some experiments heve been carried out with regard to the removal of oxigen from the fresh product.

1.2. Apple slices

Frozen sliced apples have been used for pies and other baker products. Freezing and thawing disrupt the celular structure of most plant tissue, including apple slices. It is apparent that the loss of firmness is partly due to disruption of the tissues. In frozen apple tissue there is a considerable amount of oxigen, so that oxidation catalyzed by polyphenol oxidase occurs and rapid browning results, especially during thawing. It would be desirable to have a pre-freezing process which would yield a thawed product resembling the fresh one in texture, flavour and colour.

Several treatments to improve the texture and to prevent the browning in apple slices have been reported in the bibliography. Treatments to protect the colour of these slices have usually involved a sulfite dip or ascorbic acid as anti-oxidants. Inmersion in a calcium solution was effective in preserving texture over an long storage period, as well as having a significant effect with ascorbic acid or sulfur dioxide in preventing browning. Sugars were also used in the treating solution for their effect in lowering the freeezing point of the apple tissue and in maintaining the colour and flavour.

1.3. Strawberries

Different treatments have been applied to strawberries before freezing to reach after thawing a good texture and colour, similar to the fresh product. Strawberries which were processed before freezing in water solution of 0.1 to 0.4% of various colloids showed a definite improvement in texture and in their organoleptic properties. The added colloids were either pectin, an alginate, or an agar-agar extract. Sugars with small amounts of Ca salts increase the firmness of frozen fruits.

The rate of freezing is recognized as a critical factor in tissue damage. Freezing in liquid nitrogen was best for retention of anthocyanin content and berry colour. The water holding capacity shows physical changes in the tissue and there is a significant correlation coefficient between the shear press values and percent weight loss measured by centrifugation. This suggest that the latter may be used as an objetive test for measuring textural changes in some processed foods.

2. MATERIALS AND METHODS

2.1. Preparation of samples

2.1.1. Mushrooms

The mushrooms were washed, cleaned and cut vertically into four segments. The quarters were divided into eight lots

Treatment	Packing	
	Lot	Lot
Control	(a.p) 1	(V.) 5
a) Blanching (2 min)	" 2	" 6
b) + dipping in 0.02 % $K_2S_2O_5$ (5 min)	" 3	" 7
c) Dipping in 0.2% $K_2S_2O_5$ (5 min)	" 4	" 8

The control and processed samples were frozen at -100ºC in liquid nitrogen, until the temperature was -30ºC in the core of the product, packed in polyethylene bags (200 g samples), heat-sealed at the atmospheric pressure (a.p.) and under vacuum (V.) and stored at -20ºC.

2.1.2. Apple slices

In view of the results of preliminary essays, the following treatments have been chosen:
Treatment 1. A 30 Brix degrees syrup
Treatment 2. A 30 Brix degrees syrup containing 0.28% calcium chloride
Treatment 3. A 30 Brix degrees syrup containing 0.39% sodium sulfite
Treatment 4. A 30 Brix degrees syrup containing 0.28% calcium chloride and 0.39% sodium sulfite.

Varieties of apple studied were Golden Delicious and Rennet. The apples were peeled and cored in the laboratory, then cut in slices (3-5 mm) longitudinally. The slices were divided into 4 lots and inmediately dipped in the solutions being tested. Slices were dipped for 5 min in the chosen treatment then drained, packed in polyethylene bags (150 g samples), heat sealed at the atmospheric pressure and under vacuum, frozen at -30ºC in a tunnel and stored at -20ºC.

2.1.3. Strawberries

Different Spanish strawberries varieties were investigated, Tioga and Tuff varieties from Huelva and Tioga and Cambridge varieties from Salamanca. When the strawberries reached the laboratory, they were cleaned, decaped by hand, washed and drained; afterwards they were soaked for 1 minute in the following treatments:
a) A solution containing 30% sugar, 0.3% starch and 0.3% agar-agar
b) A solution containing 0.28% $CaCl_2$ and 0.5% NaCl

The control sample received no treatment prior to freezing

Variety	Lot	Treatment
A) Tioga (Huelva)	1	a)
"	2	b)
"	3	Control
B) Tuff (Huelva)	4	a)
"	5	b)
"	6	Control
C) Tioga (Salamanca)	7	a)
"	8	b)
"	9	Control

D) Cambridge (Salamanca)	10	a)
"	11	b)
"	12	Control

The samples were frozen at -100ºC in liquid nitrogen, until the temperature was -20ºC in the core of the product. The samples were packed in poliethylene bags, sealed and stored at -20ºC.

2.2. Analytical and sensory methods

Preliminary tests of the samples with the different treatments were made before and after freezing. The methods used are described in previous reports (1, 2, 3).

2.2.1. Mushrooms

The following analysis were made: peroxidase and o-diphenoloxidase activities, texture measurements (Kramer), drip loss and organoleptic properties.

2.2.2. Apple slices

Organoleptic tests and texture measurements were made.

2.2.3. Strawberries

The following analysis were made: drip loss, soluble solids, total anthocyanins, measurements of optical density and sensory tests.

3. RESULTS AND DISCUSSION

3.1. Mushrooms

The obtained results of peroxidase and o-diphenoloxidase activities are shown in Table I

TABLE I

Peroxidase and o-diphenoloxidase (active form) activities of fresh, frozen and frozen-treated mushroom during frozen storage at -20ºC

	Peroxidase activity			O-diphenoloxidase activity			
	Initial fresh	Initial frozen	8 months -20ºC	Initial fresh	Initial frozen	2 months -20ºC	8 months -20ºC
1 (a.p.)	0.210	0.220	0.370	15.462	7.46	9.48	8.13
2 "	0.120	0.010	0.015	0.133	0.53	0	0.53
3 "		0.040	0.010		0.27	0	0.13
4 "		0.210	0.805		3.73	3.06	5.13
5 "		0.280	0.216		12.26	8.39	9.33
6 "		0.030	0.030		0.13	0	0.13
7 "		0.020	0.010		0	0	0
8 "		0.250	0.195		5.06	4.52	4.53

The peroxidase activity of mushrooms is very low. On freezing the mushrooms, peroxidase activity practically undergoes no change.

On freezing fresh mushrooms, the o-diphenoloxidase activity decreases about 50%. If the mushrooms are packed under vacuum, this decrease is about 20%. In mushrooms treated with metabisulfite before freezing (lots 4 and 8) the activity decreases more, about 75%.

Blanched mushroom (with and without treatment) after thawing have similar values of texture to those of the raw product (1). Samples without blanching show similar texture, which is about 50% lower than that of the fresh mushroom. These values are practically maintained during 8 months of frozen storage. There is no detectable difference between the samples packed at the atmospheric pressure and under vacuum.

The drip loss is high, values from 47 to 56%. There is no detectable difference between the treated and control lots.

Organoleptic properties. Frozen mushroom without treatment has a dark brown colour and bad appearance, normally and vacuum packed. In mushroom treated with metabisulfite the appearance is good and the colour slightly brown (rather good), better in vacuum packed samples. Blanched mushroom, with and without treatment, has good apppearance and a yellowish colour in all lots and during frozen storage.

The texture in cooked mushroom is firm in samples without blanching (with and without treatment) and does not change with frozen storage. Blanched mushroom has a spongly and fibrous texture.

3.2. Apple slices

Table II shows the values of shear strength (kg/g sample) in sealed samples at atmospheric pressure (a.p.) and under vacuum (v.). The value of shear strenght in Golden Delicious fresh apple slices was 1.66 kg/g and in Rennet 1.15 kg/g. Calcium treatments harden apples.

TABLE II

Shear strength. Changes in frozen apple slices with different treatments in sealed samples at the atmospheric pressure (a.p.) and under vacuum (v.)

	Storage time (months)	Shear strength (kg/g)							
		Treatment 1		Treatment 2		Treatment 3		Treatment 4	
		(v.)	(a.p.)	(v.)	(a.p.)	(v.)	(a.p.)	(v.)	(a.p.)
Golden Delicious	0	0.66	0.75	1.09	0.93	0.71	0.85	0.75	1.08
	2	0.83	0.93	1.26	0.75	1.05	0.63	0.80	0.90
	4	0.90	0.65	1.19	0.63	0.87	0.90	0.59	0.81
	6	0.72	0.64	0.82	0.79	0.65	0.59	0.73	0.77
Rennet	0	0.14	0.19	0.24	0.17	0.22	0.23	0.36	0.30
	2	0.23	0.18	0.30	0.22	0.30	0.24	0.34	0.33
	4	0.15	0.14	0.21	0.17	0.23	0.21	0.25	0.26
	6	0.16	0.14	0.19	0.19	0.22	0.20	0.19	0.20

Organoleptic properties have been evaluated. Apple slices sealed under vacuum become yellowish after freezing, the colour is more uniform than in the samples sealed at atmospheric pressure. In both varieties the slices treated only with a syrup present brown blotches in the whole product after thawing. In Rennet apple slices from the second month of frozen storage the colour was practically brown. The colour of the apple slices treated with calcium and sulfite is rather good, but the best colour is obtained with calcium-sulfite.

The flavour and odour are good in all the lots after 6 months of frozen storage.

3.3. Strawberries

The obtained results (drip loss, Brix degrees, optical density at 500 nm and 295 nm, total anthocyanins (Acy) and peroxidase activity after the freezing process in the control and treated samples are shown in Table III.

TABLE III

Inicial values of investigated strawberries varieties after the freezing process (control and treated samples)

Lot	Drip loss %	º Brix	$O.D_{500}$	$O.D_{295}$	mg/Acy/100g	Act. peroxidase / 100 g
1	50.24	7.26	0.324	0.420	56.16	0.044
2	45.47	7.00	0.233	0.383	42.13	0.046
3	56.06	6.93	0.336	0.490	58.34	0.033
4	60.92	7.51	0.250	0.376	45.57	0.049
5	50.72	6.16	0.275	0.400	47.74	0.040
6	59.96	7.2	0.315	0.456	53.57	0.045
7	53.09	9.66	0.290	0.635	58.47	0.039
8	44.57	10.06	0.358	0.349	63.53	0.033
9	54.54	8.46	0.390	0.473	64.04	0.043
10	53.53	13.86	0.207	0.437	44.92	0.006
11	48.29	11.10	0.243	0.555	53.94	0.003
12	54.10	10.76	0.177	0.642	41.26	0.003

The drip loss in the control samples ranged from 54 6o 60%; it suggest that the product structure has been seriously damaged. The Tuff variety of Huelva shows the biggest drip loss. The salt treatment reduces the drip loss.

There is a good correlation between the anthocyanins content and the values of optical density at 500 nm, as expressed by the lineal equation

$$113.42\ O.D_{500} + 20.35 = \text{mg Acy/100 g}$$

whose correlation coefficient and standard error of the estimation are c = 0.91 and e = 3.23 respectively.

According to the test panel, preference order for the fresh product is:

Cambridge variety (like best) > Tioga (Salamanca) > Tioga (Huelva) = Tuff (Huelva)

For the frozen products, preference order is:

Cambridge > Tioga (Huelva) = Tuff (Huelva) > Tioga (Salamanca)

4. CONCLUSIONS

4.1. Mushrooms

On blanching the mushrooms (cut in quarters) for 2 min in boiling water, peroxidase and o-diphenoloxidase activities are inhibited.

The treatment with metabisulfite decreases strongly the o-diphenoloxidase activity.

The samples packed under vacuum show better appearance and colour.

Cooked blanched mushroom has a fibrous and spongy texture.

Among the used treatments it is the metabisulfite treatment that gives a product with the best texture.

4.2. Apple slices

Calcium treatment harden the apples. Sulfited treatment is best for maintaining a light colour and the flavour is not adversely affected. Our taste panel confirmed this. Calcium and sulfite treatments have been found to be very effective in firming the apple slices and maintaing the colour. By using suitable proportions of calcium and sulfite in a dip, the qualities of firmness, colour and flavour can be balanced to give the most desirable product after frozen storage.

4.3. Strawberries

Among the studied varieties, Cambridge variety is considered the best for freezing; it is the that undergoes least damage in the freezing process.

The salts treatment reduces the drip loss.

According to the test panel, Cambridge is preferred to the other and of the treatments the sugar one is the best.

It is considered that the two treatments give a frozen strawberry with a good flavour and appearance, but in general the control sample likes the best.

The treatments before freezing have not great influence on the product texture.

It is not necessary to apply any treatment prior freezing.

REFERENCES

1. PRESTAMO, G. and FUSTER, C. (1982). Influence of various treatments and blanching on the quality of frozen mushroom. IIR. Meeting Com. B2, C2, D1. Sofia.
2. FUSTER, C., PREESTAMO, G. and ESPINOSA, J. (1982). Frozen apple slices: effects of different treatments and vacuum on the texture and colour. IIR-Meeting Com. B2, C2, D1. Sofia.
3. FUSTER, C., PRESTAMO, G. and CANET, W. (1982). Effects of pretreatments and freezing on different strawberry varieties. IIR-Meeting Com. B2, C2, D1. Sofia.

THE EFFECT OF BLANCHING AND FREEZING RATE ON THE TEXTURE OF POTATOES, CARROTS AND PEAS, MEASURED BY MECHANICAL TESTS

W. CANET and J. ESPINOSA
Instituto del Frío. Ciudad Universitaria. Madrid-3 (Spain)

Summary

Thermal treatment by blanching and freezing has a fundamental effect on vegetable structure and its external manifestation, that is, its texture. The effect of each of these treatments on potatoes, carrots, and peas was evaluated by different instrumental tests using an "Instron Food Testing Instrument" on samples of each of the products subjected to different forms of blanching, different rates of freezing, and various complete processes.
Conventional blanching and slow freezing rates result in irreversible losses in the rheological characteristics of the structure. In contrast, increasing the freezing rate and step-wise blanching- treatment at low temperature for a long time, followed by cooling and then conventional blanching- improve the rheological behaviour and final texture of the products. This beneficial effect of rapid freezing and the structural alterations caused by step-wise blanching are detectable by the techniques used even after cooking.

1. INTRODUCTION

The thermal treatments comprising the process for preparing frozen vegetables affect the vegetable structure, causing substantial, irreversible damage to its texture. Frozen vegetables are normally cooked prior to consumption, and it is felt that this final stage of the process minimizes the differences in texture produced by proper preparation (appropriate blanching and a rapid rate of freezing), making them imperceptible to consumers.

Instrumental methods of measuring texture are based on the study of the product's behaviour when subjected to various mechanical tests, and they offer the advantage of defining the variables measured in well known physical units and of being fast and easy to replicate /1/.

The object of the present experiment was to assess the effect of different forms of blanching -conventional and step-wise - and different rates of freezing on the final texture of potatoes, carrots, and peas, by studying their mechanical properties.

2. EXPERIMENTAL PROCEDURE

The varieties used were: the "Jaerla" potato, "Nantes" carrot, and "Negret" pea, all grown in Toledo, Spain. Each of the products was subjected to a specific preparation process, and the following sample populations were used in the study: for potatoes, tubers with a diameter of over 50 mm and a specific weight of between 1.049 and 1.059 g/cm^3; for carrots, roots 30 to 45 mm thick, with 4 cm from both the neck area and the root tip being discarded from the samples; for peas, grains ranging between 7.5 and 10.2 mm in thickness.

The following mechanical tests were used for the instrumental analysis of texture and to assess the effect of blanching, freezing, and the overall process: for potatoes, compression and shear /2/; for carrots, shear and the Kramer shear cell (KSC) /3/; and for peas, the Kramer shear cell, Back extrusion cell (BEC) /4/, and Ottawa texture measuring system (OTMS) /5/.

In the case of potatoes and carrots, the measurements were performed using specimens 25.4 mm in diameter and 5 mm thick, and for peas using 50 g of the product. Lots of 20 potato specimens, 40 carrot specimens, and 250 g of peas comprised the experimental units which, after they had been put through the different thermal treatments, were used to carry out the instrumental tests.

The mechanical tests were performed using an Instron Food Testing Instrumental (model 1140) under standard conditions of deformation speed. The variables determined from the resulting force-deformation curves were: for potatoes, maximum compression force to rupture (N), maximum shear force (N), and modulus of elasticity (N), making ten replicates per test (n = 10); for carrots, the variables measured were maximum shear force (N), with n = 10, and maximum force using the KSC (N), n = 5; for peas, the variables measured were maximum force using the KSC (N), OTMS (N), and BEC (N), n = 5 in each case.

The effect of blanching was evaluated by the mechanical testing of experimental units that had undergone conventional blanching: for potatoes, 97ºC for 2 min.; for carrots, 97ºC for 1 min.; and for peas, 97ºC for 90 sec. In step-wise blanching, the experimental units were put through a first stage at low temperature for a long time (LT-LT), carried out at 70ºC for ten minutes, followed by cooling, and then a second stage (HT-ST) equivalent to conventional blanching, in each case. Blanching took place in thermostat-controlled baths in which a constant product weight/volume of water ratio was maintained, the blanching time being established using qualitative tests of peroxidase and catalase activity.

The effect of freezing was studied by considering different rates of freezing (without prior blanching). For potatoes and peas, freezing was carried out by forced air convection at -30ºC and by forced convection in nitrogen vapor at -80ºC. For carrots, freezing was by forced air convection at -30ºC and -50ºC. The frozen products were tested mechanically after cooking in boiling water according to the recommended FAO/OMS international standard.

To study the effect of the overall process, experimental units for each of the products subjected to the different processes were tested mechanically, with combinations of conventional and step-wise blanching, the different freezing rates used, and subsequent cooking. The mechanical tests were in all cases performed using product brought to 20ºC.

The results obtained for the different variables were analyzed statistically using analysis of variance applying the complete nested model, and the LSD method was applied to compare the calculated means at the 1% significance level.

3. RESULTS AND DISCUSSION

The mean values of the mechanical parameters obtained for each of the products studied are shown in Table I for potatoes, Table II for carrots, and Table III for peas, according to the thermal treatment performed. The corresponding values for fresh and cooked control samples are likewise shown.

TABLE I

EFFECT OF BLANCHINGS, RATE OF FREEZING AND COMPLETE PROCESS IN SEVERAL MECHANICAL TESTS IN POTATOES.

Mean values and LSD (1%)

Treatments		COMPPRESION RUPTURE FORCE (N)	MODULUS OF ELASTICITY (N/cm^2)	SHEAR RUPTURE FORCE (N)
Fresh control	(A)	945.60 a	375.4 a	54.73 a
Blanching - 97ºC 2'		738.92 b	265.36 b	52.54 a
- 70ºC 10' and 97ºC 2'		746.87 b	254.13 b	54.46 a
LSD (1%)		54.71	27.51	5.09
Cooked control	(B)	82.02 a	78.69 a	9.44 a
Freezing forced air-30ºC and cqoked		83.1 a	22.29 b	5.48 b
" vapors N_2 - 80ºC " "		91.63 a	77.78 a	9.62 a
LSD		15.96	6.28	1.71
Cooked control	(C)	82.02 a	78.69 a	9.44 a
97ºC 2' and -30ºC and cooked		21.97 c	21.74 c	2.90 c
" " -80ºC " "		49.49 b	31.17 b	6.49 b
70ºC 10' and 97ºC 2',-30ºC and cooked		23.32 c	23.33 c	5.21 b
" " " ,-80ºC " "		87.31 a	70.99 a	9.68 a
LSD (1%)		15.55	7.39	1.59

Sections A in Tables I, II and III give the mean values for the different mechanical parameters found for potatoes, carrots, and peas, respectively, both for fresh produce and that subjected to conventional and step-wise blanching. The corresponding LSD (1%) values are also shown.

For all the mechanical parameters measured for the three products studied, blanching and the fresh control, when compared, show significantly different values, except for the maximum shear force, for which the values are significantly the same. Comparison of the means indicates that, for both forms of blanching, the values of the variables are significantly the same, but lower and significantly different from those for the fresh control.

These results, which coincide for the three vegetables studied, mechanically confirm the adverse effect of blanching on the vegetable structure, leading to decreased resistance to deformation and rupture of the tissues, which is interpreted to be a loss in intercellular pressure and intercellular adhesion due to the solubilization of the pectic substances and a reduction in the impermeability of the cell membrane. /6/.

The results obtained for the maximum shear force for potatoes and carrots, this variable being interpreted as the response to the degree of rupture of the cell walls, make apparent the slight effect of blanching in this regard and show a slight beneficial effect from step-wise blanching.

TABLE II

EFFECT OF BLANCHINGS, RATE OF FREEZING AND COMPLETE PROCESS ON SEVERAL MECHANICAL TESTS IN CARROTS. Mean values and LSD (1%)

Treatments		SHEAR RUPTURE FORCE (N)		KRAMER SHEAR FORCE (N)	
Fresh control	(A)	57.23	a	1593.08	a
Blanching -97°C 1'		56.50	a	1247.83	b
" -70°C 10' and 97°C 1'		59.19	a	1280.46	b
LSD (1%)		3.56		300.17	
Cooked control	(B)	34.3	a	624.55	a
Freezing forced air -30°C and cooked		20.67	c	405.03	b
" " " -50°C and "		26.34	b	388.08	b
LSD (1%)		3.72		50.56	
Cooked control	(C)	34.3	ab	624.55	a
97°C 1' and -30°C and cooked		26.36	c	333.2	c
" " -50°C " "		28.51	c	458.64	b
70°C 10' and 97°C 1', -30°C and cooked		30.87	bc	431.2	bc
" " " , -50°C " "		36.69	a	509.6	b
LSD (1%)		4.96		100.1	

Comparative analysis of the results shows that blanching affects the structure of peas to a greater extent than in does of potatoes and carrots; this latter vegetable is the least affected, and its mechanical response is less clear, because of the lower homogeneity of its structure.

TABLE III

EFFECT OF BLANCHINGS, RATE OF FREEZING AND COMPLETE PROCESS ON SEVERAL MECHANICAL TESTS IN PEAS. Mean values and LSD (1%)

Treatments		KRAMER SHEAR FORCE (N)		EXTRUSION FORCE (N)		OTTAWA CELL FORCE (N)	
Fresh control	(A)	2915.5	a	4880.4	a	4037.6	a
Blanching 97°C 90''		1832.6	b	3371.2	b	2582.3	b
" 70°C 10' and 97°C 90''		1710.1	b	2851.8	b	2577.4	b
LSD (1%)		150.58		675.4		210.25	
Cooked control	(B)	916.3	a	1274	a	1283.8	a
Freezing forced air -30°C and cooked		1004.5	a	1240.6	a	1342.6	a
" vapors N_2 -80°C " "		980	a	1332.8	a	1332.8	a
LSD (1%)		87.22		148.76		131.9	
Cooked control	(C)	916.3	a	1274	a	1283.8	a
97°C 90'' and -30°C and cooked		677.18	b	960.4	b	1004.5	b
" -80°C " "		719.32	b	1024.1	b	989.8	b
70°C 10' and 97°C 90'', -30°C and cooked		793.8	ab	1151.5	ab	1087.8	b
" " " , -80°C " "		925.12	a	1274	a	1372	a
LSD (1%)		138.15		194.56		178.41	

Sections B in Tables I, II and III show mean values for the different mechanical parameters obtained, respectively, for potatoes, carrots, and peas in the form of cooked controls and vegetables subjected to different rates of freezing (without prior blanching and then cooked. The corresponding LSD (1%) values are also shown.

Comparative analysis of the mechanical response obtained shows that the effect of the freezing rate on the structure and texture of carrots is greater than on that of potatoes and peas.

The samples of carrots frozen without blanching and later cooked show values for the variables which are significantly different from those for the cooked control. /7/. For potatoes, the values for the cooked control lots were significantly similar to those obtained for the lots that were fast frozen (-80ºC) and cooked; the values for these lots were higher and significantly different (at 1%) than those for the lots frozen more slowly (-30ºC) and then cooked. These results show the irreversible, adverse effect of slow rates of freezing on the structure and texture of both carrots and potatoes. The favourable effect of fast freezing is particularly apparent in potatoes. For peas cooked after freezing without blanching, the mechanical response obtained is completely independent of the rate of freezing, as demonstrated by the fact that all the variables measured exhibit significantly similar values for the cooked control lot and for the lots frozen at -30 and -80ºC and then cooked.

Sections C, in Tables I, II and III give the mean values for the mechanical parameters obtained for different lots of potatoes, carrots, and peas, respectively, which had undergone the complete processes consisting of the blanching and freezing temperatures mentioned above, followed by cooking. The LSD (1%) values are likewise presented, along with the corresponding values for the cooked control lots.

Comparative analysis of the results obtained demonstrates that, for the three vegetable products studied, the processes in which conventional blanching was carried out at high temperature for a short time exhibit the worst mechanical performance for structure. This adverse effect is made still worse by slow rates of freezing (-30ºC), especially for potatoes and carrots. In contrast, for all three products studied, the lots undergoing step-wise blanching showed the best mechanical performance for structure, and, in the case of fast freezing (-80ºC), the values of the variables studied are significantly equal to those for the cooked controls at the 1% level. This mechanical response is interpreted as being due to the action of pectin-methyl-esterase, PME, as a result of which pectates of Ca^{2+} and Mg^{2+} are deposited in the cell walls, which consequently become thicker; it may also be due to the gelatinization and retrogradation of starch, which increases structural resistance to the tensions undergone during freezing. /8/.

These results demonstrate instrumentally the beneficial effect of low temperature-long time (LT-LT) treatment prior to conventional blanching and fast freezing on the mechanical performance of the structure and final texture of potatoes, carrots, and peas. With the techniques used, these effects are detectable even after cooking.

REFERENCES

1. SZCZESNIAK, A.S. (1963). Objetive Measurements of Food Texture. J. Food Sci., 28, p. 410.
2. CANET, W. (1980). A study on the influence of the thermal treatments blanching, freezing and thawing on the texture and structure of potato. (Solanum tuberosum L.). Thesis. Universidad Politécnica de Madrid.

3. KRAMER, A. and AAMLID, K. (1953). The shear press, an instrument for measuring the quality of foods. III. Applications to peas. Proc. Am. Soc. Hort. Sci., 61, pp. 417-423.
4. BOURNE, M.C. and MOYER, J.C. (1968). The extrusion principle in texture measurement of fresh peas. Food Technol., 22, pp. 1013-1018.
5. VOISY, P.W. (1971). The Ottawa Texture Measuring System. J. Can. Inst. Food Technol., 4, p. 91.
6. CANET, W., ESPINOSA, J. and RUIZ, M. (1982). Effects of blanching an rate of freezing on the texture of potatoes measured by mechanical test. IIR. Proc. of Meeting. Commissions B2, C2, D1. Sofia.
7. CANET, W. and ESPINOSA, J. (1983). Influencia del proceso de congelación en la textura de vegetales. Efecto del escaldado y la velocidad de congelación en la textura de zanahoria (Daucus carota L.). Rev. Agroq. y Tecnol. de Alimentos. In press.
8. REEVE, R.M. (1977). Pectin, Starch and Texture of Potatoes: some practical and theoretical implications. J. Text. Studies, 8, p.1.

THE CONSEQUENCES OF VARYING DEGREES OF BLANCHING ON THE QUALITY OF FROZEN GREEN BEANS

K.Z.Katsaboxakis and D.N.Papanicolaou
Institute of Technology of Agricultural Products Athens
Greece

Summary

The effect of the degree of blanching (0-3 min) in boiling water (98°C ±2°C), on the quality of green beans during frozen storage at -18°C, was studied. Quality was evaluated by organoleptic tests (for colour, odour and taste) and determination of total chlorophyll, L-ascorbic acid peroxide index, colour (Hunterlab) and peroxidase activity. Samples which were frozen without previous blanching showed a rapid increase in peroxide index and a 75% loss of L-ascorbic acid during the first 3 months of frozen storage. The same samples showed an over 50% loss of total chlorophyll after 12 months and were considered organoleptically unacceptable even from the first months of frozen storage. Samples blanched from 0.5 to 3 min had a good total chlorophyll retention with no significant differences among them, for the first 6 months of frozen storage. However after that period an increased loss rate was found in samples blanched from 2 to 3 min. The organoleptic evaluation of odour and taste during frozen storage gave no significant differences for samples which had undergone blanching. However a significant green colour difference was found for mean colour scores among samples which received different degree of blanching. Samples blanched from 30 to 90 secs had a more intense colour than samples which received more prolonged heat treatment.

1. INTRODUCTION

Freezing improves the storage life of vegetables by retarding enzyme activity, chemical changes and microbial growth. However since 1930's it has been recognized that some enzymic activity occurs even at sub-zero temperatures (1) and that freezing alone does not completely prevent off flavour development, colour and texture deterioration in frozen vegetables. It has been also found many years ago that blanching, a brief heat treatment prior to freezing prevents many of the undesirable changes in frozen vegetables as a result of the inactivation of naturally occurring enzymes. Blanching improves also the visual green colour of vegetable due to the removal of gases from the vegetable surface and from the intercellular spaces and causes a drastic reduction of micro-organisms which exist in the vegetable surface and which otherwise would result in a rapid build up of infection on processing plants and the final product.

The adverse effects of blanching are mainly the solubilization and / or destruction of some nutrients and vitamins in the blanching medium and the conversion of green chlorophylls to yellow-green pheophytins (6,7,8). Colour deterioration during frozen storage is also closely related to changes in chlorophylls. It has been found (12,13) that two forms of colour deterioration occurred during storage of frozen beans. The first is associated with the conversion of chlorophyll to pheophytin and the second causes destructi of both chlorophyll and pheophytin. It was shown that both forms of colour degradation were depended upon the degree of heat-treatment prior to freezing and may be also related to the process of fat peroxidation. It was also observed (10) that some vegetables can tolerate an amount of residual peroxidase activity after blanching without significant loss of quality during frozen storage. Finally it should be pointed out that blanching is energy and time consuming process and should be limited to an optimum level.

Alternatives replacing conventional blanching have been also studied (2,9). Vacuum packaging for example was found not effective as blanching for long term frozen storage periods. Conclusively blanching is the most effective treatment prior to freezing and its optimization for each vegetable should be studied. This study was carried out to examine the consequenses of varying degrees of blanching on the quality of green beans during frozen storage. The variety studied (barbounia) is the first in consumption of fresh beans in Greece and is also partly used in the freezing industry today. The final purpose of this study is to examine the behavior of this variety to freezing and frozen storage.

2. MATERIALS AND METHODS

Processing and storage of beans. Fresh green beans (var.barbounia) were obtained from a local farm near Athens and processed as a whole on the day of harvesting. Beans were graded manually in order to separate the ones with similar stage of maturity (medium). After washing in tap-water seven experimental sub-batches were formed from the original which were subjected to different degrees of blanching from 0 to 3 min and every 30 secs. During blanching a volume of tap water was used each time and enough heat imput (steam) in order to prevent cessation of boiling due to the addition of beans. After cooling and drying, beans were frozen in a blast-type freezer of Frigoscandia (m-model) spread in large rectangular perforated aluminum trays. Finally beans were packed into high density polyethylene bags and stored at -18°C until analyzed.

Peroxidase activity. Determinations were made spectrophotometrically by using the method H_2O_2 /Guaiacol (5). The extraction of the enzyme was carried out in IM NaCI and the rate of guaiacol peroxidation was calculated by measuring the change in absorbance at 470 nm of the sample against a blank. The results are expressed as Δ.DO.470 nm. min^{-1}.

Lipid oxidation: Lipid were extracted from the frozen sample by the method of Bligh and Dyer (3) as modified by Walker (12) Peroxide index was determined by the ferric thiocyanate peroxide method. Appropriate blank solutions were prepared each time to compensate for absorbance at 480 nm due to non-peroxide materials.

L-ascorbic: Determination was carried out by potentiometric titration (Dead-stop) of the acid extract of the frozen sample, with dichlorophenol-indophenol. The dye solution was standardized each time using a fresh known solution of L-ascorbic acid.

Chlorophylls and pheophytins: They were determined spectrophotometrically in a 80% acetone extract of the vegetable tissue by measuring the absorbances at specific wave lenghts and using the equations developed by Vernon (11).

Colour by Colour Difference Meter (Model D25D2L): The instrument was standardized with a green plate (No.C2-2790 L=64.4:a_L=-14.5:b_L=5.4). Samples were measured thawed and spread in a 11cm diameter viewing dish.

Sensory evaluation: The intensity and uniformity of green colour of samples was evaluated by a taste panel immediately after thawing on a 7-point scale. The same samples were scored on a 7-point hedonic scale for their odour. Taste (eating quality) was also evaluated organoleptically on cooked samples prepared by a standard recipe and presented to panelists who asked to score on a 7-point hedonic scale.

3. RESULTS AND DISCUSSION

Peroxidase activity:Peroxidase is the most heat-stable enzyme in the majority of vegetables and is very commonly used for determined the adequacy of the blanching process. Figure 1 shows on a semilogarithmic scale the effect of blanch time at 98°C (±2°C) on peroxidase activity of green beans stored for 0,6 and 12 months at -18°C. It can be seen that 30 secs blanch treatment leaves approximately 7% of the level in the unblanched sample. For a complete or near complete enzyme inactivation 1 min blanch time is required. Measurments made during frozen storage indicated no significant peroxidase regeneration even after 12 months storage period.

Ascorbic acid retention:Figure 2 shows that over the entire storage period L-ascorbic acid levels were higher in samples which have recieved 30 to 60 secs blanch treatment,than in samples with a more prolonged blanch time. Unblanched samples showed a 75% loss in ascorbic acid content during the first 3 months of frozen storage and an almost complete loss 6 months later. Ascorbic acid content should be considered in the case of frozen vegetables as a valuable and sensitive index for the evaluation of the thermal process and of the frozen storage conditions.

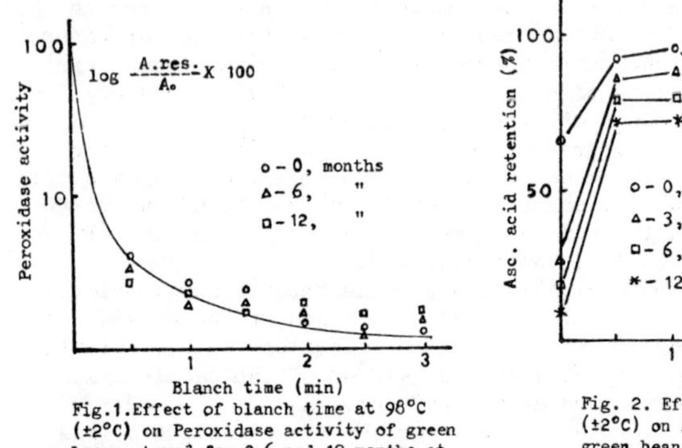

Fig.1.Effect of blanch time at 98°C (±2°C) on Peroxidase activity of green beans stored for 0,6 and 12 months at -18°C.

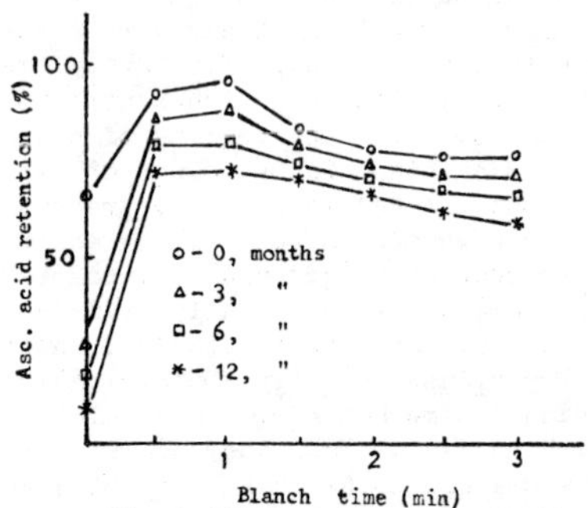

Fig. 2. Effect of Blanch time at 98°C (±2°C) on Ascorbic acid retention of green beans stored for 0.3.6.12 months at -18°C.

Changes in chlorophylls: The degree of blanch treatment influences the rate of chlorophyll deterioration during processing as well as during frozen storage. Figure 3 shows that during blanching an amount of total chlorophyll is converted to pheophytin, the conversion being proportional to the thermal treatment applied on sample. At the initial steps of blanching chlorophyll a appears to be less stable than chlorophyll b, but afte the 4th minute of blanching increased destruction rate was observed and fo chlorophyll b. Chlor.a/chlor.b ratio decreases contineously as blanch time increases from 0 to 4 minutes and then increases again due to the change in the destruction rate of chlorophyll b.

During frozen storage samples with different degree of blanching exhibited different behavior as far as chlorophyll retention is concerned as indicated in Figure 4. This figure shows the proportion of total chlorophyll as % of the initial value that remained in samples subjected to various heat treatments, as a function of storage time. In unblanched samples a constant rate of total chlorophyll loss can be observed during the entire storage period, as indicated by the almost straight line relationship between storage time and % of chlorophyll retained in samples.

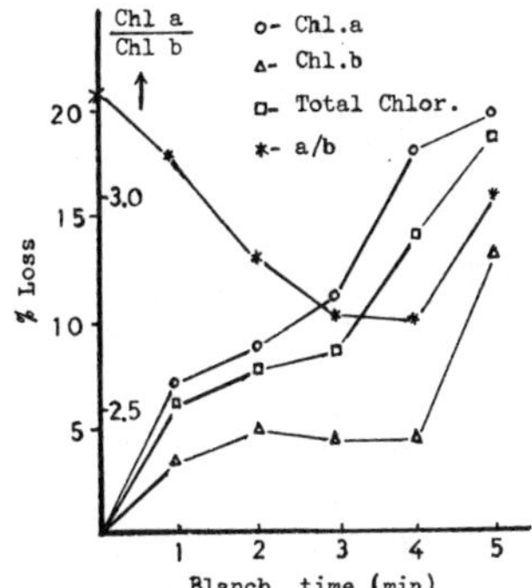

Fig.3. Effect of blanch time on % loss of chlor.a, chlor.b,total chlor. and chl.a/b ratio

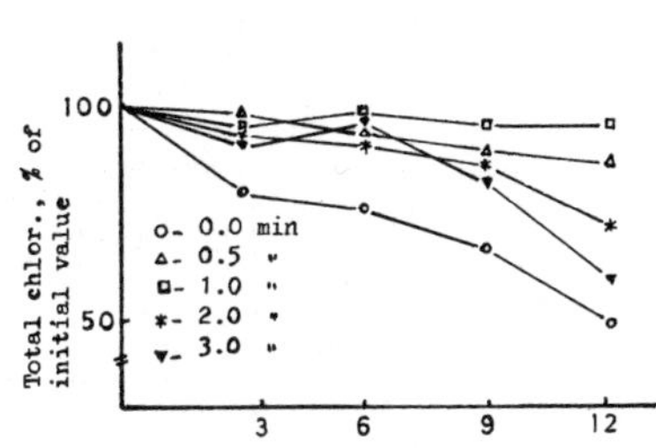

Fig.4.Total Chlorophyll, % of initial value of green beans blanched for 0,0.5, 1,2 and 3 mn during frozen storage at -18°C.

The loss of total chlorophyll diminished in samples blanched from 30 to 60 secs. Samples blanched from 2 to 3 minutes showed no significant loss in total chlorophyll for the first 6 months of frozen storage compared to samples with lower heat-trearment. However after that storage period a progressive increase was observed in the rate of chlorophyll loss. This loss rate was increased in samples of beans received more drastic thermal treatment. These results are in agreenment with those found by other workers (6,12,13) and indicate that changes in chlorophylls continue during frozen storage and are also depended or related to the amount of residual enzyme activity and / or the degree of heat-treatment given during blanching. Thermal treatment higher than that required for enzyme inactivation results in increased loss rates in chlorophylls after a certain lag storage period, apparently due to oxidation and related to the process of fat peroxidation as proved by other workers. (4,12).

Lipid oxidation: As shown in figure 5, peroxide index (O.D.480 nm) of unblanched samples increased rapidly during the first 6 months of frozen storage and then decreased slightly for the rest of the storage period. No significant differences were observed among samples with different degree of the thermal treatment. A slight increase of peroxide index is indicated during the last months of frozen storage in samples blanched for 30 and 80 secs.

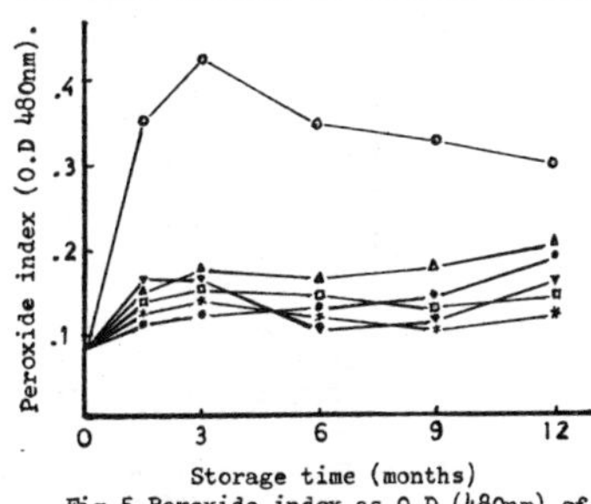

Fig.5 Peroxide index as O.D (480nm) of green beans blanched for 0,30,60,90,120, 180 secs during frozen storage at -18°C. 0-0, Δ-30, □-60, *-90, ▼-120, ●-180.

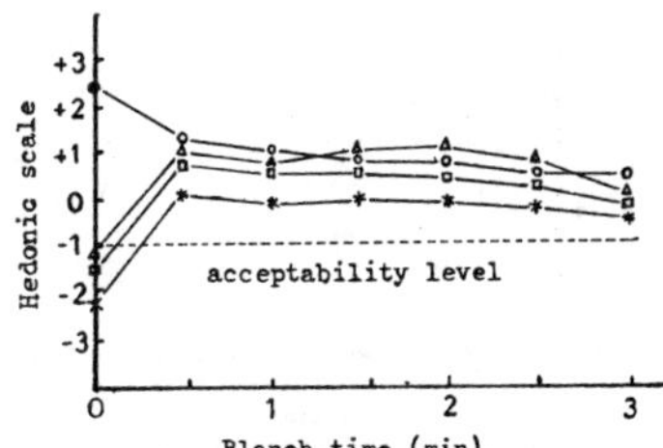

Fig.6. Effect of blanch time at 98°C on odour score of frozen-thawed green beans stored for 0,1.5,6,12 months at -18°C. o-0, Δ-1.5, □-6, *-12 months.

Table I. Hunter Colour Reflectance values* of frozen green beans blanched from 0 to 3 min and stored for 3,6 and 12 months at -18°C.

Blanch Time sec.	3 months			6 months			12 months		
	L	a_L	a_L/b_L	L	a_L	a_L/b_L	L		a_L/b_L
0	43.0	-12.0	0.64	43.3	-11.4	0.64	43.2	-11.2	0.64
30	41.8	-13.3	0.70	41.7	-12.5	0.66	42.3	-12.7	0.65
60	41.5	-13.7	0.71	39.3	-12.4	0.69	41.0	-13.1	0.68
90	40.9	-12.8	0.68	40.3	-12.6	0.65	41.8	-12.9	0.63
120	41.4	-12.2	0.62	39.8	-11.8	0.63	41.1	-12.5	0.64
150	41.3	-12.1	0.62	39.4	-10.8	0.60	41.5	-11.7	0.60
180	40.1	-11.8	0.61	39.0	-10.4	0.58	41.6	-11.2	0.59

* All Hunter values are means of triplicate determination

Colour by Hunter-Difference Meter:Table I lists the Hunter colour reflectance values for each treatment and for 0,6 and 12 months storage periods. Samples blanched from 30 to 90 secs gave increased negative a_L values and a_L/b_L ratio in all measurements made during frozen storage. Negative a_L reflectance values which measure the greeness of the samples and a_L/b_L ratio indicated a fairly good correlation with sensory colour scores for the same samples as shown in figure 8. Conclusively it can be pointed out that a_L Hunter colour reflectance values and a_L/b_L ratio differentiate well between samples with different degree of thermal treatment. This differentiation is not significant for the same sample along the frozen storage period.

Sensory evaluation: The results presented in figure 6 and 7 demonstrate that although unblanched samples got a higher score for their odour and taste at zero time they were considered organoleptically uncceptable for the same attributes even from the first months of frozen storage and at any time subsequently. As shown in the same figure 30 secs blanch treatment prevents successfully off odours development and taste deterioration during frozen storage.Odour and taste scores did not differ significantly among samples received different heat treatment. However samples blanched from 30 to 90 secs were judged slightly superior than the others which receive more drastic heat treatment.

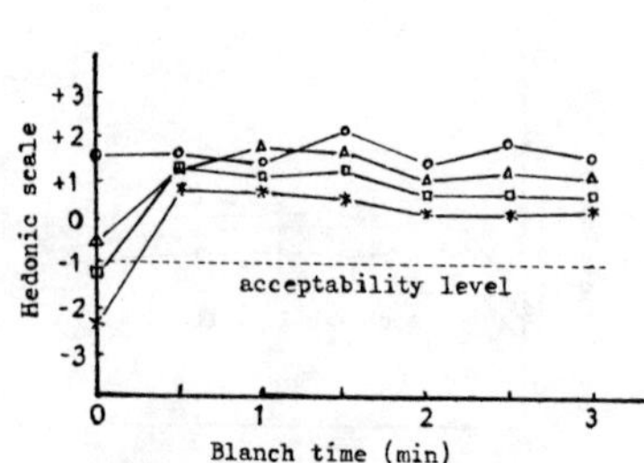

Fig.7.Effect of blanch time at 98°C on taste score of cooked green beans stored 0,1.5,6,12 months at -18°C.
0-0, ▲-1.5, □-6, ✱-12 months.

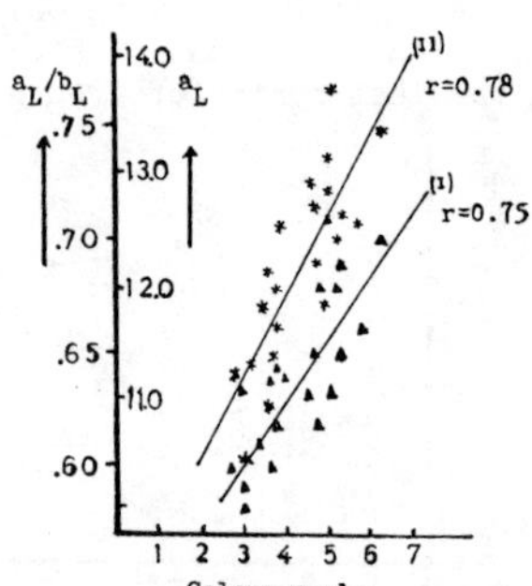

Fig.8. Relation between organoleptic scores for green colour in frozen green beans and the respective values of a_L and the a_L/b_L ratio (I) for the same samples.

Table II. Mean sensory scores for colour evaluation* of frozen green beans blanched from 0 to 3 min, stored for 0,3,6 and 12 months at -18°C.

Storage -18°C	Blanch time (sec) 0	30	60	90	120	150	180	F test	LSD P=0.05
0	3.22	6.25	4.77	5.45	4.48	4.28	3.65	* *	0.671
3	3.63	6.13	4.94	4.91	4.66	3.50	3.30	* *	0.676
6	3.56	5.76	5.13	5.20	4.93	3.50	2.93	* *	0.788
12	2.86	4.63	4.76	4.50	3.83	3.73	2.90	* *	0.783

* Green colour intensity and uniformity was scored in a 7 point scale.

As shown in table II a statistically significant difference was found for mean colour scores between samples without blanching and blanched samples. A significant also difference can be observed between samples blanched from 30 to 60 secs and samples subjected to a more prolonged thermal treatment.

4. CONCLUSIONS

Data presented here demonstrate that blanching conditions and mainly the amount of thermal treatment given influences the quality of green beans during frozen storage. Samples frozen without blanching deteriorate rapidly in all quality attributes becoming within the first months of frozen storage organoleptically unacceptable. The same samples showed an increased peroxide index, a rapid L-ascorbic loss and a gradual total chlorophyll deterioration. The absence of the above changes in blanched samples indicates that the reactions involved are principally enzymic. A blanch time of 30 secs at 98°C causes a 93% reduction on peroxide activity preventing effectivelly all the undesirable changes mentioned previously. Samples subjected to a thermal treatment which exceeds the required for the inactivation of peroxidase showed reduced L-ascorbic acid retention and after a certain storage period increased rate of chlorophyll deterioration. These results are in agreement with those found by other workers (6,12,13) and indicate that changes in chlorophylls in beans may be initiated either enzymatically or by heat. A blanch time between 30 and 60 secs at 98°C was found to be the optimum treatment in order to obtain the best quality of the tested variety of beans in the frozen state for a long time.

REFERENCES

1. BALLS, A.K. and LINEWEAVER, H. (1938). Action of enzymes at low temperatures. Food Res. 3:57.
2. BIRNBAUM, N.R. et al. (1979). Evaluation of evacuated packages as an alternative to blanching for frozen spinach. J. Food Sci. 44:404.
3. BLIGH E.G. and DYER W.J. (1959). A rapid method of total lipid extraction and purification. Can J. Biochem. Physiol. 37:911
4. BUCKLE,K.A. and EDWARDS R.A. (1970). Chlorophyll degradation and lipid oxidation in frozen unblanched peas. J.Sci. Fd. Agric. 21:307.
5. CLOCHARD A.et GUERN J. (1973). Destruction thermique de l'activité péroxydasique. Interprêtation des courbes expérimentales, Rev. Gen. Froid No. 8:860.
6. DIETRICH, W.C. et al. (1959b). Time-Temperature Tolerance of frozen foods. XVIII. Effect of blanching conditions on colour stability of frozen beans. Food Technol. 13:258.
7. FODA Y.H. et al. (1968). Effect of blanching and dehydration on the

conversion of chlorophyll to pheophytin in green beans. Food Technol. 22:119.

8. PHILIPPON J.(1969). La congélation des haricots vert. Effets combinés du temps et de la température pendant la préparation et la congélation sur la qualité finale du produit. Rev.Gen. Froid No 1:101.
9. STEINBUCH E. (1980a). Quality retention of unblanched frozen vegetables by vacuum packing II. Asparagus, parsley and celery J. Food Technolol. 15:351.
10. STEINBUCH E. (1980b). The effect of heat shocks on quality retention of green beans during frozen storage. J. Food Technol. 15:353.
11. VERNON L.P. (1960). Spectrophotometric determination of chlorophylls and pheophytins in plant extracts. Annal. Chem. 32:1144.
12. WALKER, G.C. (1964a). Color deterioration in frozen green beans (Phaseolus vulgaris). J. Food Sci. 29:383.
13. WALKER, G.C. (1964b). Color deterioration in frozen French beans (Phaseolus vulgaris). 2. The effect of blanching. J. Food Sci. 29:389.

BRUSSELS SPROUT AND CABBAGE PEROXIDASES - HEAT STABILITY AND ISOENZYME COMPOSITION

K.M. McLELLAN and D.S. ROBINSON
Procter Department of Food Science
The University of Leeds

Summary

Peroxidase in Brussels sprouts and cabbage has been found to occur in both soluble and bound forms. The soluble peroxidase shows greater stability than the bound peroxidase and also possesses the capacity for partial regeneration following heat inactivation. Isoelectric focussing has been used to demonstrate that peroxidase activity is due to a number of isoenzymes. Soluble peroxidase includes both cationic and anionic isoenzymes while bound peroxidase contains only cationic isoenzymes. Initial work indicates that the soluble anionic peroxidase isoenzymes may have greater heat stability than the soluble cationic or bound cationic isoenzymes.

Peroxidase is a highly heat stable enzyme belonging to the oxidoreductase class. Generally H_2O_2 is used as the oxidising agent in a reaction known as the peroxidatic function of peroxidase:

$$AH_2 + H_2O_2 \longrightarrow A + 2H_2O$$

AH_2, the hydrogen donor may be one of a number of organic compounds including several phenols and amines. Other reactions catalysed by peroxidase are the oxidatic and hydroxylation reactions requiring molecular oxygen as oxidising agent, and the catalatic reaction, in which peroxidase acts in the same way as catalase. While these reactions are generally believed to be of minor importance in fresh fruit and vegetables, their role may be greater in heat treated products due to the destruction of other components.

The heat stability of peroxidase and the large number of reactions catalysed by the enzyme have led to the suggestion that peroxidase may cause quality deterioration in fruit and vegetable products (1,2). While no direct evidence is available, several authors have reported a correlation between off-flavour development and peroxidase activity (3,4). The ability of peroxidase activity to regenerate following heat treatment can result in an increase in the activity of the enzyme during storage of heat treated vegetables, for example Brussels sprouts (3).

Plant peroxidases are comprised of haem prosthetic groups attached to a glycoprotein component. Inactivation is believed to involve the dissociation of the two components, while regeneration is thought to be a result of the reassociation of the haem with the glycoprotein (5,6). More severe heat treatment may result in the modification of either the haem or the glycoprotein which prevents reassociation (6).

Peroxidase consists of a variety of isoenzymes catalysing the same, or similar reactions, and containing identical haem groups but differing in the precise composition of the glycoprotein. Inactivation of peroxidase has been reported to be a biphasic process (7,8) and it is possible that this is due to peroxidase isoenzymes having different heat stabilities (5,9). In order to confirm this, however, it is necessary to isolate and characterise fully individual peroxidase isoenzymes, before studying their heat inactivation characteristics.

A large number of peroxidase assays are available. Two commonly used methods employ o-dianisidine and guaiacol as hydrogen donor and these methods can also be used for staining peroxidase activity in gels following separation of the isoenzymes by electrophoretic techniques. These assay methods measure only the peroxidatic function and provide no information concerning the other three potential peroxidase activities.

Proteins can be separated in a layer of polyacrylamide gel, according to their isoelectric points, by a technique known as isoelectric focussing. By then staining the gel specifically for peroxidase activity, it is possible to detect and compare the peroxidase isoenzymes present in various vegetable extracts. Up to 17 distinct bands can be identified on gels stained for peroxidase following isoelectric focussing of Brussels sprout extracts.

FIGURE I

Isoelectric Focussing of Soluble Peroxidase Fractions

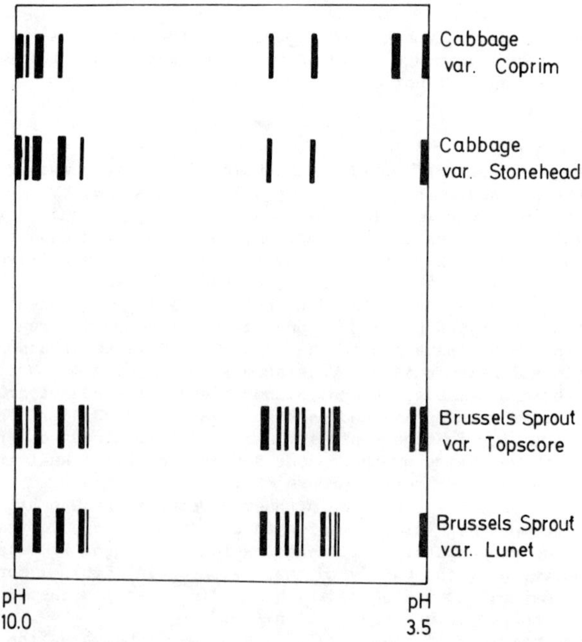

Initially column gels of approximately 5 mm diameter were used for isoelectric focussing. Peroxidase in soluble extracts of cabbage, Spring cabbage and Brussels sprouts was shown to consist of a variety of isoenzymes ranging in isoelectric point (pI) from approximately 3.7 to 10.0 (10). These isoenzymes formed three distinct groups: a cationic group with pI = 8.0-10.0, a second group with pI = 5.2-6.2, and finally, the anionic isoenzymes with pI = 3.7-4.8. Using these column gels, however, it was not possible to separate the cationic isoenzymes into distinct components. More recently a flat bed isoelectric focussing system has enabled separation on thin layer gels and improved results have been obtained. The lack of separation using column gels is believed to have been caused by the application of the sample at the cathode. Using the flat bed system, the sample can be applied at any point between the anode and cathode, and the most satisfactory results have been achieved when the sample is applied some 3 cm from the cathode. Figure I shows an example of isoenzyme profiles obtained from extracts of different varieties of Brussels sprout and cabbage.

Two types of peroxidase exist in vegetables: those occurring in the soluble portion of the cell, and those occurring in a bound form, held by ionic interactions to the cell wall (11). Both types were extracted and subjected to isoelectric focussing. The soluble peroxidases contained the three groups of isoenzymes described above, however, peroxidases occurring ionically bound to the cell wall were found to be of the cationic type with pI = 8.0-10.0.

Early work on the heat stability of soluble and bound peroxidases in crude cabbage, Spring cabbage and Brussels sprouts showed the soluble peroxidases to be slightly more heat stable than the bound peroxidase activity (12). A typical result obtained by heating crude soluble extracts to a variety of temperatures, and then determining the residual peroxidase activity is shown in Figure II. These results were obtained using cabbage soluble peroxidase extract, and demonstrate the stability of the enzyme to heating at up to 75°C. The curves seem to indicate that the inactivation process is at least biphasic with an initial period of rapid inactivation, lasting some 2 to 3 minutes, followed by a more prolonged period of much slower inactivation.

Partial purification of cabbage and Brussels sprout soluble extracts and separation of the peroxidase isoenzymes into those having a net positive charge (cationic) and those with a net negative charge (anionic) at pH 8.0, allowed comparison of the heat stabilities of anionic and cationic isoenzymes. Heating at 65°C for 10 minutes was found to cause approximately twice as much inactivation of the cationic isoenzymes as the anionic isoenzymes. Both groups of isoenzymes still showed biphasic inactivation. Extracts of cabbage and Brussels sprouts containing peroxidases which occur ionically bound to the cell wall have been found to contain only cationic isoenzymes and this is consistent with the lower heat stability of bound peroxidases compared to soluble peroxidases.

FIGURE II

Heat Inactivation of Cabbage Soluble Peroxidase

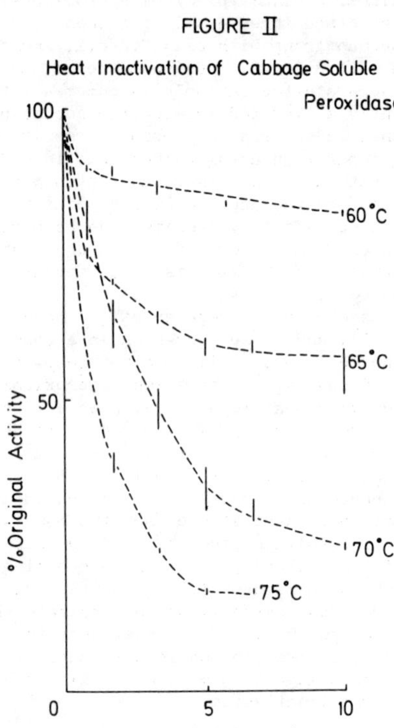

Regeneration of peroxidase activity following inactivation by heating has been shown to occur in soluble extracts from cabbage, Spring cabbage and Brussels sprouts. In the case of bound peroxidases, only those extracted from Brussels sprouts were found to regenerate, and the reason for this has not been identified. The anionic group of isoenzymes which were found to be more heat stable than the cationic isoenzymes have also been found to have a greater capacity for regeneration. Again, the apparent inability of cationic isoenzymes from soluble extracts to regenerate following heat inactivation is consistent with the lack of regeneration generally observed for bound peroxidase activity which is believed to be cationic in nature. Regeneration is a pH dependant process occurring readily at the natural pH of cabbage and Brussels sprouts (approximately pH 6). At pH 4, however, regeneration does not occur.

Present work is towards isolating and characterising individual peroxidase isoenzymes. A method has been developed using gel filtration and ion exchange chromatography which enables a cationic and an anionic isoenzyme to be obtained, uncontaminated by other isoenzymes. The purity of such isolated isoenzymes is assessed using isoelectric focussing. The use of very thin gels (0.5 mm) enables the detection of small amounts of peroxidase isoenzymes.

REFERENCES

1. BURNETTE, F.S. (1977) J. Food Sci. 42, 1-6
2. HAARD, N.F. (1977) In "Enzymes in Food and Beverage Processing" Eds. R.L. Ory and A.J. St Angelo, Amer. Chem. Soc. Symp., Series no. 7, p.143-171
3. STEINBUCH, E., HILHORST, R.A., KLOP, W., ROBBERS, J.E., ROL, W. and VAN DER VUURST DE VRIES, R.G. (1979) J. Fd. Technol., 14, 289-299
4. LEE, Y.C. and HAMMES, J.K. (1979) J. Food Sci. 44, 785-787
5. LU, A.T. and WHITAKER, J.R. (1974) J. Food Sci. 39, 1173-1178
6. TAMURA, Y. and MORITA, Y. (1975) J. Biochem. 78, 561-571
7. YAMAMOTO, H.Y., STEINBERG, M.P. and NELSON, A.I. (1962) J. Food Sci. 27, 113-119
8. DUDEN, R., FRICKER, A., HEINTZE, K., PAULUS, K. and ZOHM, H. (1975) Lebensm. Wiss u. Technol. 8, 147-150
9 WANG, S.S. and DIMARCO, G.R. (1972) J. Food Sci. 37, 574-578
10. McLELLAN, K.M. and ROBINSON, D.S. (1983) Phytochemistry 22, 645-647
11. GKINIS, A.M. and FENNEMA, O.R. (1978) J. Food Sci. 43, 527-531
12. McLELLAN, K.M. and ROBINSON, D.S. (1981) Food Chem. 7, 257-266

SUITABILITY OF SOME MATURITY INDEX TO DETERMINE OPTIMUM HARVEST TIME OF GREEN BEANS FOR FREEZING

G. CRIVELLI, G. BERTOLO, A. MAESTRELLI, E. SENESI

Instituto Sperimentale per la Valorizzazione Technologica dei Prodotti Agricoli (I.V.T.P.A.) - MILANO - Italy

Summary

The quality of frozen green beans can be improved especially by finding some suitable varieties which then are harvested at the optimum level of ripeness. The validity of the following parameters: distribution of size, seed/pod weight ratio and seed length have been examined as index for easy and quick evaluation, suitable for defining the most convenient moment of harvesting. A common index for yield and quality of frozen green beans is still not known. The optimum harvest time can only be decided with sufficient precision if the results obtained from all three parameters are considered.

1. INTRODUCTION

Green beans represent 20% of the Italian production of frozen vegetables. Over the last ten years the production has increased notably, reaching 9800 ton in 1979 with an increase of more than 50% with respect to the 6000 ton of 1970. The consumption has had almost the same increase. The possibility of having a further increase is linked to many factors, among which the high quality of the frozen products is of great importance.

The quality of the frozen product depends mainly on two kinds of factors: the first is that based on the so-called T.T.T., that links the temperature and the storage time with the product's stability, while the second one, that is based on the so-called P.P.P., refers to the influence made by Product - Processing - Packaging (P.P.P.) on the quality and storage life of the frozen product. While the storage conditions are already well known and further knowledge is particularly connected to the necessity of saving energy, the P.P.P. factors, that are often more decisive for end quality than time and temperature, need further studies especially concerning the quality of raw material.

In fact it has been seen that if the optimum storage conditions have been observed, the quality of the frozen product depends mostly on the high quality of raw material and to a less extent on processing which has in fact reached such a technological level that it guarantees the best conditions for processing. At this moment the only processing operation that could be improved is blanching, a process which is being examined very carefully by others. Technological processing can not improve the quality of the vegetabl it can only try to keep it more or less identical to the fresh product at the moment of freezing.

There are mainly two factors which determine the quality of the raw material: the suitability of variety and the best stage of maturity. The numerous studies on the suitability of the variety have contributed greatly to the excellence of the frozen product's quality, although the variety with the best yield and quality has not been discovered yet.

There is still a lot to do in order to find the index of maturity with the purpose of deciding the optimum harvest time. Normally this latter point doesn't coincide with the period of maximum productivity, as the best nutritive and organoleptic properties are found at the beginning of ripening.

The choice of an analytical parameter for the estimation of the index of maturity is conditioned mainly by its simplicity. With the aim of giving a further contribution to the research already carried out on the material index (2), a research programme has been started and the first results appear in this paper.

2. MATERIALS AND METHODS

The tests were carried out on green beans coming from an autumn harvest. The vegetables were blanched in boiling water at 98°C for 90" and then cooled. The freezing of unpackaged products was performed in air blast tunnel at -35°C for 25'. The frozen samples were packed in polythene bags and stored at -20°C. The parameters taken into consideration as possible index to estimate the maturity level were:

1) size distribution; 2) the ratio between the weight of the seeds and pods; 3) the length of the most developed seeds.

The frozen product's quality was examined after 12 months of storage by determining the sloughing and collapse and by organoleptic analysis. The sloughing grade was determined following the Van Buren method (3), but measuring the level of sediment after centrifugation, in order to make it easier. The following relation between the sediment volume and the sloughing degree was used:

$< 0,5$ ml = low ; $> 0,5 < 1$ ml = medium ; > 1 ml = high

Organoleptic properties were tested by a 8 member trained panel for colour, flavour and firmness, using an hedonic scale with 9 scores.

3. RESULTS

3.1. Size distribution

There exists a moment for each variety when it is at its highest processing suitability. This is different to the best ratio between economic yield and quality. The pod's sizing can be considered as quite a valid index to decide the optimum harvest time of the green beans, if the best percentage for each size is known for each variety. In fact, making the harvest later the size distribution tends to move towards higher size with a considerable improvement of the yield and a lowering of quality.

Table 1 - Percentage distribution of the size (A), seed/pod weight ratio (B) and seed length (C) (σ = standard deviation).

Cultivar / size	BEL AMI					PROS GITANA					VALJA				
	A	B	σ	C	σ	A	B	σ	C	σ	A	B	σ	C	σ
I <7.5mm	25,71	2,78	4,08	4,36	0,55	13,83	2,95	1,47	4,93	0,78	6,70	1,78	0,70	4,24	0,70
II 7.5–8.5mm	67,20	4,14	1,32	5,02	0,60	72,00	5,39	1,72	6,50	0,93	26,62	2,75	1,49	5,67	1,06
III 8.5–9.5mm	2,52	6,84	3,12	6,81	1,36	7,24	7,72	1,30	7,93	0,79	46,35	4,06	1,48	6,91	0,89
IV >9.5mm											14,99	6,16	2,27	7,93	1,17

An exact knowledge of the optimum values is therefore very important for deciding the harvest time. This however is not completely sufficient in order to define the quality level of the green beans because in the same size pods can be found with completely different properties.

3.2. Weight ratio seed/pod

In a high quality product, the seed/pod weight ratio is less than or equal to 5% (tab. 1). For the sizes 3 and 4 a ratio more then 5% but less than 10% can be considered valid. Some knowledge on the properties and the behaviour of the variety over some years is necessary, since even if the average values of the seed/pod weight ratio are at optimum level in each single size, there can be quite a wide variability which has to be considered when the level of maturity is estimated and a judgement is made on the freezing suitability.

3.3. Seed length

Concerning the length of the seed this must always be less than 8 mm. Seeds which are too ripe cause a lower quality even when the ratio seed/pod is less than what is considered optimum.

The latter parameters are therefore necessary to calculate in order to complete the results obtained from size distribution.

3.4. Sloughing and collapse

These properties are the defects most commonly found in frozen green beans after cooking. They don't seem to be caused by the ripening level but are variety characteristics. They can be controlled and prevented above all by breeding and screening varieties resistant to change caused by blanching and freezing.

4. CONCLUSIONS

It has been shown that:

1) the three parameters examined individually are not sufficient to decide the most convenient moment for harvesting;
2) useful information can be obtained by collecting all results with reference to the highest and lowest values for each variety;
3) for each variety it is necessary to define the percentage of the size in relation to the connection between yield and quality;
4) sloughing and collapse seems not to be connected to the level of maturity but above all to the instrinsic characteristics of the varieties.

REFERENCES

1. CRIVELLI, G., POLESELLO, A., BOGNETTI, U. (1970). Modificazioni nei fagiolini surgelati durante la conservazione. Atti XIX Congresso Nazionale del Freddo - Padova.
   SZANTO-NEMETH, E. (1975). Investigation on string bean varieties for their suitability to quick freezing. Acta Alimentaria, 4, 229.
   CRIVELLI, G., SCOZZOLI, R. (1975). Researches on green bean and pea varieties suitable to quick freezing. Proceedings of XIV International Congress of Refrigeration - Moscow.
   SENESI, E., CRIVELLI, G., BERTOLO, G. (1977). Recherches sur le comportement des légumes à la congélation rapide. Annali I.V.T.P.A., vol.III.
   CRIVELLI, G., MAESTRELLI, A., BERTOLO, G., ALLAVENA, A. (1979). Ricerche sul comportamento alla congelazione degli ortaggi. Annali I.V.T.P.A., vol. X.
2. POLESELLO, A., CRIVELLI, G., LOCATELLI, A. (1972). Primi rilievi sugli indici di raccolta del fagiolino per la surgelazione. Industrie Agrarie, 10 (3), 127.
   DURANTI, A., PROTO, D. (1981). Studio sugli indici di maturità del fagiolino. Riv. Ortoflorofrutticoltura It., 65, 433.
   STANLEY, R. (1981). Maturity index needed to up quality. Grower, July, 30, p. 32.
   PHILIPPON, J., ROUET-MAYER, M.A. (1977). Etude de developpement physiologique des haricots verts à la récolte et qualité du produit surgelé fini. Annales Technol.Agric., 23 (4), 433.
3. Van BUREN, J.P., MOYER, J.P., DILSON, D.E., ROBINSON, W.B., HAND, D.B. (1960). Influence of blanching conditions on sloughing, splitting, and firmness of canned snap beans. Food Technol. 14, 223.

STUDIES ON REDUCING SHRINKAGE AND IMPROVING SENSORY QUALITY OF FROZEN MUSHROOMS

T. R. Gormley,
An Foras Taluntais,
Kinsealy Research Centre,
Malahide Road,
Dublin, 5, Ireland.

Summary

A 1% xanthan gum solution, applied as a vacuum soak treatment, reduced shrinkage in blanched frozen mushrooms. This treatment gave a weight loss at blanching of 4.4% compared with 14.2% for samples vacuum soaked in water. The flavour of the mushrooms was not affected by the xanthan gum treatment but it did confer a softer texture to the mushrooms.

1. INTRODUCTION

Shrinkage during blanching is a major problem in the freezing of mushrooms (1-3) and techniques used to reduce weight loss include water soaking both at atmospheric (1, 2) and reduced pressure (4, 5), covering mushrooms with a batter (2), hot-air blanching (6) and soak/chill store/ soak treatments (1, 7). If left unblanched frozen mushrooms develop a brown colour unless frozen storage conditions are very good (3). Frozen mushrooms also have a tough leathery texture (1).

Of the techniques outlined above, the soaking of mushrooms in water under vacuum has been the most convenient and successful with blanching losses in the range 7.5-10% being reported (5). Singh et al. (4) reported yield increases between 10 and 15% in canned mushrooms through the use of pectin-calcium, methyl cellulose or carboxymethylcellulose, applied as a 1.5% aqueous vacuum soak prior to blanching. However, organoleptic quality was impaired. In view of these findings tests were conducted, in the present study, with "Keltrol" - a food grade xanthan gum - applied to the mushrooms as a vacuum soak as a possible method of reducing shrinkage during blanching. White strain mushrooms from different flushes were used and the mushrooms were assessed for shrinkage during blanching and for drip loss, colour, texture and flavour after freezing.

2. EXPERIMENTAL

White strain mushrooms (*Agaricus Bisporus*) were used throughout and in some cases they were stored at 2°C in a refrigerator for up to 4 days prior to use. The food grade xanthan gum used was "Keltrol" (Kelco/AIL International, Ltd., London UK) and it was applied to 200 g batches of mushrooms by soaking them in an aqueous solution of the gum under vacuum (560 mm Hg) for 30 min in an Edwards high vacuum oven at ambient temperature (22-26°C) unless stated otherwise. Samples were held in the solutions for 10 min after releasing the vacuum and were then sprayed with water and drained (2 min) prior to weighing to assess weight gain. Blanching was carried out in water at 100°C for 5 min, cooling was for 5 min in running water and samples were then drained for 2 min prior to weighing. The drained samples were blast frozen (-30°C) in plastic bags for 2 hr and were stored in cabinets (-20°C) prior to testing. Drip loss was assessed after thawing for 3.5 hr at 22-26°C; colour of the thawed product was assessed with a Hunter Colour Difference Meter (L values) using 10 mushrooms per sample and a 2.5 cm specimen port. Texture was evaluated on 100 g samples with a Kramer Shear Press fitted with a standard test cell.

2.1. Level of xanthan gum

The effect of 0, 0.25, 0.50, 0.75 and 1.00% xanthan gum solutions on blanching loss was investigated. Viscosity of the xanthan solutions was measured at 25°C with a Brookfield RVT viscometer (spindle 3) at speed 100. Samples were given to 5 tasters who were asked to comment on the flavour and mouth-feel properties of the treated mushrooms.

2.2. Effect of mushroom age post-harvest

The effect of a 1% xanthan gum solution on blanching loss in mushrooms stored for 0-4 days at 2°C was investigated in comparison with vacuum treatment in water only.

2.3. Effect of sodium metabisulphite

The effect of different levels (0, 0.1, 0.2, 0.3, 0.4%) of sodium metabisulphite (SMB) on the 1% aqueous xanthan gum system and on weight gain, blanching loss, freezing performance, colour and texture of the mushrooms was investigated.

2.4. Level and duration of vacuum treatment

The influence of different levels of vacuum (460, 560 or 660 mm Hg for 30 min), or lengths of time (10, 20, 30 min at 560 mm Hg), on the weight gain of mushrooms in 1% xanthan solution (containing 0.4% SMB) and on subsequent blanching loss was tested.

3. RESULTS AND DISCUSSION

3.1. Level of xanthan gum

The use of xanthan gum in the vacuum soaking solution had a marked reducing effect on blanching loss in the mushrooms (Table 1); weight gain after vacuum soaking was similar (59-62%) for the different treatments. In addition the xanthan solutions exerted a softening effect on mushroom texture after blanching especially at a level of 0.5% or above (Table 1). These data show that vacuum treatment with xanthan gum solutions was

Table 1: Effect of different levels of xanthan gum on shrinkage and texture in blanched mushrooms

Level of xanthan gum (%)	Weight gain (%) after vacuum soaking[1]	Blanching loss[2] (%)	Shear value[3] (kg)
0 (water control)	60	11.0	81
0.25	60	7.8	83
0.50	60	7.3	65
0.75	62	3.8	69
1.00	59	2.3	67

[1] 560 mm Hg vacuum for 30 min

[2] based on weight prior to vacuum soaking

[3] for 100 g sample of blanched mushrooms

considerably better than vacuum treatment with water in terms of mushroom shrinkage and losses recorded were considerably less than those reported in the 3S process (1) or after vacuum soaking in water (5). The xanthan gum treatment had no adverse effect on the mouth feel properties or flavour of the mushrooms as reported by 5 tasters.

3.2. Effect of mushroom age post-harvest

The 1% xanthan gum/560 mm Hg vacuum system was applied to mushrooms of different ages post-harvest. The results (Table 11) show that the xanthan/

vacuum treatment gave considerably lower blanching losses than water/vacuum.

TABLE 11: Effect of vacuum soaking in a xanthan gum solution on weight gain and blanching loss[1] in mushrooms of different ages post-harvest

Mushroom age post-harvest (days)	Weight gain (%) after vacuum soaking		Blanching loss (%)	
	Xanthan	Water	Xanthan	Water
0	48.5	47.0	14.0	23.0
1	58.0	-	6.0	-
1	55.0	59.0	6.5	15.5
2	57.0	-	6.3	-
2	62.0	60.0	2.0	11.0
3	62.0	64.0	0.0	10.3
3	56.0	60.0	2.5	11.0
4	59.0	-	+2.5	-
Mean	57.2	58.0	4.4	14.2

[1] footnotes as in Table 1

The effect of the xanthan solution was particularly noticable in the older mushrooms with a net gain in weight being recorded after blanching for 4-day old mushrooms. However, these mushrooms lost weight (water) during 4 days chill storage and so would be expected to show a low blanching loss. The highest blanching loss was recorded for freshly harvested (day 0) mushrooms (Table 11). The reduction in shrinkage after blanching observed with xanthan is a similar result to that obtained for canned mushrooms by Singh _et al._ (4) using pectin-calcium, methylcellulose, or carboxymethylcellulose. However, these authors do not quote shrinkage values at the blanching stage and so a direct comparison cannot be made.

3.3. Effect of sodium metabisulphite

While the colour of the blanched frozen mushrooms from the xanthan/vacuum system was quite acceptable, SMB was added at levels of 0 to 0.4% to study its effect on colour enhancement and on the properties of the 1% xanthan solution. The results (Table 111) show that 0.1% SMB reduced the

Table 111: Effect of sodium metabisulphite (SMB) on a 1% xanthan gum solution and on weight gain, blanching loss and quality of frozen mushrooms

SMB level (%) in xanthan solution	1% xanthan solution		Weight gain (%)[2]	Blanching loss (%)	Hunter (L)	Shear (kg)
	pH	Viscosity[1] (cps)				
0.4	5.7	659	59	2.5	68	88
0.3	5.9	643	58	3.5	68	101
0.2	6.0	628	60	3.5	65	108
0.1	6.3	600	62	3.8	65	125
0.0	7.4	418	61	2.5	63	131
Water control	-	-	64	10.3	65	193

[1] Brookfield RVT viscometer (25°C), spindle, 3, speed 100

[2] 560 mm Hg vacuum for 30 min

pH of the xanthan solution and increased its viscosity markedly; levels of SMB up to 0.4% produced further reductions in pH and increases in viscosity (Table lll). Level of SMB had no effect on weight gain, blanching loss or drip loss, but high levels gave whiter mushrooms. The magnitude of this effect was not truly reflected by the Hunter L values (Table lll). Addition of SMB also softened mushroom texture and the effect on texture of xanthan + SMB is clearly seen when compared with a shear value of 193 kg for mushrooms vacuum soaked in water without SMB. An SMB level of 0.2% in the 1% xanthan solution gave mushrooms with a highly acceptable texture (shear value 108 kg), similar to that reported for canned mushrooms (1). The blanching loss data in Table lll are based on the mushroom weight prior to vacuum soaking and the shear test was carried out on 100 g of thawed mushrooms.

3.4. Level and duration of vacuum treatment

Soaking mushrooms in 1% xanthan solution (containing 0.4% SMB) at 660 mm Hg vacuum gave a larger weight gain and a smaller blanching loss than those treated at 460 or 560 mm Hg (Table lV). This result is similar to that reported by Steinbuch (5) and McArdle *et al.* (7) for mushrooms vacuum soaked in water. Raising the level of vacuum resulted in mushrooms with a whiter colour and a softer texture (Table lV); no explanation can be given at this point in time for this effect.

Table lV: The effect of level of vacuum and time of vacuum treatment in a 1% xanthan gum solution on weight gain, blanching loss, drip loss and quality of mushrooms[1]

	Weight gain (%)	Blanching loss (%)	Drip loss (%)	Hunter (L)	Shear (kg)
Level of vacuum (30 min)					
460 mm Hg	46.5	11.5	13.2	66	102
560 mm Hg	53.3	10.0	12.8	69	85
660 mm Hg	55.8	3.5	12.3	73	72
Duration of vacuum (560 mm Hg)					
10 min	64.0	+5.0[2]	-	-	-
20 min	58.0	+3.5	-	-	-
30 min	59.0	+2.5	-	-	-

[1] footnotes as in Table lll

[2] mushrooms 4-days old post-harvest

A vacuum treatment time of 10 min was as good as 30 min in terms of blanching performance of the mushrooms (Table lV); in this part of the experiment there were no blanching losses (all gains) but this is probably due to the fact that these mushrooms were stored at 2°C for 4 days before testing.

4. CONCLUSIONS

The results show that a 1% aqueous solution of "Keltrol" (xanthan gum) containing sodium metabisulphite gave a considerable reduction in mushroom blanching loss when applied to the mushrooms as a vacuum soak treatment. A level of vacuum of 660 mm Hg and a soaking time of 0.2 hr gave the best results. The data presented are only preliminary and will be substantiated by additional tests using larger quantities of mushrooms.

5. REFERENCES

1. GORMLEY, T. R. and WALSHE, P. E. (1982). Reducing shrinkage in canned and frozen mushrooms. Ir. J. Fd Sci. Technol., 6, 165-175.
2. GORMLEY, T. R. and WALSHE, P. E. (1974). Frozen French fried mushrooms. Proc. IV Int. Congress Food Sci. and Technol., Madrid, lV, 147-154.
3. GORMLEY, T. R. (1972). Quality evaluation of frozen mushrooms. Mushr. Sci. 8, 209-219.
4. SINGH, R. P., DANG, R. L., BHATIA, A. K. and GUPTA, A. K. (1982). Water binding additives and canned mushroom yield. Indian Food Packer, 36, 39-43.
5. STEINBUCH, E. (1973). Effect of raw material and processing method on the quality and shrinkage losses of canned mushrooms. Ann. Report, Sprenger Institute, 24-26.
6. GORMLEY, T. R. (1975). Hot air blanching of mushrooms. Proc. 5th Ann. Res. Conf. Fd Sci. & Technol., Cork, 34 (abstr.).
7. McARDLE, F. J., KUHN, G. D. and BEELMAN, R. D. (1974). Influence of vacuum soaking on yield and quality of canned mushrooms, J. Fd Sci., 39, 1026-1028.

TOTAL TIME FOR FROZEN FOODS IN THE FREEZER CHAIN AND IN DISPLAY CABINETS

L. BOEGH-SOERENSEN AND K.BECH-JACOBSEN
Department of Food Preservation
The Royal Veterinary and Agricultural University, Copenhagen (Denmark)

Summary

Two surveys on residence time in the freezer chain, especially display cabinets have been carried out. Ten frozen products were included, most of them pork products. The time from day of freezing till arrival to retail store, was 2.1 month or less for 50% of the products, 3.5 months or less for 75%, and 6.4 months or less for 95%.
The total residence time in display cabinets was max. 1.2 weeks for 50% of these frozen products, max. 3.6 weeks for 75%, and max. 9.0 weeks for 95%. For two products with a high price, 95% of the packages spend about 23 weeks or less. For frozen products such as peas and hamburgers 75% of the packages spend max. 2-3 weeks, and 95% max. 6-8 weeks.
The total time in the cabinet will be divided between different positions, i.e. different temperature conditions but the mathematical/statistical work is not finalized.

1. INTRODUCTION

This work is part of COST 91: Effect of thermal processing and distribution on quality and nutritive value of food.

As nearly all frozen food packages in Denmark are labelled with date of freezing we could get very precise information about the time - but not about the temperature conditions - from freezing to display cabinet, by going to the retail stores to record the open date marking on the frozen food packages. The first part of this work, therefore, has been concentrated on display cabinets.

In display cabinets, especially the open types, frozen products in the top layers are subject to rather warm and fluctuating temperatures, which in rather short time will result in diminishing quality for many frozen products. If the product is placed in the center of the cabinet product temperature will be rather cold and uniform, meaning rather small influence on product quality. It was therefore important to get excact information about the total residence time for each package, and equally important to know where in the cabinet the package is placed.

2. METHOD

From autumn 1981 to summer 1982 (Survey A) we cooperated with 6 shops in Copenhagen, from 2 chains, all shops with open-top display cabinets. 2 shops with a quick turnover of frozen foods (no. 111 and 124), 2 shops with medium (no. 112 and 125) and 2 shops with a small turnover (no. 113 and 126). 6-10 frozen food products were selected, mainly pork products. The shops normally get delivery of frozen foods once a week, and the following morning the shop was visited.

At each visit to a shop, information was received about delivery of the selected products and freezing (packing) date, sell-by date etc. was recorded. All new packages were marked with individual self-adhesive labels, and the position of all packages of the selected products was recorded.

For each shop, weekly visits during 6-8 months, and in 2 weeks daily visits were paid.

Temperatures in display cabinets were not recorded, as it was felt that sufficient information about the temperature conditions in open-top cabinets was available, cf. Boegh-Soerensen & Bramsnaes (1977).

From June 1982 to September 1982 (Survey B) we cooperated with 6 shops north of Copenhagen, in an area with summerhouses, beaches etc., i.e. a recreation area. The same method as described above was used. These 6 shops were from different chains, but characterized by a much greater turnover during summer than during winter. As a matter of fact shop no. 136 simply is closed during the winter. From April 1983 to September 1983 (Survey C) we cooperated with 5 shops in Copenhagen.

3. RESULTS

3.1 Time in the freezer chain before display cabinet

Fig. 1 shows the distribution of time in the freezer chain for the 12 shops in Survey A + B. 50% of the frozen food products included in our experiment spend 2.1 months or less, whereas 75% spend max. 3.5 months, and 95% max. 6.4 months.

There are rather big differences between the shops and between the products. Fig. 1 also shows the distribution for shop no. 135 with the most, and shop no. 125 with the least favourable situation. It should be noted that the difference between 50% of the products is little: 1.4 and 3.8 months, while the difference between 95% is 2.9 and 12.5 months.

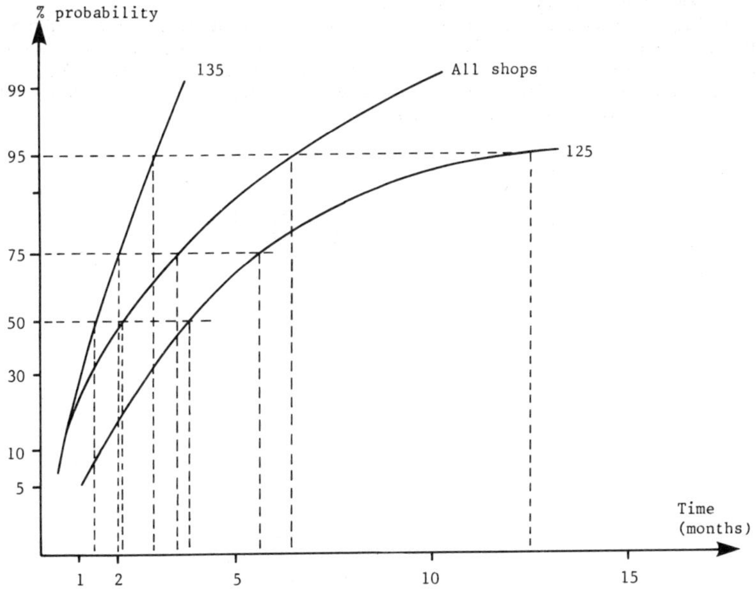

Fig. 1. Probability of time (months) from freezing to retail store.
For all shops in survey A + B, for shop no. 135 and for shop no. 125.

There are big differences between products. Table I shows the distribution of time (months) in the freezer chain for the frozen food products selected in Survey A.

TABLE I

Frozen food product	Months in the freezer chain		
	50%	75%	95%
Prepared meal (fried liver with sauce)	2.8	3.8	6.0
Raw liver	3.4	4.5	7.1
Hamburgers	1.6	2.8	6.9
Ground veal & pork	3.5	4.5	6.1
Pork sausages ("Medister")	1.7	2.6	4.8
Meat balls ("Frikadeller")	2.6	3.6	5.7
Vienna Sausages	4.0	7.2	11.7
Pork chops	1.2	2.2	5.0
Peas	7.3	10.0	13.8

It should be mentioned that retail packages with deep-frozen peas are labelled with packing date, not the day of freezing which takes place about August. We have supposed that all the peas are frozen 1st August.

3.2 Residence time in display cabinets

Fig. 2 shows the distribution of residence time in freezer display cabinets for the 12 shops in Survey A + B. The distribution was calculated for each shop and there was a marked difference between the two chains in Survey A. Fig. 2 also shows the results for shop no. 111 (the shop with the shortest residence time) and for shop no. 135 with the longest residence time.

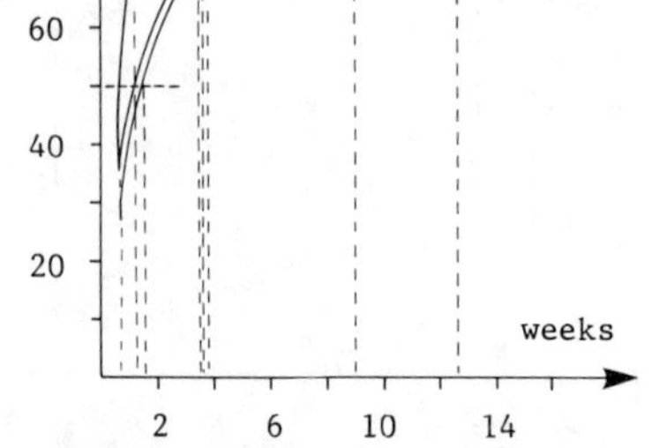

Fig. 2. Residence time in display cabinets, for all shops in Survey A + B, for the best (no. 111) and for the worst (no. 135).

Differences between products are very big, cf. Table II which shows the residence time (weeks) for the products in all shops in Survey A + B.

TABLE II

Frozen food product	Weeks in display cabinet			Average time (weeks) between delivery
	50%	75%	95%	
Prepared meal (fried liver with sauce)	7.0	15.7	27.7	11.1
Raw liver	3.5	6.2	10.7	4.5
Hamburgers	1.2	3.2	9.0	2.2
Ground veal & pork	2.0	4.8	11.2	3.1
Pork sausages	1.4	3.7	8.6	2.6
Meat balls	2.0	4.0	8.9	2.4
Vienna sausages	2.4	5.8	13.3	5.2
Pork chops	2.0	6.8	23.3	3.8
Peas	0.7	1.8	5.7	2.3

As expected, peas move much quicker than the meat products, of which only hamburgers show a reasonable turnover. Meat products with a high price, i.e. prepared meals and pork chops, may spend very long time in display cabinets.

3.3 Time between delivery to the shops

In table II is also shown average time between delivery of the different products.

Of course, there are differences between number of packages in each delivery, but normally 50% of the packages of a given product should be sold in half the time between deliveries of this product. Table II shows that there is a comparatively good agreement with this.

3.4 Stock rotation in display cabinets

In a previous paper, Boegh-Soerensen and Bech-Jacobsen (1983) described the different situations caused by placing new packages above, respectively below the existing packs. However, the mathematical/statistical treatment of the registered positions of all packages of the selected products has not been finalized.

4. DISCUSSION

This investigation should provide more data regarding residence time in display cabinets and regarding the position of the packages in the cabinet during the residence time, i.e. the product temperature.

Such information are very useful in TTT-studies where the fluctuating and rather warm product temperatures in the last components of the freezer chain should be taken into consideration. To be able to simulate the time and temperature to which frozen food products are exposed in the real freezer chain it is necessary to have rather detailed information, especially about display cabinets.

Also, it is important for scientists and for the frozen food industry to be able to calculatorily predict the quality of a product at the time of consumption. This means knowledge about the storage life of the product at various freezer temperatures, as well as knowledge about time and temperature for the product in the freezer chain. Such calculations would make it possible to use realistic storage life labelling, and indicate whether the product has sufficient storage life for the market.

Data for time and temperature in the freezer chain are often given as distribution curves, cf. Fig. 1. It has to be decided what probability you want to use. Using 50% fractiles presumably is too optimistic, and using 95% fractiles would mean that 5% of the packages should be exposed to the worst time and temperature experiences at producers, wholesale, retail, and consumer level. This will very seldom (never) be the case.

Using the 75% probability could be a sensible choice.

The authors will continue the time-temperature survey to include the other components of the freezer chain. Also, the work on display cabinets will be continued in order to obtain statistically reliable data, especially regarding stock rotation.

REFERENCES

1. BOEGH-SOERENSEN, L. AND BRAMSNAES, F.: The effect of storage in retail cabinets on frozen foods. Proc. IIR-meeting in Karlsruhe (1977), p 375-380.
2. BOEGH-SOERENSEN, L. AND BECH-JACOBSEN, K. (1983). Time-Temperature in the freezer chain in Denmark. Proceedings from 16th International Congress of Refrigeration, Paris. (In press).

THE COLOUR STABILITY OF RETAIL PACKED, FROZEN GROUND BEEF

G. BERTELSEN & L. BØGH-SØRENSEN
Danish Meat Products Laboratory, Ministry of Agriculture

Summary

The effect of two display lighting sources and fluctuating temperatures on the colour of frozen ground (comminuted, minced) beef packed in polyethylen chubs was investigated. The samples were produced at a commercial processing line, and stored at -24°C in the dark before displaying; some samples were vacuum-packed before frozen storage. The samples were evaluated visually at the beginning and at the end of the frozen storage period. During display, the visual examination was performed two or three times a week, and the surface colour was measured (objectively) with a Hunter Colour Difference Meter. Exposure to light of frozen ground beef packed in polyethylen caused appreciable colour changes in 3 days and unacceptable colour in 6 to 12 days. No difference was found in colour deterioration of the two investigated lighting sources (Free Powder Lamp and Warm White). For samples protected against light the objective colour measurings indicated a slightly better colour stability for samples at -24°C compared to samples at fluctuating temperatures in the display cabinet; visually evaluated, no difference was found. Vacuum-packaging during frozen storage at -24°C did not improve the colour stability in the display cabinet.

INTRODUCTION

The purple myoglobin is chiefly responsible for the colour of meat. The desirable bright red colour is due to oxymyoglobin (i.e. oxygenated myoglobin). The characteristics of colour degradation is the formation of brownish metmyoglobin.

The colour of frozen meat is influenced by the processing, packaging and packaging materials, freezing rate, storage temperature and time, and display conditions (temperature and light).

According to MacDougall (1982) the major colour problem during display storage is the photo-oxidation of the pigment. Lentz (1971) found that the exposure of frozen beef to light and temperature conditions normally found in display cabinets caused appreciable colour changes in 1 to 3 days and pronounced changes in 2 weeks.

According to Zachariah and Satterlee (1973) the oxidation of frozen oxymyoglobin is dependent on the temperature; the rate of oxidation increases from -5°C to -12°C and decreases at colder temperatures.

Considerable research has dealt with effects of fluctuating storage temperatures. Hustruld et al (1949) reported no significant differences between one sample of ground beef at -18.8°C constantly for six months and another sample at fluctuating temperatures between -18.8°C and -23.4°C. Townsend and Bratzler (1958) indicated decreased colour stability in meat stored at fluctuating temperatures and cyclic defrost temperatures.

This study is a preliminary work done with the main object to determine the effect of two commercial lighting sources and fluctuating temperatures on the colour stability of frozen ground (comminuted, minced) beef. Ground beef was chosen because a large part of frozen meat produced in Denmark is ground beef; most earlier investigations concerning meat colour have dealt with steaks.

MATERIALS & METHODS

Samples and experimental design

Ground beef was manufactured at a commercial processing line. The meat used for the production was fresh slaughtered forequarters, which was coarse-ground followed by a mixing with 20% hydrated soy protein (1 part of soy protein and 3 parts of water), then, the mixture was ground to the final quality. The fat content of the final product was about 21%.

The product was packed in retail-sized (1 lb.) polyethylen (PE) chubs, and was placed in cartons (24 chubs in each carton). The cartons were placed in a blast-freezer (air temperature about -30°C).

After freezing, 25 samples were transported to the Danish Meat Products Laboratory. During transport the product was packed in dry ice. At the laboratory 10 ground beef (samples packed in chubs) were vacuum-packed (five in each bag). All samples were stored three months in the dark at -24°C.

After storage, non vacuum-packed samples were divided into three batches which were allotted to the following treatments:

1. Storage in the dark at -24°C.
2. Displayed under fluorescent tube I in a display cabinet.
3. Displayed under fluorescent tube II in a display cabinet.

Vacuum-packed samples were taken out of the vacuum-packs and divided into two batches allotted to the following treatments:

4. As 2.
5. As 3.

In the display cabinet about 50% of each sample was overwrapped with black carton to prevent the effect of the light.

Packaging materials

The chub was made of 0.082 mm PE, with an oxygen permeability about 3000 cc/m^2/24 h. (measured at 26°C). For vacuum-packaging was used a 0.110 mm polyester (PETP)/PE laminate, with an oxygen permeability about 90 cc/m^2/24 h. (measured at 23°C).

Display temperature

The samples were stored in an open top frozen food display cabinet. The cabinet was defrosted automatically once a day, and the surface temperature of the product increased in that period to -9°C. In the remaining period of the day the surface temperature of the meat ranged from -16°C to -21°C, and the centre temperature from -19°C to -21°C.

Display lighting

The illumination was provided by: Tube I: Warm White, 3000K, 40W, about 2500LUX (at the surface of the meat); Tube II: Free Powder Lamp, 4000K, 36W, about 2500LUX (at the surface of the meat). The light sources were installed approximately 1 m above the product. The display cabinet was divided into two sections by a vertical flamingo plate, and the two lighting sources were installed on each side of the plate. The products were exposed to the light approximately 8 hours a day - five days a week. Night covers were used.

Colour evaluation

Subjective colour scores and objective colour measurement were registered during the three months frozen storage and during display.

The visual colour evaluation was carried out by three trained panellists. The meat colour was evaluated at the beginning and at the end of the three months of frozen storage; this evaluation was carried out in a room normally used for sensory evaluation of meat products, where the lighting was a combination of White Delux and Warm White Delux.

During display the samples were visually scored in the display cabinet under the Free Powder Lamp, two or three times a week. The samples were scored using a 0-10 scale:

10 = Very Fine
8 = Normal and Satisfactory
6 = Defects
4 = Essential Defects
2 = Distinct Defects
0 = Great Defects

A score below 6 was considered non acceptable for the consumer.

During display the surface colour was measured objectively with a Hunter Colour Difference Meter, type D25M-3. The measurements were carried out on wrapped samples. The results were expressed as lightness, hue and saturation.

RESULTS AND DISCUSSION

On arrival to the laboratory the samples were scored visually to 9, and after storage in three months the non vacuum-packed samples were scored to 8 and the vacuum-packed samples were scored to 7. This indicate rather small colour changes during frozen storage in the dark at -24°C. Lentz (1971) showed that the colour of frozen beef steaks remains attractive for three months in the dark.

The reason for the lower score of the vacuum-packed samples is presumable a deoxygenation of oxymyoglobin to myoglobin which is less attractive than the oxymyoglobin.

The changes in colour characteristics undergone by the samples during the display period are shown in Fig. 1 and 2, which shows the changes in visual scores (Fig. 1) and saturation (Fig. 2). According to MacDougall (1982) the saturation will decrease as the concentration of oxymyoglobin decreases. The Hunter values for hue and lightness are not included in the figures, but the trend was that both increased concurrently with the decrease in oxymyoglobin.

The figures shows that fluorescent light rapidly causes colour degradation of the samples. Exposure to light will cause appreciable colour changes in 3 days and the colour will get unacceptable in 6 to 12 days (colour scores <6).

In this investigation no differences in colour deterioration from the two tested light sources were found.

At Fig. 2 the colour characteristics of the exposed and the covered half of the samples from batch 2 and 3 are shown together with the characteristics of batch 1, the samples stored at constant temperature in the dark. The curves indicate only slight differences between the samples stored at constant temperature (-24°C) and the covered half of the samples stored in the freezer cabinet at warmer and highly fluctuating temperatures. At the visually evaluation no differences could be detected between storage at constant -24°C and storage in the display cabinet.

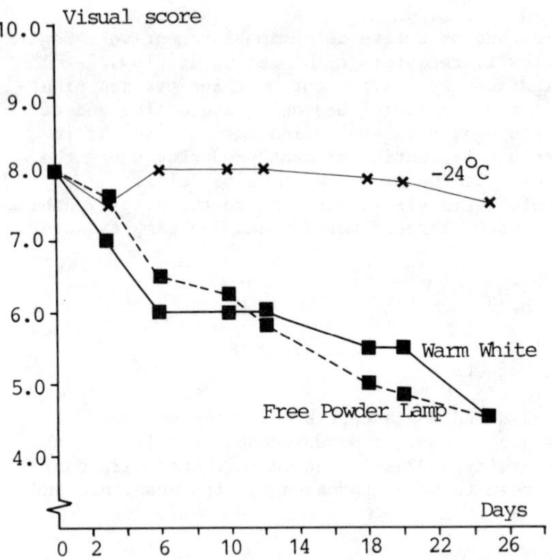

Fig. 1. Effect of display conditions on visual score for frozen ground beef packed in polyethylen chubs after storage in the dark for three months. x——x : Storage in the dark at -24°C. ■——■ : Displayed under Warm White. ■- - -■ : Displayed under Free Powder Lamp.

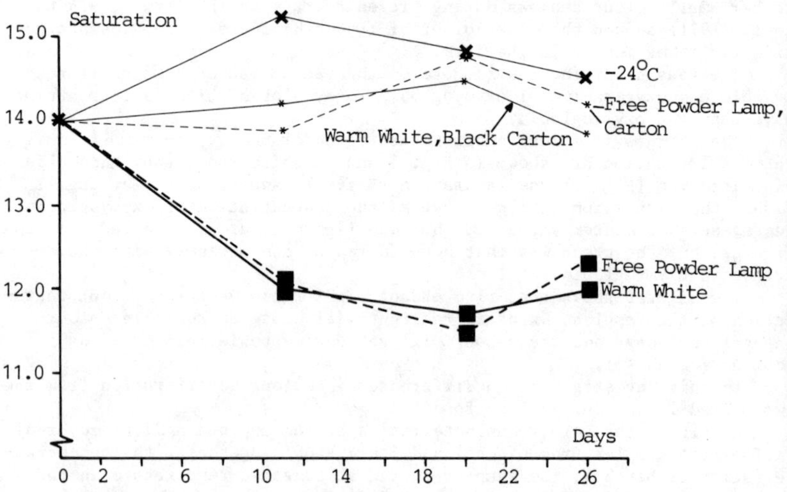

Fig. 2. Effect of display conditions on saturation for frozen ground beef packed in polyethylen chubs after storage in the dark for three months. x——x : Storage in the dark at -24°C. *——* : Displayed under Warm White. Overwrapped with black carton. *- - -* : Displayed under Free Powder Lamp. Overwrapped with black carton. ■——■ : Displayed under Warm White. ■- - -■ Displayed under Free Powder Lamp.

The results indicate that vacuum-packaging during frozen storage does not effect the colour stability in the retail cabinet. However the influence of vacuum-packaging during display has not been investigated.

CONCLUSION

The main conclusions are:

- Exposure to light of frozen ground beef (packed in polyethylen) in a display cabinet caused appreciable colour changes in 3 days and unacceptable colour in 6 to 10 days.
- No appreciable difference between the colour deterioration of the two investigated light sources (Free Powder Lamp and Warm White) was found.
- Vacuum-packaging during frozen storage does not seem to affect colour stability during display.
- For samples protected against light the objective colour measuring indicated slightly more acceptable colour of the samples stored at -24°C compared to the samples exposed to fluctuating temperatures (-16 to -21°C and -9°C at defrosting), but visually evaluated, no differences were found.

REFERENCES

Hustruld, A., Winter, J.D. and Nobel, I. 1949. How do fluctuating temperatures affect frozen foods?. Refrig. Engr. 57: 38.

Lentz, C.P. 1971. Effect of light and temperature on color and flavor of prepackaged frozen beef. J. Inst. Can. Technol. Aliment. 4: 166.

MacDougall, D.B. 1982. Changes in the colour and opacity of meat. Food Chemistry. 9: 75.

Townsend, W.E. and Bratzler, L.J. 1958. Effect of storage conditions on the color of frozen packaged retail beef cuts. Food Technol. 12: 663.

Zachariah, N.Y. and Satterlee, L.D. 1973. Effect of light, pH and buffer strenght on the autoxidation of porcine, ovine and bovine myoglobins at freezing temperatures. J. Food Sci. 38: 418.

TTT-STUDIES OF DIFFERENT RETAIL PACKED PORK PRODUCTS

K. BECH-JACOBSEN & L. BOEGH-SOERENSEN
Danish Meat Products Laboratory, Ministry of Agriculture

Summary

TTT-studies of four retail packed frozen pork products (pork chops, vienna sausages, pork sausages and ground veal & pork) were carried out, and the storage life - stability time (HQL), as well as practical storage life (PSL) - was determined at five temperatures (-40°C, -30°C, -24°C, -18°C and -12°C). Preliminary results indicate that for three of the products the quality loss followed straight lines at the coldest storage temperatures and nonlinear curves at the warmest temperatures; for one of the products, ground veal & pork, the quality loss followed cubic curves at all temperatures.
On the same time, these products were evaluated after and during a simulation of the freezer chain, i.e. the products were placed in different storage facilities and temperatures, in order to find a connection between the quality loss actually found in sensory evaluations and the shelf life loss calculated on basis of the results (PSL) from the above mentioned TTT-studie. Preliminary results indicate that the actual quality loss may be a little greater than the quality loss calculated.

1. INTRODUCTION

Retail packed frozen products are exposed to various temperatures in the freezer chain from production to consumption. Each link in the chain gives a contribution to loss of quality, especially the storage period in the retail cabinet.

One way of predicting the quality loss, is to calculate the shelf life loss in each link according to the storage time and temperature, and to add these losses, to give a percentage total loss. Here is used the general rule, that all effects are cumulative.

Another way to find the total quality loss is to expose the packages to movements similar to that in the freeser chain (i.e. do a simulation) and evaluate the product after each storage period. Both methods have been used in the present studie.

2. MATERIALS AND METHODS

Four retail packed frozen pork products (pork chops, vienna sausages, pork sausages and ground veal & pork) were sampled from a normal (commercial) production line.

Experiment A: Samples of each product were stored at constant temperatures (-40°C, -30°C, 24°C, -18°C and -12°C) in the laboratory.

Experiment B - simulation of the freezer chain: Samples of each product were stored at -24°C ("producer-store"), then in a retail cabinet, and finally stored in a chest-freezer at -18°C ("home-freezer"). Between each storage period, a transport was simulated by placing the sample at room temperature for 1½ hour. The length of each storage period was based partly on observations made in a parallel project (1), partly on previous work as summarized in Boegh-Soerensen et al, 1978 (2). Both 50%- and 95%-fractiles for each link were used.

Sensory evaluation

In experiment A and B a trained taste panel evaluated the products, using a hedonic scale (-5 to +5). Scores were given for taste, juiciness and overall impression. The first evaluation took place the day after freezing.

Products stored at constant temperatures were evaluated with regular intervals (warm temperatures: 1-2 months, cold temperatures: 2-4 months between sensory evaluations) until the average score at two successive taste sessions was -1.5 or lower.

The products in experiment B were evaluated after each storage period.

The average scores for each session were used to determine a reasonable mathematical model for quality loss with time at each storage temperature. The stability time (HQL) was defined as the time to reduce the initial score (the score the day after freezing) by 1 point, and the acceptability time (PSL) as the time to reach a score of -1.

Assuming that the quality loss in experiment B is linear with time, the quality losses were compared with the loss found by calculating the loss in each link using the results from experiment A. In the retail cabinet is was assumed the half the time were spent at -18°C and half the time at -12°C.

3. RESULTS AND DISCUSSION

Experiment A. The mathematical models for quality loss for one of the products, pork chops, are given in table I. The results for -40°C and -30°C are preliminary as the experiment is still going on. The quality loss curves are shown in figure 1, together with the actual found average scores for the samples kept at -12°C and -40°C.

TABLE I. Quality loss equations and HQL and PSL values for pork chops at different temperatures.

Temp. °C	Number of observation	Model for overall impression	R^2	HQL days	PSL days
-40	16	$y=1.32-2.40\times10^{-3}X$	0.57	405	967
-30	20	$y=0.80-2.20\times10^{-3}X$	0.24	203	816
-24	22	$y=0.92-4.40\times10^{-3}X$	0.51	129	436
-18	24	$y=1.15-6.94\times10^{-3}X$	0.64	116	310
-12	20	$y=1.35-\frac{4.57X}{162+X}$	-	46	172

The corresponding mathematical models for the other products shows straight lines at all temperatures for vienna sausages, cubic curves at all temperatures for ground veal & pork and for pork sausages straight lines at all temperatures except at -12°C, where the model is quadratic.

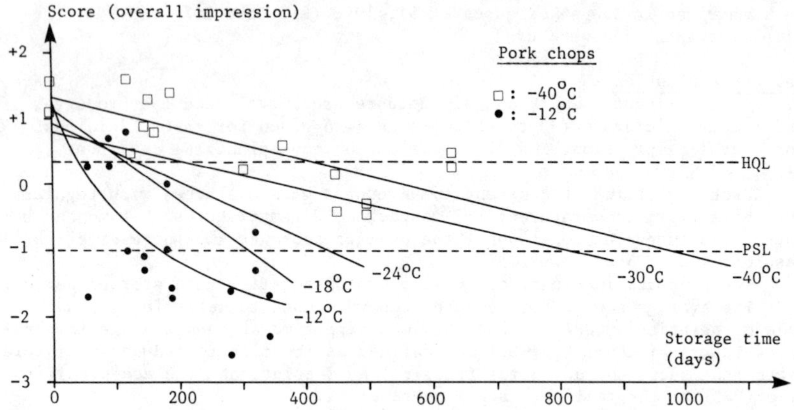

Figure 1. The changes in scores for overall impression for pork chops at different temperatures. The actual scores at -40°C and -12°C are plotted.

Figure 2 shows HQL- and PSL-curves for all four frozen pork products; HQL and PSL at each storage temperature is determined as above, see table I and figure 1. PSL-values at -30°C and -40°C are not final as the experiment has not been finalized. For vienna sausages none of the PSL-values should be regarded as final results, but the curves shows that storage life decreases at temperatures below -30°C (reverse stability). The other products shows normal stability i.e. increasing of storage life with lowering of temperature.

Experiment B. The length of each storage period in the freezer chain simulation is shown for pork chops in table II, together with the quality loss actual found, and the quality loss calculated from the PSL-values determined in experiment A, cf. figure 2.

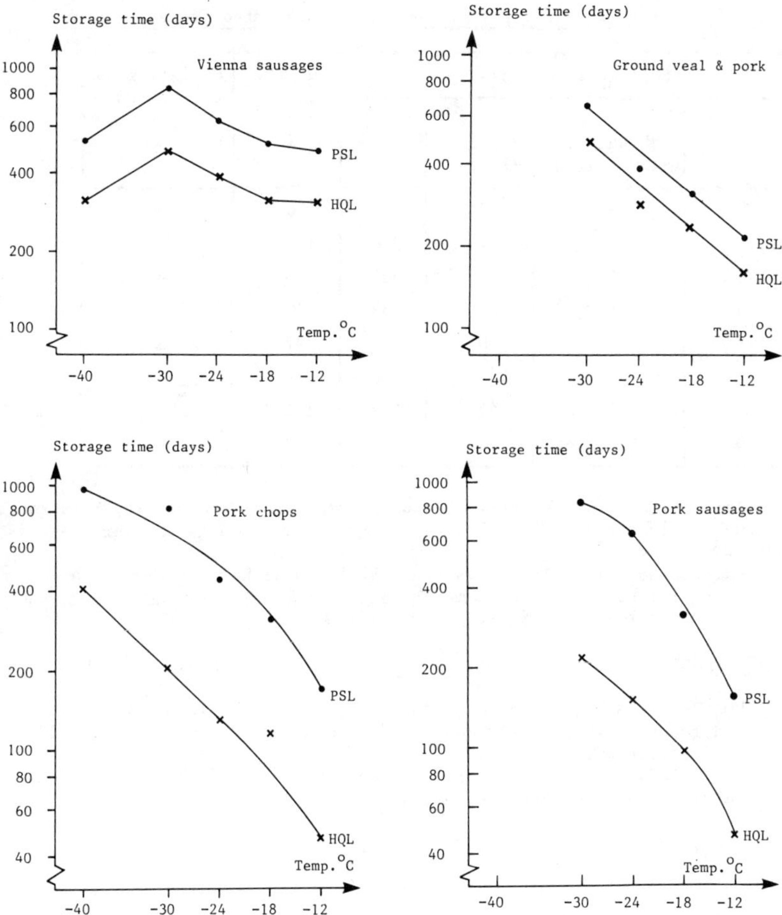

Figure 2: HQL- and PSL-curves for four frozen pork products. Vienna sausages (vacuum-packed), ground veal & pork, pork chops and pork sausages (packed in surlyn).

TABLE II. Quality loss for pork chops in the simulated freezer chain.

Storage time (days)				Quality loss found %	Quality loss calculated %
Producer store $-24^{\circ}C$	Retail Cabinet $-12^{\circ}C/-18^{\circ}C$	Home Freezer $-18^{\circ}C$	Total		
125	7	49	181	56	50
125	7	224	356	125	106
125	56	119	300	93	104
125	56	182	363	120	125
188	7	140	335	97	93
188	7	161	356	128	100
188	112	42	342	116	125
188	112	84	384	138	138

Table II shows comparatively good agreement between quality loss found and quality loss calculated. For the other three products the agreement is less pronounced.

Calculation of quality loss (shelf life loss) has not been finalized and there will presumable be some changes in the calculations regarding storage in the retail cabinet.

4. REFERENCES

1. BOEGH-SOERENSEN, L. and BECH-JACOBSEN, K. (1983). Total time for frozen foods in the freezer chain and in display cabinets. Proc. COST-91 seminar, Athen. In press.
2. BOEGH-SOERENSEN, L., JENSEN, J.H. and JUL, M. (1978): Konserveringsteknik 1, DSR-forlag, Copenhagen.

TTT-STUDIES OF CURED FISH PRODUCTS

A. TUE CHRISTENSEN & B. JESSEN
Technological Laboratory, Ministry of Fisheries
Copenhagen, Denmark

Summary

TTT-PPP experiments with frozen lightly cured fish were carried out, one of the purposes being to examine whether cured fish products showed reverse stability.
Matjes-herrings were prepared from two different raw materials, i.e. fresh herrings or frozen herrings. Matjes-herrings were covered with brine, packed in 2.5 l plastic containers.
"Gravad" halibut, prepared from frozen Greenland halibut, were either packed in PE-pouches, or vacuumpacked.
During storage of up till 6 months at -3°C, -12°C, -24°C, (and -30°C for "gravad" halibut) sensory evaluations were carried out, and the practical storage life (PSL) determined.
For Matjes-herrings, PSL was longer for the frozen raw material than for the fresh; at -24°C, 170 days against 108 days. For "gravad" Greenland halibut, vacuumpackaging resulted in considerably longer PSL; at -24°C, 174 days against 77 days for PE-packed.
In all cases, a normal stability was found.

INTRODUCTION

The storage life of frozen foods increases with colder storage temperature. However, 50 years ago it was found (Callow, 1931) that cured pork (bacon) kept better at -7°C than at -20°C. Based on several TTT-experiments with cured meat products, Lindeløv (1978) suggested three categories of frozen foods:

- Normal stability : storage life increases with colder storage temperatures
- Neutral stability : storage life is practically unaffected by the storage temperatures
- Reverse stability : storage life decreases with colder storage temperatures.

Most frozen foods have normal stability, i.e. fruits, vegetables, meat, fish, poultry, etc., although the different products (product groups) do not show the same storage life increase by lowering the storage temperature for instance from -18°C to -28°C.

Neutral stability is found in smoked pork (Lindeløv, 1978), and in vacuum-packed sliced, cured meat products (Bøgh-Sørensen, 1968).

Reverse stability is found in cured pork products, e.g. sliced bacon, which is packed in oxygen permeable plastic foil, and in unwrapped bacon.

For some products, a maximum storage life at -24°C is found, i.e. the storage life is shorter at -15°C and at -40°C (Bøgh-Sørensen, 1981).

For cured fish, very few TTT-data have been found in the literature, and as a certain amount of lightly cured fish products are frozen in Denmark, this TTT-experiment was carried out.

The purpose was to determine the practical storage life (PSL) of two commercially cured fish products at different storage temperatures from -3°C to -30°C.

MATERIALS AND METHODS

The experiment was divided in two parts. A, concerning Matjes-herring, and B, concerning "gravad" Greenland halibut.

A. Matjes-herring.

Herrings from the North Sea (Clupea harengus), were landed in Skagen (August 1982), and Matjes-herrings prepared from these unfrozen herrings (A1). As raw material was also used frozen herrings (A2), which had been stored at -28°C for 70 days, in a wooden box covered with PE; the frozen herrings were thawed in running tap water.

The herrings were gilled and dry salted with 5% salt (w/w). After 24 hours the herrings were packed in 2.5 liter plastic containers (polyethylen, PE), 2 kg in each, and filled with about 0.5 liter brine.

The containers were transferred to storage rooms at -3°C, -12°C, and -24°C, without previous freezing.

B. "Gravad" Greenland halibut.

The Greenland halibuts (Rheinhardtius hippoglossoides) were caught near Greenland and frozen on board, packed in PE-coated paperbags. The frozen fish were thawed in running tap water, filleted and cured in a brine with 10% NaCl (w/v) and 5% sugar. After 24 hours at 0°C the fillets were spiced with dill and white pepper. The fillets were packed in PE-pouches, or vacuum-packed in a Polyamid/PE-laminate.

The fillets were placed at -3°C, -12°C, -24°C, and -30°C.

Chemical analysis

The products were analyzed for % fat, % NaCl, % water.

During the storage period, analysis for TVN (Total Volatile Nitrogen), FFA (Free Fatty Acids), Peroxide number, and TBA-value, was carried out to follow the changes during frozen storage.

Sensory evaluation

A trained laboratory panel (6-7 persons) on the Technological Laboratory, Ministry of Fisheries, scored the samples for texture and taste, using a 0-10 scale; 10: excellent, 8: good, 6: no off-flavour, 4: acceptable 0: extremely bad.

RESULTS

Chemical analysis.

The results of the chemical analysis are shown in table I.

The fat content of the frozen herrings (A2) is higher than of the unfrozen herrings (A1). Salt per 100 g water was 5.5 for A1, 5.4 for A2, and 3.9 for "gravad" halibut.

The results of analysis for TVN, TBA, FFA, and peroxide number are not included, as they have not been statistically treated.

Table I. Chemical analysis, average values and standard deviation (s.d.)

Product	% Fat		% Salt		% Water	
	average	s.d.	average	s.d.	average	s.d.
Matjes-herrings, A1	15.1	2.4	3.6	0.4	63.4	3.1
Matjes-herrings, A2	21.0	1.5	3.2	0.5	59.9	1.4
"Gravad" Greenland halibut	10.8	1.8	2.8	0.3	73.1	1.7

Sensory evaluation

Fig. 1 shows the average scores for taste of Matjes-herrings (A2) during storage at -3°C, -12°C, and -24°C.

Fig. 2 shows the average scores (taste) for vacuumpacked "gravad" halibut during storage at -3°C, -12°C, -24°C, and -30°C.

From these figures, the practical storage life (PSL) at the different storage temperatures can be determined; PSL is defined as the time until the character drops below 4.

The time for the characters for taste and for texture to drop below 4 is shown in table II, also showing the resulting PSL (the shortest of the two time periods) at each storage temperature for each product.

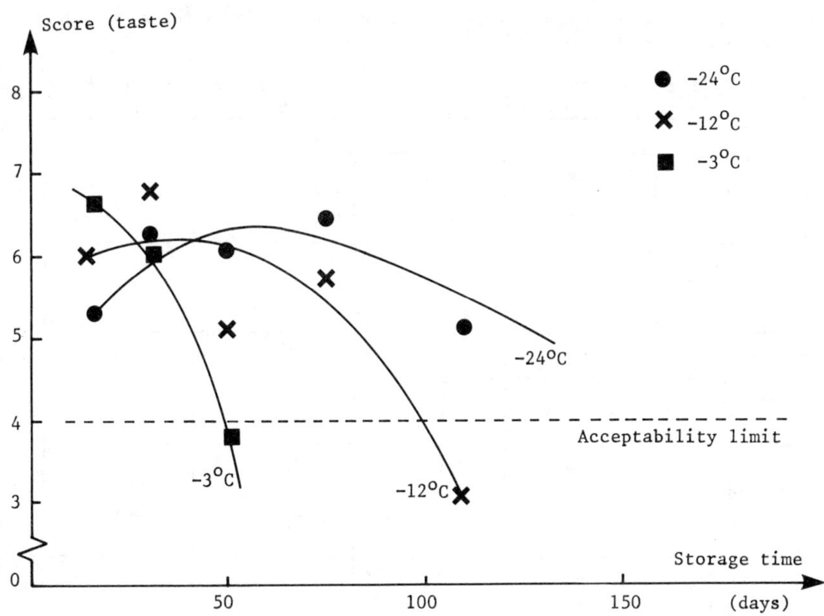

Fig. 1. Matjes-herrings, good raw material (A2). Scores for taste during storage at -3°C, -12°C, and -24°C.

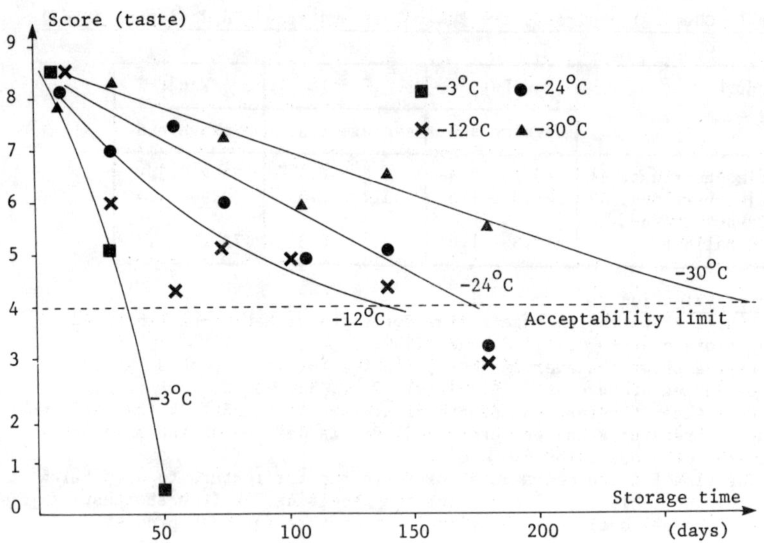

Fig. 2. "Gravad" Greenland halibut, vacuumpacked. Scores for taste during storage at -3°C, -12°C, -24°C, and -30°C.

Table II. Pratical storage life (PSL) of Matjes-herrings - two different raw materials - and of "gravad" halibut - two different packagings.

Product	Storage temperature °C	Days until taste is below 4	Days until texture is below 4	PSL days
Matjes-herring, A1	-3 -12 -24	24 65 108	52 109 123	24 65 108
Matjes-herring, A2 (good raw material)	-3 -12 -24	48 102 170	72 158 250	48 102 170
"Gravad" halibut, vac.pack.	-3 -12 -24 -30	34 137 174 288	50 133 202 273	34 133 174 273
"Gravad" halibut, PE-pouch	-3 -12 -24 -30	29 50 77 157	34 100 114 227	29 50 77 157

At $-3^{o}C$, the fish is not frozen, and the figures should not be regarded as PSL for frozen fish.

Table II shows that for Matjes-herrings, taste declines more rapidly than texture; fig. 1 indicates an improvement in taste during the first months at $-12^{o}C$ and at $-24^{o}C$. The scores for texture show the same trend. The phenomenon is presumably due to maturation, a process which normally takes place in such cured fish products when they - as have been normal practice - are put into cans and stored at max. $10^{o}C$. This maturation process is catalyzed by intestinal enzymes.

Table II shows that for "gravad" halibut, too, the scores for taste decide PSL.

For Matjes-herrings and for "gravad" halibut, PSL increases with colder storage temperatures, i.e. normal stability. There is a pronounced increase from $-24^{o}C$ to $-30^{o}C$, indicating that the often recommended storage temperature of $-30^{o}C$ for fish is really beneficial for storage life and quality of frozen fish products.

Regarding PPP, the product, the nature and quality of the raw material has a pronounced influence on PSL of Matjes-herrings. The unfrozen herring was not of very good quality, whereas the frozen herring - with a much higher fat content, see table I - was a high quality product. PSL of the good (frozen) raw material was about double of that of the less good raw material, although the frozen herrings had spent 70 days in frozen storage prior to the manufacturing of Matjes-herrings.

For "gravad" Greenland halibut, the packaging is very important. At $-12^{o}C$, vacuumpackaging increases PSL with a factor near 3, while the factor is about 2 at $-24^{o}C$ and $-30^{o}C$. This effect of vacuumpackaging is often found.

These cured fish products do not show reverse stability. If a storage temperature of $-18^{o}C$ had been included, it might have been possible to postulate neutral stability from $-12^{o}C$ to about $-24^{o}C$.

It should be noted that the panellists were not familiar with Matjes-herrings, resulting in considerable differences between the panellists, and in rather uncertain results.

REFERENCES

Bøgh-Sørensen, L. (1968): Report KFR-4. Danish Meat Products Laboratory (mimeographed).

Bøgh-Sørensen, L. and Højmark Jensen, J. (1981): International Journal of Refrigeration. 4, No. 3, p. 139-142.

Callow, E.H. (1931): Ann. Report Food Invest. Bd. (Gt. Britain).

Lindeløv, F. (1978): TTT-examinations of frozen cured meat products. Ph.D.-Thesis. Food Technology Laboratory, Technical University of Denmark.

ENERGY SAVING IN THE FROZEN FISH INDUSTRY

H. HOUWING
Institute for Fishery Products TNO, IJmuiden, the Netherlands

Summary

Are low storage temperatures necessary?
The energy consumption of one stage cooling units increases with ± 6% if the temperature decreases with 1%. If it is justified to increase the storage temperature from -25°C to -20°C 2.1 x 10^5 kWh may be saved per year for a storage room of 10,000 m^3.
Research work carried out in the Institute for Fishery Products TNO in Holland proved that these thoughts might be realizable. Mackerel, herring, cod, plaice, shrimps and mussels have been stored at different temperatures between -19°C and -25°C for one year (temperature differentiation 1°C). The quality of the raw material was good. After this storage time the quality was significantly changed. However in none of the cases the samples were rejectable. On the contrary in general the organoleptical changes were not more than 0.5 - 1.0 lower on a 9 point scale than at the beginning. The difference between -19°C and -25°C products varied between 0.2 - 0.4 points. The influence of the freezing method or the way of packaging was smaller than predicted. However glazed mussels got a better score than unglazed ones (max. 0.2 points, cooked odour). Properly packed mackerel scored higher than unpacked mackerel (max. 0.9 points, raw odour). IQF shrimps (unglazed) can be kept shorter than the block frozen, glazed product (max. 0.6 points, odour, flavour and texture).

1. INTRODUCTION

It is a known fact that energy prices are sky high at the moment. Therefore everyone, including owners of storage rooms for frozen products, tries to reduce these costs.

There are a few ways to realize a reduction. One way might be to increase the storage temperature. We have calculated that 1°C rise in temperature may result in a reduction of 6% of the costs for one-stage freezing installations within the temperature range of storage rooms. Not everyone is happy with such a suggestion because in the last decades propaganda has been made for a decrease in temperature because it would improve the shelflife; -20°C is worse than -30°C. In our Institute we have carried out some preliminary studies in order to find out wether the propagated statements are true*). These studies have been done with mackerel, herring, plaice, cod, lightly salted herring, shrimps (Crangon crangon) and mussels. The following parameters were introduced if appropriate: storage temperature, freezing rate, packaging material and glazing. Different variations have been tested after a 1, 3, 6 and 12 months' storage period.

*) This research was jointly financed by the Ministry of Economic Affairs and TNO and was coordinated by the TNO Bureau of Energy Projects.

As a quality index we used the organoleptical quality. The investigations have been supported by chemical, microbiological and instrumental analyses.

2. EXPERIMENTAL

2.1 Materials and methods

The following products have been tested:

- mackerel and herring in blocks (as such, glazed, in plastic + carton)
- matjesherring - gibbed (packed in brine in containers)
- cod and plaice fillets (consumer packs)
- boiled shrimps (block frozen, IQF and glazed)
- cooked mussels (as such in plastic cartons, IQF and glazed).

The quality of the raw material was good and generally above standard quality. The products have been stored in cabinets with temperatures of -19°C, -22°C and -25°C. The temperature fluctuation was ± 1°C.

3. ANALYTICAL

After 1, 3, 6 and 12 months samples of fish products were thawed and analyzed. Thawing took place in a chilled room at 4°C. The organoleptical assessment was done by a panel of 8 persons, utilizing a scoring system from 1.0 - 9.0 for colour, raw odour, cooked odour and flavour and texture in the cooked condition. Chemical analyses was carried out for Dimethylamine (DMA) and Extractable Protein Nitrogen (EPN). As microbiological criteria were chosen: psychrophilic plate count and Enterobacteriaceae. Occasionally other tests were carried out.

As an instrumental analysis the resistance of cooked meat was measured with the Texture Testing System of the Food Technology Corporation.

4. RESULTS AND DISCUSSION

All analytical data have been statistically processed. There has been a slight but consistent decrease of the average sensorical scores which could be related to the duration of the storage period, particularly between 6 - 12 months. The decrease was only between 0.4 - 1.5 points.

Even smaller variations in sensorical scores (0.2 - 0.4 points) could be attributed to different temperature levels (-25°C vs -19°C), but only for the texture of herring, the flavour of cod, the raw odour and cooked flavour of plaice and the cooked odour, flavour and texture of shrimps.

The influence of the packing method on the sensorical scores was sometimes significant though very small indeed. Raw odour scores for unpacked mackerel were only slightly below those of mackerel packed in plastic and carton (6.1 vs 6.4). For whole herring frozen in blocks the difference was negligible (0.1 point). Mussel colour, odour and flavour were only slightly better after glazing (0.2 point difference). Block freezing was somewhat better for shrimps than IQF but only after 12 months 0.5 points, only for flavour, -19°C vs -25°C.

The instrument texture analysis has confirmed these decreases in sensorical scores. Increases of shear resistance correlated well with lower texture scores, only herring became softer with lower sensoric scores. The temperature level has been critical for the mechanical resistance as registered by the texture recorder, but only for cod and shrimps: lower storage temperatures have led to smaller resistances.

The highest temperature level (-19°C) has had a stronger positive effect on the shear resistance of IQF unglazed shrimps than of plate-frozen glazed blocks of shrimps.

The highest temperature level (-19°) has resulted in higher DMA-values for mackerel, herring, cod and matjesherring. EPN-values were lower for herring, cod and matjes at -19°C than at -25°C. There has been a considerable decrease of EPN-values during storage of cod between 1 - 3 months (from 67.5% to 41.0%).

Bacterial numbers have not been influenced by any of the parameters studied.

5. CONCLUSIONS AND RECOMMENDATIONS

On the ground of preliminary experiments we doubt if it is necessary to store all kinds of frozen fish at relatively low temperatures. More research has to be carried out in order to be able to give justified advise to the frozen fish industry in which cases frozen fish can be kept at higher temperature, for instance -20°C instead of -30°C, thus saving energy costs.

In this further research the parameters quality of the raw material and temperature fluctuations have to be introduced.

Since it is doubtful whether consumers are as sensitive for quality changes as members of a trained panel it may be wise to pay attention also to consumer acceptability tests in this further research as well.

HARICOTS MANGE-TOUT CONGELES SANS BLANCHIMENT PREALABLE. RELATION ENTRE LA QUALITE ET LES FACTEURS TEMPERATURE ET DUREE D'ENTREPOSAGE

J. PHILIPPON , M-A. ROUET-MAYER et J. ABBAS
C.N.R.S. Laboratoire de Physiologie des Organes Végétaux après Récolte, 92190 Meudon, France.

Summary

UNBLANCHED FROZEN STRING BEANS : TEMPERATURE AND TIME OF STORAGE DEPENDANCE OF THEIR QUALITY.
Quality evolution of unblanched frozen french cut beans was studed in relation with their storage conditions ie. temperature and duration. Decrease of chlorophylls of raw beans as well as of color and flavor of cooked beans followed first order kinetics which constant increased exponentially with temperature. Q10 were found to be equal to 4.5, 3.9 and 3.0 respectively. A significant decrease of raw french beans color appeared only when 20-23% chlorophylls were lost. First perceptible color or flavor difference ($P \leqslant 0,01$) of cooked product were found to be equal at 180 days at -25°C, and 42 and 76 days, respectively, at -15°C. The essential sensorial quality patterns of unblanched French cut beans can be maintained after 6-7 month storage at -25°C.

1. INTRODUCTION

Le blanchiment préalable à l'eau bouillante ou à la vapeur d'eau à 100°C des légumes surgelés est indispensable au maintien de leur qualité lors de longues durées de conservation ou pendant leur passage dans les différents maillons de la chaine du froid. Néanmoins, cette opération entraine de par elle-même des pertes non négligeables de qualité : ramollissement de la consistance, lessivage des éléments solubles, destruction des composés thermolabiles. Corollairement, elle est responsable d'une pollution importante des eaux. C'est pourquoi il nous a semblé intéressant d'éviter le recours au blanchiment dans le cas des légumes congelés destinés à être repris ultérieurement en conserverie, notamment, pour la confection de macédoines de légumes appertisés.

Le présent travail a pour but de définir les lois générales reliant la couleur et la flaveur des haricots verts congelés non blanchis à la température et à la durée de conservation. Par sa méthode, il s'inspire des travaux américains du WESTERN REGIONAL LABORATORY D'ALBANY Ca (8). Il vient en complément des travaux récents de KOSMALA et al. (3) effectués sur des haricots verts et des pois congelés non blanchis.

2. MATERIEL ET METHODES

On a utilisé des haricots mange-tout (*Phaseolus vulgaris L.*) du cultivar VAILLANT. Aussitôt après leur cueillette, ils ont subi les opérations suivantes : calibrage et ébouttage à la main, lavage sous eau courante, essorage, emballage, congélation (de +15°C à -30°C en 3h30 environ).

Des lots de 35 à 50 paquets de 400 g ont été entreposés à 6 régimes thermiques stables : -8°C, -15°C, -20°C, -25°C, -30°C et -65°C (témoin).

La qualité sensorielle a été jugée par un jury de 12 à 14 personnes du laboratoire. Chaque séance a fait l'objet d'une répétition, ce qui a permis de recueillir de 24 à 28 informations par contrôle.

La couleur des haricots décongelés a été évaluée au moyen de tests triangulaires; le lot témoin étant stabilisé à -65°C.

La couleur et la flaveur du produit cuit a été évaluée à partir d'une échelle de notation allant de 1 (mauvais) à 5 (excellent). Les lots ont toujours été jugés par référence à un échantillon témoin qui a systématiquement reçu la note de 4 sur 5. La dérive de notation se produisant d'une séance de dégustation à l'autre a ainsi été évitée. Les durées de conservation de référence (High Quality Life) correspondent ici au moment au bout duquel 75% de réponses exactes sont recueillies lors des tests triangulaires, ou lorsque les notes attribuées à la couleur ou à la flaveur du produit cuit deviennent significativement différentes ($P < 0,01$) de celles du témoin à -65°C.

Les chlorophylles ont été dosées par la méthode de VERNON (9). Leur taux de dégradation a été exprimé à partir de la variation du rapport chlorophylle a/chlorophylle b (6). Les données numériques ainsi obtenues ont été exprimées en pourcentage de chlorophylles initiales au moyen de l'équation d'une droite d'étalonnage établie expérimentalement.

3. RESULTATS ET DISCUSSION

3.1 Couleur du produit cru décongelé

3.1.1 Seuil de perception visuelle

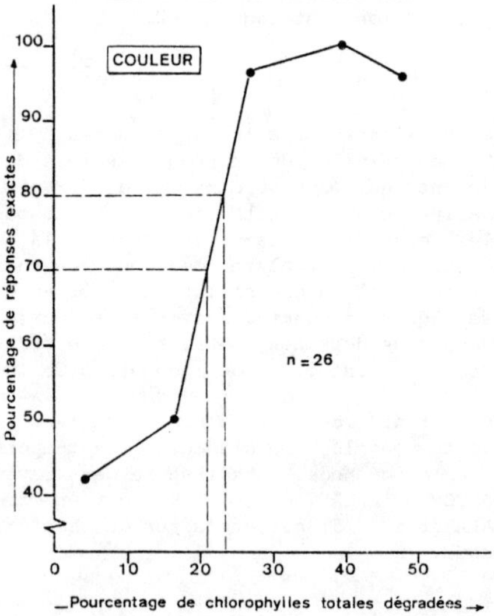

Fig. 1 - Relation entre le pourcentage de réponses exactes fournies par un jury procédant par tests triangulaires et le pourcentage de chlorophylles totales dégradées, chez des haricots mange-tout crus - Echantillons témoins à -65°C - n = nombre de réponses obtenues par points contrôlés.

La figure 1 représente la relation entre le taux de chlorophylles dégradées et le pourcentage de réponses exactes fournies par le jury lors des test triangulaires.

Comme on peut le voir les points s'alignent sur une courbe en S. On constate qu'il faut que 20 à 23% des chlorophylles soient dégradées pour que 70 à 80% des membres du jury soient en mesure de percevoir une différence de couleur par rapport au témoin stabilisé à -65°C. La difficulté rencontrée par les membres du jury pour détecter de plus faibles différences de couleur tient au fait que les haricots verts non blanchis ont un aspect très hétérogène.

3.1.2 Action de la durée d'entreposage.

La figure 2 montre, une fois encore, que la régression des chlorophylles totales évolue chez les légumes verts non blanchis, suivant une loi du premier ordre, si l'on en juge par le bon ajustement des points sur les droites de régression calculées.

La droite du lot conservé à -8°C ne démarre pas au point 100% car, pour une raison inexpliquée, la perte en chlorophylles a été beaucoup plus importante que prévu lors des tous premiers jours de l'essai.

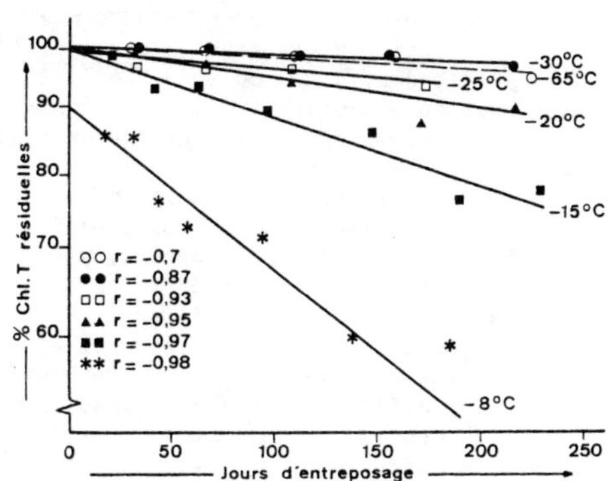

Fig. 2 - Evolution du pourcentage de chlorophylles totales résiduelles de lots de haricots non blanchis conservés à diverses températures, en fonction de la durée d'entreposage. (Chaque point représente la moyenne de 8 dosages.)

On constate que si l'altération des pigments est très importante à -8°C elle n'est pas non plus négligeable à -20°C, puisque 17% des chlorophylles disparaissent en 7 mois. Ce qui reste malgré tout en deça du seuil de perception visuelle du jury.

Il semble donc, dans le cas des haricots mange-tout non blanchis, qu'une stabilisation satisfaisante de leur coloration puisse être obtenue par une température d'entreposage de -20°C constant.

Remarquons cependant que la vitesse de dégradation des chlorophylles totales observée dans nos essais est sensiblement plus faible que celle constatée par d'autres auteurs sur le même matériel végétal non blanchi. Ainsi, MÜFTÜGIL et YÎGÎT (4) ont constaté que des haricots verts non blanchis présentaient une coloration jaune-vert après seulement 3 mois d'entreposage à -18°C. Par ailleurs, WALKER (10) a mis en évidence une perte de 40% des chlorophylles totales en 18 jours à -10°C (138 jours à -8°C dans notre cas). De même, KATSABOXAKIS et al. (2) ont observé 50% de pertes après un an à -18°C, alors que dans notre cas il aurait théoriquement fallu 3300 jours d'entreposage à -20°C pour parvenir au même résultat. Il est difficile pour le moment d'avancer une hypothèse susceptible de justifier une pareille différence.

3.1.3 Action de la température d'entreposage

La figure 3 a été établie à partir des données numériques tirées des équations des droites de régression de la figure 2. Elle représente la droite de stabilité définissant les durées de conservation de référence (High Quality Life = HQL) à chaque température.

On observe là aussi, que les points s'ajustent bien sur la droite de régression calculée (r = -0,99).

Le Q10 (rapport des vitesses de dégradation à 10 degrés d'intervalle) est une grandeur qui définit la sensibilité de la denrée aux remontées de température. Il est ici égal à 4,5. Cette valeur est notoirement plus faible que celles rapportées par OLSON et DIETRICH (5) à propos de divers légumes verts blanchis et congelés (haricots verts : 10,1 , pois : 14,2 , épinards : 15,3). Elle est très inférieure aussi au Q10 se rapportant à la dégradation des chlorophylles du persil congelé non blanchi (24,7) évalué par ROUET-MAYER et al. (7). De telles différences restent pour le moment inexpliquées.

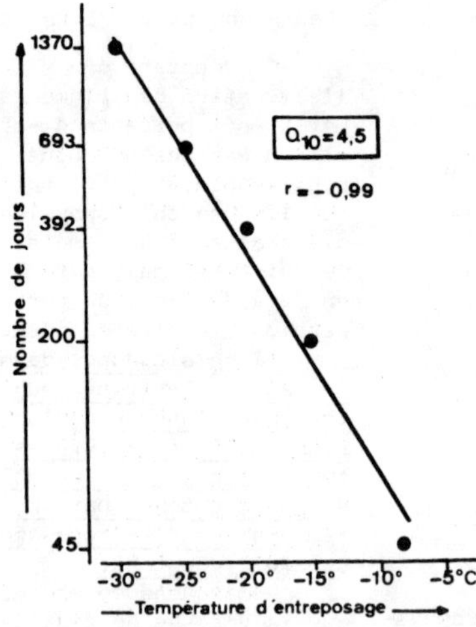

Fig. 3 - Nombre de jours à la suite desquels la perte en chlorophylles des différents lots de haricots mange-tout non blanchis s'est élevée à 21,5% de la valeur initiale, à chaque température d'entreposage expérimentée (échelle semi-logarithmique).

3.2 Qualités sensorielles du produit cuit.

Les deux graphiques de la figure 4 représentent, en fonction du temps, l'évolution de la note moyenne accordée par le jury à la couleur et à la flaveur du produit cuit dans des conditions standardisées, pour chaque régime thermique d'entreposage.

La grande irrégularité des courbes, en particulier, celles concernant les températures les plus basses, est dûe à la faible ampleur des différences par rapport au témoin stabilisé à -65°C; les dégustateurs ayant tendance à considérer systématiquement ce dernier comme meilleur que les autres.

Ces essais suffisent cependant à nous faire voir, que même à -25°C, la perte de flaveur n'est pas négligeable après 6 mois d'entreposage, et qu'au dessus de -20°C, la couleur et la flaveur du produit cuit régressent très rapidement comme on le voit encore plus clairement sur les graphiques de la figure 5. Ceux-ci montrent qu'après cuisson, les pertes de couleur et de flaveur des haricots verts non blanchis évoluent bien en fonction de la température d'entreposage suivant une loi du premier ordre, si l'on en juge par le bon ajustement des points expérimentaux sur les droites de régression calculées.

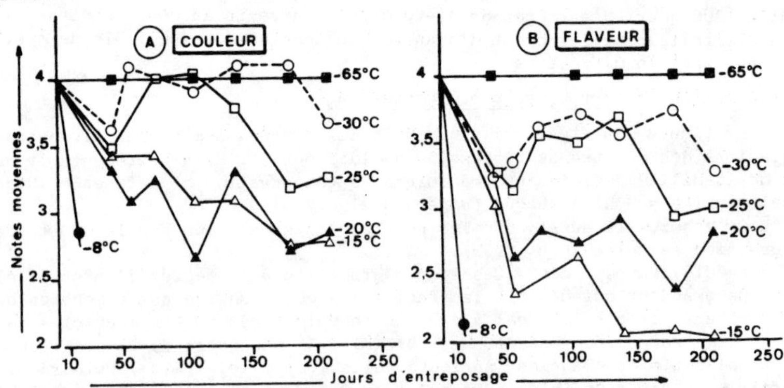

Fig. 4 - Evolution des notes moyennes attribuées par un jury de dégustateurs à la couleur (A) et à la flaveur (B) des haricots mange-tout conservés à diverses températures, en fonction de la durée d'entreposage. (Chaque note représente la moyenne de 23 à 28 informations).

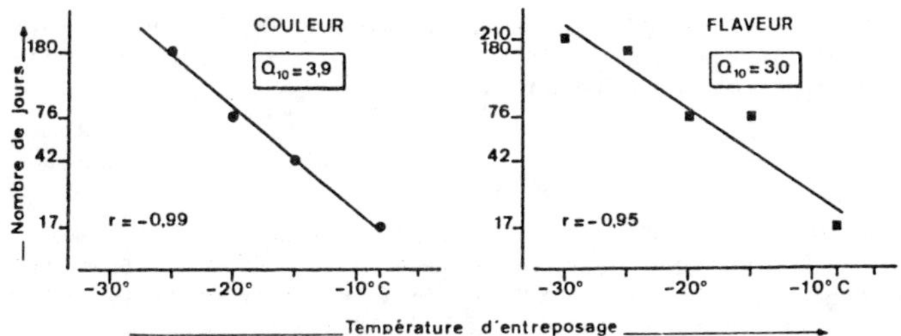

Fig. 5 - Nombre de jours à la suite desquels apparait une première différence significative (P $\leqslant$ 0,01) de couleur et de flaveur entre des haricots mange-tout cuits préalablement conservés à diverses températures stables et le lot témoin conservé à -65°C. Les différences ont été évaluées à partir des notes moyennes attribuées par un jury de dégustateurs.

Un tel résultat vient donc en confirmation des observations faites par les chercheurs américains d'Albany (8) sur différentes espèces de légumes blanchis.

Donc, l'absence de blanchiment n'entraine pas, de ce point de vue, une différence fondamentale de comportement des haricots verts congelés.

Par contre, la vitesse d'altération de la denrée est ici nettement plus rapide que dans le cas des haricots blanchis, puisque les premières différences significatives de couleur et de flaveur sont apparues en 76 jours à -20°C contre, respectivement, 101 et 296 jours à -18°C chez les haricots verts blanchis de DIETRICH et al. (1).

On remarque encore que les Q10 obtenus (couleur : 3,9; flaveur : 3,0) sont nettement plus bas que ceux auxquels sont parvenus les derniers auteurs cités (couleur: 10.1, flaveur: 7.8) - Ce qui dénote la relativement faible sensibilité du produit aux variations de la température d'entreposage.

4. RESUME ET CONCLUSIONS

Ainsi donc, tout comme dans le cas du même produit blanchi, la couleur et la flaveur des haricots mange-tout congelés sans blanchiment préalable évoluent, en fonction de la température ou de la durée d'entreposage, suivant une loi du premier ordre.

Le seuil de perception sensorielle d'une première différence de couleur a été estimé correspondre dans le cas du produit décongelé crû, à une perte d'environ 20 à 23% des chlorophylles initiales.

Les pertes de qualité se sont développées plus rapidement que dans le même produit blanchi. Les durées de conservation de référence (HQL) à -25°C du produit cuit ont été trouvées approximativement égales à 180 jours pour la couleur comme pour la flaveur.

Par contre, les Q10 de 4,5 pour la dégradation des chlorophylles et de 3,9 et 3,0 , respectivement, pour la couleur et la flaveur du produit cuit ont mis en évidence la relativement faible sensibilité des haricots verts non blanchis aux variations de températures.

Sur un plan purement pratique, compte tenu que les durées de HQL sont nettement plus courtes que les durées pratiques de conservation, on peut considérer que <u>l'abaissement de -20°C à -25°C de la température commerciale d'entreposage devrait permettre de conserver pendant une durée de 6 à 7 mois environ l'essentiel de la qualité sensorielle des haricots mange-tout non</u>

blanchis. Une telle façon d'opérer devrait éviter tous les inconvénients du blanchiment dans le cas où le produit est destiné à être repris ultérieurement dans des fabrications industrielles.

BIBLIOGRAPHIE

1. DIETRICH, W.C., NUTTING, M.D., OLSON, R.L., LINDQUIST, F.E., BOGGS, M.M., BOHART , G.S., NEUMANN, H.J., and MORRIS, H.J. (1959) . Time temperature tolerance of frozen foods-XVI. Quality retention of frozen green snap beans in retail packages - Food Technol., 13, 136-145.
2. KATSABOXAKIS, K.Z. and PAPANICOLAOU (at press). The consequences of varying degrees of blanching on the quality of frozen green beans. COST 91 Final seminar " Thermal processing and quality of foods " Athens, Nov. 14-18. 1983.
3. KOSMALA, Z., URBANIAK, M.A. and RYDZ, G.A. (at press). Some chemical and sensory properties of green peas and french beans frozen without blanching and stored in a frozen state. Proc. of 16th. Intern. Congr of Refri. PARIS, 1983, Com. C2.
4. MÜFTÜGIL, N. and YÎGÎT, V. (at press). Comparaison of quality changes in blanching and unblanched frozen vegetables. Proc. of 16th Intern. Congr. of Refri. PARIS, 1983, Com. C_2.
5. OLSON, R.L. and DIETRICH, W.C. (1969). In " Quality and stability in frozen foods " (W.B., VAN ARSDEL, M.J., COPLEY and R.L., OLSON eds.) Wiley-Interscience, 117-141.
6. ROUET-MAYER, M-A. et PHILIPPON, J. (1980). Rapport chlorophylle a / chlorophylle b. Un test de contrôle de la qualité des haricots verts mange-tout surgelés. C.R. 15ème Congr. Intern. du Froid. VENISE, 1979, 983-989.
7. ROUET-MAYER, M-A., PHILIPPON, J., FONTENAY, P. et DUMINIL, J-M. (sous presse). Altération de l'arôme et des chlorophylles du persil congelé non blanchi en fonction des facteurs température et durée d'entreposage. Lebensm.-Wiss. u-Technol.
8. VAN ARSDEL, W.B., COPLEY, M.J. and OLSON, R.L. (1969). " Quality and stability of frozen foods " Wiley-Interscience.
9. VERNON, L.P., (1960). Spectrophotométric determination of chlorophylls and pheophytins in plant extracts. Annal. Chem., 32, (9), 1144-1149.
10. WALKER, G.C. (1964). Color deterioration in frozen french beans (Phaseolus vulgaris L.)-2. The effect of blanching. J. Food Sci., 29, 389-392.

THE EFFECT OF THE STORAGE TEMPERATURE AT PROLONGED STORAGE ON THE QUALITY OF DEEPFROZEN FRENCH FRIES

J.W.LUDWIG

Institute for Storage and Processing of Agricultural Produce (IBVL)
Wageningen, the Netherlands

Summary

An investigation has been carried out with deepfrozen French fries (UK: chips) on the course of the quality (sensory/chemical/microbiological) during prolonged storage (18 months) at various low constant temperatures (-18; -15; -12; -9 and -6°C). It became apparent that:

- the sensory quality of the product deteriorates as the storage temperature is higher and the storage period advances;
- the texture evaluation is of greater importance to the overall quality as the storage temperatur is higher;
- the maximum sensory keeping quality at the 5 tested temperatures (from low to high) is resp. 24, 18, 10, 6 and 5 months;
- the course of the texture deterioration as a result of the storage conditions, becomes quite evident from the course of the moisture and fat content of the finish-fried product;
- in the fat analyses only the peroxide values of the fat give a clear picture; these increase slowly at storage and somewhat more rapidly as the storage temperature is higher;
- there is no question of systematic growth or dying-off of the yeasts, moulds and aerobic mesophilic germ count during storage; may be due to a low infection degree and inhomogeneous sampling material;
- at a maximum storage requirement of 18 months, deepfrozen French fries in fact might as well be stored at a temperature of -15°C.

1. INTRODUCTION

In deepfreeze science and deepfreeze industry people have been wondering for some time whether a storage temperature of -18°C is an absolute must for all deepfreeze products in order to maintain the quality at the right level. This also applies to deepfrozen potatoes and potato products. Particularly with a view to the recent energy crisis this question has become most topical. This report deals with the results of an investigation on the course of the quality (sensory, chemical, microbiological) of deepfrozen French fries during prolonged storage at various low temperatures.

2. EXECUTION OF THE INVESTIGATION

During an 18 months' period deepfrozen French fries were stored at 5 different temperatures, i.e. -18; -15; -12; -9 and -6°C. The deepfrozen product was collected on the day of processing at the manufacturer.
At the start and during the storage period of 18 months the following analyses were periodically (9 times) carried out:

a) sensory test 1. colour) 2. flavour/taste) 3. texture) prepared product, ready for consumption
b) moisture content (ready for cunsumption)
c) fat content (ready for consumption)

d) fat analyses (fat from product via cold extraction with methanol/chloroform)
1. free fatty acid (FFA)
2. peroxide value (PO)
3. iodine value

e) microbiological test
1. aerobic mesophilic germ count
2. germ count yeasts and moulds

To make them ready for consumption, the thawed French fries were finish-fried for 2 minutes at 180°C.
During the investigation the electricity consumption of the 5 storage cabinets was recorded during a period of 75 days with the aid of hour-counters.

3. RESULTS

3.1 Sensory test

The sensory test was carried out by a taste panel of 8 persons.
The prepared product (ready for consumption) was judged on colour, flavour/taste and texture.
Figure 1 gives an overall survey of the evaluation results of the taste panel with regard to flavour/taste and texture of the prepared products. Since it appeared that the colour is the least relevant for the overall picture, this aspect has not been included in this survey. In this overall picture both other evaluation criteria were equally important, so the lowest evaluation of both aspects has always been plotted against the storage period. In other words: if at any time the flavour/taste of a sample was judged "reasonable" and the texture as "moderate", then the overall quality at that particular time was noted as "moderate".

Fig. 1 Overall sensory evaluation

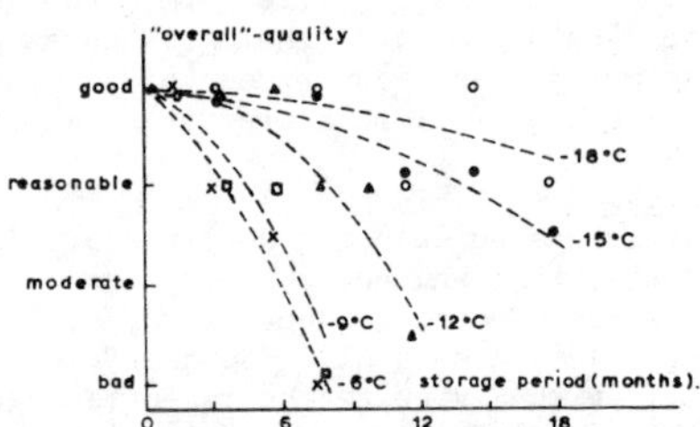

Taking a sensory judgement of reasonable/moderate for the overall quality as the limit for a still acceptable product (and this is to a certain extent arbitrary), then the maximum shelf-life at the 5 test temperatures can be derived from figure 1. This maximum sensory keeping quality is reproduced in table I.

Table I. Maximum shelf-life in months

	temperature				
	-18°C	-15°C	-12°C	-9°C	-6°C
shelf life	24	18	10	6	5

3.2 Fat and moisture content of product ready for consumption.

In the prepared (fried) "consumption ready" products, the moisture content of the French fries shows an initial decrease during the storage period, but from 6 months storage this accellerates and proceeds more quickly as the storage temperature is higher (figure 2).

Fig. 2 Moisture content finish-fried product

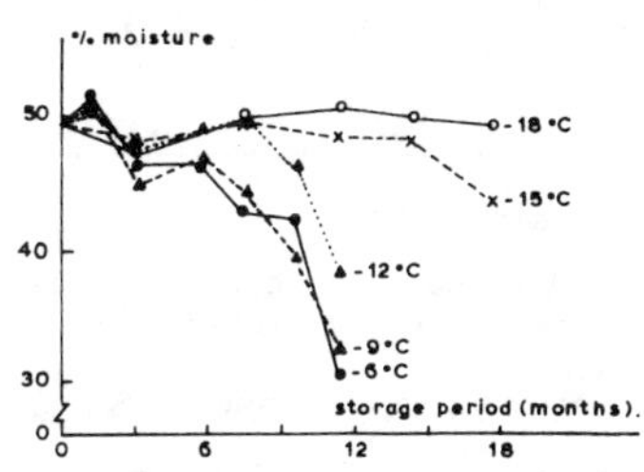

The fat content of the finish-fried French fries initially fluctuates (first 6 months) between 11 and 17%. After that it increases at further storage and more quickly as the storage temperature is higher (figure 3).

Fig. 3 Fat content finish-fried product

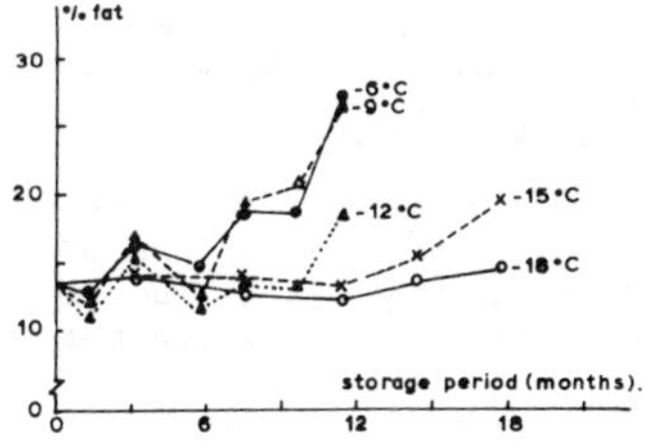

In French fries the course of texture deterioration as a result of the storage conditions, becomes evident from the course of the moisture and fat content of the finish-fried product!
At a temperature of -6 and -9°C we observed after 6 months a sharp increase of the fat content and a decrease of the moisture content; at a storage temperature of -12°C this breaking point occurs after about 10 months. These matters correspond very well with the results of table I, where the "texture" counts heavily in the overall judgment.

3.3 Fat analyses

3.3.1 Free fatty acid content (FFA)

The free fatty acid content (in % of the fat) appears to be rather variable in French fries; however, there is no increase during storage, neither at higher storage temperatures (table II).

Table II FFA in fat of French fries (in % of fat)

Storage period (weeks)	-6°C	-9°C	-12°C	-15°C	-18°C
0	0.20%	0.20	0.20	0.20	0.20
14	0.19	0.17	0.18	0.15	0.20
33	0.19	0.19	0.20	0.19	0.18
50	0.24	0.20	0.20	0.19	0.17
77	-	-	-	0.20	0.19

3.3.2 Peroxide value (PO)

The peroxide values of the French fries fat rise slowly during storage and somewhat quicker as the storage temperature is higher (table III).

Table III PO-value in fat of French fries (in meq/kg fat)

Storage period (weeks)	-6°C	-9°C	-12°C	-15°C	-18°C
0	1.9	1.9	1.9	1.9	1.9
14	1.3	1.4	1.3	1.6	1.8
33	4.2	3.2	3.6	2.8	4.4
50	7.3	6.1	4.3	4.2	5.1
77	-	-	-	-	-

3.3.3 Iodine value

The iodine value (in g J_2/100 g fat) in the fat extracted from the French fries is rather variable, but does not quite decrease as a sign of being less saturated (table IV). The variability is probably due to disturbing matters that have been extracted together with the fat.

Table IV Iodine value in fat of French fries (in g J_2/100 g fat)

Storage period (weeks)	-6°C	-9°C	-12°C	-15°C	-18°C
0	34	34	34	34	34
14	31	27	31	28	28
33	33	33	31	32	34
50	34	34	33	32	33
77	-	-	-	36	36

3.4 Microbiological tests

In the microbiological part of the investigation per gram of product, the aerobic mesophilic germ count and the germ count of yeasts and moulds were assessed. In French fries no clear growth or dying-off was observed during storage; this holds for both germ counts (tables V and VI). The exception is the germ count of yeasts and moulds at the highest storage temperature after 50 weeks storage; the growth could also be observed by microscope at that time.
At -15°C and -18°C there is no occurrence of dying-off. Dying-off at higher storage temperatures is known from literature and may not have been found in the French fries, because of a too low infection degree.

Table V Aerobic mesophilic germ count (per gram)

Storage period (weeks)	-6°C	-9°C	-12°C	-15°C	-18°C
0	100	100	100	100	100
14	60	210	70	660	14000(?)
33	200	200	220	350	140
50	160	65	75	55	95
77	-	-	-	220	450

Table VI Germ count yeasts and moulds (per gram)

Storage period (weeks)	-6°C	-9°C	-12°C	-15°C	-18°C
0	<10	<10	<10	<10	<10
14	<10	30	40	20	190
33	10	15	25	15	20
50	590	15	10	160	10
77	-	-	-	<10	30

3.5 Electricity consumption

During a period of 75 days of investigation, the electricity consumption at the 5 applied storage temperatures was recorded with the aid of hour-counters. The storage research was carried out with deep-freeze cabinets of 220 l contents and a capacity of the freezing unit of 110 W. Without attaching too much importance to the results of this small scale orienting test, it has become quite clear yet that a considerable saving of storage energy costs can be achieved by applying somewhat higher storage temperatures(-15°C instead of -18°C: saving 21%).

FREEZING PRESERVATION OF SOFT CHEESES WITH AND WITHOUT MOLD FROM GOAT'S AND SHEEP'S MILK

T. DALLES[1], G. KALATZOPOULOS[2] AND C. KEHAGIAS[1]
1. Institute of Food Technology, Lykovrissi, Attiki, Greece.
2. Agricultural College of Athens, Greece.

Summary

Cheeses were prepared from sheep's and goat's milk with or without mold (*Penicillium candidum*). Cheeses were frozen (-25°C) by slow (6h) and rapid freezing (1,5h) ten days after cheese-making. Frozen storage lasted for one year and the defrosted cheeses were held in the refrigerator (2-4°C) for 20 days. In the defrosted cheeses,pH, proteolysis and organoleptic properties were examined the first and the last day of storage. The results showed that were not observed significant: 1) changes during frozen storage of the cheeses which could affect their organoleptic properties, 2) differences between fresh and defrosted cheeses as well as between slowly and quickly frozen cheeses in the characteristics that were studied. Cheeses from sheep's and goat's milk behaved in a similar manner. It was also concluded that defrosted cheeses with or without mold were both acceptable after storage for 20 days in the refrigerator.

1. INTRODUCTION

In this country there is a unique interest for sheep's and goat's milk, because these two kinds of milk represent about 60% of the total milk production. In view of the interest for better utilization of sheep's and goat's milk, a technology was developed (2,4,5,12) for preparation of a soft cheese from sheep's and goat's milk with or without *Penicillium candidum*.Cheeses with this mold are prepared in France from goat's milk. Lactation period of sheeps and goats lasts only for about six months and it is not possible to have available products in the market for the whole year since the self life of this product is very short. It is known, that this type of cheeses deteriorate after about 20 days storage is the refrigarator because the degree of proteolysis increases intensively. In order to solve the above problem, Portman (9) in France studied the effect of frozen storage on the properties of cheese curd and found that although curd keeps its technological characteristics, however, the produced cheese was not always of good quality. Preservation of Cottage, Feta, Kaschaval cheese and Quarg was made at-18°C and-20°C by other research workers (6,11). The results of these studies showed that in some cases quality deteriorated somewhat, while soluble nitrogen content increases during frozen storage indicating an increase of the degree of ripening.

In this study, in order to solve the availability problem of soft cheeses with or without *P.candidum*,cheeses were quickly and slowly frozen and kept at -25°C for 1 year. The cheeses were then defrosted, kept in the refrigerator for the next 20 days and their characteristics were compared with those of fresh cheese which was also in the refrigerator.

2. MATERIALS AND METHODS

Seven cheese-making experiments were performed with sheep's and other seven with goat's milk according to the technology described elsewhere (2,4,5). Milk in each experiment was separated in two parts, half of it was used for preparation of cheese with P.candidum and the rest without it. Ten days later one part from the cheeses was withdrawn from the refrigerator and was used for the frozen storage, while for the other part ripening was continued. Half of the cheeses were frozen by a Frigoscandia Contracting M Model freezer (quick freezing) while the other half were placed directly in a deep freezer (slow freezing) at -25°C, which was also used for the storage of all cheeses.Before freezing, cheeses were packed in plastic poly-bags. The drop of temperature during freezing was followed with thermocouples inserted into cheeses and Figure 1 gives the temperature profile. It can be seen that, by the quick freezing process, cheeses reached storage temperature (-25°C) within $1^{1}/2$ h, while by the slow freezing process this took place in 6h. One year later cheeses were removed from the freezer and were defrosted in the refrigerator. Immediatelly after defrosting (next day)cheeses were examined for first time (1st)while next examination (2nd) took place after keeping them in the refrigerator (2-4°C) for 20 additional days. The characteristics of the defrosted cheeses were compared with those of fresh cheeses examined 10 days (1st) and 30 days (2nd) from the date of manufacture during their storage in the refrigerator.

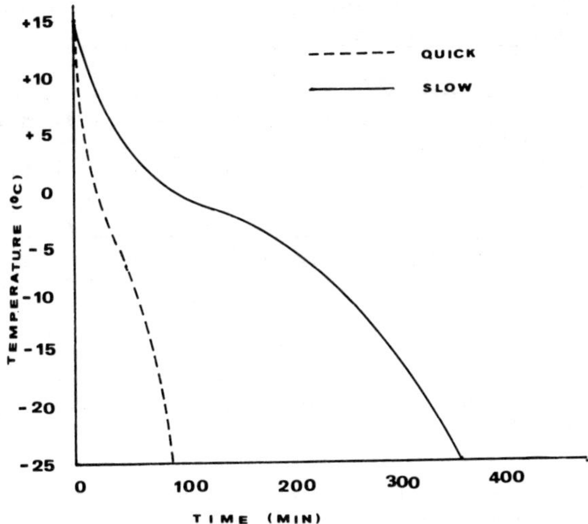

Fig. I. Change of temperature during freezing of cheeses.

Tyrosine soluble in 2% and 12% TCA was determined according to Lowry method (8) using Folin-Ciocalteu reagent. The ratio (R) of soluble tyrosine in 2% TCA to soluble tyrosine in 12% TCA was calculated because it is used as an index of the degree of proteolysis. As ripening of cheese is proceeding the value of R is approaching 1. Preparation of cheese samples for the determination of proteolysis was made according to the method described by Kalatzopoulos and Dalles (5). The values of pH were obtained with a Radiometer Copenhagen pH meter. For the sensory evaluation of the cheeses

were used more than 15 persons belonging to the staff of the Institute of Food Technology. For grading of the cheeses we took into consideration French regulations(1) and scores were awarded according to the following scales:
External appearance and shape

-Colour, flavour and texture	Taste
0-2 unacceptable	0-2 unacceptable
2-4 good	2-5 good
4-6 very good	5-8 very good

The results from the 7 experiments were averaged together and the mean values of the characteristics of fresh cheeses were compared with the values obtained for the defrosted cheeses by the statistical method,analysis of variance (7). The least significant difference was also calculated.

3. RESULTS AND DISCUSSION

Tables I and II give the mean values of the characteristics examined for the cheeses from sheep's and goat's milk respectively.

Table I. Proteolysis, pH and sensory evaluation of fresh and defrosted cheeses from sheep's milk during cold storage.

Type of cheese	Treatment	Soluble tyrosine mg/g in 2% TCA		Soluble tyrosine mg/g in 12% TCA		R		pH		Sensory evaluation	
		1st	2nd	1st	2nd	1st	2nd	1st	2nd	1st	2nd
+PC	F	2.81	4.52	1.44	2.88	1.94	1.57	4.45	4.88	5.18	5.25
	SF	2.34	4.42	1.18	2.94	1.98	1.50	4.69	4.86	4.77	4.85
	QF	2.30	4.09	1.20	2.87	1.92	1.42	4.70	4.81	5.11	4.64
LSD±(0.05)		ns	ns	ns	ns	ns	ns	ns	ns	ns	ns
-PC	F	2.42	3.18	1.24	1.66	1.95	1.94	4.34	4.68	5.29	4.87
	SF	2.00	3.09	1.03	1.52	1.94	2.01	4.68	4.86	4.88	3.78
	QF	2.05	2.81	1.09	1.52	1.88	1.85	4.61	4.80	4.92	3.95
LSD±(0.05)		ns	ns	ns	ns	ns	ns	ns	ns	ns	ns

±PC= Cheese with or without *P.candidum*, F=Fresh, SF=Slow freezing QF=Quick freezing, LSD=Least significant difference, ns=not significant, 1st and 2nd=Examinations

As it can be seen, for the cheeses from both kinds of milk there were not statistically significant differences for the soluble tyrosine and the R values of defrosted cheeses, compared with the respective values obtained for the fresh. These results indicate that proteolytic enzymes were inactivated during frozen storage at -25°C, however, when cheeses were defrosted and subsequently stored in the refrigerator, proteolytic enzymes were reactivated and soluble tyrosine was found higher in the 2nd examination. As a result to this, the R values are decreased during cold storage in both fresh and deftrosted cheeses. The freezing process, quick or slow, under the conditions that it was performed here, did not affect significally proteolysis of cheeses. As it was expected, cheeses with *P.candidum* had higher soluble tyrosine compared with those without mold. The pH of the cheeses in all cases and for the various treatments(F,QF,SF) was not found to differ statistically, with only one exception the 1st measurement of pH of the cheeses from goat's milk with P.candidum. The pH, during storage in the refrigerator of the fresh and defrosted cheeses increased, however the values were still relatively low indicating that proteolysis of the cheeses was normal. The results from the sensory evaluation also revealed that in all cases and for all treatments were not found to differ statistically. The scores awarded to the cheeses indicate that defrosted cheeses had satisfactory organoleptic properties even after one year frozen storage at -25°C.

Table II. Proteolysis, pH and sensory evaluation of fresh and defrosted cheeses from goat's milk during cold storage.

Type of cheese	Treatment	Soluble tyrosine mg/g in 2% TCA		Soluble tyrosine mg/g in 12% TCA		R		pH		Sensory evaluation	
		1st	2nd	1st	2nd	1st	2nd	1st	2nd	1st	2nd
	F	2.59	4.14	1.43	2.78	1.83	1.49	4.24	4.93	5.73	5.25
+PC	SF	2.28	4.05	1.26	2.68	1.75	1.51	4.56	4.58	5.00	4.99
	QF	2.15	3.99	1.25	2.66	1.72	1.50	4.55	4.73	4.76	5.09
LSD±(0.05)		ns	ns	ns	ns	ns	ns	0.16	ns	ns	ns
	F	2.17	2.71	1.17	1.62	1.82	1.68	4.22	4.40	5.15	4.62
-PC	SF	2.00	2.58	0.91	1.35	2.19	1.91	4.42	4.53	4.65	4.73
	QF	1.92	2.55	0.97	1.32	1.97	1.93	4.42	4.63	4.34	4.21
LSD±(0.05)		ns	ns	ns	ns	ns	ns	ns	ns	ns	ns

Abbreviations are defined in Table I.

Generally, cheeses from sheep's and goat's milk behaved in a similar manner during frozen storage followed by preservation in the refrigerator. However, it is also interesting to notice that soluble tyrosine values were higher in the cheeses from sheep's milk compared with those from goat's milk cheese. Frozen storage and type of freezing (quickly or slowly) under the conditions applied in this study didn't bring any significant changes which could have adverse effects on consumer acceptance of the cheeses. The above results are of interest since other research workers (3,10) have reported various defects in some types of cheese after defrosting.

In conclusion, soft cheeses made from sheep's and goat's milk with P.candidum or without it, stored for 1 year at -25°C,can be equally acceptable with fresh cheeses, after being defrosted and keeping them in the refriferator. Therefore, frozen storage can be used, under the conditions applied here, to make these products available during periods that sheep's and goat's milk are not produced.

REFERENCES

1. COGITORE, A. (1972). Traité pratique de reglementation laitière .Edition Sapin d'or 2e Edition.
2. DALLES, T. (1981). Preparation of soft and fresh cheeses from goat's and sheep's milk. M.Sc., Thesis, Agricultural College of Athens.
3. DONALT, K.T., CLIFORD, F., EYERS, B.S. (1957). The freezing preservation of foods AVI Publishing I.N.C.
4. KALATZOPOULOS, G., BALTADJIEVA, M., VEINOGLOU, B., DALLES,T., STAMENOVA, A., (1983). Preparation de fromage á pâte molle á partir du lait de bre-

bis, avec ou sans moisissure superficielle. Le Lait V.63:341-34-.
5. KALATZOPOULOS, G., DALLES, T. (1982). Etude comparative de l' evólution de la maturation des fromages fabriqués ā partir de laits de brebis et de chēvre avec ou sans moisissure superficielle. Revue Laitiēre Française 421:30-34.
6. KRCAL, L.Z. (1965) Stockage du fromage blanc de consommation. Prumnsl Portavin 16:118-11-. Vn Dairy Sc. Abstr. 27:303.
7. LARMOND, E. (1977). Laboratory methods for sensory evaluation of food. Canada, department of Agricultural Ottawa 33-37.
8. LOWRY, O.H., ROSEBROUGH, N.J., LEWIS FARR, A., and RANDALL, R.J. (1951). Protein measurement with the Folin phenol reagent. J. Biol.Chem. 193: 265-275.
9. PORTMAN, A. (1971). Le froid appliqué á la conservation fromage de chēvre le Technicien du lait 11:27.
10. RICHARDS, E. (1966). Cheese storage under atmospheric and deep freeze conditions. Dairy Industries 893-897.
11. SAAL, H.(1959), New methods of preserving cottage cheese. Amer. Milk Rev. and Milk Bl. 21:30-31.
12. VEINOGLOU.T., BALTADJIEVA, M., KALATZOPOULOS, G., DALLES, T., STAMENOVA, V. (1982). Preparation of soft and fresh cheeses from goat's milk. Bulletin of Agricultural studies, Agricultural Bank of Greece V.4:16-28.

QUALITY CONSIDERATIONS IN FREEZING OF DOUGH AND BAKED PRODUCTS

P. LINKO and A. KARHUNEN
Helsinki University of Technology, Department of Chemistry
Laboratory of Biochemistry and Food Technology
SF-02150 Espoo 15, Finland

Summary

The effects of freezing and thawing rates and temperatures, and frozen storage time and temperature both of unbaked doughs and of a variety of baked products on the quality of the final product were investigated. A total of over 50 products and 11 different packaging materials were included in the study. Rapid freezing and thawing minimized staling of baked goods. Quick freezing of doughs and low interior temperatures, however, resulted in an increased proof time and decreased product volume owing to adverse effects on yeast function. Many doughs could be stored at -10°C for a week without significant effects on product quality, if the interior temperature was allowed to increase to 0°C before final proofing and baking. As a result of this study the appropriate statute on frozen foods was changed to allow temporary storage of doughs in bakeries for up to one week at -10°C before final baking. Furthermore it was found that -10°C was sufficiently low for short storage of coffee bread and Danish pastry, but not for typical yeast leavened bread rolls.

1. INTRODUCTION

Refrigeration and freezing technology was introduced to the baking industry in the 1920s, first for air-conditioning and fermentation control, and soon thereafter for product freezing (1). In Finland freezing technology became increasingly popular in bakeries in the 1960s as a result of the 5-day working week and legistlation restricting night baking, to distribute work load more evenly, to reduce labour and fuel costs, and to allow deliveries of frozen products. Furthermore, freezing technology has made such products as frozen doughs, brown-and-serve goods, and ice-cream gateau possible (2). On the other hand, increasing energy costs have recently necessitated careful cost/quality optimization. The present study was undertaken in order to investigate the applicability and feasibility of freezing technology to a variety of bakery products and intermediates, with an emphasis on the effects of freezing and thawing rates and temperatures on product quality. This report summarizes results obtained with selected products. A more detailed report will be published elsewhere.

2. BAKED PRODUCTS

2.1. Freezing and thawing

The study included 54 bakery items and 11 packaging materials. Fig. 1 shows typical freezing and thawing curves of 80g coffee buns. Most product water crystallized at -12° - -14°C, although it has been shown that some water remains unfrozen at much lower temperatures (3). Time required to reach -15°C was reduced by 30% when freezer temperature was lowered from -20°C to -25°C, and by further 20% at -30°C. Time required to lower product temperature from +20°C to -15°C in liquid nitrogen freezer at -40°C varied from 8 to 160 min depending on product. Crumb staling was most noticeable organoleptically in products frozen near 0°C, and was still significant at -15°C. According to several reports (4-6) rapid staling takes place at +10° - -7°C, with a maximum rate at about 0°C. Mälkki et al. (7) studied extensively the effects of freezing and thawing conditions on rheological properties of baked products, concluding that rapid freezing minimizes staling.

Considerably less attention has been paid to the optimizing of thawing than to freezing. Nevertheless, optimal thawing conditions are particularly important with starchy materials subject to staling, owing to starch retrogradation. At +4°C as long as 2.5 h were needed for the crumb temperature to reach 0°C and, accordingly, the time to pass the temperature range characterized by rapid staling was long. Conventional compressor freezing, accompanied by thawing at +25°C resulted in staling equivalent to that caused by 12 h at room temperature, whereas rapid freezing at -40°C reduced staling to

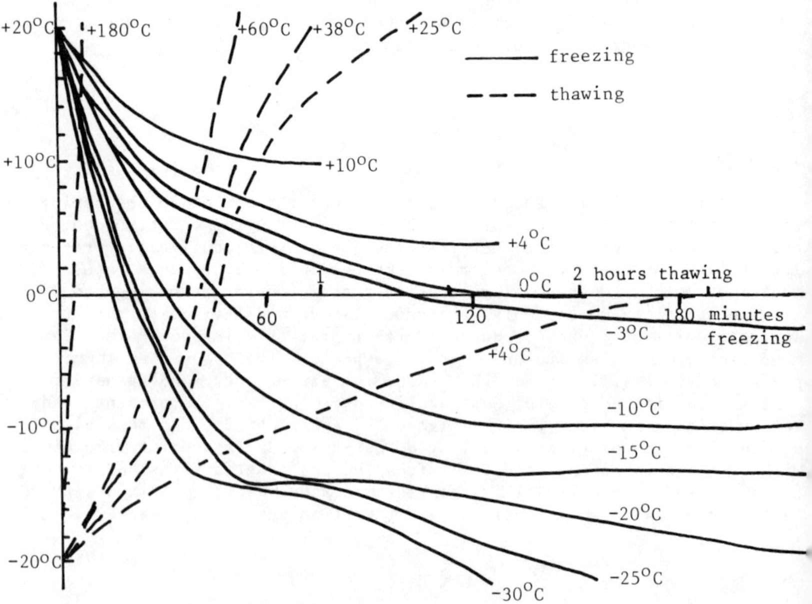

Fig. 1. Typical freezing (compressor freezer) and thawing rates of baked coffee buns (80g) at various temperatures.

the level of standing at +25°C for less than 9h. Thawing at 60°C or at higher temperatures resulted in products of organoleptic quality close to that of freshly baked unfrozen products, owing to the partial revesibility of staling (8-9).

2.3. Frozen storage

The decrease in product quality during frozen storage depends largely on improper or extended storage. Unpacked or improperly packed products may dry out significantly. In this work coffee buns stored at -18°C unpacked (air current 1 m/s) lost 1.3% moisture in 2 weeks, 2.6% in a month, and 5.1% in 3 months. Surface drying and redistribution of crumb moisture may result in the separation of crust and crumb (10-11), and/or in a light colored ring under the crust (12). Bread crumb became brittle already after one week at -18°C, whereas coffee buns could be similarly stored for about one month without noticeable staling.

3. UNBAKED PRODUCTS

3.1. Freezing

Product water crystallized in unbaked coffee buns at slightly higher temperatures of -10° - -12°C than the baked buns. Furthermore, interior temperature of unbaked products decreased faster, particularly in the liquid nitrogen freezer, probaly at least in part owing to the insulating effect of air in porous baked buns. However, the poor tolerance of yeast to quick freezing is a disadvantage when liquid nitrogen freezer is employed (13). This could clearly be seen as decreased loaf volumes in comparision with products frozen in conventional compressor freezers. The following table shows as an example that a decrease in freezing temperature from -30° to -90°C in liquid nitrogen freezer resulted in an increase of proof time by about one hour, and in about 14% decrease in loaf volume (freezing to -20°C crumb -enter, thawing for 1h at +26°C at 60% RH):

freezing temperature (°C)	proof time (min)	volume (ml)
-30 (compressor)	135	267
-30 (liq. nitrogen)	135	246
-60 "	155	232
-90 "	195	212

In addition to the effect of freezing rate in liquid nitrogen freezer, also the lowest temperature reached by the interior had an effect on final quality. The following table shows an example that decreasing of crumb temperature below -10°C increased proof time and decreased product volume, particularly if the freezing temperature was low and freezing rapid:

crumb temperature (°C)	proof time (min)			volume (ml)		
	-30°C	-60°C	-90°C	-30°C	-60°C	-90°C
- 5	125	125	180	251	249	234
-10	125	125	180	249	241	222
-15	125	155	195	246	232	212
-20	125	170	275	236	222	185
-40 (-50)		180	(300)		214	(171)

This is in accordance with the earlier finding (14) that yeast cells tolerate -10°C relatively well, but already at -15°C about 75% of cells may die. Consequently, more yeast should be used in frozen doughs stored at -15°C or lower to offset such losses (15). Taguchi et al. (16) has also shown that the addition of 0.55% of glutanic acid to dough stored at -20°C for 4 weeks resulted in no decrease in baked product loaf volume.

3.2. Thawing

Mazur (14) recommended slow thawing of frozen doughs to minimize damage to yeast. We observed that a moderate thawing time at +4°C resulted in a good quality product. Prolonged thawing, however, resulted in some fermentation already during thawing and, thus, in uneven raising and poor quality. Thawing at room temperature (+25°C) for 30 - 90 min, depending on type of product, to reach interior temperature of 0°C before proofing also resulted in good product quality.

3.3. Frozen storage

Frozen storage time of doughs significantly affected product quality. An increase in storage time increased proof time and decreased product volume, as shown in the following table for coffee bun (80g) and Danish pastry (50g) doughs (thawing for 1h at +27°C, 50% RH; proofing at +37°C, 70% RH):

frozen storage at -18°C	coffee bun (80g)		Danish pastry (50g)	
	proof time (min)	volume (ml)	proof time (min)	volume (ml)
0	75	264	50	200
1 d	80	255	55	190
2 d	105	255	60	170
1 m	120	240	60	165
2 m	135	225	75	130
3 m	150	215	90	125
4 m	200	170	120	120
6 m	215	144	120	105

Satisfactory coffee buns could also be obtained from up to one moth stored frozen doughs. Proof time could be decreased by doubling yeast quantity, but product volume remaine lower than that of the control. Although Marston (15) had recommended -18° - -25°C for the storage of frozen doughs, we found that coffee bread, Danish pastry, and breakfast rolls could be stored at -10°C for a week without any detrimental effect on quality. Storage of rolls at -10°C even slightly improved both crumb structure and loaf volume:

storage time (d) at	volume (ml) -25°C	-10°C
1	220	270
2	210	270
3	210	250
7	190	230
14	170	160

It could be calculated that about 23% savings in energy costs could be realized if the storage temperature could be decreased from -20°C to -10°C, and that the capital costs could be reduced by as much as 40%. It was also calculated that in a typical bakery conventional freezing combined with one day's frozen storage at -20°C would cost about Fmk 0.27/kg product. Consequently, based both on quality and economic considerations, the statute on frozen foods in Finland was changed to allow bakeries to store unbaked intermediate goods temporarily for a maximum of one week at -10°C before final baking.

4. REFERENCES

1. PENCE, J. W. (1968). Bread and rolls. In 'The Freezing Preservation of Foods' Vol. 4 (D. K. Tressler, W. B. Arsdel and M. J. Copley, eds) The AVI Publ. Co, Westport, Connecticut, 386.
2. DRAKE, E. (1970). Up-to-date review on freezing. Beker's Dig. 44(2),65.
3. MANNHEIM, H. C., STEINBERG, M. P., NELSON, A. I. and KENDALL, T. W. (1957). Heat content of bread. Food Technol. 11, 384.
4. KATZ, J. R. (1928). Gelatinization and retrogradation of starch in relation to the problem of bread staling. In 'A Comprehensive Survey of Starch Chemistry' Vol. 1 (R. P. Walton, ed.) Chemical Catalog Co, NY.
5. PENCE, J. W. and STANDRIDGE, N. N. (1955). Effects of storagetemperature and freezing on the firming of commercial bread. Cereal Chem. 32, 519.
6. MARANDIINI, W. and WASSERMANN, L. (1971). Brotfrischhaltung - praktische Erfahrungen. Brot u. Gebäck 25, 53.
7. MÄLKKI, Y., PUOLAKKA, L. and PAAKKANEN, J. (1974). Effects of freezing and thawing conditions on the rheological properties of bakery products. In 'DECHEMA Monographien' Band 77 'Lebensmittel - Einfluss der Rheologie' DECHEMA, Frankfurt am Main, 165.
8. SCHOCH, T. J. (1965). Starch in bakery products. Baker's Dig. 39(2),48.
9. KNJAGINICEV, M. I. (1970). Das Altbackwerden des Brotes und die Erhaltung seines Frischzustandes. Starch/Stärke 22, 435.
10. BELDEROK, K. B. and WIEBOLS, W. H. G. (1964). Studies on the defrosting of frozen bread. Food Technol. 18, 1813.
11. NEWALD, E. (1966). Krustenschäde bei tiefgekühltem Kleingebäck und ihre mögliche Verhütung. Brot u. Gebäck 20, 129.
12. PENCE, J. W., STANDRIDGE, N. N., BLACK, D. R. and JONES, F. T. (1958). White rings in frozen bread. CUreal Chem. 35, 15.
13. LORENZ, K. (1974). Frozen dough - Present trend and future outlook. Baker's Dig. 38(3), 59.
14. MAZUR, P. (1961). Physical and temporal factors involved in the death of yeast at sub-zero temperatures. Bipphys. J. 1, 247.
15. MARSTON, P. E. (1978). Frozen dough for breadmaking. Baker's Dig. 52(6), 18.
16. TAGUCHI, K., TABATA, H. and Yoshizaki, T. (1975). Stabilizers for yeast in frozen dough. U.S. Patent 3,901,975.

ACCEPTABILITY OF FROZEN WHITE SLICED BREAD

T. R. GORMLEY
An Foras Taluntais,
Kinsealy Research Centre,
Malahide Road,
Dublin, 5, Ireland.

Summary

White sliced pan bread loaves (wrapped in film) took 3.5 and 9 hr to reach -20°C using blast (-34°C) and slow freezing (-18°C) methods respectively. The frozen samples were stored at -20°C and took 9 hr to thaw (in the wrapper) to ambient temperature (17°C). Paired comparison taste panels indicated no statistically significant preference for slow or blast frozen samples; similarly no difference was found between frozen and fresh samples after 4 weeks frozen storage. Penetrometer readings confirmed these findings; mean crumb compression values were similar for blast frozen (2.6 mm), slow frozen (2.4 mm) and fresh (2.6 mm) slices. Bread tested in the different weeks was also similar, with mean compression values of 2.6, 2.4, 2.5 and 2.5 mm in weeks 0, 2, 3 and 4 respectively. As expected, crumb texture of fresh and thawed breads stored at ambient temperature for 3 days did change with a compression value of 3.1 mm on Wednesday compared with 2.3 mm on Thursday and 2.2 mm on Friday.

These data indicate that the brand of frozen white sliced bread tested was equally acceptable to fresh bread for a period of at least 4 weeks.

1. INTRODUCTION

In recent years there has been a considerable increase in the freezing of bread and other bakery products (1-9). While much of this relates to freezing at industrial level there has also been extensive freezing of bread in the home (10). It is well established that sliced bread which has been frozen is suitable for making toast and many bakers issue instructions on the wrapper for the toasting of bread directly from the frozen state. However, it is less well established that bread which has been frozen is suitable for consumption in the untoasted form, e.g. as sandwiches. There is also a lack of sensory data on the acceptability of breads that have been frozen. For this reason samples of commercially produced white sliced pan were frozen by blast and still air methods and were evaluated after thawing as untoasted slices by taste panels over a 4-week period and also using crumb compression tests.

2. EXPERIMENTAL

Thirty two loaves of commercially baked white sliced pan bread, each wrapped in a 120 gauge clear polythene bag, were frozen on Tuesday afternoon (week 0) by blast (-34°C) and cabinet (still air, -18°C) methods. The blast freezer used was a "Nu-Avon" model with air speed of 3.75 m/sec. A 24 point Honeywell temperature recorder fitted with copper-constantan thermocouples was used to monitor freezing and thawing rates. Frozen samples were tested on weeks 0, 2, 3 and 4 (after thawing); the loaves required for testing were removed from frozen storage (cabinet, -20°C) to ambient temperature each Tuesday and their acceptability and degree of staling tested on Wednesday, Thursday and Friday during ambient storage at 15-18°C. Four loaves x 4 slices were evaluated each day for each bread type using taste panels and a penetrometer. Samples of fresh bread (same brand), baked each Tuesday, were used for comparison. The bread tested had a slice thickness of 8 mm and there were 21-23 slices per loaf.

Paired comparison taste panels with 20 tasters x 2 samples were used and panellists evaluated the bread as slices (without crusts) and were asked for preference only. Samples of blast frozen bread were compared with cabinet frozen bread in the morning; the preferred sample was then compared against the fresh bread in the afternoon. The panels were carried out in weeks 2, 3 and 4 giving a total of 18 panels during the experiment.

The penetrometer used for the crumb compression test was a VEB-Feinmess Dresden model with a 40 mm diameter hemispherical probe weighing 19 g. The orientation of the bread slice in relation to the probe was the same for each measurement and the penetration of the probe into the bread (in mm) was recorded.

3. RESULTS AND DISCUSSION

3.1 Freezing and thawing

Loaves wrapped in polythene bags took about 3.5 and 9 hr respectively to reach -20°C using blast and cabinet freezing methods. The frozen samples were stored at -20°C and took 9 hr to thaw (in the bag) to ambient temperature (17°C). These times are similar to those reported by other authors (4, 6).

3.2 Taste panel evaluation

Results for weeks 2-4 indicate no statistically significant preference for blast or cabinet frozen samples (Table 1). A 38/22 break is needed for statistical significance. The data presented in Tables 1-111 are the

Table 1: Taste panel preference[1] (paired comparison) for blast and cabinet frozen white sliced pan bread over a 4-week period

Freezing method	Week			Sum of responses
	2	3	4	
Blast frozen	35	30	33	98
Cabinet frozen	25	30	27	82
Statistical significance	NS	NS	NS	NS

[1] panel responses are combined for tests on Wednesday, Thursday and Friday giving a total of 60 responses

Table 11: Taste panel preference[1] (paired comparison) for fresh and frozen bread

Bread type	Week			Sum of responses
	2	3	4	
Frozen[2]	39	24	35	98
Fresh	21	36	25	82
Statistical significance	$p < 0.05$	NS	NS	NS

[1] as in Table 1

[2] see experimental section

combined responses for 20 panellists on each of Tuesday, Wednesday and Thursday each week; this gives a total of 60 individual responses each week.

These taste panel data (Tables 1 and 11) indicate that freezing by the faster blast method seems to offer no advantage over cabinet freezing under the conditions and duration of this test. In addition, samples that had been frozen for up to 4 weeks and then thawed were as acceptable, if not more so, than samples of fresh bread that had not been frozen at all.

Preference data for frozen (thawed Tuesday) and fresh (baked Tuesday) bread during storage at ambient temperature (15-18°C) for 3 days are presented in Table 111. Again the panel showed no significant preference

Table 111: Taste panel preference[1] (paired comparison) for frozen (thawed) and fresh bread over a 3 day storage period at 15-18°C

Bread type	Day Wednesday	 Thursday	 Friday
Frozen (thawed Tuesday)	32	27	39
Fresh (baked Tuesday)	28	33	21
Statistical significance	NS	NS	$p < 0.05$

[1] panel responses are combined for tests in weeks 2, 3 and 4 giving a total of 60 responses

for fresh over frozen except on day 3 (Friday) of ambient storage when the frozen sample was preferred (Table 111). This agrees with data of Pence <u>et al.</u> (11) and Malkki <u>et al.</u> (8) who found that firmness (staleness) of bread that had been frozen develops more slowly than that of non-frozen bread.

3.3 <u>Penetrometer readings</u>

The taste panel findings were confirmed by the penetrometer readings and there was no significant difference between values for blast frozen, cabinet frozen and fresh bread except in week 4 of the experiment when cabinet frozen bread was firmer ($p < 0.05$) than the others (Table 1V).

Table 1V: Penetrometer readings (mm) for frozen and fresh breads over a 4-week period

Bread type	Week 0	 2	 3	 4	Mean
Blast frozen	2.7	2.4	2.4	2.7	2.55
Cabinet frozen	2.4	2.4	2.6	2.2	2.40
Fresh	2.5	2.2	2.8	2.9	2.60
F-test	NS	NS	NS	$p < 0.05$	-
SE	0.11	0.11	0.12	0.16	-
Mean (for frozen)	2.55	2.40	2.50	2.45	-

This overall result conflicts with that of Malkki <u>et al.</u> (7) who found that fast freezing rates were best for preserving softness and springiness in breads. The mean data (Table 1V) also show that the frozen breads showed no texture change during 4 weeks frozen storage.

Most of the staling for the three bread types took place between day 1 (Wednesday) and day 2 (Thursday) (Table V) and little further change in crumb springiness occurred between day 2 and day 3 (Friday). This may be a similar effect to that reported by Malkki et al. (8) who frequently found a disconuity phase in the elastic modulus versus storage time pattern for bread three days after baking or thawing.

Table V: Penetrometer readings (mm)[1] for frozen (thawed Tuesday) and fresh bread (baked Tuesday) over a 3 day storage period at 15-18°C

Day	Bread type			Mean
	Blast frozen	Cabinet frozen	Fresh	
Wednesday	3.2	2.9	3.2	3.10
Thursday	2.3	2.2	2.3	2.27
Friday	2.2	2.1	2.3	2.20
F-test	$p < 0.001$	$p < 0.001$	$p < 0.001$	-
SE	0.08	0.09	0.08	-

[1]data are means for 4 weeks

4. CONCLUSIONS

The data indicate that the brand of frozen white sliced bread tested was equally acceptable to fresh bread for a period of at least 4 weeks; further tests should be done to determine its long term acceptability and the study should be broadened to include other brands.

5. REFERENCES

1. WOOD, P. (1981). Products for deep freeze. Institute of British Bakers Bulletin, October, 21-36.
2. REEDICH, E. (1980). Bakers warm up to profit potential in freezing. Bakery, March, 77-80.
3. ROBB, J. and CORNFORD, S. J. (1974). The use of freezing and refrigeration by the baking industry. British Baker, March, 24-26, 29-31, 33-34.
4. BAMFORD, R. (1975). Freezing and thawing of bakery products. The Bakers Digest, June, 40-43.
5. BECHTEL, W. G. (1965). The freezing of bread. Bakers Weekly, January, 31-32.
6. BRILEY, G. C. (1975). Optimum temperatures for freezing bread. Baking Industries Journal, October, 43-45.
7. MALKKI, Y., PUALAKKA, L. and PAAKKANEN, J. (1974). Effect of freezing and thawing conditions on the rheological properties of bakery products. Dechema-Monographien, 77, 165-175.
8. MALKKI, Y., PAAKKANEN, J. and EEROLA, K. (1978). Effect of freezing and monoglycerides on staling of bread. J. Fd Proc. Preservation, 2, 101-110.
9. GORMLEY, T. R. and CORMICAN, C. (1982). Acceptability of frozen white sliced pan bread. Ir. J. Fd Sci. Technol., 6, 104 (Abstr.).
0. MUNSON, S. T., WINTER, J. D., HARME, M. L. and ALLEN, C. E. (Revised 1970). Freezing Foods for Home Use. Extension Bulletin 244, Agr. Extension Service Univ. of Minnesota, 50-55.
1. PENCE, J. W., STANDRIDGE, N. N., LUBISCH, T. M., MECHAM, D. K. and OLCOTT, H. S. (1955). Studies on the preservation of bread by freezing. Fd Technol. (Champaign), 9, 495.

STORAGE LIFE OF FISH MINCES

K J WHITTLE, K W YOUNG, P HOWGATE, A CRAIG and G L SMITH
Torry Research Station, 135 Abbey Road, Aberdeen AB9 8DG

Summary

The cold storage stability over 1 year at -15°C of fillet and mince blocks from 5 gadoid sp is compared. Assessment was by means of large panels to obtain hedonic ratings of fish fin prepared from the blocks at monthly intervals. Cod stored at -30°C provided a relatively s reference for comparison. Shelf life was determined as the point at which 75% of the asses still liked the product. Most fillet blocks had a shelf life in excess of 6 months. Most blocks had no shelf life as defined above but this definition may be too stringent. A stor temperature of -30°C is essential.

1. INTRODUCTION

Rising fuel costs and various restrictions on catching have increased th cost of catching fish substantially in the last decade, so that fish process have installed machinery to separate flesh from skin and bone and increase th yield of recoverable, edible flesh over traditional filleting methods. The availability and use of mechanically recovered, deboned, fish flesh (minced or fish mince) in fish products is such that fish mince must be considered a major raw material for the fish processing industry.

The properties and quality of minced fish in terms of colour, texture, flavour, cold storage stability and total nitrogen, bone and water contents but the range of variability is not fully known. Many factors (1) are invol (a) the type of raw material used, eg headed and gutted fish, fillet trimmin (V-cuts, lugs, belly flaps), cheeks, napes or different parts of the skeleta frame left after filleting; (b) the previous chill or cold storage history o raw material; (c) preliminary processing before flesh recovery; (d) the meth recovery; (e) the species; (f) the season and maturation state.

Growing commercial interest in developing higher value products based o minced fish has shown the need for reliable methods of assessing and predict the variety of variable quality factors. One aspect which we are studying i comparison of the stability of a number of types of minces from different ga species stored at -15°C. Assessment of the properties of the frozen minces trained sensory judges and chemical and physical tests is being carried out the usual ways. However, this paper reports on a different aspect of assess a measurement of the 'acceptability' of fish fingers made from the minced fi samples after appropriate periods of frozen storage have elapsed. Overall a ability is influenced markedly by the extent to which a person likes or disl a product, eg as measured using a 9-point hedonic scale (2). However, hedon ratings are entirely subjective and assessors must give their responses free without being influenced by a panel leader or researcher, or by previous ins tion and training. Large panels are used because of the high individual var ability in likes and dislikes. The fingers represent a familiar and popular product and some 'economy' types are made from mince commercially. They are to prepare and convenient to present to large panels for hedonic ratings. T place rates of change in perspective and aid comparison of the acceptability frozen blocks of fillets of each species were also stored at -15°C with the minces and, in each species trial, cod fillet and minced fillet were stored relatively stable conditions at -30°C to provide a bench mark or reference t gauge variability in the large panel.

This approach should help to assess the quality of mince for a consumer product more realistically and provide a more soundly based measure of shelf life. This paper is a progress report of methods, findings and conclusions to date based on the acceptability data from 5 gadoid species stored at -15°C.

2. MATERIALS AND METHODS

2.1 Fish Table I lists the origin and pre-processing history of each species. The cod for reference in each trial were caught locally, gutted, kept 2 to 3 days boxed in melting ice, processed within 1 to 3 weeks of the respective species and stored at -30°C. Two blue whiting samples covered the change in seasonal quality of the species (3,4). The hake and Alaska pollack both involved bulk shipment over long distances akin to commercial practice.

TABLE I - ORIGIN AND PRE-PROCESSING HISTORY OF GADOID SPECIES

SPECIES	CAUGHT	PRE-PROCESSING HISTORY	TRIAL BEGAN*
Blue whiting *Micromesistius poutassou*	S Faroe shelf Feb 1980	FAS 1 h whole, ST -30°C P 28 Feb	3 Mar 1980
Blue whiting	W Scotland shelf April 1980	FAS 1 h whole, ST -30°C P 14 May	19 May 1980
Whiting *Merlangius merlangus*	Locally Dec 1980	Gutted, 2-3 days BI P 16 Dec	22 Dec 1980
Haddock *Melanogrammus aeglefinus*	Locally June 1981	Gutted, 2-3 days BI P 9 June	15 June 1981
Alaska pollack *Theragra chalcogramma*	Bering Sea Unimak Pass area Dec 1981	HG, FAS 6 h, ST -30°C offloaded trawler 28 Dec, ST -30°C. 0.75 t shipped 6 Jan AF container dry ice Seattle - London. AF unrefrigerated London - Aberdeen 10 Jan. ABF ST -30°C, P 26 Jan	24 Feb 1982
South African hake *Merluccius capensis*	29°10'S 14°50'E off SW Africa Nov 1981	HG, FAS 2 h, ST -23°C offloaded trawler 28 Nov, ST -23°C, 2 Dec ST -40°C, 25 Dec. 0.5 t shipped 25 Dec by sea, refrigerated cargo -22°C, Capetown - Avonmouth 15 Jan. ST -25°C, shipped RRT Aberdeen 17 Feb. ST -30°C P 23 Feb.	24 Feb 1982

* Frozen blocks ST -30°C until trial began

FAS 1 h, frozen at sea within 1 h of catching; ST, stored; P, processed; BI, boxed with ice; HG, headed and gutted; AF, air freight; ABF, air blast frozen; RRT, refrigerated road transport.

2.2 Sample preparation Fish were stored at -30°C after receipt, processed as soon as possible (Table 1), filleted by hand and skinned by machine. Blue whiting were headed and gutted by machine and the belly cavities cleaned to remove the belly lining, swim bladder, kidney and blood tissues. Minces were produced using a Baader 694 deboner (drum perforations 5 mm diameter), the raw material being fed at random into the hopper. In each storage trial, frozen blocks (nominally 7.4 kg), packed in waxed card cartons in the normal commercial fashion (5) but without addition of polyphosphate, were made from fillets and minces of the appropriate species and from the reference cod fillets and minced fillet (Table II).

TABLE II - SAMPLE TREATMENTS FOR EACH STORAGE TRIAL

TRIAL	SPECIES	TYPES OF BLOCK:	SAMPLE TREATMENTS
1	Blue whiting (February)	A cod fillet:	skinless, trimmed, single fillets
		B cod mince:	from single fillets
		C species fillet:	skinless, single fillets
		D species mince:	from block fillets
		E species mince:	from headed and gutted fish
2	Blue whiting (April)	A to E	as 1 above
3	Whiting	A cod fillet:	as 1 above
		B cod mince:	as 1 above
		C species fillet:	skinless, single fillets
		D species mince:	from cutlet fillets
		E species mince:	from headed and gutted fish
4	Haddock	A cod fillet:	as 1 above
		B cod mince:	as 1 above
		C species fillet:	skinless, trimmed, single fillets
		D species mince:	from block fillets
		E species mince:	from headed and gutted fish
5	Alaska pollack	A cod fillet:	as 1 above
		B cod mince:	as 1 above
		C species fillet:	skinless, trimmed, single fillets
		D species mince:	from single fillets
		E species mince:	from headed and gutted fish
6	South African hake	A cod fillet:	as 1 above
		B cod mince:	as 1 above
		C species fillet:	skinless, trimmed, single fillets
		D species mince:	from single fillets
		E species mince:	from headed and gutted fish

* Assume fillets untrimmed and unskinned unless stated otherwise

2.3 Storage regime The frozen blocks were sawn longitudinally into four equa slabs and carefully wrapped in aluminium foil. Slabs of each species type w kept in a cold store room nominally at -15°C; cod slabs were kept nominally -30°C to minimise storage changes. At approximately monthly intervals for t 50 to 55 weeks of each trial, slabs of each treatment (A to E) were processe into enrobed fish fingers and kept frozen until required for tasting within 24 hours.

2.4 Taste panel assessment (acceptability of the fish fingers products) Mon assessment of samples A to E from each storage trial, was carried out by pan of 42 drawn each time from a pool of 120 of our Research Station staff, most untrained in sensory assessment. Thus, the composition of the panels varied within the pool such that members, as available, were not asked to taste mor than once a week. The panel met 6 at a time seated in individual booths. D cussion or comment was discouraged. Usually, 5 treatments from a trial were assessed hedonically (Table III) at each session.

Frozen, breaded fish fingers were deep fried (refined, deodourised vege oil; 4 min, 180°C) and presented to the panellists on paper plates divided radially into 5 equal areas, each coded with a 3-digit random number, to avo one group being influenced by another. Panels were not told which species o treatments were tested or that the cod samples were references.

TABLE III - HEDONIC ASSESSMENT SCORE SHEET

In this experiment we are interested in assessing the acceptability of fish fingers.

After tasting each sample, select from the list of like and dislike descriptions, numbers which you think fit your impression of each sampe in terms of OVERALL ACCEPTABILITY:

	Sample Code	Overall acceptability score
9 like extremely		
8 like very much		
7 like moderately		
6 like slightly		
5 neither like nor dislike		
4 dislike slightly		
3 dislike moderately		
2 dislike very much		
1 dislike extremely	Name	Date ....

2.5 Statistical analysis The usual summary statistics, mean and standard deviation, have limited use as measures of the hedonic response of a panel. They are difficult to interpret in cases of skewness or bimodality in the distributions of ratings which occur frequently. Instead, the median, a more meaningful measure of the centre of the distribution was used. Differences of 0.7 or more between medians are significant. Degree of acceptability for the purpose of estimating shelf life is given by the percentage of the panel expressing any degree of "liking", ie the number of scores 6 to 9 plus half the scores 5.

Assuming that the relationship between either median or percentage liking and time was approximately linear during the period of about 1 year, best fit lines were calculated using the method of least absolute deviations (6,7). This is the regression analogue of the median. It is less influenced than the usual "least squares" method by outlying or aberrant values. Some cautious attempts at approximate significance tests have been made, since methods for deriving tests for this type of line are not well developed.

3. RESULTS AND DISCUSSION

Figure 1 compares the lines of best fit for each species describing the changes in hedonic ratings of fish fingers prepared from each treatment during storage at -15°C for 1 year. Our assumption that the linear relationship is valid over 1 year is supported by the plot (whiting) in Figure 2 of actual median data; treatment A (cod fillets) is omitted for clarity. The pattern of results (Figure 1) is broadly similar in each trial. The order of treatment scores is important since it indicates the panel's preferences. Treatments A, B and C consistently scored higher than D and E suggesting a preference for fillets (all species) and cod fillet mince over the minces from the other species. This could be due to differences in colour and flavour affecting acceptability but space does not permit presentation of the data. The species treatments tended to be ordered C, D, E throughout the trials. After 1 year's storage acceptability tended to decrease in order A, B, C, D, E.

Fish fingers from the cod blocks (-30°C) showed no significant change in acceptability and provided relatively stable standards or benchmarks for comparison, mostly scoring > 6.5 throughout the year. The cod samples A and B had similar acceptability except in the haddock trial (June). Only in this case, is the difference in acceptability between A and B likely to be significant, although neither deteriorated much with storage time. In June, total nitrogen in the flesh of North Sea cod tends to be minimal (8), as does pH (9). Both factors could contribute to decreased acceptability being more manifest in the mince.

The species fillet samples usually scored > 6.5 initially and, in most cases, showed surprising stability at -15°C, closely matching the acceptability of the cod samples stored at -30°C. Exceptions were blue whiting (April) and hake which

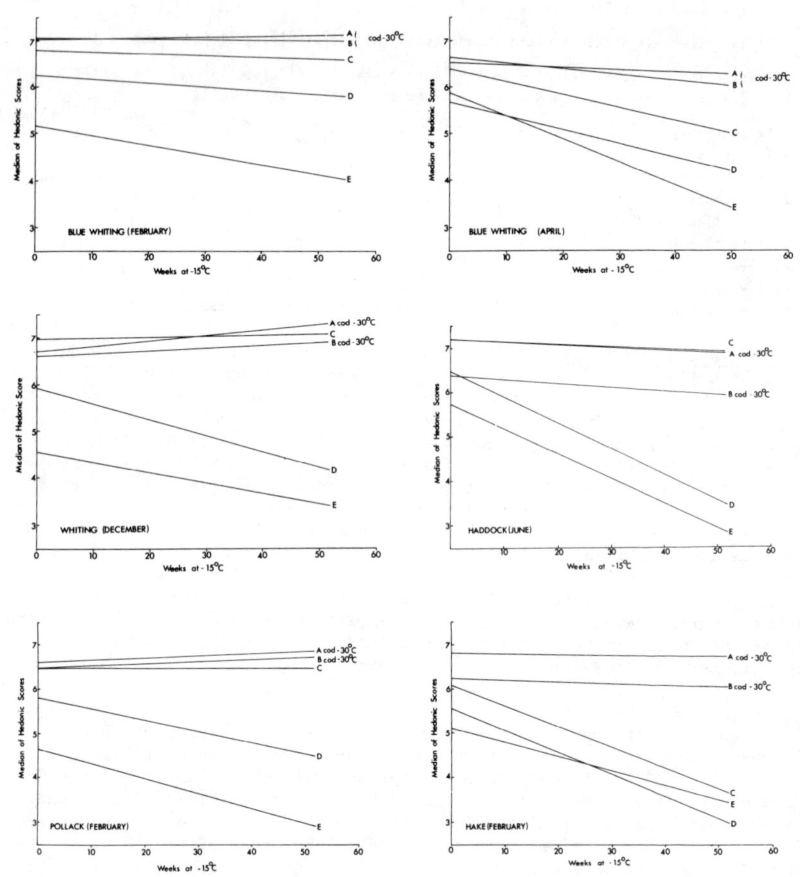

Figure 1 – COMPARISON OF CHANGES IN ACCEPTABILITY DURING COLD STORAGE OF FILLET AND MINCE BLOCKS 5 GADOID SPECIES. The best fit lines A, B, C, D, E relate to the descriptions in Ta

tended to deteriorate at similar rates to the minces. The headed and gutted (E) always scored less than 5 or 6 initially. In general, the species mince showed clear trends of declining acceptability with storage time and had dec below score 5 after 1 year, excepting block fillet mince from blue whiting (February), which was remarkably stable.

Stability of the various species treatments was compared from the slope the best fit lines expressed as the change in score units per year of storag the appropriate temperature (Table IV). Given the error about each median d point, we have assumed that the lines A and B in each trial represent the va ability within so-called stable conditions for the storage of cod blocks. T slopes of the species treatments C, D and E were compared with the range of 12 estimates of the slopes of A and B (-0.6 to +0.3) and grouped in 3 catego

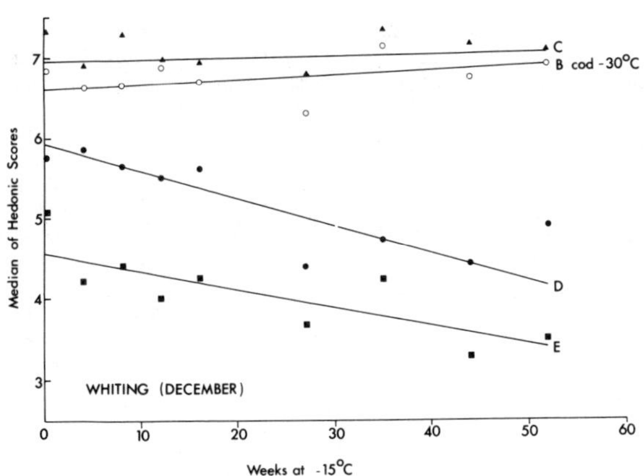

FIGURE 2 - AN EXAMPLE OF THE LINEARITY OF PANEL MEDIAN DATA VIS STORAGE TIME. Lines fitted by the method of least absolute deviations (6,7).

of deterioration as shown in Table IV: slow *; medium **; fast *** corresponding to decreases in acceptability of < 1, 1 to 2.5 and > 2.5 hedonic score units per year respectively.

TABLE IV - COMPARISON OF SLOPES OF LINES OF BEST FIT OF MEDIAN SCORES VIS STORAGE TIME

TREATMENT \ TRIAL		BLUE WHITING February	BLUE WHITING April	WHITING December	HADDOCK June	ALASKA POLLACK December	S. AFRICAN HAKE December
Cod fillet A)	-30°C	-0.1	-0.3	+0.3	-0.2	+0.2	-0.3
Cod mince B)	-30°C	-0.1	-0.6	+0.1	-0.4	+0.3	-0.2
SP fillet C)	-15°C	-0.2*	-1.4**	-0.0*	-0.2*	0.0*	-2.4**
SP mince D)	-15°C	-0.5*	-1.5**	-0.8*	-3.5***	-1.3**	-2.6***
HG mince E)	-15°C	-1.0**	-2.6***	-1.6**	-2.9***	-1.7**	-1.7**

SP, species; HG, headed and gutted fish species

A simple estimate of shelf life is obtained from the 'regressions' (Figure 1) at the median value 5.0 on the scale, ie half the respondents "like" the product to some extent and half "dislike" it. However, the median is probably not the best summary of panel response if shelf life in respect of consumer acceptability is to be measured, since a manufacturer quite rightly would expect substantially more than 50% of the population to like the product, eg 75% (10). Table V lists the shelf lives at -30°C and -15°C of the blocks assessed as fish fingers, as the points at which 75% of the assessors still like (as defined in "Statistical Analysis") the various products, ie when the lower quartile reaches a value of 5.0. We have not determined the errors of the estimates.

4. CONCLUSIONS

The different measures of acceptability, hedonic score or percentage liking, broadly agree, given the inherent variability in panel assessment. Initial acceptability and subsequent relative stability determine the comparative shelf lives of the products. The method "measures" the shelf life of frozen blocks assessed as fish fingers not the shelf life of fish fingers; further testing is being done. In practice, an allowance must be made for the further shelf lif of the product, eg fish fingers, made from the frozen block. Additionally, t shelf life measured will be affected by the pre-processing history which was noteworthy in the cases of Alaska pollack and S. African hake.

TABLE V - ESTIMATES OF SHELF LIFE OF THE FROZEN BLOCKS

TREATMENT \ TRIAL		BLUE WHITING February	BLUE WHITING April	WHITING December	HADDOCK June	ALASKA POLLACK December	S. AFRICA HAKE December
Cod fillet A)	-30°C	52	27	52	52	52	52
Cod mince B)	-30°C	52	52	52	52	52	33
SP fillet C)	-15°C	52	5	52	52	36	0
SP mince D)	-15°C	13	0	0	0	0	0
HG mince E)	-15°C	0	0	0	0	0	0

SP, species; HG, headed and gutted fish species

REFERENCES

1. WHITTLE, K.J. et al. (1981). Some factors affecting the properties of mi fish. Third National Technical Seminar on Mechanical Recovery and Utilis of Fish Flesh, Raleigh, N. Carolina, USA, Dec. 1980. Ed., R.E. Martin, p 224-264, National Fisheries Institute, Washington D.C., USA.
2. PERYAM, D.R. and PILGRIM, F.J. (1957). In "The methodology of sensory testing". IFT Symposium, Pittsburg, USA.
3. WHITTLE, K.J., ROBERTSON, I.R. and McDONALD, I. (1980). Seasonal variabi in blue whiting (Micromesistius poutassou) and its influence on processir In "Advances in Fish Science and Technology". Ed.,J.J. Connell, p 378-38 Fishing News Books Ltd, Surrey, England.
4. AFOLABI, O.A. et al. (1982). The frozen storage characteristics of blue whiting (Micromesistius poutassou). Bull. Int. Inst. Refrig, Annex 1981- p 315-322.
5. KEAY, J.N., STOREY, R.M. and WHITTLE, K.J. (1982). Other Fish Products. "Fish Handling and Processing". Eds, A. Aitken, I.M. Mackie, J.H. Merri and M.L. Windsor, p 138-151, 2nd Edition, HMSO Books, Edinburgh, Scotlan
6. KARST, O.J. (1958). Linear curve fitting using least deviations. J. Am. Statist. Soc., 53, p 118-132.
7. ARMSTRONG, R.D. and KUNG, M.T. (1978). Algorithm AS 132: Least absolute value estimates for a simple linear regression problem. Applied Statist 27, p 363-366.
8. HOWGATE, P., McLAY, R. and CROLLA, J. (1983), internal report, unpublish
9. LOVE, R.M. (1980). Biological factors affecting processing and utilizat In "Advances in Fish Science and Technology". Ed., J.J. Connell, p 130- Fishing News Books Ltd, Surrey, England.
10. HOWGATE, P. (1983). Measuring the storage lives of chilled and frozen f products. Bull. Int. Inst. Refrig. (in press).

QUALITY CHANGES IN MINCED FISH DURING COLD AND FROZEN STORAGE

P. LAJOLINNA, J. LAINE* and P. LINKO
Helsinki University of Technology, Department of Chemistry,
Laboratory of Biochemistry and Food Technology
SF-02150 Espoo 15, Finland
*University of Helsinki, Department of Food Chemistry and Technology, SF-00710 Helsinki 71, Finland

Summary

The interest in minced fish based products has recently increased as a means to improve the utilization both of by-products and of underutilized small fish species. Modern technology has made year round processing possible. The present work suggests that Baltic herring (*Clupea harengus*) could be succesfully used in minced fish production. Several methods were employed in studying quality changes during storage of minced fish at +4°, -2°, -10°, and -25°C in reference to samples stored at -40°C. Hypoxanthine and trimethylamine content appeared to be good indicators of quality before freezing. Water holding capacity ('expressive drip'), peroxide value and free fatty acids gave valuable information on quality changes during frozen storage.

1. INTRODUCTION

World wide fish harvest is of the order of 70 million tonnes. About one-half of this amount consists of pelagic species, of which about 60% is herring fish. About a quarter of the fish caught is frozen, typically as deboned cuts with estimated processing losses of as high as 70%. Minced fish can be succesfully used for a variety of products, as has been shown in Japan for centuries (1-2). With modern mechanical deboning the over-all yield may be increased to about 50% (3), and the yield from cleaned fish to as high as 70-80% (4). Minced fish has also been utlized and extensively studied in a number of other countries, particularly in the USA, Soviet Union, Norway, Denmark and Poland (1). In this work the suitablity of Baltic herring (*Clupea harengus*) in the form of frozen minced fish as a year-round raw material for further processing was investigated.

2. EXPERIMENTAL

Deponed fish was iced and minced with a Baader flesh-separator using a 3 mm sieve. The meal was immediately block-frozen at -30°C in a contact freezer. Samples were stored at +2°, -2°C, -10°, -25°, and -40°C for various periods of time. Trimethylamine content was determined according to A.O.A.C. (5), hypoxanthine according to Beuchat (6), free fatty acids by titration from chloroform extract (45g fish meal, 90 ml chloroform, 22.5g dry sodium sulfate, extraction for 1 min), peroxide value iodomaterically from the same filtrate as free fatty acids, and water holding capacity ('expressive drip') gravimetrically according to Karmas and Turk (7). Samples were also tested for organoleptic and microbiological quality.

3. RESULTS AND DISCUSSION

3.1. Chemical changes

Fish is subject to rapid chemical, physical, and microbiological changes, and long-term storage even at -20°C is known to result in major organoleptic changes (8-9). Certain portions of fish belong to most easily spoiling foodstuffs. Although rapid adverse changes may take place in improperly treated and stored minced fish, protective measures are easier that with whole or deboned fish. Oxidative changes are the limiting factor with fatty fish, whereas with other species protein denaturation may lead to a decrease in water holding capacity, emulsifying ability and other functional properties, affecting adversly further processing.

Fig. 1 shows that fish meal pH exceeded value 7 at +4°C in 3 days, and at -2°C in about 2 weeks, whereas at lower temperatures only minor changes could be observed. Similar changes have been observed during storage of minced cod (10). When fish dies, flesh adenine-5'-triphosphate decomposes enzymatically to inosine-5'-monophosphate, an important flavour enhancer in fish. During storage 5'-IMP further decomposes to bitter hypoxanthine, which accumulates and can thus be employed as an indicator of freshness (11). It was found that also with Baltic herring hypoxanthine content is a good index of the storage history. Trimethylamine oxide (TMAO) decomposes mainly by the action of bacterial enzymes to trimethylamine (TMA) and subesquently to dimethylamine and formaldehyde, and trimethylamine is largely responsible of the typical 'fishy' flavour. Fig. 2 shows that the TMA content of minced fish increased markedly in about one week or less only at storage temperatrues of +4° and -2°C. At lower temperatures TMA content remained low even up to about 2 month storage, and thus may be considered mainly as an index of quality before freezing.

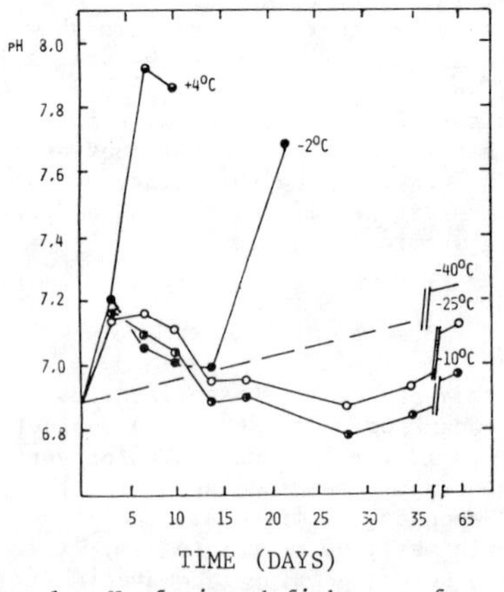

Fig. 1. pH of minced fish as a function of time.

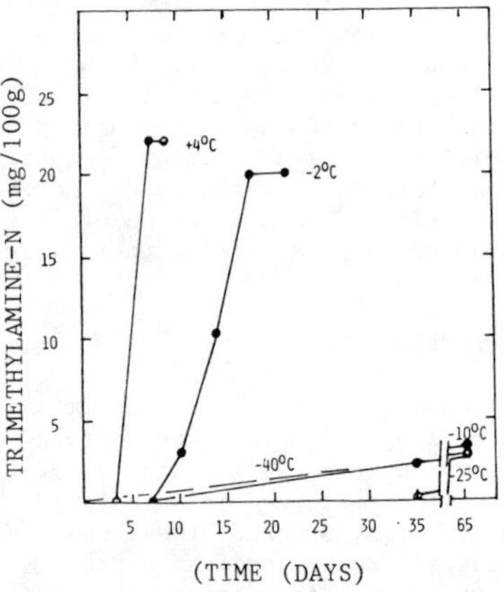

Fig. 2. Trimethylamine content of stored minced fish.

Generally 10-30 meqv O/kg has been considered as the maximum permissible peroxide value. This value was exceeded in about 2 days at +4°C and in about 10 days at -2°C. At -40°C peroxide formation was negligible (Figs 3 and 4). The formation of free fatty acids followed closely peroxide formation (Figs 4 and 5), and little free fatty acid formation could be observed during storage at -25° or -40°C.

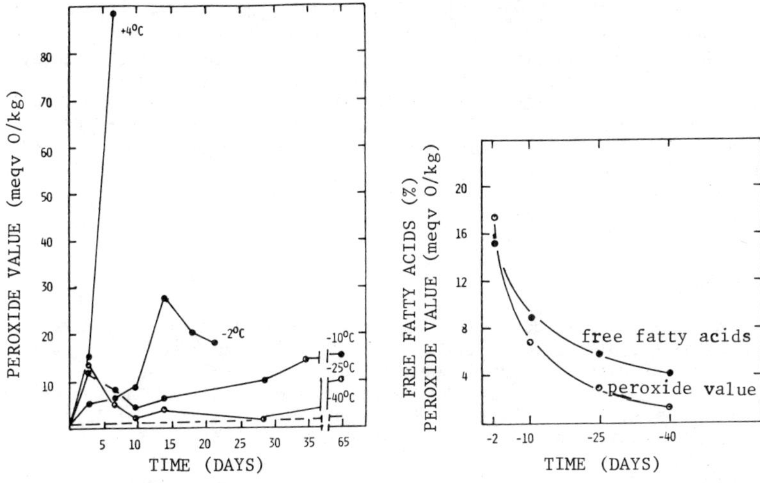

Fig. 3. Peroxide value of stored minced fish

Fig. 4. Peroxide value and free fatty acids after 3 weeks

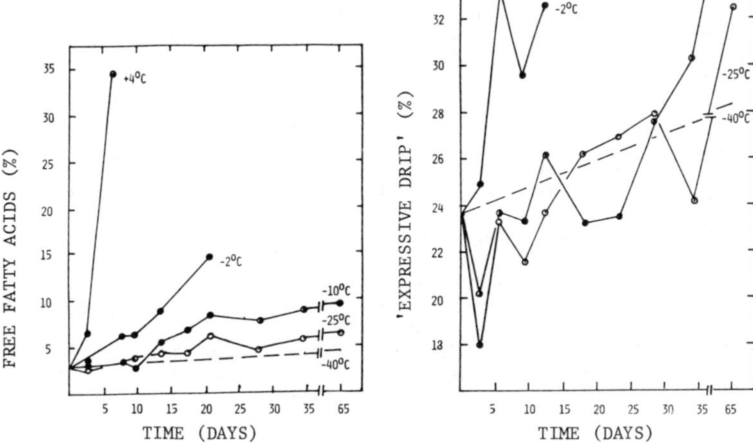

Fig. 5. Free fatty acids of stored minced fish

Fig. 6. 'Expressive drip' of stored minced fish

3.2. Water holding capacity

The denaturation of flesh proteins during processing and storage of fish results in marked changes in structure and functional proerties (8,10). Enzymatic formation of formaldehyde from TMAO has been considered as the primary reaction. However, formaldehyde formation has been observed to be insignificant in properly frozen and stored Baltic herring (12). On the other hand mechanical flesh separation and deboning results in some disintegration of cells and, thus, in increased sensitivity to spoilage. It has been reported that as little as 0.6% enzyme active kidney tissue rapidly decreases water holding capacity (10,13). Water holding capacity is one of the most important functional properties of food materials employed in further processing (7). Furthermore, poor water holding capacity results in an increased loss of water during thermal processing and, thus, in inferior product structure and colour. Fig. 6 shows that at -2°C water holding capacity decreased markedly in a few days, whereas at lower temperatures changes were slower.

4. REFERENCES

1. FAO (1975). Backgrounf paper on minced fish. FAO Fish Circ. 332, 24 pp.
2. OKADA, M., MIYAUCHI, D. and KUDO, G. (1974). "Kamaboko" - the giant among Japaneshe fishery products. Section XI, Additional Reference Papers of Interest on Flesh Separation. Second Technical Seminar on Mechanical Recovery and Utilization of Fish Flesh. Boston, June 1974, 301.
3. MARTIN, R. E. (1976). Mechanically deboned fish flesh. Food Technol. 30, 64.
4. KING, F. J. (1972). Characteristics, uses and limitations of comminuted fish from different species. Section III, Current Technical Status, Oak Brook Seminar, Sept. 1972, 15.
5. A.O.A.C. (1975). Official Methods of Analysis, 12th ed. George Banta Co. Wisconsin.
6. BEUCHAT, L. R. (1973). Hypoxanthine measurement in assessing freshness of chilled channel catfish. J. Agr. Food Chem. 21, 453.
7. KARMAS, E. and TURK, K. (1975). A gravimetric adaption of the filter paper press method for the determination of water-binding capacity. Zeit. Lebensm.-Unters. u. Forschung 158, 145.
8. SIKORSKI, Z., OLLEY, J. and KOSTUCH, S. (1976). Protein changes in frozen fish. Crit. Rev. Food Sci. Nutr. 8, 97.
9. LINKO, P., LAJOLINNA, P. and LAINE, J. (1978). On keeping quality of domestic minced fish (in Finnish). Ympäristö ja Terveys 7(3), 214.
10. JARENBÄCK, L. (1976). Frozen storage of fish mince (in Swedish; English summary). SIK-report No. 409 and 411.
11. BURT, J. R. (1977) Hypoxanthine: a biochemical index of fish quality. Process Biochem. 12(1), 32.
12. SAVOLAINEN, K., KUUSI, T. and NIKKILÄ, O. E. (1975). Formation of formaldehyde in baltic herring and cod and related changes in the fish muscle proteins during frozen storage. Scand. Refrig. 4a, 141.
13. KUDO, D. M. G. and PATASHNIK, M. (1977). Effect of processing variables on storage characteristics of frozen minced Alaska pollock. Marine Fisheries Rev. 39(5), 11.

THE INFLUENCE OF CHILLING, FREEZING AND FROZEN STORAGE ON POULTRY MEAT MAINTENANCE

T.T.PETRAK, D.ROSEG, S.KOŠMERL[x], A.HRASTE, A.JELIĆ
Centre of Animal Products Technology, Faculty of Veterinary Sciences, University of Zagreb and Koka Poultry Industry, Varaždin[x]

ABSTRACT

Maintenance of poultry meat selected as carcass, breasts, breast fillet, legs, and wings, unpacked and packed in foodtainers (fizzines polystyrene) and stretch-wrapped in PVC wrapper, evaluated in iced water chilling conditions, dry chilling (air chiller), and a combination of these two methods of chilling and freezing. Significant differences were established in maintenance of poultry carcasses and cut-up pieces. Cut-up pieces and carcasse showed a shorter maintenance period packed in foodtainers with PVC foil, than unpacked meat at the same temperature (+4°C and +8°C). These results were the opposite of those for frozen poultry carcasses and cut-up pieces. Histo-morphological investigation showed that mm. pectoralis superficialis (white meat) are more prone to destruction when frozen than mm. ilio tibialis lateralis (dark meat).

INTRODUCTION

The significance of poultry meat processing in human nutrition increases each year, and therefore a major part of research is directed towards improvement of hygienic and technological conditions, as primary factors which influence the maintenance of fresh or frozen poultry meat. Time-temperature tolerance of retail-level poultry meat depends directly on how it is processed, particularly on the cooling and freezing process.

These are all delicate problems (Stadelman,1974), particularly a combination of iced water immersion chilling with air chill wind tunnel. The organoleptic properties of poultry meat resulting from bacteriological and autolitical processes at cooling and freezing, and particularly in thawing i.e. maintenance, is a subject which has been of interest to many authors (Roseg,1966., Berner,1969., Ristić,1982).

In this study examinations were carried out to give relevant indicators on carcass and cut-up poultry meat maintenance, in respect of organoleptic, chemical, bacteriological and histomorphological changes (properties).

MATERIALS & METHODS

Cooling

After carcasses (Hybro line) are removed from the eviscerating line, they are chilled for 30 minutes in a spin-chiller with water at 16°C (slush ice connected), and then cooled in an air chill wind tunnel (AC) at a temperature of -5°C, relative moisture 96% and air velocity 3.6 m/sec for 100 minutes (KOKA,Varaždin). The alternative method was to chill the carcasses only in the air chill wind tunnel for 120 minutes. The cooled carcasses (A) were cut up and prepared for the market as: breasts (B), breast fillet (C). legs (D), and wings (E). Poultry meat selected in this way was stored in cooling chambers at +4°C and +8°C. Part of the samples were unpacked, and the other part packed in foodtainers and stretch-wrapped in PVC wrapping (P).

Freezing

After spin-chilling, some of the samples were frozen (F) at -40°C for two hours (two-phase freezing) and some were frozen at the same temperature as above, directly from the air wind chill tunnel. The samples were stored in cooling chambers at -18°C. Some samples were unpacked and others packed, as described above.

Methods of Analysis

Organoleptic analysis was adapted on the basis of triangular tests (Goodall,1967). Chemical analysis was based on NH_3 mg % examination Conway,1962). Results were interpreted as follows: less than 8.33 NH_3 mg % - organoleptically irreproachable poultry meat; 8.33 - 8.84 - organoleptically unchanged; 8.84 - 9.35 - organoleptically unchanged, close to spoiling point. Bacteriological analysis used GSP- agar (Kielwein, 1972) for Pseudomonas sp. In histo-morphological analysis we used mm. ilio tibialis lateralis and mm. pectoralis superficialis. Samples were taken of fresh meat, frozen meat and thawed meat. Thawing was carried out at +22°C for 4 hours. Pieces of muscle were placed in cool +4°C 10 % formalin (frozen samples only) and the others in normal formalin. After fixation, the muscles were placed in varying concentrations of alcohol aethylicus pa. and imbedded in paraffin, and preparations 10 microns in thickness were made (Reichert Spencer) and stained in hemalaun eozin by the Majer method (Romeis 1968). Slides were examined by Zetopan Reihert microsc Ten samples of each party were taken for examination. Findings of more than 80 % - positive or negative - were taken as findings for interpretation.

RESULTS

Organoleptic property results presented in Table 1 demonstrated the maintenance of poultry meat processing S+AC,AC,S+AC+P, AC+P and stored at +4°C and +8°C. Ivestigation of organoleptic properties of frozen poultry meat also presented in Table 1, processing S+F,AC+F,S+F+P,AC+F+P and stored at -18°C.
NH_3 mg % chemical analysis results on poultry meat processing as above, stored at +4°C, +8°C and -18°C are given in Table 2.

ORGANOLEPTIC ANALYSIS OF CHILLED AND FROZEN POULTRY MEAT

Table 1

Temp. Processing	ABCDE	ABCDE	A B C D E	A B C D E	A B C D E	A B C D E	ABCDE
+ 4°C Hour	24	48	72	96	12o	144	168
S + AC	−	−	−	− − + − −	− + + + −	− + + +	+
AC	−	−	−	− − − − −	− − + − −	− + + +	−
S+AC+P	−	−	−	− + + + −	+ + + + +	+ + + +	+
AC + P	−	−	−	− − + − −	− + + + −	+ + + +	+
+ 8°C							
S + AC	−	−	− − + − −	− + + − +	+ + + +	+	
AC	−	−	− − − − −	− − + − −	− + + +	+	
S+AC+P	−	−	− + + − −	+ + + + +	+ + + +	+	
AC + P	−	−	− − + − −	− + + + +	+ + + +	+	
− 18°C Month	1	3	6	9	12	15	18
S + F	−	−	− − + − −	− + + − −	− + − −	− + + +	− ++
AC + F	−	−	− − − − −	− − + − −	− − − −	− + − −	− ++
S + F + P	−	−	− − − − −	− − − − −	− − + − −	− − − −	−+ ++
AC+F+P	−	−	− − − − −	− − − − −	− − + − −	− − − −	−+ −−

CHEMICAL ANALYSIS (NH_3) OF CHILLED AND FROZEN POULTRY MEAT

Table 2

Temp.	Processing	ABCDE	A	B	C	D	E	A	B	C	D	E
+4 °C	Hour	24			48					72		
	S + AC	< 8,33			< 8,33			< 8,33	8,50	8,84	8,50	8,67
	AC								8,33	8,50	8,33	8,33
	S + AC + P								8,67	9,18	8,67	8,67
	AC + P								8,50	8,67	8,33	8,67
+8 °C	S + AC	< 8,33	< 8,33	8,33	8,50			8,50	9,18	9,01	8,67	8,67
	AC			8,33	8,33			8,33	8,67	8,67	8,50	8,50
	S + AC + P			8,50	8,50	< 8,33	< 8,33	8,67	9,69	9,86	9,01	9,01
	AC + P			8,50	8,50			8,50	9,01	9,01	9,01	8,67
-18 °C	Month	1			3					6		
	S + F	< 8,33			< 8,33			< 8,33	8,67	8,84	8,67	8,50
	AC + F								8,50	8,67	8,50	8,33
	S + F + P								8,50	8,67	8,33	8,50
	AC + F + P								8,50	8,67	8,33	8,50
	AC + F + P								8,33	8,50	8,33	8,67

Table 2 (cont.)

Temp.	Processing		A	B	C	D	E	A	B	C	D	E
+4°C		Hour			96					12o		
	S + AC		8,67	8,84	9,69	8,84	9,ol	9,ol	9,35	-	9,69	9,69
	AC		8,5o	8,67	9,ol	8,67	8,5o	8,84	9,ol	9,52	8,84	9,ol
	S + AC + P		8,5o	8,84	lo,2o	9,18	9,69	8,84	lo,o3	-	lo,o3	lo,2o
	AC + P		8,5o	8,67	9,69	8,84	9,ol	9,ol	9,35	-	9,ol	9,18
+8°C												
	S + AC		8,84	lo,o3	lo,o3	9,18	8,84	9,35	-	-	lo,o3	9,86
	AC		8,67	9,69	9,52	8,84	8,67	9.17	-	-	9,35	9,35
	S + AC + P		8,84	lo,88	11,22	9,86	lo,37	9,35	-	-	-	-
	AC + P		8,84	9,86	lo,2o	9,18	9,86	9,35	-	-	9,67	-
-18°C		Month			9					12		
	S + F		9,67	9,ol	9,35	9,ol	8,67	8,84	9,35	9,69	9,35	9,ol
	AC + F		8,5o	8,5o	9,ol	8,84	8,5o	8,5o	9,ol	9,52	9,ol	8,84
	S + F + P		8,5o	8,5o	9,52	8,84	8,5o	8,5o	9,ol	-	9,ol	8,84
	AC + F + P		8,5o	8,5o	8,67	8,5o	8,5o	8,5o	8,84	9,01	8,67	8,84

Table 2 (cont.)

Temp.	Processing	A	B	C	D	E	A	B	C	D	E
+4 ^{o}C	Hour			144					168		
	S + AC	9,35	1c,2o	-	9,86	1o,o3	9,69	-	-	-	-
	AC	9,o1	9,35	-	9,35	9,35	9,35	9,52	-	9,35	9,35
	S + AC + P	9,18	-	-	1o,54	11,o5	9,52	9,86	-	-	-
	AC + P	9,18	9,52	-	9,52	9,86	9,69	-	-	-	-
+8 ^{o}C											
	S + AC	9,86	-	-	-	-					
	AC	9,52	1o,o1	-	-	-					
	S + AC + P	9,86	-	-	-	-					
	AC + P	1o,o3	-	-	-	-					
	Month			15					18		
-18 ^{o}C											
	S + F	9,18	9,52	-	9,52	9,18	9,35	-	-	-	9,86
	AC + F	8,84	9,52	-	9,o1	9,o1	9,18	-	-	9,35	9,18
	S + F + P	8,67	9,35	-	9,35	9,38	9,78	9,52	-	9,52	9,35
	AC + F + P	8,67	9,o1	9,52	9,o1	9,18	8,84	9,35			

Bacteriological analysis presented in Table 3 showed that there was no difference in the total contamination count of Pseudomonas sp. between carcasses and cut-up meat stored at +4°C and +8°C. We isolated the highest number of Ps. fluorescens, Ps. putrefaciens, Ps. putida and Ps. aeruginosa. At freezing, we found that the total count of bacteria Pseudomonas sp. was less than 10^2 in the first month, to 10^4 after 18 months. Histo-morphological examinations were carried out on mm. ilio tibialis lateralis and mm. pectoralis superficialis. "Fresh" poultry carcasses after cooling showed us (Photograph 1) that mm. pectoralis superficialis had normal histological structure as did mm. ilio tibialis lateralis (Photograph 2). Frozen poultry carcasses samples after freezing and storage (3 months) were taken for trial directly from frozen materials (tissue) and placed in cooled formalin. Mm. pectoralis (Photograph 3) demonstrated that certain muscle fibres had intracellular destruction shown as longitudinal striation - rarely vertical. Some cells showed normal structure. Mm. ilio tibialis lateralis (Photograph 4) demonstrated normal histological structure. Samples from thawed poultry carcasses, after 4 hours thawing at 22°C showed us that the greater number of mm. pectoralis cells demonstrated normal structure but in some places tissue with longitudinal intracellular destruction could be seen (Photograph 5). Mm. ilio tibialis lateralis demonstrated normal histological structure.

Table 3

BACTERIOLOGICAL CURVES FOR

PSEUDOMONAS SP.

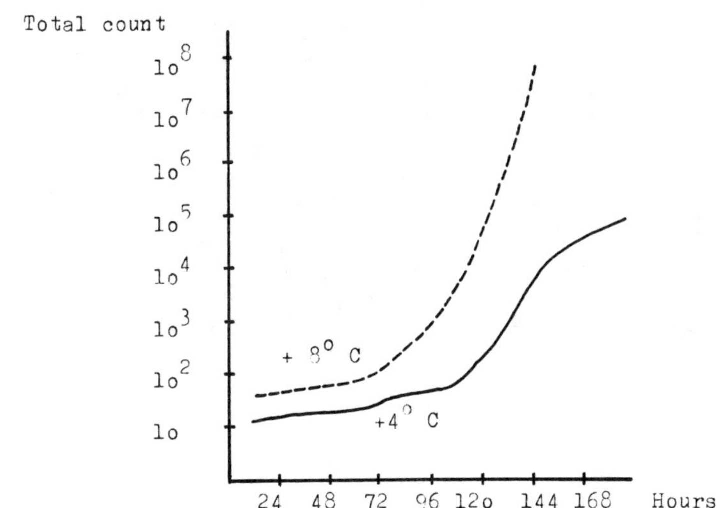

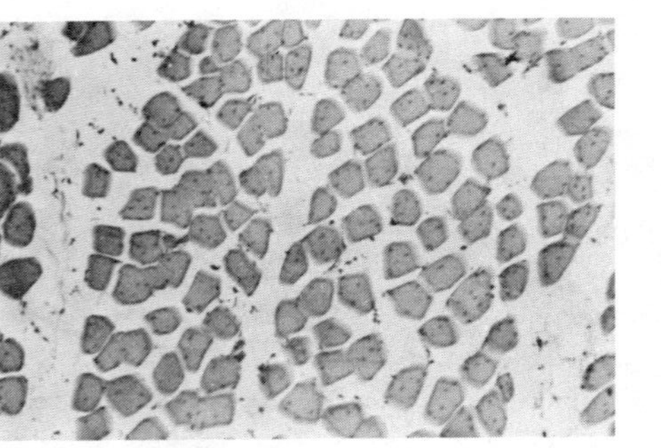

Photograph 1.

Mm.pectoralis superficialis,
"fresh"poultry carcasses
Enlargement loxlo

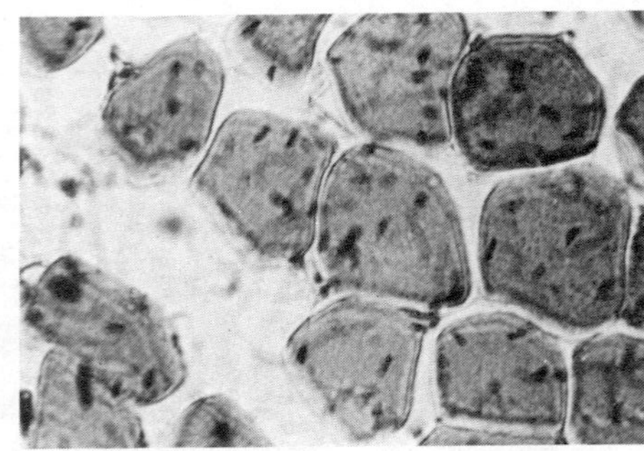

Photograph la.

Mm.pectoralis superficialis
"fresh" poultry carcasses

Enlargement lox4o

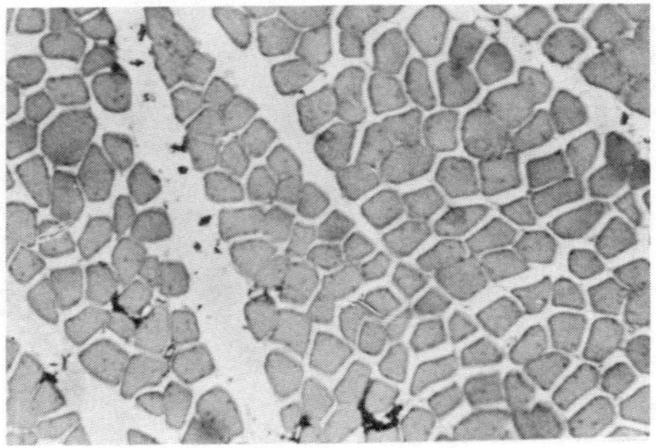

Photograph 2.

Mm. ilio tibialis lateralis

"fresh" poultry carcasses

Enlargement loxlo

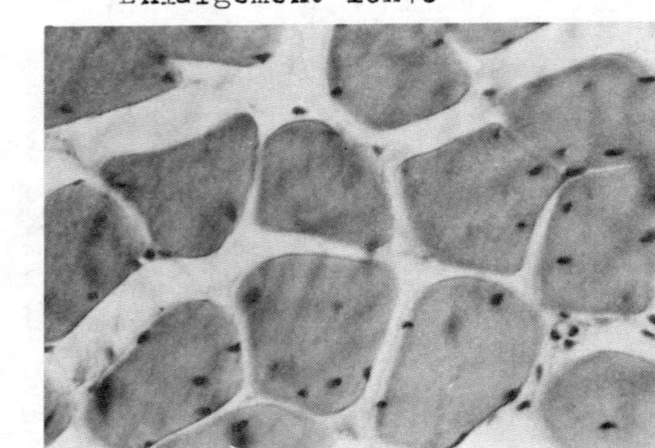

Photograph 2a.

Mm.ilio tibialis lateralis

"fresh" poultry carcasses

Enlargement lox4o

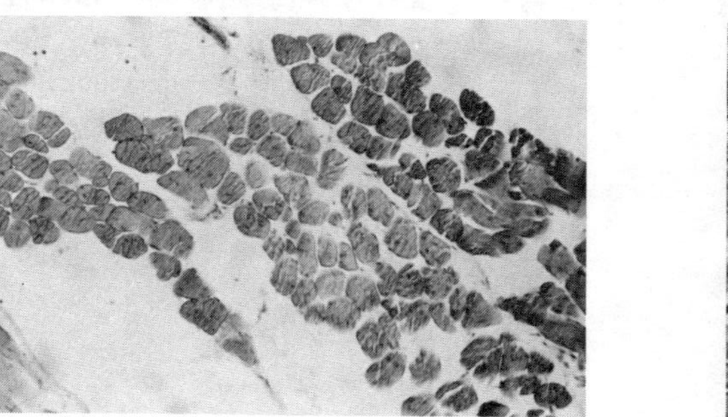

Photograph 3.
Mm. pectoralis superficialis
frozen poultry, enlargement loxlo

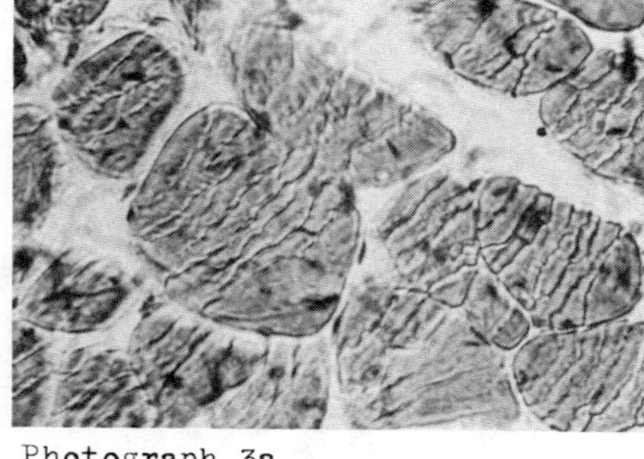

Photograph 3a.
Mm. pectoralis superficialis
frozen poultry,enlargement lox4o

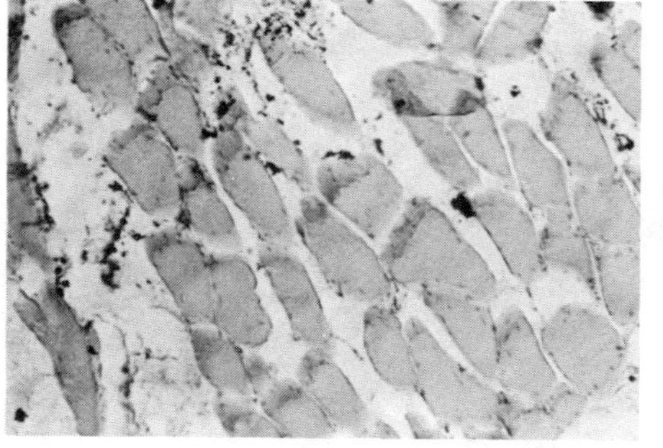

Photograph.4
Mm. ilio tibialis lateralis
frozen poultry, enlargement loxlo

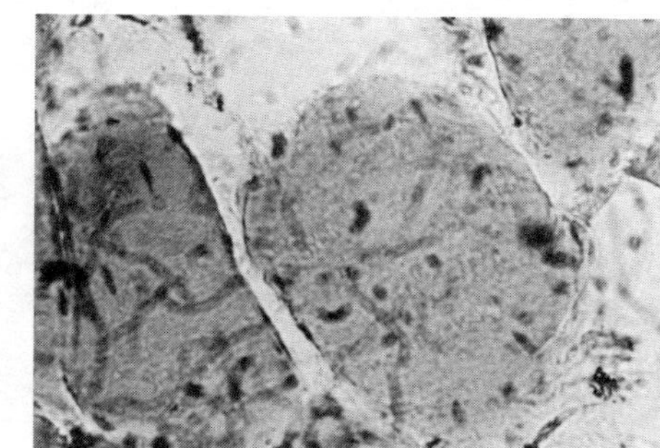

Photograph 4a.
Mm. ilio tibialis lateralis
frozen poultry, enlargement lox4o

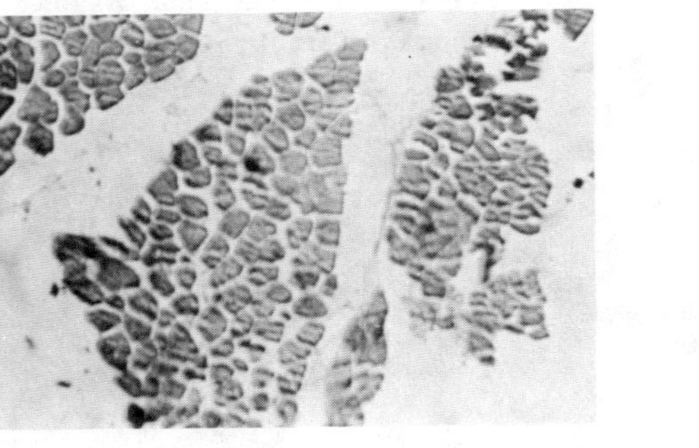

Photograph 5.
Mm. pectoralis superficialis
thawed poultry,enlargement loxlo

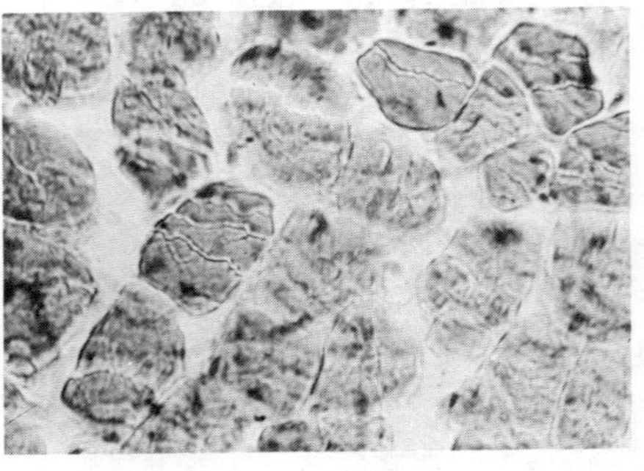

Photograph 5a.
Mm. pectoralis superficialis
thawed poultry, enlargement lox4o

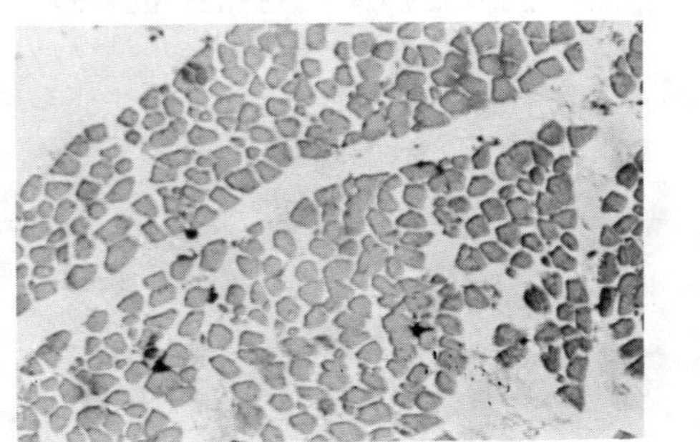

Photograph 6.
Mm. ilio tibialis lateralis
thawed poultry,enlargement loxlo

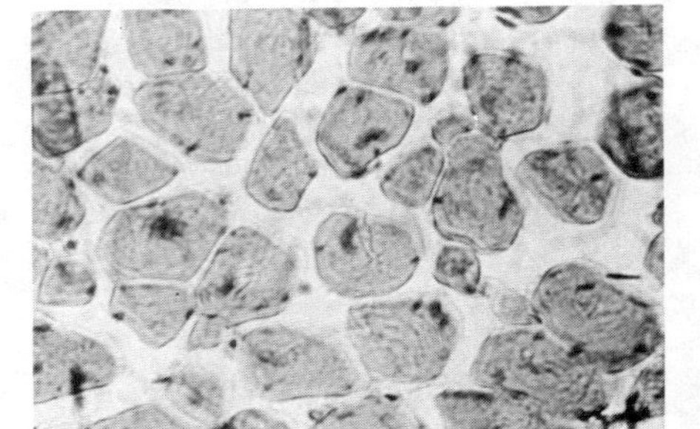

Photograph 6a.
Mm. ilio tibialis lateralis
thawed poultry,enlargement lox4o

CONCLUSION

A. Fresh poultry meat

1. Maintenance

The results of organoleptical examination of poultry meat prepared as carcasses and retail level pieces, showed that so-called "fresh meat" technologically processed by direct cooling in the air chill wind tunnel (dry cooling) and stored at +4°C, could be maintained as follows: carcasses-168 hours, wings, legs and breasts 120 hours, and breast fillet 96 hours. When sorage is at +8°C maintenance results are on average lower by 24 hours.
Analysis of the amount of ammoniac in the tissue of the meat showed that spoiling point of the carcasses and breast fillet stored at +4°C was identical with that shown by organoleptic examination, while with wings, legs and breasts spoiling point appeared 24 hours later. Storage at +8°C gave maintenance results for carcasses and fillet identical to those shown by organoleptical examination, while wings and legs could be maintained for an additional 24 hours, and unboned breasts and breast fillet had similar maintenance properties i.e. the maintenance period was shorter by 24 hours.

2. Technological processing

AC and S+AC processing of fresh meat gave a longer period of maintenance of poultry meat than that of samples wrapped in PVC.

B. Frozen meat

1. Maintenance

Organoleptic analysis demonstrated that maintenance of poultry meat stored at -18°C was 18 months for carcasses, 15 for breats, wings and legs, and 9 for breast fillet.
Tests for ammoniac content showed maintenance of 18 months for carcasses and breasts, 15 months for wings and legs and 12 months for fillet.

2. Technological processing

AC+F+P processing gave the best maintenance results as compared with the other methods applied.

Bacteriological screening showed that a relatively low initial number of psychrophillic micro-flora at the beginning of the cooling process, but this increased rapidly at cold storage temperatures. The total number og Pseudomonas sp. in frozen poultry meat did not increase proportionally to the number of months storage.

Histo-morphological study demonstrated that frozen white poultry meat had a more marked tendency to destruction of the muscle tissue.

Note:
The subject matter of this research requires further histo-morphological, histo-chemical, and stereological analysis of frozen poultry meat.

REFERENCES

1. BERNER,H.,KLEEBERGER,A. und BUSSE,M. (1969): Untersuchungen über neues Kühlverfahren für Schlachtgeflügel. Fleischwirtschaft 49,1617-1620,1623.
2. CONWAY,E.J. (1962): Microdifusion analysis and volumetric error. Crosby Lockwood and Son.Ltd.,London.
3. GOODALL,H.,COLQUHOUN,J.M. (1967): Sensory Testing of Flavou and Aroma. Scientific and Technical surveys 49.
4. KIELWEIN,G.(1971): Die Isolierung und Differenzierung von Pseudomonaden aus Lebensmitteln. Arch.Mikrobiol. 34,270.
5. ROMEIS,B. (1968): Mikroskopishe Technik. München, Wien.
6. ROSEG,D.,(1966): Influence of autolytic processes of freezing and defrostation on poultry meat. Dissertacion Work.
7. RISTIĆ,M.(1982): Einflüsse der Kühlung auf die Qualität von Broiler-Schlachttierkörpern. Tehnologija mesa, 9,255.
8. STADELMAN,W.J. (1974): Chilling poultry meat. Poultry Sci. 53, 1267-1268.

RETENTION OF VITAMIN B_1, B_2, AND B_6 IN FROZEN MEATS

K. MIKKELSEN*, E.L. RASMUSSEN* and O. ZINCK**
*Department of Food Preservation, Royal Veterinary and Agricultural University, Copenhagen
**National Food Institute, Soeborg (Copenhagen), Denmark

Summary

Four of the most common meat products in Denmark were used in this study: pork loin, ground meat, vienna and "medister" sausages. The products were frozen and stored at -24°C or -12°C. Samples were analysed for B_1, B_2 and B_6 every two month for one year, both before and after the final preparation. Retention (%) after preparation and one year storage at -12°C or -24°C showed the following range: For pork loin B_1 50-85, B_2 90-112 and B_6 48-88. For ground meat B_1 77-78, B_2 84-85 and B_6 107-122. For "medister" sausage B_1 83-87, B_2 90-93 and B_6 98-115. For vienna sausage B_1 79-82, B_2 66-67 and B_6 115-133.

1. INTRODUCTION

Meat is an important source of B-vitamin. On average meats supply from 25 to 40% of the recommended intake of vitamin B_1, B_2 and B_6 in Denmark. Freezing is usually considered one of the most favourable preservation methods. There is, however, no garantee that the result will be acceptable. For the consumer it is the intake that counts. One have to consider the entire process i.e. freezing, storage, thawing and preparation and not only the freezing process itself. A substantial part of the meat consumed has been frozen at some stage. It is estimated that up to a third has been frozen either in the home or by the industry. The results for retention of nutrients during freezing of animal tissue are confusing especially regarding loss of B-vitamins. Sometimes there appears to be considerable increases in the vitamin content which are difficult to explain. The bulk of work concerning retention of B-vitamins in frozen meat was done in the fifties. Different studies of the retenion of B-vitamins in the same muscle stored at the same temperature in the same period has produced different results. (1), (2) and (3) found a retention of 85-112% for thiamine, 56-106% for riboflavine in pig longissimus dorsi stored at -18°C for 180 days.

2. EXPERIMENTAL

Four of the most common meat products in Denmark were used in this study: pork loin (longissimus dorsi), vienna sausages, "medister" sausages (coarsely ground pork sausage) and ground meat (veal and pig). These four products also represent different ways of preparation. Pork loin is sliced and fried as whole pieces of meat, vienna sausage is a processed meat product which is prepared by boiling, "medister" sausage is also a processed meat product, which is prepared by boiling and frying. Ground meat (veal and pig) is just ground, formed as hamburgers and fried. The products for this study were taken from a normal processing line, frozen and stored at -24°C or -12°C. Samples were analysed for B_1, B_2 and B_6 prior to freezing, every two month for one year, and both before and after the final preparation. On pork loin analyses were also made after one week storage. The vitamin analyses were done using the standard technics, i.e.

B_1 using the chemical thiochrom method, B_2 microbiologically with Lactobacillus casei, and B_6 also microbiologically, with Saccharomyces calsbergensis. Calculations of retentions were based on dry matter.

Twelve pork loins with a normal pH_{24} (5.5-5.7) were selected, sliced, vacuum packed and stored. Six loins at each temperature. Vitamin content in the drip was determined. Emphasise was given to make samples representative for the whole loin.

The ground meat was vacuum packed in 400 grams packs and stored. The drip was to small to be of importance. For preparation, a hamburger of 85-100 grams was formed and fried for 5.5 minutes on each side.

"Medister" and vienna sausages were vacuum packed, frozen and stored. The "medister" sausages were boiled for 10 minutes in water and then fried 4.5 minutes on each side. The vienna sausages were boiled for 10 minutes, then rested for 5 minutes and finally boiled for an additional 5 minutes.

3. RESULTS

3.1 Vitamin B in meat products

Table I shows the vitamin content before freezing. As comparison the figures from the Danish food composition tables are also listed. The differences are within normal range.

Table I. Vitamin B_1, B_2 and B_6 content in pork loin, ground meat, "medister" and vienna sausages before freezing.

Vitamin µg/g	Pork loin* Initial	Pork loin* FCT**	Ground meat Initial	Medister Initial	Medister FCT**	Vienna Initial	Vienna FCT**
B_1	6.0	7.5	3.8	2.9	2.0	1.8	1.4
B_2	1.6	2.3	2.6	1.8	1.6	2.5	2.0
B_6	3.5	3.6	1.9	1.4	2.0	1.2	1.5

* Figures of B_1 and B_2 in pork loin are average of 12 loins (S.D. = 0.84 and 0.15) and the figures for B_6 are based on 6 loins (S.D. = 0.42). The figures for pork loin from FCT are for lean pork meat.

** FCT = Danish Food Compositions Tables, National Food Institute, publication no. 75. 1983.

3.2 Effect of freezing

The effect of freezing and one week of storage on the retention of B-vitamins in pork loin is shown in table II. Very limited effects are found. Due to problems with the analyse for content of vitamin B_6 the data for the initial contents in the 6 loins stored at -24°C had to be disregarded.

Table II. Retention of B_1, B_2 and B_6 after one week storage of pork loin*

	Vitamin B_1 x	Vitamin B_1 S.D.	Vitamin B_2 x	Vitamin B_2 S.D.	Vitamin B_6 x	Vitamin B_6 S.D.
-12°C	101	9.0	101	6.0	112	11.8
-24°C	98	11.6	114	19.5	-	-

* Based on six loins at each temperature.

3.3 Effect of storage temperature and time

A covariance analyses on the effect of storage temperature showed that only for vitamin B_1 (raw tx/raw t one week) were there significantly

(1% level) differens between the two temperatures.

In pork loin the retention of vitamin B_1 at -12°C is higher than the retention at -24°C. The mean retention of all observations is 110 at -12°C and 88 at -24°C (S.D. 11.4 and 6.6) respectively. A regression analysis was made on the effect of temperature, it showed no significant effect of time for both samples.

3.4 Effect of thawing and preparation

Drip and cooking losses in pork loin are shown in table III. The concentration of B_1 and B_6 in the drip is at the same level whereas for B_2 is only half.

Table III. Drip loss, cooking loss, and drip content of vitamins in pork loin.

	Drip %	Vitamin in drip % B_1	B_2	B_6	Cooking loss %
-12°C average	7.6	8.3	4.3	8.0	22.5
-12°C S.D.	1.1	1.5	0.7	1.3	2.7
-24°C average	6.6	7.2	3.4	7.3	24.9
-24°C S.D.	1.4	1.4	0.8	1.5	5.7

The vitamin content in the juice lost during cooking is not known. There is good evidence that the distribution of the various vitamins is at the same, as found in the drip. The overall retention of vitamins after preparation is shown in table IV.

Table IV. Average retention and standard diviation (S.D.) in pork loin (prepared tx/raw tx).

	-12°C			-24°C		
	B_1	B_2	B_6	B_1	B_2	B_6
Average	84	96	88	75	94	88
S.D.	8.9	9.6	3.9	17.4	5.1	7.9

B_2 seems only to be moderately affected by the cooking procedure.

3.5 The overall retention in prepared pork loin

Retention of B-vitamins in pork loin after storage and preparation compared to base line figures are shown in fig. I. Regarding B_1 it is seen that generally, the retention is highest if stored at -12°C. Mean values found for the whole period are 95 and 68 (S.D. 11.6 and 18.7) for -12°C and -24°C respectively. A significant difference at 1% level. For neither of the temperature levels any significant effect of storage time was found.

The findings on B_2 is opposite to B_1. Here the retention is generally higher if the samples are stored at -24°C. The mean values for samples stored at -12°C and -24°C are 90 and 107 (S.D. 4.6 and 17.3) respectively. The retention at the two storage temperatures was found to be significantly different (5% level), but no significant effect of storage time.

Because the base line data is missing for B_6 at -24°C, the comparison of the overall effect on retention of this vitamin is based on findings after one week of storage. From this it is found that the retention is higher when the products are stored at -24°C (mean 97, S.D. 23.3 at -24 versus mean 86 S.D. 8.5 at -12°C). No significant effect of storage temperature and time could be found.

Figure I. Retention of B-vitamin in prepared pork loin during one year of frozen storage at -12°C and -24°C. (Base line data: B_1 and B_2 analyses made before freezing B_6 analyses made after one week of frozen storage).

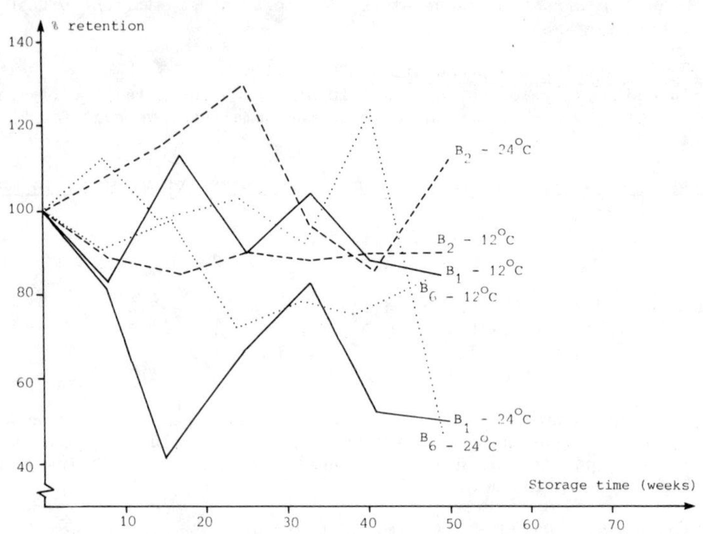

3.6 Retention of B vitamins after storage and preparation

The overall effect of one year of storage followed by preparation, on the retention of B vitamins is shown in table V. The table indicate that the loss of B_1 generally is approximately 20% at -12°C whereas a higher loss is seen from pork loin stored at -24°C.

Table V. Retention of B vitamins in pork loin, ground meat, "medister" and vienna sausages after one year storage and subsequent preparation.

Product	Temperature °C	B_1	B_2	B_6	Week of storage
Pork Loin	-12	85	90	88	49
	-24	50	112	48	50
Ground meat	-12	77	84	122	52
	-24	78	85	107	52
"Medister"	-12	87	93	98	50
	-24	83	90	115	50
Vienna	-12	79	67	115	51
	-24	82	66	133	51

B_2 shows large differences between the products. Approximately 1/3 is lost during storage of vienna sausage whereas for pork loin a loss may or may not occur when stored at -12°C and -24°C respectively.

The results obtained on B_6 after one year of storage are even more confusing. The B_6 content is reduced to one half in pork loin stored at -24°C whereas for vienna stored at the same temperature there is an increase.

4. DISCUSSION AND CONCLUSION

In this presentation the emphasize has been given to pork loin. The results from the other products are necessary to determine whether the retention is typical for both processed and unprocessed meat products.

In pork loin the loss of B_1 from the unfrozen product to time of serving depends on the storage temperature. Storing at -12°C gave a significantly higher retention at time of serving than at -24°C. The drip loss during thawing in pork loin is 7-8% and it correspond with the same loss of B_1 in the drip.

Regarding B_2 a significant difference in retention in pork loin was found between storage at -12°C and -24°C, with a higher retention at -24°C. This difference is not due to one of the single effects of either storage temperature, thawing or preparation. The loss of B_2 in the drip in pork loin is only half of what was found for B_1 and B_6.

For B_6 there was problems with the analytical method for determination of the initial content in the loins that were subsequently stored at -24°C. Therefore it is impossible to compare with the unfrozen products. Calculations based on findings after one week of storage show no significant difference in retention rate due to either time or temperature.

Finally it is difficult to explain the changes from one result to another over the time period. This may, as other workers have suggested, be due to changes during frozen storage that makes the vitamins easier to analyse with the present methods. Especially increases in vitamin content for some vitamins during storage are difficult to understand. Also the present methods of analyses make this kind of work very resource consuming, it is only possible for one technician to make 4 double analyses of B_1, B_2 and B_6 during one week. This obstacle, of course, limits the number of samples it is possible to analyse.

REFERENCES

1. LEHRER, W.P., Jr., WIESE, A.C., HARVEY, W.R. and MOORE, P.R. (1951). Effect of frozen Storage and subsequent cooking on the thiamine, riboflavine, and nicotinic acid content of pork chops. Food Res. 16, 485-491.
2. LEE, F.A., BROOKS, R.F., PEARSON, A.M., MILLER, J.I. and WANDERSTOCK, J.J. (1954). Effect of rate of freezing on pork quality. J. Am. Dietet. Assoc. 30, 351-354.
3. WESTERMAN, B.D., OLIVER, B. and MACKINTOSH, D.L. (1955). Influence of Chilling Rate and Frozen Storage on B-complex Vitamin Content of Pork. Agricultural and Food Chemistry 3(7), 603-605.

TT-INTEGRATORS - SOME EXPERIMENTS IN THE FREEZER CHAIN

P. Olsson
SIK - The Swedish Food Institute

Summary

As a basis for the development of time/temperature integrating monitors for use in the freezer chain several studies have been performed. The aim has been to investigate conditions in the freezer chain especially of interest for the proper use of integrating devices. In this paper three such studies are briefly described namely field tests with indicators, exposure tests in freezer cabinets and laboratory simulations giving the background for a computer programme.

1. INTRODUCTION

The problem of quality deterioration during frozen food distribution is well known and has been dealt with since the beginning of the frozen food trade.

The quality losses during storage and distribution can be related both to temperature and to time. This concept was first thoroughly investigated in the U.S. during the 1950s by van Arsdel and Guadagni (1). These studies resulted in the Time Temperature Tolerance-hypothesis, which has become the basic guidelines in question of frozen food stability. In some special cases the hypothesis does not seem valid.

The main weakness with today's most common control system, the open dating, is that it reacts only on Time and not on the intregrated Time-Temperature history.

2. BACKGROUND

For several years Time-Temperature monitors have been patented. Normally they are too expensive for use in a larger scale.

Recently indicators which are unexpensive and effective have been developed. Such an indicator is the i-point TTM which is the result of more than 10 years R&D work. It can be described as a self-adhering label, 25 x 60 mm in size, containing a biochemical process, that has the ability to integrate and accumulate time/temperature exposures.

When predetermined time/temperature limits have been reached, the TTM reacts by changing colour, clearly and irreversibly.

During the development of this device SIK has performed several investigations. This paper will give some examples from this work which started already in 1972 with literature surveys regarding product shelf-life (2) and conditions in the Scandinavian freezer chain (3).

Among the studies performed we can mention field tests, exposure tests in freezer cabinets, simulated studies in the laboratory regarding the temperature history of the indicator versus the food temperature.

These studies also founded the background for a computer programme, developed in our laboratories for foods in rectangular shape, from individual consumer packs to entire pallet loads (4).

In this paper there will be given some examples from the vast unpublished work which has been performed. The paper will give some details from three areas; time-temperature studies in short freezer chains in Sweden, an example from time-temperature measurements in freezer cabinets and an example from the experimental verification of our computer programme.

3. EXPERIMENTAL

3.1. Field tests

The field tests were carried out jointly by i-point TTM, two Swedish food companies and SIK - The Swedish Food Institute.

3.1.1. Background

SIK assisted in two field tests on a consultancy basis. In each field test approximately 200 each of 5 indicators with different time-temperature sensitivities were distributed over 40-50 institutional packages at the packing plant, directly after deep freezing. The packages were then placed on pallets and distributed in the normal manner from the packer to a distributor and then to the staff kitchen of a large industry in Malmö, where a check of the colour change frequency and degree of colour change was carried out.

A summary account of the tests, along with comparative comments, is given below.

3.1.2. Test objective

The intention of both tests was only to check the general reaction capability and mechanical functioning of the indicator under realistic handling conditions and to obtain some idea of the general sensitivity range required of the indicators during normal handling. The test was not intended to attempt to correlate the colour change interval and product quality. The time-temperature history during distribution was to be registered on temperature recorders in order to check the reactions of the indicators.

3.1.3. Indicators and raw material

After being activated, "blocks" of indicators were placed in different positions on the outer shipping cartons and inner packages, according to a schedule prepared in advance. Two temperature recorders for registering the actual time-temperature history of the shipment were placed in cartons in the top and next to top pallet layers. Institutional packs of IQF-frozen meat balls were used in test 1 and pork loins in test 2.

3.1.4 Results

The duration of the test from activation of indicators to reading of the indicator colour change at the end station was 27 and 26 days, respectively, for the two tests.

3.1.4.1. Colour change

At the time of final reading, in test 1, all indicators of type 1-3 with a few exceptions changed colour, while indicators of type 4 were in the process of changing. Indicators of type 5 were unchanged. In test 2, types 1-2 changed colour, while type 3 was in a state of changing, and types 4 and 5 were unchanged.

3.1.4.2. Comparisons between colour change of indicators placed on individual outer cartons and inner packages

In test 2 colour change had progressed slightly more on the outer cartons than on the inner package, with a high degree of statistical significance,while only a tendency in the same direction was obtained in test 1.

3.1.4.3. Temperature history in storage and during transportation

In Figures 1 and 2, the temperature history for the two upper pallet layers as registered by the temperature recorders, is shown.

In test 1, a higher temperature in the lower pallet layer was registered during the first days, which might have been the result of the product not being frozen down to the temperature of the distributor's warehouse. As expected, the temperature recorder in the upper pallet layer registered greater temperature variations than the other in both tests. In test 2, a relatively steep rise in the temperature up to approximately -10°C in the upper layer was obtained during transportation from the intermediate storage to the customer's kitchen, in which an unrefrigerated van was used, a rise of temperature up to approximately -10°C was obtained in both tests. It should be pointed out, that it is not possible to draw any general conclusions about the normal temperature conditions in the distribution chain from only two separate measurements, and this was not the intention of the tests.

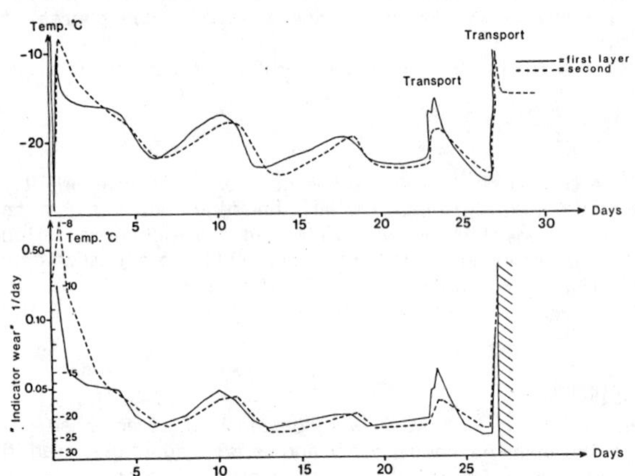

Figure 1.
Field test 1. Temperature recording in first and second layer of a pallet load and the corresponding "indicator wear" through distribution.

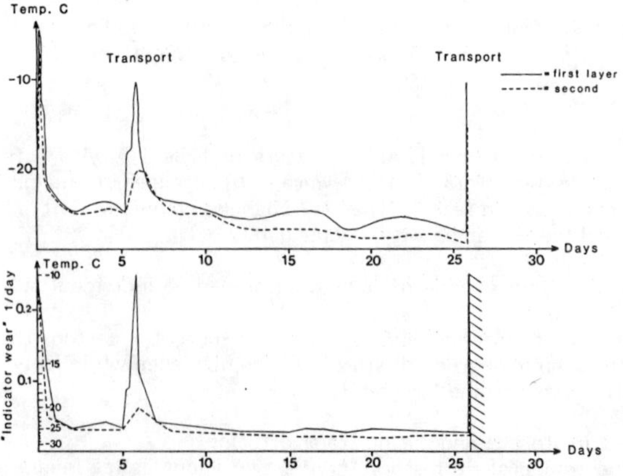

Figure 2.
Field test 2. Temperature recording in first and second layer of a pallet load and the corresponding "indicator wear" through distribution.

3.1.4.4. Calculated and observed time-temperature reaction of the indicators

In test 1 an average colour change up to and including indicator 4 was obtained and in test 2 up to and including indicator 3. With the aid of time-temperature reaction specifications for the indicators (determined by i-point AB) and the time-temperature histories in Figures 1 and 2, it was possible to estimate the theoretical time-temperature reaction for a given indicator, expressed, for example, as the number of equivalent days at a certain temperature in relation to the time for colour change at this temperature, the indicator "wear".

The result of such calculations for the indicators 4 and 3 in tests 1 and 2 respectively give a good correlation between calculated theoretical indicator "wear" and the average colour change actually obtained.

In order to obtain an approximation of the actual quality deterioration for a food product which would result from the brief temperature histories observed in these tests, an example for pre-cooked food was calculated. Average values for product acceptability collected from a literature review made at SIK (2) were used. In this example the calculated quality deterioration was insignificant, lying in the area of 3-6% of the "acceptability" time. The calculations showed differences of only a few percent between pallet layers and between the two tests. It must be emphasized that the indicators used were chosen to change colour during the relatively short frozen food chain.

3.2. Freezer cabinets

3.2.1. Background

It is well known from the literature (5, 6, 7) that the temperature in freezer cabinets are not uniform through the stored food articles. There are also variations due to type of cabinets, load, ambient conditions, maintenance.

3.2.2. Test objective

Differences in temperature in different layers in a freezer cabinet and also the fluctuating temperatures during defrosting periods give differences in temperature history outside and inside food packages. This must be considered when setting the time-temperature feature for an integrating indicator.

3.2.3. Test performance

An example from such a test is the following performed in an open top display cabinet (Figure 3). (Dimensions: length 8 m, width 0.7 m, height 0.30 m from bottom to load mark).

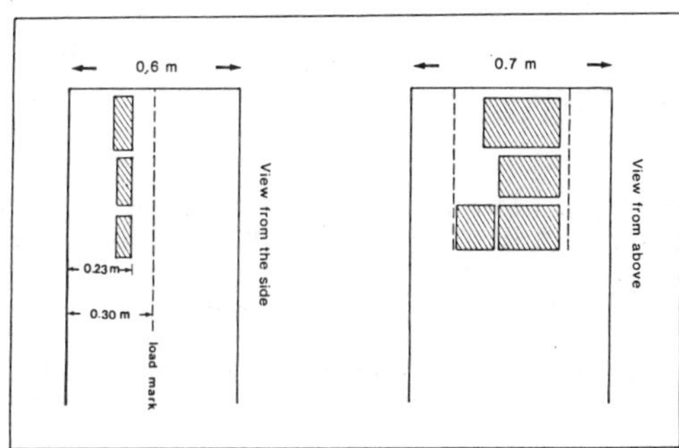

Figure 3.
Position of consumer packages in a retail cabinet (top layer).

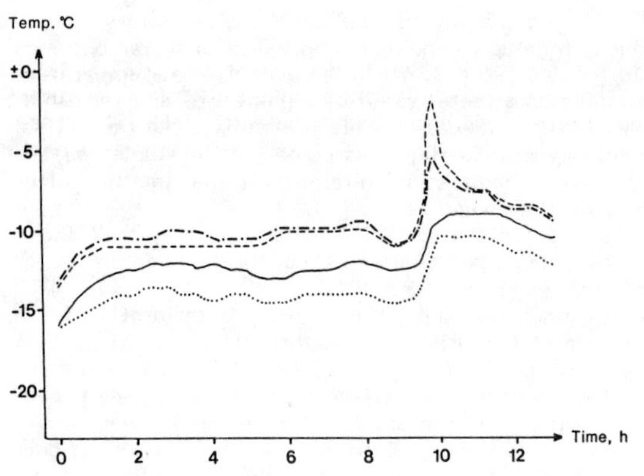

Figure 4.
Seafood salad in aluminium foil pouch (250 g).
_ . _ surface outside package (top)
---- product surface temperature (top)
—— product centre temperature
····· surface outside package (bottom)

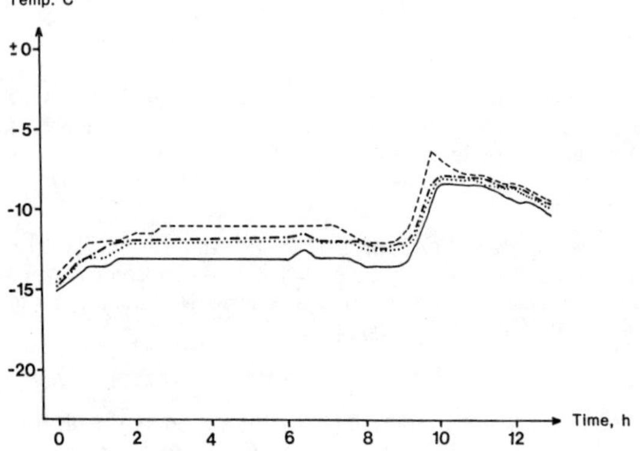

Figure 5.
Blockfrozen cod fish in laminated paperboard (450 g).
---- surface outside package (top)
— · — product surface temperature (top)
····· product centre temperature
—— surface outside package (bottom)

3.2.4. Results

Some of the results are shown in Figure 4 and Figure 5.

3.2.5. Discussion

In the original study a wide variation of food packages and products were investigated. During the defrosting periods the temperature on the upper area rose considerably for some combinations of package and product.

However, calculations showed that this did not affect the remaining practical storage life to any greater extent.

For the products shown in the figure the differences in quality deteriorating between the food and the assumed indicator with the same characteristica during assumed 15 days storage:

Seafood salad in al-foil pouch (0.25 kg) 2% and block frozen cod (0.45 kg) 1%.

3.3. Computer simulation

As experimental time-temperature studies in the frozen food chain are very time-consuming, it was felt desirable to develop a mathematical model, so that the influence of different time-temperature exposures could be investigated by computer simulation instead. An explicit finite difference technique with integral transformation of the thermophysical properties has been used to solve the transient heat conduction equation with temperature dependent thermophysical properties. The programme assumes food packages, cases or pallet loads to be a homogenous body of the food material, with exterior heat transfer surfaces only. It is limited to rectangular and homogenous blocks of food in symmetrical loads. The number of measuring cells or grid-points is limited to between 500 and 1000. A further limitation is of course that the thermal data of food and packaging material as a function of composition and temperature have to be known with fair approximation.

The output from the programme comprises temperature distributions and quality wear distributions as well as cumulative temperature and quality wear diagrams, the latter function being derived by integration over the time-temperature exposure curve for the known time-temperature-tolerance of the food. In the course of the development work on the programme a considerable number of experimental exposure studies were made on actual foods and model foods in consumer packages as well as shipping cases and entire pallet loads. Temperatures in these exposure studies were measured by thermocouples at different locations on the surfaces and interior of packages. A datalogging unit (Schlumberger-Solartron-Compact 33) with 50 channels, and with direct computer processing of the data was used. The agreement between simulated time-temperature exposures and experimental exposures has been very acceptable.

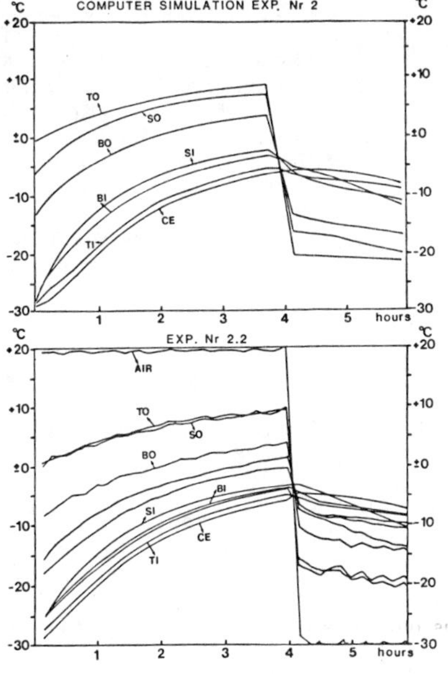

Figure 6.
Comparison between computer simulation and experimental exposure for a shipping case of frozen fish during a combined heating-recooling exposure (from 4).

TO = Top-outside shipping case
TI = Top-inside retail package
SO = Side-outside shipping case
SI = Side-inside retail package
BO = Bottom-inside retail package
BI = Bottom-outside shipping case
CE = Centre

Figure 6 shows such a comparison for a shipping case of frozen fish during a combined heating-recooling exposure. The output from the programme may also be presented as plots of temperature and quality wear.

It is apparent that such a simulation programme may be very useful as a tool in efforts to optimize a new frozen food chain or to point out the weak points of an existing one, even if it is limited in its present scope to symmetrical rectangular loads of homogenous packaged foods.

4. GENERAL DISCUSSION

It is clear that the time-temperature response curve of a food and the time-temperature exposure determine the actual accumulated quality loss. However, it is very difficult to determine this in practice for two main reasons: the great variability in time-temperature sensitivity of foods due to the PPP-factors, and the practical impossibility to measure the time-temperature history of individual packages in the freezer chain, at least on a regular basis. For this reason many attempts have been made to utilize temperature dependent reactions with known response curves to design simple devices to be fixed to food packages and which will integrate the accumulated exposure, the response curve of the indicator hopefully coinciding with that of the food materials. The first technically successful indicator or integrator was presented by the Honeywell Company many years ago, but was too expensive for practical use. Today one of the companies which has developed low-cost time-temperature integrating indicators is the I-POINT Company. Their indicator, described in the background chapter, is based on a biochemical reaction. The response curve of the reaction may be chosen close to that of actual foods.

Early field tests in shipments of frozen products showed overall good agreement between colour change and indicator "wear" calculated from temperature recordings and the predetermined indicator response curve. The indicator has since then been further developed and is now an effective time-temperature integrating device. However, as the quality stability of a given food has been shown to be very variable, it is not likely that any TTT-integrator (irrespective of its own precision and accuracy) could be used to pin-point the quality of a sample of frozen food. Instead it may become very useful for indicating how the food sample has been handled in the freezer chain, so that corrective measures may be taken. For such an application it is not necessary that the response curve of the indicator matches that of the individual frozen food very closely, and it is not necessary that the difference is very small between the time-temperature exposure of an indicator on the outside of the package or case and the exposure of the food inside. The computer simulation programme previously outlined can be used for comparing the exposure of indicators and the exposure of the food, and to integrate the resulting accumulated indicator "wear" and quality wear.

REFERENCES

1. VAN ARSDEL, W.B. and GUADAGNI, D.G. (1959). Time-temperature tolerance of frozen foods. XV. Food Technol. 13(1):14
2. BENGTSSON, N. et al. (1972). An attempt to systemize time-temperature tolerance data as a basis for the development of time-temperature indicators. IIR, Commission C1 and C2, Warsaw.
3. OLSSON, P. and BENGTSSON, N. (1972). Time-temperature conditions in the freezer chain - a critical literature review primarily directed towards conditions in Scandinavia. SIK-Rapport 309.
4. DAGERSKOG, M. (1978). Computer simulation of time-temperature distribution and quality changes in the frozen food chain. Scand. Refrig. 7:1.
5. LORENTZEN, G. (1971). Present status of the freezer chain. Scand. Refrig., Aug/Sept, p. 5.
6. BOEGH/SOERENSEN, L., BECH-JACOBSEN, K. (1983). Time-temperature in the freezer chain in Denmark. IIF-IIR, Paris 1983, C2, p. 455.
7. ALRIC, P. (1983). Le controle en France des temperatures des denrees surgelees dans les meubles de vente. IIF-IIR, Paris, D2, p. 234.

THE I-POINT TTM - A VERSATILE BIOCHEMICAL TIME-TEMPERATURE INTEGRATOR

K. BLIXT, I-POINT AB, MALMOE, SWEDEN

All over the world there is in our days an enormous waste of perishable products. This waste is caused by temperature abuse in the distribution.

Some sources of information claim, that the value of such spoilage in US only, amounts to many billions of dollars.

Therefore, it is easy to be a believer in the need for better handling systems of perishable products, than obviously is at hand today. As the temperature treatment seems to be one of the most crucial factors, this is the area, where the best results of improvements can be expected.

Here, temperature treatment means time-temperature handling, where it is well known that for most perishables, the combined effect of both time and temperature has to be considered.

The main weakness with today's most common control system, the Date-Stamping, is that it does not react to more than one of these two factors- the Time - and does not tell anything about the exposure to the other crucial factor - the Temperature.

One has to admit, that the Date-Stamping was the only realistic possibility at hand at the time, when the Health Authorities, Trade Organizations and others involved started working out practices and regulations for the handling of perishable products.

However, things are changing in this world, and the statement made by one of the members of an open dating committee at that time "the only way to complete quality control of perishable products is to have one inspector following each shipment" was some years ago accepted as the ironic joke it was meant to be, but is today no more a joke - it is now quite realistic.

The I-POINT TTM (Time-Temperature Monitor) shown in figure 1, can be said to be able to act as such an inspector and can be used not only on each shipment, but also on each carton.

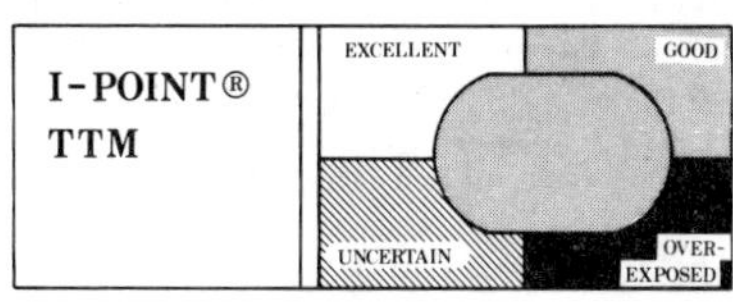

Fig. 1. I-point TTM

I-POINT TTM is the result of more than 10 years R&D work, and the first generation of products is now offering its service. It can be described as a self-adhering label, 25 x 60 mm in size, containing a biochemical process, that has the ability to integrate and accumulate time-temperature exposures.

When predetermined Time-Temperature limits have been reached, the TTM reacts be changing color, clearly and irreversibly.

The colors, that can be defined, are green, yellow, brownish and red, and the meaning of these colors can be expressed either in words or in figures.

Our presently recommended interpretations of the color signals are:

When color 1 (GOOD) is reached, 80% of the preset Time-Temperature Tolerance is used up.

When color 2 (UNCERTAIN) is reached, 100% of preset Time-Temperature Tolerance is used up.

When color 3 (OVEREXPOSED) is reached, 130% of the preset Time-Temperature Tolerance is used up.

As regards text or figures on the label, as well as the meaning of the different color signals, it is of course possible to meet individual requirements.

The "heart" of the TTM is an inner ampoule, which before activation consists of two parts, one with an enzyme solution together with a liquid pH-indicator, and one with a substrate solution.

By applying pressure on the TTM the seal between the two parts is broken, the two solutions are mixed, and the reaction is started. By influence of the pH-indicator, the color of the solution becomes green, which is shown in the window.

A process of pH-degradation caused be enzymatic activity is now going on, and the development thereof is reflected by the reaction of the pH-indicator, changing color in accordance with the color printed on the label.

The velocity of this reaction is Time-Temperature dependent, and the device reacts in accordance with the "Time-Temperature Tolerance Hypothesis" of which the main conclusions are:

1. Quality loss is proportional to the integrated, accumulated effects of both Time and Temperature.

2. The rate of quality loss will increase, as storage temperature increases, and decrease, as temperature decreases.

3. All quality losses are cumulative and irreversible.

4. All temperature effects are commutative - they have the same result on product quality, no matter when they occur during the life of a product.

By developing the ability to control the two parameters

1. <u>Change of the speed of the reaction</u>

2. <u>Parallel displacement of all TT-curves</u>

an enormous flexibility of I-point TTM has been developed which makes it possible to meet most TTT-requirements.

The present Standard Assortment consists of these basic curves:

Type	Q10	Temperature area
1	1.9	$+20^{o}C$ - $+10^{o}C$
	2.6	$+10^{o}C$ - $\pm 0^{o}C$
	3.3	$\pm 0^{o}C$ - $-10^{o}C$
	42	$-10^{o}C$ - $-20^{o}C$

Type	Q10	Temperature area
2	2.2	+20°C - +10°C
	2.4	+10°C - ± 0°C
	2.7	± 0°C - -10°C
	2.85	-10°C - -20°C
3	2.1	+20°C - +10°C
	5.2	+10°C - ± 0°C
	5.2	± 0°C - -10°C
	5.2	-10°C - -20°C
4	4	+30°C - +20°C
	13	+20°C - +10°C
	15	+10°C - ± 0°C
	>15	± 0°C - -20°C
5	30	+30°C - +20°C
	30	+20°C - +10°C
	>40	+10°C - ± 0°C

All types can be parallelly moved, and by using the feed back from market contacts, a detailed standard assortment is being built up.

Thus, what is needed to go from here to the common use of a monitoring system for perishable products, that, correctly designed and utilized, will show to meet any reasonable requirements of feasibility and cost, is storage life (stability) data for the perishables.

The supplier of such data must be the food producers, if convenient in cooperation with research institutes. What is required, is nothing more than should be at hand today. Otherwise, it could not be possible to set a meaningful expiration date.

CORRELATION OF SENSORY AND INSTRUMENTAL METHODS IN THE DETERMINATION OF COLOUR IN FROZEN GREEN BEANS

J.B. Adams and A. Robertson

The Campden Food Preservation Research Association
Chipping Campden, Glos. GL55 6LD

ABSTRACT

A range of colour quality was produced experimentally in green beans (*Phaseolus vulgaris L.*) by varying the blanch time from 0-60 seconds prior tc frozen storage. Sensory evaluations of colour were correlated with measureme using a tristimulus colorimeter and with pigment composition as determined by high performace liquid chromatography. The data was treated by principal components analysis. High correlation coefficients were obtained, particular between the major sensory principal component and the colorimeter 'a' value (the major colorimeter principal component. Chlorophyll a was shown to corre with good colour whilst pheophytin a correlated with poor colour in the unbl and underblanched samples.

1. INTRODUCTION

The colour of frozen vegetables, after cooking, has a major influence o the initial perception of product quality. The chlorophylls play a major ro in the colour of green vegetables, and although changes which occur during a after processing, or differences between varieties, may not be solely the result of these pigments, nevertheless, it is important that all pigments should be analysed to provide an understanding of their effects on sensory colour measurements. Ultimately, this knowledge can be used as a basis for optimum preservation processes and cooking treatments.

The work described herein, involved the assessment of the colour of gre beans (*Phaseolus vulgaris L.*) by sensory methods and by tristimulus colorime Measurements of the individual chlorophyll pigments and their degradation products were made and β-carotene was also measured. Relationships between sensory evaluations of colour and colorimetric data were examined as were relationships between colour and pigment data.

2. EXPERIMENTAL PROCEDURES

2.1. Processing conditions: The variety of beans was Cascade (mean ler of 10 of the most mature seeds = 135mm). Samples (approx 6kg) were blan for 10, 20 and 60 seconds at 98°C in a ratio of 1 part green beans to 12 parts tap water. The samples were cooled for 5 minutes in cold tap wate drained, sliced and then frozen using a fluidised-bed freezer. After freezing, 250g sub-samples were weighed into polythene bags and stored a -18°C. An unblanched sample was also frozen, after slicing, and then su sampled and stored as above. Fresh samples were retained for immediate analysis. The storage intervals for colour and pigment analyses were 0, 7, 14 and 28 days.

2.2. Sensory evaluations of colour: Sensory assessments of colour were carried out both on the raw and on the frozen beans at zero time and aft 3, 7, 14 and 28 days storage. The sensory team consisted of 3 trained panallists who evaluated the samples, after cooking, using the methods described in CFPRA Technical Memorandum No.278 (Adams and Bedford, revis by Geering, 1981).

2.3. Instrumental measurements of colour.

2.3.1. Tristimulus colorimetry: The beans were cooked as for sensory evaluations but cooled in tap water immediately afterwards to maintain the colour. Measurements were carried out using the Gardner Colour Difference Meter as described by Springett et al (1980). The instrument was standardised on a light green tile having the specification L-70.0, a=16.0, b=6.0. Ten sets of measurements were taken for each sample, the whole sample being thoroughly mixed between each set.

2.3.2. High Pressure Liquid Chromatography (HPLC): 25g samples of green beans were extracted three times with cold ethyl acetate. The final volume of extract was 30ml and 10µl of this was injected into the High Pressure Liquid Chromatograph (Pye Unicam XP modular system). The pigments were separated on a 100mm x 4.6mm ID column containing Hypersil ODS (Shandon) stationary phase using the linear gradient mobile phase described previously (Adams et al, 1982). Sodium dihydrogen orthophosphate was added to the aqueous methanol at 1mM final concentration to maintain the pH at 7.0.

RESULTS AND DISCUSSION

3.1. Sensory changes in colour: Large losses in colour quality of the green beans were found at zero time in the unblanched samples. Reductions in greenness and increases in khaki/brown were particularly noticeable. These changes were noted within one hour of freezing and may have been due to very high enzyme activities during slicing or during the freezing process, or perhaps as the material warmed to cooking temperature, prior to analysis. Concomitant increases in yellow and grey and reductions in brightness also occurred. Similar changes, although not so large were observed in the beans blanched for 10 seconds. The observed changes in all colours stabilised within 14 days and remained relatively constant thereafter.

Principal components analysis on the sensory colour scores of all samples and treatments is shown in the Figure (A). 93.6% of the variance was explained in two principal components of which PC1 accounted for 88.4%. The positive side of the PC1 axis was composed of khaki/brown, grey and yellow (poor colour) while bright green and uniform parameters contributed to the negative side (good colour). In the case of the minor principal component, PC2, which explained only 5.2% of the total variance, it is not possible to assess which terms are contributing to good colour and poor colour as the points are equally distributed about the PC1 axis.

3.2. Changes in tristimulus colorimeter values: Reductions in the Gardner -a values were particularly obvious in the unblanched green beans and in the beans blanched for 10 seconds while the -a value of the two samples blanched for longer times was relatively unaffected. The changes in L and b values were small in comparison with those in the -a values; the L values of the unblanched beans were slightly higher than any of the blanch treatments while the b values of the beans blanched for 60 seconds were generally somewhat higher than the other treatments.

As shown in the Figure (B), principal components analysis gave 99.1% of the data variance in two components, of which PC1 accounted for 95.3%. The PC1 axis was mainly due to changes in the 'a' value although 'L' and 'b' also made significant contributions. By comparison with the principal components analysis of the sensory data, PC1 may be considered as the 'hue' axis where +a may be interpreted as the khaki/brown side of the axis and -a as the greenness side. The distribution of the various blanched treatments along the PC1 axis corresponds well with this interpretation. PC2 accounted for only 3.8% of the Gardner variance and was made up of L and b changes predominantly.

A

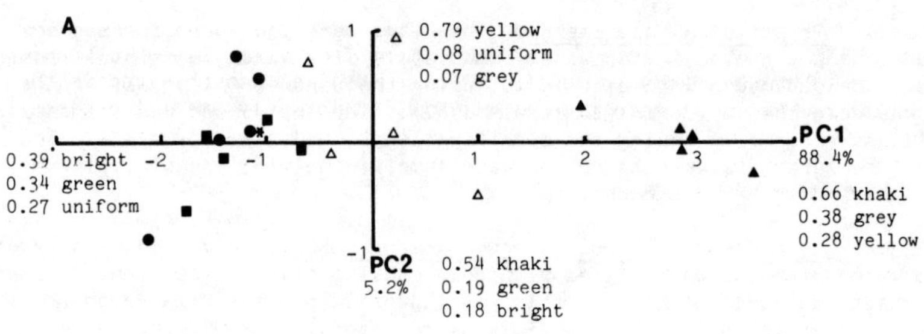

B

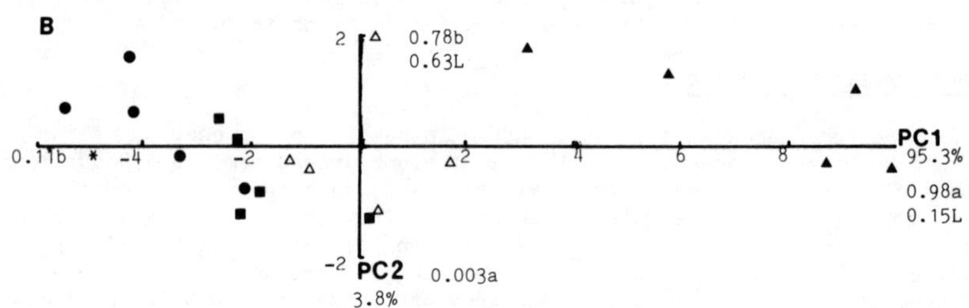

C

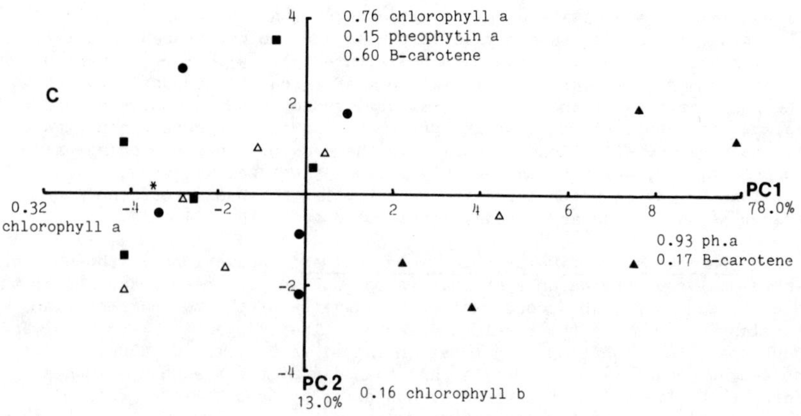

Figure. Principal components analysis on sensory (A) Gardner (B) and HPLC pigmen (C) data. ▲ unblanched; △ 10 second blanch; ■ 20 second blanch; ● 60 second bla * fresh. Data variance explained by each PC is given in percentage. Factor Loadi for each variate contributing to PC's is also given. The origin of the PC axes represents the mean. Units are those from the raw data and scaled by an arbitary factor.

3.3. Changes in pigments: Large losses in chlorophyll a were found at zero time of frozen storage, in the unblanched beans and in the beans blanched for 10 seconds. During frozen storage, chlorophyll a degraded at a relatively slow rate for all treatments. A large increase inpheophytin a, one of the breakdown products of chlorophyll a, was observed in the unblanched beans at zero time of storage. Relatively small increases in pheophytin a were found in the blanched beans and these did not vary significantly with the blanch treatment. The pheophytin a was stable during frozen storage in all samples. As isolated, this pigment is greyish in colour and the changes correlated quite well with greyness observed by the sensory panel (r=0.75). Chlorophyll b, although present in smaller amounts than chlorophyll a, was found to be relatively stable, whatever the blanch treatment, and would thus be expected to contribute to the observed colour. B-carotene was the only carotenoid to be quantified in this work and showed a rapid degradation during frozen storage at a rate which was independent of the blanch treatment.

As with the sensory assessments and the measurements made using the Gardner colorimeter, a large proportion of the data variance (91.0%) was accounted for by two principal components, PC1=78.0% and PC2=13.0%. Positive PC1 containing mainly pheophytin a and a little B-carotene would be expected to represent poor colour whereas negative PC1, with only chlorophyll a contributing, would represent good colour. PC2 contained chlorophyll a, B-carotene and a little pheophytin a on the positive side with chlorophyll b on the negative side. Further work is required to determine whether PC2 has any significance in terms of sensory colour.

The correlations of all data and principal components are shown in the table. Every coefficient is highly significant ($P<0.001$).

TABLE: Correlation coefficients obtained from linear regression analysis of the raw data and of the principal components.

Correlation coefficients (19 degrees of freedom)				
Sensory	Gardner		HPLC	
	a value	PC1	chl. a	PC1
PC1	0.92	0.92	0.78	0.79
Greenness	0.87	0.86	0.72	0.75

CONCLUSIONS

In this work, it has been shown by principal components analysis that most of the information from the original data can be incorporated in a single variate (PC1). In the case of sensory assessment, samples of good colour (20 and 60 second blanches) which are bright, green and uniform, have negative PC1 values whereas, generally, samples of poor colour (unblanched and 10 second blanch) which are khaki, grey and yellow, have positive PC1 values. The colorimeter and pigment PC1s were interpreted similarly. It was evident that, of the principal components, only the colorimeter PC1 is useful if it is required to differentiate precisely between the colours arising from all of the blanch treatments.

5. REFERENCES

Adams, M.J. and Bedford, L.V., revised by Geering, J. (1981) QAV - A method for the sensory appraisal of quality of processed vegetable varieties. CFPRA Technical Memorandum No.278.

Adams, J.B., Robertson, A. and Springett, M.B. (1982) Instrumental methods of quality assessment. CFPRA Technical Memorandum No.307.

Springett, M.B., Adams, M.J., Baker, A., Armstrong, K. and Butler, R.C. (1980) The sensory and instrumental methods used in blanching research, 1977-79. CFPRA Technical Memorandum No.241.

6. ACKNOWLEDGEMENTS

The authors thank the Ministry of Agriculture, Fisheries and Food for fun the work and the Director of Campden Food Preservation Research Association f permission to publish the results. They also thank the following staff at Campden R.A.:- Mr H Davies for technical assistance on High Performance Liqu Chromatography, Mr K Jewell for statistical analysis, the Quality Assessment Team for sensory analysis, the Experimental Processing Team for assistance in blanching and freezing and the Department of Agriculture for obtaining the gr beans used in this work.

PHASE TRANSITIONS IN FROZEN FOODS EXAMINED BY CALORIMETRY

K. Porsdal POULSEN
The Technical University of Denmark, Bygning 221, DK2800 Lyngby

SUMMARY

In order to forecast neutral and reverse stability concentration of reactants is followed by means of calorimetry. From changes in enthalpy the specific heat is calculated. Plots of specific heat versus temperature illustrate changes in concentration. It is concluded that large changes in concentration leads to reverse and neutral stability but that these phenomena also exist when such changes not have been detected.

1. INTRODUCTION

Freezing foods such as meat and fish down to $-10^{\circ}C$ brings them to a state in which they are rock hard and apparently solid. There is, therefore, a tendency to overlook that part of the product may be in a liquid phase and the term "frozen" is often used synonymously with "solid".

From phase diagrams of binary systems the existence of an equilibrium between a frozen (pure) solvent and a given concentration of solutes in the liquid region can be observed. This mixture of two phases exists between the initial freezing point and the eutectic temperature of the system. The concentration in the liquid phase is independent of the amount of solutes in the solution before freezing, but the volume of liquid phase is related to the total amount of solute.

E. Pincock and T.E. Kiovsky describe the relationship between reactions in liquids and frozen solutions of bimolecular reactions. The rate of a second order reaction ($A + B \longrightarrow P$) in the liquid phase of a frozen solution is given by equation (1)

$$\frac{d\,[A_s]}{d\,t} = -\,k_2\,[A_s]\;[B_s]\,\frac{V_s}{V_f} = -\,k_2\,[A_s]\;[B_s]\,\frac{C_f}{C_s} \qquad (1)$$

k_2 is the second order rate constant which depends only on temperature.

A_s and B_s are the molar concentrations of the reactants A and B in the thawed solution.

V_s and V_f are the volumes of the thawed solution and the volume of liquid regions in frozen solutions respectively.

C_s and C_f are the total concentrations of solutes with similar meaning of the indices.

From the second part of equation (1) it can be concluded that the product k_2C_f has a maximum at a temperature below the freezing point of the solution.

The mathematical solution shows that k_2C_f has a maximum at a temperature normally lower than the intersection of the exponential k_2 line and the C_f line (Figure 1). The use of mathematics is, however, of little relevance, because the organoleptic limitation for foods is caused by chain reactions of a high order.

The ratio $\frac{C_f}{C_s}$ is never less than one and consequently the "frozen" rate is never less than the "unfrozen" rate at the same temperature.

In the systematic considerations given by O.R. Fennema it is taken into account that $C_2 = A_s + B_s + P_s + I_s$, where P_s and I_s are concentrations of the resulting products and the concentration of inert or inhibiting components. The final influence may be:

1: Normal stability: A temperature decrease results in a slower reaction rate (and better stability when foods are stored).
2: Neutral stability: The temperature has no influence on reaction rate.
3: Reverse stability: A temperature decrease results in increased reaction rate.

To-day more and more frozen foods are sold as "ready-to-eat-dishes" with different ingredients, salt, and spices in the mixture. Such foods must be assumed to be complex in their behaviour during frozen storage at different temperatures. When storage life (log.scale) is plotted against temperature figures appear which are very far from the most frequently depictured straight "Arrhenius" lines (Figure 2).

2. EXPERIMENTAL

A number of different product were examined by the use of a Calvet microcalorimeter and by a Perkin-Elmer differential scanning calorimeter, model DSC2C. As examples the following bacon sample:

Water	50.96 %
Protein	13.63 %
Fat	31.4 %
NaCl	3.8 %
Carbohydrate	0
$NaNO_2$	229 mg/kg
KNO_3	63 mg/kg

and the following Danish salami:

Water	28.19 %
Protein	13.81 %
Fat	52.8 %
Salt	5.4 %
$NaNO_2$	4 mg/kg
KNO_3	61 mg/kg
Carbohydrate	0.2 %

were examined by use of the Calvet microcalorimeter. For each experiment at least one cooling and one heating cycle was undertaken. When C_p is plotted versus temperature, the phase transitions are indicated as peaks at the curves. Phase transitions, e.g. water freezing into ice, are followed by (large) changes in concentrations of solutes/reactants and consequently (large) variations in shelf life can be expected.

The bacon cooling curve (no. 1 in Figure 3) shows a drastic change at about -16°C indicating a supercooling of the water phase. Some correction for lag of signals from the sample must also be considered. Curve 1 is also giving a relative maximum at about -42°C. Two heating curves, one just after the 16 hours cooling and one after 100 hours storage at -86°C, are

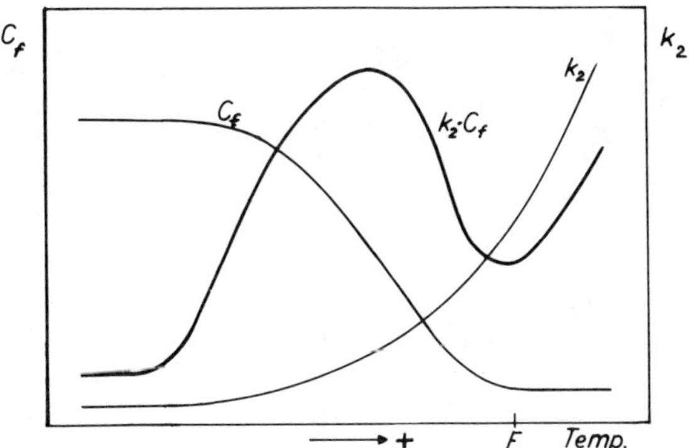

Figure 1. Relation of k_2C_f to temperature. k_2 is the rate constant of a second order reaction and C_f is concentration of solutes in the liquid portion of the product. F is the freezing point.

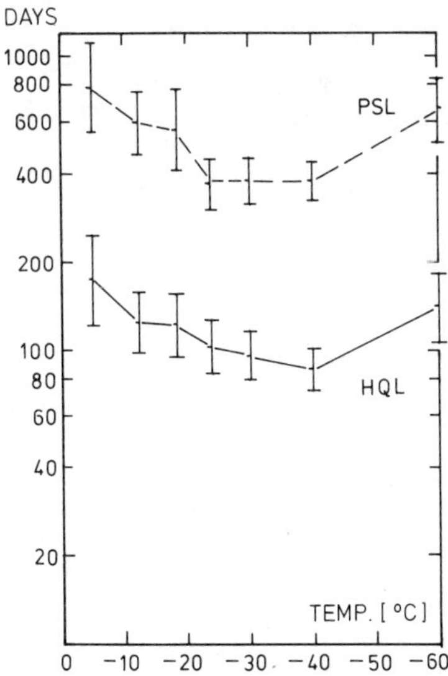

Figure 2. High-Quality-Life and Practical-Storage-Life for smoked bacon (6.92 g salt/100 g H_2O) as a function of temperature.

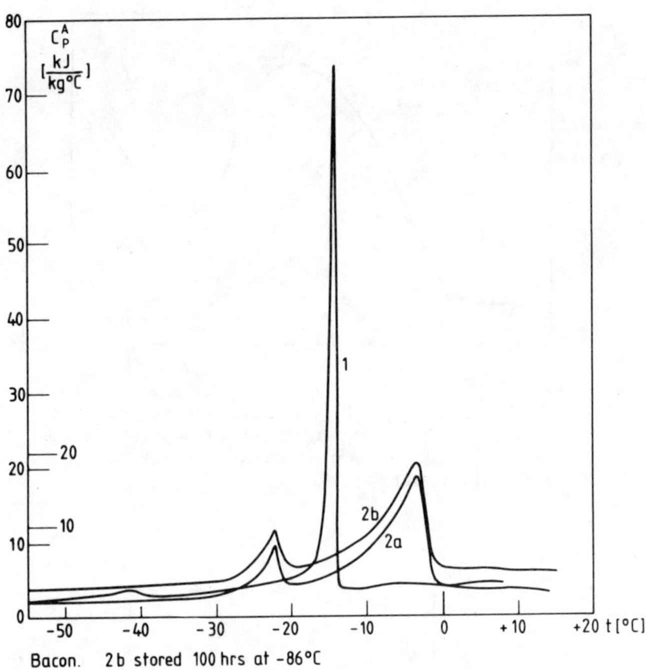

Bacon. 2 b stored 100 hrs at -86°C

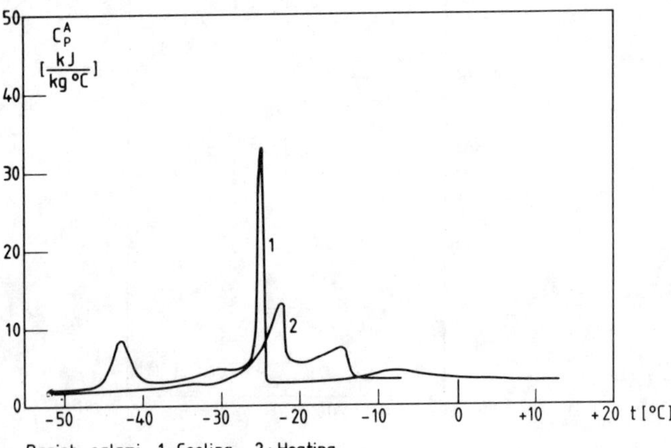

Danish salami. 1: Cooling. 2: Heating

Figures 3 and 4. Specific heat as a function of Temperature measured by a Calvet microcalorimeter.

to be seen. These two curves are practically identical but clearly different from the cooling curve. The heating curves show a phase change at -23°C indicating the eutectic solution of sodium chloride.

Cooling of Danish salami (Figure 4) shows a distinct peak at -26°C and another peak at -43°C. Heating of salami shows the eutectic mixture primarily consisting of sodium chloride and melting of ice at about -15°C.

3. DISCUSSION AND CONCLUSION

The two examples presented show considerable phase transitions in the temperature interval between -18°C and -30°C and therefore we can not expect normal stability for these products.

However, products known to have neutral or reverse stability e.g. vienna sausages, ref. 3 could not be recognized by the technique used in our experiments. So it is concluded that:

1. Products showing distinct phase transitions in or close to the commercial temperature area will not have normal stability.

2. The sensibility of the present calorimetric technique is not sufficient to detect all important phase transitions and/or stability of foods as a function of temperature in the frozen state is so complex that more parameters have to be followed.

REFERENCES

1. PINCOCK, R.E. and KIOVSKY, T.E. (1966). J. Am. Chem. Soc., 88, 4455.
2. POULSEN, K. Porsdal and LINDELOV, F. (1975). Stability of frozen systems and frozen foods. Scand. Ref. 4, 4, 165-169.
3. POULSEN, K. Porsdal and LINDELØV, F. (1981). Acceleration of chemical reactions due to freezing. In Water activity: Influence on food quality. Edited by: L.B. ROCKLAND and G.F. STEWART. Academic Press, New York.

ENGINEERING PARAMETERS FOR FREEZING AND THAWING EQUIPMENT

C.H. VEERKAMP
Spelderholt Centre for Poultry Research and Extension
Ministry of Agriculture and Fisheries
Spelderholt 9, 7361 DA Beekbergen, The Netherlands

Summary

The selection of the method for freezing or thawing and the optimization of the design of the equipment are nowadays the result of tradition and experience of the food manufacturer and the equipment company. The parameters used in these procedures are analyzed. The main parameters are the state of the product, the capacity, quality changes, processing time and capital and operating costs. The specification of the product and related quality variables do not generally present a problem. The thermal properties can be calculated or measured. The methods for calculating the processing time numerical or based on an empirical analytical method are accurate within 10%. Little is known about the influences of the various parameters related to the capital and operating cost of freezing and thawing methods. Selection and optimization of procedures for freezing and thawing processes should be developed.

1. INTRODUCTION

Several methods are known for freezing and thawing food products. In freezing, heat is extracted from the food by five different methods: (a) air blast freezing, (b) contact plate freezing, (c) immersion freezing, (d) cryogenic freezing and (e) liquid freon freezing. The purpose of the freezing process is to decrease the temperature below the initial freezing point of the product so as to get a longer shelf life during storage. Methods (a) to (d) are those most applied in the food industry. The commercial application of liquid freon freezing is limited by a number of practical problems (e.g. loss of freon).

In thawing, heat is transported to the food by convection or generated internally by the conversion of electrical energy via the dielectric and resistive properties. Thawing processes make the products suitable for further processing or selling and for preparation for use in the kitchen. The commercial application is nearly always restricted to the convective heat transfer methods of thawing: (a) vacuum steam, (b) immersion in water, (c) spraying with water and (d) air blast. Microwave processes are practicable in tempering processes for heating frozen blocks of fish or meat to a temperature just below the initial freezing point. At this temperature cutting and mincing is preferred, so there is no need for complete thawing these products.

The end-product quality and the cost are the main factors in the selection of the freezing or thawing method and the equipment for the process. No examples are found in the literature which make use of available selection or optimization procedures to choose the most suitable method and to optimize the design of the equipment. Until now it was the combination of tradition and experience of the food manufacturer and the equipment company. In the future more sophisticated optimization procedures for the freezing and thawing method and the design of the equipment should be developed. The scope of this presentation is to summarize and classify the parameters involved for this selection procedure.

2. PRODUCT SPECIFICATION

The product variables which are important for the selection and design of freezing and thawing equipment are summarized in Table I.

Table I   Required product specification.

Name	Massflow	Average initial temperature
Dimensions	Continuous-batch	Initial freezing point
Composition	Quality-variables	Storage temperature
Density		Thermal properties
Liquid-solid		
Homogeneous-bulk		
Package		

In practice many of these variables are easy to specify, others can be estimated.

The thermal properties (enthalpy, specific heat and the thermal conductivity) depend on the temperature in the freezing and thawing temperatures range. The COST 90 project (9) studied extensively the availability of reliable thermal properties data of food in literature, methods of measurement and predicting models. As a result of this study a computer program is developed which calculates the thermal properties of all kind of foodstuffs by using the composition, density and initial freezing point (1). For this purpose the composition should contain only the main components: water, proteins, fat, carbohydrates and minerals.

The quality of the product after freezing or thawing depends mainly on the initial quality before processing. However, some products (e.g. lettuce) are not suitable for freezing. It is concluded from a survey of the literature that the main quality variables involved in freezing and thawing processes are: the dehydration, the texture, the colour and the driplosses after thawing (Table II). The dehydration of the product depends on the package, surface area, initial temperature, ambient temperature and freezing time. It is well known that the weight loss of a product inside a package, in which there is an air space with poor heat transfer between package and product, is greater than freezing the same product without the package in an air blast tunnel. Dehydration may account for as much as 5-6 % weight loss in a poorly designed freezer (11).

The texture of food is influenced by the rate of solidification of water in the product. Soft fruits are especially susceptible to textural changes during freezing and thawing. The colour of e.g. frozen poultry products can easily be influenced by the rate of solidification of the ice in the layer near the outside surface. The colour of the thawed product is not influenced by the rate of solidification of the water or melting of the ice. Drip losses after thawing can be influenced by the rate of solidification and the thawing time of the product (2). The rate of solidification of the water or melting the ice at any point in the product is related to the processing time (freezing or thawing time). The freezing time is defined as the time required to decrease the average temperature of the product to an average temperature below the initial freezing point equal to the storage temperature.

The thawing time is the time required to increase the average temperature of the product from the storage temperature to the temperature of the initial freezing point.

Table II Quality variables important in freezing and thawing processes.

Quality variable	Related variables
Dehydration	art/state product T_i, T_a, t_p
Texture	art/state product solidification rate (t_p)
Colour	art/state product solidification outside layer (t_p)
Drip loss	art/state product solidification melting rate (t_p)

T_i = initial temperature
T_a = ambient temperature
t_p = processing time

All effects on the quality of any product can be related to the processing time required to freeze or thaw the product. For optimization purposes it is suggested that the effects on the perceived quality due to changes in the quality variables should be calculated using predictive models (8).

3. PROCESSING TIME CALCULATIONS

Numerical solutions of the heat transfer equations in freezing and thawing and computer programs to predict processing time, heat flow and rate of solidification or melting of any product in relation to the process conditions have existed since 1970 (2, 5, 6, 7, 12).
The use of these research efforts for engineering purposes is limited by the availability of the computer programs or their accessibility. Other calculation methods are based on Plank's formula which are elaborated for various shapes of the product (4, 10).

The accuracy of all predictions is the result of approximations of the physical behaviour of the food and errors in experimentally estimated parameters. An accurate estimation of the heat transfer coefficient is of paramount importance. There is no reason to expect that one of the above mentioned methods should predict the processing time more accurately than another. The accuracy of both calculation methods are compared with experiments generally within 10%. In selecting the freezing or thawing method calculations of the processing time and heat load in relation to ambient temperature and heat transfer coefficient are necessary.

4. SELECTION OF FREEZING OR THAWING METHOD

Before starting the design of the equipment a choice should be made between the methods for freezing or thawing. Usually experience and general information available about the methods are the base for making a decision. The factors affecting this selection are listed in Table III.

Table III Factors influencing the selection of the method for freezing and thawing.

Art/state of product
Capacity
Processing time
Indications of capital cost
Indications of operating cost

The art and state of the product and the capacity are determined by the product specification. The processing time depends on all quality variables involved. Indications about capital and operating cost of the available methods are required in relation to processing time and capacity. The information about all the above mentioned factors should be arranged in a flow chart or a program to facilitate the selection of the method of freezing or thawing. Until now too little information has been available for this purpose.

5. DESIGN OF THE EQUIPMENT

The capital cost and the operating cost are the ultimate factors in the design of the equipment. The parameters influencing these factors are listed in Table IV.

Table IV Parameters important for the design of freezing and thawing equipment.

Capital cost	Operating cost
Batch/continuous	Heating/cooling medium
Capacity	Energy
Internal transport	Fuel
Refrigerating/heating	Labour
Control and instrumentation	Water
Building	Maintenance
Installation	Waste handling
Interest	
Depreciation	

In many cases the parameters mentioned in this Table are depending on each other. Little information about the influences of all these parameters on the capital or operating costs is generally available. However, for selecting and designing the equipment it is essential that the relationship between the total cost per unit product and the processing time is known. The optimization of the processes of freezing or thawing should however not only keep the cost per unit product as low as possible, but also take into account the effects on the perceived quality due to the changes in the quality variables.

REFERENCES

1. BEEK, G.van and VEERKAMP, C.H. (1982). A computer program for the calculation of thermal physical properties of foodstuffs. Voedingsm. Tech. 15,(19), 63-66.

2. CALVELO, A. (1981). Recent studies on meat freezing. In: Developments in meat science. Ed. R.Lawrie, Vol.2, 125-128, Appl.Sci.Publ., London.
3. CLELAND, A.C. and EARLE, R.L. (1977). A comparison of analytical and numerical methods for predicting the freezing times of foods. J.Food Sci. 42, 1390-1395.
4. CLELAND, A.C. and EARLE, R.L. (1982). Freezing time prediction for foods - a simplified procedure. Int.J.Refrig. 5, 134-140.
5. CLELAND, A.C., EARLE, R.L. and CLELAND, D.C. (1982). The effect of freezing rate on the accuracy of numerical freezing calculations. Int.J.Refrig. 5, 294-301.
6. COMINI, C., GUIDICE, S.del, LEWIS, R.W. and ZIENKIEWICZ, O.C. (1974). Finite element solution of non linear heat conduction problems with special reference to phase change. Int.J.Num.Meth.Eng. 8, 613-624.
7. CREED, P.G. and JAMES, S.J. (1981). Predicting thawing times of frozen boneless beef blocks. Int.J.Refrig. 4, 355-358.
8. FARKAS, D.F. (1980). Optimizing unit operations in food processing. In: Food Process Engineering. Ed. Linko, P. et al. Vol.I, 103-115. Appl.Sci.Publ., London.
9. JOWITT, R. et al. (1983). Physical properties of foods. Appl.Sci.Publ. London, ch.15-18.
10. Kouwenhoven, H.L.J. (1972). The evaluation of heat removal during the freezing of poultry in air blast freezers. Bull. IIR Annexe 2, 195-207.
11. PERSSON, P. and ASTRÖM (1969). Selection of freezing equipment, its influence on productivity and profits. Food Proc.Market. 38(3), 93-98.
12. VEERKAMP, C.H. (1970). Theoretical and experimental freezing and thawing of poultry. Proc. XIV World Poultry Conf., Madrid, Part III, 861-872.

THE EFFECT OF THE RATE OF FREEZING AND THAWING ON THE DRIP LOSS FROM FROZEN BEEF

S.J. James, C. Nair and C. Bailey
Agricultural Research Council, Meat Research Institute
Langford, Bristol, BS18 7DY, U.K.

Summary

These experiments have not found any effect of freezing rate on drip loss from frozen beef. Trials indicate that very short (less than 5 minutes) and very long (greater than 2000 minutes) thawing times produce greater than average drip losses with minimum loss occurring at a thawing time of approximately 400 minutes.

1. INTRODUCTION

The main detrimental effect of freezing and thawing meat is the large increase in the amount of proteinaceous fluid (drip) released on final cutting, yet the influence of the rate of freezing and thawing on drip production has long been a source of contradictory claims. A review of relevant literature shows that considerable research has been carried out on the effect of freezing during the last half century, but it is difficult to compare results since different materials have been used, freezing rates have not been defined in the same way and initial and final temperatures have varied.

Some authors (1,2,3) have concluded that freezing rate has no influence on drip production, whilst work at Oklahoma State University (4) on beef steaks concluded that the use of freezing temperatures below -18°C increased the drip loss. Other authors (5), also using beef steaks, reported that increased freezing rates lessened drip, although the extent to which they did so, depended on the ratio of cut surface to volume. Further investigations (6,7) reached similar conclusions.

Several researchers maintained that other factors were more important. Some (1,8) considered that pH had far more effect on drip than freezing rate, while others (9) stated that storage conditions, especially temperature, were far more important.

Far less work has been carried out on the effect of thawing rate. Investigations using beef (1) and both beef and pork (10), showed that there was no significant effect of thawing rate on drip loss. It has been stated (11,12,13) that a faster thawing rate would result in a greater drip loss due to the shorter time allowed for the water to re-establish equilibrium with muscle proteins. In contrast to the above statements studies (14,15) have shown that a fast thawing rate resulted in a product with significantly less drip, and poultry carcasses (16) thawed slowly in a refrigerator at 3-5°C lost more drip than those thawed faster in air at 27°C, or in water at 21°C or 55°C.

Further evidence (17) indicated that high rates of protein denaturation occurred when the temperature of beef muscle remained in the temperature range -2°C to -5°C. Fast freezing and fast thawing should consequently minimise drip loss.

The results are therefore very conflicting and provide no useful design data for optimising freezing and thawing systems.

All the previous experiments that have been quoted had either used samples of a substantial size in each of which there was a range of freezing rates, or very small samples. While the use of small samples overcame the problems of differing freezing rates within the samples, they did not simulate the conditions found inside large pieces of meat. Anon and Calvelo (18) overcame both problems by careful design of their experimental equipment and produced a relationship between the time taken to freeze from 0° to -7°C and the difference between the percentage drip of the fresh and frozen sample. The relationship showed that minimum drip was obtained at very short freezing times of 5 minutes or less, maximum drip was produced by times from 15 to 20 minutes and drip loss was a constant independent of freezing rate at freezing times of 30 minutes and over. In commercial practice there are very few air-based freezing systems capable of achieving freezing times below 30 minutes, consequently freezing rate would have no effect on drip production. The implication that a freezing time in the region of 17 minutes could increase the drip production by approximately 30% was, however, worrying since a number of the individual portion freezing systems and plate freezers achieve freezing times of this order.

The authors therefore decided to check the validity of the results before looking for similar patterns, both during thawing and in other muscles and or species. Since a number of experimental details were missing from Anon and Calvelo's paper and initial attempts to obtain the required data by personal communication failed, the experiments could not be exactly duplicated. This paper details the results of two sets of experiments carried out using similar, but not identical methods and equipment to study the effect of both freezing and thawing rate on drip production.

2. EXPERIMENTAL

2.1 Material

M. semimembranosus from thirteen (7 for freezing studies, 6 for thawing) beef animals that had been stunned using a captive bolt pistol, slaughtered, dressed and split were used in the experiments. The sides were hung for 6 hours at approximately 12°C before being chilled in a room at a nominal 0°C, 1 m/s. At 24 hours post mortem the muscles were removed and aged in covered trays for a further 24 hours at 0°C.

2.2 Method

2.2.1 For effect of freezing rate

Cylindrical samples 70 mm long (diameter 25 mm for the determination of drip from fresh and diameter 50 mm for freezing), with their longitudinal axis parallel to the fibre direction, were removed from each muscle, the number of samples being dependant upon the size and shape of the muscle. The 25 mm diameter cylinders were immediately divided into eight discs 5 mm thick. The amount of available drip was then measured using the centrifugation method described by Penny (19), centrifugation being carried out for 1 hour at 3000 rm (1700 g).

The remaining cylinders of muscle were inserted into 2 mm polyvinyl chloride (PVC) tubes, 70 mm long, 50 mm internal diameter, insulated at one end and around the curved surface with 30 mm of expanded polyurethane foam (Figure 1). Copper-constantan thermocouples inside stainless steel sheaths were then inserted through previously drilled holes and positioned at 10 mm intervals along the longitudinal axis of the cylinder. A 1 mm thick stainless steel plate was clamped to the exposed end of the cylinder and a block of solid carbon dioxide placed on the plate.

The thermocouples were monitored continuously to ±0.25°C during the freezing process using a Solartron data-logging system until all the temperatures were below -7°C. The sample complete with thermocouples was then removed from the holder and placed in a room at -30°C for 24 hours to equalise. After equalisation the thermocouples were removed and six discs 5 mm thick, 50 mm diameter sawn from the cylinder, each one centred on a thermocouple position. A 25 mm diameter disc was removed from the centre of each 50 mm diameter disc, placed in previously weighed centrifuge tubes with inserts and allowed to thaw at 20°C for 2 hours. The drip was then measured by the method previously described.

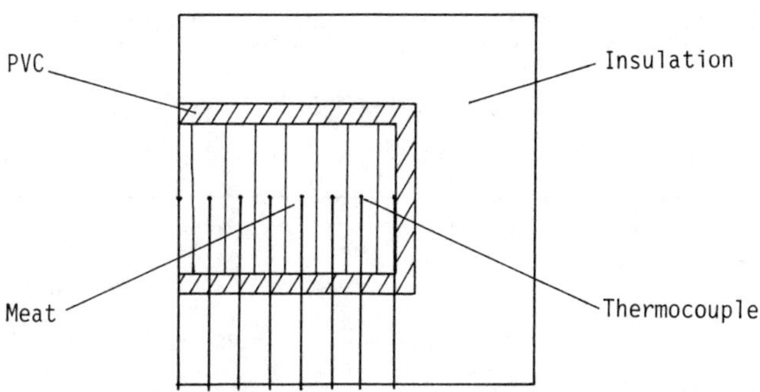

Figure I. Configuration during freezing.

2.2.2. For effect of thawing rate

One cylindrical sample (diameter 25 mm and 70 mm long), and two cylindrical samples (diameter 50 mm and 70 mm long), with their longitudinal axis parallel to the fibre direction, were removed from each muscle. The 25 mm diameter cylinders were immediately divided into six discs 10 mm thick (previous tests having revealed problems in cutting 5 mm discs from thawed cylinders) and the amount of available drip from each disc measured as previously described.

The two 50 mm diameter cylinders were inserted into PVC tubes insulated with 30 mm of expanded polyurethane foam on the flat surfaces and thermocouples inserted in the same manner as described before. Samples were then frozen in air at a nominal -30°C, 0.75 m/s, with the direction of air movement being perpendicular to the longitudinal axis of the cylinders. The freezing process was terminated when all thermocouples registered -28°C or below.

After freezing the meat cylinder complete with thermocouples was transferred into a second PVC tube of the same dimensions that had been insulated in the same way as those used in the freezing determinations. A hollow steel plate heat exchanger was then clamped to the exposed end and a water/glycol mixture at a controlled temperature between 1° and 30°C passed through the exchanger.

Very long thawing times were obtained by leaving frozen cylinders in the uninsulated PVC tubes in still air (<0.2 m/s) at 1° or 18°C. Very short thawing times were obtained by tahwing one cylinder in a microwave oven. The temperatures along the microwaved sample were taken every 15 seconds and the temperatures in all the other samples were monitored continuously until they were all above 0°C.

After thawing six 10 mm thick slices were cut from the cylinders, each slice centred on a thermocouple position. Two 25 mm diameter discs were then removed from each slice and the drip potential of the discs measured using the method previously described.

2.2.3 pH

A sample (approximate weight 2g) was removed from each muscle at 48 hours post mortem and the pH determined using the method described by Bendall (20).

3. RESULTS

Table I Mean weight of discs, mean % drip loss and variance of % drip loss from unfrozen cylindrical sample.

(a) Fresh control for freezing determination (9 replicates)

Animal No.	Mean weight discs (g)	Mean % drip	Variance
1	2.652a,b,c	21.2a	11.7
	2.230a,c	21.9a	12.7
2	3.163b	18.4a,c	31.5
3	2.486a,c	19.7a,c	19.9
	2.176a,c	20.5a	23.3
4	2.780b,c	10.0b	5.7
5	2.434a,c	13.4b,c	21.7
6	2.115a	24.1a	6.6
7	2.381a,c	19.3a	1.9

Means within column with same subscript do not differ significantly ($p<0.01$).

(b) Fresh control for thawing determination (6 replicates)

Animal No.	Mean weight discs (g)	Mean % drip	Variance
8	5.294a	11.2a,b,c	3.8
9	4.303b,c	8.6 a,b,c	0.7
10	4.609b	15.5b	8.1
11	4.138c	15.4b	9.1
12	4.138c	7.5c	3.3
13	4.350b,c	9.4c	8.3

Means within column with same subscript do not differ significantly ($p<0.01$)

It can be seen in Table I that there is a considerable variation in percentage drip loss from different cylinders. The mean percentage drip loss is smaller and the variation less, both between and within vylinders, with the larger sample size used for the determination of the effect of thawing rate.

The position of the disc along the axis of the cylinder that produced the maximum percentage drip loss varied between samples. An analysis of the data from all 15 cylinders produced a very low correlation (correlation coefficient 0.15) between percentage drip and position along cylinder.

The time in minutes for the centre to cool from 0° to -7°C is plotted against the percentage drip loss from each disc in Figure II(a). Figure II(b) shows the difference in percentage drip loss between frozen discs and the mean of unfrozen discs of the same muscles against freezing time. There is a large scatter in both the results for percentage drip and the difference in percentage drip at any freezing time. There was no effect of freezing time on percentage drip loss in either case.

The data on thawing time and drip loss have been displayed in a similar manner in Figures III(a) and III(b), thawing time being defined as the time taken (minutes) for the temperature at the centre of each sample to pass between -7°C and 0°C. There is again a large scatter in the results and no clear relationships. Figure III(b) does however indicate a tendency for very fast thawing (thawing times of less than 1 minute) and very slow thawing (thawing times of greater than 2000 minutes) to produce greater than average drip losses with a minimum drip loss at approximately 400 minutes.

Ultimate pH's varied between 5.3 and 5.7 within the normal range for beef.

5. DISCUSSIONS AND CONCLUSIONS

The variation of drip loss within a particular muscle found in this experiment does not seem to have been reported in any detail in the literature. However considerable differences in drip loss between muscles on the same carcass have been reported (21). Penny (3), concluded that drip loss is affected by the collagen and fat content of the material and its temperature/pH history. Any muscle that is allowed to cool while still on the carcass will experience a range of cooling rates within it. One would therefore expect to find variation in drip potential within the muscle due to differences in temperature/pH history. A tacit assumption of Anon and Calvelo's analysis is that there is little or no variation in drip potential along the length of the unfrozen cylinder of muscle. They subtracted the drip loss from unfrozen samples from that of the corresponding frozen samples to remove the effect of animal and pre-freezing treatment. The present authors have found that in some samples the drip loss after freezing and thawing is less than the mean loss from the corresponding fresh control and negative differences are produced. Recent contact with Anon has revealed that their unfrozen loss was the mean loss from 2 to 3 samples and posssible variation along the cylinder was not considered. The magnitude of the variations found in the present experiment is very high and might be a result of the particular centrifugation method used to determine the drip potential. This method has been well tested for pork, but when used with beef great care was taken to avoid localised areas of high collagen content (22). If such areas are present in the small samples used in this method they could lead to increased variability in the drip results. Experiments are therefore being carried out using other methods of drip determination to check the validity of these results.

No obvious relationship was found between freezing rate and drip loss over the range of freezing times (from 2 to 70 minutes) used in this experiment. This range covers the majority of individual portion freezing systems and the plate freezing of meat slabs less than 50 mm thick. Many blast air freezing operations for beef quarters or cartoned meat achieve freezing times of between 24 and 72 hours and neither Anon and Calvelo's

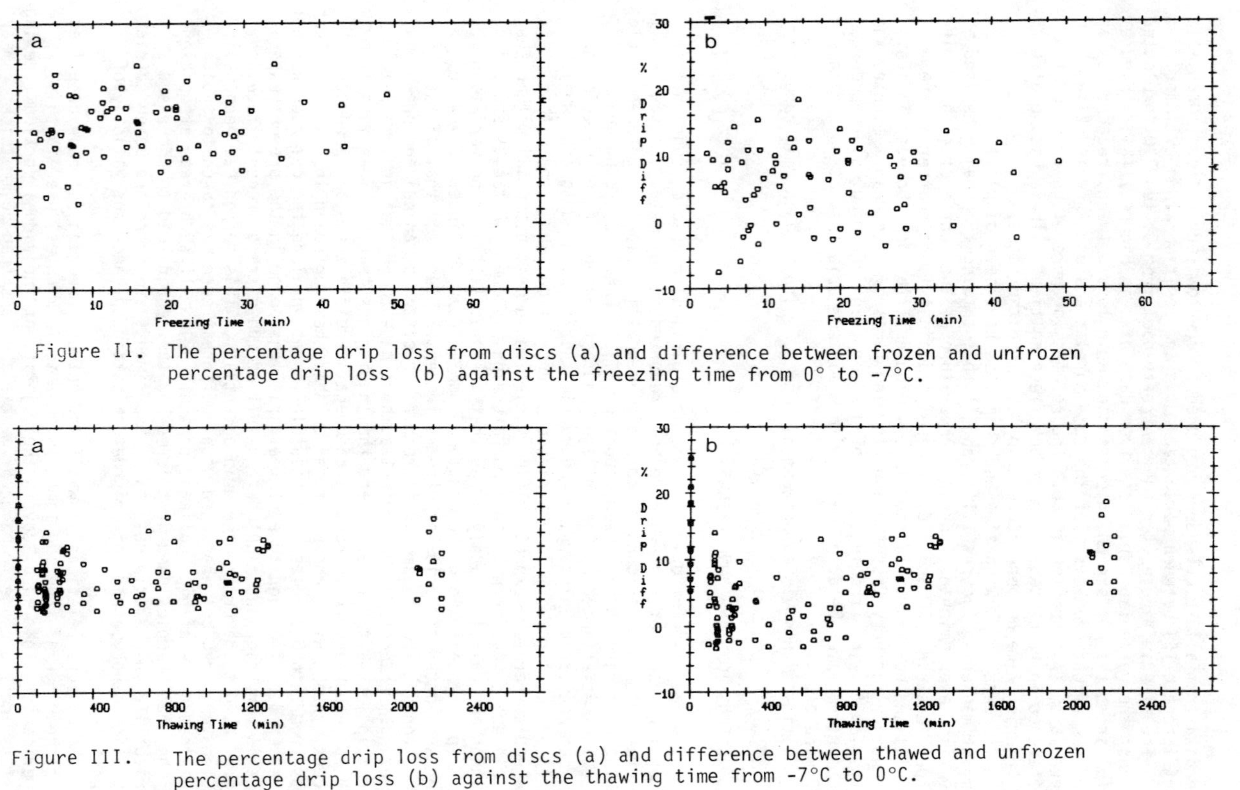

Figure II. The percentage drip loss from discs (a) and difference between frozen and unfrozen percentage drip loss (b) against the freezing time from 0° to -7°C.

Figure III. The percentage drip loss from discs (a) and difference between thawed and unfrozen percentage drip loss (b) against the thawing time from -7°C to 0°C.

or this report covers that range. We must conclude from the results of this experiment that any relationship between freezing rate and drip was masked by variability in the material before it was frozen.

The range of thawing times, from 1 minute to 40 hours, considered in the thawing experiments would cover the majority of commercial processes. Again no clear relationship between thawing rate and drip production was revealed. The results do indicate that fast thawing using microwaves and very long thawing times can increase drip production.

REFERENCES

1. EMPEY, W.A. (1933). J. Soc. Chem. Ind., 52,p235.
2. BAILEY, C. (1972). Meat Research Institute Memorandum No. 19.
3. PENNY, I.F. (1974). Meat Freezing: Why and How? Meat Research Institute Symp. No. 3. p. 8.1.
4. ANON. (1963). Oklahoma State University Report No. 56.
5. RAMSBOTTOM, J.M. and KOONZ, C.H. (1939). Fd. Res., 5, p423.
6. COOK, C.A., LOVE, F.J., VICKERY, J.R. and YOUNG, W.J. (1926). Aust. J. Biol. Med., 3, p15.
7. JAKOBSSON, B. and BENGTSSON, N.E. (1969). 18th Eur. Meeting Mt. Res.Wkrs. p.482
8. SAIR, L. and COOK, W.H. (1938). Canad. J. Res., 6, p139.
9. MORAN, T. and HOLE, H.P. (1932). J. Soc. Chem. Ind., 51, p16.
10. CIOBANU, A. (1972). Lucrari Cercetare R., 9, p63.
11. CUTTING, C.L. (1974). Refrig. Air Con. 77, p45.
12. LOVE, R.M. (1966). In Cryobiology H.T. Academic Press, London.
13. RICHARDSON, A.S. and SCHRUBEL, R. (1908). Cited in Cryobiology.
14. DEATHERAGE, F.E. and HAMM, R. (1960). Fd. Res. 25, p623.
15. FINN, D.B. (1932). Proc. Royal Soc. 111, p396.
16. SINGH, S.P. and ESSARY, E.O. (1971). Poultry Sci. 50, p364.
17. REAY, G.A. (1934). J. Soc. Chem. Ind. 53, p413.
18. ANON, M.C. and CALVELO A. (1980). Meat Sci. 4, p1.
19. PENNY, I.F. (1977). J. Sci. Fd. Agric. 26, p1593.
20. BENDALL, J.R. (1975). J. Sci. Fd. Agric. 26, p55.
21. TAYLOR, A.A. and DANT, S.J. (1971). J. Fd. Technol. 6, p131.
22. PENNY, I.F. (1982). Personal communication.

NUTRITION

INTRODUCTION AND CONCLUSION

A E BENDER

Department of Nutrition, Queen Elizabeth College, University of London, England

There is a great deal more to nutrition than simply measuring the nutrient content of foods. Nutritionists need to know which sections of the population are consuming which foods in what amounts, what factors may influence their dietary habits, whether any are unusually heavily dependent on a limited choice compared with the average and, with respect to changes that take place in processing, the role played by the food in question in the diet as a whole.

Special attention must be paid to the more vulnerable groups of the population.

For these reasons the role of sub-group IV has been somewhat restricted since most of these problems do not fall within the area of thermal processing. Consequently the number of projects undertaken was fewer than the sub-group would have wished and problems that they would have liked to attempt to solve could not be undertaken. However, several of the projects of the other sub-groups included nutritional input.

When considering changes that take place during processing it is essential to view the subject in perspective. The question is whether a change, beneficial or detrimental, will have any significant effect on the nutritional status of any section of the population. If not then other quality considerations must overshadow nutritional qualities.

Such perspective can be illustrated by an investigation, one of the few of its kind, carried out by Hellendoorn et al on meals canned for the Dutch Army (J.amer.diet.Ass. 58, 434-441, 1971). The findings are particularly valuable because (1) they involve meals rather than the more usual single foods or model systems; (2) the effects of both the process, canning and subsequent 5 years storage were investigated compared with the usual analyses immediately after processing, and (3) a range of nutrients were examined.

The results show that vitamin A, as an example, suffered 50% destruction during processing and was completely destroyed after 1½ years storage. However, since the original raw foods supplied less than 2% of the recommended daily allowance of vitamin A such a loss is of no importance. Even had none been lost the meal would not have made any useful contribution to the vitamin A intake. Thiamin, on the contrary, was present in such large amounts, namely 9mg, that even after the loss of 75% due to processing and 5 years storage, the meal still supplied more than the RDA.

The work also showed that several vitamins, including riboflavin, and vitamins B6 and B12 were completely stable even after 5 years.

Another illustration of perspective is provided by the potato. This supplies about one-third of the vitamin C and one-sixth of the thiamin of the average European diet so that losses would be nutritionally significant. It is well established that vitamin C can be damaged during cooking, and particularly when kept hot after cooking, and the loss is so great in dehydration that these products are often fortified with added vitamin C. It is not so well known that treatment with sulphite to inhibit polyphenol oxidase damages the thiamin.

Another problem area is that of analysis. Several nutrients can undergo chemical changes during processing which may not be detected by the usual chemical and physical methods of assay but are of biological importance. For example, carotenoids isomerise to forms of lower potency during canning of vegetables, vitamin B6 exists in three forms which can interchange, protein quality and the availability of carbohydrates and minerals may be changed. While bioassay would reveal changes in bioavailability such assays are lengthy and expensive and, apart from being beyond the resources of most control laboratories, they are notoriously variable.

Within the area of analysis the sub-group established a working party on vitamin assay and their report is now ready for publication, as will be described by Dr. Brubacher. So far as is possible for a limited number of vitamins methods have been standardised for use in different laboratories - a factor which may become of special importance if nutritional labelling develops.

In this whole area we need to decide which nutrients are worthy of consideration, how they can be measured and, particularly if novel foods and novel processes are introduced, how reliable are the commonly used methods as an index of usefulness for the human consumer. In the same vein, although not part of COST 91 project, the safety of novel processes must be assured, and since extrusion cooking can apparently increase the amount of undigestible materials in the food it is necessary to consider the importance of fibre in the diet.

Finally, domestic food preparation must be included as a food process so it is necessary to compare industrial processing with the domestic equivalent rather than to consider industrial processing in isolation.

Overview

A. MARIANI-COSTANTINI
Istituto Nazionale della Nutrizione
Via Ardeatina, 546-00179 Roma-Italy

The problem of livelihood, in other words the problem of providing nutritional adequacy for healthful life, goes back to the most ancient times of human history.

Having the honour of speaking in Athens, birthplace of the Western civilization, I would like to introduce my talk with a citation from Hesiod (line 42 "Works and days").

Κρύψαντες γὰρ ἔχουσι θεοὶ βίον ἀνθρώποισι

"keep in fact the gods hidden what is necessary for men to live".

The intuition of the poet is perhaps oversimplifying given that - as recites the following line - "otherwise it would be sufficient to work a day to cover the needs of a year".

However the two lines powerfully express the painful difficulty of having what is indispensable for survival, in other words "enough food". And such pressing need, strictly associated with the development and evolution of human communities, is still dramatically afflicting the great majority of the world's population living in developing countries.

In spite of the increases in total food production, in developing countries (Fig. 1), in fact, energy supplies remain submarginal, as opposed to the wellknown excesses of developed countries. Furthermore, (Fig. 2) the gap in protein and fat supply, particularly of animal origin, between developed and developing countries tends to increase. (1)

Consequently, while protein-energy malnutrition continues to represent the main cause of suffering and death in the 3° world, at the same time industrialized countries are faced with new challenges in the fields of nutrition and safety of foods.

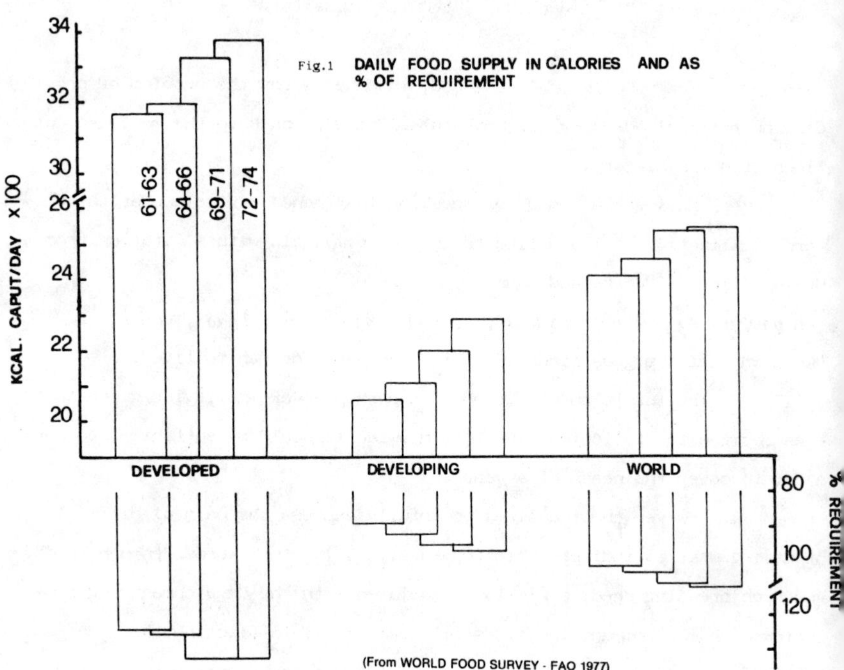
Fig.1 DAILY FOOD SUPPLY IN CALORIES AND AS % OF REQUIREMENT
KCAL. CAPUT/DAY x100
34
32
30
26
24
22
20
61-63
64-66
69-71
72-74
DEVELOPED
DEVELOPING
WORLD
% REQUIREMENT
80
100
120
(From WORLD FOOD SURVEY - FAO 1977)

Fig. 2 **PROTEIN AND FAT SUPPLY IN DEVELOPED AND DEVELOPING COUNTRIES : 1961-63 AND 1972-74**
(Grams per caput per day)

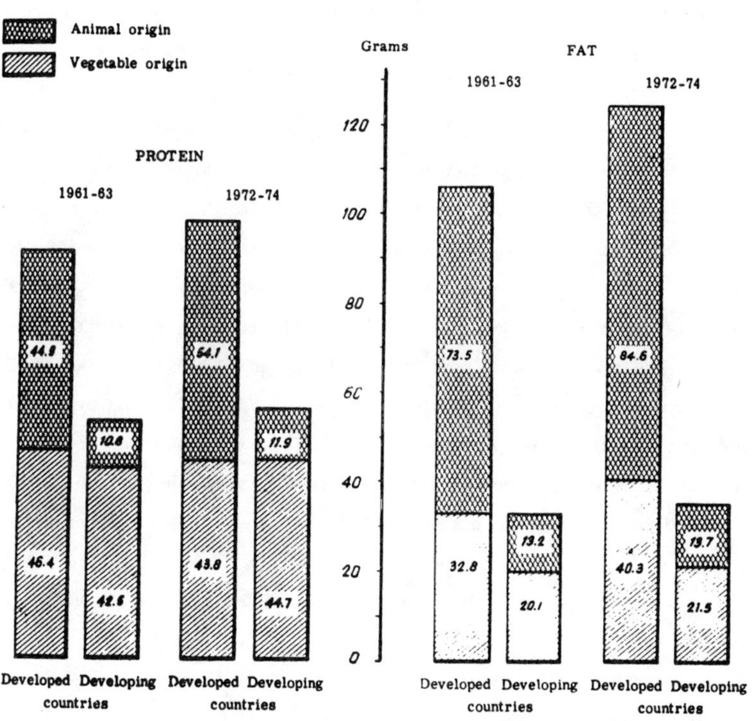

(From WORLD FOOD SURVEY - FAO 1977)

On the occasion of the concluding seminar of COST 91, example of collaboration open to all OECD European States,and prior to the session wich will deal with the effects of thermal processing on the quality and nutritive value of foods, it may be appropriate a general review on some of those aspects that will be of a prominent relevance in the future.

As a matter of fact, the excesses and/or unbalances of consumption in the developed countries can be attributed mainly to 3 orders of factors: increase in total food production, improvement of modern techniques of food storage and distribution, progressive and parallel increase in the population's income .

In any case, it has to be emphasized that the unbalance, established particularly with regard to energy, protein and animal fat is definitely more relevant than it would appear from a simple comparison of figures.

In fact, the average per caput energy supply should be at present reduced considering the decline of energy expenditure for work and physical activity and the modification of the structure of the population.

The trend towards reversal of the pyramid of the italian population between 1951 (Fig. 3) and 1980 (Fig. 4) is an explanatory example of the last aspect of the problem, in that it shows a consistent reduction of the fractions with higher energy requirements (young people) in the face of a marked increase of the segments with lower energy needs (older people, over 60).

A wealth of epidemiologic evidence on the effects of overconsumption has been provided (2).

The need of rationalizing and correcting the model of eating behaviour established in industrialized countries appears therefore evident. This would also contribute to save resources wich could be used for exchanges and/or food aids to developing countries.

In industrialized countries (3-4-5-6-7) however the classical deficiency syndromes have been replaced by more complex forms of malnutrition.

In this context it seems appropriate to stress, as shown in Fig. 5, the impact, which may be harmful or beneficial, of the nutrients

Fig. 3 - PYRAMID OF THE STRUCTURE OF ITALIAN POPULATION IN 1951

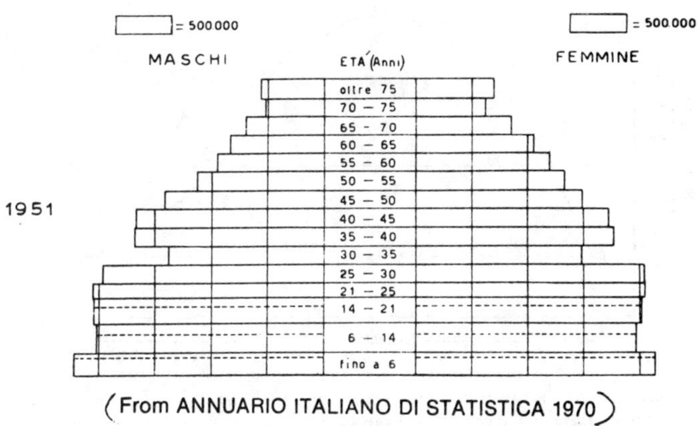

(From ANNUARIO ITALIANO DI STATISTICA 1970)

Fig. 4 - PYRAMID OF THE STRUCTURE OF ITALIAN POPULATION IN 1980

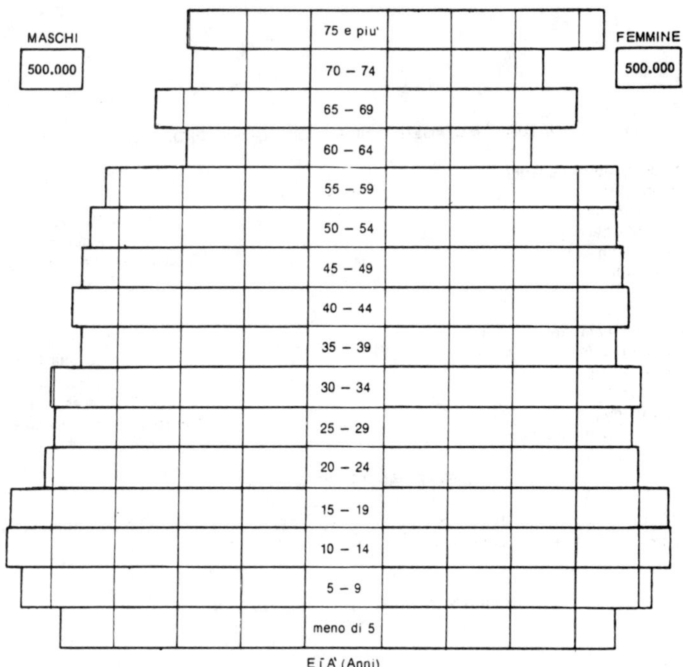

(From ANNUARIO ITALIANO DI STATISTICA 1980)

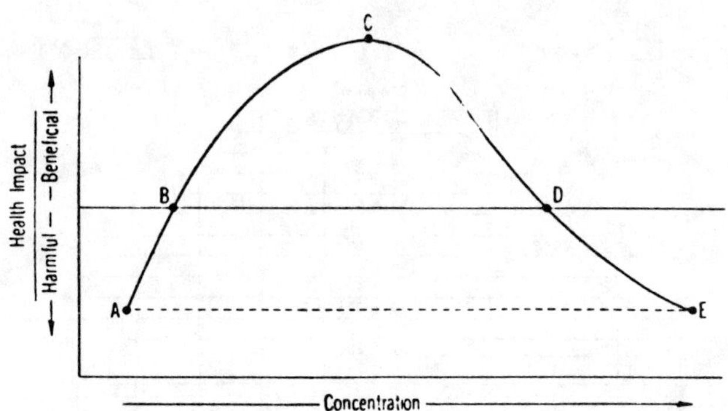

Fig. 5 Impact of nutrient concentration on health. The concentration referred to may be in diet or tissues. A, lethally low concentration; B, minimum concentration compatible with good health; C, concentration for optimal health; D, maximum concentration compatible with good health; E, lethally high concentration. (from Hathcock, 1982).

Fig.6 NATURAL MUTAGENS AND CARCINOGENS IN FOOD

Safrole, Estragole, Methyleugenol
Hydrazines
Furocoumarins
Solanine and Chaconine
Quercetin
Quinones

Sesquiterpene Lactones
Phorbol Esters
Canavanine
Alcohol
Mold carcinogens
Nitrite, Nitrate and Nitrosamines

Theobromine
Pyrrolizidine
Vicine and Convicine
Gossypol
Sterculic acid and malvalic acid
Anagyrine

Rancid fat
Peroxisomes
Trans fatty acid
Cooked Food as a Source of
Ingested Burnt and Browned Material

DIETARY ANTICARCINOGENS

Vitamin E
Ascorbic Acid

B-Carotene
Uric Acid

Selenium
Phenols

Glutathione

(from B.N. Ames Science 221, 1256, 1983)

concentration on health (8).

A clear demonstration of the contradictory effects of the western model of nutritional behaviour is provided by a recent survey of young men and women in Heidelberg, Arab et al., in 1982 (9). The study shows that unsatisfactory intakes of calcium, iron, thiamine, pyridoxine and folate are associated with more than adequate protein intake (11-13% of the energy) and with excesses in fat (41-43%) and alcohol (26-30%) levels. Consequently, factors of risk by excess and defect coexist. On the other hand, as concerns the elderly, dietary calcium deficiency is emerging as a widespread problem (Rivlin, 1983 (10); Albanese, 1978 (11)).

Besides the problems of safety raised by additives and contaminants (on this matter we'll come back later), an issue is now receiving growing attention. It is well known that comparison of data from different countries reveals wide differences in the rates of many types of cancer, while, at the same time, epidemiological studies have indicated that dietary factors may be the most promising area to explore (12-13-14-15).

Ames (16) has recently desumed the role of natural mutagens and carcinogens in food. He reported many examples (Fig. 6) and speculated on the biochemical mechanisms, particularly on the role of oxygen radicals and their inhibitors, also naturally contained in the diet, in relation to cancer promotion and anticancerogenesis.

It is also worth mentioning that according to Ames the human dietary intake of "nature's pesticides" is likely to be several grams per day - at least 10,000 times higher than the dietary intake of "man-made pesticides".

Now it is evident that a whole series of priority areas for research and interventions in nutrition policy and education can be recognized. In the limits of this presentation, hoever, I will restrict myself to the subjects of a more immediate practical interest.

The problem of nutritional guidelines

Various authors have analyzed the different definitions and interpretations of dietary standards. Waterlow (17-18) emphasized the need to distinguish between recommandations (intake, recommended allowances) and estimates of requirements or needs or safety levels. Recommandations refer to specific situations (population groups) and are affected by various non dietary factors (mainly socioeconomic and cultural).

It is clear however that while ensuring the intake of the essential nutrients represents the primary need for developing countries, developed countries are faced with the problem of elaborating guidelines to avoid excesses or unbalances. In spite of the discussions raised in the USA by the formulation of the socalled "Dietary goals" (19-20-21-22-23), a new approach to establishing an upper or prescriptive level of recommended intakes of nutrients was recently devised for use in the U.K. by Passmore,Hollingsworth and Robertson (24). It was based on present supplies and suggested opportune modifications of food habits, such as a consistent reduction in the intake of fat, sugar, meat and ancohol, a proportional increase of cereals, potatoes, other vegetables and fruits, while the intakes of milk, eggs, fish, pulses and nuts were to remain unchanged. This suggestion does not seem to be fully accepted, particularly in the non-mediterranean countries.

Just in the U.K., in this same year, the official acceptance of the nutritional guidelines formulated by an ad hoc Committee chaired by Prof. James, was received with difficulties and reservations by the industy, although the results of a special study (the Laing Special Report (Fig. 7) - aimed at verifying the effective health improvement deriving to the population from a correct application of the suggestions of the report- showed the benefits in therms of prevention and potential savings to the N.H.S. (25) that could be achieved.

The contribution of technology to the evolution of food consumption

Certainly, in industrialized countries, the possibilities offered by modern technologies in the storage and processing of foods, from the more traditional to the latest (Fig. 8), determinantly contributed to changes in the nutritional behaviour. In developed countries, urbanization, work rhythms and life styles induced a progressive reduction in the consum

Fig. 7 - EVALUATION OF BENEFITS IN TERMS OF PREVENTION AND SAVINGS TO NHS IN THE U.K.

DISEASE	INFLUENCE OF PRESENT DIET		POTENTIAL SAVINGS TO NHS 1983/84	TOTAL DEATHS 1981	
	Dietary factor	% Disease caused by diet		Under 65s	65 and over
Allergies	Fat, Sugar, Salt, Fibre	5-40	n/a	n/a	n/a
Anorexia	Fat, Sugar, Salt, Fibre	n/a	n/a	n/a	n/a
Cancer of large bowel and colon	Fibre	20-80	£8m-£31m	2,702	8,845
Constipation	Fibre	100	£36m	3	24
Diabetes	Fat, Sugar, Fibre	65-80	£78m-£96m	1,079	4,199
Diverticular disease	Fibre	100	£16m	128	1,307
Gallbladder disease	Fat, Sugar, Fibre	75-80	£28m-£30m	54	441
Heart disease	Fat	30-90	£80m-£239m	38,746	134,903
High blood pressure	Fat, Salt	20-90	£32m-£146m	1,097	4,800
Irritable bowels	Fibre	40-80	n/a	n/a	n/a
Overweight and obesity	Fat, Sugar, Salt, Fibre	100	£21m	69	96
Strokes	Fat, Salt	20-80	£114m-£458m	9,257	69,584
Tooth decay	Sugar	100	£452m	0	1
Vitamin and mineral deficiences	Fat, Sugar, Salt	100	£16m	7	46
References are too much fat, sugar, salt or too little fibre			£881m-£1,541m	53,142	224,246

(From LAING REPORT 1983)

Fig. 8 - HISTORICAL EVOLUTION TOWARDS THE IMPROVEMENT OF PROCESSED FOOD QUALITY

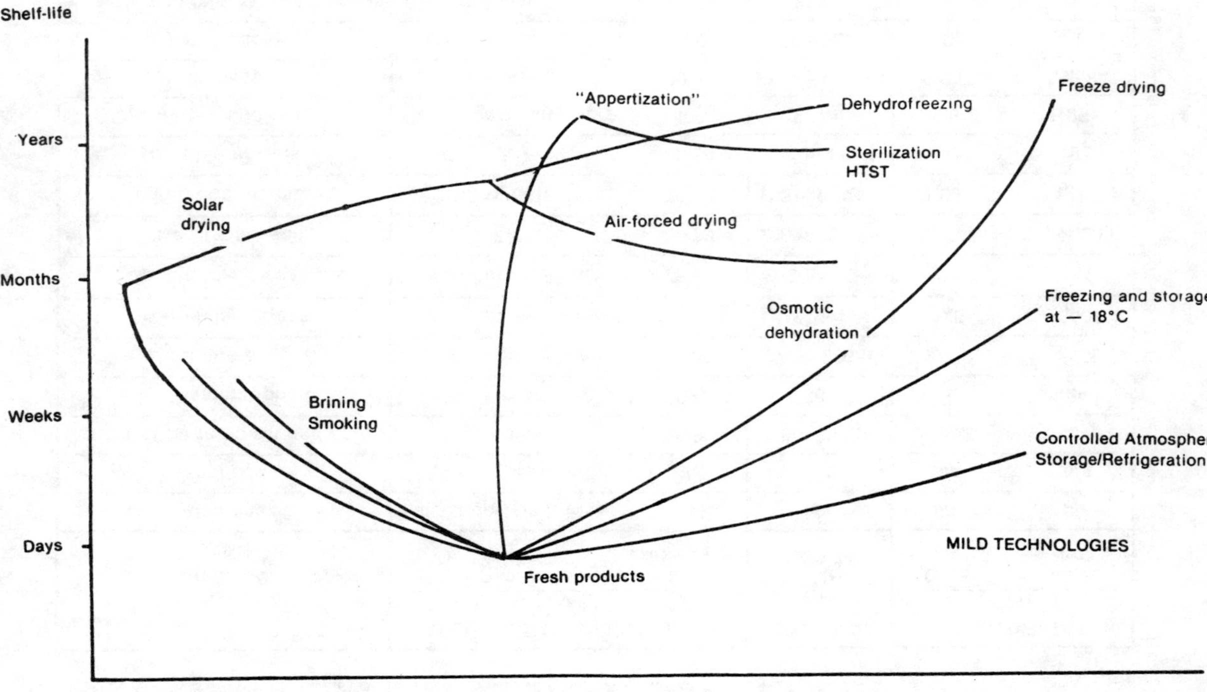

(From C. PERI, 1983)

ption of fresh products.
Obviously, the choice and/or success of the different technologies is primarily dictated by economical and/or practical criteria (storage time and conditions, distribution benefits, ready for use meals, etc.). The historical evolution towards the improvement of the quality requirements of processed foods appears evident, as shown by Fig. 8. The application of "mild technologies", i.e. "minimally processed foods", designed to guarantee the preservation of the organoleptic and nutritional characteristics of fresh foods, represent the most advanced solutions (26).

Optimizing food quality

It appears evident that safety, meant to indicate in a broad sense that a food is free of any chemical or microbial contamination at the time of its consumption, represents the first requisite in food production. Actually, growing confusion and concern about the safety of food additives and chemicals in our environment have led several food regulatory agencies to issue guidelines for testing and assessing chemicals in food which reflect both the advances of experimental toxicology and the degree of ability to interpret the experimental findings in terms of extrapolating them to man. (27)

A model of safety evaluation in the field of chemicals in foods, evolved through the activities of the Joint FAO/WHO Expert Committee on Food Additives(JECFA) is illustrated in Fig. 9. In the light of the information provided by the diagram, assessment of toxicity of food chemicals can be visualized a s a complex process having a dynamic rather than a static character, since new scientific findings may at any time challenge the results of previous evaluations. The two-way arrows at (a) and (b) in the diagram are intended to represent this aspect.

Following the circulation of a document prepared by Vettorazzi (International Program on Chemical Safety, IPCS, Internal Technical Report, 1982) joint WHO Secretary to the JECFA, a strategy meeting on updating principles and methodologies used by JECFA in testing and assessing chemicals in food was held in Oxford (U.K.) last September. According to the views expressed in the document and accepted at the meeting in Oxford, such updating should primarily be concerned with general principles for testing classes of food chemicals, interpretation of experimental findings and general guidelines and opinions about the capability and usefulness of various types of test programmes discussed in various JECFA reports. In other words, the methods and procedures for the toxicity testing and assessing of food additives and contaminants dealt with during the years of the activity of JECFA should be comprehensively reviewed and brought into line with advances in toxicology and related disciplines.

Once met the fundamental requisite of safety, it is possible to develop and adopt optimal conditions for the treatment of food directed to the preservation of its quality attributes, namely texture, flavour, colour and nutritive value. To that end, as underlined by Tannenbaum (29), it is important to know those properties that are important characteristics of safety and high-quality of foods, as well as those chemical and biochemical reactions that have important effects on their quality and/or wholesomeness (Fig. 10).

With particular regard to the preservation of the nutritive value, attention should be paid not only to the reactions leading to the loss or degradation of vitamins, minerals, proteins and lipids but also to those leading to their bio-unavailability. Extreme treatment conditions (high temperature, elevated pH and oxygen concentration, etc.) may induce the formation of artifacts that perhaps are not toxic in themselves but cause a more or less relevant reduction in the nutritive quality of foods, so contributing to the increase of the non-nutrient fraction. This relationship between nutrients and non-nutrients in food remains in fact the key factor in the preservation of the nutritional integrity of a processed food. This is particularly true when considering nutrition, as proposed by McLaren (30), in the context of an Agent-Host-Environment interaction system (Fig. 11), where the human body is the host, that part of the environment we normally consume is food and the agent in nutrition is that part of food represented by nutrients.

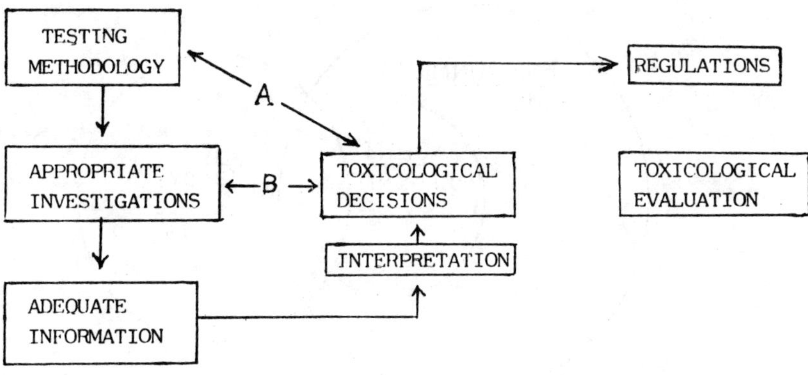

Fig. 9 A model of safety evaluation of chemicals in food.

(from G.Vettorazzi,1982)

Fig. 10 ANALYTICAL APPROACH TO QUALITY

1) Determining those properties that are important characteristics of safe, high-quality foods

2) Determining those chemical and biochemical reactions that have important influences on loss of quality and/or wholesomeness of foods

3) Integrating the first two points so that one understands how the key chemical and biochemical reactions influence quality and safety

4) Applying this understanding to the various situations encountered during storage and processing of food

(from S.T. Tannenbaum, 1976)

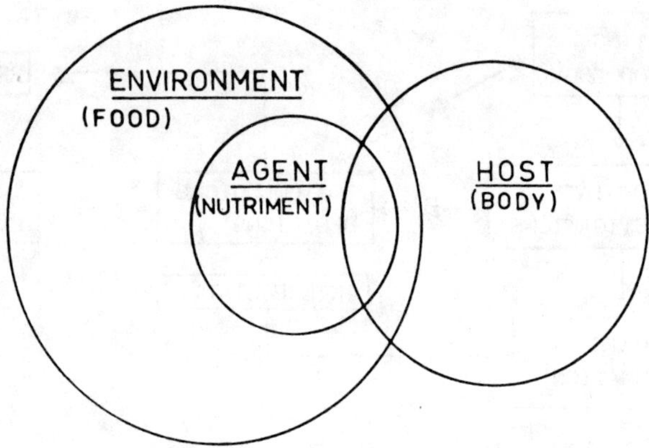

Fig. 11 Agent-Host-Environment concept applied to nutrition (from McLaren,1981)

Conclusions

Without any doubt, the solution of the world's malnutrition problems will be achieved only through the improvement of our basic and specific knowledge in the field of food production and distribution and in that of human nutrition. Although not clearly recognized, in the past, these two aspects are strictly interrelated, particularly at the operational and practical level. Consequently, the initiative of the COST group, taking into consideration both aspects, can only lead to positive outcomes. And this will be particularly true when the project will be able to count on a more effective and concrete recognition from the interested parties and/or countries, both at the political and economical level.

REFERENCES

1. The fourth World Food Survey - F.A.O. - Roma, 1977

2. Farinaro E., Stamler G., Mancini M. - Proceedings of Int.Symp. on Epide miology and Prevention of Atherosclerotic Disease,Prev.Med.12,1, 1983

3. Mariani A. - Nutritional needs and the situation in the European Community in Symposium on nutrition, food technology and nutritional informa tion - Report EUR 70805, EN., pag.5

4. Srinisasan T.N. - Development, poverty and basic human needs: some issues. Food Res.Inst.Studies 1 6:11, 1977

5. The Lancet; Editorial - A radical approach to Zinc.Lancet 1:191, 1978

6. Farrel P.M., Levine S.L., Murphy M.D., Adams A.Y. - Plasma tocopherol levels and tocopherol-lipid relationships in a normal population of children as compared to healthy adults - Am.J.Clin.Nutr.31:1720, 1978

7. Combs G.F. - Assessment of vitamin E status in animals and men - Proc.Nutr.Soc. 40:187, 1981

8. Hathcock J.N. - Nutritional Toxicology - Ac.Press 1982 N.Y.

9. Arab L. et Al. - Nutrition and Health - A survey of young Men and Women in Heidelberg Ann. Nutr. a.Met.Suppl. 26/S1/82, 1982

10. Rivlin R.S. - Nutrition and the health of the elderly - A growing concern for all ages - Arch.Int.Med. 143, 1200, 1983

11. Albanese A.A. - Calcium nutrition in the elderly - Postgrad Med. 63, 167, 1978

12. Doll R. and Peto R. - J. Natl. Cancer Inst. 66, 1192, 1981

13. Peto R. and Schneiderman,Eds. - Banbury Report 9. Quantification of Occupational Cancer (Cold Spring Harbor Laboratory, Cold Spring Harbor, N.Y., 1981)

14. Cancer Facts and Figures. 1983 (American Cancer Society,New York,1982)

15. National Research Council - Diet, Nutrition and Cancer (National Acade my Press, Washington D.C. 1982)

16. Ames B. - Dietary Carcinogens and Anticarcinogens - Science 221,1256, 1983

17. Waterlow J.C. - But et utilization des apports et allocations Alimentaires recommendés - Cah.Nutr.Diet. 13,253, 1978

18. Waterlow J.C. - Use of recommended intake and the purpose of dietary recommandations, Food Policy 1979

19. Select Committee on Nutrition and Human Needs - U.S. Senate - Dietary Goals for the United States, Washington, D.C.: U.S. Government Printing Office (1st. ed.) february 1977

20. Select Committee on Nutrition and Human Needs - U.S. Senate - Dietary Goals for the United States, Washington, D.C. : U.S. Government Printing Office (2nd ed.) december 1977

21. Royal Norwegian Ministry of Agriculture - Report no. 32 to the Storting on Norwegian nutrition and food policy (1975/1976)

22. Harper A.E. - Dietary Goals - a skeptical view, A.J.Clin.Nutr., 31,310, 1978

23. Hegsted D.M. - Dietary Goals - a progressive view, Am.J.Clin. Nutr., 31, 1504, 1978

24. Passmore R., Hollingswort D.F., Robertson J. - Prescription for a better British diet - Brit.Med.J. 1, 527, 1979

25. James P.W. personal communication

26. Peri C. - Il contributo della tecnologia alimentare alla evoluzione dei consumi ed al miglioramento della qualità dei cibi - XIX Congresso Naz. di Medicina Sociale, Verona, ott. 1983

27. U.S.F.D.A. - Toxicological Principles for the safety Assessment of direct Additives and Color Additives used in Food - Bureau of Food, 1982

28. Vettorazzi G. - International Program of Chemical Safety, IPCS, Internal Report, 1983

29. Tannenbaum S.T. - Integration of chemical and biological changes in foods and their influence on quality, Principles of food science, ed by O.R. Fennema Part. I Food Chemistry, Marcel Dekker, Inc. New York and Basel 1976

30. Mc Laren D.S.: Nutrition and its disorders: 3° ed. Churcill Livingstone. Edinburgh (1981)

NUTRIENTS CONSIDERED TO BE WORTHY OF EXAMINATION IN PROCESSED FOOD

K. PIETRZIK
Institute of Nutritional Science
Dpt. Pathophysiology of Human Nutrition
University Bonn

Summary

The recommendations regarding critical nutrients considered to be worthy of examination in processed food are preceded by some explanations for an objective evaluation of the nutritional status to provide our industrial colleagues the reasons why some of the nutrients are of importance.

Basing on the nutritional supply situation in Europe the recommendations are made especially for vitamins (thiamine, riboflavin, pyridoxine, folate, ascorbic acid, vitamins A and D), minerals (iodine, iron, calcium) and dietary fibre to be worthy of examination. As all the nutrients are of importance, there is no possibility to place them in order of priority. It must be taken into account whether or not methods exist for determination of the nutrients under consideration, whether or not a diminished intake of the nutrient in question may create serious problems in a population group.Additionally the examination of vitamin content depends whether or not a food item contributes substantially to the vitamin supply of the population group. Food items are listed according to their vitamin contents and to their contribution to the daily vitamin intake.

If a choice has to be made, unstable vitamins have to be considered in first instance. Finally some special aspects of bioavailability and analytical difficulties are discussed.

1. Introduction

Classical nutrient deficiency diseases resulting in clinical symptoms are rare among people in European countries. The relatively few cases found are usually associated with clinical diseases. There are, however, "modern" nutrition problems associated with overconsumption of food and alcohol and there is a high risk in coronary heart diseases combined with hyperlipemia, high blood pressure, diabetes etc..Cirrhosis of the liver, gallstones, pancreatitis, constipation and varicosis are as well common nutrition related health problems.

Though the major nutrition problems are caused by overconsumption, lack of nutrients is also found. Especially the vitamin and mineral intake is often inadequate -even in industrialized countries- and alterations of specific biochemical and hematological parameters indicate the existence of subclinical deficiency symptoms.

Furthermore, there are special population groups that might be at risk with regard to specific nutrients. The elderly, for example, have a reduced food intake and a consequent need for foods higher in nutrient density. They depend to a great extent on the preservation of nutrients in such manufactured foods. There are not only changes in the nutrient content of processed foods but there are changes in consumption patterns as well. Products that were previously seasonal may now be obtained all over the year and new products appear on the market. Our food consumption changes with our changing life-style. These changes may be modest from one year to the next, but over a longer period they become drastic and result in new nutritional problems.

2. Nutrients considered to be worthy of examination

The diet of European countries is characterized in several ways. On the one hand we have overconsumption on the other hand we have a low need of energy in case of decreasing physical activity and a relatively high intake of refined foods. According to Brubacher (1) an average individual may eat with daily diet of 2400 kcal (10 MJ) about 90 g sugar, 30 - 40 g white flour, rice and so on and drink about half a litre of beer or two or three glasses of wine. Under these conditions there are only 1100 - 1300 kcal (4,6 - 5,4 MJ) left in which all necessary nutrients besides energy have to be incorporated and it will be difficult to provide a nutritionally adequate diet.

On the occasion of a congress entitled "Undernutrition in Middle Europe?" in 1981 in Augsburg an expert group (2) reported the following situation (figure 1).

Nutritional Deficiencies in Middle Europe					
	well proved	probable	possible●	affected group	preventive measure
iron			+	young people women reproduct. age pregnant women	appropriate food " ~ med. substitution
iodine	+			whole population	iodized salt
calcium (in relation to P-intake)			+	young people and old	milk and milkproducts
dietary fibre		+		food naturally rich in dietary fibre generally desired	
vitamin A			+	young and old men pregnant women	appropriate food
vitamin B1			+		"
vitamin B6			+	women (pill?)	"
folate			+		"
vitamin C			+	old people, smokers	"

●further information see text

Figure 1. Results from round table discussion on the occasion of the congress "Mangelernährung in Mitteleuropa?" (2)

The most severe deficiency situation concerns iodine and goitre is a problem all over Europe except in those countries where iodized salt is generally consumed by the population.

Because of the relatively high intake of refined foods there is fibre deficiency in the diet and due to this, up to 30 % of the population suffer from constipation (3).

Also mineral and vitamin intake is often inadequate and deficiency might be possible in different groups of people.

Increasing evidence suggests that iron deficiency even in the absence of obvious anemia may affect health (4, 5). Special groups of population have increased needs for iron, for example women, because of additional requirements due to reproduction and children due to rapid growth. According to a WHO report (6) also work capacity is impaired by low levels of hemoglobin. The amount of iron in the diet is usually higher than man's nutritional requirements, but the poor absorption and bioavailability of most of the iron in present day diets is responsible for the insufficient iron status of large segments of the population (6, 7, 8).

In this context ascorbic acid is a powerful promoter of iron absorption. Vitamin C acts not only as a reductant but also binds non-heme iron in the diet in equimolar concentrations to form a readily absorbed complex. To prevent iron deficiency and anemia we have to look for both the nutrients as the promoting effect of ascorbic acid on the iron absorption is dose-dependent (9, 10, 11).

In contrast to this, the heme iron absorption is much better and independent from ascorbic acid. The better absorption can be ascribed to the fact that heme is taken up into the mucosal cells as such. This means that ascorbic acid plays a particularly critical role in diets in which little or no meat is present. In considering the iron nutritive value of those diets it is therefore essential to have a knowledge of their ascorbic acid contents(6).

The reported incidence of vitamin deficiency varies widely and absolutely contrary results concerning the frequency of vitamin deficiency are found in the literature (● fig. 1). The reason for these differences in estimating the vitamin status is to be found in the lack of objective criterions. Since it is well recognized that some period of depletion of body stores precedes the manifestation of classical clinical symptoms of nutrient deficiency, nutritionists have elaborated sensitive diagnostic tests which indicate the different stages of deficiency viz.: 1.latent vitamin deficiency, 2. the sub-clinical level, 3. the marginal vitamin deficiency level and 4. the clinical deficiency stage.

The latent vitamin deficiency is no problem as long as the vitamin supply and requirement remain constant. But from the health point of view the subclinical level should no longer be ignored as in this case vitamin dependent enzymes or hormones are reduced. The marginal vitamin deficiency status is characterized by morphological or functional disturbances before the well-known classical symptoms of vitamin deficiency appear leading finally to irreversible damage of the tissues.

Basing on the criteria to objectify vitamin deficiency on the subclinical level and the marginal vitamin deficiency level we have numerous reports that thiamine intake as well as pyridoxine and folate intake is often insufficient in European countries. According to the last German nutrition report (12) there is a frequency of thiamine deficiency (α ETK 1,15) between 7,2 % and 32,3 % depending on the group tested (young pupils 7,2 %, industrial workers 13 %, women aged 18 to 46, 22,6 % and finally a representative population group from Heidelberg 32,3 %).

The pyridoxine status (α EGOT 2,0) is in 10 to 30 % of the tested groups on a subclinical deficiency level. And folate deficiency seems to be the most common hypovitaminosis even in industrialized countries (ranging between 12 and 66,5 % (12)). According to our own findings about 10 to 15 % show morphological alterations in the blood picture (hypersegmentation of granulocytes) (13) corresponding to folate deficiency which indicates a marginal vitamin deficiency as the last stage before the classical folate deficiency symptom in form of megaloblastic anemia will develope.

The data given in the nutrition report of FRG are comparable to those published in other European countries. Fidanza, for example, investigated the nutritional status of an elderly Italian group of people and published similar results. Additionally he pointed out that nearly 50 % of very old people (70 years and over) have no adequate vitamin A intake (14) compared with the data for younger people given in the above mentioned nutrition report (1-3 % deficiency based on serum retinol level 30 µg/dl, and 12-20 % deficiency based on serum retinol level 35 µg/dl).

In figure 1 calcium is mentioned possibly to be inadequate in the daily diet especially in young people and in the elderly. Additionally "Sub-group 4 - Nutrition" recommends to look for vitamin D in the diet as there is a significant correlation between ingested cholecalciferols (CC) and serum levels of its metabolite 25-hydroxycholecalciferol (25-OH-CC) (15). Though there is also a significant correlation between serum levels of 25-OH-CC and the amount of sunlight , there are several reports of rickets in different European countries indicating thereby that the exposure to sunlight often might not be quite sufficient. Especially in those countries where a basic enrichment of baby food with vitamin D is not performed the problem is more severe than in other parts of Europe(16).

Another problem of an inadequate provision of vitamins and minerals for man is in case of calorie reduced diet as a lot of people have to diminish their calorie intake in accordance with their reduced physical activities to prevent obesity. In this context Mareschi et al. (17) tested menues on different caloric levels (e.g. 1500, 2500, 3000 kcal). The vitamin content of the menues was calculated on the basis of food consumption tables considering nutritional losses resulting from usual cooking methods.

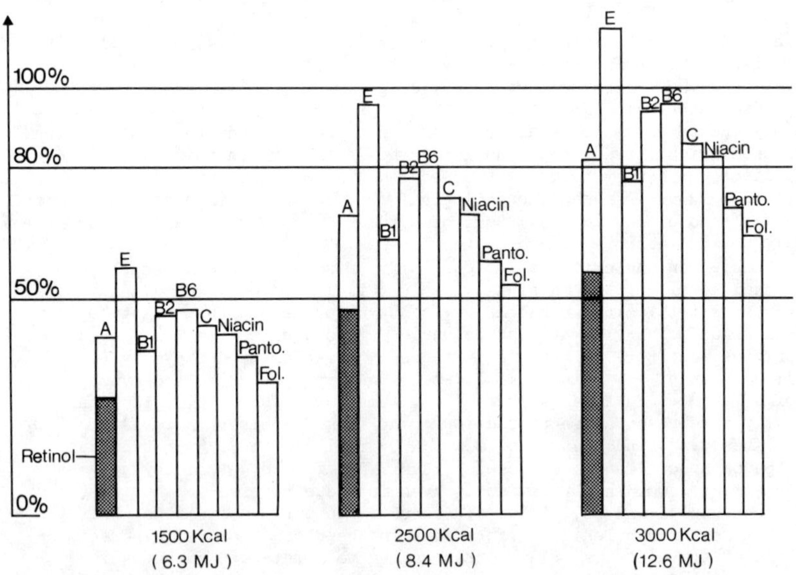

Figure 2. Vitamin intake on three caloric levels based on french food habits(100 % level = Recommended Dietary Allowances) modif. from Mareschi et al. (17)

The results are given in figure 2. They show that some vitamins as folic acid and the vitamins A, B$_1$ and C had a high risk of inadequacy in food intake. The results of epidemiologic surveys, made in other European countries are in agreement with the results of Mareschi's study. The threshold of 80 % of the recommended allowances is not reached with a caloric level of 2500 kcal (10,5 MJ). For an allowance of 1500 kcal (6,3 MJ) most of the vitamins do not or just reach 50 % of the recommendations.

In face of these findings "COST 91-Sub-group 4-Nutrition" recommends the following suggestions concerning critical nutrients.

3. Suggestions concerning critical nutrients

Taking into account

- whether or not food items contribute substantially to the vitamin supply of the general population or of a population group at risk,
- whether or not methods exist for determination of the nutrients under consideration,
- whether or not diminished intake of the nutrients in question may create serious problems in a given population group,

the following recommendation is made by "COST 91-Sub-group 4-Nutrition":

Main emphases should be given to vitamins, minerals and dietary fibre, according to the following list.

- Vitamins: (thiamine, riboflavin , pyridoxine, folate, ascorbic acid, vitamin D, and vitamin A including β-carotene)
- Minerals and trace elements: (iron including iron/ascorbic acid relationship, iodine and calcium)
- Dietary fibre

In table 1 food items and their contribution to the daily intake of vitamins, minerals and dietary fibre are listed.

Daily Intake of Calories and Nutrients (%) from Different Foods (Table 1)

	Cereals	Roots and Tubers	Vegetables	Fruit	Meat and Offals	Fish and Seafood	Milk
Kcal	25.86	4.56	1.93	2.87	14.96	0.92	8.98
iron	22.53	8.24	12.64	4.95	28.02	2.75	2.75
vitamin A*	–	–	–	–	38.24	1.31	28.76
vitamin B1	28.41	10.80	9.66	3.98	28.98	1.14	8.52
vitamin B2	7.82	3.35	8.38	3.35	19.55	1.68	42.46
vitamin B6	14.12	14.88	7.34	7.38	24.45	1.57	5.98
ascorbic acid	–	26.15	45.39	23.85	–	–	–
folate	19.99	9.08	20.58	9.21	7.55	0.37	5.54
calcium	5.60	1.98	7.70	3.38	1.98	2.45	68.26
dietary fibre	36.42	11.74	13.88	17.61	2.85	–	–

* retinol + β-carotene – insignificant intake

The per caput intake is taken from the FAO Food Balance Sheets of Western Europe from 1977 (18) except the data for pyridoxine, folate and dietary fibre which are taken from the German Nutrition Report (3). For the determination of folate in food there are no good methods available (microbiological assay). In so far we don´t know the real intake of folate.

Average Daily Nutrient Intake from Different Foods in %

vit A	daily intake: 1.3 mg
1. meat products	26.58
2. meat	18.03
3. fresh vegetables	10.75
4. butter	9.51 %
5. oils and fats	9.34
6. eggs	6.79
etc.	19.00

vit B1	daily intake: 1.2 mg
1. meat	29.41
2. cereals	18.85
3. potatoes	10.07
4. meat products	8.31 %
5. flour products*	7.34
6. milk	6.61
etc.	19.41

vit B2	daily intake: 1.6 mg
1. milk	24.07
2. meat	17.42
3. meat products	10.01
4. eggs	8.48 %
5. cereals	7.72
6. cheese and curds	7.11
etc.	25.19

vit C	daily intake: 76.5 mg
1. fresh vegetables	24.28
2. potatoes	23.31
3. tropical fruit	15.25
4. refresh. beverages	11.99 %
5. fresh native fruit	11.21
6. vegetable products	6.12
etc.	7.84

iron	daily intake: 11.40 mg
1. cereals	26.58
2. meat products	15.70
3. meat	10.18
4. flour products*	8.64 %
5. potatoes	6.54
6. eggs	6.36
etc.	26.00

calcium	daily intake: 652.2 mg
1. milk	40.82
2. cheese and curds	17.79
3 cereals	8.54
4. sweets	4.49 %
5. fresh vegetables	3.23
6. refresh. beverages	3.16
etc.	21.97

dietary fibre	daily intake: 18.23 g
1. cereals	36.42
2. flour products*	14.95
3. potatoes	11.71
4. fresh native fruit	10.26 %
5. fresh vegetables	8.17
6. vegetable products	5.68
etc.	12.81

table II

* flour products = Nährmittel

Additionally folate in food is characterized as a mixture of pteroylpolyglutamic and pteroylmonoglutamic compounds, and absorption is mainly from pteroylmonoglutamate (80 %); polyglutamic compounds remaining nearly unabsorbed. As long as we don't know the proportion of the different compounds there is no reliable basis for the calculation of intake (absorption). Nevertheless it can be derived from the listed food items potentially rich in folate.

Though data for iodine are available (3, 19), they are not being mentioned, as recent findings indicate that these data are often not to be relied upon (20).

As the major source of human vitamin D is through the conversion of 7-dehydrocholesterol by ultraviolet radiation from sunlight, the daily intake of adults is not specified. To prevent rickets in children is recommended to enrich baby food with vitamin D.

In table 2 only those nutrients are listed which guarantee reliable data, as good methods exist. In this case the food items are given according to their contribution to the daily intake of the respective nutrient. The percentage of nutrient intake in table 2 is based on data from the german nutrition report (3). Calculations on the basis of the food balance sheets of western Europe (FAO) sometimes show different results but the sequence of the listed food items in general is the same.

If a choice has to be made, unstable vitamins have to be considered in first instance. The potentially deleterious agents are given in figure three.

The most sensitive vitamins are ascorbic acid and folate. Their maximal losses will be about 100 % depending on the exposure to the different agents. Cooking for two minutes, for example, destroys folate to 80 % in vegetables (21). During longer cooking and warmholding of meals there is a complete destruction of folate.

Ascorbic acid as well might be destructed completely because it is unstable at higher temperatures and in alkaline medium. The presence of heavy metals and the exposure to oxygen contribute in destructing the vitamins.

The other vitamins of importance are vitamin B_1 and B_2 which are also very sensitive, as maximum losses will reach to 80 resp. 75 %.

	pH7	<pH7	>pH7	oxygen	light	temperature	maximum losses(%)
vitamin A	●	↓	●	↓	↓	↓	40
vitamin B1	↓	●	↓	↓	●	↓	80
vitamin B2	●	●	↓	●	↓	↓	75
vitamin B6	●	●	●	●	↓	↓	40
vitamin B 12	●	●	●	↓	↓	●	10
vitamin C	↓	●	↓	↓	↓	↓	100
vitamin D	●		↓	↓	↓	↓	40
vitamin E	●	●	●	↓	↓	↓	55
vitamin K	●	↓	↓	●	↓	●	5
biotin	●	●	●	●	●	↓	60
folate	↓	↓	●	↓	↓	↓	100
pantothenic acid	●	↓	↓	●	●	↓	50

● stable ↓ unstable

Figure 3. Potentially deleterious agents for vitamins and maximum losses

In case of thiamine besides of SO_2 an alkaline medium is regarded to be the most deleterious agent.

Riboflavin as well as pyridoxine are especially unstable in light,in case of pyridoxalphosphat it reacts with free amino-groups. Vitamin A is unstabile in an acid medium and by exposure to oxygen as well as in presence of heavy metals.

Thermal processing, exposure to light and oxygen in home cooking as well as in **catering** must be considered as the most deleterious agents for vitamins.

From these findings it should become understandable that in different groups of people -as shown above- the vitamin supply often must be diagnosed as inadequate.

In case of new products, which are likely to replace existing foods, those should be examined for their nutrient contribution to the diet compared with foods they may replace.

If a drastic change occurs in the nutrient content in a food, which is an important source of a specific nutrient, it might affect the nutritional status of large segments of the whole population. In many European countries potatoes, for instance, supply 20 - 40 % of the vitamine in the diets and if a new variety with a low vitamin C content will be introduced, it might seriously affect the vitamin C intake of the population.

In general food technologists developing new strains of potatoes, tomatoes etc. for canning purpose, for example, have to look for changes in the nutrient content, too.

4. Special problems

To objektify the nutrient changes due to cooking and thermal processing there seems to be an urgent need for improving some of the commonly used analytical procedures. The procedures for the determination of some of the vitamins as e.g. folate and pyridoxine are so inaccurate that comparisons among different investigations in various countries are questionable. To calculate the contribution of vitamin in food to the daily vitamin supply, it is essential to know the different natural forms of vitamins occuring in the food. As mentioned above the bioavailability of folate, for instance, depends to a great extend on the binding with glutamates, as polyglutamates nearly remain unabsorbed.

A similar problem is the bioavailability of iron. Though there is enough iron in the diet, the absorption of food iron may be as low as 5 % and only under favourable circumstances -such as the presence of sufficient ascorbic acid in the diet- it exceeds 10 %. In view of modern eating habits including much processed food and in case of reduced energy intake accompanied by low content of vitamins, it often becomes difficult to cover the physiological requirements for iron from diet, especially for women. From the nutritional point of view it is important to enhance the bioavailability of the non-heme iron which is the main component of the dietary iron. Knowing the iron/ascorbic acid relationship in food, a better calculation of the supply situation might be possible.

It is also generally agreedthat the assessment of dietary fibre is not very accurate. And finally there seems to be an obvious need to reach an agreement on the methods of assessment of iodine content in food. Different food tables give absolutely different iodine concentrations (milk, fish, beverages incl. alcohol (19, 22, 23).

In fish, for example, there are only few data available concerning the iodine losses during processing (cooking, canning, etc.)(23).

On the basis of insufficient data a reliable calculation of the nutrient intake in some cases is still impossible.

Therefore, the developments in the whole food production should not only be the concern of the food technologists and food chemists but of the nutritionists as well.

References

1. BRUBACHER, G.B. (1978), Vitamin B_1 supply in industrialized countries. in: "The importance of vitamins to human health" (1978) Proc. 4th Kellog Nutr. Symp. London
2. SCHLIERF, G., WOLFRAM, G. (1982), Mangelernährung in Mitteleuropa? Wiss. Verlagsgesellschaft, Stuttgart
3. Material zum Ernährungsbericht 1980, Deutsche Gesellschaft für Ernährung e.V., Frankfurt/M.
4. JACOBS, A. (1977), The nonhematological effects of iron deficiency. Clinical Science and Molecular Medicine 53, 105-9
5. BHASKARAM, P., SIVAPRASAD, J., KRISHNAMACHARI KAVR. (1977) Anemia and immune response. Lancet, i: 1000
6. WORLD HEALTH ORGANISATION (1975), Control of nutritional anemia with special reference to iron deficiency, Technical Report Series No. 580. Report of an IAEA USAID WHO Joint Meeting
7. INTERNATIONAL NUTRITIONAL ANEMIA CONSULTATIVE GROUP (INACG) (1977), The Nutrition Foundation, New York and Washington, 1-29
8. INTERNATIONAL NUTRITIONAL ANEMIA CONSULTATIVE GROUP (INACG) (1981), Iron deficiency in women. A report prepared for INACG by TM. Bothwell and RW. Charlton. The Nutrition Foundation, 1-68
9. BJÖRN-RASSMUSSEN, E., HALLBERG, L., ISAKSSON, B. and ARVIDSSON, B.(1974) Food iron absorption in man. Applications of the two-pool extrinsic tag method to measure heme and non-heme iron absorption from whole diet. J. Clin. Invest. 53:247
10. DERMAN, D.P., BOTHWELL, T.H., MACPHAIL, A.P., TORRANCE, J.D., BEZWODA, W.R., CHARLTON, R.W. and MAYET, F.G.H. (1980), Importance of ascorbic acid in the absorption of iron from infant foods. Scand. J. Hemat. 25:193
11. SAYERS, M.H., LYNCH, S.R., JACOBS, P., CHARLTON, R.W., BOTHWELL, T.H., WALKER, R.B., and MAYET, F. (1973), The effects of ascorbic acid supplementation on the absorption of iron in maize, wheat and soya Brit. J. Hemat. 24:209
12. Ernährungsbericht 1980, Deutsche Gesellschaft für Ernährung e.V., Frankfurt/M.
13. PIETRZIK, K., URBAN, G., HÖTZEL, D. (1978) Biochemische und haematologische Maßstäbe zur Beurteilung des Folatstatus beim Menschen, Intern. Z. Vit. Ern. Forschung 48, 391-396
14. FIDANZA, F., LOSITO, G. (1981) Nutritional status of the elderly, Bibliotheca Nutr. Dieta, 30, 70-80
15. HADDAD, J.G., HAHN, T.J. (1973), Natural and synthetic sources of circulating 25-hydroxyvitamin D in man. Nature, London, 244:515
16. WIDHALM, K. (1982) "Mangelernährung im Kindesalter" in: "Mangelernährung in Mitteleuropa?" Wiss. Verlagsges.Stuttgart,
17. MARESCHI, J.P., COUSIN, F., DE LA VILLEON, B., BRUBACHER, G.B. (1982) Caloric value of food and the adequate provision of vitamins for man. Unpublished
18. Food Balance Sheets of Western Europe, FAO (1977)
19. SOUCI, S.W., FACHMANN, W., KRAUT, H. (1981) Food composition and nutrition tables 1981/82, ed. by Deutsche Forschungsanstalt für Lebensmittelchemie, Garching b. München, Wiss. Verlagsges. Stuttgart
20. TOLLE, A. (1983) "Colloquy on iodine" chaired by A. Tolle, Institut für Hygiene, Bundesanstalt für Milchforschung, D 2300 Kiel

21. ROETZ, R., MERKER, H.-J. (1972) Die Bedeutung der Folsäure für Schwangerschaft und Entwicklung des Menschen, Therapiewoche 22,4,255
22. HAENEL, H. (1979) Energie- und Nährstoffgehalt von Lebensmitteln, Verlag Volk und Gesundheit, Berlin,GDR
23. MONTAG, A., GROTE, B., (1982) Untersuchungen zur Jod-Brom-Relation in Lebensmitteln, Z. Lebensm.Unters.Forsch. 172:123-128

LABORATORY ANALYSIS AND BIOLOGICAL AVAILABILITY OF NUTRIENTS

A.E. Bender
Department of Nutrition, Queen Elizabeth College, University of London, England.

Summary

The nutrients present in foods are usually determined in the laboratory by relatively rapid and reproducible chemical and physical methods. These, however, are responses to specific parts of the molecule of the nutrient and may be enhanced or repressed by interfering substances. The 'true' assay is a biological one but apart from its lack of precision and high cost the results may not easily be extrapolated to man. In the past such difficulties have not presented practical problems of any importance but the introduction of novel foods and novel processes which may change our traditional diets will require investigation before current laboratory methods can be accepted as useable indexes of the bioavailability of nutrients for human beings.

1. Introduction

We are continually developing new processes, making use of different ingredients and seeking new varieties of foods with higher yields and other advantages. Public health authorities need to monitor such changes in case the nutritive value of traditional diets deteriorates.

When analysing foods the information sought is the nutritional contribution they make to the human diet, sometimes termed bioavailability. This, in practise, is usually impossible to ascertain and, instead, we have to use one of a variety of chemical, physical, microbiological and animal assays which can serve only as an index of the usefulness of the nutrient for man.

The 'true' assay of a nutrient is the biological one, in which the defined response of the test animal is measured. If vitamin E is defined as that substance which prevents the resorption of the fetus (in an experimental animal) then the amount of food required to carry out this function (compared with a standard) is, by definition, a true measure of the vitamin E content of the food. Such assays are long, laborious and subject to the enormous errors due to biological variation. For example in an investigation to determine the potency of diastereoisomeric forms of alpha-tocopherol, Weiser and Vecchi (1) used six groups of rats each at a different dose level of standard, and another six groups, each on a different dose of the material under test. A preliminary stage of vitamin E-depletion took 3 months verified by mating at 4-weekly intervals. When no living embryos could be detected after day-20 of pregnancy then the animals were considered prepared for the assay. This involved mating them and sacrificing them after 20-days of pregnancy in order to examine the fetuses. The dose-response relation of test substance compared with standard, using protection against resorption as the criterion, provides the basis for the results.(The complexity is emphasised by the fact that apart from their isomers, there are four tocopherols and four tocotrienols which are vitamin-E active).

Compared with such bioassays chemical and physical measurements are precise, rapid and reproducible but the presence of other substances may enhance or interfere with the results since the procedure only measures a colour reaction with a particular chemical grouping or the physical effect of a particular portion of the molecule. Even when partial purification of an extract of a food is carried out we are still attempting to measure one part of the nutrient in many hundreds or thousands of parts of a wide variety of other substances.

However, over the years we have come to accept these chemical/physical methods as providing a useful index of nutrient content, holding the expensive and inaccurate bioassay in reserve for confirmation in a few, special instances. Even when the lengthy biological assay has been carried out we are still faced with the almost insurmountable drawback that we have to extrapolate from the animal to man.

2. Examples of problems

2.1. Iron.

Iron serves as an example of some of the problems. The availability of iron in foods is a well-established problem, although its extent is not always realised(2). The WHO Report, 1972 (3) and standard text books (4), corrected later (5) provide a figure showing the proportion of total iron in various foods found to be available from human experiments using doubly-labelled iron (Figure 1). This shows, by a divided rectangle, the proportion of iron absorbed and implies that the rectangle indicates the mean and standard error or deviation. The original publications (6) show that they are, in fact, the antilogs of the log standard error on each side of the geometric mean - a statistical device to deal with the enormous variation in the results. The mean absorption of iron ranges from 1% from rice, 3% from beans, to 8% from soya and 20% from veal muscle. However, the individual values cover the enormous ranges shown in Figure 2.

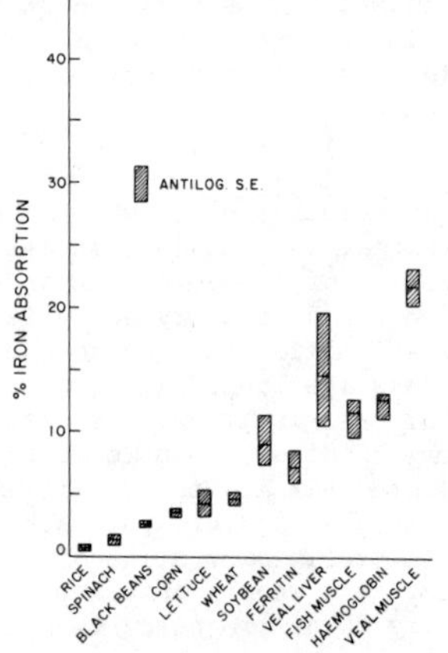

Fig. 1: % Iron Absorption from Foods.

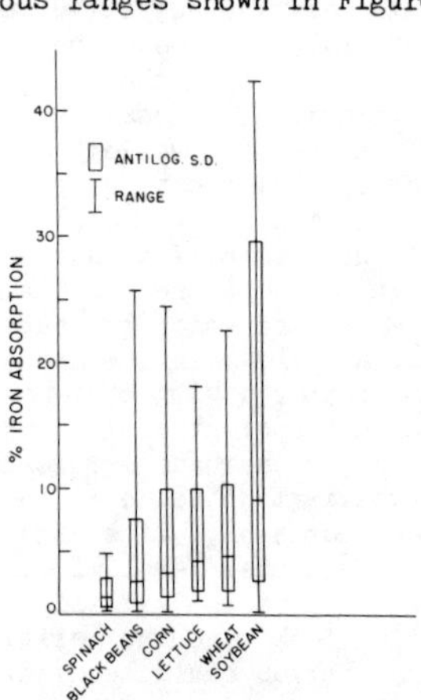

Fig. 2: % Iron Absorption from Foods.

The absorption of iron depends on three groups of factors. 1) The chemical form of the iron in the food and the amount of iron in the diet. 2) Other foods consumed at the same time. Absorption is enhanced by ascorbic acid, an acid medium, and by fructose and amino acids which chelate with the iron,and decreased by phytates, oxalates and phosphates. 3) The physiological state of the subject, including iron reserves, concentration of acid in the stomach and intestinal motility. Measurements carried out directly on human subjects, even were they possible under routine conditions, show such enormous variations as to render them virtually valueless. Layrisse and Matinez-Torres (6) used labelled iron and found that in a group of 28 individuals under controlled conditions, the amount of iron absorbed from soybeans ranged from 2% to 42.2%!

The problems of iron availability, which are widely appreciated, are in fact, not very different from those for vitamin A which are not generally appreciated. This also exists in foods in a variety of chemical forms, its absorption is influenced by the presence of other nutrients in the diet and the physiological state of the individual, but the figure for retinol equivalents, calculated from chemical estimation of retinol and carotenoids stated in food composition tables, is generally accepted as the amount that is available. For vitamin A we tend to regard dietary and physiological factors quite separately from the chemical forms of the vitamin, whereas for iron, we combine all the factors.

The value of the chemical and physical measurements of vitamin A has become established over many years of comparison with bioassays and is generally accepted (but there are still problems as mentioned later). However, if novel foods replace traditional foods to any large extent then these established procedures will have to be re-examined.

2.2 Energy

There is even a problem in estimating the energy available from foods. The procedure usually employed involves determination of fat, carbohydrate and protein and then calculation of the energy by using the appropriate conversion factors.

Carbohydrate can be measured 'by difference', as reducing sugars, or by subtracting dietary fibre from the total. Fat presents known problems and the question of nitrogen conversion factors for protein is ever-present.

The energy conversion factors used are usually those of Rubner (4.1, 4.1 and 9.3 respectively) or Atwater (4,4 and 9 which allow for incomplete absorption). However, the Tables of Food Composition of the United States make use of digestibility values for groups of foods. Table IA indicates the resulting confusion.

The change to SI system made a small difference, of legislative rather than scientific importance, when figures in kcal were rounded off to kjoules as shown in Table IB.

3. Current problems

We are faced with two practical problems; (1) we need to know what nutritional changes take place in current food processes and, (2) what effects the replacement of traditional foods by novel preparations may have.

Although it is the source of much comment and criticism losses during food processing cannot be detrimental since, apart from poor selection, the wide variety of processed foods has made nutrient deficiency a thing of the past. So when we look on established processes such as canning, dehydration, freezing and so on, although much effort is devoted to determining and possibly reducing processing losses such processes, judging from results, must be regarded as being nutritionally accep table; the investigations are rather to reassure the consumer and to satisfy nutritional labelling requirements than to investigate nutritional damage.

In any case the problem appears, superficially, to be relatively simple because the nutrients are measured before and after treatment and so any of the methods, chemical, physical or animal assay, should provide the required comparative information. However, recent work indicates that such is not always the case. For example, prior to 1971 it was accepted that canning of fruits and vegetables had no effect on vitamin A potency because colorimetric assay of the extracted carotenoids showed no change. It was then shown (7) that heat isomerises the carotenoids and changes their biological potencies - a change not revealed by total carotenoid determination. It is now concluded that green vegetables, which contain largely beta-carotene, lose about 15-20% of their vitamin A potency, while red and yellow vegetables, which contain largely alpha-carotene, lose 30-35% of their potency. Bioassay would reveal this (with low precision) chemical and physical assay require separation and measurement of all the isomers (assuming we already know their relative potencies).

Even more intractable problems arise with changes in protein quality since processing may render unavailable some of the essential amino acids and the only one that can be measured (with some still outstanding problems), is available lysine. If any other amino acids are rendered unavailable this would be revealed only by multiple bioassays after supplementation with a range of amino acids (again,with the imprecision of biological work).

The second problem, that of replacement of traditional foods by novel foods, and traditional foods processed by novel methods, is even more difficult to overcome. While it is unlikely that changes would be of sufficient magnitude to give rise to real concern, if there were any major change then the only true assay would be on human beings, if such is possible.

In the earlier days of measuring human requirements, such as those for vitamin A and vitamin C in Sheffield in 1942-44 (8,9) subjects were made deficient and the therapeutic dose (of the food as well as the pure vitamin) determined. Modern techniques tend to measure turnover rates of pools of labelled vitamins and would not provide this information.

Indeed the Sheffield experiments revealed the extent of the problem under discussion since they showed for vitamin A, for example, that the physical form of the food (i.e. the state of subdivision of carrots) and the presence of fat greatly influenced the amount of carotene absorbed. It has even been shown in experimental animals that vitamin E serves to protect the vitamin A after ingestion.

Consequently there is a need for continuous monitoring of the health of the communities even of well-fed western countries. A specific example has recently arisen in modern Japan, not among poor peasants in the countryside but among middle class adolescent high-school athletes. A diet based too heavily on sugar, sweetened soft drinks and polished white rice gave rise to several hundred cases of clinical beri-beri between 1975 and 1980 (10).

This should reinforce the warnings that came to light when early samples of dehydrated potatoes and a modified preparation of infant food were marketed. The potato supplies up to half the average intake of vitamin C in winter in many European countries and some of the earlier (early 1960's) preparations of instant potato were low in, or devoid of vitamin C. The only change was drying the cooked potato - hardly a novel process.

In the case of the infant formula, treatment at a higher temperature, again hardly a novel process, damaged so much of the vitamin B6 that a number of cases of convulsions in infants were reported.

Clearly, while there may be no cause for immediate concern there is always a danger of complacency.

Table 1A Comparison of Energy Content of Foods Calculated by Three Methods (Kcal)

Food	4.1;9.3;3.75[a]	4;9;4.2[b]	FAO 1947 Merrill and Watts (1955)[c]
Bread, brown	242	245	251
Wholemeal flour	339	336	327
Butter	793	768	748
Milk	66	65	68
Beef steak	177	172	177
Apple	45	45	55
Bananas	77	76	103
Currants, black	29	28	79
Oranges	35	35	49
Beans, raw butter	266	264	350
Potatoes	87	86	92
Groundnuts	603	586	576

a - protein x 4.1; fat x 9.3; available carbohydrate as monosaccharides x 3.75
b - protein x 4; fat x 9; available carbohydrate as starch x 4.2
c - specific factors for different foods (Food and Agriculture Organisation).

Energy-yielding components of foods and computation of calorific values. 1947. UN, FAO, Washington DC; Merrill, A.L. and Watts, B.K. (1955) Energy Value of Foods - Basis and Derivation. US Dept. of Agriculture, Agr. Handbook No. 74. Washington D.C.

Quoted from Paul, A.A. and Southgate D.A.T. 1978. McCance and Widdowson's The Composition of Foods. 4th Ed. MRC Spec. Rpt. Ser. No. 297. HMSO. London.

Table IB Atwater Energy Conversion Factors Currently is Use (kJ)

	Germany	United Kingdom
Fat	38	37
Middle-chain length triglycerides	34	-
Protein	17	17
Carbohydrate	17	16
Ethanol	30	29
Organic acids	13	-

REFERENCES

1. WEISER, H. and VECCHI, M. (1981) Stereoisomers of alpha-tocopherol acetate; characterisation of the samples by physico-chemical methods and determination of the biological activities in the rat resorption-gestation test. Internat. J. Vit. Nutr. Res. 51. 100-113.
2. LOCK, S. and BENDER A.E. (1980) Measurement of chemically-available iron in foods by incubation with human gastric juice in vitro. Br. J. Nutr. 40. 413-420.
3. WORLD HEALTH ORGANISATION (1972) Tech. Rep. Ser. No 503. Geneva.
4. DAVIDSON S. et al. (1975) Human Nutrition and Dietetics, 6th Edn. Churchill-Livingstone.
5. DAVIDSON, S. et al (1979) Human Nutrition and Dietetics, 7th Edn.
6. LAYRISSE, M. and MARTINEZ-TORRES, C. (1971) Food iron absorption; iron supplementation of foods. Prog. Haematol. 7. 137-160.
7. SWEENEY, J.P. and MARSH, A.C. (1971) Effect of processing on provitamin A in vegetables. J. Amer. Dietet. Assoc. 59. 238-243.
8. HUME, E.M. and KREBS, H.A. (1949) Vitamin A requirements. Spec. Rep. Ser. Med. Res. Coun. London. No. 264.
9. BARTLEY, W., KREBS, H.A. and O'BRIEN, J.R.P. (1953) Vitamin C requirements. Spec. Rep. Ser. Med. Res. Coun. London. No. 280.
10. ABE, T. (1982) A relapse of beri-beri in Japan. Appendix in Methodologies for Human Population Studies in Nutrition Related to Health. NIH Publ. No. 82-2462 USA.

METHODS OF EVALUATING PROTEIN QUALITY

F. FIDANZA
Istituto di Scienza dell'Alimentazione
Università degli Studi di Perugia

Summary

The most suitable methods in food technology for protein nutritional quality evaluation are critically reviewed. The advantages and drawbacks of the following procedures are examined: In vivo methods: a) single level assays (protein efficiency ratio, net protein ratio, net protein utilization); b) multi-level assays (relative nutritive value, relative protein value). In vitro methods: a) amino acid profile alone or in combination with digestibility (chemical score, computed protein efficiency ratio); b) enzymatic methods (enzymatic ultrafiltrate digest); c) microbiological assays; colorimetric methods for available amino acids and dye-binding procedures. In synthesis the criteria for the selection of an appropriate protein assay procedure are also provided.

A primary function of dietary protein is to supply the organism with a mixture of amino acids in the correct proportions for the synthesis and maintenance of tissue proteins. Consequently the nutritional quality of the protein depends on how closely it can supply these amino acids in accordance with the specific needs of the organism. These needs may change according to age and physiological function, implying, therefore, that the quality of the protein is dependent upon such different physiological states (4). In addition to the nutritional quality of the protein, the amount is obviously important.

The preferred method of assessing protein quality depends on the purpose of the determination, such as for planning a food and nutrition policy, agricultural research (particularly plant breeding), regulatory requirements, and to determine the effects of food processing and storage. But in the present paper mostly those suitable in food technology will be considered.

The evolution of the methodology is fully described in the several reviews and books published in the last few years (2; 3; 10; 12; 13; 15) and so will not be further discussed.

There are two main groups of methods of evaluating protein quality: in vivo or biological, based on animal models including man, and in vitro- chemical, microbiological and enzymic.

Those based on animal models can be classified into single-level assays and further sub-divided mainly into techniques using body weight change and nitrogen retention.

Of the techniques using body weight changes the most common are:
1.1) Protein efficiency ratio (PER), which represents the ratio of the body weight gain over a four-week period to the protein intake. This method is the official procedure for regulatory purposes in some countries (USA and Canada), but presents fundamental drawbacks. The results are dose-dependent and the values increase with the increasing intake of food, plateauing at the maximum growth rate and then falling slowly at higher intakes. Because it is based on weight gain, PER does not take into consideration the requirement of protein for maintenance, and consequently values are not proportional; i.e., a protein with a PER of 1 does not have a quality half of that of a protein with value of 2. The variations in food intake, as in other factors related to it, increase the variability of PER estimates, decreasing the precision of the method particularly for poor quality proteins. The inter-laboratory variability is considerable and even when the correction factor is applied with an assumed PER value of 2.5 for casein (an internal standard) the results of inter-laboratory comparisons are not improved. (Samonds and Hegsted, unpublished data). The variation in body composition is not taken into account, because PER is based on weight changes only. This can be an important factor that has to be verified further. This method is considered simple, but time consuming and of moderate cost.
1.2) Net protein ratio (NPR), is an improvement of PER by the inclusion of a group of animals consuming a non-protein diet for a similar period of time (now in general 10 days), in order to take into account the requirement of protein for maintenance (1). NPR is based on the tacit assumption that the response is linear between the zero intercept and the point at which curvature begins to occur. This is not valid for lysine-limiting proteins that will have a higher nutritive value. The laboratory-to-laboratory variability is slightly better than for PER and the result does not vary with the amount of protein eaten; the simplicity is the same, less time-consuming and the cost lower. A recent comparison with PER showed a very good agreement particularly for values of leguminous seeds (19).

Of the techniques depending on nitrogen retention the most widely used is:
2) Net protein utilization (NPU) that measures the amount of the fed protein retained in the body of the rat. In order to consider the requirements of protein for maintenance, a non-protein group of rats is included. The difference between the protein in the bodies (NPU carcass) of the non-protein group and the fed group is a measure of the amount retained. If digestibility (D) is also measured then Biological Value can

be calculated from NPU/D.

In order to ensure that protein quality is the limiting factor, the level of the protein in the diet is fixed at about 10%. At higher levels quantity can compensate for quality. Similar results were obtained measuring body water instead of N, if the ratio N/H_2O is known.

The principle of this method is similar to the NPR. In NPR the weight change is measured, with NPU, the nitrogen retention. To determine NPU nitrogen balance can also be measured, and retention is estimated indirectly via excretion.A distinction between NPU standardized (NPU st) and NPU operative (NPU op) has been proposed. The NPU st is determined at low level of protein, usually 10%; while NPU op is obtained at the level of protein occurring in the food or diet in question and is thus an overall measure of the food as consumed (13). The same drawbacks observed for NPR apply to NPU. The cost, in experienced laboratories, is lower than for PER determination.

Other methods of the same classes as above or based on specific responses (plasma or liver protein regeneration, liver protein utilization, liver enzymes, blood urea levels, etc.) are very seldom used and they need much further experimentation.

Because in the above methods the linearity from zero intake to some preselected level in certain cases has not been demonstrated, in recent years multi-level assays have been proposed (14). The most useful techniques are:

3.1) Relative nutritive value (RNV) in which three levels of protein are selected which fall along the relatively linear portion of the curve where protein is limiting. In addition, a zero-protein group is included. RNV is then calculated by dividing the slope of a test protein by the slope of the reference protein, lactalbumin (14). Because of lack of linearity from zero intake of protein limiting in lysine and to a lesser degree in methionine (for conservation and reutilization of lysine when intake is low), modification of this method has been proposed in which the zero intake point is omitted.

3.2) Relative protein value (RPV) in which the slope-ratio is calculated at graded levels of protein intake in the range of linearity, omitting the zero protein intake. The analytical part of this method, shortened to 14 days, has recently been simplified by autoclaving the carcass and then homogenizing it in a Waring blender. If faeces are collected for the last five days, digestibility can also be determined (6). The laboratory-to-laboratory variability was the smallest found in an inter-laboratory comparison of four biological methods (Data of Samonds and Hegsted, unpublished).

In a recent comparison RPV and NPU were found to agree very well if the body nitrogen is the parameter of response. Minor differences were observed with gluten and faba bean protein concentrate (6).

From the experience of various laboratories, RVP can be considered at the moment to be the most reliable method of measuring the nutritional quality of a given protein. But problems remain, such as that of selecting the reference protein, of determining the effect on the response of the non-protein energy content of the diet and of the effect of the second limiting amino acid (6).

Table I , taken from Pellet and Young (13), provides comparative information on some criteria to be used for the selection of an appropriate rat assay procedure.

Further limitations of all biological methods on rats are that they are affected by other interfering materials, and that a single index is used for multiplicity of purposes, while extrapolation of data from animals to man (if man is the final consumer) is always questionable.

For this last reason human experimentation has been recommended in recent years. Most of the methods used on rats can be used with man, but the response is evaluated with nitrogen balance or weight change. Apart from the cumbersomeness of the procedure (in fact human experimentation can be carried out only in experienced and specialized laboratories), ethical consideration (particularly with children) and difficulties of interpretation, there are many technical problems that need further investigation (studies on man have not reached the same degree of standardization as animal procedures). Last, but not least, is the problem of time and cost. More practical are some biochemical indices potentially sensitive to protein quality like amino acid levels in plasma muscle, metabolites in plasma or urine, and enzyme activity. Further studies are needed on all these potential methods.

The *in vitro* methods have the advantage of being rapid, accurate in some respects, widely-applicable, relatively simple and inexpensive.

At the moment they can be grouped into the five following classes.

4.1) *Assays utilizing the amino acid profile alone or in combination with digestibility*. The chemical score is the most widely used. The content of each essential amino acid in a protein (after chemical hydrolysis) is expressed in proportion to the content of corresponding amino acids in a reference protein or amino acid pattern. The amino acid with the lowest proportion is the limiting amino acid and the ratio is the chemical score. The limitation is of course referred to the synthesis of body proteins and other metabolically important nitrogenous metabolites. Various modifications of this method have been suggested but the original one is still valid. As reference protein, instead of the egg protein firstly proposed (and the result termed protein score) it is preferable to use the provisional amino acid pattern recommended by the joint FAO/WHO expert group (7) (termed chemical score). Because the availability of amino acids is not taken into account, recently the chemical score has been corrected for digestibility.

In line with this approach is the recent method of the Computed PER (C-PER), proposed particularly in countries where the PER is still an official method (16). This requires determination of the essential amino acid (EAA) profile of the sample protein and of a reference casein and their *in-vitro* digestibilities.

TABLE I

SOME CRITERIA FOR THE SELECTION OF AN APPROPRIATE RAT ASSAY PROCEDURE

Criterion	PER	NPR	NPU (carcass)	NPU (balance)	RNV	RPV
A direct measure of digestibility	no	no	no	yes	no	no
Rank protein	poor	moderate	moderate	moderate*	good	good
Ability to discriminate	moderate	moderate	moderate	moderate*	good	good
Reproducibility in other labs	poor	moderate	moderate	N/A**	good	good
Applicability to proteins very low in lysine	poor	poor	poor	poor	poor	moderate
Proportionality	poor	poor	moderate	moderate	moderate	poor
Test for linearity	no	no	no	no	yes	yes
Simplicity	simple	simple	simple	complex data collection	complex statistical analysis***	moderate
Time required (days)	28	10-14	10	9	14-21	14-21
Relative cost	moderate	low	moderate	moderate	high	high

* Good if corrected to a standard NPU value for casein plus methionine.

** N/A - Insufficient information available on collaborative assays using NPU (balance).

*** Requires multiple regression analysis.

The *in-vitro* digestibility is measured on an aqueous suspension with a combination of trypsin-chymotrypsin-peptidase at 37°C, pH 8. After exactly 10 min a bacterial protease is added, the solution is transferred to a 55°C water bath for nine minutes; then the fall in pH is recorded at 20 minutes with the temperature adjusted to 37°C.

For the C-PER each EAA is expressed as a percentage of the FAO/WHO (7) standard and corrected for the *in-vitro* digestibility. The percentages less than 100% of the FAO/WHO standard are multiplied by a constant increasing in value as the sample's EAA percentage of the FAO/WHO standard decrease. This calculation then forms a vector from which a harmonic mean is used to obtain the sample's average EAA profile expressed as a percentage of the FAO/WHO standard EAA profile. This harmonic mean is then compared to a single, similarly -computed value for the control casein, giving the sample value as a percent of casein (arbitrarily adjusted to PER 2.5). Then the C-PER is calculated using a quadratic equation.

The C-PER model was constructed from samples having corrected rat PERs ranging from 0.67 to 3.22. The standard error of estimate of the C-PER is about ±0.36 PER units. The C-PER assay can be completed in 72 hr; provides a reasonably accurate prediction of PER (the C-PER values are very close to PER values obtained from the rat assay); is independent of the level of additive spice, carbohydrate and/or fat in the food being tested; is sensitive to trypsin inhibitors and changes in protein structure, which occur during processing; is much less expensive than the rat PER assay.

4.2) *Enzymatic methods*. The evaluation of this method has been fully described by Mauron (11) and only the more recent developments will be reported here; the Enzymatic Ultrafiltrate Digest (EUD) amino acid index, (8).

The sample, diluted to contain 1 g of protein is incubated and stirred at 37°C for 6 hr, with pepsin at pH 1.8. After adjustment of pH to 7.8 and volume to 50 ml, an aliquot of 10 ml is transferred to an Amicon stirred ultrafiltration cell, thermoregulated at 37°C, with UM2 Diaflo Membrane. After addition of 0.2 ml of a 1% solution of trypsin grade A and 2 ml each of pancreatin and erepsin extracts the ultrafiltration is carried out for 6 hr at a speed of 8 ml/hr. The collected ultrafiltrate is adjusted to 50 ml with water. After hydrolysis of glutamate and asparagine with HCl, the available amino acids are determined chromatographically according to Spackman et al. (18) with an amino acid analyser. Total amino acids determined after hydrolysis are used to calculate an integrated index index, using egg protein as reference, according to the method of Sheffner (17) for the Pepsin Digest Residue amino acid index (PDR), in which the most important factors are the pattern of essential amino acids in the enzymatic and residue stages and the amount of amino acids as a group containing the particular pattern. The release of amino acids with ultrafiltration *in vitro* has been checked with that *in vivo* in the rat and good agreement obtained.

Comparison of this method with RPV using 19 protein preparation has shown a correlation coefficient of the order of 0.8.

With this method, which allows results in about two working days, we have obtained very good results on line during bread, biscuits and alimentary pastes technology, with raw and cooked foods and with frozen dishes.

4.3) Microbiological assays. The most suitable microbiological method seems to be that using the ciliated protozoan requiring the same ten amino acids as the rat: Tetrahymena pyriformis. More recently Tetrahymena termophila is preferred, shortening the incubation time from 96 to 66 hr and determining the growth with a Coulter counter. This protozoan is sensitive to food additives and spices, so limiting its application.

Microbiological methods have been used for availability of selected essential amino acids, a method particularly useful for the plant breeder (9).

4.4) Colorimetric methods for available amino acids. Processing renders some essential amino acids unavailable, particularly lysine and sulfur-containing amino acids.

The available lysine can be determined with fluorodinitrobenzene (5) but this method is not applicable to products rich in carbohydrates.

The procedures proposed for determination of cysteine and methionine and their oxidized forms after enzymatic hydrolysis or in the intact protein need further studies, particularly in comparison with biological assay values. (For details see 13).

4.5) Dye-Binding Procedures. These procedures are rapid and inexpensive and can be useful for damaged lysine amino groups. Very promising in this regard are azo dyes and Acid Orange 10 or 12. There are technical problems and much further experimentation is needed.

In addition to particular drawbacks for each assay examined above, the *in vitro* methods present additional problems such as the rate of release of amino acids during digestion, possible different utilization of amino acids and proteins, role of non-specific nitrogen,role of toxic materials, and amino acid imbalance.

Table II from the highly recommended publication of Pellet and Young (13) gives some criteria for the selection of an appropriate protein nutritional quality assay procedure. No comments are needed; the only modification lies in the simplicity of the technique for the multi-level assays, which is less complex than some of the other modifications previously described.

In agreement with these authors and other experts in this field the practical suggestion that can be given for protein nutritional quality evaluation in food processing and handling is to start with *in vitro* assays, only measuring amino acid availability if this is expected to be involved, or preferably with amino acid scoring that include digestibi-

TABLE II

SOME CRITERIA FOR THE SELECTION OF AN APPROPRIATE PROTEIN ASSAY PROCEDURE

Criterion	Amino acid scoring	Amino acid scoring and digestibility	Micro-biological	Human and rat assays	
				single-level	multi-level
Suitability for screening	yes	yes	yes	no	no
Sample size required	small	small	small	moderate	large
Expense	low	low	low	moderate (rat) high (human)	moderate (rat) high (human)
Simplicity of technique	simple	simple	simple	moderate	moderate
Can, by itself, give information on limiting amino acid and complementary potential	yes	yes	no	no	no
Value obtained is inclusive of digestibility and/or availability	no	yes	yes	yes	yes
Affected by toxins, stimulants, and/or inhibitors	no	no	yes	yes	yes
Ability to discriminate between proteins	moderate	moderate	moderate	moderate	high

lity or enzymatic methods. Then if time, money and sample size are not limiting factors or there are some specific reasons the biological assays represent the final step.

REFERENCES

1. BENDER A.E. and DOELL B.H. (1957). Biological evaluation: a new aspect. Brit. J. Nutr. 11: 140.
2. BODWELL C.E., ed. (1977). Evaluation of proteins for humans. Avi Publishing, Westport, USA.
3. BODWELL C.E., ADKINS J.S. and HOPKINS D.T., eds. (1980). Protein Quality in Humans Assessment and in Vitro Estimation. Avi Publishing Co., Westport, USA.
4. BRESSANI R. (1977). Human Assays and Application, in BODWELL C.E., ed.: Evaluation of Proteins for Humans, Avi Publishing Co., Westport, USA, pp. 81-118.
5. CARPENTER K.J. (1960). The Estimation of Available Lysine in Animal Protein Foods. Biochem. J. 77: 604.
6. CORCOS BENEDETTI P., TAGLIAMONTE B., BISES G., GENTILI V. and SPADONI M.A. (1981). Evaluation of Protein Quality: Comparison of Single-Point and Multi-Point Assays. Ann. Nutr. Metab. 25: 48.
7. FAO/WHO Ad Hoc Expert Committee (1973). Energy and Protein Requirements. FAO Nutrition Meetings Report Series No. 52, Rome.
8. FLORIDI A. and FIDANZA F. (1975). Food Protein Quality. III Enzymatic Ultrafiltrate Digest (EUD) Amino Acid Index. Riv. Sci. Tech. Alim. Nutr. Um. 5: 13.
9. FORD J. (1980). Microbiological Assays. In BODWELL C.E.,ADKINS J.S. and HOPKINS D.T., eds.: Protein Quality in Humans: Assessment and in Vitro Estimation. Avi Publishing Co., Westport, USA.
10. FRIEDMAN M., ed. (1977). Protein nutritional quality of foods and feeds. M. Dekker, Inc., New York.
11. MAURON J. (1970). Nutritional evaluation of proteins by enzymatic methods. In BENDER A.E. ed.: Evaluation of Novel Protein Products. Pergamon Press, Oxford.
12. PELLET P.L. (1978). Protein Quality Evaluation Revisited. Fd. Technol.32 (5): 60.
13. PELLET P.L. and YOUNG V.R. (1980). Nutritional Evaluation of Protein Foods. United Nation University, Tokyo.
14. SAMONDS K.W. and HEGSTED D.M. (1977). Animal Bioassays: A Critical Evaluation with Specific Reference to Assessing Nutritive Value for the Human. In Bodwell C.E. ed.: Evaluation of Proteins for Humans. Avi Publishing Co., Westport, USA.
15. SATTERLEE L.D., ed. (1977). The Midland Conference - New Concepts for the Rapid Determination of Protein Quality. Fd. Technol. 31 (6) 69.
16. SATTERLEE L.D., MARSHALL H.F. and TENNYSON J.M. (1979). Measuring Protein Quality. J. Am. Oil Chemists' Soc. 56: 193.

17. SHEFFNER A.L. (1967). In Vitro Protein Evaluation. In ALBANESE A.A. ed.: Never methods of nutritional biochemistry. Vol. 3, Academic Press, New York.
18. SPACKMAN D.H., STEIN W.H. and MOORE S. (1958). Automatic Recording Apparatus for Use in the Chromatography of Amino Acids. Anal. Chem. 30: 1190.
19. WOLZAK A., ELIAS L.G. and BRESSANI R. (1981). Protein Quality of Vegetables Proteins as Determined by Traditional Biological Methods and Rapid Chemical Assays. J. Agric. Food Chem. 29: 1063.

THE DETERMINATION OF VITAMINS IN FOODS

A report of the working group
on vitamin assay in foodstuffs

G.B. BRUBACHER AND W. MÜLLER-MULOT
Department of Vitamin and Nutrition Research
F. Hoffmann-La Roche & Co. Limited Company, Basle, Switzerland

Summary

Vitamin content may be a criterion of food quality. An approved collection of suitable methods for determining vitamin content is therefore highly desirable, yet no such collection exists. One approach to this problem is described below. In line with suggestions elaborated by a group of vitamin experts in the course of a workshop, a manual has been prepared which describes in detail the recommended, state-of-the-art methods (vitamins A, B_1, C and E and ß-carotene). The manual also provides information on methods not yet thoroughly tested or applicable only in certain cases (vitamins B_2 and D). Finally, for three vitamins only some references are given, since for the time being no recommendations can be made (vitamins B_6, PP and folic acid). Vitamin K, pantothenic acid, vitamin B_{12} and biotin have been omitted altogether since these four vitamins are of minor interest in connection with food quality. It is hoped that the methods described will be used by nutritionists and food chemists. Comparative, multi-centre trials involving other methods should be inaugurated so that the manual can be revised in one or two years. This will pave the way for approval of the methods by official or scientific bodies.

Vitamin content may be a criterion of food quality and is a function of internal and external factors such as genetic capacity and environmental conditions during growth, on the one hand, and storage and processing conditions on the other in the case of vegetables. Approved methods for vitamin determination are therefore highly desirable in all cases in which the results of such investigations are to be compared. As no such collection of approved methods exists, a group of vitamin experts met in March 1981 to set the requirements which such methods must meet. The working group consisted of the following members: A. Blumenthal, Institut für Ernährungsforschung, Rüschlikon, Switzerland; F. Brawand, Schweizerisches Vitamininstitut, Basle, Switzerland; G. Brubacher, F. Hoffmann-La Roche & Co. Ltd., Basle, Switzerland; F. Fidanza, Istituto di Scienza dell'Alimentazione, Perugia, Italia; U. Manz, F. Hoffmann-La Roche & Co. Ltd., Basle, Switzerland; W. Schüep, F. Hoffmann-La Roche & Co. Ltd., Basle, Switzerland; D. Scuffam, Laboratory of the Government Chemist. London, U.K.; P. Scheffeldt, Institut für Ernährungsforschung, Rüschlikon, Switzerland; D.A.T. Southgate, Food Research Institute, Norwich, U.K.; J.P. Vuilleumier, F. Hoffmann-La Roche & Co. Ltd., Basle, Switzerland; P. Walter, Schweizerisches Vitamininstitut, Basle, Switzerland.

In vitamin assay of foodstuffs two points must be considered:

1) Each group of vitamins consists of a number of different chemical compounds, sometimes called vitamers. These chemical compounds can be divided into two classes. There are compounds which can be interconverted by simple chemical or biochemical reactions, e.g. vitamin A palmitate may be saponified to retinol or all-trans-vitamin A may be isomerized to cis-isomers. On the other hand there are compounds which cannot be interconverted by simple means, e.g. α, β and γ tocopherol, or α and β carotene. A comprehensive method would allow determination of each chemical compound separately. In this way the fate of each compound during food processing and storage could be followed. Unfortunately, no feasible comprehensive methods exist. The existing methods can be used only in special investigations but are not suitable for large-scale screening.

For the first class of compounds, a pragmatic approach is to convert all compounds to the same compound and determine only the concentration of the latter. For the second group, either all vitamers would have to be determined together (disregarding the individual vitamers) or the main vitamer would have to be determined alone.

2) In most cases the analytical value provides only limited information on the biological efficacy of a given vitamin, since bioavailability is also a factor to be taken into account. Bioavailability depends not only on the chemical nature of the vitamers considered but also on the composition of the meal. The working group therefore felt that bioavailability should be disregarded in this context.

The group of experts decided to prepare a manual describing recommended, state-of-the-art methods. The manual contains information on sampling, preservation and transport of the sample to the laboratory. Practical procedures are described in detail in order to ensure reproducibility of the method in the lab. The methods were judged according to the following criteria of reliability:

a) Reproducibility (intra-laboratory repeatability and inter-laboratory comparability)
b) Correctness (comparison with other methods)
c) Sensitivity
d) Specificity
e) Robustness (Influence of small changes of the laboratory procedure).

A practical approach was chosen. In other words, even where not all aspects were investigated, the method was approved on the basis of the committee's own experience. Information on preservation and transport was too scanty to include these points in the description of the methods. Therefore, only general instructions are given in the introduction.

The manual has been prepared on these lines. It contains recommended methods for determination of vitamins A, B_1, C, E and β-carotene. These methods can be used in a variety of foodstuffs. Sufficient knowledge has been accumulated on the criteria mentioned above.

Methods for determination of vitamins B_2 and D have also been described but must be regarded with certain reservations because our knowledge is insufficient for accurate judgment.

Lastly, for vitamins B_6, PP and folic acid, only references are given since for the time being there is no recommended method.

More data on reproducibility, correctness, sensitivity, specificity and robustness are required. Nutritionists and food chemists are therefore invited to compare the methods described in the manual with their own techniques, to organize multi-centre trials and to refer their results to the working group on vitamins. It is hoped that with these new data the manual can be revised in one or two years and that official or scientific bodies will approve the final version.

The methods described in the manual should be employed in investigations where the results are to be compared with the results of similar investigations. Alternatively, they may be used as reference methods, where simpler methods are available or where the laboratory's own method will be retained for one reason or another. The methods described in the manual will be of value in two main fields: establishing food composite tables and providing data for nutritional labelling of food.

A COMPARISON OF NUTRITIONAL CHANGES TAKING PLACE DURING MANUFACTURE AND STORAGE OF FOODS COMPARED WITH CHANGES TAKING PLACE IN DOMESTIC PREPARATION AND CATERING

G. Varela
Instituto de Nutrición
Facultad de Farmacia
Madrid -3. SPAIN

Summary

Food undergoes changes in nutritive value (NV) after production until consumed and metabolized by man. Cooking is in current fashion, and this is important for nutrition. As an example, three of the four working subgroups of COST-91 deal with cooking. We shall study here the nutritional changes taking place during manufacture and storage of foods as compared with those taking place during industrial or home cooking. We will start by seeing why cooking is popular today and the arguments in favor of and against home cooking and catering. The fact is that, as a rule, we do not like to eat in the collective dining room (CDR) and we prefer the home. Sometimes the aims of home cooking and catering have been shown as opposite although they actually are complementary and coinciding in many aspects. Manufacturing and storage of food can complement catering and home cooking. In our view, home cooking is "industrializing" while the aim of catering is to make prepared food look "home made".

1. INTRODUCTION

Cooking is in current fashion in the developed countries. This fact is obviously important for nutrition. In this paper we intend to point out the reasons for this popularity.

As an example of this interest, of the four working subgroups of the COST-91 Project, the 1st deals with extrusion cooking, the 2nd with industrial cooking and the 4th, on nutrition, has devoted special interest to home cooking.

This Project has shown along the years that, even though catering and home cooking have their own specific problems, coordination among the various working groups would be very useful. This was the theme of the Hudderfield Meeting in 1982 (2).

This work is aimed at comparing nutritional changes in manufacture and storage of food with those taking place in both catering and home cooking processes.

This difficult taks is facilitated by the excellent and modern literature on the subject, such as the reviews by G. GLEW (4), P.A. LACHANCE (5) and R. ZACHARIAS (9) and the papers of the 1981 CEE London Symposium (3). As to the general effects of processes on the NV of food, A.BENDER's book (1) is a valuable and well known reference.

2. CHANGES TAKING PLACE DURING MANUFACTURE AND STORAGE OF FOODS

We have recently (6) dealt with the influence of industrial, culinary, digestive and metabolic processes on the nutritive value of food after their obtention from earth or water (Potential Nutritive Value) until they are metabolized (True Nutritive Value).

This diagram shows the conventional division of such changes in two phases, separated by the moment of intake. Here we will try to place the culinary and industrial processes within the general scheme.

With regard to dietetics and its position in the scheme, we shall note that in the developed countries eating has three facets: health, pleasure and food habits. Cooking has a relevant place in the latter two.

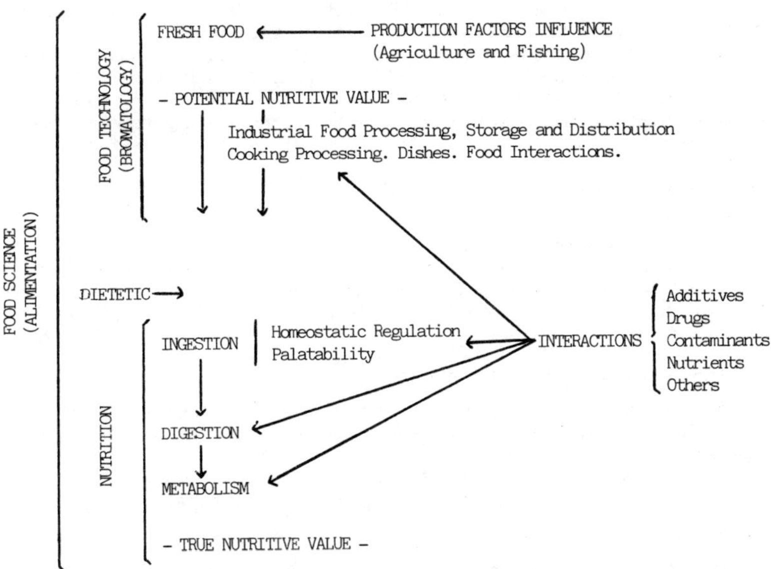

There is an extensive literature on the changes in nutritive value (NV) of food during manufacture and storage, very well summarized in the above mentioned work by A. BENDER (1). Two general conclusions may be drawn from these papers: 1) Such processes, if well performed, should not practically affect the NV of food; and 2) The advantages offered by such processes offset by far the possible drawbacks.

3. CURRENT INTEREST IN COOKING IN GENERAL

This interest is possibly due to the following reasons:

1) Cooking food is a biologic feature of man.
2) Culinary techniques are a social and cultural heritage.
3) The high value reached by gastronomy in the developed world. The pleasure of eating.
4) Need of a cooking designed for collective feeding. Are we going towards a "novissima" cooking?.
5) Week-end home cooking.
6) The need to know the nutritional changes of food during cooking.

As all processes, cooking has advantages and drawbacks. From a practical point of view, the benefits outweigh the disadvantages.

Benefits of culinary processes in general

1) Destruction of most of the toxines and pathogenic germs.
2) Increase of nutritional use of food as to palatability, digestibility and metabolicity.
3) Combination of foods favor covering dietary recommendations of energy and nutrients.
4) Possibility of manipulating the diet with specific purposes.

One of the most important aims of cooking is to make possible and increase the nutritive value of food in the phases relevant for its utilization: palatability, digestibility and metabolicity.

Culinary treatments are essential for acceptance of foods. Some examples of this effect are presented in this paper. Palatability is substantially increased by frying. This is probably the reason why deep-frying techniques are being applied to new areas where it is not normally used or even considered more harmful than other culinary techniques. Reference is made in this connection to previous papers of the author (7) showing how frying of fod enhances palatability without affecting in practice its digestive and metabolic utilization.

Disadvantages of culinary processes in general

1) Possibility of formation of toxic substances.
2) Loss of nutritive value.

When discussing these disadvantages, we can draw the general conclusion that negative effects have no practical bearing if the cooking processes are carried out in a correct manner and, in any case, the overall result is truly favorable and beneficial to man.

4. COLLECTIVE FEEDING AND INDUSTRIAL COOKING

We are witnessing a revolution in food habits, because are shifting from the home to collective dining rooms. This fact raises very complex problems, not only from the nutritional point of view.

As a rule, we do not like to eat in collective dining rooms and we prefer to do it at home. Each type of behavior has however its advantages and drawbacks.

a) Nutritionals

1. Better possibilities of estimating RDA's in energy and nutrients.
2. " " " knowing true intakes.
3. " " " complementing the home diet.
4. " " " follow-up of the group's nutritional state.
5. " " " controlling food quality.
6. Better control of culinary processes (temperature, time, food/water ratio, etc.).
7. Easy analysis of NV changes in compound dishes.
8. In many instances better NV than in home cooking.
9. Better possibilities of correctly preparing therapeutical diets.
10. Policy aimed at changing food habits, if appropriate (new food products of technologies).

b) Toxicological

1. Increased safety in raw materials, processes and control of finished products.

c) Social and economic

1. Increased labor productivity
2. Reduced cost per meal.
3. Essential for non-split working hours.

Disadvantages of the collective dining room

1. Nutritional } The risk of toxicity or malnutrition is graver since it would affect
2. Toxicological } many people.
3. Palatability: acceptance is one of the main problems in collective feeding.
4. Psychological: probably the most difficult to solve.
5. Technological: those resulting from manipulation of large amounts of food and the time elapsed between preparation and consuming.

These advantages and disadvantages should be taken into account by the nutritionist in charge of the management and programming of a collective dining room. Some nutritional criteria for collective feeding, as summarized below, should also be taken into account:

a) Groups receiving the whole daily diet (homes for the elderly, military units, hospitals, etc.).
b) Groups receiving only a part of the daily diet (schools, factories, etc.).

1. It is neccesary to know the nutritional status of the population to which the group belongs, so as to be able to:
   1.1. Complement nutritionally the food consumed at home.
   1.2. Follow the food habits of the group in planning menus.
2. Even though RDA's are given by day, remember that it is not necessary to adjust the daily intake to those values. Balancing the amounts over larger periods enlarges diet planning alternatives.
3. Try to determine the actual and optimal amounts of collective dining as a percentage of the total daily food intake,
4. When assessing the possible losses in NV due to the culinary processes, the following factors should be taken into account:
   4.1. Error causes in NV determinations.
   4.2. Nutritional status of the group for each nutrient.
   4.3. Importance of the contribution of nutrients of a certain food product for the total of the diet.

5. HOME COOKING VERSUS INDUSTRIAL COOKING?

The aims of home cooking and industrial cooking have sometimes been termed as opposite. We do not share this approach. In our view, they have many things in common and are complementary in all cases.

We think that the main differences that can be found between the two ways of cooking are the following three processes: reheating, warm-holding and repeated frying (RF). Special mention is made of RF in the paper, since reheating and warm-holding are subjects dealt with by subgroup 2 of COST-91.

The diffusion of frying techniques has brought about the need of industrializing them. One of the major problems is the so-called "useful life" of fats, related to the number of fryings possible without toxicity problems or a marked decrease in palatability. This subject has been studied by us (8).

It should be recalled that the extensive information on RF refers to frying performed with industrial or home fryers, but there is very little information of frying in a frying pan as is usually done at home. Therefore we conduct our work in the laboratory and with families controlled by us using frying pans.

6. FINAL REMARKS

It is true that home cooking is a traditional cooking. But this does not mean that there have been no changes, many of which have come from procedures used in industrial cooking.

All this has led to a deep revolution in home cooking, which becomes "industrialized". On the other hand, one of the aims of catering is to make the meals as similar as possible to those prepared at home. For this reason, there are many features common to both types of cooking and this makes it necessary to have a cooperation and coordination among nutritionists and technicians in these areas, without forgetting what they owe to the houswife, who is the star performer of the kitchen.

Let us recall finally that the common objective is to make people enjoy eating, in all circumstances of their life and work and with the available food sources.

7. POSSIBLE RESEARCH PATHS

In accordance with what has been stated, we suggest as useful a more deep investigation in the following areas:

1) Search for appropriate culinary treatments for the new catering systems.
2) Search for appropriate culinary processes for the so-called "new foods", particularly single cell foods which will not be readily accepted by man without adequate cooking, in spite of their high nutritive value.
3) Deepening the scientific knowledge of traditional dishes of the various countries. This knowledge could lead to identifying the factor of acceptance with practical results applicable to food technology.

These suggestions refer to industrial cooking as well as to the modern home cooking.

REFERENCES

1. BENDER, A.E. (1978). "Food Processing and Nutrition". Academic Press. London.
2. CEE. COST-91. (1982). "Nutrient Control in Catering. Available and Missing Data. Minutes of the workshop". Huddersfield (UK).
3. COMMISSION DES COMMUNAUTES EUROPEENNES. (1981). Symposium sur la Nutrition, la Technologie des Produits alimentaires et l'information nutritionelles". London (1980). Rapport EUR-7085 FR.
4. GLEW, W. (1980). "The contribution of Large-Scale Feeding Operations to Nutrition". Wld. Rev. Nutr. Diet., 34, 1.
5. LACHANCE, P.A. (1975). "Effect of Food Procedures on Nutrient Retention with Emphasis upon Food Services Practices". In: R.S. Harris y E. Karmas. "Nutritional evaluation of Food Proceesing". Pub. Westport. Cap. 16.
6. VARELA, G. (1982). "Influencia de los procesos industriales, culinarios, digestivos y metabólicos sobre el valor nutritivo de los alimentos". Libro del II Congreso Nacional de la Sociedad Española de Nutrición. Granada.
7. VARELA, G. (1982). "Nutritional aspects of home frying". In. CEE. COST-91. "Home Cooking nutrient changes and emerging problems". Proceeding of workshop. Roma.
8. VARELA, G., MOREIRAS-VARELA, O. y RUIZ ROSO, B. (1983). "Utilización de algunos aceites en frituras repetidas. Cambios en la grasa y análisis sensorial de los alimentos fritos" Grasas y Aceites, 34, 101.
9. ZACHARIAS, R. (1977). "Effects of domestic and large scale cooking on the quality and - nutritive value of vegetables and fruits". In: W.L. Downey. "Food Quality and Nutrition". Proceeding of COST-91 Seminar. Dublin. Applied Sciences Publ. London.

THE ROLE OF DIETARY FIBER IN THE DIET OF INDUSTRIALIZED COMMUNITIES

A.E. HARMUTH-HOENE
Federal Research Center for Nutrition, Karlsruhe
Federal Republic of Germany

Summary

Based on observations that many of the non-infectious diseases prevalent in industrialized societies are rarely known in countries of the Third World it has been suggested, about ten years ago, that a liberal intake of indigestible dietary fiber, present in cereal products, vegetables and fruits typical for developing societies, has a protective effect against such diseases. To test this hypothesis a rapidly increasing number of investigations on the various physiological and clinical effects of dietary fiber have been conducted and are still under way. Today, the postulated protective effect of fiber against chronic constipation, diabetes and perhaps diverticulosis of the colon is well documented. The role of fiber in the development of colorectal cancer has still to be clarified. Much information has been gathered on the influence of dietary fiber on colonic function and lipid metabolism, as well as interaction with essential nutrients.

1. INTRODUCTION

During the last decade hardly any other subject has received as much attention from research workers in Nutrition and related disciplines in many countries, as the role of dietary fiber in health and disease. This development was initiated through a hypothesis advanced by Burkitt and Trowell (1) and Trowell (2) that a diet rich in dietary fiber affords a protection against diseases which prevail in affluent western communities such as obesity, chronic constipation, diverticular disease, large bowel cancer, coronary heart disease and diabetes. This also implies that highly refined diets containing small amounts of indigestible material are either the causative factor or will provide favorable conditions for the development of so-called "western diseases".

This assumption was largely based on the observation that many of these diseases are rare among Africans accustomed to a diet high in dietary fiber.

With the development of modern food processing technologies in the industrialized countries, the indigestible portion of flour and other carbohydrate containing food was deliberately removed because it was considered to be nutritionally inactive and possibly irritable to the gastro-intestinal tract. At the same time food of plant origin was increasingly replaced by animal products. This resulted in a drastic decline in the daily intake of dietary fiber, for instance, in Germany from 103 g/day in 1879/81 to 25 g/day in 1972/75 (4), or in Denmark from 40 g/day in 1927 to 20 g/day in 1978. Comparable figures could be cited for other industrialized countries.

With much expectation to find an effective means to prevent or even to cure some of the diseases prevalent in industrialized communities, numerous attempts have been made. Many different approaches were used to clarify the mechanism(s) by which dietary fiber exerts its assumed protective effects. During the course of time a number of difficulties have become evident which must be taken into consideration.

a) Dietary fiber, defined by Trowell et al. (3) as the sum of lignin and the plant polysaccharides that resist digestion by endogenous secretions of the human gastro-intestinal tract, includes a wide array of polymeric substances with divergent chemical and physical properties. This not only presents difficulties in designing a simple routine analytical method but implies wide differences in physiological effects depending on source and composition of the particular dietary fiber under investigation.
b) Diets rich in dietary fiber usually have a lower energy density, and contain less fat and refined sugar, and more foods derived from plants, than diets low in dietary fiber. In epidemiological studies the effects of these associated characteristics cannot be separated from the effects of dietary fiber.
c) Many of the diseases typical for industrialized communities are developing slowly before becoming evident and are believed to be caused by several factors in addition to the lack of dietary fiber.

In view of the difficulties listed above, it is not surprising that some aspects concerning the mode of action of dietary fiber still await clarification.

In the following, some of the physiological and clinical effects of dietary fiber will be discussed.

2. OVERWEIGHT

It is widely believed that because of its low caloric availability dietary fiber can be effectively used to lower the energy density in reducing diets. Due to a high water-binding capacity and the chewing required during ingestion of dietary fiber, an increased satiety effect has been observed in human subjects in short term experiments (6,7,8). Experimental evidence on the long-term effect of dietary fiber on satiety, however, is not unanimous (8,9,10,11). The crucial question whether an increased consumption of dietary fiber is effective in the prevention or treatment of obesity has not been sufficiently clarified. To simply add some form of isolated fiber to our typical diet high in fat and refined carbohydrates, will not be sufficient. A more promising approach is a change of our whole dietary pattern to include more foods rich in natural dietary fiber, i.e. unrefined cereal products, vegetables and fruit.

3. CHRONIC CONSTIPATION

Chronic constipation is predominant among western populations, especially women, and has led to a widespread use of laxatives. There is general agreement that an increased intake of dietary fiber increases stool volume to a varying extent depending on the source, as well as chemical and physical properties of the fiber. Measuring this bulk-forming effect of 20 g of fiber from various sources Cummings et al. (12) observed an increase in fecal weight between 20 % (guar) and 127 % (bran) in 5-6 healthy volunteers.

The daily addition of 22,5 g agar-agar or guar to a diet of 12 healthy young women, increased stool volume by an average of 58 % (agar-agar) and 21 % (guar), respectively (13). Although the high water-binding capacity of fiber offered a ready explanation, comparison of the bulk-forming property of various fiber sources with their ability to bind water, did not confirm this theory (14).

Transit time of ingested substances through the entire digestive tract tends to be shortened by dietary fiber, showing an inverse relationship to stool volume (15). This effect, however, is less well documented, due to inherent variation in individual transit time.

Evidence is accumulating that the observed changes in colonic functions induced by dietary fiber are related to its rather extensive degradation by the colonic microflora. The extent of this fermentative breakdown varies with the source and properties (molecular structure, solubility, particle size) of the fiber and differs considerably between individual persons. In general cellulosic polysaccharides are digested to 50 % and non-cellulosic fiber to an even greater extent (16). The end-products of bacterial fermentation are short chain fatty acids which can be absorbed from the colon (17) and gases, mainly hydrogen and carbon dioxide. The concomitant increase in microbial cell mass is thought to contribute significantly to the increase in stool volume, following the ingestion of a fiber rich diet (14).

4. DIVERTICULAR DISEASE

Diverticulosis of the colon, a disease prevalent among industrial populations but rare in rural Africa, has been linked to a fiber-depleted diet. This disease is characterized by hypersegmentation of the bowel and high intracolonic pressure caused by excessive colonic muscular contraction. Serious complications are known to arise from this condition. Studies comparing the dietary fiber intake of patients with diverticulosis with that of non-affected individuals of the same age and sex revealed a significantly lower fiber consumption in the patient group (18,19). The value of fiber in the treatment of diverticulosis has not been clarified sufficiently. In some studies relief of the symptoms could be demonstrated (18,20) while in another double-blind investigation the effect of bran was equal to that of placebo (21).

5. LIPID METABOLISM

From numerous epidemiological studies and experimental investigations on man and small rodents evidence is accumulating that some indigestible polysaccharides have a pronounced effect on lipid metabolism. These include a lowering of serum cholesterol by pectin, guar and carubin (23,24,11), but not by wheat bran, cellulose and fiber from various vegetables, increased fecal losses of bile acids by pectin (24,25), and changes in bile salt metabolism (26). It has been suggested that this effect is mediated via bile acid-binding by certain gel-forming types of fiber, as well as an increased rate of transit which will reduce the accessibility of bile acids to enzymic degradation and consequently their reabsorption. It has been argued that altered bile acid reabsorption could be responsible for a reduction in cholesterol synthesis.

The implication of hypercholesterolemia in the development of coronary heart disease has initiated a number of studies comparing the dietary fiber of various population groups with the incidence of coronary heart disease and high serum cholesterol levels. Not surprisingly the evidence from these studies is not very convincing because of the multiplicity of factors involved in the development of coronary heart disease (27,28).

6. DIABETES

One of the important aspects of dietary fiber was the observation that the postprandial hyperglycemia in diabetics and normal subjects is reduced by guar and pectin (29,30) and, to a lesser extent, by wheat bran (31). This indicates the role of dietary fiber in the therapy of diabetes, and possibly its prevention. This has promoted numerous studies in man and experimental animals to confirm the original findings, and several theories have been proposed to explain the underlying mechanism of action. It has been suggested that dietary fiber retards gastric emptying and consequently may slow down intestinal absorption and subsequent glucose and insulin responses. The greatly increased viscosity of the intestinal contents after ingestion of gel-forming fibers was thought to hinder the diffusion of glucose to the sites of absorption (32). The findings that gum guar reduces the level of gastro-intestinal hormones, in particular gastric inhibitory peptide (GIP) which activates the release of insulin (33), offers yet another explanation for the effect of dietary fiber on the postprandial blood glucose elevation. It is quite possible that all these mechanisms might be involved. Despite these uncertainties the short-term beneficial effects of fiber supplemented diets on glucose tolerance and insulin requirements are well documented. Long-term therapeutic treatment of diabetics with high fiber diets has been successful, provided the patients strictly adhered to the diet.

7. CANCER OF THE COLON

Burkitt (34) was the first to suggest that the incidence of colorectal cancer is related to dietary factors, characteristic of industrialized societies, that may contribute to the mutagenic potential of the feces. Since then several epidemiological and case control studies have indicated that diets high in dietary fiber and low in fat and meat may play an important role in the prevention of cancer of the large bowel (35,36,37). Various theories have been advanced to explain this protective effect of dietary fiber.

a) Fecal bile concentration, which shows a striking direct relationship to the incidence of colorectal cancer may serve as a substrate for the production of carcinogens or co-carcinogens by certain bacterial enzymes in the colon. Although a number of dietary fibers have been shown to increase fecal bile acid excretion the overall concentration of bile acids in the colon is considerably reduced as a result of the increased fecal bulk.
b) There are indications that a high concentration of ammonia in the large intestine may increase the risk of tumor development. The ingestion of a fiber rich diet greatly increases microbial cell growth and subsequently stimulates conversion of ammonia into microbial protein (38). At the same time colonic pH will be lowered which may possibly inhibit nitrite production by intestinal bacteria, and thus the endogenous synthesis of carcinogenic N-nitroso compounds.
c) The increased production of short chain fatty acids, especially butyrate, by bacterial fermentation of indigestible polysaccharides has also been mentioned as a protective factor against colorectal cancer. Experimental evidence to support this proposition is still rather scanty.

Although there are several indications for the protective effects of dietary fiber against colorectal cancer the exact mechanism(s) of this effect still needs to be clarified by future work.

8. ABSORPTION OF NUTRIENTS

In view of such strong evidence for the beneficial action of dietary fiber possible negative side effects, like the interaction of dietary fiber with essential nutrients, should be considered. Due to the ion-exchange capacity and gel-forming property of most dietary fibers, a reduction in the bioavailability of essential minerals and trace elements in fiber rich diets may pose a problem. In addition, a shortened transit time through the small intestine could have a similar effect. Although a number of reports indicate decreased or negative balances of calcium, magnesium, iron and zinc caused by an increased intake of various dietary fibers, there is no general agreement, even when the same source of fiber has been used (39,40). This is not surprising because, besides level and source of dietary fiber, several other factors are known to influence mineral metabolism in human subjects. These are a) composition of the basal diet, in particular source and level of the minerals and of protein, as well as the presence or absence of food items which will inhibit or enhance mineral absorption b) the mineral status of the subjects prior to the study and c) the length of the study period. The absorption of minerals and trace elements is regulated by a complex system of homeostatic control which is quite adaptable to dietary changes. For these reasons, a severe deficiency in essential minerals or trace elements is not likely to occur, except at an extremely high intake of dietary fiber over an extended time period, combined with a low supply with essential minerals.

There is little information on the effect of fiber on vitamin availability in human subjects. It is quite conceivable, that some vitamins such as carotene and nicotinic acid and possibly others will be bound to certain fibers and thus become less available (39).

9. CONCLUSION

Evaluation of the dietary fiber hypothesis by an increasing number of research workers has broadened present day knowledge of the role of fiber in the diet of industrial communities considerably. Although there are still many open questions to be solved there is good evidence that the number of western diseases linked to the fiber content of diets is much larger than the number of those in which any other single dietary component is involved. This has changed the status of dietary fiber from an inert and useless food constituent to that of an essential nutrient.

REFERENCES

1. BURKITT, D.P. and TROWELL, H.C. (1975). Refined Carbohydrate Foods and Disease. Some Implications of Dietary Fibre. Academic Press, London.
2. TROWELL, H. (1976). Definition of dietary fiber and hypotheses that it is a protective factor in certain diseases. Am. J. Clin. Nutr. 29: 417-427.
3. TROWELL, H., SOUTHGATE, D.A.T., WOLEVER, T.M.S., LEEDS, A.R., GASSUL, M.A. and JENKINS, D.A. (1976). Dietary fibre redefined. Lancet 1: 967.
4. ROTTKA, H. (1980). Der Verzehr von Pflanzenfaserballaststoffen in der Bundesrepublik Deutschland, in: Pflanzenfasern-Ballaststoffe in der menschlichen Ernährung. H. Rottka, Ed. Georg Thieme Verlag, Stuttgart, S. 63-72.

5. HELMS, P. (1980). Der Verzehr von Pflanzenfaserballaststoffen in Dänemark, in: Pflanzenfasern-Ballaststoffe in der menschlichen Ernährung. H. Rottka, Ed. Georg Thieme Verlag, Stuttgart, S. 72-76.
6. WILMHURST, P. and CRAWLEY, J.C.W. (1980). The measurement of gastric transit time in obese subjects using 24-Na and the effects of energy content and gum guar on gastric emptying and satiety. Brit. J. Nutr. 44: 1-6.
7. BOLTON, R.P., HEATON, K.W. and BORROUGHS, L.F. (1981). The role of dietary fiber in satiety, glucose and insulin: Studies with fruit juices. Am. J. Clin. Nutr. 34: 211-217.
8. GRIMES, D.S. and GORDON, C. (1978). Satiety value of wholemeal and white bread. Lancet 2: 106.
9. EVANS, E. and MILLER, D.S. (1975). Bulking agents in the treatment of obesity. Nutr. Metabol. 18: 199-203.
10. BRYSON, E., DORE, C. and GARROW, J.S. (1979). Wholemeal bread and satiety. Lancet 2: 260-264.
11. LEITZMANN, C. (1980). Der Einfluß von Pflanzenfaserballaststoffen auf den Energiehaushalt, in: Pflanzenfasern-Ballaststoffe in der menschlichen Ernährung. H. Rottka, Ed., Georg Thieme Verlag, Stuttgart, S. 113-116.
12. CUMMINGS, J.H., SOUTHGATE, D.A.T., BRANCH, W.J., HOUSTON, H., JENKINS, D.J.A. and JAMES, W.P.I. (1978). Colonic response to dietary fibre from carrot, cabbage, apple, bran and guar gum. Lancet 1: 5-8.
13. HARMUTH-HOENE, A.E. (1980). Der Einfluß von Guarmehl und Agar-Agar auf die Stickstoffbilanz, die Resorption von Mineralstoffen und Spurenelementen und die verdauliche Energie beim Menschen. Berichte der Bundesforschungsanstalt für Ernährung, Karlsruhe 1980/5.
14. STEPHEN, A.M. and CUMMINGS, J.H. (1980). Mechanism of action of dietary fibre in the human colon. Nature 284: 283-284.
15. CUMMINGS, J.H. (1978). Physiological effects of dietary fibre in man, in: "Topics in Gastroenterology 6", Truelove, S.C. and Hayworth, M.F. eds., Blackwell, Oxford, pp. 49-62.
16. CUMMINGS, J.H. (1982). Consequences of the metabolism of fiber in the human large intestine, in: Dietary Fiber in Health and Disease. G.V. Vahouny and D. Kritchewsky, eds., Plenum Press, New York and London, pp. 9-22.
17. McNEIL, N.J., CUMMINGS, J.H. and JAMES, W.P.T. (1978). Short chain fatty acid absorption by the human large intestine. Gut 19: 819-822.
18. BRODRIBB, A.J.M. and HUMPHREYS, D.M. (1976). Diverticular disease: three studies. Brit. Med. J. 1: 424-430.
19. GEAR, J.S.S. (1979). Dietary fibre and asymptomatic diverticular disease of the colon, in: Dietary Fibre: Current Developments of Importance to Health, K.W. Heaton, ed., Technomic Publish. Co. Inc. Westport, U.K., pp. 57-62.
20. PAINTER, N.S., ALMEIDA, A.I. and COLEBOURNE, K.W. (1972). Unprocessed bran in the treatment of diverticular disease of the colon. Brit. Med. J. 2: 137-140.
21. ORNSTEIN, M.H., LITTLEWOOD, E.R., BAIRD, I.M., FOWLER, J., NORTH, W.R. and COX, A.G. (1981). Are fibre supplements really necessary in diverticular disease of the colon? A controlled clinical trial. Brit. Med. J. 282: 1353-1356.
22. KEYS, A., GRANDE, F. and ANDERSON, J.T. (1961). Fiber and pectin in the diet and serum cholesterol concentration in man. Proc. Soc. exp. Biol. (N.Y.) 106: 555-558.
23. JENKINS, P.J.A., LEEDS, A.R., NEWTON, C. and CUMMINGS, J.H. (1975). Effect of pectin, guar gum and wheat fibre on serum cholesterol. Lancet 1: 1116-1117.

24. KAY, R.M. and TRUSWELL, A.S. (1977). Effect of citrus pectin on blood lipids and fecal steroid excretion in man. Am. J. Clin. Nutr. 30: 171-175.
25. MIETTINEN, T.A. and TARPILA, S. (1977). Effect of pectin on serum cholesterol, fecal bile acids and biliary lipids in normolipedemic and hyperlipidemic individuals. Clin. Chim. Acta 79: 471-477.
26. POMARE, E.W. and HEATON, K.W. (1973). Alteration of bile salt metabolism by dietary fibre (bran). Brit. Med. J. 4: 262-264.
27. McCULLAGH, E.P. and LEWIS, L.A. (1960). A study of diet, blood lipids and vascular disease in Trappist monks. New Engl. J. Med. 263: 569-574.
28. MORRIS, J.N., MARR, J.W. and CLAYTON, D.G. (1979). Dietary fiber from cereals and the incidence of coronary heart disease, in: Dietary fibre: Current Developments of Importance to Health. K.W. Heaton, ed., Technomic Publish. Co. Inc. Westport, U.K., pp. 45-56.
29. JENKINS, D.J.A., LEEDS, A.R., GASSUL, M.A., WOLEVER, T.M.S., GOFF, D.V., ALBERTI, K.G.M.M. and HOCKADAY, T.D.R. (1976). Unabsorbable carbohydrate and diabetes: Decreased postprandial hyperglucaemia. Lancet 2: 172-174.
30. JENKINS, D.J.A., LEEDS, A.R., GASSUL, M.A., COCHET, B. and ALBERTI, K.G.M.M. (1977). Decrease in postprandial insulin and glucose concentrations by guar and pectin. Ann. Intern. Med. 86: 20-23.
31. JEFFREYS, D.B. (1974). The effect of dietary fibre on the response to orally administered glucose. Proc. Nutr. Soc. 33: 11 A.
32. ELSENHANS, B., SUFKE, U., BLUME, R. and CASPARY, W.F. (1980). The influence of carbohydrate gelling agents on rat intestinal transport of monosaccharides and neutral amino acids in vitro. Clin. Sci. 59: 373-380.
33. MORGAN, L.M., GOULDER, J.J., TSIOLAKIS, D., MARKS, V. and ALBERTI, K.G.M.M. (1979). The effect of unabsorbable carbohydrate on gut hormones. Diabetologia 17: 85-89.
34. BURKITT, D.P. (1971). Epidemiology of cancer of the colon and rectum. Cancer 28: 3-13.
35. IARC (International Agency for Research on Cancer) Intestinal Microecology Group (1977). Dietary fibre, transit time, fecal bacteria, steroids and colon cancer in two Scandinavian populations. Lancet 2: 207-211.
36. DALES, L.G., FRIEDMANN, G.D., WRY, H.K., GROSSMAN, S. and WILLIAMS, S.R. (1979). A case control study of relationship of diet and other traits to colorectal cancer in American blacks. Am. J. Epidemiol. 109: 132-144.
37. MODAN, B., BARELL, V., LUBIN, F. and MODAN, M. (1975). Dietary factors and cancer in Israel. Cancer Res. 35: 3503-3506.
38. CUMMINGS, J.H., STEPHEN, A.M. and BRANCH, W.J. (1981). Implications of dietary fiber breakdown in the colon, in: Banbury Report No. 7, Gastrointestinal cancer: Endogenous factors, R. Bruce, S. Tannenbaum, and P. Correa, eds., Cold Spring Harbor, New York, pp. 71-81.
39. KELSAY, J.L. (1982). Effects of fiber on mineral and vitamin bioavailability, in: Dietary Fiber in Health and Disease, G. Vahouny and D. Kritschewsky, eds., Plenum Press, New York and London, pp. 91-103.
40. KELSAY, J.L. (1981). Effect of diet fiber level on bowel function and trace mineral balances of human subjects. Cereal Chemistry 58: 2-5.

STABILITY OF VITAMINS A AND D IN VEGETABLE COOKING FAT DURING STORAGE

F. BRAWAND, U. OLBRECHT, R. SCHULTHESS and P. WALTER
Swiss Vitamin Institute, University of Basel
Vesalianum, Basel, Switzerland

The stability of vitamin D in foodstuffs, especially in margarines and fats is not well known because of the difficulties of its determination. Due to these analytical problems, the stability of vitamin A which can easily be determined has in many instances been taken as a monitor for the stability of vitamin D.

We have compared the stabilities of the two vitamins in a commercially available vitamin vegetable fat supplemented with 9 IU vitamin D, 110 IU vitamin A per g over a period of 42 months stored at 4°C in the dark. Vitamin A was determined according to Carr-Price (1) and vitamin D by the curative test with rachitic rats according to Bourdillon (2). No changes in vitamin content were observed after 3 and 6 months. After 11 months, vitamin D had dropped by about 40 %, whereas vitamin A still showed no decay. After 42 months, a drop of 10 % of vitamin A as compared to 50 % of vitamin D was observed. It is concluded that the stabilities of vitamins A and D are non identical in the vegetable oil tested.

REFERENCES

1. CARR, F.H. and PRICE, E.A. (1926). Biochem. J. 20, 497
2. BOURDILLON, R.B., BRUCE, H.M., FISCHMANN, C. and WEBSTER; T.A. (1931). Special Report Series Medical Research Council No. 158 (H.M. Stationery Office, London)

DETERMINATION OF ADDED VITAMIN D TO FOODSTUFFS BY HIGH PERFORMANCE LIQUID CHROMATOGRAPHY

F. BRAWAND, L. GANZONI, R. HEIZMANN, U. OLBRECHT and P. WALTER
Swiss Vitamin Institute, University of Basel, Vesalianum, Basel, Switzerland

Summary

Samples of various foodstuffs or pharmaceutics were saponified in alcoholic KOH and extracted with ether. Pre-purification of vitamin D was achieved by fractionating the ether-extracted material on a reverse phase column or by thin layer chromatography. The purified vitamin D probes were analyzed by HPLC and UV-detection at 260 nm. These results are compared to those from the determination in rachitic rats by the curative test or to the colorimetric method using antimontrichloride. Good agreement of the vitamin D values was observed between these methods in margarine, vegetable fat, tablets or capsules.

1. INTRODUCTION

In foodstuffs vitamin D is found in extremely low concentrations and usually accompanied by interfering material. As a consequence the determination of this vitamin is still a matter of debate among vitamin analysts. They key to successful analysis, however, seems to be indicated in the literature and comprises the elaboration of procedure for saponification, extraction, cleanup of the sample and final analysis by HPLC. The present paper reports a method for the assay of vitamin D based on these points.

2. ASSAY PROCEDURE

2.1 Saponification and extraction

Throughout the assay procedure samples were protected from direct light. For each sample, a parallel analysis was carried out which contained sample + defined amount of vitamin D (internal standard).

The sample (10-20 g fat, margarine; 100 g milk; 1 capsule; 1 tablet) was refluxed in a 500 ml round bottle flask for 20 min in the presence of 120 ml ethanol and 40 ml 50 % KOH. 250 mg vitamin C (or hydroquinone) was added to protect vitamin D. After cooling the mixture was extracted 3 times with 100 ml ethylether. The amount of ethylether has to be defined. The ether phases were combined and washed neutral. Sodium chloride or ammonium chloride is added if emulsions occur during the washing steps. The ether extract was evaporated in the presence of 20 mg 2,6-Di-butyl-4-methyphenol (BHT) and dried by repeated additions of ethanol. Finally the residue was brought to 400 IU D/5 ml with methanol.

2.2 Cleanup of extracts

10 IU were used in the cleanup step on a preparative HPLC column under the following reverse phase conditions:

Stationary phase:	Lichrosorb RP-18, 10 μm ; 250 mm length ; 8 mm ID
Mobile phase:	Methanol : H_2O : THF (83 : 7 : 10)
Flow:	4.0 ml/min
Temperature:	40°C
Detection wavelength:	260 nm
Retention time:	~45 min

2.3 Quantification of vitamin D

The fraction between the 40th and 50th min was collected and used for HPLC analysis on a column under the following isocratic conditions:

Stationary phase:	Lichrosorb Si-60, 7 μm ; 250 mm length ; 4,6 mm ID
Mobile phase:	n-Hexane : Isopropanol (98.5 : 1.5)
Flow:	2.0 ml/min
Temperature:	40°C
Detection wavelength:	260 nm
Retention time:	~6 min
Calculation:	The vitamin D content of each sample was calculated on the basis of the recovered vitamin D in the corresponding internal standard probe.

3. STABILITY OF VITAMIN D SOLUTIONS

Analytical samples as well as analytical samples plus vitamin D added (internal standard) were frozen and stored in the dark during 17 days. No significant decrease of vitamin D could be detected during this time. It is therefore possible to store several precleaned samples and analyze them together later on with HPLC. This procedure is only recommended if the samples are sufficiently washed to neutral before freezing.

4. RECOVERY AND REPRODUCIBILITY

Defined amounts of vitamin D were added to vitamin D_3-free margarine (gift from Dr. Manz, Roche, Basel) and analyzed according to the described method. The recovery was 95.5 ± 3 % with n = 4.

5. RESULTS AND DISCUSSION

The results with the described method were compared to those obtained by antimontrichloride (1) or by the curative rat test (2). On the whole the results (table) obtained by the different methods show satisfactory agreement. It has to be considered that vitamin D is thermally isomerized during the saponification step. The isomer (previtamin D) is separated from vitamin D on the second HPLC column and subsequently not measured. Based on the use of internal standard it appears that the amount of isomerization does not exceed 5 %.

Desintegration of material without saponification proved to be laborious and ineffective to remove lipids. In favorable cases (table**) cleanup could be achieved after saponification by thinlayer chromatography (3).

TABLE
DETERMINATION OF VITAMIN D (IU)
COMPARISON OF DIFFERENT METHODS

Sample		HPLC	biol./$SbCl_3$	
Margarine	(50 g)	526*	485	
Fat	(50 g)	422*	420	
Milk	(500 ml)	170*	135	
Capsules	(1)	235*	300	
		339*		359
Tablets	(1)	583**		629

*indicates cleanup by means of HPLC (RP-18)
**indicates cleanup by means of TLC (SiO_2)

REFERENCES

1. BROCKMANN, H. and CHEN, Y.H. (1936). Ueber eine Methode zur quantitativen Bestimmung von Vitamin D. Hoppe-Seyler's Z. Physiol. Chem. 241, 129-133
2. BOURDILLON, R.B., BRUCE, H.M., FISCHMANN, C. and WEBSTER, T.A. (1931). Special Report Series Medical Research Council No. 158 (H.M. Stationery Office, London)
3. BOLLIGER, H.R. and KOENIG, A. (1965). Quantitative Bestimmung von Vitamin D in Konzentraten, Arzneimitteln und weiteren Kombinationspräparaten mittels Dünnschicht-Chromatographie. Z. Analyt.Chem. 214, 1-23

FOLIC ACID: ASSAY AND STABILITY

A.E. Bender and N.I. Nik-Daud
Department of Nutrition, Queen Elizabeth College, University of London.

Summary

The assay of folic acid in foods has been standardised using Lactobacillus casei and results appear to be reproducible. Assays indicate that the vitamin is relatively stable to heat in several vegetables although a large proportion is leached into the cooking water.

Introduction

Folic acid presents two major problems; it exists in foods in a large number of different forms which makes analysis difficult, and there is little evidence of the relative potencies of these forms for human beings. The joint FAO/WHO group (1) recommends that only folates consisting of fewer than four glutamate residues should be taken into account, so-called 'free folate', although a variable amount of the higher complexes are broken down during digestion and may become available.

The limited evidence on requirements is based on pteroyl glutamic acid, which is the synthetic, unnatural form.

Consequently, when problems of possible deficiency arise it is difficult to evaluate diets. It is currently believed that spina bifida may be related to a relative deficiency of folate in the mother at the appropriate time of development of the fetus. Moreover, there appears to be an enhanced need for folate during the time of increased cell division so that the possibility of relative deficiency duting pregnancy has long been suspected.

Information is lacking of the potential contibution from food, and the stability of the vitamin and its losses during processing, cooking and storage.

A microbiological assay for folates in blood has been standardised and is in general use but only a few forms of folate are present in blood compared with the many forms in food. The present work has made use of that methodology and attempts to standardise the many factors that affect the assay. Since the method depends on the response of a micro-organism there is still no evidence of the availability of the vitamin for human beings but at least it now becomes possible to make comparisons between foods and between different laboratories, and to investigate the effects of processing.

2. Materials used.Difco folic acid medium (0822-15) and microinoculum broth (0320-02); lactobacillus agar AOAC (0900-15) were used; also partly purified Difco chicken pancreas extract (0459-12). The organism was Lactobacillus casei NCIB 10463 (chloramphenicol-resistant strain). full details of the procedure will be published elsewhere.

3. Procedure. The assay consists of the following stages and each requires rigid control:-

1. The preparation of the inoculum; 2. extraction of the vitamin from the food; 3. hydrolysis of the extracted folates to their simple forms; 4. 'setting-up' of the assay; 5. incubation of the vitamin with the organisms; and 6. measurement of their growth reponse.

3.1. Preparation of the inoculum. Since L.casei responds to more forms of folate than other tested organisms it is the most commonly used and was used in the present work, (while RDA's are based on assays with L. casei, AOAC (1980) recommends S. faecalis (2))

It is essential to introduce into the medium an inoculum of constant size and growth phase. The culture must be maintained continuously and transferred to stab culture at regular intervals. The usual procedure required that 1 16-hour culture is further subcultured for 6-8 h in medium containing folic acid; the harvested organisms are freed from folate by repeated washing and centrifuging during which time they must be maintained in sterile conditions. The inoculum must be prepared on the day of the assay and it is necessary to commence centrifuging about 9 h before the start of the assay- a difficult timing procedure.

This problem was overcome by storing the inoculum at a low temperature. The dried culture was regenerated in broth containg 300 ug/ml chloramphenicol and incubated at 37'C for 72 h to reach suitable density, then re-subcultured for further periods, washed free from traces of folate and stored at temperatures of -20'C, -60'C or in liquid nitrogen. The viability of the culture after storage was tested at various times by withdrawing an aliquot and inoculating the assay medium to obtain a standard curve.

Results showed that the culture could be stored at -20'C for periods up to 28 days; when stored at -60'C it was still fully active after 90 days; and when stored under liquid nitrogen it was still active after one year.

Thus the harvested inoculum can be frozen for subsequent use so obviating the need to maintain stab cultures and permitting better control of inoculum size and growth phase between assays as well as avoiding the 9 h preparation period.

3.2. Extraction of the vitamin. It is difficult to extract substances from plant materials. In the instance of folate the problem is complicated by its occurrence both as 'free folate' and higher complexes which can be degraded to 'free folate' by conjugase systems naturally present in the food. In order to determine free and combined folates it is essential to inactivate this enzyme system. This is usually achieved by autocalving the food before extraction. While this procedure yields higher results it is not clear whether this is due to more complete extraction or to destruction of the naturally-occurring conjugase.

In order to ascertain whether autoclaving destroys any of the folate, samples in phosphate-ascorbate buffer were heated for 5 min at 5, 10 and 15 psi, and the results compared with untreated samples. All samples were homogenised in a blender, centifuged and assayed for total folate content.

In a sample of fresh lettuce (*Lactuca sativa*) the value for the raw, autocalved material was 33 ug/100 g compared with 59 ug after autoclaving at 5 psi, so demonstrating the need for the heat process. After autoclaving for 10 and 15 psi similar values, namely 60 ug/100 g were obtained so demonstrating no destruction of folate had occurred by the more svere heat treatment and indicating that the normal heating at 5 psi did not damage the vitamin.

In three additional samples of food, namely red pepper, green cabbage and commercially frozen Brussels sprouts, the values after autoclaving were respectively increased by 29%, 22% and 22%; that of lettuce was increased by 51%. These figures indicate the relative inefficiency of extraction from homogenised and unautoclaved foods.

Ascorbic acid is usually added to the phosphate buffer to protect the folate from oxidation. The optimal concentration of ascorbic acid was determined by adding amounts reanging from zero to 1000 mg/100 ml buffer solution.

The values for free and total folate were as follows:-

Ascorbic acid (mg/100 ml buffer)	free folate (ug/100 G)	total folate (ug/100 g)
0	7	99
50	11	150
100	32	196
250	34	203
500	25	189
1000	26	208

100 mg ascorbic acid per 100 ml thus appears to be adequate in agreement with the findings of other authors (3, 4), who used 150 mg/100 ml: to provide a margin safety 250 mg ascorbic acid was used in subsequent work,

3.3. Deconjugation process. *L. casei* responds only to free folate so that determination of the total amount requires liberation of the complex forms.This is achieved by incubation with the enzyme preparation (from chick pancreas or hog kidney) for periods of 1½-16 h and there are reports of conflicting values with the different systems. Moreover, the pH must be strictly controlled and the process carried out under sterile conditions to avoid contamination with folate-producing organisms.

The variable factors in this stage include the use of an adequate amount of enzyme, maintenance of the optimum pH and control of sterility.

Based on the work of Hurdle et al (2) three levels of chicken pancreas extract were used, namely 25, 50 and 100 mg per g sample of Brussels sprout extract. After incubation for 16 h at 37'C the assays at all three levels of conjugase yielded values of 214-220 ug/100 g total folate, so indicating the adequacy of the concentration of 25 mg/g sample. 50 mg/g was used to provide a safety margin.

The optimum pH was found to be between 7.0 and 8.0 with a fall of 20% in the results when pH fell below 6.5 or rose above 8.5. The free folate values were unchanged when pH was varied between 6.0 and 9.0.

The results agree with those of Malin (4) using unpurified chicken pancreas extract. In subsequent work the pH was maintained at 7.0.

Sterility was maintained by using chloramphenicol. When this was not used 4 out of 5 tubes showed bacterial growth. The use of toluene as preservative was not satisfactory since 2 out of 5 tubes became contaminated; the results also indicated that the addition of the antibiotic did not interfere with the activity of the pancreatic extract.

3.4. 'Setting-up' the assay. This is a lengthy procedure and requires strict sterility. It consists of dispensing folate standards and samples of extracted foods into a large number of individual test-tubes, followed by the addition of a specified volume of assay medium, autoclaving and inoculating aseptically with the organism at the correct stage of growth and using an inoculum of constant size.

Storage of the inoculum at the temperature of liquid nitrogen obviated the need for repeated subculturing and washing, as well as the need to autoclave the assay medium containing the samples and standards. The use of chloramphenicol permitted the use of an automatic inoculation device Ithe Fison autodiluter, LFA 100) with which 0.03 ml of standards or samples and 6 ml of inoculated assay medium were added to each tube.

This permitted better control and was far more rapid than the earlier procedures.

3.5. Measurement of growth of the organisms. There is conflicting evidence with regard to the effect of incubation time on values obtained. Samples of Brussels sprouts were assayed after 16, 20, 24 and 40 h incubation. After 16 h there was little difference in growth at levels of standard pteroyl glutamic acid between 6 and 18 ng/ml. A good growth curve was obtained after 20, 24 and 40 h incubation with no apparent difference between the values after these three time periods. Although 40 h incubation gave the greatest response the growth was so heavy that it was necessary to dilute in order to read the turbidity so 24 h incubation period was selected.

The growth is usually measured by nephelometer or turbidity. The former requires optically-matched tubes so that the method of choice was a spectrophotometer fitted with a flow cell. Readings were carried out at 630 nm wave length in order to compensate for the colour of the medium.

4. Reliability of the assay. The reliability of the method was tested by assaying a sample of Oxoid yeast extract for free and total folate on 8 occasions over period of 4 weeks. Values for free folate varied between 0.6 and 0.8 ug/g - S.D. 0.7, C.V. 10%: for total folate, 24-27 ug/g, S.D. 1.7, C.V. 6.5%.

5. Folate content of foods. The results indicate that (a) folate is relatively stable to heat; and (b) it is leached into the processing water,

(Values are mean of 4 samples):-

vegetable	free folate		total folate	
	ug/100 g	%raw food	ug/100g	% raw food
French beans				
raw	20		76	
cooked	12	58	50	70
cooking water	7	35	22	29
cauliflower				
raw	23		175	
cooked	10	42	100	57
water	8	34	83	48
broccoli				
raw	45		320	
cooked	18	39	180	55
water	11	25	105	32
leeks				
raw	21		90	
cooked	13	61	50	53
water	7	32	45	50
green cabbage				
raw	12		64	
cooked	4	33	37	60
water	4	33	24	37
Brussels sprouts				
raw	23		175	
cooked	12	54	110	64
water	8	35	70	39

5.1 Loss from fresh vegetables. Brussels sprouts (*Brassica oleracea*, var. *gemmifera*) lost no folate when stored at -20'C up to 180 days.

At 4'C there was a steady loss from a mid-season variety (Rampart) of 40% over 14 days accompanied by an increase of 20% in free folate. corresponding figures for a late variety (Glentora) were 30% and 20%.

These changes indicate a breakdown of complex folate by naturally-occurring conjugases accompanying some degree of destruction of the free form.

Freshly harvested endive (*Cichorium endiva*) contained 390 ug/100 g dry wt free folate and 1845 ug total. Total folate fell at a rate of 15% per day (45% in 72 h) when stored at 23'C, while free folate was unchanged over this period.

When stored at 4'C the fall was only 5% per day with no change in free folate.

REFERENCE

1. W.H.O. JOINT FAO/WHO EXPERT GROUP (1970) Technical Report Series No. 452, Geneva.
2. AOAC (1980). Official Methods of Analysis, Association of Official Analytical Chemists. 13th Ed, pp 759-763. Washington, DC.
3. Hurdle, A.D.F., Barton, D. and Searles, I.H. (1968) Amer. J. Clin. Nutr. 21. 1202-1207.
4. Malin, J.D. (1975) Ph.D. Thesis, University of Strathclyde.

ETUDE PAR MICROANALYSE EN SELECTION D'ENERGIE DE LA MIGRATION DES MINERAUX DANS LA VIANDE AU COURS DU CHAUFFAGE

(MINERAL MIGRATION IN MEAT DURING COOKING STUDIED BY E.D.S.)

M. LAROCHE*, B. BOUCHET+, I. BRONNEC*, D.J. GALLANT+
Institut National de la Recherche Agronomique
*Laboratoire des Aliments d'Origine Animale
+Laboratoire de Technologie des Aliments des Animaux
Rue de la Géraudière - 44072 NANTES Cedex - FRANCE

Résumé

La migration des matières minérales dans la viande au cours du chauffage a été étudiée dans un microscope électonique à balayage par microanalyse en sélection d'énergie de coupes longitudinales ou transversales de muscles lyophilisés ou de leurs cendres. Le taux de phosphore reste pratiquement constant (échantillons lyophilisés ou cendres) au cours du chauffage. Il a donc servi de base de référence aux calculs de concentrations pour lesquels on a retenu les résultats de LAWRIE (1981) avec P = 200 mg/100 g de viande fraiche. La perte du soufre est négligeable, mais cet élément disparaît presque totalement au cours de la calcination. La perte relative des différents éléments est supérieure aux pertes de jus, ce qui indique que la plus grande partie des éléments dosés se trouve dans la phase liquide de la viande. L'addition de chlorure de sodium n'a pas d'influence significative sur les teneurs des autres éléments.

Summary

The mineral migration in the meat was studied under a scanning electron microscope by an energy dispersive system in function of heating time on, either longitudinal or cross sections of freeze-dried muscles, either on their ashes. The phosphorus yield being constant during heating time, mineral concentrations were calculated in mg/100 g of fresh meat according to LAWRIE (1980) with a phosphorus basis P=200. It was found that sulfur loss was negligeable during heating and the relative loss of mineral elements higher than the loss of juice. It was concluded that analysed minerals were particularly concentrated into the aqueous phase of the meat. Addition of salt (NaCl) had no significant influence on the mineral content of the meat during heating.

1. INTRODUCTION

Les pertes au chauffage ne sont pas exclusivement de l'eau, mais elles incluent également une partie de tous les constituants de la viande. Parmi ceux-ci, les minéraux sont faibles quantitativement, mais ils présentent une importance nutritionnelle non négligeable.

Après avoir envisagé la répartition des cendres entre la viande hachée et le jus (LAROCHE et NICOLAS, 1983) et celle de différents éléments entre la viande et le jus par microanalyse des cendres (LAROCHE et al., 1983) en fonction des conditions de traitement thermique, nous avons voulu comparer dans cette étude la migration des minéraux dans la viande cuite avec ou sans sel. L'effet du type d'échantillon (coupes transversales ou longitudinales et utilisation de cendres) sur ces déterminations a également été envisagé.

2. MATERIELS ET METHODES

2.1. Préparation des échantillons

Des échantillons de muscles Pectoralis profundus de bovins ont été placés dans des boites de conserves dans lesquelles était rajoutée la moitié du poids de la viande d'eau distillée ou salée (NaCl) à 3 %. Les boites serties ont été immergées dans de l'eau bouillante et maintenues pendant 4 h à 100°C. Un morceau de chaque échantillon chauffé a été gardé intact et le reste a été broyé. Après lyophilisation de l'ensemble, une partie de la viande hachée a été utilisée pour préparer les cendres par calcination à 550-600°C pendant environ 40 h.

2.2. Microanalyse en sélection d'énergie

L'analyse des échantillons a été effectuée à l'aide d'une diode Si(Li) sur microanalyseur de type EDAX 711 couplée à un microscope à balayage JEOL 50A. La sonde excitatrice, soumise à une tension de 20 KeV et à une intensité de 5.10^{-11} A, analyse une surface de 5.000 microns carrés. Les différents éléments analysés sont détectés simultanément en choisissant des surfaces les plus planes possibles sur les échantillons disposés dans des cavités creusées dans des cylindres de carbone pur, cet élément n'étant pas détecté par le système. L'analyse d'un cristal de $BaCl_2$ excité dans les mêmes conditions permet de calculer les concentrations élémentaires sur 8 éléments maximum après correction ZAF (nombre atomique, absorption, fluorescence) en mg/g de matière sèche présente.

Les analyses ont été effectuées sur les cendres (C) et sur des coupes transversales (T) et longitudinales (L) de viandes crue (0), cuite sans sel (-) ou avec sel (+). Pour chacun de ces 9 types d'échantillon nous avons effectué 7 analyses.

2.3. Traitement statistique

Pour comparer les différents résultats obtenus, nous avons considéré l'analyse de variance sur les moyennes de chaque type d'échantillon, ces informations étant complétées par la comparaison individuelle des moyennes obtenues. Pour les tableaux, nous avons représenté : -, non significatif ; (+), significatif au seuil de 10 % ; + significatif au seuil de 5 % ; ++, significatif au seuil de 1 %.

3. RESULTATS ET DISCUSSION

3.1. Résultats bruts

Les résultats bruts obtenus sont entrêmement dispersés (moyenne des coefficients de variation 80 %, le tiers de ces coefficients étant supérieur à 100 %). Cette dispersion peut être liée à des problèmes d'irrégularité de topographie sur la surface de l'échantillon.

La réponse des cendres est évidemment très supérieure à celle des viandes du fait d'une diminution considérable du bruit de fond liée à la disparition des matières organiques. Pour pouvoir comparer les valeurs obtenues, tous les résultats sont rapportés à la quantité de viande dont nous connaissons la teneur en cendres.

L'analyse de variance sur les moyennes n'indique aucun effet significatif ni du mode de détermination (T,L,C), ni du type de traitement (O,-,+). Il apparaît cependant un certain nombre de différences significatives entre moyennes, pour le sodium et le chlore, compte tenu du chauffage en présence de sel, et pour le soufre, du fait de l'utilisation de cendres.

Pour pouvoir comparer les concentrations observées avec les différents échantillons, nous devons rechercher une expression des résultats nous permettant d'éliminer l'effet des irrégularités de surfaces et les variations importantes mais d'origine connue pour certains éléments.

3.2. <u>Résultats rapportés au phosphore</u>

Nous avons observé précédemment (LAROCHE et al., 1983) que la teneur en phosphore restait constante dans la viande au cours du chauffage, et c'est l'élément pour lequel les variations observées dans cette étude sont les plus faibles. De ce fait, nous avons recalculé toutes les valeurs obtenues en prenant pour base P = 200, valeur correspondant approximativement aux teneurs en phosphore rapportées par LAWRIE (1981) en mg/100 g de viande fraiche.

L'analyse de variance sur les moyennes donne :

	Cl	Na	Mg	S	K	Ca	Fe	Total
O,-,+	(+)	-	-	-	(+)	-	-	++
T,L,C	-	-	+	+	-	-	+	++

Les variations du total peuvent être reliées aux variations du soufre et du chlorure de sodium. En ce qui concerne le fer, sa présence en plus grande quantité dans les échantillons de viande traduit une pollution des échantillons lors des coupes, nous ne considérerons donc plus cet élément par la suite.

La comparaison des moyennes nous fournit le tableau suivant :

Effet du traitement							Effet du mode de détermination						
	Cl	Na	Mg	S	K	Ca		Cl	Na	Mg	S	K	Ca
TO/T-	-	-	-	-	++	-	LO/CO	++	++	++	++	-	++
LO/T-	-	-	-	++	++	-	L-/C-	++	-	++	++	-	++
CO/C-	+	++	++	+	++	++	L+/C+	+	++	++	++	-	++
TO/T+	++	++	-	+	++	-	TO/CO	++	+	++	++	+	++
LO/L+	++	++	-	+	++	-	T-/C-	-	-	+	++	-	++
CO/C+	++	++	++	-	-	++	T+/C+	++	-	++	++	-	(+)
T+/T-	++	++	-	-	-	-	TO/LO	-	-	-	-	-	-
L+/L-	++	++	-	-	-	-	T-/L-	-	-	-	-	-	-
C+/C-	++	++	-	+	-	-	T+/L+	-	-	-	-	-	-

En ce qui concerne le traitement, l'effet du chauffage se traduit par une diminution de la teneur en potassium dans la viande chauffée, et celui du sel par une augmentation des teneurs relatives en chlore et en sodium. Le chauffage entraînerait d'autre part une augmentation de la teneur relative en soufre dans le produit obtenu.

Pour aucun des éléments considérés il n'apparaît de différences entre les coupes transversales et longitudinales, mais sauf pour le potassium, et à un degré moindre pour le sodium, les différences entre les coupes et les cendres sont systématiques. Dans les cendres, le chlore et le soufre sont dosés en quantités moins importantes tandis que le magnésium et le calcium sont dosés en quantités supérieures que dans le matériel brut, ce qui confirme les résultats obtenus par ailleurs (MASTEAU et al., 1980).

3.3. Evolution au cours du chauffage

Afin de suivre l'évolution des différents éléments au cours du chauffage, les valeurs obtenues pour les viandes cuites ont été rapportées au poids de viande crue initiale à partir des rendements observés. Pour cette comparaison, nous avons regroupé les valeurs obtenues sur les coupes transversales et longitudinales. Nous reprenons également dans le tableau, les valeurs rapportées par LAWRIE (1981) en mg/100 g de viande fraiche ainsi que les résultats des dosages de chlorure de sodium, effectués parallèlement sur les viandes par méthode coulométrique.

	Cl	Na	Mg	P	S	K	Ca
LAWRIE (1980)	69	84	23	207	205	358	8
Viande crue coupes ...	35	8	6	(200)	204	631	3
cendres ..	4	33	19	(200)	5	476	16
sans sel coupes	12	11	2	116	177	205	0,3
cendres	0,1	11	7	116	0,1	192	3
avec sel coupes	186	53	2	122	193	225	0,6
cendres	96	121	6	120	12	187	3
dosage NaCl	210	140					

Sauf pour le soufre, les valeurs obtenues avec la viande crue ne correspondent pas aux données de la littérature. Les valeurs de Na et de Cl sur la viande ou sur les cendres ne correspondent pas aux dosages du chlorure de sodium.

La perte de soufre au cours du chauffage n'est pas significative ; cet élément serait donc lié à d'autres constituants restant dans la viande.

4. CONCLUSIONS

L'utilisation de la microanalyse sur des sections transversales ou longitudinales de fibres musculaires ne fournit pas de résultats statistiquement différents. Le volume d'émission de fluorescence X par l'échantillon est suffisant pour obtenir une valeur "moyenne" ne distinguant pas les éléments principalement extra- ou intra-cellulaires.

L'analyse des cendres ne permet pas de doser le soufre, mais elle fournit des valeurs moins discordantes, par rapport à celles habituellement rencontrées dans la littérature, que celles obtenues sur matériel non calciné. Il semblerait que le chlore disparaisse partiellement lors de la préparation des cendres, alors que la calcination du chlorure de sodium dans les mêmes conditions s'effectue sans changement de masse.

Si l'on excepte le soufre pour lequel la perte est pratiquement nulle, la perte relative des différents éléments est supérieure aux pertes de jus au cours du chauffage, ce qui indiquerait que la plus grande partie des éléments dosés se trouve dans la phase liquide de la viande, résultat que nous avions obtenu précédemment.

L'addition de sel dans l'eau de cuisson modifie bien entendu les teneurs en chlore et en sodium de la viande chauffée, mais elle n'a pas d'influence significative sur les teneurs des autres éléments considérés.

Ces travaux ont été réalisés dans le cadre d'un contrat de programme du Ministère de la Recherche et de l'Industrie : "Effet des traitements thermiques sur les aliments".

BIBLIOGRAPHIE

LAROCHE M., BOUCHET B., BRONNEC I., GALLANT D.J., 1983. Utilisation de la microanalyse pour l'étude des traitements thermiques de la viande. Réunion Viandes et Produits Carnés, PARIS, 3-4 mars.

LAROCHE M., NICOLAS N., 1983. Influence des conditions de chauffage sur les teneurs en matière sèche et en cendres d'échantillons de viande hachée. Réunion Viandes et Produits Carnés, PARIS, 3-4 mars.

LAWRIE R.A., 1981. Nutrient variability due to species and production practice. In : Meat in nutrition and health, (proceedings), ed. Interstate Printers and Publishers inc., p. 7-17.

MASTEAU Y., BOUCHET B., GALLANT D.J., 1980. Migrations minérales dans les tissus du caryopse de blé mises en évidence par microanalyse en sélection d'énergie. Congrès S.F.M.E., POITIERS, 4-6 juin.

THE EFFECT OF THERMAL PROCESSING ON THE PHYSIOLOGICAL VALUE OF MILK

B. BLANC
Federal Dairy Research Institute, CH-3097 Liebefeld-Berne

Summary

Studies were carried out with the collaboration of student volunteers, in good health, in order to see if there existed measureable differences in physiological behaviour after the ingestion of raw milk or milk treated at ultra high temperature. This was done by measuring blood leucocytes.

In order to evaluate the results it was necessary to carry out a preliminary study to measure the values of leucocytes in fasting volunteers or, in order to simulate volume intake, mineral water was given as a control. No significant difference was found between these two situations.

Based on the control group it was possible to show significant differences between raw and UHT milk as regards total leucocytes, monocytes and the ratio of monocytes to total leucocytes at 30 and 90 minutes and also for the latter at 210 minutes. Some significant differences were found in the same group at different times of observation. The significance of these changes in the physiological state of the organism under the influence of aliments is discussed.

1. INTRODUCTION

Even though there are numerous studies on the utilization of aliments, their digestability and the biological value of their nutriments, it is only recently that the intimate relations which possibly exist between the constituents of the diet and the physiological reactions which they provoke have been taken into consideration. Postprandial leucocytosis is of particular interest (1).

The importance of these physiological effects (which are not only nutritional) of the aliments is very great. If we recall to mind the frequent medical observations concerning the role of the diet on the state of health and the resistance to infections it is surprising that so few people have carried out research in this field. However, it is probably because this field necessitates delicate experimentation with human subjects in a healthy and stable physiological and psychological state. As well as this, the metabolic controls and the impulses from the nervous system lead to complex modulations of the global body reaction after the ingestion of aliments of defined volume and composition.

Nothing has been published on the measurement of postprandial physiological reaction of the variation of leucocytosis since the work of Kouchakoff (5) and Kollath (4).

During our primary studies carried out on leucocytosis after the ingestion of either raw milk or milk having undergone thermal treatment we realised that there was a lack of literature on the subject of the variations of total leucocytes and their various categories.

This modest presentation gives some information about the modulation of leucocytosis during fasting compared to that of the same subjects having absorbed mineral water, raw milk or UHT milk (ultra high temperature treated).

2. METHODS OF STUDY

The volunteers were three students aged between 20 and 30 and in good health. They arrived at the laboratory early in the morning, while fasting, on several consecutive days having neither smoked nor run. They had absorbed nothing, except perhaps a little mineral water, since nine o'clock the previous evening. With the help of a qualified nurse a first venous blood sample was taken and the volunteers received nothing or 250 ml of mineral water or milk. Other blood samples were taken at 30 and 90 minutes after ingestion. The relative evaluations were obtained by calculating the difference existing between the parameters being considered the day of fasting and those obtained after the ingestion of the products tested the following day or the day after. The light mineral water chosen comes from Henniez-Santé spring (Vaud, Switzerland) and has the following composition (expressed in mg/l): Calcium 110, Magnesium 20, Sodium 8, Strontium 0.2, Bicarbonate 375, Nitrate 25, Chloride 17, Sulphate 14, Metasilicic acid 15, Orthoboric acid 0.15, Carbonate 6500.

The raw milk, taken at the latest 24 hours after milking, contained a total of 30 - 40 aerobic mesophiles and had a protein and fat concentration of 3.2 g/l and 3.8 g/l respectively. Indirect UHT treatment permitted the sterilization of milk by increasing the temperature to 145°C for 2.4 seconds. After cooling it was introduced aseptically into rectangular cartons covered with a polymer and coated with a sheet of aluminium according to the Tetra-Brik process of Tetrapack.

The leucocytes were counted with the help of a Coulter Counter for counting cells, as well as the special smears for reading in the abbot automat for leucocyte distribution.

3. RESULTS

A comparison of the values found in volunteers, fasting and after ingestion of mineral water on the following day is presented in table I. Statistical analysis of the results showed no significant difference between the volunteers in the two situations. However, some highly significant differences between individual volunteers were found.

When the values from both situations are pooled we can define a domain of normal values for our group of students. Table II contains the approximate regions of the group medians found, and the average standard deviation (per group of 18 - 22 values).

The results of the investigations with different milks are to be found in table III. A significant difference between raw and UHT milk groups was found for the total number of leucocytes at 30 and 90 minutes when compared to the situation just before ingestion of the aliment. One notes a higher leucocyte value in the situation where raw milk was drunk instead of UHT milk. As regards the monocytes the most significant difference noted was between the median values obtained in the same individuals drinking raw or UHT milk. Even at time zero the physiological state of the

Table I

Categories/time	fasting median values · $10^9/1$	fasting standard deviation	mineral water median values · $10^9/1$	mineral water standard deviation
Total leucocytes				
0 min	6.482	1.653	6.768	0.909
30 min	6.511	1.006	6.784	0.906
90 min	6.501	0.630	7.102	0.689
Neutrophiles (Segmented)				
0 min	3.888	1.114	3.861	0.521
30 min	3.897	0.956	3.764	0.509
90 min	3.854	0.632	3.962	0.465
Eosinophiles				
0 min	0.126	0.119	0.119	0.098
30 min	0.097	0.092	0.066	0.113
90 min	0.073	0.069	0.095	0.082
Basophiles				
0 min	0.050	0.038	0.037	0.052
30 min	0.032	0.035	0.040	0.039
90 min	0.039	0.048	0.039	0.021
Monocytes				
0 min	0.432	0.129	0.452	0.120
30 min	0.402	0.163	0.408	0.081
90 min	0.400	0.146	0.478	0.124
Lymphocytes				
0 min	2.509	0.603	2.535	0.508
30 min	2.373	0.428	2.586	0.472
90 min	2.102	0.246	2.402	0.392

18 - 22 values were available for calculating each presented result.

Table II

Median values

	$\cdot 10^9/l$	$\cdot 10^9/l$
Total leucocytes	6.50 - 7.10	(1.00)

	$\cdot 10^9/l$	$\cdot 10^9/l$
Leucocyte fractions, absolute:		
Neutrophiles (Segmented)	3.80 - 4.00	(± 0.70)
Eosinophiles	0.07 - 0.12	(± 0.10)
Basophiles	0.03 - 0.05	(± 0.04)
Monocytes	0.40 - 0.48	(± 0.13)

	$\cdot 10^9/l$	$\cdot 10^9/l$
Lymphocytes	2.10 - 2.60	(± 0.40)
Leucocyte fractions, percent:		
Neutrophiles (Segmented)	55 - 60	(± 5)
Eosinophiles	1.0 - 1.8	(± 1.3)
Basophiles	0.5	(± 0.5)
Monocytes	5.6 - 6.6	(± 1.7)
Lymphocytes	33 - 38	(± 5)

Table III

HL 6 - Median values and median increases

Total leucocytes:

(± 1.20)	contr. · $10^9/l$	raw	UHT	(± 0.40)	contr. · $10^9/l$	raw	UHT
0	5.20	5.30	5.00				
0.03	5.00	4.90	4.80	0-0.03	-0.20	0.20[x]	-0.10[x]
0.09	5.20	5.10	4.90	0-0.09	0	0.10[x]	-0.40[x]
0.21	5.80	5.30	5.20	0-0.21	0.40	0	0

Neutrophiles (Segmented), absolute:

(± 1.30)	contr. · $10^9/l$	raw	UHT	(± 0.40)	contr. · $10^9/l$	raw	UHT
0	2.30	2.30	2.90				
0.03	2.40	2.40	2.70	0-0.03	0	0.30	0
0.09	2.70	2.60	2.70	0-0.09	0.20	0	-0.10
0.21	3.00	3.00	2.80	0-0.21	0.70	0.40	0.10

Eosinophiles, absolute:

(± 0.06)	contr. · $10^9/l$	raw	UHT	(± 0.07)	contr. · $10^9/l$	raw	UHT
0	0.08	0.11	0.09				
0.03	0.06	0.12	0.09	0-0.03	-0.02	0.01	0.02
0.09	0.08	0.07	0.11	0-0.09	-0.06	-0.01	0.04
0.21	0.10	0.07	0.06	0-0.21	-0.02	0.02	-0.04

Basophiles, absolute:

(± 0.04)	contr. · $10^9/1$	raw	UHT	(± 0.04)	contr. · $10^9/1$	raw	UHT
0	0.03	0.04	0.02				
0.03	0.04	0.04	0.04	0-0.03	0	0	0.02
0.09	0.05	0.04	0	0-0.09	0	0	0
0.21	0.04	0.02	0.04	0-0.21	-0.02	-0.04	0.02

Monocytes, absolute:

(± 0.14)	contr. · $10^9/1$	raw	UHT	(± 0.12)	contr. · $10^9/1$	raw	UHT
0	0.34	0.25^x	0.43^x				
0.03	0.24	0.33	0.25	0-0.03	-0.04	0.02^x	-0.13^x
0.09	0.35	0.27	0.33	0-0.09	0	0.05^x	-0.09^x
0.21	0.26	0.27	0.39	0-0.21	-0.06	0.01^x	-0.13^x

Lymphocytes, absolute:

(± 0.30)	contr. · $10^9/1$	raw	UHT	(± 0.50)	contr. · $10^9/1$	raw	UHT
0	2.40	2.20	2.30				
0.03	2.30	2.20	2.10	0-0.03	-0.10	-0.10	-0.10
0.09	2.20	2.20	2.20	0-0.09	-0.20	0	0
0.21	2.40	2.00	2.30	0-0.21	-0.10	-0.20	0.40

Neutrophiles (Segmented), % of leucocytes:

(± 10)	contr.	raw	UHT	(± 7)	contr.	raw	UHT
0	42	43	55				
30	47	47	53	0-30	4	3	1
90	53	51	53	0-90	9	-2	2
210	51	55	50	0-210	12	2	-3

Eosinophiles, % of leucocytes:

(± 1.6)	contr.	raw	UHT	(± 1.3)	contr.	raw	UHT
0	1.5	2.0	1.8				
30	1.0	1.5	1.5	0-30	-0.5	0.0	0.5
90	1.5	0.5	2.0	0-90	-1.0	0.0	0.8
210	1.8	1.0	1.0	0-210	-0.5	0.0[x]	-0.5[x]

Basophiles, % of leucocytes:

(± 0.8)	contr.	raw	UHT	(± 0.8)	contr.	raw	UHT
0	0.5	1.0	0.5				
30	0.7	0.7	0.7	0-30	0.0	0.0	0.2
90	1.0	0.5	0.0	0-90	0.0	0.0	0.0
210	0.8	0.5	0.5	0-210	-0.2	-0.5	0.2

Monocytes, % of leucocytes:

(± 2)	contr.	raw	UHT	(± 2.4)	contr.	raw	UHT
0	5.3	5.0[x]	8.1[x]				
30	5.0	5.2	5.0	0-30	-0.4	0.1[x]	-2.2[x]
90	5.4	5.5	6.3	0-90	0.0	1.4[x]	-2.1[x]
210	4.1	5.9	6.0	0-210	-1.5[x]	0.9[xy]	-1.5[y]

Lymphocytes, % of leucocytes:

(± 9)	contr.	raw	UHT	(± 7)	contr.	raw	UHT
0	48	48	34				
30	43	44	35	0-30	-3	-3	-2
90	41	33	34	0-90	-4	0	0
210	42	34	40	0-210	-6	-4[x]	6[x]

volunteers showed a significant difference between the raw ($0.250 \cdot 10^9$) and UHT milk ($0.430 \cdot 10^9$) variations, however, this does not diminish the importance of the significant differences repeatedly obtained at all observation times for the raw and UHT groups.

After calculating the monocytes as a percentage of total leucocytes, it was possible to confirm the differences mentioned above as regards the effect of milks on monocytosis. One notes that the relative increase in monocytes is greater after the ingestion of raw milk. Also after 210 minutes a significant difference between the control and raw milk variation is detected.

Changes in values were also detected during the observation period in the same variation. The median values differing significantly from time zero are encircled in table III. Thus the total leucocyte value in the control variation was higher at 210 minutes than at time zero and at 30 and 90 minutes when the volunteer had absorbed UHT milk. It is also true for segmented neutrophiles at 210 minutes in the variation with raw milk; for the monocytes at 90 minutes with raw milk and at 30 minutes for UHT milk; for lymphocytes at 30 minutes in the control group. Other increases were also noted when either neutrophiles, eosinophiles or monocytes were expressed as a percentage of total leucocytes (see table III).

3. DISCUSSION

Experiments on human subjects is becoming more and more difficult especially when it involves taking blood samples. One must resign oneself to not being able to present a large number of cases and figures which are normally demanded of in modern biological experimental proceedures. This presentation has no other goal but to give some basic results which will allow one to appreciate future work and to stimulate intrest in a still forsaken field.

However, it would already seem possible to say that there is no marked effect of light mineral water (Henniez-Santé) on the equilibrium of serum leucocytes. This confirms the observations of Kouchakoff (5).
What seems to be more important than the ingestion of a volume of neutral liquid is the circadian rythm and its nyctohemeral variations. During the three and a half hours observation period it is already possible to detect certain variations. These stand out even more if one takes into consideration, not only the medians of the group of subjects being studied, but the values of each individual subject over the three and a half hours.

When we consider the effect of raw or UHT milk on subjects with known leucocyte variations, either fasting or after ingestion of mineral water, we rejoin the preocupations of Kouchakoff (5) and Kollath (4) concerning modifications caused by thermal treatment of food. Fruit and vegetables were their main interest. Using a different approach and using another aliment, milk, it would seem possible to demonstrate a convincing effect statistically assured.

The greatest variations detected in our experiments were for monocytes. Monocytes, even if they are relatively few in number and only in transit in the blood, play an important role in reactions concerning the defence of the organism. Their extravascular lifespan after their transformation to macrophage may be as long as several months or even years (Hoffbrand and Pettit, 1980). Pinocytosis and phagocytosis are properties which are well developed in the monocyte-macrophage group which come into play not only in the antimicrobial anti-infections protection but also in the elimination of organic and inorganic particles (Kleinhauser 1978) and in conjunction with lymphocytes in immunological reactions.

Kouchakoff (5) and Kollath (4), even though they apparently agreed on some of the results of their observations, had contradictory interpretations. After resumption of the studies on the influence of food on the variations of human leucocytes, the facts appeared to be even more complex. Any ordinary explanation would be premature and haphazardous for the moment.

It is not possible at present to draw conclusions as regards how alimentary products, such as milk having undergone thermal treatment, induce reactions less favourable than the corresponding crude products. However, it would seem necessary, in studies with aliments, not to limit oneself to the nutritional contribution alone. One should also concern oneself with the entirity of the physiological, biochemical and immunochemical reactions set in motion and whose influence could well effect the state of health and resistance of the organism to infection.

4. ACKNOWLEDGEMENTS

I would like to thank Dr. P. Rüst for the statistical evaluation of the results and Dr. M. Casey for the english presentation of the text.

REFERENCES

1. BLANC, B. (1980). Einfluss der thermischen Behandlung auf die wichtigsten Milchinhaltsstoffe und auf den ernährungsphysiologischen Wert der Milch. Alimenta-Sonderausgabe, 5-25
2. HOFFBRAND, A.V. and PETTIT, Y.E. (1980). Essential haemetology. Blackwell Scientific Publications, 93.
3. KLEINHAUER, E. (1978). Hämatologie: Physiologie, Pathologie, Klinik. Springer-Verlag, Berlin-Heidelberg, 253-4
4. KOLLATH, W. (1939). Von Nahrungseinwirkungen vor der Resorption durch den Darm. Ein Beitrag zur Frage: gekocht oder roh? Klin. Wschr. 18 (16) 557-563
5. KOUCHAKOFF, P. (1937). Nouvelles lois de l'alimentation humaine basées sur la leucocytose digestive. Mémoires de la Société vaudoise des Sciences naturelles 5 (8) no 39, 319-348.

INFLUENCE OF THE CFR (COOKING-FREEZING-REHEATING) SYSTEM ON THE NUTRITIVE QUALITY OF FOOD PROTEIN

O. Moreiras-Varela, B. Ruiz-Roso and G. Varela
Instituto de Nutrición
Facultad de Farmacia
Madrid -3. Spain

Summary

We have studied the influence of the CFR System on the nutritive value of the protein of some food products of animal origin: meat (beef and pork), fish (hake and sardine) and eggs, as well as two compound dishes: meat hamburger and fish hamburger.

Intake, Diet Growth Index (DGI) (weight increase/intake), Net Protein Utilization (NPU) and Biological Value (BV) were calculated. MILLER and BENDER (4) technic was used.

1. INTRODUCTION

Food habits have drastically changed in the last ten years and an increasing number of people eat in collective dining rooms.

The fact that home cooking has become industrial cooking makes it necessary to study this new feeding systems, which has three phases: cooking preparation, conservation and distribution to the consumer after reheating. The sum of these three phases is known as CFR (Cooking-Freezing-Reheating) System (2) (3).

On the other hand, the effect of frying in an oil bath (deep-frying) on the nutritive value of food (7) is not very well know. This is a form of cooking used in the Mediterranean countries which has enormously spread to other countries not familiar with this technique. Our laboratory has been working for a long time on this issue and Doctoral Thesis (5) (6) and research papers have been published. Recently were reviewed and updated by VARELA (7) (8).

2. PURPOSES

In this work we study the influence of the CFR System on the nutritive value of the protein of some food products of animal origin: fish (hake and sardine), meat (beef and pork) and eggs, as well as two compound dishes: meat hamburger and fish hamburger, containing carbohydrates, which could affect the nutritive value of protein.

Each food is studied raw; fried in virgin olive oil fixing fat/food ratio, temperature and frying time; frozen at -18ºC and keeped it at this temperature for 30 days; and reheated by macrowaves oven.

In raw food as in each step of the CFR System: fried, frozen and reheated, the following parameters are studied: Intake, Diet Growth Index (DGI) (weight increase/intake), True Digestibility (TD), Net Protein Utilization (NPU) and Biological Value (BV) were calculated. MILLER and BENDER (4) technic was used.

Figure I. Nutritive value of raw and processed beef hamburger protein

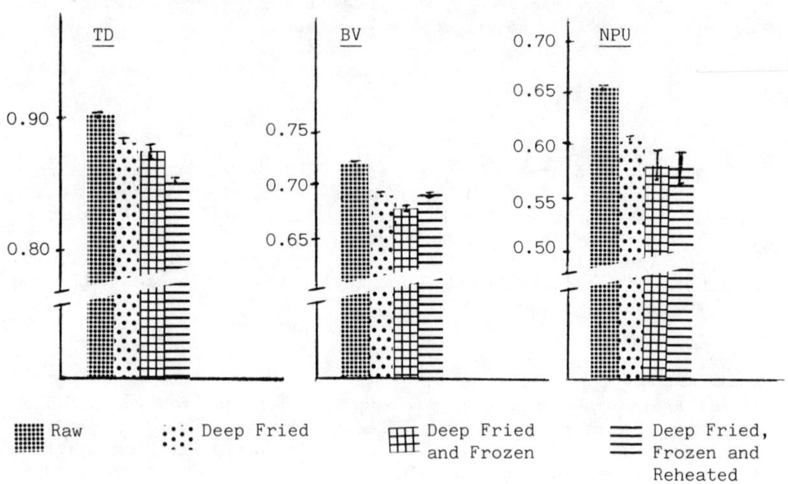

Figure II. Nutritive value of beef and beef hamburger protein

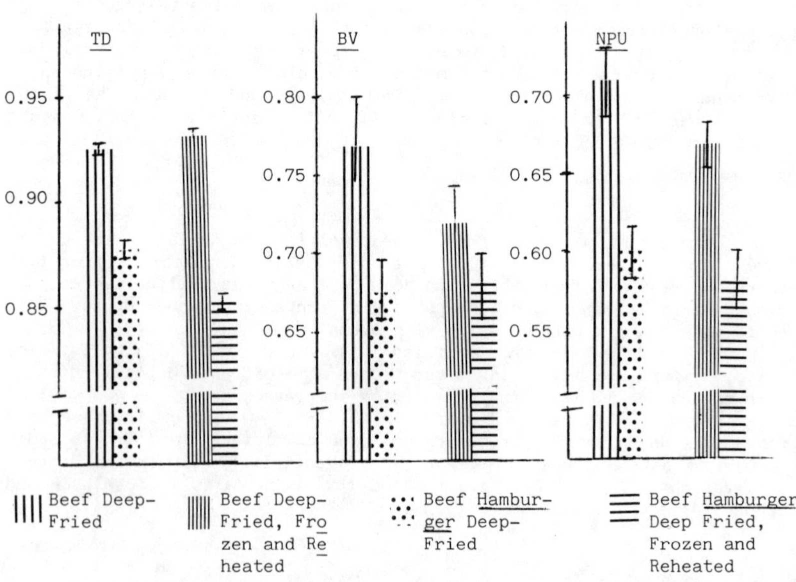

3. RESULTS

3.1. Intake: The immediate effect of increasing intake by frying was observed in almost all cases. Such increases were significant in various instances, for example fish hamburgers (8.3 ± 0.2 to 9.0 ± 0.2 d/day) and meat hamburgers (8.2 ± 0.2 to 9.1 ± 0.2 g/day).

This confirms that frying improves the palatability of food. However, when this food is frozen, intake generally decreases to values near to those of raw food.

3.2. DGI: The processes the studied foods haven been submitted to, did not affect this index. This shows that animals weight increase is only a function of intake.

3.3. TD: Digestibility of protein does not change. We only have found alterations in this index in both types of hamburgers and sardines. In hamburgers it decreases mainly as a result of frying (0.92 ± 0.004 raw fish hamburger and 0.89 ± 0.003 fried), but there are also losses during freezing and reheating. In sardines, the high content of polyinsaturated fat (more than 50% of fat in dry matter), become rancid, and animals fed with such diets showed a negative growth. This was solved by repeating the experiments with sardines to wich a antioxidant had been added. With such a treatment, a significant improvement was recorded.

3.4. BV: This index practically does not change as a result of submitting these food products to the CFR treatment, with the exception of the cases when temperature increase takes place in the presence of carbohydrates (0.72 ± 0.02 raw meat hamburger, 0.68 ± 0.02 fried) (Figure I).

3.5. NPU: No changes in this parameter were observed for single processed food as compared with the same raw food. There is the obvious exception, as for BV, of hamburgers containing carbohydrates. Processing of meat hamburgers leads to significant digestibility losses with consequent significant decreases in NPU. As for BV, even though there are losses due to frying, they are not significant (Figure II).

As a general rule, we can draw the conclusion that the CFR system in our experimental conditions is suitable for its use in collective feeding. Our results agree with those of BODWELL and WOMACK (1).

Frying is the phase of the CFR system having the highest incidence, but only in compound food products. There are no practical differences in the nutritive value of protein of food before and after frying. The nutritive value of protein of single and compound food is practically not affected by freezing or reheating.

REFERENCES

1. BODWELL, C. and WOMACK, M. (1978). "Effects of heating methods on protein nutritional value of five fresh or frozen prepared food products". J. Food Science, 43, p 1543.
2. GLEW, G. (1980). "The contribution of large scale feeding operations to nutrition". World Rev. Nutr. Dietet., 34, p 354.
3. GLEW, G. (1981). "Further treatment for catering". In: CEE COST-91 "Industrial Cooking". Minutes of the Workshop. Versailles. France.

4. MILLER, D.S. and BENDER, A.E. (1955). "The determination of net utilization of proteins by shortened method". Brit. J. Nutr. 9, p 382.
5. RODRIGUEZ, A. (1982). "Estudio comparativo de las alteraciones en la lipidemia y lipoproteinemia en ratas por la ingestión de grasas crudas o procedentes de fritura". Tesis Doctoral. Universidad Complutense. Madrid.
6. RUIZ ROSO, B. (1982). "Influencia del sistema CFR (fritura-congelación-recalentamiento) sobre la calidad nutritiva de la proteína de algunos alimentos de origen animal". Tesis Doctoral. Universidad Complutense. Madrid.
7. VARELA, G., MOREIRAS-VARELA, O. and RUIZ-ROSO, B. (1980). "Utilization of some oils in repeated domestic fries". Proceedings of the III[d] International Congress on the Biological Value of Olive Oil, Chania, Creta. Grecia.
8. VARELA, G. (1982). "Nutritional aspects of home frying". In: CEE COST-91. "Home cooking nutrient changes and emerging problems". Proceedings of workshop. Rome.

CHANGES OF PHYTIC ACID CONTENT DURING BREADMAKING

P. SCHEFFELDT, R. SCHOENHAUSER and A. BLUMENTHAL
Institute for Nutrition Research
Seestrasse 72, CH-8803 Rueschlikon, Switzerland

Summary

During the production of 9 different types of bread in commercial bakeries, the changes of the phytic acid content were followed from ingredients, to sponge or sour dough, to final dough, and to bread. The sponge or sour dough led to high phytic acid reductions (40-100%), depending on the use of flour or meal (particle size), starting culture and time of development. The total dough fermentation gave very different extents of phytic acid degradation (18-99%), depending on the type of flour/meal used (fine/coarse, wheat/rye), the use of a sponge or sour dough, and the time. The baking process resulted in additional phytic acid reductions (13-52%). Pumpernickel,which requires a low temperature baking, showed 52% degradation. Overall phytic acid reduction was determined for white wheat bread to be 99%, for brown wheat bread 54-63%, for wheat whole meal bread 33%, for rye-wheat flour bread with sour dough 83%, and for rye whole meal bread with sour dough 49-78%. The use of sponge or sour dough method increased degradation of phytic acid. As long as the dietary role of phytic acid is not clear, these breadmaking procedures would be advantageous.

1. INTRODUCTION

Today bread is still considered a staple food although only 130 g is consumed per adult and day in Switzerland (1). Nutritionist and new dietary guidelines recommend increased consumption of bread, particularly bread made of high extraction flours or whole meals. These breads provide more nutrients and dietary fibre per unit weight than white bread. On the other hand, they also contain more phytic acid and/or its salts. In a previous publication we have reported the phytic acid content of different types of commercial bread available in Switzerland (2). Based on this we estimated that the average daily intake through bread was 107 mg of phytic acid per adult.

The relevance of phytic acid in our diet is not fully understood, but nutritionist agree that phytic acid may reduce the bioavailability of minerals (Ca, Mg) and trace elements (Fe, Zn). It is known that during breadmaking phytic acid is partially hydrolyzed by phytase or phosphatase to myo-inositol and inorganic phosphorus. Both of these degradation products do not interfere with bioavailability of minerals or trace elements.

As long as the dietary role of phytic acid is not clear, it may be appropriate to develop breadmaking procedures aiming at maximum phytic acid reduction. Thus the aim of our work was to monitor the phytic acid degradation occurring during commonly used breadmaking procedures as applied in Switzerland. Investigations were carried out in several commercial bakeries and we determined the extent of the phytic acid degradation during the single breadmaking steps.

2. MATERIALS AND METHODS

2.1 Materials

Two industrial and four small bakeries cooperated in this study. Nine different types of bread were chosen in order to include bread made by straight-, sponge-or sour dough method as well as bread made from wheat, rye, soybean, and blend of them. All samples, including baking ingredients (flour(s) and/or meal(s), yeast, salt), the pre-fermented dough (sponge or sour dough), the dough after final fermentation (i.e. after floor and proof time), and bread were collected in the participating bakeries.

2.2 Methods

The phytic acid was determined according to the method of Wheeler and Ferrel (3). The extraction procedure used for flours and meals was slightly modified. The extraction solution used was 15% TCA instead of 3% TCA and the mixing ratio was 1 part material to 10 parts of 15% TCA solution. Extraction was carried out for 2 hours by shaking. Dough samples were taken and immediately homogenized in 3.75% TCA solution at a ratio of 1:2 (w/w), the bread samples were homogenized in 3% TCA solution at a ratio of 1:3 (w/w) and subsequently stored at -20°C until further analysis. After thawing the samples were diluted with 3 % TCA solution (w/w) to obtain thin slurries which facilitate the extraction. The aliquots (weight) were taken to contain 1-6 mg phytic acid-phosphorus.

The dry weight of all samples was determined by drying at 105°C for 24 hours. Water homogenates were prepared of dough (1:2, w/w) and bread samples (1:3, w/w) prior to drying.

3. RESULTS AND DISCUSSION

Bread varieties investigated are characterized according to ingredients and breadmaking procedures used (Table I). Four breads were made using the straight-dough method, two the sponge-dough method, and four the sour dough method. The phytic acid content of the flours and/or meals used for the pre-fermented doughs and of the developed sponges and sour doughs are compiled in Table II. It is known that phytic acid degradation is the result of cereal phytase and microbial enzymes (phytase, phosphatase). For the interpretation of the extent of phytic acid hydrolysis, one has to consider duration and temperature of the fermentation as well as other factors such as water content, pH and size of flour or meal particles. It is evident that the pre-fermentation led to reductions of 40-100% of phytic acid (Table II). In sponge b (bread No. 5), which was not inoculated but underwent a spontaneous fermentation, the phytic acid was reduced by 40% only, whereas in sponge a and sour doughs c,d,e,f, which were inoculated, the phytic acid was reduced from 61-100%. It appears that the higher reduction of phytic acid in inoculated samples compared to the not inoculated one, is the result of microbial enzymes. These findings are in agreement with those of REINHOLD (4) and HARLAND and HARLAND (5). Phytic acid in sour dough c was completely hydrolyzed, inspite of the fact that wheat is considered to have less phytase activity than rye and furthermore that the temperature was 4°C during fermentation. The sour dough c used in this case was held for seven years. It is likely

Table I Breadmaking procedures

Bread variety (bread weight in g) (% wheat and/or rye)	Dough handling -straight-dough -sponge-dough -sour dough method	Pre-ferment.of dough			Final dough fermentation*				Baking	
		percentage of total flour/meal used	fermentation time hours	fermentation temp. °C	yeast addit.	time min.	temp. °C	final pH	time min.	temp. °C
1 White (500) (100% wheat)	straight	-	-	-	+	115	28	5.4	50	240-200
2 Brown (500) (100% wheat)	straight	-	-	-	+	110	28	5.6	50	240-200
3 Brown (500) (100% wheat)	sponge a	33	8	26	+	90	26	5.7	65	230-210
4 Whole meal (500) (90% wheat, 10% rye)	straight	-	-	-	+	115	28	5.6	50	240-200
5 Wheat-Rye whole meal (500) (75% wheat, 25% rye)	sponge b	48	13	30	+	65	30	4.3	40	280-240
6 Rye-Wheat flour (500) (55%rye, 45% wheat)	sour dough c	28	24	4	+	85	30	4.0	50	240-230
7 Rye whole meal/Wheat flour (500) (75 % rye, 25% wheat)	sour dough d	16	16	23	+	105	23	4.7	50	240-200
8 Rye whole meal - (100% rye)	sour dough e	25	6	27	-	65	27	3.9	480	140-110
9 Pumpernickel (whole meal) (100% rye) -	sour dough f	25	6	27	-	65	27	3.9	960	140-110
10 Soybean meal/Wheat flour (33% soyb.,67% wheat) (250)	straight	-	-	-	+	90	32	5.7	36	250-230

*floor and proof time

Table II Changes of phytic acid content during sponge and sour dough development

Bread No.	Sponge or sour dough	Flour and/or meal used (% wheat and/ or rye)	Phytic acid content mg/100 g dry weight		Phytic acid reduction %	Fermentation				
			flour/meal	sponge/ sour dough		time hours	temp. °C	water content %	final pH	starter (part)
3	sponge a	Brown flour (100% wheat)	546	142	74.0	8	26	50.4	5.6	yeast (1%)
5	sponge b	Wheat aleurone (coarse) and rye whole meal (50% wheat, 50% rye)	669	404	39.6	13	30	45.2	4.1	none
6	sour dough (1 stage)c	Brown flour (100% wheat)	493	1	99.8	24	4	55.3	3.7	yes (30%)
7	sour dough (1 stage)d	Flour and meal (fine) (34% wheat, 66% rye)	471	25	94.7	16	23	50.1	3.5	yes (1%)
8/9	sour dough e/f (2 stage)	Whole meal (coarse) (100% rye)	482	188	61.0	6	27	44.1	3.9	yes, basic sour (12%)

that special microorganisms had been selected which must have high phytase activity even at this low temperature. A possible reason for this may be that these organisms use more phytic acid as a phosphorus source to cover their high phosphorus requirement.

The effects of the entire breadmaking processes applied in this study on the phytic acid level are summarized in Table III. The total reduction of phytic acid during breadmaking is the result of the fermentation (pre-fermentation, floor and proof time) and the baking. The reduction during fermentation ranged from 18-99% and during baking from 13-52%. The initial phytic acid content of the white flour used for bread No. 1 was the lowest and nearly all phytic acid was hydrolyzed. The advantages of sponge or sour dough pre-fermentation, as previously shown in Table II, carry over into the final dough fermentation resulting in higher phytic acid reduction (during total dough fermentation of bread No. 3 versus 2, and No. 5,6,7,8,9 versus 4,10). Data compiled in Table III also indicate that the particle size of flours and meals used affected the phytic acid degradation. With one exception (bread No. 9), breads made from flours or fine ground meals (No. 2,3,6,7) had higher phytic acid reductions than breads made from coarse meals (No. 4,5,8,10). The particle size appears to determine the access of enzyme to substrate. The finer the particle, the higher the phytic acid reduction during a given period.

The phytic acid hydrolysis which we determined during the baking process is not surprising. Reports in the literature (6, 7) indicate that phytase is quite heat stable. Since the temperature within a loaf does not exceed 98°C and gradually approaches this temperature during the baking process, phytase will be active for part of the baking time. Support for this hypothesis was demonstrated on the example of Pumpernickel (No. 9). This bread was baked for 2 hours at 140°C followed by 14 hours at 110°C. We determined phytic acid level after 2, 12 and 16 hours and found reductions of 18.4%,45.6% and 52.2% respectively. This shows that phytase was still present after 12 hours baking. The rye whole meal bread (No. 8) was made from the same ingredients as the Pumpernickel but was baked for 4 hours at 140°C and then 4 hours at 110°C. The phytic acid reduction was 23%. This suggests that the phytic acid degrading enzymes were inactivated earlier.

The overall phytic acid reduction during the breadmaking processes analysed ranged from 33-99%. High hydrolysis was found in breads made from white (99%) and brown wheat flour (54-63%) by the straight-dough or sponge-dough method. Breads made from whole meal (wheat and/or rye), that means from ingredients with a high to a very high phytic acid content, demonstrated the favorable effect of sponge- or sour dough method on phytic acid degradation. It is known, that the sponge-dough and the sour dough method will produce bread of high acceptance (8, 9). The long fermentation of dough applied with these methods leads to the formation of more flavour compounds, and, as we have shown, to a higher phytic acid reduction. As long as the dietary relevance of phytic acid is not clear, this degradation is advantageous and makes the consumption of nutritious breads made from high extraction flours or whole meals a recommendable proposition.

Table III Changes of phytic acid content during breadmaking

Bread variety (%wheat and/or rye)	(bread weight in g)	Phytic acid content mg/100 g dry weight			Phytic acid reduction %		
		Ingredients	Dough after final fermentation	Bread	during dough fermentation	during baking	total
1 White (100% wheat)	(500)	91	<1	<1	99	<1	99
2 Brown (100% wheat)	(500)	460	327	168	28.9	34.6	63.5
3 Brown (100% wheat)	(500)	525	351	240	33.2	21.1	54.3
4 Whole meal (90% wheat, 10% rye)	(500)	1145	941	760	17.8	15.8	33.6
5 Wheat-Rye whole meal (75% wheat, 25% rye)	(500)	585	445	339	23.9	18.1	42.0
6 Rye-Wheat flour (55% rye, 45% wheat)	(500)	388	115	65	70.4	12.9	83.3
7 Rye whole meal/Wheat flour (75% rye, 25% wheat)	(500)	416	235	171	43.5	15.4	58.9
8 Rye whole meal (100% rye)	-	456	338	233	25.9	23.0	48.9
9 Pumpernickel (whole meal) (100% rye)	-	456	338	100	25.9	52.2	78.1
10 Soybean meal/Wheat flour (33% soyb.,67% wheat)	(250)	668	550	438	17.7	16.7	34.4

Acknowledgements: This work has been supported by the Swiss Ministry for Education and Science and constitutes a part of the E.E.C. Scientific Programme COST 91, Subgroup 2 (industrial cooking).

REFERENCES

1. BLUMENTHAL, A., SCHEFFELDT, P. and SCHOENHAUSER, R. (1983). Zum Nährstoffgehalt schweizerischer Brote und deren Beitrag zur Bedarfsdekkung der Bevölkerung. Mitt.Gebiete Lebensm. Hyg. 74, 80-92.
2. BLUMENTHAL, A. and SCHEFFELDT, P. (1983). Zum Phytinsäuregehalt schweizerischer Brote. Mitt.Gebiete Lebensm.Hyg. (in press).
3. WHEELER, E.L. and FERREL, R.E. (1971). A method for phytic acid determination in wheat and wheat fractions. Cereal Chem. 48, 312-320.
4. REINHOLD, J.G. (1975). Phytate destruction by yeast fermentation in whole wheat meals. J.Am.Diet.Ass. 66, 38-41.
5. HARLAND, B.F. and HARLAND, J. (1980). Fermentative reduction of phytate in rye, white, and whole wheat breads. Cereal Chem. 57, 226-229.
6. PEERS, G.F. (1953). The phytase of wheat. Biochem.J. 53, 102.
7. RANHOTRA, G.S. and LOEWE, R.J. (1975). Effect of wheat phytase on dietary phytic acid. J.Food Sci. 40, 940-942.
8. SPICHER, G. and STEPHAN, H. (1982). Handbuch Sauerteig. BBV Wirtschaftsinformationen GmbH, Averhoffstrasse 10, 2000 Hamburg 76.
9. ROTHE, M and RUTTLOFF, H. (1983). Aroma retention in modern bread production. Nahrung 27, 505-512.

A QUANTITATIVE BASIS FOR THE CLASSIFICATION OF CONVENIENCE FOODS

J. RYLEY Procter Department of Food Science,
Leeds University, U.K.

Summary

Methods of nutritional classification of foods for educational purposes and meal planning are often commodity biased and insufficiently flexible to incorporate "recipe" products. Following an analytical study of a wide range of convenience foods, a modification of Hansen's "Index of nutritional quality" was devised and used to assess the results.

1. INTRODUCTION

The messages and methods of nutrition education are both controversial and topical issues, and reflect both scientific progress and social and economic changes. In the post World War II years, the message concentrated on adequacy. Food intake was limited by rationing so that information about the dangers of overconsumption was not required. Rationing in the United Kingdom provided a diet very close to the "prudent" diet nutritionists and dietitians are currently advocating. Consumption of fat and sugar presented no problems because their supply was limited. Although dietary fibre was not emphasised, the need to rely on home-grown vegetables and the "national loaf" undoubtedly ensured a generally higher level of fibre in the average diet than in the post-rationing decades.

The current proliferation of high fat, high sugar containing foods and the whole range of convenience and snack foods which are the consequences of an affluent society has outdated the nutritional message of the post war era. However, replacing that message with one which takes into account increased consumer spending, the wide choice available, the change in the nation's food supplies and advances in nutritional knowledge and food technology is a problem of considerable concern to those involved in nutrition education.

Classifying foods into groups still appears to be a desirable objective, the aim being to give guidance for food selection and meal planning. The "post war three" ("protection" by minerals and vitamins, "growth" from protein containing foods and "energy" from fats and sugars) is now totally out of date with its positive requirement for foods containing fat and sugar. The American "four" (meat, fish, beans; milk group; fruit and vegetables; cereals) has a more modern image since no specific mention is made of fats or sugar which are simply assumed to be present as a result of food preparation. Neither of these methods has really considered the problems associated with classifying the "recipe" foods and snack foods characteristic of the diets of technologically advanced societies. The Department of Health and Social Security of the United Kingdom, in its publication "Eating for Health" (1979) (1) ignored the issue with its advice to "include cereal foods, protein foods, some fat and fruit and vegetables so that energy needs are met".

De Belcher and Ter Haar (1980) (2) of the Dutch Nutrition Education Bureau have described a "Meal Guide" which is a breakthrough in three respects - it embraces current nutritional thought, it allows for the inclusion of "modern" food items and it attempts to classify foods on a quantitative basis.

In the United Kingdom, the source of nutrient data in Paul and Southgate's "McCance and Widdowson's Composition of Foods" (1978) (3). The tables are compiled from analytical data on retail food supplies, but convenience foods are only included if they have a well established market record. As this meant that very few such products are included, Ryley, Klesko, Turner and Glew (1780) (4) published a report of an analytical study on a wide range of convenience foods, the results of which have been summarised by Ryley (1984) (5). This paper is concerned with the classification of results of that study.

2. METHOD

Eighty convenience menu items prepared to the "ready-to-eat" stage including the final heating process were compared with similar types of product made by traditional methods in institutional kitchens. Immediately after completion of the preparation stages, the menu items were frozen and stored at $-20^{\circ}C$ for the assay of protein, total fat, moisture, calcium, iron, vitamin A, vitamins B_2, B_6 and B_{12}. Vitamins B_1 and C were assayed immediately after sample collection because of their labile characteristics. The menu items were mainly entrée and dessert items. The majority of the convenience products were either frozen or "dry mix" (4)(5).

3. RESULTS

Initially, because one aim of the project was to examine micronutrient levels in relation to energy levels, the concept of "index of nutritional quality" as developed by Hansen (1973) (6) and co-workers was examined for its applicability for assessing the results.

The concept of "index of nutritional quality" or INQ was originally introduced as a means of quantitatively evaluating the nutritional profile of whole diets. The term relates the nutrient content of a food to the energy content (both expressed as a % of the Recommended Daily Amount), and is independent of moisture content. The rationale behind the term is that when energy requirements are satisfied, all other nutrient requirements should also be satisfied to achieve a balanced diet. The derived value, which is a simple ratio, therefore examines the rate of supply of nutrients relative to the rate of supply of energy. An INQ of 1.0 for each nutrient in a total daily diet is a significant target.

The INQ of a food is calculated for each nutrient by the formula:

$$\frac{\text{\% RDA of nutrient supplied by the food.}}{\text{\% RDA for energy supplied by the same quantity of food.}}$$

Thus it is independent of moisture content or portion size and allows all nutrients, whether g, mg or ug quantities to be expressed on the same scale.

When judged by the index of nutritional quality, comparable convenience and traditional products had similar nutrient profiles. In 105 comparisons, traditional and convenience INQ values were similar on 51 occasions. Of the remaining 54 comparisons, traditional product INQ values were higher on 41 occasions. Convenience product INQ values were higher on 13 occasions. Traditional products tended to be higher in protein, calcium, thiamine, riboflavin and nicotinic acid. Convenience values were higher for ascorbic acid (mainly from frozen vegetable and fortified instant potato). INQ values were similar for iron and vitamin A. Some of the differences (protein, calcium, riboflavin) could be attributed directly to the extensive use of skimmed dried milk in the traditional school meal.

4. DISCUSSION

The Society for Nutrition Education (1976)(7), in trying to assess the nutritional quality of "fast foods" proposed that a food should be considered nutritious if the INQ is greater than one for four or more nutrients. When this approach was applied to the results of the study described in this paper, most main courses would have been classified as nutritious but fish cakes, pies and flans were marginal. Only desserts containing egg or milk as major ingredients qualified as nutritious. A drawback of this approach is that the number of nutrients assayed becomes a critical factor - the concept cannot be justifiably applied to incomplete data.

Witwer et al. (1977)(8) suggested that for most nutrients INQ values can be directly translated into adjectives which are commonly used to describe the status of food items. Their proposals link INQ values to five terms (poor, fair, adequate, good and excellent sources of particular nutrients), but as can be seen in the examples for protein (Table 1), these proposals can be very misleading. Roast beef is sensibly classified but defining sprouts or cottage pie as good sources of protein is disputable. Thus INQ only offers a sound basis for defining the nutritional status of foods if the energy density is also considered.

Table 1

Product	MJ/100g	Protein g/100g	INQ value for protein	Definition according to Wittwer
Roast beaf in gravy	.43	14.6	5.6	Excellent
Cottage pie	.64	6.5	1.7	Good
Sprouts	.13	3.6	4.6	Good

The mean energy density of foods in the U.K. is .6 - .8 MJ/100 g(9). Wittwer proposals can be applied directly to foods of approximately average energy density, but the scales need adjusting pro rata for foods with both higher and lower energy densities.

To provide a more reliable basis for describing the attributes of food items the following scale linking the INQ data to the commonly used adjectives is proposed.

Table II Proposed classification of foods according to Index of Nutritional Quality and energy density

Energy density MJ/100g	Index of Nutritional Quality				
	Poor	Fair	Adequate	Good	Excellent
1.25 MJ	0.2	0.4	.8	2.0	4.0
1.00 MJ	0.25	0.5	1.0	2.6	5.0
0.75 MJ	0.35	0.7	1.4	3.5	7.0
0.50 MJ	0.50	1.0	2.5	5.0	10.0
0.25 MJ	1.00	2.0	5.0	10.0	20.0
0.10 MJ	2.50	4.5	12.5	25.0	50.0

The five columns poor, fair, adequate, good and excellent are characterised by the contribution per 100 g of 2.5%, 5%, 10%, 25% and 50% of the R.D.A., respectively. Roast beef in gravy, cottage pie and sprouts are now sensibly classifed as good, adequate and fair sources of protein, respectively. If this table is used in menu planning, energy intake can be controlled while maintaining nutritional quality. Thus 10 menu items per day of "adequate" classification for all nutrients is a minimal target. If some items fall below "adequate" they can be balanced by the inclusion of some items in the "good" and "excellent" categories. Such a table provides more guidance for menu planning than for instance, the list of "good sources" formerly published for the planning of school meals (10).

Table III shows the results of using the proposed classification on the results of the analytical study previously described. Generally, the differences between convenience and traditionally prepared products was not great enough to cause the items to be differently classified. Exceptions were caused either by obvious compositional differences (e.g. the proportion of potato in cottage pie) or the inclusion of a high proportion of dried skimmed milk in the traditional product (e.g. custard and custard tart). The use of fortified instant potato in the convenience cottage pie and fish cakes causes the product to be highly rated with respect to vitamin C.

This technique could be modified to include fat and sodium by utilising "maximum daily amounts" instead of "recommended daily amounts" and classifying into high, medium and low food groups

Table III Nutrient classification of convenience foods

Key: 1 = poor 2 = fair 3 = adequate 4 = good 5 = excellent

Product	Source	Protein	Ca	Fe	A	B_1	B_2	C	B_6	B_{12}	niacin
					Vitamins						
Roast beef in gravy	Frozen	4	-	3	-	1	3	-	3	4-5	4
	Trad	4	-	3	-	2	3	-	3	4-5	4
Beef stew	Frozen	3	-	3	-	1	3	2	2	4	3
	Trad	3	-	3	-	1	3	-	2	4	3
Steak and kidney pie	Chilled	4	2	3-4	2-3	3	4	-	2	5	3
	Trad	4	2	3-4	-	3	4	-	-	-	4
Sausage and egg pie	Frozen	3	3	3	-	2	4	-	3	-	3
	Trad	3	3	3	1	2	4	-	-	-	2
Sausage rolls	Frozen	3	3	3	-	2	2	-	1-2	3	3
	Trad	3	3	3	-	2	2	-	-	-	3
Cottage pie	Frozen	2	1	2	-	2	2	5	2	3	2
	Trad	3	2	3	-	2	2	-	3	4	3
Chicken in sauce	Dehyd	3	1	2	-	1	2	-	1	2	3
	Trad	3	2	2	-	1	2	-	2	-	4
Cod mornay	Frozen	4	3	1	-	1	3	-	2	4	2
	Trad	4	3-4	1	-	1	3	-	2	4	3
Fish cakes	Frozen	3-4	2	2	-	2	2	3	2-3	4	3
	Trad	4	2	3	-	2	3	-	2	4	3
Sponge cake	Dry mix	2-3	2-3	2-3	-	2	1-2	-	<1	-	2
	Trad	2-3	3-4	2-3	-	2	3	-	-	-	2
Custard	Dry mix	1	2	<1	1	1-2	2	-	<1	2-3	<1
	Trad	2-3	4	1	2	1	4	-	-	-	<1
Egg custard tart	Frozen	3	3	2-3	-	2	3	-	1	3	1
	Trad	3	3-4	2-3	2-3	2-3	4	-	1	3	1
Bakewell tart	Frozen	2-3	2-3	2-3	-	3	2	-	1	-	1
	Trad	2-3	2-3	2-3	-	3	2	-	1	-	1
Apple pie	Frozen	2	2-3	3	-	2	1	-	<1	-	1
	Trad	2	2	3	-	2	1	-	-	-	1

REFERENCES

1. Department of Health and Social Security (1979). "Eating for Health", HMSO, 79
2. De BEKKER, G.J.P.M. and TER HAAR, G.I. (1980). "The Meal Guide", Voorlichtings bureau voor de Voeding, The Hague, Netherlands
3. PAUL, A.A. and SOUTHGATE, D. (1978). "McCance and Widdowson's Composition of Foods", HMSO
4. RYLEY, J., KLEZKO, K., TURNER, M. and GLEW, G. (1978). "A Nutritional Evaluation of Convenience Foods", Leeds University
5. RYLEY, J. Human Nutrition: Applied Nutrition (in press)
6. HANSEN, R.G. (1973). Nutr. Rev. 31, 1
7. ANON. (1976). "Nutritional claims for foods", Society of Nutrition Education, Berkeley, California
8. WITTWER, A.J., SORENSON, A.W., WYSE, B.W. and HANSEN, R.G. (1977). J. Nutr. Ed., 9, 26
9. Ministry of Agriculture, Fisheries and Food (1981). Annual Report of the National Food Survey Committee, HMSO
10. Department of Education and Science (1975). "Nutrition in Schools", HMSO, 27

RAPID METHODS FOR PROTEIN NUTRITIONAL QUALITY EVALUATION: EUD AND C-PER

F. Fidanza and A. Maurizi Coli*
Istituto di Scienza dell'Alimentazione
Università degli Studi di Perugia

Summary

The protein nutritional quality of 50 samples of novel protein sources and 6 types of pasta was assessed by two rapid methods: the Enzymatic Ultrafiltrate Digest (EUD) and the Computed Protein Efficiency Ratio (C-PER). For both methods the ability to discriminate among proteins is good and changes in protein nutritional quality due to preparation and processing can be detected. For the EUD, the validity was assessed by the Relative Protein Value method, with very good results being obtained. Both EUD and C-PER can be used with success for regulatory purposes and particularly for the monitoring of variables introduced by food processing.

Rapid methods of protein nutritional quality evaluation are needed by the food industry in many areas -- monitoring raw materials and ingredients, processing conditions, final formulations and products, as well as shelf life stability. For such methods not only rapidity should be attained but also accuracy, validity and wide applicability.

In a previous paper (1) we examined the advantages and drawbacks of the most suitable methods for nutritional quality evaluation in food technology.

With the use of two of them, (the Enzymatic Ultrafiltrate Digest (EUD) aminoacid index and the Computed Protein Efficiency Ratio (C-PER), the results obtained on 50 samples of novel protein sources and 6 types of pasta will be shown here.

Materials and methods

The novel protein sources were in most cases protein isolates or concentrates produced with different methods by other operative units as part of a project for novel protein sources sponsored by the National Research Council.

In addition, the pasta was examined before and after technological preparations with different time and thermal conditions.

The methods for EUD and C-PER are described in detail elsewhere (2,3).

Results and conclusions

In table I the results on 50 samples of novel protein sources are shown. In general, for both methods the ability to discriminate among

---

* With the technical assistance of M. Buono, R. Coli and G. Parretta

TABLE I

EUD AMINOACID INDEX AND C-PER OF SOME NOVEL PROTEIN SOURCES

FOODITEM	No. OF SAMPLES	EUD MEAN	EUD RANGE	C-PER MEAN	C-PER RANGE
BEEF BLOOD PROTEIN CONC.	1	78.3		2.46	
CANDY BARS	1	72.6		2.63	
CHICK-PEA WHOLE FLOUR	1	66.8		2.22	
CHICK-PEA PROTEIN CONC.	2	58.5	58.4-58.7	1.83	1.82-1.84
DEFATTED CHICK-PEA PROTEIN ISOLATE	1	55.8		1.72	
CORN GERM FLOUR	1	61.6		2.07	
BOLTED CORN GERM	1	66.6		2.15	
BOLTED CORN GERM, EXTRUDED	2	58.1	57.1-59.1	1.85	1.80-1.90
FABA BEAN PROTEIN CONC.	1	58.5		1.95	
FABA BEAN PROTEIN ISOLATE	1	50.1		0.91	
FABA BEAN (MINOR) PROTEIN CONC.	1	46.5		1.51	
FABA BEAN (MINOR) AND WHEAT MEAL EXTRUDED	2	48.7	48.0-49.4	1.38	1.24-1.52
GRAPE SEEDS PROTEIN CONC.	1	39.2		0.36	
LUPIN PROTEIN CONC.	2	50.4	48.8-51.9	1.86	1.73-1.98
PEA PROTEIN CONC.	2	57.6	55.8-59.4	1.61	1.59-1.63
PEAS	10	57.8	54.8-61.3	1.87	1.76-2.02
SPIRULINES (MAXIMA AND PLATENSIS, DRYED)	8	63.8	56.7-70.5	2.16	1.82-2.49
SUNFLOWER SEED PROTEIN FLOUR	2	57.9	57.0-58.8	1.80	1.78-1.83
SUNFLOWER SEED PROTEIN CONC., FREEZEDRYED	2	55.5	54.8-56.2	2.09	2.08-2.10
DEFATTED SUNFLOWER SEED PROTEIN CONC.	3	45.7	40.2-50.5	1.60	1.22-1.71
SUNFLOWER SEED, HULLED AND GROUND	1	49.8		1.87	
DEFATTED SUNFLOWER SEED, HULLED AND LAMINATED	1	56.2		1.85	
TOBACCO PROTEIN CONC.	2	61.3	61.1-61.3	2.50	2.45-2.56
WHITE EGG, FREEZED DRYED	1	79.2		2.66	

TABLE II

TECHNOLOGICAL CHARACTERISTICS AND PROTEIN NUTRITIONAL QUALITY OF SEMOLINA AND PASTA

	DRYING		PACKAGING		EUD	C-PER
	temperature (°C)	time (h)	temperature (°C)	time (min)		
SEMOLINA	-	-	-	-	50.6	0.88
MACARONI	55	33	38-40	90	48.2	0.96
SEMOLINA	-	-	-	-	50.2	0.91
SPAGHETTI	71-72	11	55-62	80	48.5	0.77
SEMOLINA	-	-	-	-	50.5	0.98
SHELLS (small)	60	6 1/3	58	40	49.9	1.17
SEMOLINA	-	-	-	-	50.9	1.02
SHELLS (large)	70	4 1/2	80	30	48.9	0.94
EGG-DOUGH (4 eggs)	-	-	-	-	57.6	1.51
EGG-NOODLES (4 eggs)	51	11	49	25	52.9	1.41
EGG-DOUGH (6 eggs)	-	-	-	-	62.6	-
EGG-NOODLES (6 eggs)	51	11	49	25	55.8	-

protein is good. Only in a few cases there some disagreements. The procedure for the preparation of concentrates or isolates causes some changes in protein nutritional quality.

In table II the technological characteristics and the protein nutritional quality of the pasta are summarized. The temperature and time of drying and packaging influence the protein nutritional quality of the products. The addition of eggs, as expected, increases the protein nutritional quality. The C-PER values are, in two cases, (macaroni and small shells) not consistent with EUD and their scatter is larger than with EUD.

We do not consider it correct to correlate the EUD data with C-PER data because the values with the second method are not proportional to the protein nutritional quality.

The validity of EUD has been assessed by others (4), using on some of the same samples, the relative protein value method, which is considered at the moment as the most reliable.

The correlation coefficient between the two variables was highly significant (r=0.8). The ability to discriminate among proteins was rather similar; only with proteins of low nutritional quality (faba beans and grape seed concentrate) the EUD showed a slightly higher value than RPV. The standard error of measurement for 50 duplicate sample was $\pm$1.8.

Considering the fact that with EUD one can have the results in about 24 hours and one can obtain in addition the relative release of the various aminoacids in a protein, this method can be suggested for routine regulatory purposes and particularly for the monitoring of variables introduced by food processing. In countries where PER is the official method for regulatory purposes, the C-PER can be recommended.

REFERENCES

1. FIDANZA F. (1983). Methods of evaluating protein quality. This publication.
2. FLORIDI A. and FIDANZA F. (1975). Food Protein Quality. III Enzymatic Ultrafiltrate Digest (EUD). Amino Acid Index. Riv. Sci. Tech. Alim. Nutr. Um. 5: 13.
3. PELLET P.L. and YOUNG V.R. (1980). Nutritional Evaluation of Protein Foods. United Nation University, Tokyo.
4. SPADONI M.A. and CORCOS BENEDETTI P. Personal communication.

Participants

Adenier, H., Department de Génie Biochimique, Université de Technologie, Creusot-Loire, France.

Ahlström, A., Department of Nutrition, University of Helsinki, Viiki, SF-00710, Helsinki 71, Finland.

Ahvenainen, R., Food Research Laboratory, Technical Research Centre of Finland, Biologinkuja 1, 02150 Espoo 15, Finland.

Andersson, Y. SIK - The Swedish Food Institute, Box 5401, S-402 29 Gothenburg, Sweden.

Anifantakis, E., Agricultural College of Athens, Iera Odos 75, Votanikos, Athens, Greece.

Antila, J., Technical Research Centre of Finland, Biologinkuja 1, SF-02150 Espoo 15, Finland.

Arkoudilos, J., Food Technology Institute, Lykovrysi, Attiki, Greece.

Asp, N. G., Department of Food Chemistry, Chemical Centre, University of Lund, Box 740, S-222 07 Lund, Sweden.

Aspridis, J., Ministry of Agriculture, Lambrou Katsoni 47, Athens, Greece.

Athanasiou, N., Food Industries Department, Agricultural Bank of Greece, Omirou 8, Athens, Greece.

Athanasopoulos, P., Food Industries Department, Agricultural Bank of Greece, Omirou 8, Athens, Greece.

Auckland, J. N., Department of Catering Studies, The Polytechnic of Huddersfield, Queensgate, Huddersfield, HD1 3DH, England.

Auclair, J., I.N.R.A., Domaine de Vilvert, F-78350 Jouy-En-Josas, France.

Babatzimpoulou, M., Higher Technical and Professional Center, Sindos, Thessaloniki, Greece.

Bech-Jacobsen, K., Department of Food Preservation, Danish Meat Products Laboratory, Howitzvej 13, DK-2000 Copenhagen F, Denmark.

Ben-Gera, I. Wenger International Incorporated, Franklin Roosevelt Place 12, B-2008 Antwerp, Belgium.

Bender, A. E., 2 Willow Vale, Fetcham, Leatherhead, Surrey, BT22 9TE, England.

Berset, C., E.N.S.I.A., Avenue des Olympiades, F-91305 Massy, France.

Bertelsen, G., Department of Food Preservation, Danish Meat Products Laboratory, Howitzvej 13, DK-2000 Copenhagen F, Denmark.

Betzios, V., Agricultural College of Athens, Iera Odos 75, Votanikos, Athens, Greece.

Billon, M., Department Materials pour Industries Alimentaires, Creusot-Loire, BP No. 34, 42701 Firminy, France.

Blanc, B., Swiss Dairy Research Institute, CH-3097 Liebefeld-Berne, Switzerland.

Blanchot, G., Centre Technique Raymond Dellamagne, Biscuits Belin SA, 15-17 Avenue de l'Europe, 02400 Château Thierry, France.

Blixt, K., I-Point A.B., Nya Agnesfridsvägen 181, S-213 75 Malmo, Sweden.

Björck, I., Department of Food Chemistry, Chemical Centre, University of Lund, Box 740, S-220 07 Lund, Sweden.

Boegh-Soerensen, L., Department of Food Preservation, Royal Veterinary and Agricultural University, Howitzvej 13, DK-2000 Copenhagen F, Denmark.

Bognar, A., Bundesforschungsanstalt für Ernährung, Institut für Hauswirtschaft, Garbenstrasse 13, D-7000 Stuttgart, Federal Republic of Germany.

Bounie, D., Université des Sciences et Techniques du Languedoc, Laboratoire de Biochimie et Technologie Alimentaires, Place E. Bataillon, 34060 Montpellier Cedex, France

Boyazoglu, E., Food Industries Department, Agricultural Bank of Greece, Omirou 8, Athens, Greece.

Brawand, F., Swiss Vitamin Institute, University of Basel, Vesalgasse 1, CH-4051 Basel, Switzerland.

Brubacher, G., Department of Vitamin and Nutritional Research, c/o Hoffmann-La Roche & Co. AG., Grenzacherstrasse 124, CH-4002 Basel, Switzerland.

Brümmer, M., Federal Research Centre of Grain and Potato Processing, Institute of Baking Technology, Schützenberg 12, Postfach 23, D-4930 Detmold, Federal Republic of Germany.

Buss, D. H., Ministry of Agriculture, Fisheries and Food, Great Westminster House, Horseferry Road, London SW1P 2AE, England.

Canet, W., Instituto del Frio, Ciudad Universitaria, Madrid 3, Spain.

Carballido, A., Department de Investigaciones Bromotologicas, Edificiofacultad de Farmacia, Ciudad Universitaria, Madrid 3, Spain.

Cheftel, J. C., Université des Sciences et Techniques du Languedoc, Place E. Bataillon, F-34060 Montpellier Cedex, France.

Chouvel, H. SA des Biscuits Belin, Centre Technique, 15-17 Avenue de l'Europe, 02400 Château Thierry, France.

Codounis, M., Ministere de l'Agriculture, Institut de Technologie des Produits Aliment, Lykovrysi, Athens, Greece.

Corrieu, G., Institut National de la Recherche Agronomique, Laboratoire de Génie Industriel Alimentaire, 369 Rue Jules Guesde, 59650 Villeneuve D'Asq, France.

Couveliotis, C., Ministry of Agriculture, Acharnon 2, Athens, Greece.

Crawshaw, A., Research and Technology Centre, Dalgety Spillers Limited, Station Road, Cambridge, CB1 2JN, England.

Crivelli, C., I.V.T.P.A., Via Venezian 26, I 20133 Milano, Italy.

Cuq, J. L., Laboratoire de Biochimie et Technologie Alimentaires, Université des Sciences et Techniques du Languedoc, Place E. Bataillon, 34060 Montpellier Cedex, France.

Dagerskog, M., Nordreco, Box 500, S-267 00 BJUV, Sweden.

Dale, K. J., Ministry of Agriculture, Fisheries and Food, Great Westminster House, Horseferry Road, London SW1P 2AE, England.

Dalgleish, J. McNair, Ministry of Agriculture, Fisheries and Food, Great Westminster House, Horseferry Road, London, SW1P 2AE, England.

Dalles, T., Food Technology Institute, Lykovrysi, Attiki, Athens, Greece.

Dalezios, J., Agricultural Bank of Greece, Food Industries Department, Omirou 8, Athens, Greece.

De Coninck, V., CPC Nederland BV, Zetmeel en - derivaten groep, Westkade 119, 4551 LA Sas van Gent, Netherlands.

Doxanakis, V., E.V.G.A., SA, Iera Odos 88, Votanikos, Athens, Greece.

Eriksson, C., SIK - The Swedish Food Institute, Box 5401, S-402 29 Gothenburg, Sweden.

Espinosa, J., Instituto del Frio, Ciudad Universitaria, Madrid 3, Spain.

Fidanza, F., Università degli studi Perugia, Dipartmento di Science e Technologie, Casella Postale 333, 06100 Perugia, Italy.

Fiellettaz-Goethart, R., Instituto CIVO Technology TNO, Utrechtseweg 48, 3700 AJ Zeist, The Netherlands.

Fitton, M. G., CPC Europe R & D Centre, Havenstraat 84, B-1800 Vilvoorde, Belgium.

Frazier, P. J., Dalgety-Spillers Limited, Group Research Laboratories, Research and Technology Centre, Station Road, Cambridge, CB1 2JN, England.

Fuster, C., Instituto del Frio, Ciudad Universitaria, Madrid 3, Spain.

Galiotou, M., Agricultural College of Athens, Iera Odos 75, Votanikos, Athens, Greece.

Garcia-Matamoros, E., Instituto del Frio, Ciudad Universitaria, Madrid 3, Spain.

Garcia-Olmedo, O., Instituto del Frio, Ciudad Universitaria, Madrid 3, Spain.

Gerakopoulos, D., Food Technology Institute, Lykovrysi, Attiki, Greece.

Glew, G., Department of Catering Studies, The Polytechnic of Huddersfield, Queensgate, Huddersfield, HD1 3HD, England.

Gormley, R., Kinsealy Research Centre, Malahide Road, Dublin 5, Ireland.

Goutefongea, R., Institute National de la Recherche Agronomique, Laboratoire des Aliments d'Origine Aminale, La Géraudière, 44072 Nantes Cedex, France.

Gray, P. S., Head of Division III/A/2, Commission of the European Communities, Brussels, Belgium.

Griffin, M., National Ice and Cold Storage Co. Ltd., Cockstown Estate, Belgard Road, Tallaght, Co. Dublin, Ireland.

Guilbert, S., Laboratoire de Biochimie et Technologie Alimentaires, Université des Sciences et Techniques du Languedoc, Place E. Bataillon, 34060 Montpellier Cedex, France.

Hadjiantoniou, D., Hellenic Sugar Industry, Larisa Sugar Factory, Larisa, Greece.

Hakulin, S., Helsinki University of Technology, Department of Chemistry, SF-02150 ESPOO 15, Finland.

Harmuth-Hoene, A. E., Bundesforschungsanstalt für Ernährung, Institut für Biochimie, Engesserstrasse 20, 7500 Karlsruhe 1, Federal Republic of Germany.

Harper, J. M., Vice President for Research, Colorado State University, Fort Collins, Collorado 80523, USA.

Härrod, M., SIK - The Swedish Food Institute, Box 5401, S-402 29 Gothenburg, Sweden.

Hawthorn, J., University of Strathclyde, Department of Food Science and Nutrition, James P. Todd Building, 131 Albion Street, Glasgow, G1 1SD, Scotland.

Helsing, E., WHO Regional Office for Europe, 8 Scherfigsvej, 2100 Copenhagen Ø, Denmark.

Hirons, J., Ministry of Agriculture, Fisheries and Food, Great Westminster House, Horseferry Road, London SW1P 2AE, England.

Holm, F., Jutland Technological Institute, Marselis Boulevard 135, 8000 Arhus C, Denmark.

Holtz, E., University of Lund, Division of Food Engineering, PO Box 50, S-230 53 Alnarp, Sweden.

Houwing, H., Institute for Fishery Products TNO, Dokweg 37, PO Box 183, NL-1970 CA Ijmuiden, The Netherlands.

Iconomou, D., Food Technology Institute, Lykovrysi, Attiki, Greece.

Ioannides, E., Agricultural College of Athens, Iera Odos 75, Votanikos, Athens, Greece.

Iordanou, P., E.L.S.A. SA., PO Box 80107, Pireas 18510, Greece.

Jacobson, E., The National Swedish Board for Technical Development, Box 43200, S-100 72 Stockholm, Sweden.

Jägerstad, M., Department of Nutrition, University of Lund, PO Box 740, S-220 07 Lund 7, Sweden.

James, S. C., Agricultural Research Council, Meat Research Institute, Langford, Bristol, BS18 7DY, England.

Janssen, L. P., Delft University of Technology, Laboratory for Physical Technology, Prins Berhardlaan 6, 2628 BW Delft, The Netherlands.

Javier-Borderias, A., Instituto del Frio, Ciudad Universitaria, Madrid 3, Spain.

Jensen, J. H., National Food Institute, Mørkhoj Bygade 19, 2860 Søborg, Denmark.

Jeunink, J., CREALIS, Centre de Recherches et d'Etudes Alimentaires, Rue Frédéric Sauvage, 19100 Brive, France.

Jimenez Colmenero, F., Instituto del Frio (CSIC) Ciudad Universitaria, Madrid 3, Spain.

Johannsmann, H., Bundesministerium für Ernährung, Landwirtschaft ünd Forsten, D-53 Bonn, Federal Republic of Germany.

Jowitt, R., Presently at: The Catholic University of Leuven, c/o Laboratory of Food Preservation, Kardinaal Mercierlaan 92, B-3030 Heverlee-Leuven, Belgium.

Jul, M., Danish Meat Products Laboratory, Department of Food Preservation, Howitzvej 13, DK-2000 Copenhagen, Denmark.

Katsaboxakis, K., Institute of Agricultural Products of Athens, Lykovrissi, Amaroussion, Athens, Greece.

Kazantinou, S., Food Technology Institute, Lykovrisi, Attiki, Greece.

Kazazis, I., Higher School of Food Technology, Ag. Spyridonos & Palikaridi, Egaleo, Attiki, Greece.

Karaoulanis, G., Food Technology Institute, Lykovrysi, Attiki, Greece.

Karazanos, G., Agricultural Bank of Greece, Aristotelous 13, Thessaloniki, Greece.

Kehagias, C., Food Technology Institute, Lykovrysi, Attiki, Greece.

Kervinen, R. Helsinki University of Technology, Department of Chemistry, Otakaari 1, 02150 Espoo 15, Finland.

Kim, J. C., Institute for Cereals, Flour and Bread TNO, Lawickse Allee 15, PO Box 15, 6700 Wageningen, The Netherlands.

Kiritsakis, A., K.A.T.E.E. Thessaonikis, Sindos, Greece.

Koivistoinen, P., University of Helsinki, Food Chemistry Division, Department of Nutritional Chemistry, Helsinki 71, Finland

Kondylis, T., Vegetable Research Institute, Korinos Ilias, Castouni, Greece.

Kristensen, K. H., Jutland Technological Institute, Marselis Boulevard 135, 8000 Arhus C, Denmark

Laroche, M., Institut National de la Recherche Agronomique, Laboratoire des Aliments d'Origine Aminale, I N R A Nantes, Rue de la Géraudière, 44072 Nantes Cedex, France.

Launay, B., E N S I A, Avenue des Olympiades, F-91305 Massy, France.

Laustsen, K., Arhus Oliefabrik A/S, LMT Afdeling, 8100 Arhus C, Denmark.

Lazaridis, S., Agricultural Bank of Greece, Food Industries Department, Omirou 8, Athens, Greece.

Leniger, H., Agricultural University Biotechnion, De Dreijen 12, NL-6703 BC Wageningen, The Netherlands.

Linko, P., Helsinki University of Technology, Department of Chemistry, Otakaari 1, SF-02150 Espoo 15, Finland.

Linko, Y. Y., Helsinki University of Technology, Department of Chemistry, Otakaari 1, 02150 Espoo 15, Finland.

Londahl, G., Frigoscandia AB, Box 912, S-251 09 Helsingborg, Sweden.

L'Orient, D., Ecole Nationale Supérieure de Biologie, Appliquée à la Nutrition et à l'Alimentation (ENS.BANA), Département de Biochimie Alimentaire, Campus Universitaire Montmuzard, 21100 Dijon, France.

Ludwig, I. W., Institute for Research on Storage and Agricultural Produce (IBVL), Bornsesteeg 59, PO Box 18, NL-6700 AA Wageningen, The Netherlands.

Lund, D., Department of Food Science, Balcock Hall, 1605 Linden Drive, Madison, Wisconsin 53706, USA

McLellen, K., Procter Department of Food Science, The University of Leeds, Woodhouse Lane, Leeds, LS2 9JT, West Yorkshire, England.

Maingonnat, J., INRA, Centre de Recherches de Lille, Laboratoire de Genie Industriel Alimentaire, 360 Rue Jules Goesde, 59650 Villeneuve D'Asq, France.

Mallidis, C., Food Technology Institute, Lykovrysi, Attiki, Greece.

Malkki, Y., Technical Research Centre of Finland, Food Research Laboratory, Biologinkuja 1, SF-02150 Espoo 15, Finland.

Mariani, N., Istituto della Nutrizione, 00179 Rome, Italy.

Matalas, L., Agricultural Bank of Greece, Food Industries Department, Omirou 8, Athens, Greece.

Mathioudakis, B., Division III/A/2, Commission of the European Communities, Brussels, Belgium.

Megaloeconomou, S., Ministry of Agriculture, Halkokondyli 3I, Athens, Greece.

Mercier, C., INRA, Laboratoire de Biochimie des Glucides, Chemin de la Géraudière, 44072 Nantes Cedex, France.

Metaxopoulos, J., Agricultural Bank of Greece, Food Industries Department, Omirou 8, Athens, Greece.

Meuser, F., Technische Universitat Berlin, Institut für Lebensmitteltechnologie - Getreidetechnologie, Berlin, Germany.

Mildner, P., University of Zagreb, Laboratory of Biochemistry, Zagreb, Yugoslavia.

Moreiras-Varela, O., Istituto de Nutrition, Facultad de Farmacia, Madrid 3, Spain.

Moscicki, L., Institute of Food Engineering, Agricultural University in Lublin, PKWN 28, 20-612 Lublin, Poland.

Muller, H. R., Nestlé, Case Postale 88, CH-1814 La Tour de Peilz, Switzerland.

Munoz-Delgado, J. A., Instituto del Frio, Ciudad Universitaria, Madrid 3, Spain.

Newman, M., Ross Foods Limited, Ross House, Grimsby, Lincs DN31 3SW, England

Nielsen, S., Irish Board of Science and Technology, Shellbourne Road, Dublin 4, Ireland.

Nitzke, J., Institut für Lebensmitteltechnology, Universität Hohenheim, Garbegstr. 25, 7000 Stuttgart 70, Federal Republic of Germany.

Ohlsson, T., SIK - the Swedish Food Institute, Box 5401, S-402 29 Gothenburg, Sweden.

Olsson, P., SIK - The Swedish Food Institute, Box 5401, S-402 29 Gothenburg, Sweden.

Öste, R., University of Lund, PO Box 740, 222-07 Lund, Sweden.

Papanikolaou, D., Food Technology Institute, Lykovryssi, Attiki, Greece.

Paneras, E., Laboratory of Agricultural Technology, Aristotelion University of Thessaloniki, Thessaloniki, Greece.

Papotto, G., Mapimpianti S.p.a., Via Europa 25, I-35015 Galliera Veneta, Padova, Italy.

Park, J. R., Ministry of Agriculture, Fisheries and Food, Great Westminster House, Horseferry Road, London, SW1P 2AE, England.

Paulus, K. O., Federal Research Centre for Nutrition, Engesserstrasse 20, D-7500 Karlsruhe, Federal Republic of Germany

Petrak, T., Institute of Animal Products Technology, Veterinary Faculty, Heinzlova 55, YU-41000 Zagreb, Yugoslavia.

Philippon, J., Laboratoire de Physiologie des Organes Végétaux après Recolte, Station du Froid, 4ter Route des Gardes, F-92190 Meudon, France.

Pietzrik, K., Institute für Ernährungswissenschaft, Universitat Bonn, Endenicher Allee 11-13, D-53 Bonn, Federal Republic of Germany.

Pikoulas, E., Alimenta Ltd., Lagoumitzi 24, Athens, Greece.

Plastourgos, S., Agricultural Bank of Greece, Food Industries Department, Omirou 8, Athens. Greece.

Pion, R., Institute Nationale de la Recherche Agronomique, Laboratoire d'Etude du Métabolisme Azoté. Thiex - Saint Genes Champanelle, 63122 Ceyrat, France.

Poulsen, K. P., Technical University, Lundtoftevej 100, Building 221, DK-2800 Lyngby, Denmark.

Prozenz, K., Buss AB Basel, CH-4133 Pratteln 1, Switzerland.

Rasmussen, E. L., Department of Food Preservation, The Royal Veterinary and Agricultural University, Howitzvej 13, DK-3000 Copenhagen F, Denmark.

Remy, R., Test-Achat, 13 Rue de Hollande, Bruxelles, Belgium.

Rask, O., SIK - The Swedish Food Institute, Box 5401, S-402 29 Gothenburg, Sweden.

Ristic, M., Bundesanstalt für Fleischforschung, E C Baumannstrasse 20, 8650 Kulmbach, Federal Republic of Germany.

Rizos, A., Ministry of Agriculture, Acharnon 2, Athens, Greece.

Robertson, A., Campden Food Preservation Research, Chipping Campden, Glos. England.

Rodis, P., Agricultural College of Athens, Iera Odos 75, Votanikos, Athens, Greece.

Roseg, D., Veterinary Faculty, Zagreb University, 41000 Zagreb, Heinzelova 55, Yugoslavia.

Rosset, R., C N E R P A C, 5 Rue Mazet, F-75006 Paris, France.

Rouet-Mayer, M-A., Laboratoire de Physiologie des Organes Végétaux après Recolte, Station du Froid, 4ter Route des Gardes, F-92190 Meudon, France.

Ruiz-Roso, B., Instituto de Nutrition, Facultad de Farmacia, Madrid 3, Spain.

Ryley, J., Procter Department of Food Science, The University of Leeds, Woodhouse Lane, Leeds, LS2 9JT, West Yorkshire, England.

Salmon, J., 8 Cherrydale Road, The Maultway, Camberley, England.

Sanderson-Walker, M., Birds Eye Walls Limited, Station Avenue, Walton on Thames, Surrey, KT12 1NT, England.

Samaras, F., Food Technology Institute, Lykovrysi, Attiki, Greece.

Scheffeldt, P., Institute for Nutrition Research, Seestrasse 72, CH-8802 Ruschlikon, Switzerland.

Seibel, W., Federal Research Centre of Grain and Potato Processing, Institute of Baking Technology, Schutzenberg 12, Postfach 23, 4930 Detmold, Federal Republic of Germany.

Seiler, H., Institute für Anorganische Chemie, Universitat 51, 4056 Basel, Switzerland.

Seiler, K., Federal Research Centre of Grain and Potato Processing, Instituto of Baking Technology, Postfach 23, Schutzenberg 12, D-4930 Detmold, Federal Republic of Germany.

Shearer, G., Ministry of Agriculture, Fisheries and Food, Food Laboratory, Colney Lane, Norwich, NR4 7UA, Norfolk, England.

Skjöldebrand, C., Division of Food Engineering, University of Lund, PO Box 50, S-230 53 Alnarp, Sweden.

Sluimer, P., Institute for Cereals, Flour and Bread TNO, Lawickse Allee 15, PO Box 15, 6700 AA Wageningen, The Netherlands.

Smith, A. C., ARC Food Research Institute, Colney Lane, Norwich, NR4 7UA, England.

Solms, J., Swiss Federal Institute of Technology, Universitätstrasse 2, CH-8092 Zurich, Switzerland.

Southgate, D. A. T., ARC Food Research Institute, Colney Lane, Norwich NR4 7UA, England.

Spiess, W. E. L., Bundesforschungsanstalt für Ernährung, Engessertsrasse 20, D-7500 Karlsruhe 1, Federal Republic of Germany.

Srebrnik-Friszman, S., Ministere de la Sante Publique et de la Famille, Institut d'Hygiène et d'Epidémiologie, Rue Juliette Wytsman 14, 1050 Bruxelles, Belgium.

Stavrinos, P., Agricultural Bank of Greece, Food Industries Department, Omirou 8, Athens, Greece.

Steinbuch, E., Sprenger Institute, Haagsteeg 7, NL-6808 PM Wageningen, the Netherlands.

Stollman, U., SIK - The Swedish Food Institute, PO Box 5401, S-402 29 Gothenburg, Sweden.

Stolp, W., Agricultural University, Department of Food Processing, Biotechnion, De Dreijen 12, NL-6703 BC Wageningen, the Netherlands.

Storey, R. M., Humber Laboratory of Torry Research Station, Ministry of Agriculture, Fisheries and Food, Wassand Street, Hull HU3 4AR, England.

Tejada, M., Istituto del Frio, Ciudad Universitaria, Madrid 3, Spain.

Theander, O., Department of Chemistry, Swedish University of Agricultural Sciences, Uppsala, Sweden.

Tolboe, O., Jutland Technological Institute, Marselis Boulevard 135, 8000 Arhus C, Denmark.

Trichopoulou, A., Athens School of Hygiene, Vas Sofias 90, 11528 Athens, Greece.

Tsaktani, E., Agricultural Bank of Greece, Food Industries Department, Omirou 8, Athens, Greece.

Tsouroufllis, S., Hellas Can S.A., El. Venizelou & Solonos 135, Kallithea, Athens, Greece.

Turner, M., Catering Studies Department, The Polytechnic of Huddersfield, Queensgate, Huddersfield, England.

Upton, P. K., National Board for Science and Technology, Shelbourne Road, Dublin 4, Ireland.

Valassopoulos, U., Agricultural Bank of Greece, Food Industries Department, Omirou 8, Athens, Greece.

Valdehita, M. T., Department de Investigaciones Bromotologicas, Edificio Facultad de Farmacia, Ciudad Universitaria, Madrid 3, Spain.

Van Boxtel, L., Instituto CIVO Technology TNO, Utrechtsweg 48, Box 3700 AJ Zeist, The Netherlands.

Vanderpoorten, R., Administration de la Recherche Agronomique, Chaussée d'Ixelles 29-31, B-1050 Brussels, Belgium.

Van Hoof, J., Faculteit van de Diergeeneskunde, R.U.G., Wolterslaan, B-9000 Gent, Belgium.

Van Lengerich, B., Technosche Universität Berlin, Institut für Lebelsmitteltechnologie - Getreidetechnologie, Seestrasse 11, D-1000 Berlin 65, Germany.

Van Zuilichem, D. J., Department of Food Processing and Engineering, The Agricultural University, De Dreijen 12, NL-6703 Wageningen, The Netherlands.

Varela, G., Instituto de Nutrition, Facultad de Farmacia, Ciudad Universitaria, Madrid 3, Spain.

Vartis, G., Sevath S.A., M. Karaoli 79, Greece.

Veerkamp, C. H., Spelderholt Centre for Poultry Research and Extension, Het Spelderholt 9, 7361 DA Beekbergen, The Netherlands.

Verbeke, R., Rijksuniversiteit Gent, Laboratorium voor Hygiene en Technologie van Eetwaren van Dierlijke Oorsprong, Wolterslaan 16, B-9000 Gent, Belgium.

Vincent, M. W., Vincent Processes Limited, Turnpike Road Industrial Estate, Shaw, Newbury, Berkshire, RG13 2NT, England.

Von Kozlowski, Langnese-Iglo, Entwicklung Tieflühlkost, Luthersweg 50, D-3050 Wunstorf 1, Federal Republic of Germany.

Voutsinas, L., National Dairy Committee, Athens, Greece.

Wade, P., United Biscuits (UK) Limited, St. Peter's Road, Furze Plaat, Maidenhead, Berkshire, SL6 7QU, England.

Wainwright, A., United Biscuits UK Limited, Group Research and Development Centre, St. Peter's Road, Furze Plaat, Maidenhead, Berkshire, SL6 7QU, England.

Weidmann, W., Werner und Plfeiderer, Maschinenfabrik, PO Box 301220, D-7000 Stuttgart, Federal Republic of Germany.

Winger, R. J., Meat Industry Research Institute of New Zealand Inc., PO Box 617, Hamilton, New Zealand.

Yacu, W. A., RHM Research Limited, The Lord Rank Research Centre, Lincoln Road, Cressex, High Wycombe, Bucks, HP12 3QR, England.

Yanniotis, S., Agricultural College of Athens, Iera Odos 75, Votanikos, Athens, Greece.

Zeuthen, P., Department of Food Preservation, Danish Meat Products Laboratory, Howitzvej 13, DK-2000 Copenhagen F, Denmark.

Zinck, O., Nutrition Unit, National Food Institute, Mørkhoj Bygade 19, 2860 Søborg, Denmark.

AUTHOR INDEX